AN AMERICAN ASTRONAUTICAL SOCIETY PUBLICATION

eighth annual meeting

Volume 11

ADVANCES IN THE ASTRONAUTICAL SCIENCES

Edited by Horace Jacobs

Proceedings of the
Eighth Annual Meeting
of the American Astronautical Society
16-18 January 1962
Washington, D.C.

Printed and Distributed by Western Periodicals Co., 13000 Raymer St., North Hollywood, California

THE AMERICAN ASTRONAUTICAL SOCIETY
516 FIFTH AVE.
NEW YORK 36, N.Y.

Affiliated with the American Association for the Advancement of Science

Member of the International Astronautical Federation

Printed and Bound in the U.S.A.

FOREWORD

This volume of "Advances in the Astronautical Sciences" is a record of the proceedings of the American Astronautical Society's Eighth Annual Meeting, held 16-18 January 1962, in Washington, D.C. The welcoming address was delivered by Mr. Alfred M. Mayo, President of the Society. Special events included a luncheon address by Hon. George P. Miller, Chairman, Committee on Science and Astronautics, U.S. House of Representatives, and an Honors' Night Banquet address entitled "First Place in theSpace Race!?" by Dr. Edward C. Welsh, Executive Secretary, National Aeronautics and Space Council. The 1961 Flight Achievement Award was presented to Commander Alan B. Shepard, Jr., USN, and Captain Virgil Grissom, USAF. The 1961 Melbourne W. Boynton Award for Space Medicine was presented to Captain Ashton Graybiel, USN (MC). Recipient of the 1961 Space Flight Award was Dr. Fred L. Whipple.

The Eighth Annual National Meeting was sponsored and organized by the Washington, D.C. Section of the American Astronautical Society. The General Chairman was Mr. Willson H. Hunter, National Aeronautics and Space Administration, and the Technical Chairman was Dr. George R. Arthur, General Electric Missile and Space Vehicle Department. The technical program consisted of the following eight sessions:

1. Basic Research. Chairman: Dr. Robert Jastrow, Director, NASA Institute for Space Studies
2. Guidance and Control. Chairman: Dr. Robert E. Roberson, Consultant
3. Bioastronautics. Chairman: Dr. William Helvey, Republic Aviation
4. Communications. Chairman: Mr. R.A. Glaser, Boeing, Aero-Space Division
5. Lunar and Planetary Exploration. Chairman: Dr. Eliot T. Benedikt, Norair, Northrop Corporation
6. Astrodynamics. Chairman: Mr. Hans Lieske, General Electric, Missile and Space Vehicle Department
7. Applications of Astronautical Systems. Chairman: Mr. G.L. Hollingsworth, Boeing Scientific Laboratories
8. Space Vehicle Design. Chairman: Mr. S.B. Kramer, Lockheed Missile and Space Company

The technical papers in this volume appear basically in the order of presentation at the meeting. They are published essentially in the form submitted by the authors.

Committee chairmen who were responsible for the various aspects of this meeting include the following:

Finance	James S. Troutman Analytic Services, Inc.
Public Affairs	Haig Kafafian
Arrangements	John L. Crone Advanced Research Projects Agency
Registration	John A. Curtis ACF Electronics
Exhibition Management	F.G. McGavock

Special thanks are due to those who served on these committees and helped to make this meeting a success. Thanks are also extended to the editorial and art staff and to many others who assisted in preparing this volume for publication.

Horace Jacobs
Director of Publications and Editor

CONTENTS

Page

Page

PART VIII SPACE VEHICLE DESIGN

eighth
annual
meeting

AMERICAN ASTRONAUTICAL SOCIETY

INTRODUCTION

THE FIRST YEAR OF MANNED SPACE FLIGHT*

Alfred M. Mayo +

The first year of manned space flight has just ended. The American Astronautical Society, despite increased attention to the qualifications of its members, has grown by more than thirty-four percent. Its new members have ranged in technical disciplines from environmental biologists to solid state physicists, from mechanical engineers to bioengineers. Many of these people, outstanding specialists in their own field, also belong to societies contributing to their specialized interest. This broadening of AAS membership is not an accident. At the beginning of the year, your Board of Directors decided that the American Astronautical Society could make an important contribution to Astronautics by concentrating a major effort toward eliminating the knowledge and communication barriers restricting optimum integration of efforts by the broad spectrum of scientists and engineers who must cooperate to explore space.

Just as the "art" of medical practice is being augmented by the application of quantitative scientific principles, the "art" of space vehicle design must be supplemented by the application of more scientific methods to permit the quantitative blending of the diverse skills and materials necessary to produce the equipment and vehicles for the effective exploration of space.

In past aircraft design, sophisticated techniques were used to determine detail aerodynamic shapes, to design structural members and in the development of electronic circuits. The overall design and the important integration of these parts, though, depended to a major degree on the skill and judgment of a "key" designer. The "cockpit" where the man and the machine came together, was often arranged as result of the composite opinions of a group of experts convened as a "mockup board". The absence of a developed science of system interactions forced this procedure. Fortunately the average aircraft could carry between one-tenth and one-half of its weight as useful load. The cost penalty of non-optimum system interactions was large but not prohibitive. Early

---

* Opening statement for AAS National Meeting, January 16-18, 1962 - Washington, D. C.

+ President of the American Astronautical Society

lunar space flight systems, on the other hand, may return only between a hundredth and a thousandth of their initial launch weight to Earth at the mission's end. The economic incentive to improve the design is very great indeed. Research in many scientific disciplines is building up an impressive body of principles permitting improved objectivity in the engineering of many of the components and functions required for space flight. Comparable progress is not yet being made in the science of interactions.

Over a period of time, beautifully integrated systems have evolved in nature. The coding of information represented by the arrangement of atoms in the DNA molecules of genes and the smooth coordination of motion in the body of a cat pursuing a mouse are examples of integrated solutions to complex requirements in nature. Man is just beginning to learn to quantize interactions. Fluid mechanics, magnetics, and electrodynamics are being integrally considered in magnetohydrodynamics. It is clear that magnetohydrodynamic processes may also involve mechanical, chemical and nuclear effects. These in turn have a direct bearing on the entire dynamic interaction.

In spacecraft structures the use of single-purpose materials and components is becoming more and more costly. In an integrated spacecraft structure the same material might serve as a radiation shield, a heat shield, a meteor barrier, a load carrier and an atmosphere container. In addition, it could be configured to meet the equipment support, restraint, and esthetic needs of the crew. Future structures may also serve some of the functions of present electronic equipment by using circuits integrated into their design. Food, water, and life support equipments are being configured for use as radiation shielding. The water may be pumped from place to place in accordance with the mission need to serve variously as shielding, for the transfer of heat, or as an adjunct to the control of the vehicle.

To provide optimum consideration of these complex variables and interactions, it is important to strive for an understanding of the value of each function related to mission gains or objectives. Principles might then be developed to permit a quantitative integration of the variables in a manner to maximize the gains from space exploration per unit of cost.

President Kennedy, in his second state of the union address of May 25, 1961, stated:

> "Now is the time to act, to take longer strides, time for a great new American enterprise, time for this Nation to take a clearly leading role in space achievement. I believe that the nation should commit itself to achieving the goal before the decade is out, of landing a man on the moon and returning him safely to earth."

Some people have raised questions concerning the amount of emphasis given a single mission in space. Others have questioned the high cost of the project against the gain of "merely landing a man on the moon and returning him to earth".

If one considers only the mission statement and the surface gains, both objections would be unquestionably valid. By selecting a broadly challenging objective capable of being accomplished within a reasonable time, the President has offered the Nation far more than the achievement of the difficult stunt. He has provided a "system of challenges" which can force development of new knowledge and new technical capability in essentially every field of science and technology. At the same time, progress in the frontier of space technology will serve notice to all the world's people of a dynamic and vigorous United States. Practical applications of the technological gains from space operations undoubtedly will have a major impact on our economy during this decade.

In concert with the directed effort toward a manned lunar landing and return to Earth, major programs are being projected within the NASA, in Space Sciences and in Advanced Research and Technology. An Office of Applications has been instituted to ensure improved "feedback" of knowledge and technology to help bolster our economy and to speed the utilization of knowledge gained from the exploration of space.

Other government agencies are already active in the exploitation of space. The Department of Defense is by far the most active, but others such as the Atomic Energy Commission, the Weather Bureau, and the Federal Communications Commission have major activities in preparing for or in utilizing space to better serve their functions.

The weather and communication satellites are well known, and their advantages are better appreciated than are many of the other space technology benefits. Orbiting spacecraft are assuming roles of increasing importance to the defense of our nation. Technology, in the form of new knowledge, materials, equipment, and processes stimulated by early space exploration, is beginning to be utilized.

In summary, the first year of manned space flight has brought dramatic growth to activity in Astronautics. The scientific satellites have gained much information on the environment of space. Weather and communication satellites have begun to produce and indicate great future importance. Defense satellites are becoming important to our safety. Universities and industries are beginning to utilize knowledge and technology developed from efforts in the space program. Man, for the first time in recorded history, has flown in space.

The President has given our Nation a broad challenge to advance our technology and world prestige by a dynamic program of space exploration. If we are to rise to this challenge, progress must be made in practically every field of science and engineering.

This alone is not enough! Major gains are needed in blending the knowledge and skills of our varied disciplines to produce more effective and better integrated space exploration systems. It is in the dissemination of this kind of knowledge and the stimulation of integrative effort that the American Astronautical Society might perform a major service toward the advancement of Astronautics.

SPACE PROGRAMS AND THE BENEFITS OF SPACE RESEARCH*

George P. Miller +

As chairman of the House Committee on Science and Astronautics, I am pleased to have this opportunity of speaking to a group of outstanding specialists in a field which today has become essential for survival and in the never-ending quest of mankind for a brighter tomorrow.

As a boy, I was fascinated with the fiction then published, dealing with fanciful adventures in the universe which surrounds us. Despite my close association with the nation's astronautical program, it is often startling to realize that today, these flights of fancy have become hard, practical realities of vital concern to you, to me, and to every citizen of the world on both sides of the Iron Curtain.

Our committee is the Committee on Science as well as Astronautics. While we may have spent more time in the Committee on Astronautics, our jurisdiction and responsibility include the scientific work and endeavors of the Government. We appreciate, despite the cost of research, development and production of space vehicles, supporting instrumentation, and ground facilities, that the by-products of this effort will have economic, social, and defense values far in excess of the original cost.

Our need to lead in space exploration is not merely a matter of survival; it is not simply the result of a normal impulse to surpass our competitors, nor is it the result of a selfish desire for the yet undreamed-of conveniences and luxuries which the mastering of space technology can create. It arises from a broader and nobler purpose which has existed in the hearts and minds of men since the first human thought occurred -- the need to know, the need to grow, the desire for fulfillment of the ultimate destiny of mankind.

---

* Luncheon address

+ Chairman, Committee on Science and Astronautics, U. S. House of Representatives

We know even at this early stage of development and research that space exploration is of such immense importance to man's total knowledge that it will benefit and alter the course of his existence in ways no more foreseeable today than those which resulted from the invention of the wheel. And yet, despite the long-term goals which underlie the space program, and the fundamental desire of our nation to develop and share with the world the benefits arising from the exploration of space for peaceful purposes, many Americans do not yet fully comprehend its significance. Citizens who have no difficulty in understanding the significance of the ICBM, or other defense aspects of the program, ask me, "Why do we want to go to the Moon?" Considering the fact that the National Aeronautics and Space Administration will spend more than 1.6 billion dollars this year, and that the estimated cost of putting men on the moon will amount to 20 billion dollars or more, they certainly deserve an answer to their question.

In his State of the Union message on May 25, last year, President Kennedy ignited the fuse of what may become the world's fourth great scientific revolution. He said:

> "Now it is time to take longer strides -- time for a great new American enterprise -- time for this nation to take a clearly leading role in space achievement...
>
> "...I believe that this nation should commit itself to achieving the goal, before this decade is out, of landing a man on the moon and returning him safely to the earth."

The Congress, American industry, and American scientific and technical personnel, within and outside the Government, responded promptly to the President's challenge. Today, although less than a year has elapsed, those in charge of our space program are confidently predicting that this goal will be met, possibly in advance of the deadline which the President set.

Great progress has already been made in Project Mercury -- initial steps along the path of space exploration. Last September 13, the United States launched an unmanned spacecraft into a single orbit, returned it from space, and recovered it from the Atlantic Ocean. A little more than two months later, on November 29, the chimpanzee, Enos, earned his place in history by orbiting the earth twice before he was recovered from the Atlantic. Today, we are on the eve of the first manned orbital flight in the Mercury program. From these flights our scientists are accumulating basic knowledge about the achievements possible for man in space -- how well he can operate controls and instruments, what he can learn that cannot be obtained from electronic devices, how the human body reacts to prolonged weightlessness. Initially, manned orbital flights will be limited to three trips around the earth, but, as the program develops, they will be extended up to 18 orbits -- 27 hours in space.

Utilizing the knowledge gleaned from these manned orbital flights, the lunar exploration will be undertaken, as a part of Project Apollo, in a three-man capsule for which development contracts have already been let. As in the case of its less refined predecessors, the Apollo capsule will first be flown in orbits around the earth, to test its components, train its crew, and develop operational techniques. Gradually, the distances from the earth and the duration of flights will be increased, culminating in manned flight around the Moon which will accumulate information and experience of value in subsequent landings on the moon itself. We are beginning to gain experience with the larger boosters which will be required in Project Apollo. On October 27, NASA launched the first Saturn vehicle -- a 460-ton structure, ten feet taller than the Statue of Liberty, with eight engines which generated 1,300,000 lb. of thrust. It is the most powerful rocket stage known to exist in the world. In later flights this thrust will be increased to 1,500,000 lb., and in orbital missions a second stage will be added.

Less than ten days ago, I was at the George C. Marshall Space Flight Center in Huntsville, Alabama, where the scientific and technical development of the Saturn is being carried on. On that trip I also visited the Michoud plant in New Orleans where several stages of the Saturn will be manufactured by private industry for the National Aeronautics and Space Administration.

Flight around the Moon will be accomplished with a three-stage vehicle generating 7.5 million pounds of thrust in the first stage, capable of placing 175,000 lb. in orbit around the earth and 70,000 lb. around the moon. A lunar landing may be accomplished by bringing together two of these vehicles in space. The potentialities of such a "rendezvous" approach are being carefully considered. This would require two launches of the advanced Saturn. The first would place into orbit a large, fully-fueled rocket stage. Once its orbit had been established, a second Saturn would launch the Apollo spacecraft and its auxiliary rockets into the same orbit and the crew would join it with the first orbiting stage. The largest rocket would then launch it into a flight path to the moon.

Contracts have already been awarded for the development of a two-man spacecraft, to be called Gemini, to experiment with the rendezvous technique. Launched by a modified Titan II missile, it will be capable of remaining in orbit for a week or more. Eventually it will rendezvous with an Atlas-boosted Agena stage. Two-man flights will precede Project Apollo, which will begin with orbital flights, advance to circumlunar flights, and culminate with actual landings on the Moon.

A direct flight to the Moon will require an even larger launch vehicle with nearly double the thrust of the advanced Saturn. About 350 feet in height, it is called the Nova.

One of the major benefits being gleaned from this vast effort to conquer space is the stimulus which it is providing for scientific research in new and uncharted areas. The scope of space research may be judged from the resources which it involves. Other industrial revolutions have been based upon efforts by relatively few individuals. For example, when Edison invented the first high-voltage electric bulb, there were only 2,000 industrial scientists and engineers in the United States. Today, however, it is estimated that there are more than 700,000 such individuals contributing to progress in the age of space. This work is not related solely to our progress on the path to the Moon. Much of it derives from the new knowledge of materials and techniques which has application in our daily lives. Our accelerated scientific research in space already produced useful knowledge which otherwise might not have been discovered in a decade, or even a century. These advances are being experienced by the housewife in her kitchen, the farmer in his field, the doctor in his operating room. Let me give you some examples.

One of man's great concerns has long been to know more about weather, whether merely to plan a picnic, or to aid in planting a crop. Despite great advances in the process of gathering meteorological information, forecasters are still handicapped by the fact that only one-fifth of the globe is covered by land-based weather reporting systems. Since the launching of the Tiros satellites, a wealth of new information regarding weather conditions around the world has been made available to meteorologists. Tiros I, launched April 1, 1960, provided television coverage of the earth's cloud patterns. Subsequently, Tiros III, orbited in July 1961, to coincide with the hurricane season, is still gathering information on the origin, development, and movement of these massive tropical storms. In fact, the Weather Bureau used photographs taken from Tiros III to analyze and track Storm Eliza in the Pacific and Hurricanes Carla and Esther in the Gulf and the Atlantic. In the case of Esther, the Tiros satellite provided two precious days of advance warning before the giant wind was observed by aircraft. The Tiros series will be succeeded by the Nimbus satellite, scheduled for launching late this year. An advanced design, which will be placed in polar orbit, Nimbus will be able to view every area on earth twice a day. Meteorologists believe this new information will ultimately permit accurate long- and short-range weather forecasting for periods as long as a month or even a season.

A second area of great progress is that of cummunications. Citizens of almost every nation have already sighted Echo I, the huge passive communications satellite launched by NASA in 1960. Echo I demonstrated conclusively that, used as communications relays or reflectors, satellites can extend line-of-sight transmissions to intercontinental ranges. With the world's growing need for communication facilities, and for cheaper and more reliable methods of transmission, the use of satellites holds forth enormous promise. During the past decade, the number of overseas telephone calls made or received in the United States grew from one million to nearly four million a year. It is estimated that some 5,000 or more circuits will be needed in the next 5 years. An industry expert has estimated that a single communications satellite, costing about $40 million and placed in a 22,300-mile equatorial orbit, could accommodate as much traffic as a $500 million cable system. A number of private companies have shown interest in commercial satellite communications. For example, the American Telephone and Telegraph Company is negotiating with NASA for use of a rocket to launch a satellite which would relay signals across the Atlantic and amplify them with power from the sun. The company has announced its willingness to spend up to $25 million this year on experiments leading to an earth-orbiting communications system. With satellites in orbit, the postman can anticipate less difficulty with "snow, rain, heat and gloom of night." Only last November, NASA assisted the Post Office Department in an experiment with Speedmail. A letter, urging the public to send Christmas presents early, was bounced off the satellite Echo from Stump Neck, Maryland, to Newark, New Jersey -- the first letter to be processed through outer space.

Research in space may also some day be responsible for saving your life. For example, people with some types of cardiac condition already are wearing a rhythm-control device to keep their hearts functioning while they live a normal life. This device, miniaturized by space scientists, is powered by tiny mercury batteries so small that it can be sewn into the patient's body. Space research has led to miniature instruments which continuously measure a hospital patient's heart action, brain waves, blood pressure, and breathing, permitting him to be "watched" by remote control. Any change in condition is instantly recorded and a warning flashed. One drug, developed from a missile propellant, is being used to treat mental ills. The problems involved in maintaining astronauts aboard spacecraft have led to a new method of rapidly lowering blood pressure during operations.

The search for new materials of lighter weight, greater strength and resistance to high temperatures and radiation -- all necessary for use in space vehicles -- is leading to countless new or improved products which will add to the pleasure of life in the future. Pyroceram, a material originally devised for nose cones, now provides the housewife with pots and pans which can be taken from the freezer and immediately put in the hottest flame without damage. A fluxless aluminum solder can now be used to repair eavespouts, flashings, and utensils which formerly would have been discarded. Miniaturization has produced radios as small as a lump of sugar. Research in electronics for space has brought remote control television switches. It is even possible to buy a wristwatch powered by the same mercury batteries that operate a timing mechanism developed for the Explorer satellite.

The by-products of space research seem to be almost limitless, and are certain to change the course of our lives. Most important, however, is the fact that because of the impetus of space requirements, these new materials and new products will be available for use in our own lifetimes, not decades hence. Other space research of more basic nature will certainly produce major changes in the way we live, and offer solutions to the many problems which the world and its growing population. Future spacecraft will require new sources of power, and the research now being undertaken to find these sources may very well change the methods by which we heat our homes, power our automobiles, or cook our food. Among those being studied are the fuel cell, which converts fuel directly into electrical power without machinery or moving parts. Another is plasma power, generated through use of hot ionized gas. Still another potential source is photoelectric power, to say nothing of atomic energy. Space research may also produce new sources and uses of water. It will lead to new discoveries with respect to noise -- a vital factor in human engineering. The highly specialized computers required for launching and guiding satellites have already produced electronic mechanisms capable of functions which are beyond the limitations of a human being.

All of these efforts will lead to vast changes in our economy. They already are producing new employment and new leisure opportunities. As the program advances, these opportunities will grow. We can count ourselves fortunate that space research will pay such large dividends in scientific knowledge and practical applications, since it would probably be necessary even without them. Man must go to the moon, and getting him there offers a challenge to every nation. In our position as leader of the Free World, the United States cannot fail to meet that challenge, for failure could only mean victory for those who seek to change our way of life. All of us share a heavy responsibility to make certain that this is not our fate.

FIRST PLACE IN THE SPACE RACE!?*

Edward C. Welsh +

It is indeed a privilege and a welcomed opportunity to express some ideas on space policy, space thinking and space doing to this audience of dedicated authorities on the subject. Much of what I plan to say will undoubtedly seem commonplace, or at least rather obvious, to most of you. However, we are so busy with the complex, the obscure, and the unknown these days that an occasional exposure to the simple or elementary may well be useful. At least, I have attempted to insure against any harmful or excessive dosage by injecting that rare ingredient of brevity.

A NATIONAL SPACE PROGRAM

The space business is a national business and the space program is a national program. It is not an expensive pastime for a few gifted scientists and engineers. Rather, it is everybody's business, just as democracy, education, and freedom are everybody's business. To a continuing and increasing degree, each and every sector of the economy will be affected by how much and how well we do in the space race. People want to know, and have the right to know, how the space effort is going to help them, their community, their country. They must be kept informed.

SPACE PRODUCTIVITY

Impressive is the mounting space output in electronics, employment, and engineering; in medicine, metallurgy, and meteorology; in national defense, navigation, and negotiations; or in pedagogy, prestige, and propulsion. Yet, it is not enough for you and me to know of this space productivity. Moreover, it is not enough for us to know that this space output will more than pay for itself in gross national product, in greater standards of living and in government

---

* Address delivered as part of the annual Honors Night Program

\+ Executive Secretary, National Aeronautics and Space Council

revenue. Everyone should know. Everyone should understand. Everyone should get the encouragement, the inspiration, and the sense of pride from these national accomplishments. This is a job for us all, particularly for you who know the story so well. Sell the story of space and you will not be selling your country short.

ECONOMY THROUGH SPEED

Most of us have, I am sure, been asked where we stand in this race. Are we behind? Are we ahead? And, I do not need to point out that those are not easy questions to answer. However, one thing we can be reasonably sure of is that it is a long race and a lagging frontrunner will quickly find himself overtaken and passed. Moreover, let us not deceive ourselves into thinking that there is any payoff for coming in second. The only thing worse than coming in second is not having run at all. Knowing that, the course has now been charted: the United States has stopped strolling, with occasional pauses to glance at the moonlight, and has started running with steady progress closer to the moon.

Haste, it is often said, makes waste. Such time-honored sayings often contain some grains of truth, but more often they are just easily voiced generalizations used as a substitute for thinking or are rationalizations for sheer laziness. Certainly, I can say with no fear of effective counter-argument, that in the space business we already know that slow, intermittent efforts are infinitely more wasteful, unless the goal is to go no place expensively. The one resource we can least afford to misuse is time and that is the resource wasted the most when we move slowly. We have tried that, to our Nation's loss. Indeed, we would relish the question as to where we stand relatively in the space race, if we had made more haste and wasted less time during the past decade.

PROGRESS IN 1961

Fortunately, the year 1961 was the United States' biggest year in space, primarily because it was in that year that a sense of urgency was combined with a willingness to make the tough decisions. That fact vastly outweighs the individual accomplishments of manned flights, or any other of the numerous magnificent accomplishments in space, during 1961.

Despite certain notable events, the U. S. achieved a substantially greater rate of space progress in 1961 than did the Soviet Union. There is no reason why we cannot do that every year. If those assumptions are correct, and I believe they are, then the conclusion is inescapable: We will not only lead in space but we will annually increase that margin of advantage. Let me hasten to add, however, that I am speaking of the possible, not the inevitable. I am also speaking of the probable, but only if we keep working harder every day and every year. We can be certain that the Russians are going to speed up rather than slow down. But they can match us, only if we place short-run convenience ahead of our Nation's future. The general public and its representatives in the Congress will make the funds and other resources available in quantities and qualities sufficient for us to win the race, but they will not continue the flow of such wealth into space unless they see the necessary speed, the necessary progress, and are kept fully informed as to what is being done and why.

We must make it clear that space exploration is probably the most difficult and certainly the most complex effort undertaken by man. I do not say that it is the most important of man's efforts, but it is one of the most important. In broad general terms, it is more important to attain permanent world peace, to encourage and protect freedom, and to raise the standard of living, especially in those areas where it is so very low. However, space activity can, directly or indirectly, contribute materially toward the attainment of all those objectives.

RISKS IN SPACE PROGRAM

Space activity is a risky business, but it is important to understand the nature of the risks in order that man may proceed to profit from space undertakings. When I refer to risks, I am not talking about the risks to human life, although of course, they do exist, and the courage required of space pioneers is great. There are some intrepid souls, however, who say that the human risks involved in space exploration are no greater than the hazards of being a pedestrian.

It is important to understand that, contrary to the truisms expounded by some economists and some businessmen, profits do not come from the assumption of risks. Rather, profits come to those who take more effective action than others in the elimination of risks or at least in making adequate provision for handling those risks which cannot be eliminated. So it is in the space business.

Let me identify and categorize briefly some of the risks which do exist in a space program and which we therefore must either eliminate or provide against, if we are to be truly successful. The biggest risk is one that has already been eliminated this year. I refer to the risk that this country would not make a firm and prompt decision to go forward into space, with definite objectives and with accelerated speed. The smallest risk is that we cannot do the job. The most unnecessary risk is that we will exaggerate the expense of the space program compared with its potential productivity. Corollary to that risk is the hazard that we will understate the benefits which can flow from space exploration. One of the most troublesome risks is that we will not learn to talk sense and give the public facts about space, including space successes and failures. One of the most stupid risks comes from misevaluation of the space competences and accomplishments of the Russians. These hazards arise from varying extremes of opinion, such as those who believe that the Soviets will wait until we catch up to them in large rocket development and others who believe that their primary accomplishment is in borrowing scientific and engineering knowledge from others. The most discouraging and possibly the most unwarranted risk flows from the belief on the part of some people that this country cannot afford an all-out space effort. There are those who seem to take for granted that a country like Soviet Russia, with less than half the per capita income of the United States, can afford a major successful space program, while we can not. That is, at best, ridiculous and, at worst, deliberate sabotage.

As previously mentioned, the space business is a risky one, but, as we look over the hazards, it is impressive that many of them are not technological in nature. Rather, they are emotional, irrational, or psychological. I hasten to point out, however, that such labels do not minimize the significance of the risks involved.

PEACEFUL USES: MILITARY AND NON-MILITARY

I now turn briefly to a specific anomaly in our space efforts, one which can cause us continuing difficulty unless it is thoroughly understood, and, through such understanding, dissolved. There are some well intentioned people who would draw a line - in fact, a barrier - between what they call peaceful uses and non-peaceful uses of outer space. As I have done so many times in the past, I again emphasize that it is our national policy that all our space efforts be peaceful. We do not have a division between peaceful and non-peaceful objectives for space. Rather, we have space missions to help keep the peace and space missions to improve our ability to live well in peace. There is no one agency in the government that has full responsibility for peaceful missions, just as there is no agency that has responsibility for non-peaceful missions.

In the National Aeronautics and Space Act, which established the National Aeronautics and Space Council and the NASA, the Congress made it explicit that all activities in space should be devoted to peaceful purposes. The law affirms that some of those peaceful purposes should be the responsibility of the NASA and some should be the responsibility of the Defense Department, with the President determining which such agency shall have the responsibility for and the direction of each of those peaceful space activities. I add that the act recognizes that other agencies also have space responsiblities. Those responsibilities are likewise peaceful in intent.

FRAGMENTATION AND COORDINATION

As one readily notices, there exists some fragmentation within the Government as to the handling of space responsibilities. It has been suggested that such fragmentation could be solved and that the space program could be made more effective and efficient, if all the space activities were put into one agency, such as a Department of Space.

This idea has some appeal, but it would not solve the fragmentation problem. In fact, it could create other fragmentation of an even more serious nature. For example, it might slice off part of the missions for keeping the peace and put them in a new agency, leaving other such missions in the Defense Department. It might slice off part of the missions in the nuclear field and put them in a new agency, leaving other nuclear functions in the AEC. It might divide the weather functions as between the Weather Bureau and the new agency. And so, the fragmentation would exist in another, and possibly even more, divisive form. I do not suggest that there may not come a time when a Space Department would not be a useful organization. Certainly, the mission is sufficiently important. However, I do suggest that such a new agency in the Government would not necessarily accomplish what is sometimes claimed for it, and, more important, at this time it could cause an ill-afforded delay while we were slicing and patching, organizing and reorganizing.

Consequently, the President has attempted to meet the coordination issue by activating the National Aeronautics and Space Council. Under the positive leadership of the Vice President, this mechanism has already shown results.

MORE PROGRESS IN 1962

As previously emphasized, this past year was one of decision - decision to accelerate, decision to put our resources to work more effectively, and particularly decision to lead rather than follow. You have seen results flowing from those decisions, in more launches, bigger contracts, new installations, more coordinated planning, and larger budgets.

It is essential, however, that the momentum thus gained be maintained. There is no place for "business as usual" in space. Notable as it has been, our recent progress furnishes no justification for satisfaction or complacency. During this year, we expect more sophisticated manned and unmanned space flights, progress in conventional as well as more modern rockets and rocket engines, lengthy strides in meteorological and communications satellites, greater international cooperation in space activities, increases in our relative national security through space, and progress in the rendezvous technique for lunar and other missions. As you know, these activities will call for larger budgets, more careful planning, and less duplication. We will strive to eliminate duplication, such as unnecessary duplication of effort, dangerous duplication of lack of effort, and all duplication which comes from following rather than leading. The space vitamins of 1961 will produce the space strength of 1962. It is up to all of us to use that strength to push the frontiers of space into the backyards of the planets.

eighth
annual
meeting

AMERICAN ASTRONAUTICAL SOCIETY

FORMATION OF THE SOLAR SYSTEM

A. G. W. Cameron*

Information bearing on the early history of the solar system can be deduced from a study of the anomalies in the isotopic composition of certain elements extracted from meteorites. In stone meteorites, xenon is sometimes enriched in Xe^{129}; and in iron meteorites, silver is sometimes enriched in Ag^{107}. If these anomalies are attributed to the decay of the extinct radioactivities I^{129} and Pd^{107}, then we may deduce the approximate time intervals between the cessation of nucleo-synthesis in the interstellar gas, and the formation and cooling of the meteorite parent bodies to the point where further fractionation of the elements involved ceases. This time interval is about 1.5×10^8 years for the xenon anomaly and about 2 to 4×10^7 years for the silver anomaly. The atmosphere of the Earth contains xenon that has been subjected to additional enrichment processes; the Earth apparently did not start retaining any xenon until about 10^8 years following the retention of xenon by the meteorites.

One of the principal scientific objectives of the space program is the exploration of the Moon and planets, and one of the principal goals of this exploration will be a determination of the histories of these bodies. Many highly refined techniques will undoubtedly be used in the exploration program; among these will be the radioactive dating of certain events in the early history of the Moon and planets. This can be done by studying anomalies in the isotopic composition of certain elements produced by the decay of extinct radioactivities. It is the purpose of this discussion to show how such anomalies are already giving information about the early history of the solar system from a study of meteorites.

The first evidence for an effect due to an extinct radioactivity was obtained by Reynolds[1] at the end of 1959. He found that xenon gas extracted from certain stone meteorites had large excess abundances of the isotope Xe^{129}. The excess abundance of this isotope is attributed to the decay of the extinct radioactivity

---

* Institute for Space Studies, NASA Goddard Space Flight Center, New York, N. Y.

of I^{129}, which has a half-life of 1.7×10^7 years. Measurable amounts of this isotope must have been present at the time the meteorite parent body was formed, but they would have decayed shortly thereafter.

Actually, this anomaly in the isotope Xe^{129} is only one of many that have since been discovered in xenon extracted from various sources. Perhaps we should regard primordial xenon to be that which is extracted from bodies having the greatest amount of xenon per gram of material present. On this basis, the meteorites in general have much more xenon per gram than does the Earth in its atmosphere, and certain of the stone meteorites called carbonaceous chondrites contain much more xenon than do most of the ordinary stone meteorites. Therefore, we might expect the xenon in the carbonaceous chondrites to resemble primordial xenon most closely.

The isotopic composition of xenon extracted from three carbonaceous chondrites (all of the measurements due to Reynolds) is shown in Fig. 1. The histogram at the base of the figure shows the relative abundances of the various isotopes of xenon in the Murray carbonaceous chondrite. At the top of the figure are shown the differences between the composition of xenon from two other carbonaceous chondrites and from Murray (regarded as a standard), with the

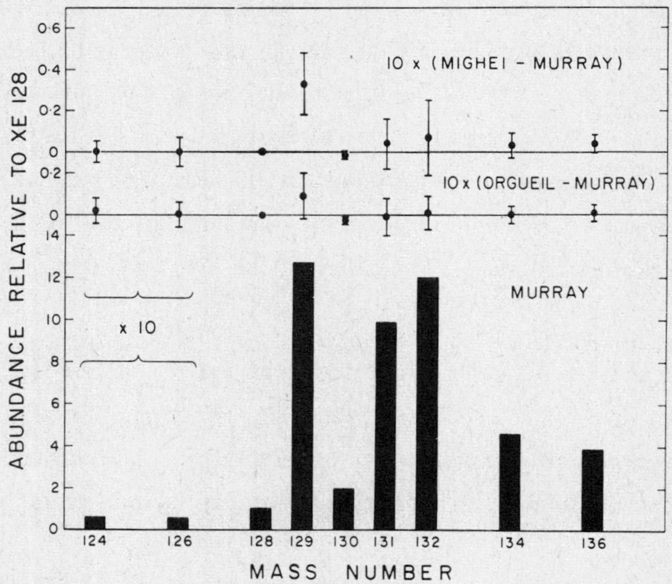

Fig. 1 Comparison of the Isotopic Compositions of Xenon Extracted from Three Carbonaceous Chondrites

ordinate expanded by a factor of ten. It may be seen that there is no essential difference between the composition of xenon in these meteorites, with the possible exception of slight differences in Xe^{129}. This constancy supports our use of the Murray composition as a primordial standard.

In Fig. 2 is shown the difference between the isotopic composition of xenon extracted from a more ordinary chondritic stone meteorite, Richardton, and from the Murray standard, again shown at the bottom of the figure. In this

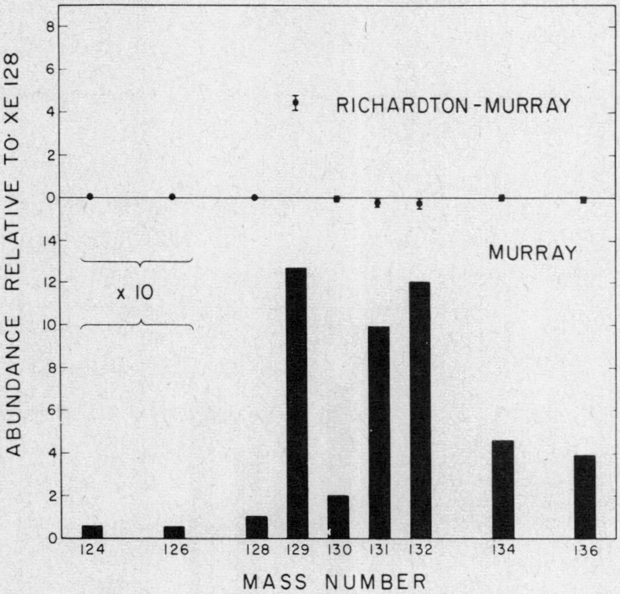

Fig. 2 Comparison of the Isotopic Compositions of Xenon in the Richardton and Murray Chondrites

case the ordinate has not been expanded by a factor of ten. There is no essential difference between the isotopic composition of any of the isotopes of xenon, with the exception of Xe^{129}, which is seen to be greatly overabundant in the Richardton meteorite. Many other examples are now known in which a similar overabundance of Xe^{129} occurs. As indicated above, we attribute this overabundance to the effects of the incorporation of some I^{129} in the meteorite parent body. Because xenon is a volatile element, there is a large fractionation between the abundances of iodine and xenon, and therefore only a small amount of I^{129} need be incorporated in the meteorite parent body in order to show up as an isotopic anomaly in the xenon daughter element.

When we turn to the body having the least fractional weight of xenon, the Earth (including its atmosphere), we find an entirely different situation. The comparison between the composition of xenon in the atmosphere and that of the Murray standard is shown in Fig. 3. It may be seen that Xe^{129} is overabundant in the Earth's atmosphere, but not by an amount that is large in comparison with similar overabundances of the isotopes Xe^{131}, Xe^{132}, Xe^{134}, and Xe^{136}. It may also be seen that there are smaller but nevertheless real abundance differences in the lighter isotopes Xe^{124}, Xe^{126}, and Xe^{130}. This comparison has been made by normalizing on the isotope Xe^{128}; it is likely that there is an anomaly in Xe^{128} comparable to that in Xe^{130}.

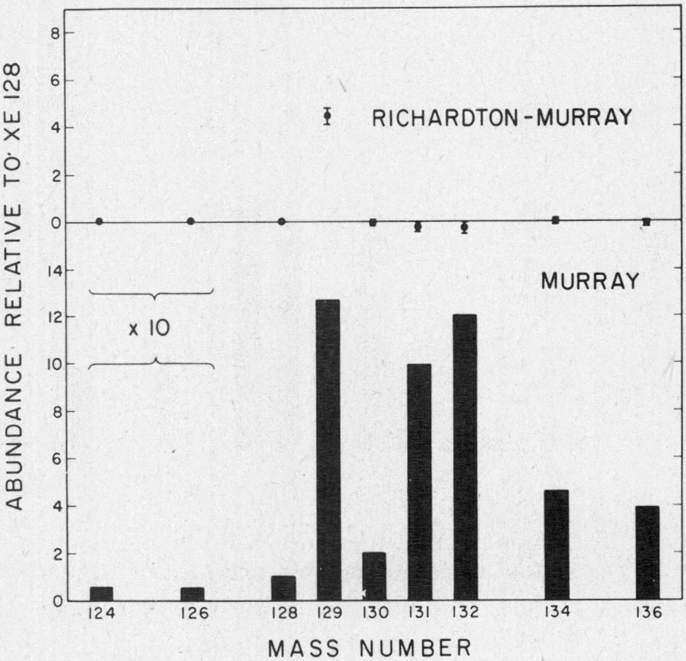

Fig. 3 Comparison of the Isotopic Compositions of Xenon in the Terrestrial Atmosphere and in the Murray Carbonaceous Chondrite

The isotopes shown in Fig. 3 with large overabundances, Xe^{129}, Xe^{131}, Xe^{132}, Xe^{134}, and Xe^{136}, are precisely those isotopes of xenon that can be formed from the beta decay of fission products. This suggests that some of the xenon in the Earth's atmosphere is fissiogenic in origin. However, the amount of this fissiogenic xenon in the Earth's atmosphere is much greater than can be expected to be produced by the spontaneous fission of uranium and thorium isotopes or by the neutron-induced fission of U^{235}. Also, the mass-yield curve of these types of fission do not agree with the overabundance pattern shown in Fig. 3. Kuroda[2] has suggested that the extinct radioactivity Pu^{244} has been responsible for the production of these fission products. Pu^{244} has a half-life of 7.6×10^{7} years, and we must suppose that some of it was present in the Earth at the time the Earth was formed. However, since Pu^{244} should also be present in the meteorite parent bodies at the time they were formed, and because the xenon extracted from meteorites does not show the presence of these fission products, we must ask why the Earth does not contain a very much larger overabundance of Xe^{129} than is found. Because there is so much less xenon per gram of Earth than per gram of meteorite, we might expect the majority of the xenon in the Earth's atmosphere to be Xe^{129}. Since this is not so, it seems necessary to conclude that the Earth did not retain xenon in its atmosphere for some considerable time after xenon was retained in the meteorite parent bodies.

It does not seem possible to explain the small anomalies effecting the isotopes Xe^{124}, Xe^{126}, Xe^{128}, and Xe^{130} as resulting from the effects of extinct radioactivities. This pattern of anomaly can, however, be produced if xenon is subjected to a large flux of neutrons in the presence of comparable amounts of neighboring elements. This exposure may have occurred during the early history of the Sun, when it was contracting from the interstellar medium, and when it burned up its deuterium in thermonuclear reactions. Some of the deuterium thermonuclear reactions produce neutrons, and the capture of these can produce small changes in the isotopic composition of the heavier elements. I estimate that about thirty percent of the atmospheric xenon was once in the Sun where it was subjected to a large neutron flux presumably accompanying deuterium burning. Perhaps this xenon has been subsequently captured into the atmosphere from the solar wind.

Effects due to one other extinct radioactivity have been discovered by Murthy.[3] He examined the isotopic composition of silver extracted from various phases of iron meteorites. Silver contains two stable isotopes of comparable abundance, Ag^{107} and Ag^{109}. Murthy discovered that the Ag^{107} in silver meteorites was enriched relative to Ag^{109} by amounts of the order of 2 to 4 percent. It happens that Ag^{107} is the daughter product produced by the decay of

Pd^{107} which has a half-life of 7×10^6 years. It also happens that the abundance of silver in iron meteorites is very low, while that of palladium is relatively high. Hence, if the meteorite parent body is formed with appreciable amounts of Pd^{107}, this can produce an enrichment of Ag^{107} under conditions where silver is greatly fractionated relative to palladium, as in iron meteorites.

These radioactivities give us clues about the early chronology of the solar system. However, in order to make a proper interpretation of these extinct readioactivity effects, we must also learn a great deal about the history of the galaxy preceding the formation of the solar system. Current determinations of the ages of old galactic clusters of stars indicate that the galaxy is probably about three times as old as the solar system, which itself is about 4.5×10^9 years old. The oldest stars contain a much smaller ratio of heavy elements to hydrogen than do the Sun and the newly formed stars. Hence, we believe that the galaxy was once nearly pure hydrogen gas. As the galaxy developed, stars were formed, and the more massive ones evolved quite quickly. The nuclear reaction products that were produced during this evolution were then thrown forth into the interstellar medium by supernova explosions, thus gradually building up a larger abundance of the heavier elements in this medium. Because this is a gradual process, radioactivities with half-lives of a few millions, or tens of millions, of years can be expected to be present in the interstellar medium to a small extent at any given time. It is the task of astrophysicists and nuclear physicists to determine what the appropriate model of element formation during galactic history has been, and therefore to determine the level of abundances of these radioactivities at any time in the interstellar medium.

At some point, the solar system will start to condense out of the interstellar medium, probably as part of the formation of a cluster or association of stars. Once this contraction has started, we may expect that the solar system material will be isolated from the injection of any further radioactivities due to supernova explosions in nearby space. Hence, the existing level of the extinct radioactivities starts dying away. If we can estimate the abundance level in the interstellar medium before a contraction started, and if we can also measure the abundance level of the extinct radioactivities in various chemical systems at the time chemical fractionation ceases, then we can determine the time at which the final chemical fractionations occur relative to the start of the contraction. In this way, we should be able to determine from the anomalous silver content of the iron meteorites the time from the start of the contraction through the collection, thermal insulation, and heating of the meteorite parent bodies, and their cooling to the point where iron solidifies. From the amounts of anomalous Xe^{129} trapped in the stone meteorites, we can determine the time interval from the start of the contraction through the heating and cooling of the meteorite parent bodies to the point where xenon can no longer diffuse through them, which would be at a temperature of about 200°K. We can similarly date the retention of xenon in the Earth's atmosphere.

I have made these time-interval estimates using the model of galactic history shown in Fig. 4. The philosophy behind the construction of this history is as follows. If we expect that the galaxy was once composed entirely of gas, then star formation would occur much more rapidly than it does at present.

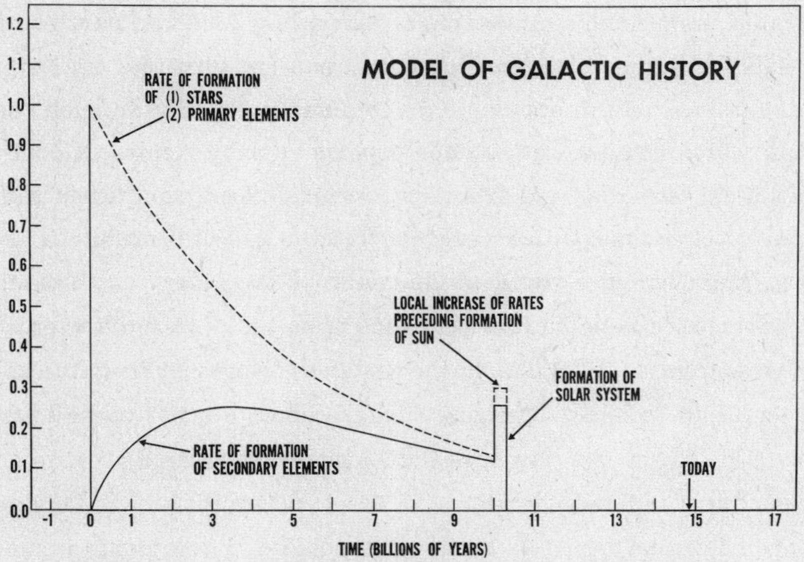

Fig. 4 The Model of Galactic History used for Calculating the Abundances of Radioactive Nuclei in the Interstellar Medium at the Start of the Condensation that Forms the Solar System

I have therefore assumed that star formation falls off exponentially with time from the beginning of galactic history. This beginning of galactic history is taken to be about 1.5 x 10^{10} years ago. A study of the nuclear reactions responsible for element formation leads to a classification of the elements into two groups. First, we have the elements that can be formed directly by a sequence of nuclear reactions starting from a pure hydrogen gas. These include all the abundant lighter and medium-weight elements through iron. We may call these the primary elements. Secondly, there are the elements that can only be built by using as raw material the lighter and medium-weight elements that are produced as primary products. We may call these the secondary elements, and they include all the extinct radioactivities. It is assumed in the model shown in Fig. 4 that the primary elements are formed at the same rate as star formation, and the secondary elements are formed at this rate multiplied by the amount of build-up of the primary elements toward their saturation level. The rate of fall-off of the primary element production was determined in Fig. 4 from the relative abundances of Th^{232} and U^{238}, for the assumed galactic age.

Many different kinds of observations show that star formation is strongly concentrated toward the spiral arms of our galaxy, and, in particular, that it is associated with the very dense clouds of gas and dust that can be found in such spiral arms. Stars appear to be continually forming in such regions, and those that evolve rapidly and produce supernova explosions will contribute fresh radioactivities, especially to this region where star formation is going on. Furthermore, the observations suggest that the gas in the plane of the galaxy is streaming radially outwards from the center of the galaxy at a rate that amounts to some 50 km/sec at about 3 kiloparsecs from the center of the galaxy, and at a rate of the order of 8 km/sec at the distance of the Sun from the center of the galaxy, or 8 kiloparsecs. The gas will take about 3×10^8 years to traverse this distance. These observations thus suggest that the gas that is going to form fresh stars is locally enriched in fresh radioactivities. This accounts for the feature shown in Fig. 4 as the local production of elements immediately prior to the formation of the solar system. The length of time assumed for this local production is 3×10^8 years, and the increased rate of the production was determined from the ratio of U^{235} to U^{238} in the solar system. We assume that, following this increased local rate of element production, condensation started and no further contribution of fresh radioactivities took place to the material contracting to form the solar system. There is then an unknown but, as we shall see, fairly short interval of time required for the condensing material to get into the form of solid bodies circling about the Sun. We know from the variations in the relative abundances of the lead isotopes in the Earth and in the meteorites that 4.5×10^9 years has elapsed since uranium was chemically isolated from lead in such bodies.

The chronology of the events that can thus be deduced for the early history of the solar system is shown in Fig. 5. The time intervals shown in this figure have been calculated for the abundances of fresh radioactivity in the interstellar medium corresponding to the model of galactic history shown in Fig. 4; the ends of these intervals are deduced from the levels to which the radioactivity has fallen at the time that chemical isolation of the appropriate system has occurred.

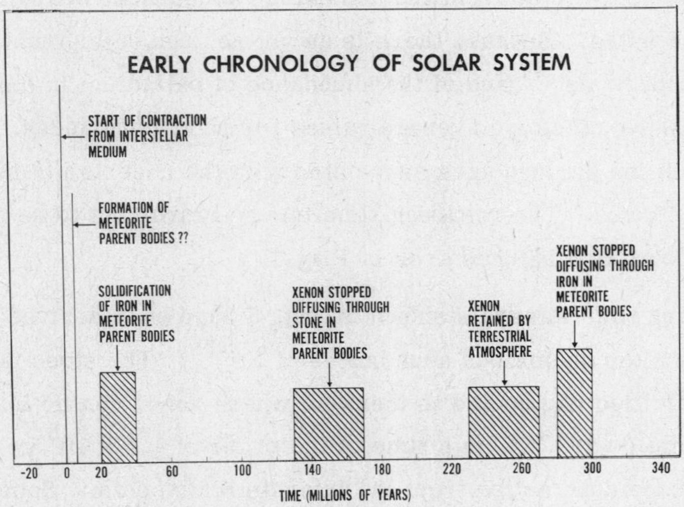

Fig. 5 The Early Chronology of the Solar System as Deduced from the Abundances of the Decay Products of Extinct Radioactivities in Various Systems

The first of the intervals shown in Fig. 5 has not been deduced from any existing evidence for an extinct radioactivity, but is a possibility corresponding to one hypothesis concerning the heating of the meteorite parent bodies. It is rather difficult to see where the heat necessary to melt iron in such meteorite parent bodies can arise, since the evidence indicates that the bodies cooled quickly and were therefore rather small. The gravitational potential energy is much less than that required to provide necessary heating. Fish, Goles, and Anders[4] have suggested that the extinct radioactivity Al^{26} was responsible for this heating. If that is the case, then we can set a lower limit to the amount of Al^{26} that must have been present at the time the meteorite parent bodies were collected together and became thermally insulated. The resulting time interval from the start of the contraction to the thermal insulation of the meteorite parent bodies turns out to be less than 6×10^6 years. This is a rather short time for the formation of the solar system by contraction out of the interstellar medium, but it seems by no means excluded. However, it must be emphasized that we are not restricted to this time interval if another method can be found for heating the meteorite parent bodies.

The next interval determined by the extinct radioactivities is that associated with the cooling of the meteorite parent bodies to the point where iron solidifies. This is determined from the anomalous abundances of the isotope Ag^{107} in iron meteorites. Because there is as yet no good determination of the anomalous abundance of Ag^{107} and of the abundance of palladium in the same meteorite samples, we must use average values for such abundances. This leads to a spread in the derived ages associated with the uncertainty in the silver to palladium ratio. The resulting time interval turns out to be 2 to 4×10^7 years. This is shown by a hatched area in Fig. 5.

The remaining time intervals shown in Fig. 5 have been derived from the evidence concerning the anomalous abundances of Xe^{129}. The stone meteorites appear to have solidified and cooled to the point where xenon can no longer diffuse through them ($\sim 200\,°K$) in a time of the order of 1.5×10^8 years following the start of the contraction from the interstellar medium. Somewhat different results are obtained for different samples, as might be expected, since different samples may come from different places in the meteorite parent bodies which have cooled to a sufficiently low temperature at different times. The resulting range of uncertainties is indicated in Fig. 5.

We have already discussed the evidence indicating that the Earth did not retain xenon in its atmosphere for some considerable time after the meteorite parent bodies cooled to this low temperature. From the overabundance of Xe^{129} in the atmosphere, it can be estimated that the time interval associated with the retention of xenon in the atmosphere seems to be about 10^8 years longer than that associated with the retention of xenon in the stone meteorites. In principle, we can obtain a similar estimate of the time at which the Earth started retaining its xenon from the abundances of the xenon fission products from the decay of Pu^{244}. Because of the various uncertainties involved, this time is not very well determined but is quite consistent with the interval shown in Fig. 5.

Reynolds, Merrihue, and Pepin[5] found primordial xenon in the iron meteorite Sardis. This has an anomalous abundance of Xe^{129}. The evidence shows that xenon was able to diffuse out of this iron component of the meteorite parent bodies until a time slightly greater than that at which the Earth started retaining xenon in its atmosphere. This time is also shown in Fig. 5. It is perhaps not too surprising that this turns out to be considerably greater than that at which the stone parts of the meteorite parent bodies cooled sufficiently to retain xenon, since the iron component of the meteorite parent bodies has obviously been heated much more than the stone component.

These results show the great power of the radioactive dating method in casting light on the sequence of events in the early history of the solar system. It will be very interesting to determine the isotopic composition of xenon in the atmospheres of the planets and of the Moon, if the Moon contains any xenon in its atmosphere and, hence, to determine the dates at which such atmospheres have started to retain some xenon. We shall also be interested in learning whether the xenon in the rocks of the Moon, the planets, and the asteroids contains xenon of abnormal composition, which would allow the thermal history of such bodies to be determined. This indicates one very desirable direction in which future experiments in space research can be designed. At the same time, we have seen how much the proper interpretation of these results depends upon a knowledge of the preceding history of the galaxy. This is another indication of the importance of an interdisciplinary approach in space science, and shows the need for vigorous investigations in many branches of astrophysics at the same time as the exploration of the Moon and planets is being carried out.

The analyses and results summarized in this paper are being published in more detail elsewhere.

REFERENCES

1. J. H. Reynolds, "Determination of the Age of the Elements," Phys. Rev. Letters, vol. 4, No. 1, Jan 1, 1960, pp. 8-10

2. P. K. Kuroda, "Nuclear Fission in the Early History of the Earth," Nature, vol. 187, No. 4731, July 2, 1960, pp. 36-38

3. V. R. Murthy, "Isotopic Composition of Silver in an Iron Meteorite," Phys. Rev. Letters, vol. 5, No. 12, Dec 15, 1960, pp. 539-541

4. R. A. Fish, G. G. Goles, and E. Anders, "The Record in the Meteorites. III. Or the Development of Meteorites in Asteroidal Bodies," Astrophys. J., vol. 132, No. 1, July, 1960, pp. 243-258.

5. J. H. Reynolds, C. M. Merrihue, and R. O. Pepin, "Extinct Radioactivity and Primordial Rare Gases in Iron Meteorites," Bull. Am. Phys. Soc., vol. 7, No. 1, Jan 1962, p. 35

SUNLIGHT PRESSURE PERTURBATIONS ON SATELLITE ORBITS

Irwin I. Shapiro**

This paper develops semiquantitatively the theory of sunlight pressure perturbations on orbits of satellites with large area-to-mass ratios. Particular emphasis is given to resonant and to stable orbits. Deductions from the effects of sunlight pressure on the Echo balloon are discussed, as are some implications for passive satellite communication systems. Finally, this paper considers the influences of sunlight pressure on the dust belt that may be orbiting the earth.

INTRODUCTION

Light pressure has a long and venerable history in astronomy, having been recognized as an important consideration in celestial mechanics as far back as 1746, when Euler first suggested that the pressure exerted by sunlight waves was responsible for comets' tails.[1] Since this pressure would affect smaller particles more than larger ones, he reasoned that the smaller ones would form a tail. However, Euler's demonstration that light waves could exert pressure was not very convincing, and his suggestion lay dormant and unappreciated for almost 160 years. It wasn't until early in the 20th century* that solar radiation was again considered seriously in the celestial mechanical problems of the solar system. The dominant personality in this work was that of Poynting.[3] He deduced the remarkable fact that sunlight reflected (or absorbed and isotropically reradiated) from spherical orbiting objects would exert a drag force; and, therefore, that the small dust particles in the solar system would be swept into the sun --emptying the space between the planets. Poynting's theory was both brilliant and startling; however, his quantitative calculations

** Staff member, M. I. T. Lincoln Laboratory, Lexington, Massachusetts, operated with support from the United States Air Force.

* In 1872 Maxwell first showed that, as a consequence of his electromagnetic theory, light waves would exert pressure;[2] but the experimental verification (by Lebedev in Russia and by Nichols and Hull in the United States) was not achieved until the beginning of the 20th century.

were in error. In fact, a great controversy raged for the next several decades, with scientists such as Larmor,[4] Plummer,[5] and Poynting all obtaining different (numerical) results. Their work had, however, one thing in common: All of them deduced incorrect coefficients for the drag force. In fact, each at different times suggested two different results, yielding five distinct incorrect sets of coefficients in all. The controversy was not finally settled until 1937 when the late H. P. Robertson formulated the problem correctly from a relativistic viewpoint.[6]

Except for occasional articles on solar sailing, and one by Wyatt and Whipple[7] extending Robertson's work, not much attention was devoted to the orbital effects of sunlight pressure. Indeed, in the early days of the artificial satellite age it was not generally considered to be an important perturbation. However, in late 1958, while investigating the orbital properties of the West Force dipoles,[8] Parkinson, Jones, and the present author discovered that these very sunlight pressure perturbations were perhaps the most important of all for such satellites. We therefore developed a detailed quantitative theory of the effects.[9]

Since their magnitude is proportional to the area-to-mass ratio of the satellite, it was obvious that large, lightweight satellites, like the then-contemplated Echo balloon, would also be severely perturbed - actually "blown about" - by sunlight pressure. In particular, our findings indicated that sunlight pressure perturbations would have to be considered in estimating air densities from the changes in period of Echo's orbit. Reports of our work stimulated further interest in these rather startling perturbations and various people began to investigate them, including Levin,[10] Geyling,[11] Musen,[12] Kozai,[13] Wyatt,[14] and Bryant.[15]

During that period (1959) there were no satellites aloft with a sufficiently high area-to-mass ratio for sunlight pressure to have produced noticeable perturbations. We tried to discern the effect in the eccentricity variations of the Vanguard I orbit, but the data available to us were not precise enough to prove its existence. On the other hand, Musen, Bryant and Bailie,[16] working with more accurate data, were able to show that inclusion of the sunlight pressure effect reduced the discrepancy between the theoretical and observed variations of the perigee height of the Vanguard orbit. A truly quantitative verification of the theory, however, had to await the launching of the Echo I satellite.

SUNLIGHT PRESSURE PERTURBATIONS

To understand qualitatively the effects of sunlight pressure, consider the sun to lie in the orbit plane (Fig. 1). Near A, the satellite's velocity is increased by sunlight pressure, tending to produce perigee at this point, whereas near B, the satellite's motion is retarded, tending to produce apogee there. A similar analysis can be made for all other points along the orbit. On balance, we find a net displacement of the orbit perpendicular to the direction of the earth-sun line. For a positively charged particle moving in a similar (circular) orbit in a uniform magnetic field, we find that a perturbing electric field in its orbit plane produces the same type of perturbation as discussed above. The analogy breaks down, however, when the perturbing field vectors are perpendicular to the orbit plane. This latter case produces essentially no perturbations of the satellite orbit by sunlight pressure, but causes an acceleration of the charged particle along the magnetic field line in an ever-increasing spiral.

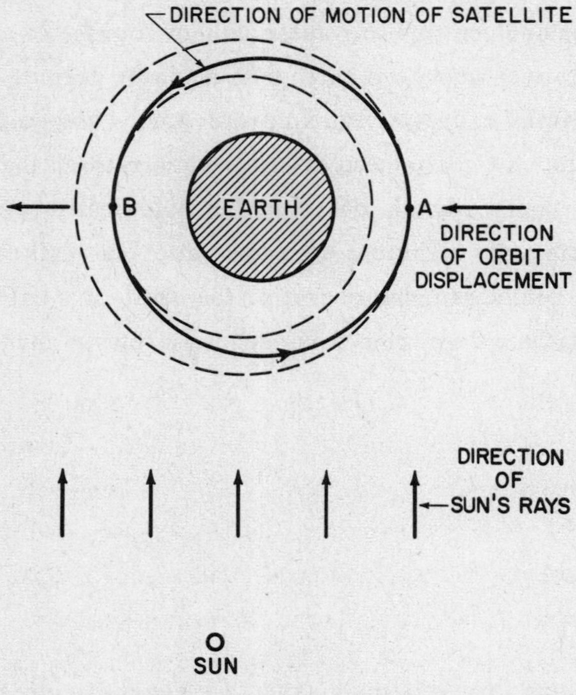

Fig. 1 Displacement of a Satellite Orbit by Sunlight Pressure

We assumed in our illustration that the satellite was spherical*; if it is not, the perturbations could be quite different. For example, the orbit of a specularly reflecting flat plate, whose normal is almost perpendicular to the direction of sunlight, will be displaced along the direction of the earth-sun line, not perpendicular to it. Further, a small change in the orientation of this plate could lead to a 180° change in the direction of displacement of its orbit.

The acceleration caused by sunlight pressure is proportional to KA/M, where A is the cross-sectional area of the satellite (as projected on a plane perpendicular to the earth-sun line), and M is its mass. The reflection function, K, is generally a tensor quantity, reflecting the fact that the force exerted on the satellite by the radiation is not always directed parallel to the impinging solar rays. The magnitude of K can vary with time (because of changing surface conditions and changing orientation of the satellite), but is constrained to lie between zero and two. For example, $K = 0$ for a transparent object; $K = 2$ for a suitably oriented specularly reflecting plate; and $K = 1$ for an absorbing object or a specularly reflecting sphere.

The orbit is not displaced continuously in the same direction by sunlight pressure because the second harmonic of the earth's gravitational field causes the perigee position to rotate; and, of course, in addition, the direction of sunlight revolves about the earth with a yearly period. Hence, in general, perigee height will oscillate with a certain period and amplitude which depend on the orbit and on the effective area-to-mass ratio of the satellite. For certain initial orbits, perigee height decreases monotonically, provided that the satellite has a sufficiently high area-to-mass ratio: We call these "resonant orbits". When perigee height remains constant, the orbits are called stable. In the following sections we examine these special orbits in some detail.

---

* Or, equivalently for this discussion, we could assume that the satellite was a perfect absorber and reradiated isotropically.

Resonant Orbits

In describing the characteristics of sunlight pressure resonances, let us restrict the discussion initially to equatorial orbits. The main effect of this pressure is to displace orbits in a direction perpendicular to that of the in-plane component of the earth-sun line. Hence, if the major axis were maintained perpendicular to this component, perigee height would decrease monotonically (Fig. 2a)*. This result can be obtained (to first order) if the second harmonic of the earth's gravitational field causes perigee to rotate at the same rate and in the same direction as the in-plane component of the earth-sun line, thereby maintaining the desired relative orientation. In this orientation, sunlight pressure causes the maximum change in eccentricity, but no change in the argument of perigee. Unfortunately (from this point of view), the motion of perigee caused by the second harmonic ($\dot{\Omega}_{2h} + \dot{\omega}_{2h}$) is a function of the orbit parameters and does not remain constant as the orbit is displaced. An initial equality of angular rates can therefore not be maintained. To show this

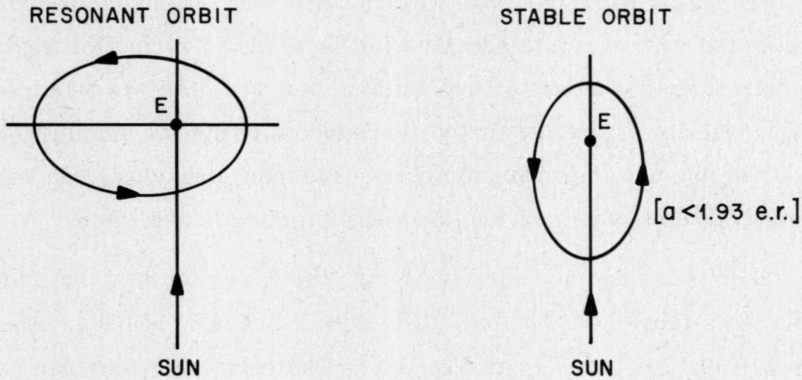

Fig. 2 Relative Orientation of Orbit Major Axis and Earth-Sun Line for Resonant and Stable Equatorial Orbits

---

* If the orbit were initially rotated in its plane by 180°, perigee height would first increase (until circularity was reached) and then decrease.

quantitatively, we note that the resonance conditions (i.e., the equality of angular rates) for an equatorial orbit is

$$\dot{\Omega}_{2h} + \dot{\omega}_{2h} \approx 5 \frac{(5 \cos^2 i - 2 \cos i - 1)}{a^{7/2} (1 - e^2)^2} \text{ deg/day}$$

$$= \dot{\beta} \approx 1 \text{ deg/day} \tag{1}$$

where $\dot{\beta}$ is the angular rate of rotation of the earth about the sun, and where the orbital elements have their usual definitions with $\underline{a}$ being expressed in earth radii. We see that, as the eccentricity increases, so does $\dot{\Omega}_{2h} + \dot{\omega}_{2h}$ and, hence, the angle between the earth-sun line and the orbit major axis increases from its initial 90° value.[+] As long as the total increase remains less than 90°, perigee height will continue to decrease. Therefore, for sufficiently high values of KA/M, perigee height will decrease monotonically until the orbit intersects the earth's surface. On the other hand, for lower values of KA/M, the orbit behavior throughout the satellite lifetime will not be resonant; that is, perigee height will oscillate and air drag is needed to bring the satellite down from orbit. In other words, a rather long period of time is required for the (small) inequality in angular rates to produce a change in angular positions of the order of 90° and therefore to destroy resonant behavior. Only for satellites with relatively low values of KA/M is sufficient time available.

For polar and near-polar orbits, the description of sunlight pressure resonances is more complicated. The added difficulty is caused by the importance of the change in inclination angle.* The relevant resonance condition is given approximately by

$$\dot{\Omega}_{2h} - \dot{\omega}_{2h} = \dot{\beta} \tag{2}$$

[+] Even if air drag is neglected, $\underline{a}$ will decrease (due to the effect of the earth's shadow) and will therefore also contribute to the increase in $\dot{\Omega}_{2h} + \dot{\omega}_{2h}$.

* The change in inclination has a negligible effect for equatorial (and near-equatorial) orbits because in these latter cases the dependence of the resonance condition on inclination is only of second order in $\underline{i}$.

i.e., by

$$a^{7/2} (1 - e^2)^2 \approx 5(1 - 2 \cos i - 5 \cos^2 i) \tag{3}$$

where a is again expressed in earth radii. Thus, for $\Delta i = i - \pi/2 \ll 1$, we have

$$a^{7/2} (1 - e^2)^2 \approx 5(1 + 2\Delta i) \tag{4}$$

as an approximate resonance condition. The important change in i is due to radiation pressure. Neglecting the effects of the earth's shadow, we find to first order that the change in i per revolution is

$$\delta i \approx - \frac{3\pi\lambda e}{(1 - e^2)^{1/2}} W \cos \omega \tag{5}$$

where λ is the ratio of the magnitude of the acceleration due to sunlight pressure to that due to the central force field of the earth at a distance a. W is the cosine of the angle between the radiation pressure vector and the normal to the orbit plane. (The positive direction of this normal is that of $\underline{r} \times \underline{v}$, where $\underline{r}$ is the position vector and $\underline{v}$ the velocity vector of the satellite.) We see that the magnitude of the change in i increases approximately linearly with e, but that the sign of the change depends on the relative signs of cos ω and W. The precise relationship between the signs of these latter two quantities is rather complicated, especially over long periods of time. As an illustration, we explain this relationship for an initially circular polar orbit satisfying the resonance conditions, Eq. (2). When the sun passes the longitude of the ascending node, the argument of perigee lies approximately within ± 23 1/2° [+] of the value $\omega = 90°$. (Remember that if the resonance condition is satisfied for an initially circular orbit, with the sun in the orbit plane, perigee appears 90° from the earth-sun line at the position at which the satellite is moving in the

---

[+] This possible variation represents the effect of the inclination of the equator to the ecliptic.

same direction as the sunlight.) From its definition, we see that W remains positive as the sun passes from the longitude of the ascending node to that of the descending node. During that time, ω decreases (through zero) from within $\pm$ 23 1/2° of ω = 90° to within approximately the same angular distance from ω = 270°. For this half-year, then, the product W cos ω is predominantly positive. The same result is easily seen to obtain for the other half-year, when both W and cos ω change sign. Consequently, while e increases from zero, i and Δi decrease secularly. The decrease in e on the left side of Eq. (4) is therefore partially compensated for by the decrease in Δi on the right side. As a numerical example [with KA/M = 50 cm^2/g and an initially circular polar orbit], we note that when the eccentricity reaches 0.25, $(1-e^2)^2 \approx 0.88$ while $1+2\,\Delta i \approx 0.90$; and Eq. (4) seems to remain nearly satisfied. However, by considering the decrease in a caused by radiation pressure, we find that the left and right sides of Eq. (4) differ by about 5% when e = 0.25. The qualitative implications of these results for resonant behavior are the same as those discussed above for equatorial orbits.*

It is clear that resonant orbits can be useful in limiting the lifetime of balloon satellites, and for investigating air densities over a wide range of perigee altitudes in a relatively short time.

Stable Orbits

Let us consider a noncircular orbit whose major axis coincides with the earth-sun line (Fig. 2b). Radiation pressure will then cause no change in either eccentricity or perigee height, but rather a maximum change in the position of perigee. Thus, by choosing an orbit for which the combined effect of sunlight pressure and the second harmonic causes the position of perigee to rotate at the same rate and in the same direction as the earth-sun line, we would maintain this coincidence. The result is a stable orbit.

On the basis of a rather complicated analysis, we can show that for each satellite there exist between zero and four such stable equatorial orbits

---

* For a complementary discussion of polar orbit resonances, see Ref. 17.

for a given initial mean altitude. For direct stable orbits, we find that the eccentricity satisfies

$$e_s \approx - \lambda(1 - e_s^2)^{1/2} \cos^2 (\Psi/2) \frac{3(\pi - \eta) + \sin\eta \cos\eta}{|\dot{\Omega}_{2h} + \dot{\omega}_{2h} - \dot{\beta}|} \tag{6}$$

where, as mentioned above,

$$\lambda = (KA/M)(I/c)(a^2/\mu), \tag{7}$$

where Ψ is the inclination of the equator to the ecliptic, and where 2η is the angular extent of the shadow region. [$I/c \approx 4.65 \times 10^{-5}$ dynes/cm^2 is the value of sunlight pressure found at a distance of one astronomical unit from the sun.] Depending on the values of a and KA/M* there can exist zero, one, two, or three solutions to Eq. (6) satisfying $0 < e_s < 1$. For retrograde stable equatorial orbits, the eccentricity satisfies the same equation, but with $\dot{\omega}_{2h}$ replaced by $-\dot{\omega}_{2h}$. There is at most one solution to this latter equation; in total, we have between zero and four stable orbits for every set of values of a and KA/M.

In Figs. 3 and 4, we show the relationships between e_s and KA/M for various values of a. There is an upper limit on e_s, for each value of a, caused by the physical constraint that perigee height must exceed the radius of the earth. In fact, for these stable orbits to be physically meaningful, we should restrict perigee height to lie above about 1.25 e.r., thus minimizing the effects of the (neglected) air drag. The solid portions of the curves in Figs. 3 and 4 represent perigee heights greater than 1.25 e.r., while the dotted portions represent perigee heights between 1 e.r. and 1.25 e.r. For each $a \leq 1.93$ e.r. there is at most one direct stable orbit. By confining ourselves to stable orbits with perigee heights greater than 1.25 e.r., we find that there are at most three direct stable orbits for each a satisfying

---

* We assume for simplicity that for each satellite, KA/M is a scalar constant.

1.93 e.r. $\leq a \leq$ 2.20 e.r., at most two for each a satisfying 2.20 e.r. $\leq a \leq$ 3.5 e.r., and at most one for $a \geq 3.5$ e.r. There are no direct stable orbits, for any value of a, if KA/M exceeds 2200 cm^2/g.+

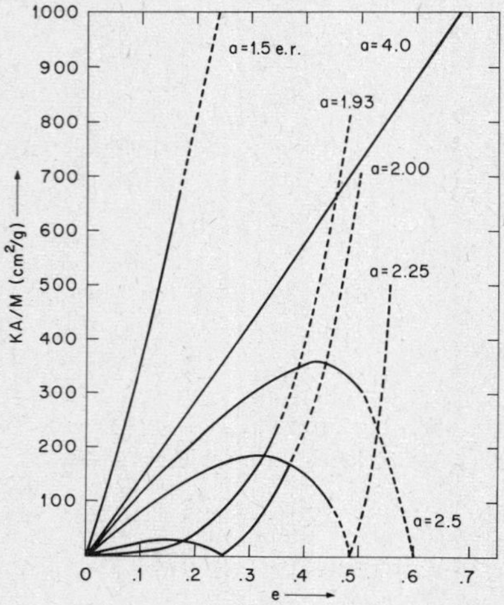

Fig. 3 Characteristics of Direct Stable Equatorial Orbits

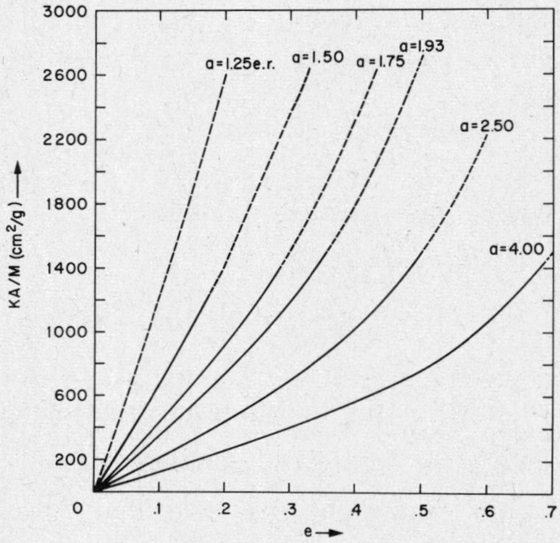

Fig. 4 Characteristics of Retrograde Stable Equatorial Orbits

For direct stable orbits, perigee is always in shadow if $a \leq 1.93$ e.r., whereas for $a > 1.93$ e.r., the perigee is in sunlight except for those stable orbits corresponding to the third solution; i.e., the one with the largest eccentricity when three exist (Fig. 3). The perigees of retrograde stable orbits are all in sunlight.

Even when neglecting air drag and other minor perturbations, the above described orbits are still not rigorously stable. We have neglected small periodic perturbations caused by sunlight pressure; these are proportional to $\sin^2 (\psi/2)$. To understand qualitatively a physical source for such oscillations, note that because the sun moves in the ecliptic and not in the earth's equatorial plane, the projection of the earth-sun line on this latter plane moves at a varying rate. A relatively small semiannual oscillation is thereby produced in the eccentricities of these (almost) stable orbits, since

---

+ This limiting condition, as a function of a, can be derived easily from Eq. (6). A similar analysis can be carried out for retrograde orbits.

the component of the earth-sun line in the orbit plane oscillates about the position of the orbit major axis with this period.

There are no polar orbits which are stable in the above sense. However, there are classes of near-polar orbits for which the amplitude of oscillation of perigee height is reasonably small, but still larger than for the corresponding equatorial or near-equatorial orbits. Their properties are quite complicated and will not be discussed further.

Near-Stable Orbits

Since it is quite difficult to attain precisely an orbit with a predetermined eccentricity and perigee position, we might ask about the behavior of orbits near those that are stable.* The center of such a near-stable orbit rotates about the center of the corresponding stable orbit (i.e., one for the same value of KA/M and with the same mean altitude), as shown in Fig. 5. Thus, the center of the stable orbit rotates about the earth with angular velocity $\dot{\beta}$ whereas the center of the near-stable orbit rotates about that of the stable orbit with an angular velocity $(\dot{\Omega}_{2h} + \dot{\omega}_{2h})_s - \dot{\beta}$, which is equal to the difference of the rate of motion of the earth-sun line and that of the perigee of the stable orbit which is caused by the second harmonic. The maximum variation of the perigee height for near-stable orbits is $2ae_s$ if $\rho \geq ae_s$ and 2ρ if $\rho \leq ae_s$ (Fig. 5).

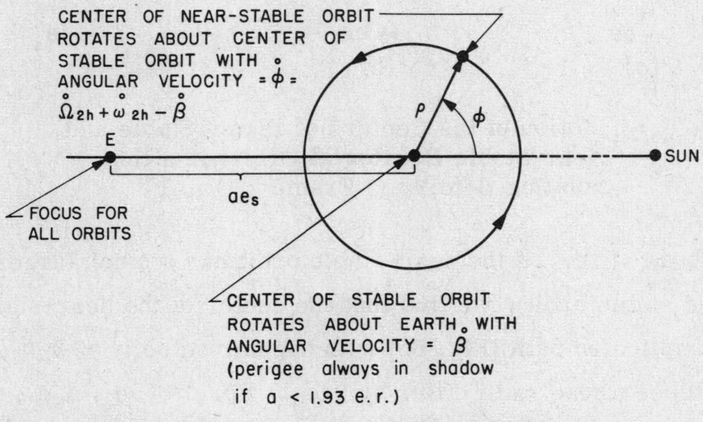

Fig. 5 Motion of the Centers of Direct Near-Stable Equatorial Orbits in a Rotating Reference Frame

---

* We shall discuss only direct near-stable orbits; the theory of the corresponding retrograde orbits can be developed analogously.

There is one value of the mean altitude ($\approx$ 2500 km) for which the small oscillation (discussed earlier) of the center of the equatorial stable orbit is particularly simple. We show in Fig. 6 the motion of the center of this orbit and some typical near-stable orbits. First of all, we consider the near-stable orbit whose center is the same distance from the center of the stable orbit as the latter is from the center of the earth (Fig. 6a). For this orbit, therefore, $\rho = ae_s$, and the center of the near-stable orbit passes through the center of the earth periodically. As can be seen, it traces out a cardioid in a geocentric non-rotating reference frame. The angular velocity of the orbit center along this path is approximately $2\dot{\beta}$, and we note that the corresponding variation in perigee height will exhibit cusps. This is true if, and only if, the center of the near-stable orbit passes through the center of the earth.

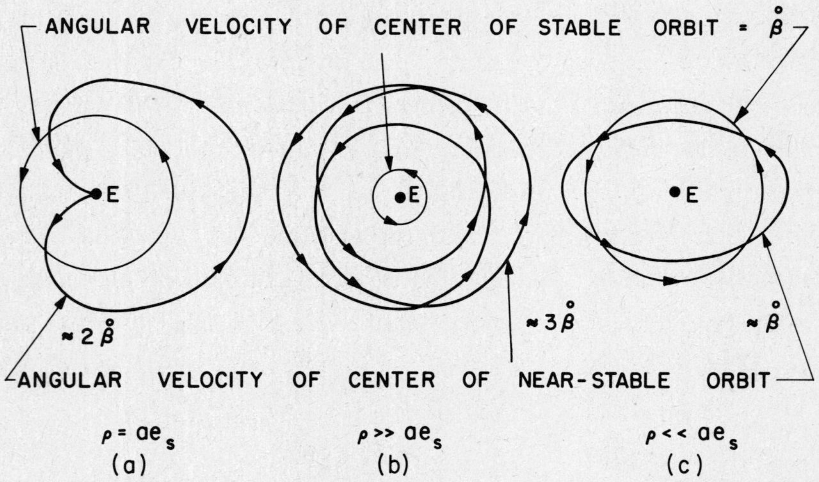

Fig. 6 Motion of the Centers of Direct Stable and Near-Stable Equatorial Orbits in a Non-rotating Reference Frame

If $\rho >> ae_s$ (i.e., if the near-stable orbit has a much larger eccentricity than that of the stable orbit), we find that the center of the near-stable orbit moves in a complicated path (Fig. 6b) with angular velocity $\approx 3\dot{\beta}$ and between circular envelopes whose radii differ by $2ae_s$. For $\rho << ae_s$, i.e., for an orbit very close to the stable one, the center follows an elliptical path (Fig. 6c) with semimajor axis $ae_s + \rho$ and semiminor axis $ae_s - \rho$, at an angular velocity $\approx \dot{\beta}$.

In this analysis, at least, it would seem that we have returned to the original epicycle descriptions of Ptolemy. And, in fact, we can develop the motion of near-stable orbits quite accurately in terms of a quickly convergent series of epicyclical motions.

Effects of the Earth's Shadow

Were it not for the earth's shadow, the solar radiation field would always act like a conservative force field. That is, a satellite with a constant area-to-mass ratio moving in a closed orbit about the earth would neither gain nor lose energy from this field after a complete revolution. However, if the orbit is noncircular and the shadow region asymmetric with respect to the orbit major axis, the orbital energy of the satellite will be changed (Fig. 7). If the satellite leaves the shadow closer to the sun than when it entered, the satellite gains energy: It is not being retarded during that distance. Similarly, the satellite loses energy if it leaves the shadow region at a greater distance from the sun than when it entered. The change in energy is simply proportional to the difference in the projected distances of the radial positions of the satellite along the component of the earth-sun line in the orbit plane. With the potential energy of a satellite in the solar radiation field defined as

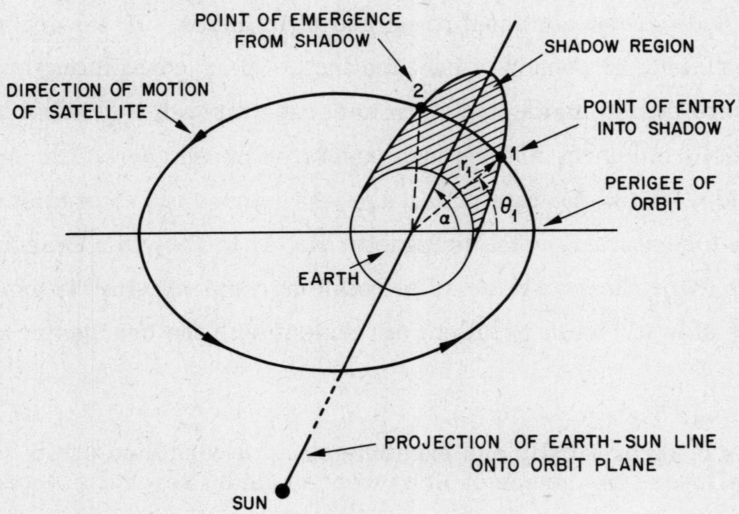

Fig. 7 Passage of a Satellite Through the Earth's Shadow

$$V(x) = (KA/M)(I/c)\, x \tag{8}$$

where x is the distance of the satellite from the sun, we find (Fig. 7)

$$\Delta E = (KA/M)(I/c)\left[1 - W^2\right]^{1/2}\left[r_1 \cos(\alpha - \theta_1) - r_2 \cos(\alpha - \theta_2)\right] \tag{9}$$

where ΔE is the change in energy per revolution. For $0 < \alpha < \pi$, point 2 will always be further from the sun than point 1. Similarly, for $0 < \alpha < \pi$, sunlight pressure will tend to decrease perigee height. From this it follows generally that whenever radiation pressure causes perigee height to decrease, the orbital energy will also decrease (if part of the orbit is in shadow), and vice-versa for perigee height increasing because of sunlight pressure.

COMPARISONS WITH THE ORBIT OF ECHO I

The Echo balloon, with an initial area-to-mass ratio of about 100 cm^2/g, more than 300 times larger than that of Vanguard I, provided an opportunity to test quantitatively the theory of sunlight pressure perturbations.* Unfortunately it was not just a simple matter of comparing the experimental orbit with a theoretical one deduced solely from the initial values of the orbital elements. The possible variations in one additional parameter had to be known precisely, namely, those of KA/M. Were Echo a perfectly smooth sphere, K would be unit; A, its cross-sectional area; and M, its mass. However, in order to keep Echo inflated, 33 pounds of sublimating powder, constituting about 20% of its mass, were dusted inside, and were to escape through tiny holes introduced in the balloon before launch, and through punctures caused by collisions with micrometeoroids. Since the rate of leakage was impossible to predict accurately, we were forced to allow the parameter KA/M to vary in a reasonable manner+ in our theoretical work so as to obtain agreement with the experimental data. We were able to obtain excellent agreement with the changes in all the

---

* Comparisons of theoretically and experimentally determined orbits for the first few weeks following the launch of Echo were made by several groups.[18,19,20]

\+ That is, in a manner consistent with our *a priori* knowledge of the balloon's properties.

orbital elements over the first six months by allowing KA/M to vary as if K and A were fixed at their nominal values and with M decreasing from its initial value to a value 20% lower by the middle of January 1961. In other words, a model consistent with our results indicated that the balloon maintained its shape and specular reflection characteristics while the sublimating powders leaked out slowly, having essentially escaped completely by the middle of January. This rate of gas leakage is about an order of magnitude slower than that predicted by NASA.

We tested the sensitivity of our results to the rate of gas leakage by comparing the observed maximum value of the eccentricity ($e_{max} = 0.0788$) reached on December 27, 1960, with the following theoretical calculations:

1. By assuming that KA/M maintained its initial value of 102 cm^2/g, we found $e_{max} = 0.0705$.
2. By assuming that the satellite lost all of its gas soon after launching, whereas K and A retained their assumed initial values (KA/M = 130 cm^2/g), we obtained $e_{max} = 0.0885$.

There are, of course, other models consistent with these results. For example, it is conceivable that M decreased very rapidly, and that A simultaneously decreased giving rise to only a small net effect on A/M, but that K kept increasing due to a gradual roughening of the balloon's surface.* On the basis of orbital information alone, it is difficult to distinguish among these various hypotheses.+

Our results were published last year,[21] without further comment, since we were unaware of any independent supporting data. However, recently we saw a study of the radar measurements made on Echo I at Millstone Hill between launch and October, 1961.[22] The characteristics of the amplitude vs. time curves for the received signals showed a rather dramatic change between

---

* Note that for diffuse reflection from a sphere, $K \approx 1.44$.

\+ It is also possible that the variation found in KA/M had no direct physical significance, but was merely the result of "covering up" for the effects of some neglected perturbations and/or for the accumulation of errors in our numerical integration procedure.

January 11 and 18, 1961 (Fig. 8). Lines 1 and 2 of this figure are reasonably typical of the returns from the period from launch until January 11, 1961. From January 18, 1961 onward, as exemplified by lines 3 and 4, a gradual deterioration of the signal quality was found. These results, of course, can be understood on the basis of our postulated model: After essentially all the gas escaped, the internal pressure inside Echo was sufficiently reduced so that the plastic surface tended to reform the folds which were introduced when the balloon was packed initially in its small launch container.*

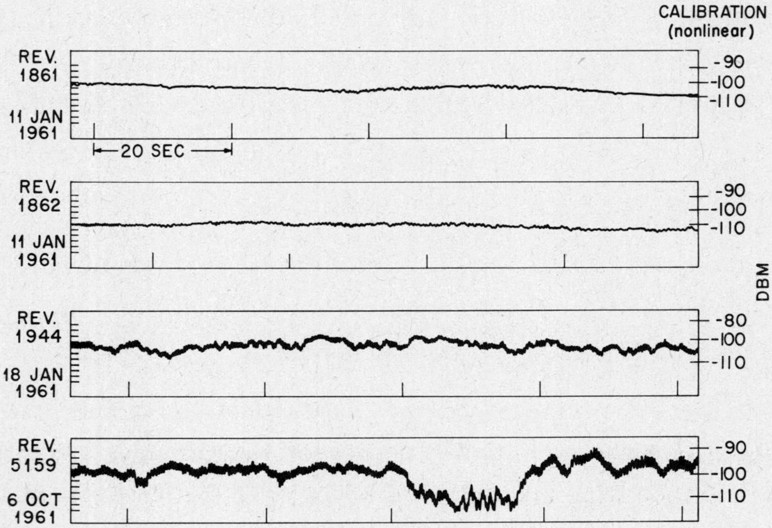

Fig. 8 Amplitude vs. Time of Radar Returns from Echo I Received by Millstone Hill

While quite pleased with such seeming consistency with our orbital analysis, we felt it prudent to determine the results on Echo obtained by other radar groups. We discovered that observations made under the direction of W. Jakes at Bell Labs, Holmdel, did not indicate any such dramatic change. Jakes felt, however, that the envelope of their observed amplitude oscillations did begin to increase slowly starting in about January, 1961 but not so drastically as shown in Fig. 8. The major difference between the two sets of

---

* Here again other interpretations are possible. For example, this deterioration in radar reflectivity could have been caused by meteor-produced holes and/or tears in the balloon.

observations is that Bell operated at 960 mc, and Millstone at 440 mc. The average width of the gores in the balloon is about halfway between the two wavelengths. Therefore, we may perhaps assume that the Bell radar was sensitive to scintillations caused by the unevenness of the gores, whereas Millstone Hill could detect only larger scale irregularities; hence the change in signal quality in the latter case was more dramatic. However, we have not undertaken any quantitative investigation of this rather complicated problem of the scattering of radio waves by an almost spherical body composed of 82 metallic gores, partially insulated from each other, and no firm conclusions can therefore be drawn.[+]

Were this model for the gas leakage rate correct, we would like to make meaningful deductions about the micrometeoroid puncture rate. However, the difficulties would be at least twofold: We would have to know the temperature distribution over the balloon's surface accurately in order to calculate the vapor pressure, since the leakage rate depends on it linearly; and we would also have to know the size distribution of the holes made by impinging micrometeoroids which, in turn, depend on their velocities, shapes, and densities. About all we could say, without much further study, is either that the temperature had been somewhat lower than calculated (only about a 20°C change in temperature produces an order of magnitude change in vapor pressure), or that the puncture sizes and rates of impact were considerably less than expected.

In Fig. 9, we have sketched our theoretical predictions, made about a year ago,[23] of the future variations of Echo's perigee height. We note first that after January 1, 1961, the perigee was increasing; and hence, Echo gained orbital energy from the solar radiation field. In fact, over a period of several weeks in February, 1961, it actually gained more energy from the radiation field than was lost due to air drag. This marked the first time that an increase had ever been observed in the period of an artificial satellite. On the figure, this is indicated by an increase in the satellite's mean altitude.

There are two major difficulties in making accurate long-term predictions of the Echo orbit. First, now that the gas has escaped, the shape of

---

[+] To shed more light on this problem, data has been solicited from two other radar groups that observed Echo.

567R

Echo can change in unpredictable ways which will affect radiation pressure and air drag perturbations.* Second, the density of the atmosphere changes significantly and somewhat unpredictably. In fact, because of the approach of the quiet sun period, upper atmosphere densities are expected to decrease markedly over the next few years. The various curves shown at the end of Fig. 9 indicate possible behaviors of perigee height during the balloon's final year in orbit. The path by which Echo will descend will be very sensitively dependent on the air density variations during that time.

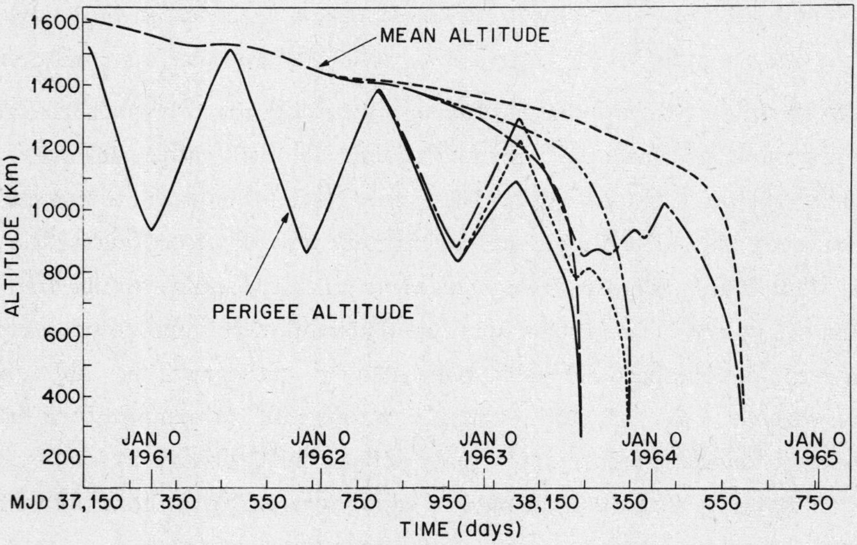

Fig. 9 Time Variations of Perigee Altitude and Mean Altitude of Echo I for Various Air Density Models

We calculated the solid curve using the atmospheric density data obtained from an analysis of the Echo orbit during the first four months after launch, and assuming no change in its shape, we therefore might predict that

---

* These effects are partially mitigated by the averaging, resulting from the rapid rotation of the balloon. According to estimates made by Gehrels[24] from interferometric measurements, the period of rotation of Echo in October, 1961 was about 75 sec. (Such a rapid rotation can be accounted for by assuming that a small net torque is caused by sunlight pressure acting on the uneven surface of Echo.) The resultant centrifugal pressure at the equator of the balloon is about 0.02 dyne /cm^2, which is two orders of magnitude smaller than the predicted initial internal pressure but almost three orders of magnitude greater than sunlight pressure. The balloon should therefore resist shape changes caused by most external forces.

Echo would cease to orbit in the summer of 1963. However, by making allowances for a general decrease in atmospheric densities, it is clear that Echo will remain in orbit at least until 1964. In fact, because of a (possible) decrease in size and, more importantly, a decrease in air density, Echo's mean altitude did not decrease so much in 1961 as predicted on the basis of the earlier shape and air density values: The second minimum perigee height was about 30 km higher than is shown on the figure.*

PASSIVE COMMUNICATION SATELLITES

Let us now consider the implications of solar pressure perturbations for communication satellites of the balloon type. Although Echo was shown to be quite useful for communication purposes, it was very difficult to track, and coverage conditions varied widely with time. Communication scientists therefore prefer not to have to cope with large differences in apogee and perigee height. On Echo these reached 1200 km. Hence, we wish to choose orbits which minimize the maximum apogee-perigee altitude difference. Our earlier discussion showed that (neglecting air drag) a stable orbit has a constant apogee-perigee altitude difference which is, in fact, half that of the maximum difference for a nonstable orbit with the same mean altitude and direction of motion. In Table I, we show the differences in apogee and perigee altitudes

Table I

EQUATORIAL STABLE ORBITS

a	Apogee-Perigee Altitude Difference*	
	Direct Orbit	Retrograde Orbit
1.25 e.r.	215 km	135 km
1.50	705	295
1.75	3,080	520
2.00	4,400	820

** KA/M = 100 cm^2/g

---

* The orbit calculated at the Smithsonian Astrophysical Observatory[25] shows that perigee height reached its second minimum on December 7, 1961, in precise accord with Fig. 9.

for stable equatorial orbits of satellites with area-to-mass ratios of 100 cm^2/g and with various mean altitudes. The rather striking increase for direct orbits is caused by the existence of a resonance at a mean altitude of 0.93 e.r. Stable orbits in this vicinity therefore have large eccentricities (Fig. 3). In general, the apogee-perigee altitude difference is much smaller for retrograde than for direct stable orbits. Had Echo been launched at the same mean altitude of 1600 km, but in a stable retrograde equatorial orbit, the apogee-perigee altitude difference would have been an order of magnitude smaller. In addition, its lifetime would have been far greater because the satellite would not as soon have dipped into the lower regions of the atmosphere.

When we increase the mean altitude (as is usually desired for communication satellites to increase coverage), the apogee-perigee spread increases. At a one-earth-radius altitude it is clearly better to use retrograde stable orbits for which the spread is relatively small - only 820 km. Of course there is also a cogent reason for not using retrograde orbits: Launching against the direction of rotation of the earth requires a more powerful rocket for a given payload. The only other available alternative to inhibit large apogee-perigee height differences is to increase appreciably the mass of the satellite. But this latter approach would require far greater rocket capabilities. Hopefully, then, these considerations on stable orbits will be of use in planning communication systems based on sets of Echo-like passive satellites.

THE EARTH'S DUST BELT

We wish, finally, to discuss the implications of radiation pressure for the dust belt that may surround the earth. Over the past few years, starting in 1955 with an article by de Jager,[26] many people have suggested that the concentration of micrometeoroid matter about the earth is considerably greater than in deep space at comparable distances from the sun. Experimental data, while still subject to some doubt, seems to support this idea. In fact, Hibbs,[27] after analyzing Explorer I micrometeoroid data, concluded that dust particles were orbiting the earth. Others, however, feel that these rather meager data warrant only a conclusion of an increased concentration of dust near the earth.[28] Whipple[29] has suggested that this increase might be due in part to trapping in orbit about the earth of matter liberated from the moon during meteor impacts on that body. Singer[30] and others have suggested that charge drag would serve to trap dust particles into orbits about the earth. At the very least, the earth's gravitational attraction will cause somewhat of a concentration in the vicinity of the earth. We have long felt that, since the attraction of the sun on the dust is significantly reduced (because of sunlight pressure), the earth should more easily be able to capture these particles into circumterrestial orbits. As a consequence of such a capture process, we would expect the concentration of smaller dust particles to be relatively larger* and this is in accord with recent data.[31 +] However, to our knowledge, none of these various qualitative ideas has been subjected to any significant quantitative analysis.

For the sake of discussion, let us assume that dust particles are in closed orbits about the earth. What could we expect of their orbital behavior? The most important perturbing forces would probably be those of sunlight pressure and charge drag. We can compare their magnitudes for spherical particles, as follows:

$$\frac{F_{rp}}{\pi r_m^2} \approx 4.65 \times 10^{-5} \text{ dynes/cm}^2 \tag{10}$$

$$\frac{F_{cd}}{\pi r_m^2} \approx -2 \times 10^{-10} \, NV^2 \text{ dynes/cm}^2 \tag{11}$$

where r_m is the particle radius, N is the number of charged particles per cm^3, and V is the electrostatic potential on the particle in volts. The right side of Eq. (11) is, unfortunately, only a crude approximation; without a detailed knowledge of the plasma characteristics and the surface conditions of the particle, it is not possible to calculate the charge drag force accurately. On the other hand, the uncertainty in Eq. (10) is less than a factor of two, and is caused

---

* This conclusion on relative concentrations would, of course, also follow, were charge drag the dominant trapping mechanism.

+ For the comparison to be meaningful, the dust particle size distribution in deep space must also be considered.

mainly by the uncertainty in the reflection characteristics of the dust particle.* Radiation pressure will probably dominate in the altitude regions of about one earth radius (where the ion density is between 10^2 and 10^3 particles/cm^3) unless V is 10v or above, which, on the basis of present knowledge, seems unlikely.+ If the potential is only of the order of 1 v, the magnitude of the solar radiation force will be several hundred times greater. Of course, we must realize that the actions of these two forces are quite different: Sunlight pressure will cause a displacement of the orbit as a whole, whereas charge drag will cause a shrinkage of the orbit.

Singer[30] has argued that dust in circumterrestrial orbits would be concentrated at about an altitude of one earth radius because he expected the electrostatic potential on the dust particles to be very low there. However, he did not consider the effect of sunlight pressure. We find that at precisely this altitude over the equator there exists a resonance which would bring dust particles down from orbit in an extremely short time - on the order of several days for the very small particles.

On the other hand, any dust particles which are trapped in stable orbits might have very long orbital lifetimes, provided that charge drag is not too large. In other words, if some of the dust does get trapped into circumterrestrial orbits and if the *a priori* probability of capture into orbit is a sufficiently smooth function of the orbit parameters, we would expect to find relatively concentrated dust bands in the vicinities of the stable orbits. Since the parameters of a stable orbit depend on the area-to-mass ratio of the orbiting particle, each concentrated region might contain particles with a noticeably different mean value of area-to-mass ratios. Crudely speaking, we might find the dust segregated into bands depending on their size. These deductions are,

---

* We assume that the dust particle is sufficiently large compared to the relevant wavelengths of sunlight so that geometrical optics applies.

+ Although measurements of the electrostatic potential on a satellite at $\approx$ 1500 km, the highest altitude for which such measurements have so far been published, indicate potentials of a few tenths of a volt,[32] the potentials on dust particles can conceivably be much higher, especially since the plasma characteristics at higher altitudes may be radically different.

of course, highly speculative. In view of this, it seems premature to describe in any greater detail the spatial density distributions which could result from dust particles with various area-to-mass ratios being trapped in stable orbits with different inclinations, eccentricities, and mean altitudes. Much more work, both experimental and theoretical, must be carried out before even tentative conclusions can be drawn on the existence, origin, and characteristics of a dust belt orbiting about the earth.

In recent studies of the exosphere, some attention has been devoted to the component of neutral hydrogen which may be in orbit about the earth.[33, 34] We wish to point out that sunlight pressure would severely limit the lifetime of such particles. In fact, it is unlikely that a hydrogen atom could complete even one revolution.*

ACKNOWLEDGMENT

The author wishes to express his gratitude to Dr. Harrison M. Jones with whom he collaborated on most of the analyses described in this paper. He also wishes to thank C. W. Perkins for several useful discussions.

---

* Estimates based on the approach of Baade and Pauli,[35] but corrected to account for recent solar spectral measurements, indicate that sunlight pressure on neutral hydrogen is a significant fraction of the earth's gravitational attraction. Hence, the ballistic trajectories[33, 34] of exospheric hydrogen atoms may also be subject to substantial sunlight pressure perturbations.

REFERENCES

1. L. Euler, Hist. de l'Acad. Roy. des Sciences et Belles Lettres, Berlin 1746, p. 117

2. J. C. Maxwell, A Treatise on Electricity and Magnetism, Clarendon Press, Oxford, vol. 2, 3rd edition, 1892, p. 441

3. J. H. Poynting, "Radiation Pressure in the Solar System: Its Effect on Temperature and Its Pressure on Small Bodies," Phil. Trans. A, vol. 202, June 1903, pp. 525-552. See also, J. H. Poynting, Collected Scientific Papers, University Press, Cambridge, 1920

4. J. Larmor , "On the Dynamics of Radiation," Fifth Intern. Cong. Math., vol. 1, Aug. 1912, p. 19

5. H. C. Plummer, "On the Possible Effects of Radiation on the Motion of Comets, with Special Reference to Encke's Comet," Monthly Notices, R.A.S., vol. 65, Jan. 1905, p. 229

6. H. P. Robertson, "Dynamical Effects of Radiation in the Solar System," Monthly Notices, R.A.S., vol. 97, No. 6, April 1937, pp. 423-438

7. S. P. Wyatt and F. L. Whipple, "The Poynting-Robertson Effect on Meteor Orbits," Astrophys. J., vol. 111, No. 1, Jan. 1950, pp. 134-141

8. W. E. Morrow, Jr., "Orbital Scatter Communication," presented at the XIII th General Assembly of the International Scientific Radio Union, London, England, Sept. 1960

9. R. W. Parkinson, H. M. Jones, and I. I. Shapiro, "Effects of Solar Radiation Pressure on Earth Satellite Orbits," Science, vol. 131, No. 3404, March 1960, pp. 920-921

10. E. Levin, Rand Corporation, unpublished notes

11. F. T. Geyling, "Satellite Perturbations from Extra-Terrestrial Gravitation and Radiation Pressure," J. of Franklin Inst., vol. 269, No. 5, May 1960, pp. 375-407

12. P. Musen, "The Influence of the Solar Radiation Pressure on the Motion of an Artificial Satellite," J. Geophys. Res., vol. 65, No. 5, May 1960, pp. 1391-1396

13. Y. Kozai, "Effects of Solar Radiation Pressure on the Motion of an Artificial Satellite," Smithsonian Astrophys Obs., Spec. Rep. No. 56, Jan 1961, pp. 25-33

14. S. P. Wyatt, "The Effect of Radiation Pressure on the Secular Acceleration of Satellites," Smithsonian Astrophys. Obs., Spec. Rep. No. 60, Mar 1961, pp. 1-16

15. R. W. Bryant, "The Effect of Solar Radiation Pressure on the Motion of an Artificial Satellite," Astronomical J., vol. 66, No. 8, Oct 1961, pp. 430-432

16. P. Musen, R. Bryant and A. Bailie, "Perturbations in Perigee Height of Vanguard I," Science, vol. 131, No. 3404, Mar 1960, pp. 935-936

17. I. I. Shapiro and H. M. Jones, "Lifetimes of Orbiting Dipoles," Science, vol. 134, No. 3484, Oct 1961, pp. 973-979

18. I. I. Shapiro and H. M. Jones, "Perturbations of the Orbit of the Echo Balloon," Science, vol. 132, No. 3439, Nov 1960, pp. 1484-1486

19. D. O. Muhleman, R. H. Hudson, D. B. Holdridge, R. L. Carpenter and K. C. Oslund, "Observed Solar Pressure Perturbations of Echo I," Science, vol. 132, No. 3439, Nov 1960, p. 1487

20. R. Jastrow and R. Bryant, "Variations in the Orbit of the Echo Satellite," J. Geophys. Res., vol. 65, No. 10, Oct 1960, pp. 3512-3513

21. P. E. Zadunaisky, I. I. Shapiro, and H. M. Jones, "Experimental and Theoretical Results on the Orbit of Echo I," Smithsonian Astrophys. Obs., Spec. Rep. No. 61, Mar 1961, pp. 1-22

22. D. Hynek, M.I.T. Lincoln Laboratory, private communication

23. H. M. Jones, I. I. Shapiro, and P. E. Zadunaisky, "Solar Radiation Pressure Effects, Gas Leakage Rates, and Air Densities Inferred from the Orbit of Echo I," Proc. Cospar Space Symposium, North-Holland Publishing Co., Amsterdam (in press)

24. E. Gehrels, M.I.T. Lincoln Laboratory, private communication

25. B. Miller, Smithsonian Astrophys. Obs., private communication

26. C. de Jager, "The Capture of Zodiacal Dust by the Earth," Mem. Soc. Roy. Sci. Liege, vol. 15, 1955, pp. 174-182

27. A. R. Hibbs, "The Distribution of Micrometeorites Near the Earth," J. Geophys. Res., vol. 66, No. 2, Feb 1961, pp. 371-377

28. M. Dubin, "Remarks on the Article by A. R. Hibbs, 'The Distribution of Micrometeorites Near the Earth'," J. Geophys. Res., vol. 66, No. 8, Aug 1961, pp. 2592-2594

29. F. L. Whipple, "The Dust Cloud Around the Earth," Nature, vol. 189, No. 4759, Jan 1961, pp. 127-128

30. S. F. Singer, "Distribution of Dust in Cislunar Space - Possible Existence of a Terrestrial Dust Belt," Am. Astronaut. Soc. Lunar Flight Symp., MacMillan Co., New York, 1961

31. C. W. McCracken, W. M. Alexander, and M. Dubin, "Direct Measurement of Interplanetary Dust Particles in the Vicinity of Earth," Nature, vol. 192, No. 4801, Nov 1961, pp. 441-442

32. E. C. Whipple, Jr., Goddard Space Flight Center, NASA, private communication. See also, R. E. Bourdeau, J. L. Donley, G. P. Serbu and E. C. Shipple, Jr., "Measurements of Sheath Currents and Equilibrium Potential on the Explorer VIII Satellite (1960)," NASA, Technical Note D-164, Jul 1961, pp. 1-25

33. J. C. Brandt and J. W. Chamberlain, "Density of Neutral Gas in a Planetary Exosphere," Phys. Fluids, vol. 3, No. 3, May-Jun 1960, pp. 485-486

34. E. J. Öpik and S. F. Singer, "Distribution of Density in a Planetary Exosphere," Phys. Fluids, vol. 3, No. 3, May-Jun 1960, pp. 486-488

35. W. Baade and W. Pauli, Jr., "Über den auf die Teilchen in den Kometenschweifen ausgeübten Strahlungsdruck," Naturwiss., vol. 15, No. 2, Jan 1927, pp. 49-51

FIRST RESULTS IN NASA TOPSIDE SOUNDER SATELLITE PROGRAM

J. E. Jackson,* R. W Knecht,** and S. Russell†

The first radio-pulse soundings of the topside of the ionosphere were obtained during two recent rocket flights at the NASA Wallops Island facility. These tests were performed as part of the NASA S-27/S-48 Topside Sounder Satellite Program. The primary objective of this program is the investigation of the ionosphere above the height of maximum ionization of the F2 region. On 24 June 1961, a two-frequency radio-pulse sounder was carried to an altitude of about 1,060 km using a four-stage solid-propellent rocket. Reflections from the topside of the ionosphere were obtained on one or both frequencies during 13 of the 15 minutes that the payload was above the maximum electron-density level in the ionosphere. Unexpected effects were observed as the sounder passed through the levels in the ionosphere where the plasma frequency was equal to the sounding frequency. Some evidence of the presence of ionization irregularities at altitudes between 700 and 950 km was obtained. A second rocketborne sounder was launched during disturbed ionospheric conditions on 13 October 1961 using the same type of vehicle. "Spread F" returns were obtained during the first portion of the flight, in addition to the normal topside echoes. This paper presents the results of the rocket tests and reviews the overall NASA S-27/S-48 Topside Sounder Satellite Program.

---

* Goddard Space Flight Center, National Aeronautics and Space Administration, Greenbelt, Maryland.

** Central Radio Propagation Laboratory, National Bureau of Standards, Boulder, Colorado.

† Airborne Instruments Laboratory, a Division of Cutler-Hammer, Inc., Deer Park, Long Island, New York.

INTRODUCTION

The tenuous gas in interplanetary space and in the vicinity of the planets is substantially ionized under the influence of radiation from the sun. Immediate objectives in the ionosphere program of the National Aeronautics and Space Administration (NASA) are concerned with ionization in the vicinity of our planet. The NASA is presently supporting six projects designed to study the nature, dynamic behavior, and distribution in time and space of charged particles in the ionosphere. These studies are aimed at a more complete understanding of the upper atmosphere and, in particular, of the influence exerted by the sun on the earth's ionosphere. Four of the projects will use satellites, the fifth will use space probes, and the sixth will use sounding rockets. Two of the projects are topside ionospheric sounders that will examine the structure of the ionosphere from above, since the region above hmaxF2 cannot be explored by conventional ground-based stations. This paper reviews the overall Topside Sounder Satellite Program and presents the results of preliminary rocket tests of an ionospheric sounder.

The most important measurement that can be made above hmaxF2 is that of electron-density distribution as a function of altitude; a density profile of this type can provide many clues concerning various ionospheric parameters. For example, the density profile shown in Fig. 1 can be used to determine the temperature of the electron-ion gas and the composition and relative abundance of the ion species (O^+, H^+, or He^+). The electron-density data is also important in the study of the diffusion and recombination process.

Correlation of electron-density distribution with data from other disciplines can provide information of great value. By comparing the temperature of the ionized particles with the neutral-gas temperature, the extent of thermal equilibrium in

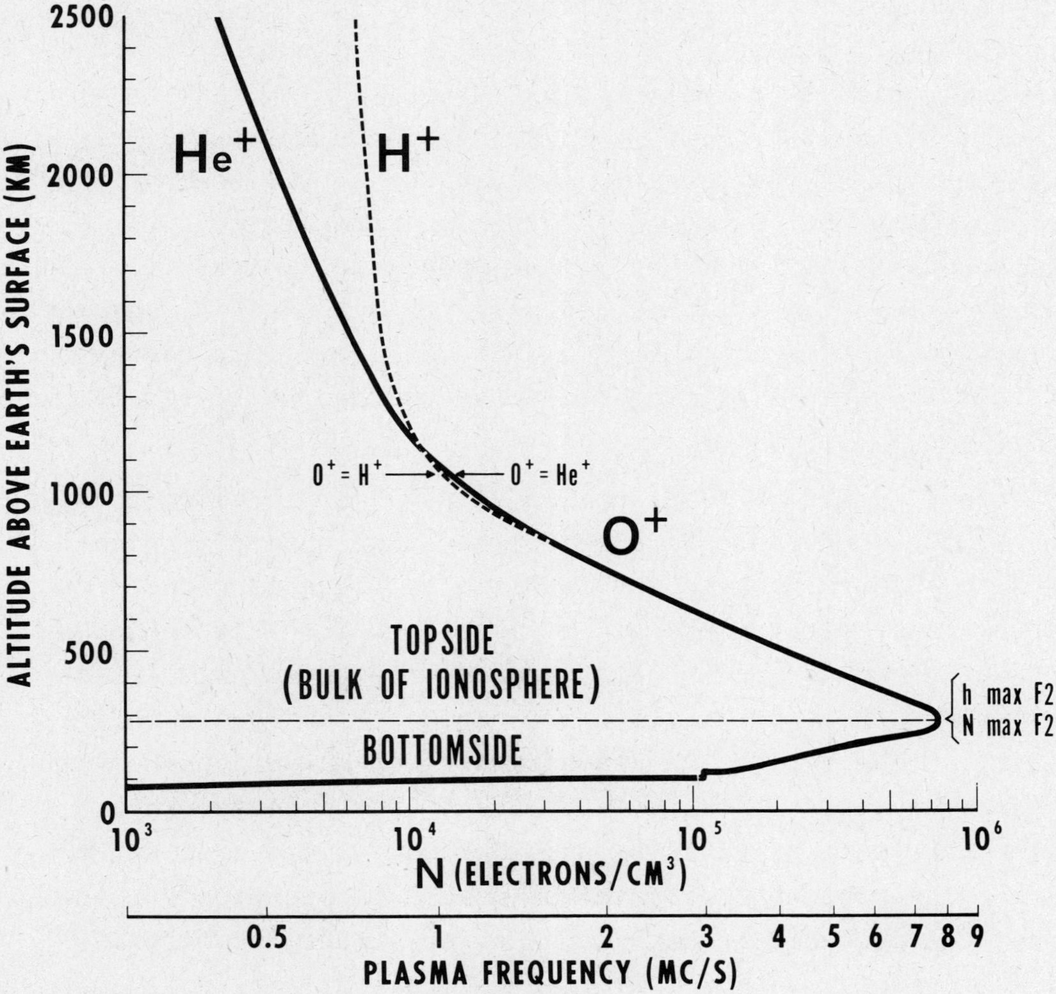

Fig. 1 Average Daytime Electron-Density Distribution in the Ionosphere as a Function of Altitude for Temperature of 1350°K

the ionosphere can be determined. By correlating observed ionization with the flux of incoming ionizing radiation, the ionization process itself can be investigated.

The topside sounder technique is the only technique known that can provide electron-density profiles synoptically above hmaxF2. Topside soundings should permit the investigation of the previously mentioned parameters as a function of time and geographical location. The anticipated data should be particularly valuable since the planned orbital inclination of 80 degrees should provide data on the little known and very complex high-latitude regions (polar and auroral zones). The profiles obtained in the equatorial regions will be of interest from another standpoint, since the distribution near the equator should be materially affected by the earth's magnetic field. (Since the magnetic field at the equator is nearly horizontal, vertical diffusion of the charged particles is significantly inhibited.)

In addition to its scientific value, the increased knowledge of the ionosphere is directly applicable to communications, tracking, and guidance. The importance of the ionosphere to terrestrial radio communication is well-known. Predictions of ionospheric storms and disturbances are often unsatisfactory, since these predictions are based on inadequate information. Since a thorough understanding of natural phenomena is a prerequisite to their intelligent use, a better knowledge of the entire mechanism should lead to more precise forecasts. The prediction of maximum usable frequencies for communication purposes is based upon observations from ground-based ionospheric stations. This information was considered of sufficient importance to justify the establishment of about 150 ground-based ionospheric sounding stations throughout the world. Even this number of stations is not sufficient for accurate world-wide mapping of the bottomside ionosphere. Two of the most important observations obtained by these stations are the height and density of maximum ionization in the ionosphere. In principle, several topside sounders could provide this information synoptically and with better geographical coverage.

It is important to realize that the ionosphere and the lower atmosphere present severe problems in designing precision tracking systems. Since the disturbing effects of the lower atmosphere (both refraction and absorption) become significant before a frequency is reached where the ionosphere is completely transparent, the ionospheric problems cannot be bypassed by raising the operating frequencies. High frequencies have other disadvantages, such as necessitating complex antennas to maintain a large capture area. It is important to note that, in problems associated with guidance and tracking of space vehicles, the entire ionosphere (both bottom and topside) must be taken into account. In this respect, an adequate topside observational program is crucial.

NASA TOPSIDE SOUNDER PROGRAM

Consideration of satelliteborne topside sounding experiments was under way early in 1958. On 3 July 1958, Dr. Lloyd V. Berkner, Chairman of the Space Science Board of the National Academy of Sciences, sent out a request for ideas on satellite experiments; several organizations replied with proposals for topside sounding experiments. Further support was provided by the Space Science Board's Working Group on Satellite Ionosphere Measurements meeting in Boulder, Colorado in September 1958. At a special meeting in October 1958 in Ithaca, New York, called by Dr. Henry Booker to consider the topside sounder experiment in detail, at least seven groups expressed interest in this type of experiment. In particular, this meeting stimulated a proposal from the Canadian Defence Research Telecommunications Establishment (DRTE), which came to NASA at the end of 1958. The NASA was pleased to accept DRTE's cooperation on the Topside Sounder program, and it was agreed that this scientific undertaking would be a joint effort between Canada and the United States; each country would pay its own costs of the project. This was the first international arrangement for cooperative research in a space vehicle. A joint announcement of this arrangement was made by both countries on 20 April 1959.

Concurrently, NASA had requested that the Central Radio Propagation Laboratory (CRPL) of the National Bureau of Standards examine the topside sounder proposals received by NASA for scientific merit and engineering feasibility and recommend immediate and long-range approaches for this area of research.

In June 1959, a CRPL study report recommended the fixed-frequency system as a first-generation experiment and suggested that the DRTE of Ottawa, Canada be encouraged to develop its swept-frequency system as a second-generation experiment. This second recommendation was, in fact, a concurrence by CRPL with the decision already reached between NASA and DRTE. The topside sounder effort, which began in Canada early in 1959, is currently known as the S-27 project. NASA and CRPL discussed the development of the simpler fixed-frequency system as a parallel effort, in view of its complementary features. A joint proposal from CRPL and Airborne Instruments Laboratory (AIL) was received by NASA in January 1960. CRPL was to provide scientific supervision of a fixed-frequency experiment and analyze the resulting data, and AIL was to design, construct, and test the satellite payload. The CRPL/AIL topside sounder experiment is now known as the S-48 project. Work began at AIL on 9 May 1960 under a contract letter of intent that provided partial funding. About a year later, the AIL contract was finalized and fully funded. For this program, the expenses of CRPL are paid annually by NASA through a routine interdepartmental transfer of funds.

Since the S-27 and S-48 projects are under the same project manager at Goddard Space Flight Center and in view of the similarity of objectives and techniques in the two projects, a joint working group was set up early in 1960. The United Kingdom then expressed an interest in participating in the Topside Sounder program. Under the present agreement, the United Kingdom supports the United States program by operating a telemetry station in the South Atlantic. In return for the use of this station, the United States will provide topside soundings over the United Kingdom stations at Winkfield (England) and Singapore. Data from all three United Kingdom stations will be used in ionospheric research in England. Fig. 2 is a flow chart of the responsibility and data distribution for ionospheric topside sounding.

Project S-48 is one of six NASA projects designed to study the nature, dynamic behavior, and distribution in time and space of charged particles in the vicinity of the earth. The objectives of this experiment are similar to those of the S-27. The S-48, however, will emphasize the study of meridional cross sections through the ionosphere using telemetry stations along the 75°W meridian. The S-27 and S-48 telemetry stations are shown in Fig. 3. The United States system has a low resolution and a fast profile acquisition rate; whereas, the Canadian experiment has a high resolution and a slow profile acquistion rate. In the S-48, six fixed frequencies will provide downward pulse transmissions and echo receptions in the 3 to 9 Mc range. The S-48 payload, weighing about 105 pounds, is scheduled for launch in 1962 by a Scout vehicle into a near circular orbit at an altitude of about 1,000 km (620 miles) and with an 80-degree inclination.

ROCKET TESTS

Since previous rocket experimentation provided no precedents for the Topside Sounder program, an important part

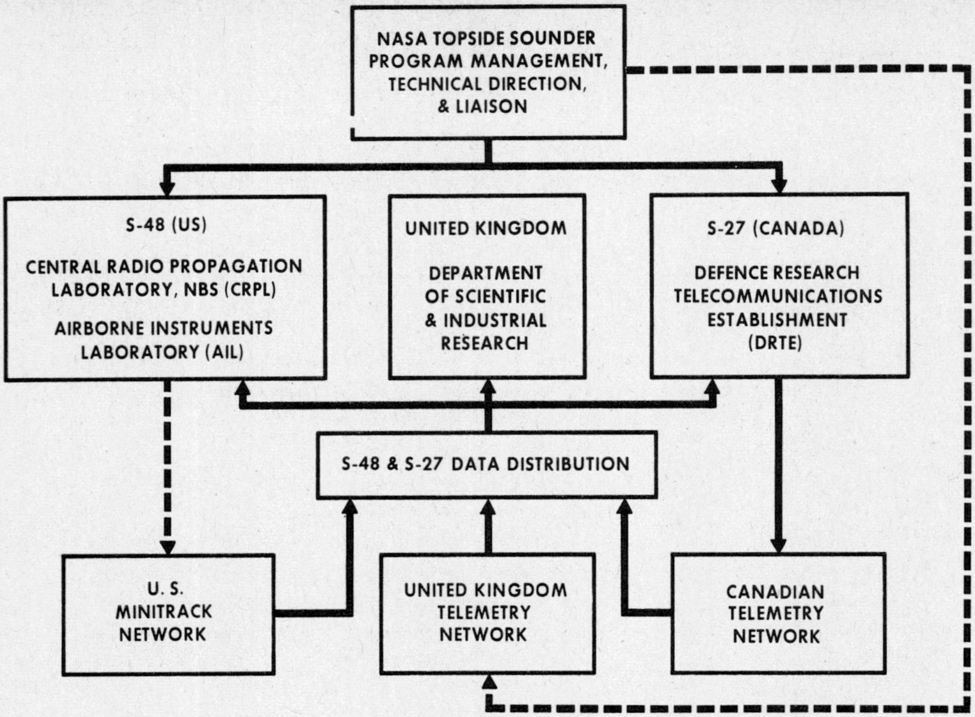

Fig. 2 Flow Chart of Responsibility and Data Distribution for Ionospheric Topside Sounding

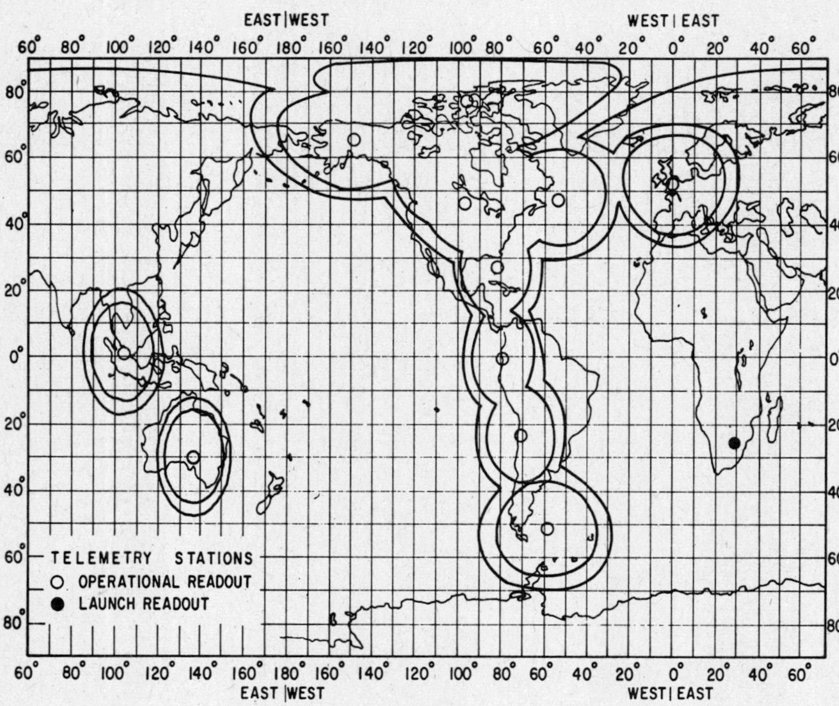

Fig. 3 S-27 and S-48 Telemetry Stations Showing Area of Coverage

of the CRPL/AIL effort has been rocket testing of a simplified topside sounder. The orginal plan, which called for a single daytime firing to a height of about 1,000 km during undisturbed ionospheric conditions, had the following principal objectives:

1. To obtain a measure of the effective reflection coefficient of the topside of the F-layer and to provide an indication of the required transmitter power and the appropriate level of receiver gain necessary to obtain optimum ionospheric echoes with the satellite sounder
2. To determine whether the topside of the F-layer is rough or smooth under quiet daytime conditions. For instance, it is important to know at an early stage whether blobs of ionization are encountered on the topside of the F-layer, whether large-scale ripples are common, or whether the topside is predominantly horizontally stratified
3. To obtain an experimental determination of the separation of the ordinary and extraordinary components of the reflected echo under sample experimental conditions,
4. To obtain an indication of the possible importance of the field-ducted echoes as have been discussed, for instance, by Gallet and Booker. This is crucial as preparation for complex and possibly confusing echo returns in the satellite experiment
5. To gain information on the type and degree of antenna detuning that will take place
6. To test the adequacy of the data telemetry link, including provisions for synchronization, bandwidth, etc.
7. To obtain a qualitative measurement of the level of the cosmic-noise background at two frequencies. This will be a useful repetition and supplement to the results of the Canadian DRTE group at 3.85 Mc in Transit IIA.

At a meeting of the Working Group in June 1960, it was decided that it would be scientifically inadvisable to finally adjust the parameters of the satellite topside sounders on the basis of a single "quiet" daytime rocket observation and that a nighttime firing under disturbed conditions was highly desirable.

The final plan, therefore, was to include a second rocket test to study topside reflection characteristics at a time when spread F echoes indicative of disturbed ionospheric conditions were being observed on bottomside soundings. Spread F is the name given to a phenomenon involving the reflection of radio waves from the F region of the ionosphere. Under certain conditions, for periods lasting from a few minutes up to several hours, the radio-wave echo ceases to be well-defined over a relatively narrow range (time) interval and instead, a large number of scattered returns at various ranges are observed. Because of the prevalence of spread F echoes at high latitudes, a foreknowledge of the degree to which topside echoes will be degraded during this condition is of considerable importance in the planning of the DRTE experiment. Since a variety of spread F is regularly observed in the equatorial region, the results will be very useful in the final planning of the CRPL/AIL satellite experiment.

EXPERIMENT-DESIGN CONSIDERATIONS

Selection of Sounding Frequencies

On the basis of predictions of the F2-layer maximum electron density and the likely shape of the topside profile, sounding frequencies near 4 and 6 Mc were chosen for the rocket experiment. Limitations on antenna size prevented the use of frequencies less than about 4 Mc. To minimize interference from ground broadcast stations, lists of frequency allocations were studied and frequencies of 4.070 and 5.970 Mc were selected as the exact sounding frequencies to be used in the first rocket sounder. These frequencies and several others nearby were monitored continuously at the National Bureau of Standards field station at Fort Belvoir, Virginia for about one month to confirm their suitability.

It was later decided to include only the 4.07-Mc sounding frequency in the second rocket payload. For a night firing during disturbed conditions, it was estimated that there would be less than a 10-percent chance of obtaining echoes at 5.97 Mc. To facilitate interpretation of the results and to evaluate the success of the test, it was decided to include an ion trap in place of the 5.97-Mc transmitter to provide a local measurement of ion density during the ascent portion of the rocket trajectory. In a previous flight an ion trap designed and built by the Lockheed Missiles and Space Company, measured ion densities that were in excellent agreement with electron densities deduced from bottomside ionograms taken at the same time and place (Hanson and McKibbin, 1961).

Power Radiated and Transmission Loss

Consideration of the amount of peak pulse power required for sounding was based upon the calculation of the probable transmission loss and the best available information on the level of the cosmic noise at the two sounding frequencies.

The transmission loss is principally caused by two factors:

1. Path attenuation,
2. Magnetoionic splitting into the ordinary (o) and extraordinary (x) components.

Path attenuation, α, depends on the effective distance to the reflecting layer where the electron density corresponds to a plasma frequency equal to the radiated frequency.*

---

* The transmitted signals reflect from the ionosphere when they reach an electron density, N, in electrons per cubic centimeter given by

$$N = \frac{f^2}{80.5}$$

where f is the frequency in kilocycles per second. (This frequency is known as the plasma frequency corresponding to the density N.)

$$\alpha = 33 + 20 \log f_{Mc} + 20 \log 2d_{km} \qquad (1)$$

where

α = path attenuation in db,

f_{Mc} = frequency radiated in Mc,

d_{km} = distance to the reflecting layer in kilometers.

Note that d_{km} is not the true distance to the layer where reflection takes place because the ionosphere acts as a dielectric producing a divergence of the radiation. This can be considered as an additional transmission loss due to an increased distance to a reflecting layer. Assuming an altitude of 800 km for the sounder and a representative topside electron-density profile, the values of d_{km} for 4.07 and 5.97 Mc were calculated to be 760 and 810 km.

Magnetoionic splitting results in up to a 6-db loss, 3 db in splitting and 3 db because the returned wave is elliptically or circularly polarized whereas the antenna is linearly polarized.

Antenna orientation loss was not considered because a favorable orientation of the sounding antenna with respect to the ionosphere is provided by the spin stabilized vehicle.

The total losses are listed in Table I. Cosmic noise was computed from the empirical relation $T_{eff} = 500(100/f)^{2.3}$. Values for a 30-kc RF bandwidth are listed in Table I. The 30-kc bandwidth was determined on the basis that a range resolution of 5 to 8 km was desirable. The minimum detectable signal was considered to be 10 db above noise and is also listed in Table I. From this the required radiated power is determined and listed in the same table. The signal-to-noise ratio and minimum detectable signal do not take into account beneficial integration from film presentation of data.

TABLE I

POWER CALCULATIONS

Frequency (Mc)	Maximum Total Loss (db)	Cosmic Noise (dbm)	Minimum Detectable Signal (dbm)	Required Radiated Power (watts)
4.07	114.6	-92.0	-82.0	1.8
5.97	118.8	-96.4	-86.4	1.75

ROCKET PAYLOAD

The sounder payload was designed to mate with an Argo D-4 (Javelin) four-stage solid-propellent vehicle. To obtain a trajectory that would exceed an altitude of 1000 km, it was necessary to limit the payload weight to about 50 pounds.

The radio sounder consisted primarily of two pulsed transmitters, a single receiver for both frequencies, two 16-foot telescoping antennas, antenna matching networks, and an electronic programmer. Fig. 4 is a block diagram of the payload. Pulsed RF transmissions have a duration of 100 microseconds and have a 15-millisecond pulse-repetition period, the longest anticipated delay time for an ionospheric return. The sounder is programmed for a transmit and receive interval first at 4.07 Mc, then at 5.97 Mc, and followed by a period used for synchronization purposes. The cycle is repetitive at a 22.2-sec^{-1} rate.

An RF peak pulse power of 40 watts was obtained from the completely transistorized transceiver. Losses in the dual-frequency reactive-matching networks operating into a 32-foot dipole antenna reduce the radiated power to between 2 and 3 watts for each frequency.

The received echo signals at the output of the transmitter combined with suitable synchronization signals were telemetered via a standard FM/FM VHF radio link to several ground stations.

Fig. 5 shows the first flight payload. The telescoping sounder antennas are shown collapsed and folded forward in the position that permits a protective nose cone to surround the payload. Shortly after ejection of the nose cone, the antennas are unfolded into a plane normal to the vehicle axis. As the antennas approach this position, explosive charges in the base of each antenna pole are fired causing the antennas to extend to a 16-foot length. When the antennas completely unfold and lock into position, power is applied to the ionospheric sounder and the sounding experiment begins.

QUIET IONOSPHERIC TEST

The first of the two scheduled firings took place at 1817 EST on 24 June 1961 at Wallops Island. Topside echoes on one or both frequencies were received during 13 of the 15 minutes that the payload was above the maximum of the F region. Except for the fact that the 4-Mc echoes were somewhat weaker than expected, the experiment proceeded as planned and was completely successful. The preliminary results are:

1. It was confirmed that a peak radiated power of about 2 to 3 watts is sufficient for topside sounding of the quiet ionosphere from altitudes of about 1000 km.
2. The fixed-frequency sounding system proposed for S-48 has been shown to operate successfully.
3. Unexpected effects occur as the sounder exits and re-enters the overdense ionosphere.

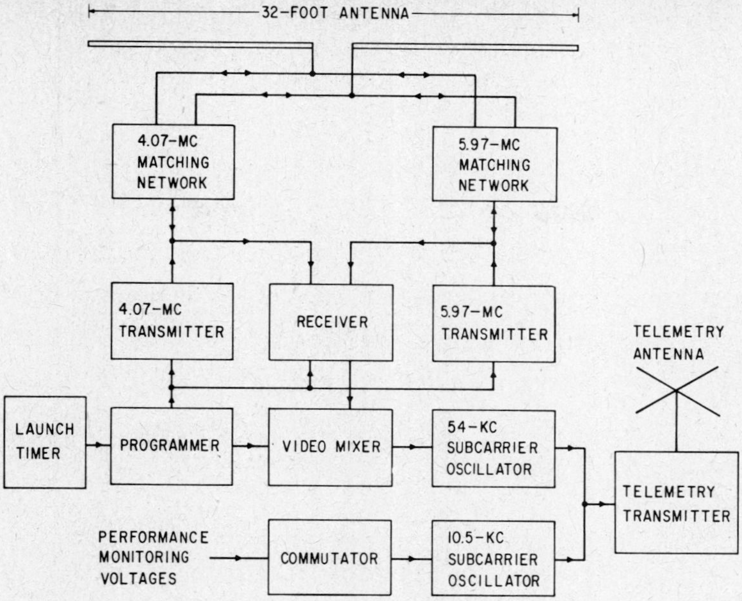

Fig. 4 Block Diagram of S-48 Rocket Payload

Fig. 5 Payload for NASA Flight 8.15

4. Some evidence was obtained for the presence of ionization irregularities in the 700 to 1,000 km height range (maximum effect in the 850 to 950 km range).

The flight was affected somewhat by the vehicle despin operation, which reduced the spin rate to less than 6 rpm rather than the desired 60 to 90 rpm. Consequently, the vehicle was no longer spin stabilized and began to tumble. The indication of this was a periodic fading of both the ionospheric echc returns and the telemetry signal.

A possible effect on the experiment, which cannot be verified, was that the last sections of one or both sounder antennas, lacking the aid of centrifugal force, may have failed to extend fully. This would cause a slight detuning of the antenna system. The effect is suggested by the noticeably lower echo signal strengths that were observed at 4.07 Mc as compared with the 5.97-Mc echoes. Since the loaded Q of the antenna system at 4.07 Mc was about three times higher than at 5.97 Mc, a given change in antenna pole length would degrade the efficiency of the antenna much more at 4.07 Mc.

Film records of the 4- and 6-Mc data are reproduced in Fig. 6. The ordinate on the film records is the "virtual depth" of the reflection level below the rocket sounder; this is obtained by multiplying one-half of the echo delay time by the speed of light in a vacuum. The virtual depths would be expected to be zero at the points where the sounder emerges from the reflection levels for a particular frequency and magnetoionic component and then increase to a maximum value at the time when the sounder is farthest from the reflection level--this occurs near the apogee of the trajectory. The virtual depths should be symmetrical on the down-leg. On the best film records, the topside echoes conform rather well to the expected pattern. But on the films reproduced in Fig. 6, which have a compressed time scale, the weaker 4-Mc trace is only barely discernible. A better presentation is obtained using an expanded time scale, Fig. 7. The double return on the 4-Mc record is due to the presence of two magnetoionic components.

Unusual effects on both frequencies ("splashes," which may be a kind of plasma resonance effect) are observed as the sounder emerges from the topside of the ionosphere and passes through levels corresponding to the three reflection levels--the ordinary level ($X = 1$), the extraordinary level ($X = 1 - Y$), and the z level ($X = 1 + Y$).* Also, on the 6-Mc x-component (and to a much lesser extent on the 4-Mc z-component), several traces that are almost exactly harmonically related in virtual depth are observed during the first 10 seconds after topside exit. Considering the relatively small radar cross section of the antenna, it is remarkable if these are, in fact, multiple echoes (indicating up to three round trips between the sounder and the ionospheric reflection level). Enlarged views of these topside "exit" phenomena are shown in Fig. 8.

The echoes were weak or missing entirely during the time when the sounder was within 100 to 150 km of the reflection level. This is undoubtedly due to detuning of the system by the relatively high electron density in this region.

---

* $$X = \left(\frac{f_N}{f}\right)^2 ; \quad Y = \frac{f_H}{f}$$

where

f_N = plasma frequency,
f_H = gyro frequency,
f = sampling frequency.

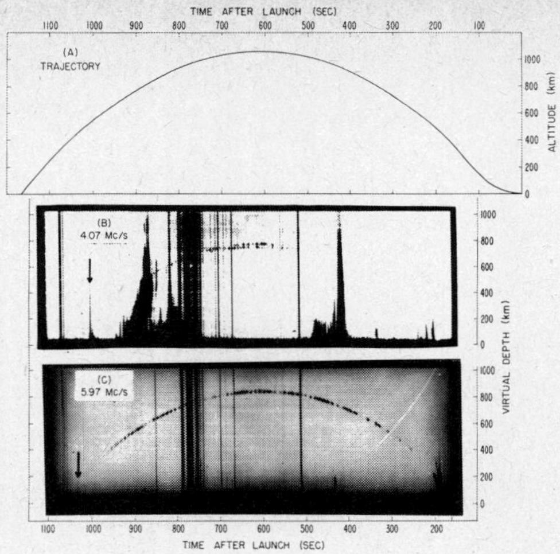

Fig. 6 Sounding Records of Quiet-Ionosphere Test Obtained During Flight NASA 8.15, Wallops Island, Virginia, 24 June 1961

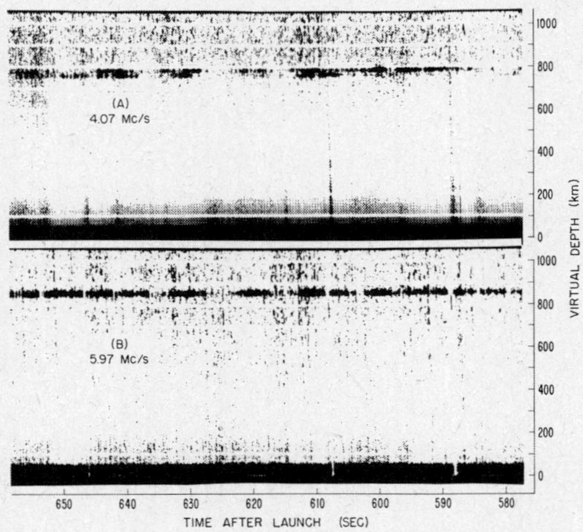

Fig. 7 Expanded Record of Apogee Portion of NASA 8.15 Sounding Experiment

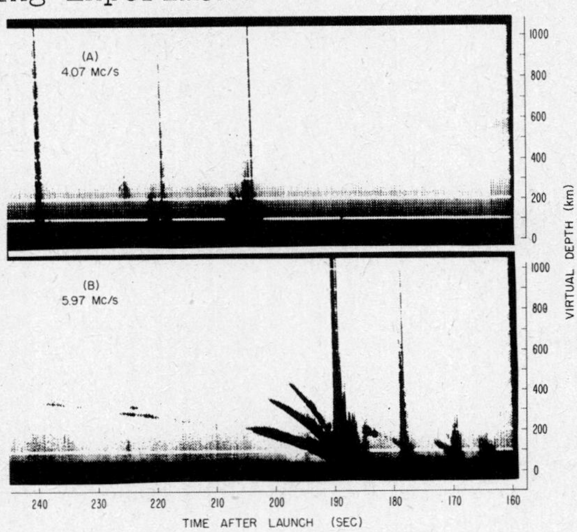

Fig. 8 Topside Exit Effects Observed During Flight of NASA 8.15

Prominent effects are observed on the 4-Mc record between about 400 to 480 seconds (910 to 1000 km) and between about 780 and 940 seconds (960 to 700 km). It can be seen that large effects occurred in the 910- and 950-km regions on both up- and down-legs; however, the "disturbance" lasted over a much wider height range on the downgoing portion and a second large effect occurred as the sounder passed through the 840-km level. The causes of these effects are not understood, but it seems likely that they indicate the presence of scattering irregularities in the 700 to 1000 km height range. The occasional vertical gaps in the records are due to telemetry outages. An increase in noise background, probably due to cosmic noise, is also evident during the higher parts of the flight.

Shortly after the topside echoes cease due to detuning on the downward leg of the flight, splashes, which are believed to indicate ionospheric re-entry, are observed on both frequencies--arrows in Figs. 6(B) and 6(C). It is not yet clear as to why the three magnetoionic components are not observed individually as in the topside exit. After an interval during which the payload was passing through the over-dense region, it breaks out of the bottomside of the F-layer with sharp increases in interference observed almost simultaneously on both frequencies. Prior to bottomside exit, the ionosphere shielded the sounder from interfering terrestrial signals. The relatively large time difference between 4- and 6-Mc topside re-entry together with the almost simultaneous bottomside exit is a graphic indication of the substantial difference between the topside and bottomside electron-density gradients. On the basis of the exit heights alone, the topside electron-density distribution was nearly exponential and the neutral atmosphere scale height between 400 and 600 km was about 72 km at the time of the experiment, Fig. 9 . This indicates a temperature of about 1,200°K assuming an atomic-oxygen atmosphere. At the time of the firing, the F2 layer maximum penetration frequency (foF2) was 7.8 Mc and the height of the F2 peak electron-density level about 305 km as determined by the Wallops Island bottomside sounder.

DISTURBED IONOSPHERE TEST

In order to satisfy the objectives of the experiment, the second firing was delayed until observations made using the Wallops Island bottomside sounder indicated that suitable spread F was present. In temperate latitudes, spread F tends to be associated with magnetic disturbance. Unfortunately, however, during a magnetic storm, the F2 penetration frequency is usually decreased very substantially. Therefore, though a considerable amount of spread F is often present, the maximum electron density in the ionosphere is frequently less than that required for topside reflections on the 4-Mc sounding frequency. It was found that either periods following magnetic storms or nights during which moderate magnetic bay-type disturbances were observed were most likely to meet both of the required conditions--spread F present and sufficiently high F2 layer penetration frequency. About 4-1/2 weeks of readiness was necessary until the ionospheric conditions were satisfied and, at the same time, the proper weather conditions and a clear firing range existed. Finally, on the night of 13 October 1961, the conditions were favorable. Mild spread F had begun at about 2240 EST during the occurrence of a small magnetic bay. By 2326 EST, the "spreading" was deemed sufficient, and the rocket was fired.

A trajectory very similar to the earlier firing was flown during the second experiment. A peak altitude of about 1070 km was reached in 610 seconds. The 4-Mc sounder was automatically turned on 132 seconds after launch, when the payload was at an altitude of about 270 km. The film record

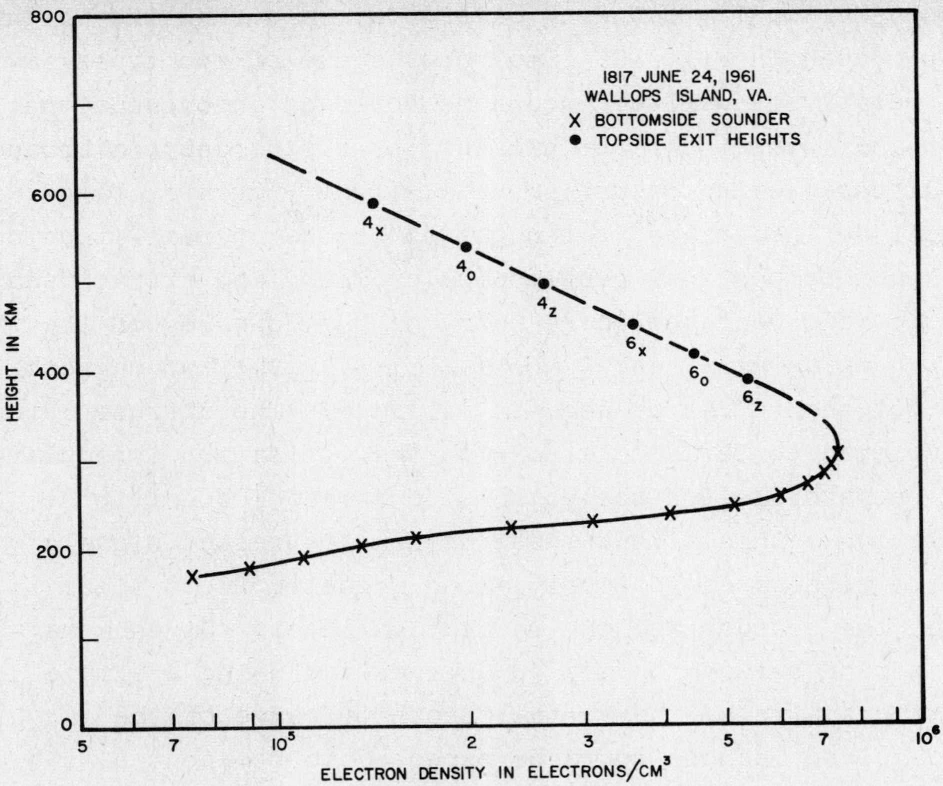

Fig. 9 Topside Profile Obtained From Exit Points

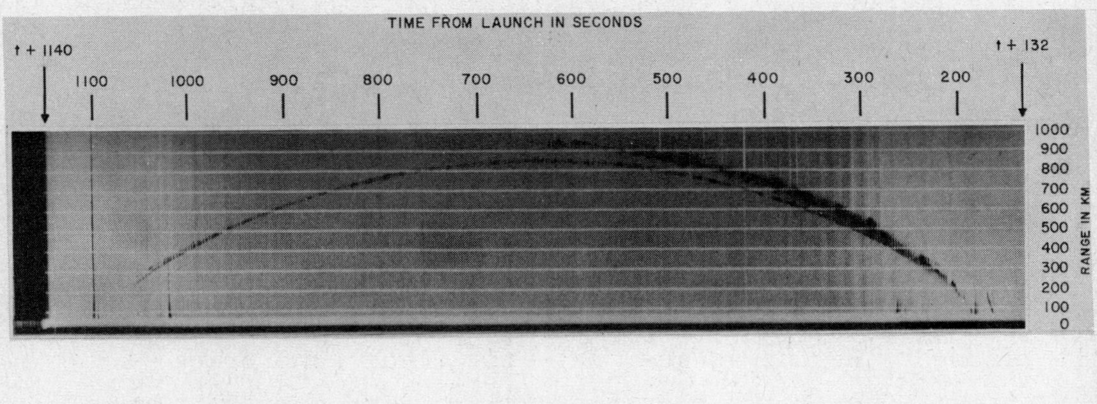

Fig. 10 Sounding Records of Disturbed-Ionosphere Test Obtained During Flight NASA 8.17, Wallops Island, Virginia, 13 October 1961

of the 4-Mc topside sounding data obtained during the flight is reproduced in Fig. 10. The returns are of two types--a clean relatively well-defined echo observed throughout the flight and, over the first half of the flight only, a broader more-diffuse return occurring at a slightly greater range. The well-defined trace is thought to be the "normal" topside reflection composed of overlapping ordinary and extraordinary echoes. Near ionospheric re-entry on the down-leg of the flight (beginning at about 1040 seconds), the two magneto-ionic components can be seen to separate. The diffuse returns are believed to be associated with the ionization irregularities responsible for the bottomside spread F conditions. Calculations suggest that these returns represent signals that are propagated obliquely along magnetic field lines between the sounder and the reflection level. Given a magnetic dip of between 68 and 70 degrees and using a simple geometric analysis, the virtual depth or range of the "field-guided" echoes would be expected to be about 6 to 8 percent greater than vertically incident topside echoes. At apogee, the observed separation appears to be about 60 km, the range of the diffuse returns being about 7 percent greater than the well-defined trace. Also, at times during the flight, the spread echo signal strength was great enough to saturate the receiver (40 to 50 db above cosmic noise). Signal strengths of these magnitudes were as much as 30 db above expected levels (calculated on the basis of isotropic radiation).

These results seem to suggest that field-guided ducting along magnetic field-aligned irregularities is taking place from the time that the sounder emerges from the topside of the ionosphere in the vicinity of Wallops Island until apogee is reached. Beyond apogee, only the clean echo trace is observed. One of the possible explanations for the termination of the diffuse returns at the midpoint of the flight involves the azimuth on which the flight took place. Because an azimuth of about 8 degrees south of east was flown, the sounder traveled progressively into regions of lower magnetic dip. Consequently, it would not be unreasonable to assume that the region in which the ionization irregularities existed did not extend beyond the midpoint of the trajectory, since bottomside sounding stations in the Bahamas did not observe any spread F during the night of the firing.

As in the first flight, plasma-resonance effects appear to be observed as the sounder passes through the levels in the ionosphere corresponding to several reflection conditions. However, in this experiment, only resonances associated with the extraordinary and z levels seem to be observed on topside "exit," with the resonance corresponding to passage through the ordinary level very weak or missing. Effects similar to the earlier flight also occurred on topside re-entry and bottomside exit from the ionosphere.

On the basis of the heights of the resonance effects, the topside electron-density distribution was approximately exponential with a neutral atmosphere scale height of about 50 km. Values of virtual depth computed on the basis of a 50-km scale height for both the normal and ducted echoes are in substantial agreement with the observations. Assuming an atomic-oxygen atmosphere, this scale height would imply a temperature of about $800^{\circ}K$.

CONCLUSIONS

The successful tests of the topside ionospheric sounders are significant since they demonstrate the feasibility of an important technique for ionospheric research in the future. At the same time, they have provided valuable and, in some cases, unexpected geophysical data.

Specific examples of the impact of the rocket-test results upon the satellite program are:

1. Weak echoes received during the first rocket topside sounding showed that the receiver gain characteristic in the satellite should be revised to provide a better quality of telemetry data, when weak ionospheric echoes are being received by the satellite and the received telemetry signal-to-noise ratio is poor.
2. The second rocket topside sounding showed a completely unexpected situation: echoes, possibly field-ducted, were received that have signal strengths much greater than expected. These caused saturation and blocking of the receiver. This required further re-examination of the satellite receiver design to provide protection against saturating signal.
3. Quite unexpected and yet not entirely understood phenomena have been observed in both cases, indicating that the data analysis will require considerable personal judgment and hand processing prior to automatic machine reduction.

The information gained from the two experiments provides a substantial background for the extensive topside sounding program that will begin in the latter part of 1962 with the launching of the Canadian S-27 and United States S-48 Topside Sounder Satellites.

REFERENCES

1. "Worldwide Radio Noise Levels Expected in the Frequency Band 10 Kc to 100 Mc," National Bureau of Standards Circular 557, Superintendent of Documents, 1955.
2. R. M. Gallet and W. F. Utlaut, "Evidence on the Laminar Nature of the Exosphere Obtained by Means of Guided High-Frequency Wave Propagation," Phys Rev Letters, Vol 6, p 591-595, 1961.

3. W. B. Hanson and D. D. McKibbin, "An Ion Trap Measurement of Ion Concentration Profile Above the F2 Peak," *Journal of Geophys Res*, Vol 66, p 1667-1671, 1961.

4. R. W. Knecht and S. Russell, "Pulsed Radio Soundings of the Topside of the Ionosphere in the Presence of Spread F," to be submitted to the *Journal of Geophys Res*.

5. R. W. Knecht, T. E. Van Zandt, and S. Russell, "First Pulsed Radio Soundings of the Topside of the Ionosphere," *Journal of Geophys Res*, Vol 66, p 3078-3081, 1961.

RUBY OPTICAL MASERS FOR IN-FLIGHT MEASUREMENT OF HYPERSONIC FLOW FIELD PARAMETERS

G. K. Wessel*, H. C. Rothenberg**, B. Zendle**

The use of a ruby optical maser is proposed for measuring plasma parameters when conditions render impractical the use of more conventional instruments like Langmuir type probes and microwave interferometers. The instrument to be discussed would, for example, be useful in measuring the electron density distribution and electron velocity distribution in hypersonic flow fields of vehicles in flight. As the instrument would have a spatial resolution of approximately 1 cubic centimeter, it could determine plasma electron density distributions directly. This would be a considerable improvement over the microwave interferometer method for determining electron density profiles in laboratory plasmas by comparing the total phase shift calculated by integration over an assumed electron density distribution with the measured total phase shift.

This paper presents in detail the physical principles upon which the design of the new instrument is based, and discusses two applications - one to a hypersonic flow field situation and the other to a high temperature laboratory plasma situation.

INTRODUCTION

In this paper we propose a new method for direct measurement of electron density distribution and electron velocity distribution in plasma. The method utilizes a ruby optical maser in the manner shown in Fig. 1. The intensity of light scattered in the plasma volume common to the maser incident beam and the receiver beam pattern can be related to the number of free electrons per unit volume in the common region. As the common volume can be made approximately 1 cm^3, or smaller depending on

---

* Syracuse University Physics Dept., Syracuse, N.Y.

** General Electric Co., Defense Electronics Div., Syracuse, N.Y.

the optics, much better spatial resolution can be obtained by the instrument to be proposed here than has been obtained by other techniques found useful in some current areas of plasma research.

In diagnosing high temperature plasmas established for controlled fusion research where physical probes of the Langmuir type are impractical, Wharton[1/] and others have exploited a microwave interferometer technique in which the total phase shift for a microwave beam passing through the plasma is measured. An electron density distribution is assumed which is substituted into the expression for the phase factor for a plasma medium. The resulting phase factor is then integrated over the extent of the plasma along the microwave path and the calculated total phase shift is compared with the measured total phase shift. Aside from being a determination by assumption, the method usually has the additional disadvantages of the assumption of plane wave and infinite plane plasma boundaries in its related analysis, as well as the assumption of slowly varying dielectric constant over distances comparable with a wavelength. These assumptions are difficult to comply with in situations where the technique is often applied.

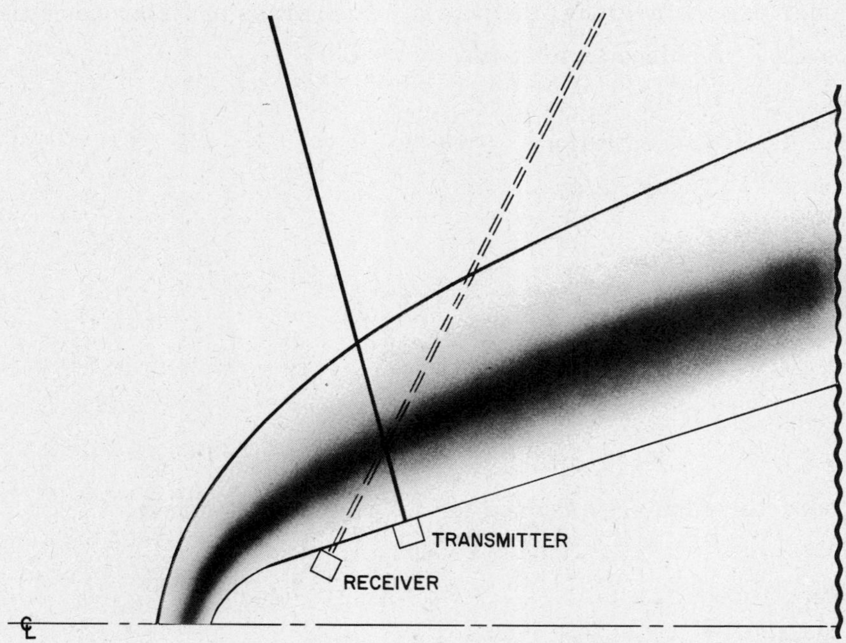

Fig. 1 Measurement Scheme for Optical Maser Flow Field Probe

A current research area presently handicapped by the lack of a good research tool for measuring electron density distributions is hypersonic aerodynamics. To date no instrument has been developed that can be used to measure directly the electron density and electron velocity distributions in flow fields of models in hypersonic shock tunnels and flow fields of vehicles moving through the atmosphere at hypersonic speed. Current flow field theories are in need of experimental evidence as to the state of the plasma in these flow fields, that is, whether equilibrium, 'frozen', or nonequilibrium. The ruby optical maser system to be described would enable direct measurements on real flow fields to be made, thereby providing extremely useful data for analyzing the state of the plasma.[2/]

The spatial resolution of approximately 1 cm^3, or even smaller depending on the optics, would also permit experimental studies of turbulent mass motion in plasma.

OPERATIONAL THEORY AND APPLICATION TO HYPERSONIC FLOW FIELD

In the following we investigate the scattering of light from electrons, atoms, molecules, and ions. A numerical example is given for the case of a vehicle moving through the atmosphere at an altitude of about 200,000 feet with an approximate hypersonic velocity corresponding to Mach 25 and an accompanying stagnation temperature of about 8000°K. Under these conditions, the population densities of electrons, ions, and neutral particles are determined as follows:

Air at 10^{-3} atm., 8000°K:[3/]

Neutral atom (O, N) Concentration.... ≈100%

Electron, ion (N^+) Concentration .... ≈ 1%

Molecules at 200,000 ft.

($P \approx 10^{-4}$ atm.) .... $N_1 \approx 2.8 \times 10^{15}$ cm^{-3}

Atoms (O, N) in test volume

($d/d_o \sim 10$) .... $N_2 \approx 5.6 \times 10^{16}$ cm^{-3}

Electrons, ions (N^+) in test volume .... $N_3 \approx 5.6 \times 10^{14}$ cm^{-3}

The following characteristics are assumed for the ruby optical maser:

Wavelength .... $\lambda = 6943$ Å

Corresponding frequency............ $f = 4.32 \times 10^{14}$ sec^{-1}

Spectral line width .... $\Delta f = 10^9$ sec^{-1}

Peak optical power .... $P = 500$ K watt/cm^2

For electromagnetic radiation at optical frequencies, the index of refraction is essentially unchanged by the presence of free electrons for densities being considered here. The effect due to atoms, molecules, and ions is also very small, as can be seen below from the tabulated experimental values of refractive index for some gases of interest and densities approximately equal to our value of N_2.

INDEX OF REFRACTION, n, FOR GASES AT 10^{-3} ATM. AND 0°C (CORRESPONDING TO $\frac{P}{KT} = 2.6 \times 10^{16}$ $cm^{-3} \sim N_2$)

GAS	$(n-1) \times 10^6$	at wavelength Å
AIR	0.29	6709
O	0.27	6709
N	0.44	6709
NO	0.30	whitelight

From the tables we see that $n-1 \approx 3 \times 10^{-7}$. Therefore reflection and refraction effects in the plasma sheath are negligible.

SCATTERING FROM FREE ELECTRONS

In Thomson scattering from free electrons the power scattered from a unit volume of plasma to a detector of unit area at distance, r, from the scattering volume is[4/]

$$S_f = \frac{\left(\frac{e^2}{mc^2}\right)^2}{r^2} \sin^2 \theta \, N_3 \, P \, \frac{\text{watts}}{cm^2}$$

where θ is the angle between the direction of oscillation of the electron* and the direction of observation. For r = 10 cm, $\theta = 90^o$,* $e = 4.8 \times 10^{-10}$ esu, $m = 9.1 \times 10^{-28}$ gm., $c = 3 \times 10^{10}$ cm/sec , and using the previously given values for the other parameters we get for the peak power scattered into the detector

$$S_f = 2.2 \times 10^{-7} \text{ watts/cm}^2$$

---

* We assume the optical maser output is plane polarized in the direction perpendicular to the plane formed by the direction of incidence and the direction of observation.

Thomson scattering by nuclei will be negligible as the nuclear masses are much larger than the electron mass.

SCATTERING FROM BOUND ELECTRONS

For light scattering from bound electrons of atoms, molecules or ions, it is assumed that the individual electrons are bound to an equilibrium position around which they can oscillate with a characteristic frequency. Assuming that electrons of one particular molecule, atom, or ion oscillate coherently, whereas electrons of different molecules oscillate incoherently, and that the incident radiation frequency is remote from any resonant frequency, a calculation[4/] yields for the scattered power entering the detector,

$$S_b = \frac{\pi^2 f^4}{c^4 r^2} \sin^2\theta \, \frac{(n^2-1)^2}{N_2} P$$

where n = index of refraction of the gas at the particular pressure. Using $n^2-1 = (n+1)(n-1) \cong 2(3 \times 10^{-7})$ and the values already specified for the other parameters, we get for the peak power scattered into the detector

$$S_b = 2.65 \times 10^{-8} \text{ watts/cm}^2$$

Thus, the scattered radiation from the bound electrons of 1 cm^3 of atoms, molecules, and ions is about 12 percent that of the scattered radiation from the free electrons, in spite of the much smaller density of the latter.

Further discrimination between scattered and bound electrons should be possible as a result of the difference in Doppler broadening of the scattered radiation (see Fig. 2). This will be discussed later.

BACKGROUND RADIATION

In addition to the internal noise of the receiver (e.g., photomultiplier tube) the following sources of background noise have to be considered:

1. radiation from the plasma
2. radiation from the sun and the earth
3. scattered sunlight from the atmosphere.

The amount of radiation from these sources is roughly estimated here and appears to be smaller than the scattered signal light. In addition, the use of a narrow band-pass filter and modulation of the light from the optical maser in connection with a phase-locked amplifier in the detection system should provide good discrimination between signal and background radiation.

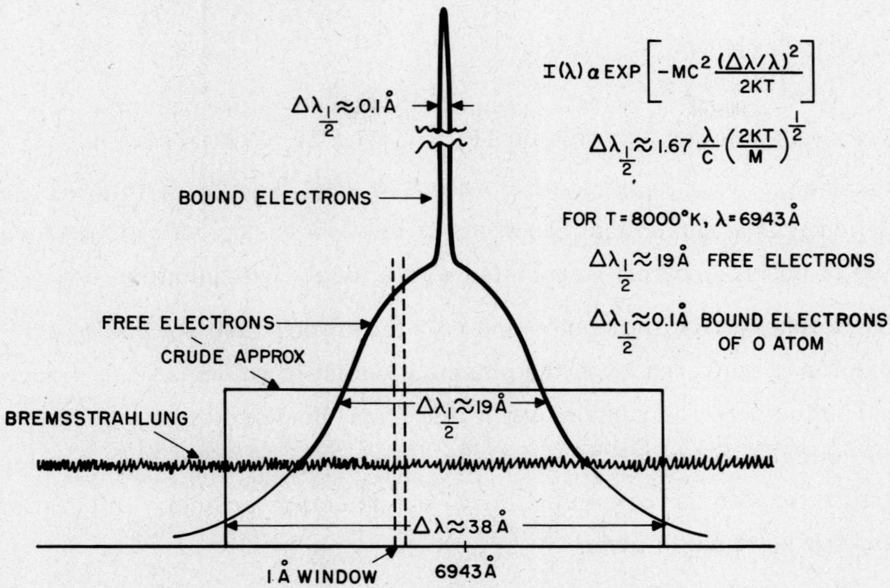

Fig. 2 Spectral Distribution of Scattered Power

Radiation From the Plasma

For the calculation of radiation from the plasma it is necessary to know whether the plasma is transparent or opaque for the frequency of the optical maser ($\lambda = 6943 \mathring{A}$). In the first case only the bremsstrahlung from the collisions between electrons and ions or neutral atoms has to be considered. In the latter case the whole plasma must be treated as a black body.

A measure of the transparency of the atoms and molecules in the plasma region is the optical layer thickness K(f) l where K(f) is the absorption coefficient and l the thickness of the layer. The layer is considered transparent or thin if $K(f)\, l \ll 1$. This will be the case if l is ~1/2 meter and the assumption holds that was made previously in connection with the validity of the Rayleigh scattering formula, namely, that the maser frequency is remote from any resonance.

The bremsstrahlung from electron-neutral and electron-ion collisions can be calculated using the next formula expressing the radiated power per unit volume:

$$P_N = \frac{N_3^{\,2} Z^2 e^6}{6hc^3 \epsilon_0^{\,3} m \pi^2} \left(\frac{2\pi k\, T}{m}\right)^{1/2} e^{-\frac{hf}{2kT}} K_o\left(\frac{hf}{2kT}\right) \frac{h}{2kT} \Delta f \; \frac{\text{watts}}{m^3}$$

For $N_3 = 5.6 \times 10^{14}$ cm^{-3}

$Z = 1$

K_o (1.283) = 0.442 (modified Bessel function, second kind)

$T = 8000^oK$

$\Delta f = 6.23 \times 10^{10}$ sec^{-1} $\rightarrow \Delta\lambda = 1\mathring{A}$

we get for the power radiated per unit volume, $P_N = 1.57 \times 10^{-7}$ watts/cm^3 $\mathring{A}$.

Z has been taken equal to unity because electrons at 8000°K have an average energy of 1.03 eV and it is assumed that electrons of this low energy will not penetrate the negative 'cloud' of bound electrons associated with a singly ionized atom.

This radiation is spherically distributed and only the fraction $\Delta V\ P_N \cos\phi/4\pi r^2$ is entering the detector of unit area from the plasma element of volume ΔV (ϕ = angle between the normal to the detector surface and a line from the detector to a point on a ring-shaped volume element $\Delta V = 2\pi r^2 \sin\phi\, d\phi\, dr$. The ring-shaped volume element, ΔV, is axially symmetric with the perpendicular to the detector surface). Integration of P_N over the detector solid angle gives

$$S_{br} = P_N \int_0^{20\text{ cm}} \int_0^{2.5^o} \frac{\cos\phi}{4\pi r^2}\, 2\pi r^2 \sin\phi\, d\phi\, dr$$

which reduces to $S_{br} = 2.98 \times 10^{-9}$ watts/cm^2 $\mathring{A}$.

Previously we found that the scattered power at the detector from the free electrons is 2.2×10^{-7} watts/cm^2. Depending on the magnitudes of electron density and wavelength of incident light, this scattered energy may be distributed over a spectral distribution corresponding to Doppler broadening by free electrons at 8000°K. In a subsequent section (see also Fig. 2) it is shown that the half intensity width would be about 19 $\mathring{A}$. For a crude comparison of signal-to-noise we assume a uniform (flat) spectral distribution of twice this half-intensity width. On a per angstrom basis we have as signal power

$$S_f/A = \frac{2.2 \times 10^{-7}}{38} = 5.78 \times 10^{-9} \text{ watts/cm}^2\ \mathring{A}$$

The signal-to-noise ratio under these circumstances is ~2

$$\frac{S_f}{S_{br}} = \frac{5.78 \times 10^{-9}}{2.98 \times 10^{-9}} = 1.92$$

As mentioned earlier, it may be possible to improve the signal-to-noise ratio by modulation of the light from the optical maser and the use of a phase-locked amplifier in the detection system.

Bremsstrahlung from electron-neutral collisions has been neglected because at these conditions of temperature and constituent densities the electron collision frequency with ions is 5 times greater than with neutrals.[6/] Time limitations precluded evaluation of other radiation emitting processes that occur in plasmas. For example, the power in the detector passband due to radiative recombination has not been estimated.

Radiation From the Sun and the Earth

The detector looking directly at the sun or directly at the scattered sunlight from the earth will have to be avoided as the power from these sources in the receiver pass band would be intolerably high.

Scattered Sunlight From the Atmosphere

The sunlight power, in the wavelength region of interest for a 1 Å band, is 4 x 10^{-11} times the power in the maser incident beam. The sunlight power, Rayleigh-scattered in the flow field test volume, is therefore not more than 4 x 10^{-11} of the Rayleigh scattered maser power. Integration over the detector solid angle, assuming an exponential atmosphere, shows that the Rayleigh scattered sunlight power incident on the detector would be considerably less than the maser power that has been Thomson scattered by the free electrons in the test volume.

An estimate was made of the sunlight Thomson scattered into the detector by the free electrons in the ionosphere. Assuming a constant electron density of 10^6 cm^{-3} over the altitude range 80-500 km and a sunlight power flux density of 5 x 10^{-5} watts/cm^2/°A, we arrived at 1.6 x 10^{-13} watts/cm^2 °A into the detector.

ELECTRON VELOCITY DISTRIBUTION

The velocity distribution of the free electrons may be obtained from the measurement of the Doppler line width if each electron acts as a distinct scattering center. The degree of scattering incoherence will depend on the magnitudes of electron density and the wave length of incident light. The Doppler distribution for the scattered light based on a Maxwellian distribution for electron velocities is given by

$$I(\lambda) \;\alpha \exp\left(-mc^2 (\Delta\lambda/\lambda)^2/2kT\right)$$

The half-intensity width of the line is found to be

$$\Delta\lambda_{1/2} \cong 1.67 \frac{\lambda}{c} \left(\frac{2kT}{m}\right)^{1/2}$$

For $T = 8000^{\circ}K$, $\lambda = 6.943 \times 10^{-5}$ cm

$$\Delta\lambda_{1/2} \cong 19 \text{ Å for free electrons}$$

$$\Delta\lambda_{1/2} \cong 0.1 \text{ Å for bound electrons of 0 atom}$$

APPLICATION TO A HIGH TEMPERATURE LABORATORY PLASMA

Another interesting application area of the ruby maser plasma probe is controlled fusion research. As an example, we will consider Model B-65 Stellarator[7/] in which the electron density is of the order of 10^{12} electrons per cm^3 and the electron temperature is 80 ev or 6.18×10^5 degrees Kevlin. The tube diameter of this device is 10.2 cm. If we consider deuterium plasma completely filling the tube, the bremsstrahlung flux density at the probe detector is 1.19×10^{-14} watts/cm^2 Å whereas the flux density of ruby light scattered from 1 cm^3 of free electrons into the detector 10 cm away is 3.94×10^{-10} watts/cm^2 over $\Delta\lambda_{1/2} \approx 167$ Å. Again making the crude assumption of a flat spectral spreadover twice $\Delta\lambda_{1/2}$, we have $1\ 18 \times 10^{-12}$ watts/cm^2 Å and a signal to bremsstrahlung noise ratio of 100. As the ion temperature in the B-65 stellarator is also 80 ev, the plasma is essentially fully ionized and there is no scattering from bound electrons.

CONCLUSIONS

On the basis of the foregoing discussions, it would seem that a detection system, based on a ruby optical maser, 1 P21 photo multiplier tubes, and 1 Å windows could be used to measure electron density and velocity distributions in hypersonic flow fields and laboratory plasmas by observing incoherently scattered light from an optical maser beam. A detector such as this could make flow-field measurements on vehicles in flight and could greatly assist in our understanding of flow-field phenomena.

Doppler broadening of the light scattered incoherently by the free electrons is large compared to the line width of the maser. In addition to providing a measure of the electron velocity distribution (temperature distribution for equilibrium in the flow field), the different Doppler spreads for bound and free electrons could provide the basis for discrimination between the two. It may also be possible to provide improved discrimination between signal and background radiation noise by modulating the primary light beam and using a phase-locked amplifier in the detection system. Time limitations precluded evaluation of all radiation emitting processes occurring in plasmas (e.g. radiative recombination). Therefore, knowledge of the sensitivity limit for detecting free electrons in plasma with an optical maser will require further investigation.

REFERENCES

1. C. B. Wharton and D. M. Slager, "Microwave Determination of Plasma Density Profiles," J. Appl. Phys. Vol. 31, No. 2, page 428, Feb. 1960.

2. H. T. Nagamatsu, Research Laboratory, General Electric Co., private communication, July 1961.

3. R. Smelt: "Frontiers of Aerodynamic Research," Chapter 15, second series, Mod. Phys. for Eng., page 352, McGraw-Hill, 1961.

4. A. Sommerfeld, Atombau und Spektrallinien, I. Band, Friedr. Vieweg & Son, Braunschweig.

5. J. E. Drummon, Plasma Physics, McGraw-Hill Book Co., Inc., 1961.

6. J. A. Ratcliffe: The Magneto-ionic Theory, page 33, Cambridge University Press, 1959.

7. Data supplied by M. B. Gottlieb, Project Matterhorn, Princeton, N.J. to D.J. Rose and M. Clark, Jr., "Plasmas and Controlled Fusion," M.I.T. Press, 1961.

eighth
annual
meeting

AMERICAN ASTRONAUTICAL SOCIETY

ATTITUDE STABILITY AND MOTIONS OF PASSIVE, GRAVITY-ORIENTED SATELLITES

D. B. DeBra*

The stability of a gravity-oriented satellite is investigated as a function of its orbit. For an inverse-square-law central-force field the large-motion (non-linear) attitude behavior is numerically investigated as a function of the orbit eccentricity and vehicle shape. It is shown that a major mechanism of tumbling is the excitation of motion in the yaw-roll plane by the forced motion in pitch through the second order coupling terms which are neglected in linearized analyses. The eccentricity for marginal stability and the sensitivity of this stability boundry to damping are also discussed.

INTRODUCTION

The study of the angular motions of a rigid body on orbit under the influence of gravity torques dates to the eighteenth century. The early work, of course, was done in analyzing librations of the moon. Cassini was one of the first to systematically observe the lunar librations and postulate a theorem based on his observations. The theorem stated that the equator of the moon and the lunar orbit plane always intersect the plane of the ecliptic along the same line. These observations were later confirmed by Mayer in the year 1749. By 1754 D'Alembert had already attempted an analytical explanation for this behavior. Successively, Lagrange, Laplace, and finally Poisson extended this analytical work, a summary of which is given by Routh[1] and Plummer[2]. In their work they tried to deduce the moments of inertia of the moon based on observational data, but the calculated values are still uncertain due to the difficulties of measuring small librations using observational data that contains

* Senior Scientist, Lockheed Missiles and Space Company, Sunnyvale, Calif.

very large apparent librations due to the eccentricity of the moon's orbit around the earth.

At the beginning of the twentieth century, pioneers in astronautics began thinking of the attitude control problems of space vehicles. However, it was not until after World War II, and the idea of developing satellites for peaceful and military observation of the earth, motivated examining methods of passive stabilization, that the space vehicle attitude control problem and the classical lunar libration problem were brought together. Rather complete discussions of the recent literature up to 1960 have been given by Roberson[3] and Frye and Stearns[4]. Among the works described, several discuss passive gravity stabilization. In these works derivations of the gravity torque are given, and justification of linearizing the equations of attitude motion are presented. The stability of a satellite as a function of its principal moments of inertia has also been discussed for the case of a circular orbit[5]. When the orbit is not circular, however, there is more than the natural motions to consider since eccentricity and/or oblateness effects of the attracting body introduce forcing terms into the equations and make the coefficients of the differential equations time varying.

This paper will report some research that was done in examining the stability of a rigid body in orbit without the restricting assumptions of linearization or a circular orbit. The equations used will be described and the numerical results presented with a discussion of their implications.

EQUATIONS OF MOTION

The dynamical behavior of a rigid body is described by Newton's 2nd law. The conventional form of this law for attitude motion about the body's center of mass can be written.

$$\dot{H} = L \tag{1}$$

where the symbols are described in the section on notation at the end of the paper. Quadrature methods (numerical integration) are most easily applied to a set of first order differential equations written in canonical form, that is, written so the first derivatives of each variable appears on the left side of an equation as a function of the variables on the right hand side. Eq. (1) is in this form for derivatives taken with respect to an inertial space. The angular momentum, $H = I \cdot \omega^b$, is not always the variable needed. If we wish to work with the angular velocity it is more convenient to write the differential equation in body coordinates since the elements of the coordinatized moment of inertia dyadic of the vehicle are constant in that coordinate frame. This can be done by taking into account the angular velocity of the coordinate frame in addition to the time derivative taken with respect to that frame. After multiplying by the inverse moment of inertia dyadic, Eq. (1) can be rewritten as a canonical differential equation in the angular velocity of the body:

$$\overset{\circ}{\omega}{}^{b}_{b} = I^{-1} \cdot (-\omega^{b} \times I \cdot \omega^{b} + L) \tag{2}$$

The circle above the ω is used to denote time differentiation with respect to a non-inertial frame which is specified by the subscript, in this case b for the body. This notation is described in more detail in an earlier paper[6].

The only external torque considered will be that due to the gravity field of the attracting body. The gravity torque about the satellite's center of mass due to an oblate spheroid can be derived in vector form[6] as the expression:

$$L^g = \frac{3k}{|R|^3} \hat{R} \times I \cdot \hat{R} + \frac{kJ(R^e)^2}{|R|^5} \left\{ \begin{array}{l} \hat{R} \times I \cdot \hat{R} \left[5 - 35(\hat{R} \cdot N)^2 \right] \\ + (\hat{R} \times I \cdot N + N \times I \cdot \hat{R})(10 \hat{R} \cdot N) \\ + N \times I \cdot N \, (-2) \end{array} \right\} \tag{3}$$

It can be seen in Eq. (3) that the external torque depends upon the attitude of the body with respect to the radial direction $\hat{R}$ and the polar axis N of the spheroid. Hence the attitude must be known in order to solve for the torque from which the angular velocity can be found in Eq. (2).

It is well known that a vehicle has zero torque positions for J = 0 when any one of its principal axes are aligned with the radial direction. This can be seen from the first term in Eq. (3): if **R** is an eigenvector of the moment of inertia dyadic, **I**, the torque is zero. For proper vehicle configurations in a circular orbit some of these positions may be stable equilibrium positions as shown in Fig. 1 (taken from Ref. 5). If these stable zero torque attitudes become unstable when the orbit is made eccentric, it is helpful in visualizing the way in which the instability develops to use attitude coordinates which describe the motion relative to a reference such as the zero torque orientation. For this reason, the attitude is described with respect to a reference which may be the zero torque position or a position which approximates it.

There are a number of ways that attitude can be described. The most common are Euler angles and the direction cosines of the body axes. The direction cosines have many advantages for numerical calculations; however, they are not as easy to use for visualizing the attitude as a set of Euler angles. Another set of coordinates that combine the advantages of each are the Euler parameters. For small angles, the first three Euler parameters are equivalent to the Euler angles if the Euler angles are rotations about each of the three orthogonal axes. For large angles, the Euler parameters are more convenient than the Euler angles for visualizing attitude.

The important advantages are for numerical work. The time derivatives of the Euler parameters do not contain singularities, do not require evaluation of trigonometric functions, and are very well suited for evaluation by digital computer. The canonical form of the differential equations relating the Euler parameters to the relative angular velocity can be written

$$\overset{o}{W}_b = W_4\, \omega_b^{b-r}/2 - \omega_b^{b-r} \times W/2 \qquad (4)$$

$$\dot{W}_4 = -\, \omega^{b-r} \cdot W/2 \qquad (5)$$

where the significance of the four Euler parameters are described in more detail in the section on notation. The angular velocity of the body with respect to the reference is given by

$$\boldsymbol{\omega}^{b-r} = \boldsymbol{\omega}^{b} - \boldsymbol{\omega}^{r} \tag{6}$$

where the reference frame and, hence, its angular velocity is specified as a function of time or of the translational motion but does not depend on the attitude motion.

The canonical form of the differential equation relating the time derivatives of a set of Euler angles to the angular velocity of the body is

$$\overset{\circ}{\boldsymbol{\phi}} = B^{-1} \cdot \boldsymbol{\omega}_{b}^{b-r} \tag{7}$$

where matrix B is a function of the Euler angles themselves[7]. The Euler angles used here are defined in Table I. The reference axes from which attitude is measured is an orthonormal triad. All references have their axes in approximately the same direction as the local reference which has its $\hat{1}$ axis radially outward, and the $\hat{3}$ axis in the direction of the orbital angular momentum vector. The $\hat{2}$ axis completes a right hand set. The numerical results presented here are motions measured with respect to a local reference.

In addition to the rigid body, a sphere was included which was coupled to the rigid body through a viscous fluid. The sphere was included to simulate the effect of a damping control system. It was chosen for two reasons. First, when the sphere is small, it can be thought of as a passive device that might be considered for the damping of librations about the vehicle's zero torque position for which its potential is a minimum. For a given size sphere the optimum value of viscosity can be chosen. Second, when the sphere is very large, the viscous torques have a negligible effect on its motion so the vehicle is damped with respect to a reference that rotates uniformly with the mean motion of the orbit if the sphere is given those initial conditions. Therefore, it is mathematically equivalent to an active control system supplying rate damping.

TABLE I
EULER ANGLES

Angle	Axis	Frame Name
		r reference
ϕ_1	1r	
		r'
ϕ_2	2r'	
		b'
ϕ_3	3b	
		b body

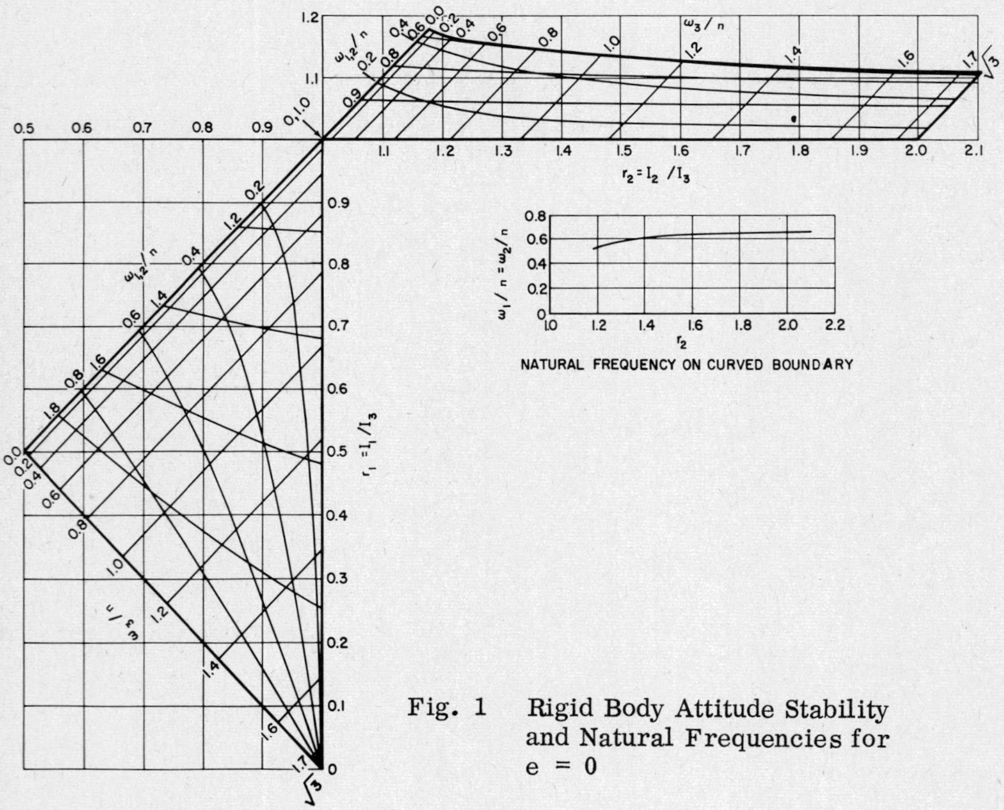

Fig. 1 Rigid Body Attitude Stability and Natural Frequencies for e = 0

Since the dynamics of the sphere is independent of the body motion the angular velocity of a large sphere could also be made equal to the angular velocity of any reference frame with respect to which rates are measured. For any reference, the large sphere is a simple analytical artifice for introducing rate damping terms of the type that one might use with an active damping system.

When the sphere is included, the equations which describe the motion of the body are only altered by the addition of a torque in Eq. (2) which can be written as

$$L^d = - c(\boldsymbol{\omega}^b - \boldsymbol{\omega}^s) \tag{8}$$

But the motion of the sphere must be described too. The torque on the sphere is just the negative of Eq. (8), so only the angular velocity and not the attitude of the sphere is needed which can be obtained from an equation similar to Eq. (2). Since the moments of inertia, I_o, of a sphere are equal about all axes through its center of mass the equation is simply

$$\overset{o}{\boldsymbol{\omega}}{}^s_b = - \boldsymbol{\omega}^b \times \boldsymbol{\omega}^s - L^d/I_o \tag{9}$$

In addition to the exact equations, a set of linear equations with constant coefficients is given below for the body alone. It was found during the study that including the time variations in the coefficients of the equations did not improve the ability of the linear equations to describe the actual attitude motion. (A comparison of the exact and the linear equations with the constant coefficients will be given later). The homogeneous part of these equations has been given many times in the literature[1,2,8,] and does not depend upon the way in which the reference attitude is defined as long as the reference approximates to first order the average equilibrium position normally associated with a circular orbit. The derivation of the forcing terms due to oblateness[9] is too lengthly for inclusion in this paper and will be presented in a later paper in which the forced motion due to oblateness will be discussed. The linear equations are

$$I_1 \ddot{\phi}_1 + (I_3 - I_2)n^2 \phi_1 - (I_1 + I_2 - I_3)n \dot{\phi}_2 = - I_1 \dot{\omega}^r_{1r}$$

$$- (I_3 - I_2)\, \omega^r_{3r} \omega^r_{2r} - 2 \left[kJ(R^e)^2/|R|^5\right](I_3 - I_2)N_{3r}N_{2r}$$

$$(I_1 + I_2 - I_3)n\dot{\phi}_1 + I_2\ddot{\phi}_2 - (I_1 - I_3)(4 n^2)\phi_2 = \tag{10}$$

$$-I_2\dot{\omega}^r_{2r} - (I_1 - I_3)\, \omega^r_{3r}\, \omega^r_{1r} + (I_1 - I_3)(3k/|R|^3)\hat{R}_{3r}$$

$$+ 8(I_1 - I_3)\left[kJ(R^e)^2/|R|^5\right] N_{1r} N_{3r}$$

$$I_3 \ddot{\phi}_3 + (I_2 - I_1)(3n^2)\phi_3 = -I_3 \dot{\omega}^r_{3r} + (I_2 - I_1)(3k/|R|^3)\hat{R}_{2r}$$

$$+ 8 (I_2 - I_1)\left[kJ(R^e)^2/|R|^5\right] N_{2r} N_{1r}$$

Fig. 1 is developed from the secular equation of the homogeneous part of Eq. (10). The reader will find it helpful to refer to Fig. 1 frequently as he follows the discussion of the numerical study of stability which follows. It will help illustrate how the stable regions (implied by the linear equations) shrink as the eccentricity is increased, and it helps correlate the many discrete runs which will be presented later in Tables II and III.

TABLE II
MARGINAL STABILITY
WITHOUT DAMPING

Run	r_1	r_2	Critical Eccentricity	Critical Axis
1	1.05	1.3	.065	YP
2	.2	.9	.065	Y
3	.5	.833	.0085	Y
4	.6	.933	.059	Y
5	1.05	1.1	.143	P
6	.5	.6	.0879	P
7	.4	.7	.062	P
8	.5	.7	.0958	Y
9	.6	.7	.0879	P
10	.3	.8	.048	Y
11	.4	.8	.0369	Y
12	.5	.8	.0261	Y
13	.6	.8	.105	YP
14	.7	.8	.0621	P
15	.2	.9	.0621	Y
16	.3	.9	.0439	Y
17	.4	.9	.0439	Y
18	.5	.9	.0155	Y
19	.6	.9	.0284	Y
20	.7	.9	.192	P
21	.8	.9	.114	R
22	1.01	1.18	.161	P
23	1.02	1.18	.161	YP
24	1.05	1.18	.0958	YP
25	1.1	1.18	.0621	YP
26	1.15	1.18	.136	P
27	1.17	1.18	.114	YP
28	1.18	1.18	.067	P
29	1.1	1.3	$<10^{-4}$	P
30	1.1	1.433	$<10^{-4}$	RP
31	0.	1.0	.215	P
32	.1	.9	.145	Y
33	.1	.92	.117	Y
34	.1	.94	.0611	Y
35	.1	.96	.1026	Y
36	.1	.98	.112	Y
37	.1	1.0	.215	P
38	.12	.92	.0638	Y
39	.14	.94	.086	Y
40	.16	.96	.083	Y
41	.18	.98	.166	Y
42	.2	1.0	.173	P
43	.6	.61	.265	P
44	.7	.71	.18	P
45	.8	.81	.165	P
46	.9	.91	$>.4$	
47	1.0	1.333	$<10^{-4}$	R
48	1.03	1.363	.0287	R
49	1.07	1.403	.0115	R
50	1.12	1.453	.000128	YP
51	1.0	1.7	$<10^{-4}$	R
52	1.03	1.730	.0465	R
53	1.07	1.77	.018	Y
54	1.1	1.8	$<10^{-4}$	R
55	.7	.9	.108	Y

NOTE: Initial Offset about Yaw Axis: $W_1 = 0.1$

TABLE III
MARGINAL STABILITY
WITH DAMPING

Run	r_1	r_2	r_0	c	Critical Eccentricity	Critical Axis
1	.5	.833	.01	.014	.0104	Y
2	.5	.6	.01	.014	.088	P
3	.4	.7	.01	.014	.062	Y
4	.5	.7	.01	.014	.096	Y
5	.6	.7	.01	.014	.096	P
6	.3	.8	.01	.014	.052	P
7	.4	.8	.01	.014	.0403	Y
8	.5	.8	.01	.014	.0285	Y
9	.6	.8	.01	.014	.0522	Y
10	.7	.8	.01	.014	.0628	P
11	.2	.9	.01	.014	.0739	Y
12	.3	.9	.01	.014	.044	Y
13	.4	.9	.01	.014	.044	Y
14	.5	.9	.01	.014	.017	Y
15	.6	.9	.01	.014	.037	Y
16	.7	.9	.01	.014	.191	P
17	.8	.9	.01	.014	.114	Y
18	1.01	1.18	.01	.014	.161	P
19	1.02	1.18	.01	.014	.161	YP
20	1.05	1.18	.01	.014	.088	P
21	1.1	1.18	.01	.014	.062	YP
22	1.15	1.18	.01	.014	.148	YP
23	1.17	1.18	.01	.014	.114	P
24	1.18	1.18	.01	.014	.096	P
25	1.1	1.3	.01	.014	.00082	P
26	1.1	1.433	.01	.014	$<10^{-4}$	RP
27	.6	.8	0.	0.	.032	Y
28	.6	.8	.01	.0014	.0515	Y
29	.6	.8	.01	.014	.034	Y
30	.6	.8	.1	.14	.083	Y
31	.6	.8	10^5	.0014	.0515	YR
32	.6	.8	10^5	.014	.0562	Y
33	.6	.8	10^5	.14	.215	P
34	.6	.8	.01	.0003	.043	Y
35	.6	.8	.01	.0008	.038	Y
36	.6	.8	.01	.003	.038	R
37	.6	.8	.01	.01	.040	Y
38	.6	.8	.01	.1	.037	YP

NOTE: Initial Offset about Yaw Axis: $W_1 = 0.1$

NUMERICAL CALCULATIONS

With nonlinear behavior it is necessary to define what is meant by stability, and this definition may be quite arbitrary. In the present numerical study the attitude motion is said to be unstable if the vehicle rotates with respect to its reference more than 90° within the 20 orbital revolutions over which the calculation is carried out. For Euler angles the criterion was applied to each of the angles, while for the Euler parameters it was applied to the net angle of rotation. This large bound was placed on the motions so that there would be a better opportunity to see the effect of nonlinearities and therefore understand the way in which the motions become unstable. It may also have physical importance in extreme cases. Passive systems are usually proposed for missions which do not have tight attitude control requirements, for example, the pointing of an antenna with a very large beamwidth. Such a large bound is an upper limit for the specification that might be given for this mission.

Any study of nonlinear equations depends not only on the parameters of the problem but also on the initial conditions. Since the eccentricity e, the two normalized vehicle shape parameters $r_1 = I_1/I_3$ and $r_2 = I_2/I_3$, the damping sphere size $r_o = I_o/I_3$, and the viscous coupling c already imply a five-parameter family, it was decided to limit the generality of initial conditions considered. It was desired to have initial conditions which would allow all of the motions of the vehicle that could be excited without intentionally introducing perturbations that would cause instabilities. (Refer to Fig. 2.)

The initial angular velocity of the vehicle and the damping sphere were chosen normal to the orbit plane with magnitude equal to the mean motion of the orbit. In this way, if the vehicle were isoinertial, it would continue to rotate uniformly without secular (increasing) terms in its deviation from the radial direction.

When there is no oblateness in the attracting body there is no forced motion of the yaw-roll plane, and unless there is an offset in the yaw or roll axes there is no coupling between the pitch motion and the yaw-roll plane. Therefore, in nearly all cases studied an initial offset about the yaw axis was introduced.

Oblateness is normally small. A value as large as .03 was used without introducing behavior not predicted by the linear equations. No oblateness is included in any of the runs discussed below.

Numerical Results

Results presented in graphical form were plotted by a digital-to-analog converter directly from the output tape of an IBM 7090 digital computer.

The nonlinear coupling between the pitch axis and yaw-roll plane was a major mechanism for the development of instabilities. The inadequacy of the linear equations in predicting this type of motion can be seen in Figs. 3, 4, and 5. It can also be seen that the difference between the two linear cases in Figs. 3 and 4 is negligible; therefore, the time variation of the coefficients are not responsible for the excitation of the oscillations in the yaw-roll plane. Of several hundred cases that were run, the nonlinear coupling between pitch and the yaw-roll plane was responsible for tumbling in over 50% of the cases that were unstable. This behavior is relatively insensitive to the initial conditions about the yaw axis. In fact, when reduced by a factor of 10 to an initial yaw offset of .01 radians, the behavior shown in Fig. 6 resulted. The latter is unstable, while the case with the larger initial condition is stable by the present definition, but neither is calculated to warm the control engineer's heart.

A second nonlinear effect which is less interesting and of less importance can be predicted from the behavior of a pendulum excited at resonance. When there are zero initial conditions in the yaw-roll plane, the vehicle motion about the pitch axis builds up at first as in a linear system. As the motions become large, however, the natural frequency changes and the exciting force

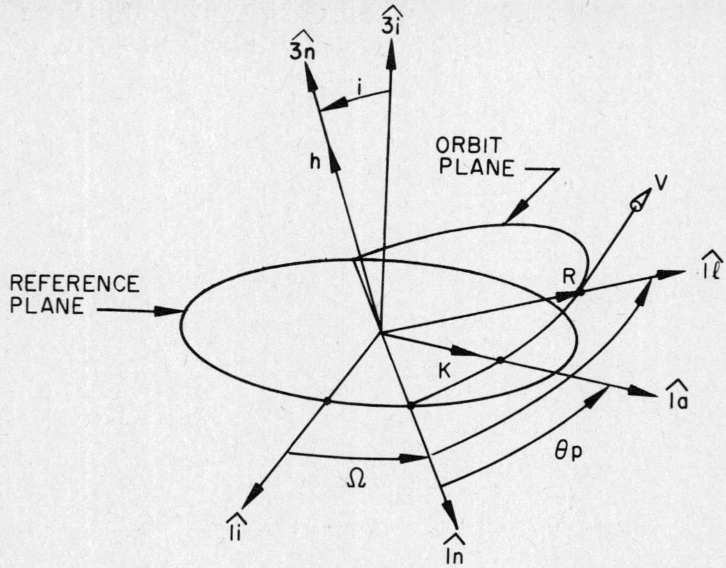

Fig. 2 Orbit Parameters and Coordinates

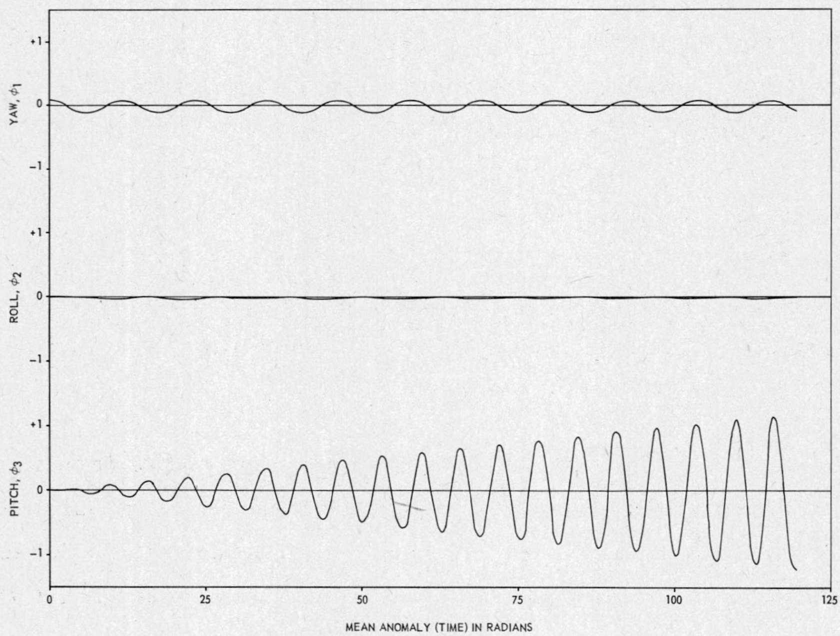

Fig. 3 Attitude Motions from Linear Equations with Constant Coefficients $r_1 = 0.5$, $r_2 = 0.833$, $e = 0.01$, $\phi_1(0) = 0.1$

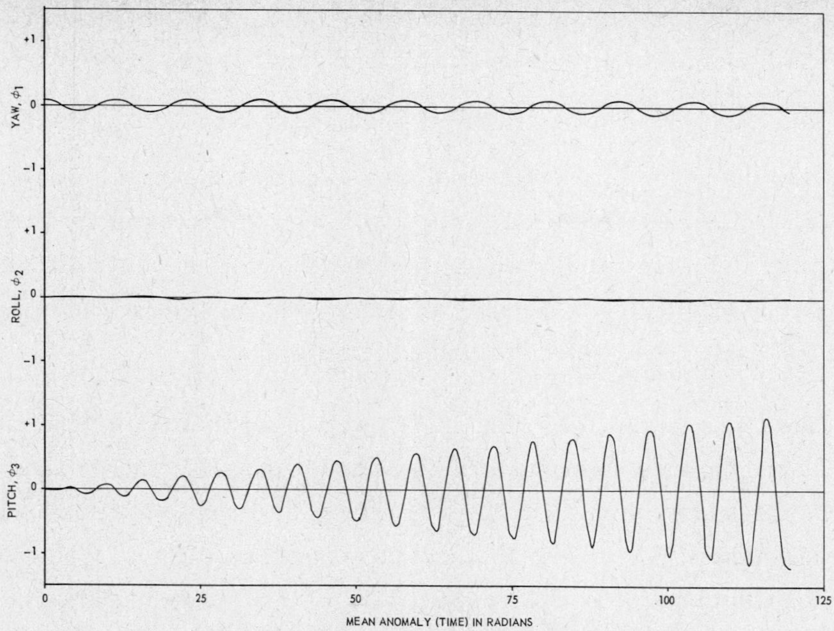

Fig. 4 Attitude Motions from Linear Equations
$r_1 = 0.5$, $r_2 = 0.833$, $e = 0.01$, $\phi_1(0) = 0.1$

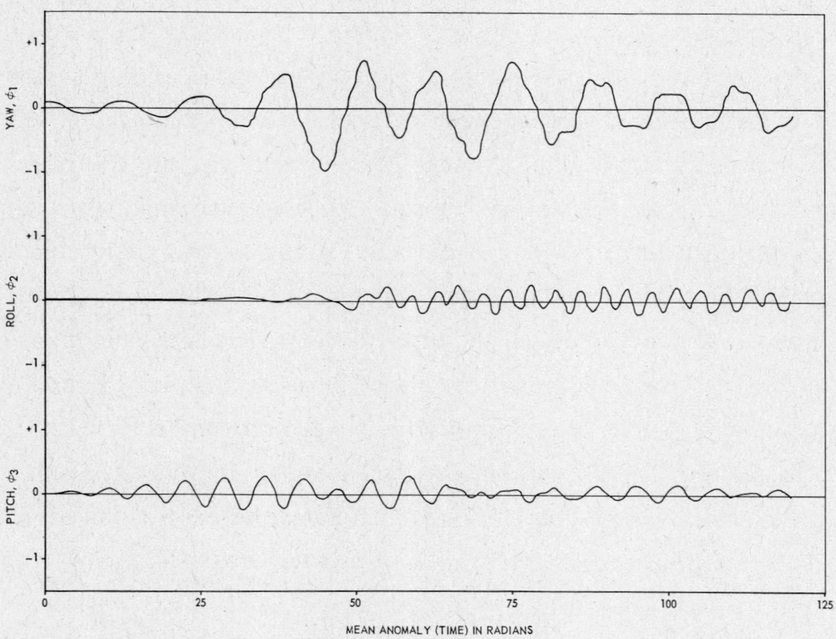

Fig. 5 Attitude Motions
$r_1 = 0.5$, $r_2 = 0.833$, $e = 0.01$, $\phi_1(0) = 0.1$

(which has the small amplitude natural frequency) will go out of phase and remove energy. This effect is shown in Fig. 7. Though stable by the present definition, it would hardly be a wise choice in an actual design.

Approximately one hundred cases were plotted to study the general behavior. Later a search mode was developed to automatically change the eccentricity until a value for the eccentricity for marginal stability was established. The values determined for this, without damping and with damping are given respectively in Table II and III. The results of Table II are shown graphically in Fig. 9.

In the solution of the Mathieu Equation, isolated regions of stability and instability can occur as one changes a parameter. In an analogous way, though stability is defined quite differently, run 13 in Table II does not appear to agree with run 27 in Table III. The stability search in the first case started with $e = .1$ and was stable. Successive calculations for $e = .2$, .141, .119, .109 all were unstable. They were followed by a run of .031 which was stable. The first case, $e = .1$, is therefore an isolated stability.

There are two stable regions for the case of a circular orbit[5] as shown in Fig. 1. In the conventional region, the axis of minimum moment of inertia is aligned with the radial direction, and the axis of maximum moment of inertia is normal to the orbit plane, i.e. $r_1 < r_2 < 1.0$. In the second region, the principal axis of intermediate moment of inertia is aligned with the vertical, and the axis of minimum moment of inertia is normal to the orbit plane. In the conventional region, vehicles have their minimum potential energy at their zero torque position and, hence, when damping is introduced, natural oscillations tend to die out. In the second region, the motion is stable as a result of gyroscopic coupling. The potential at the zero torque position is a saddle point. The motion about the pitch axis increases the potential energy but rotation about the roll axis decreases it; hence, internal damping of vehicles in the second region makes them go unstable just as a top or a body rotating about its axis of minimum moment of inertia without external torques drifts away from its equilibrium position due to energy dissipation. The undamped motions of a vehicle in this region are typical of that shown in Fig. 8. By contrast with the motions in the

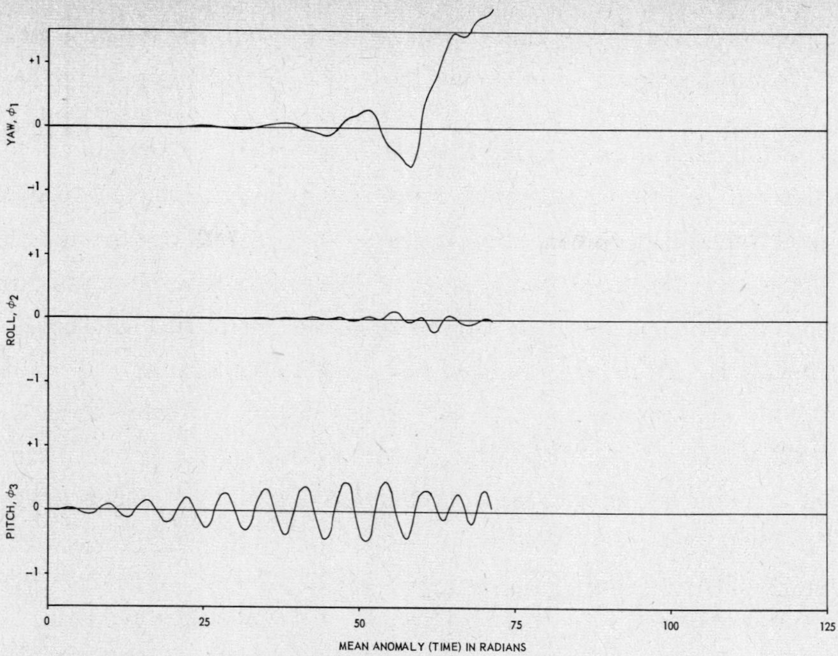

Fig. 6 Attitude Motions
$r_1 = 0.5$, $r_2 = 0.833$, $e = 0.01$, $\phi_1(0) = 0.01$

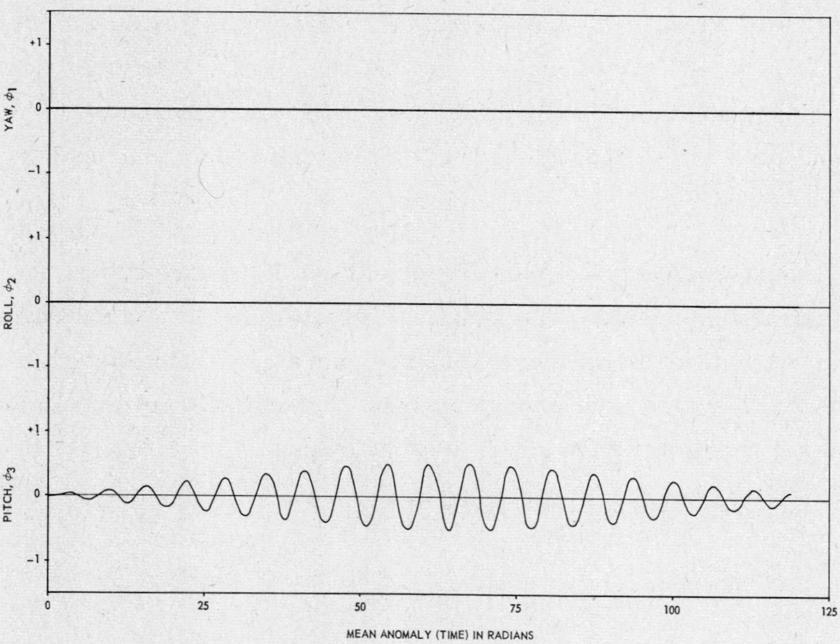

Fig. 7 Attitude Motions
$r_1 = 0.5$, $r_2 = 0.833$, $e = 0.01$, $\phi_1(0) = 0$

conventional region, the oscillations tend to diverge from the initial conditions until an angular velocity has built up to supply the gyroscopic torques which stabilize it

The energy of the oscillations in the yaw-roll plane can be affected both by damping and the eccentricity; however the damping always tends to reduce energy and therefore causes an instability for vehicles in the second region. On the other hand, if the motion of the vehicle in the second region is primarily about the pitch axis, damping tends to stabilize it. These effects can be seen in several runs in Tables II and III. Undamped run 24 is critical in yaw, and the marginal eccentricity is reduced by damping, as shown in damped run 20. On the other hand, undamped run 28 shows an increase in the marginal eccentricity over damped run 24 where the pitch axis is critical, but damping tends to bring out the beast in yaw just enough to have it share the blame for instability. The marginal eccentricity increases, however, since the motion is dominantly about the pitch axis.

CONCLUSION

Satellites which have a stable attitude when in a circular orbit are subject to excitations which can make them tumble when the orbit is eccentric or distorted due to oblateness. The eccentricity above which tumbling occurs varies with the vehicle shape, but is between .05 and .2 for most vehicles, Fig. 9. In the majority of unstable cases studied, the forcing torques about the pitch axis cause the vehicle to tumble about its yaw axis through second order small coupling terms between the pitch and yaw axes.

When gravity external torques alone are considered, linearized equations with constant coefficients make a good mathematical model of the system as long as the angles stay small and there is no chance of instability developing. Including the time variation in the coefficients of the linear equations does not help describe the motions more accurately or predict the stability of the system.

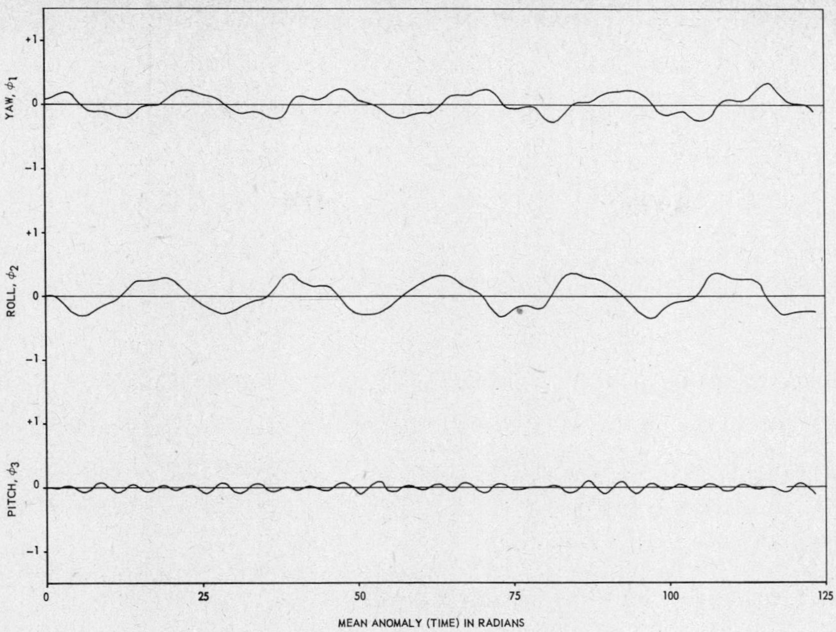

Fig. 8 Attitude Motions
$r_1 = 1.05$, $r_2 = 1.6$, $e = 0.01$, $\phi_1(0) = 0.1$

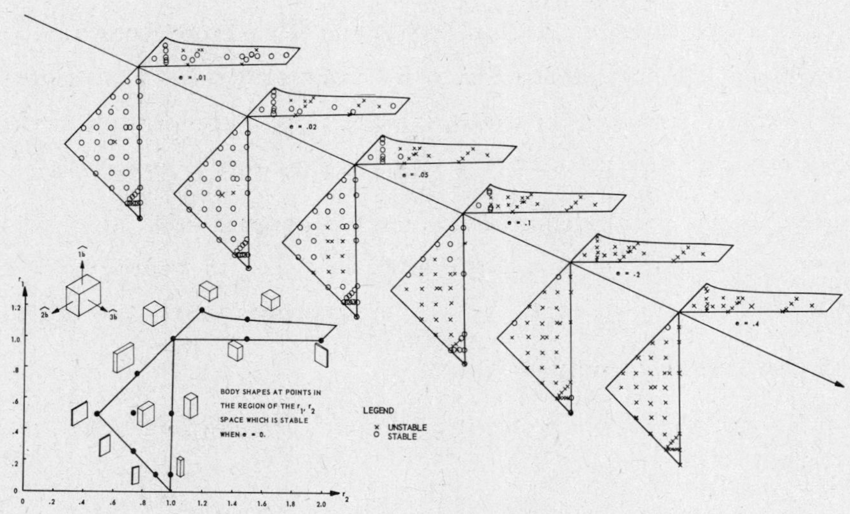

Fig. 9 Attitude Stability as a Function of Vehicle Shape (r_1 and r_2) and Orbital Eccentricity

The results presented are for a satellite subject only to gravitational external torques. Though stability in the presence of gravity torques may be necessary, it is not necessarily sufficient to assure a stable orientation. The effect of other external and internal torques must be evaluated.

NOTATION

Symbols

B	dyadic relating the time derivatives of set of Euler angles to the angular velocity vector of the orientation they specify
I	the moment of inertia dyadic
$\overline{\overline{1}}$	the idemfactor
h	orbital angular momentum vector
H	angular momentum of vehicle about its center of mass
L	torque vector
N	unit vector along the pole of an oblate spheroid
R	radius vector from center of spheroid
W	a vector with magnitude 2 sin $\psi/2$ and direction along the axis about which a rotation of ψ carries the reference triad into the body triad. The components are the first three Euler parameters. The fourth Euler parameter, W_4, is a scalar 2 cos $\psi/2$
ϕ	a generic term including the three Euler angles ϕ_1, ϕ_2 and ϕ_3. Its time derivative, $\overset{o}{\phi}$, is the angular velocity vector expressed on a basis of the axes about which each performs its rotation
ω	angular velocity vector
c	coefficient of viscous coupling between sphere and body
e	orbital eccentricity
J	oblateness coefficient

k	gravity field constant of the spheroid
n	orbital mean motion
r_o, r_1, r_2	ratios of moments of inertia to I_3 of the body
W_4	fourth Euler parameter, see W
ϕ_1, ϕ_2, ϕ_3	Euler angles specifying body orientation with respect to a reference
$\omega_1, \omega_2, \omega_3$	natural frequencies of vehicle

Sub and superscripts

b	body
d	damping
e	equatorial
g	gravitational
r	reference
s, o	sphere
1, 2, 3,	components along the 1, 2 and 3 axes

Operations

$\cdot$, $\times$	dot and cross products
$\wedge$	indicates a unit vector e.g.: $\hat{R} = R/\lvert R \rvert$
$\lvert \; \rvert$	the magnitude of a vector
V^b	the vector quantity V for b, the body
V^{b-r}	the vector quantity V for b, with respect to r
V_b, V_{1b}	the vector V expressed in b coordinates, the first component of V expressed in b coordinates

REFERENCES

1. Routh, E. J., The Advanced Part of a Treatise on the Dynamics of a System of Rigid Bodies, Dover, 1955, Chapter XII

2. Plummer, H. C., An Introductory Treatise on Dynamical Astronomy, Dover, 1960, Chapter XXII

3. Roberson, R. E., "A Review of the Current Status of Satellite Attitude Control," presented at the IAS-ARS Regional Meeting on Space Exploration, San Diego, 5 Aug. 1958

4. Frye, W. E., and Stearns, E. V., "Stabilization and Attitude Control of Satellite Vehicles," American Rocket Society Journal, vol. 28, No. 12, Dec. 1959, p. 927

5. DeBra, D. B., and Delp, R. H., "Satellite Stability and Natural Frequencies in a Circular Orbit," Journal Astronautical Sciences, vol. VIII, No. 1, Spring 1961

6. DeBra, D. B., "Vectors and Dyadics: The Natural Tools of Space Vehicle Mechanics", AAS preprint 61-76, presented San Francisco, Calif., Aug. 1961

7. Goldstein, H., Classical Mechanics, Addision-Wesley, Mar. 1956, p. 134

8. Roberson, R. E., "Attitude Control of Satellite Vehicles - An Outline of the Problem," Proc. of VIII Intern. Astro, Fed. Congress, Barcellonna, 1957, p. 338, Eqs. (24), (25), and (26)

9. DeBra, D. B., The Attitude Large Motions and Stability, due to Gravity, of a Satellite with Passive Damping in an Orbit of Arbitrary Eccentricity about an Oblate Body, Ph. D. Dissertation, Stanford University, Jan. 1962

ATTITUDE CONTROL OF SATELLITES USING INTEGRATING GYROSCOPES

Joseph E. DeLisle,* Glenn Ogletree,+
and Benjamin M. Hildebrant+

Vertically oriented satellites instrumenting the orbital plane exhibit certain gyroscopic properties. If one or more integrating gyroscopes are properly installed, this characteristic may be used to accomplish semi-passive damping of vehicle librations. Vehicle oscillations cause gimbal precession in the gyroscopes. Damping occurs because of energy dissipation as heat in the viscous restraint of each gyro. The systems described comprise only the gyro or gyros, properly supplied with constant power. Neither external error sensing nor on-board computation is required. Of the various systems discussed, the simplest uses a single gyroscope to accomplish three-axis damping.

* Assistant Director, Instrumentation Laboratory, Massachusetts Institute of Technology

+ Group Leader, Research Staff, Instrumentation Laboratory, Massachusetts Institute of Technology

INTRODUCTION

In the utilization of a satellite of the earth to accomplish a specific mission, one prime requisite is generally the accurate orientation of the entire vehicle or a specific portion of it according to some predetermined plan. Such orientation is accomplished by applying the (relatively old) fundamentals of the relatively new field of satellite attitude control. The stabilization or control of a satellite vehicle reduces ultimately to a three dimensional problem in rigid body (or non-rigid body) dynamics. Nevertheless there is more involved than conceptual grasp of spinning tops or elementary gyroscopic theory in understanding the dynamics of satellites. Visualization of a satellite moving along its path about the earth is usually not difficult, but the dynamics that are the consequence of the simultaneous rotation of a vertically-oriented satellite about its center of mass seem less well understood.

Aside from purely gyroscopic phenomena associated with vertically oriented satellites, certain other factors derived from the vehicle's orbital environment are vital to its operation but sometimes misunderstood. For example, far from being in a "gravity-free environment" as is sometimes suggested, the satellite's motion is dramatically influenced by gravity. The specific force of gravity is the vector resultant of the gravitational acceleration and the negative of the centripetal acceleration[1*]. The gravity force acting on the portion of the satellite nearest the earth is downward along the geocentric radius toward the earth's center, while that acting on the portion away from the earth is directed outward along the radius. This is because the gravitational acceleration varies as the inverse square of the distance from the mass center of the earth to the point in question, while the centripetal acceleration is proportional to the

---

[*] Superscripts refer to similarly numbered references at the end of the paper.

radius to the point in question divided by the cube of the radius to the mass center of the vehicle. Thus, gravitation is dominant on the downward end of the satellite and the negative of centripetal acceleration is dominant on the upper end. This gravity variation - from "downward" to "upward" over the length of the satellite - provides the necessary opposing forces of a restoring couple that is generally referred to as the Gravitational Gradient Torque acting on the satellite about its mass center. This varies as three halves of the sine of twice the angle of deflection of the axis of least inertia from the geocentric radius, and has received extensive treatment in the literature (e.g., Refs. 2, 3) It will be evident from subsequent remarks that it provides the only restoring moment influencing pitch axis motions of the satellites studied here and contributes to the total roll axis restoring torque, but has no component about the yaw axis since the opposing forces described above can have no moment about the geocentric radius.

The present paper is concerned with a general technique of achieving three-axis control of the attitude motions of a specific class of earth satellites, those that are both vertically - and orbital plane-oriented. The technique involves the use of one or more integrating gyroscopes in the multiple (and semi-passive) role of motion sensors and control actuators. An attempt will be made to further the general understanding of gyroscopic phenomena in satellite attitude motions, after which a series of gyroscopic attitude control systems will be illustrated and discussed.

TYPES OF EARTH SATELLITES

Satellites of the earth may be divided into several categories that depend upon the type of attitude control required. For example, satellites may serve as platforms for astronomical observation or recording instruments, in which case the vehicle must be held rotationally-fixed relative to an inertial frame of reference.* On the other hand a satellite similarly equipped for telescopic or photographic observation of space objects might be used to study other bodies in our solar system such as the moon, the sun, and other planets or their moons. In that application, the vehicle - or at least the telescopic apparatus - must be caused to rotate relative to the inertial frame such that the optical line of sight tracks the body being observed as the satellite moves along its orbital path. A third broad group of satellites are those that are to be held in a prescribed orientation relative to the earth. Within this category are included satellites that track the local horizontal line in or normal to the orbital plane, those that track a non-vertical position vector from the satellite to a specified (fixed) or non-fixed) point on the earth's surface, satellites that track other near-earth vehicles such as satellites or missiles, and satellites that track the local vertical direction. In the latter category, of vertically-oriented vehicles, it might be required that the axes of the satellite instrument a local geographic reference frame such as the northerly, easterly and vertical directions. Or, more typically, the vehicle axes might be required to instrument a local orbital reference frame, namely, the local vertical and horizontal directions in the orbital plane and the normal to the orbital plane. In the present paper, the discussion is specifically restricted to this last situation.

COORDINATE REFERENCE FRAMES

To further clarify the orbital situation studied here, it will be instructive to define certain sets of cartesian coordinate reference frames that will be useful in the remarks to follow. These are the:

---

* A reference coordinate frame that is at rest relative to any set of extra-solar star lines at the satellite may be considered an inertial reference frame for purposes of discussing earth-oriented satellites.

Geocentric Inertial Reference Frame	I	(Fig. 1*)
Geocentric Earth Reference Frame	E	(Fig. 1)
Geocentric Orbital Plane Reference Frame	K	(Fig. 2)
Geocentric Orbital Position Reference Frame	P	(Fig. 2)
Vehicle-Centered Orbital Reference Frame	O	(Fig. 3)
Vehicle-Centered Vehicle Reference Frame	V	(Fig. 4)
Vehicle-Centered Principal Axis Reference Frame . .	A	(Fig. 5)
Vehicle-Centered Gyro Case Axis Reference Frame . .	GU	(Fig. 6)
Vehicle-Centered Gyro Gimbal Axis Reference Frame .	GIM	(Fig. 7)

Complete coordinate transformations between these frames are set forth in Ref. 3 They may also be readily developed from a study of the above figures. These reference frames are adequate for discussion of the gyroscopic attitude control systems to be described in this paper. In the remarks to follow, it is presumed that the V frame is exactly aligned with the A frame, except where otherwise noted in connection with skewed vehicle configurations.

THE SATELLITE AS A GYRO ELEMENT

The term gyro element as used here refers to a body having finite, non-zero mass, and which is characterized by a non-zero angular velocity about some axis through its mass center**. The motions of such a body have been shown[4, 5, 6] to be governed by the law of Conservation of Angular Momentum. Specifically, if, as seen by an observer fixed in the I

* The figures and notation are generally consistent with Ref. 3.

** Actually, satellites of this type have certain of the dynamic characteristics of the gyro element as more rigorously defined in Derivation Summary No. 1 of Ref. 6.

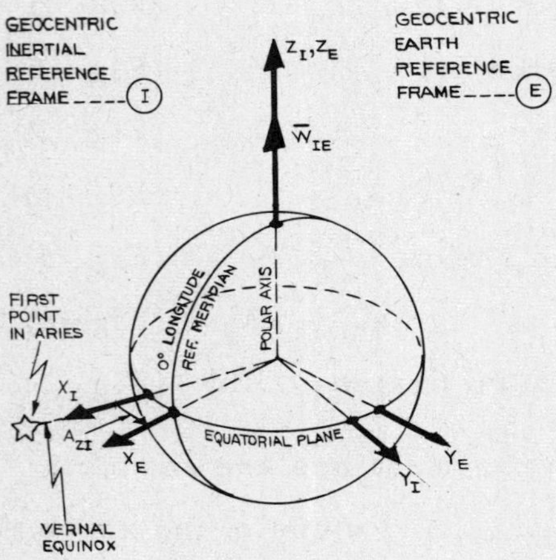

Fig. 1 Geocentric Inertial (I) and Earth (E) Reference Coordinate Frames

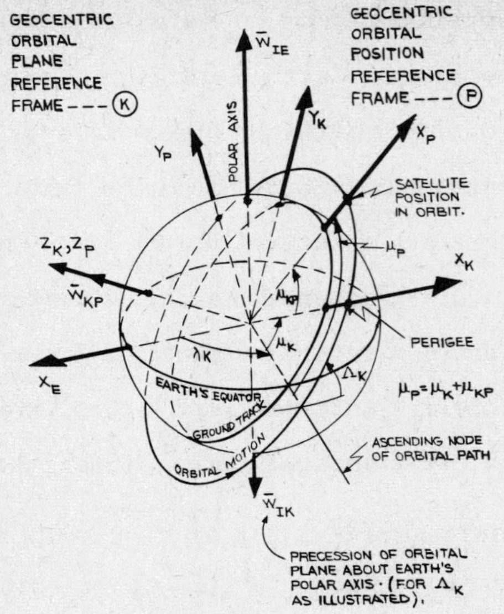

Fig. 2 Geocentric Orbital Plane (K) and Orbital Position (P) Reference Coordinate Frames

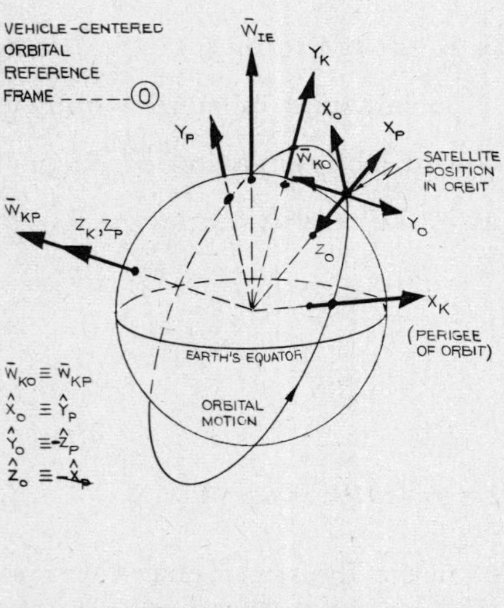

Fig. 3 Vehicle-Centered Orbital (O) Reference Coordinate Frame

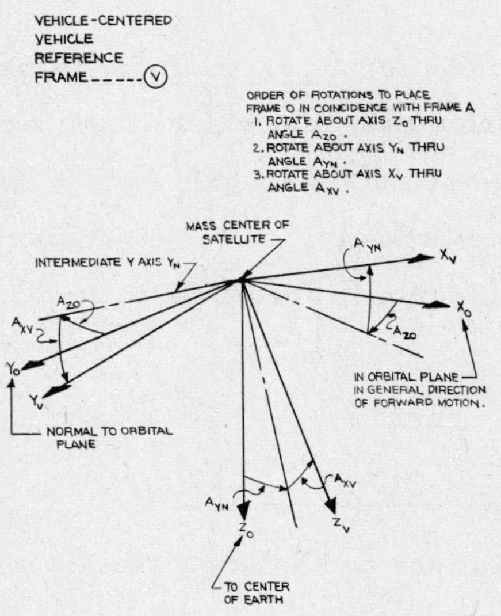

Fig. 4 Vehicle-Centered Vehicle (V) Reference Coordinate Frame

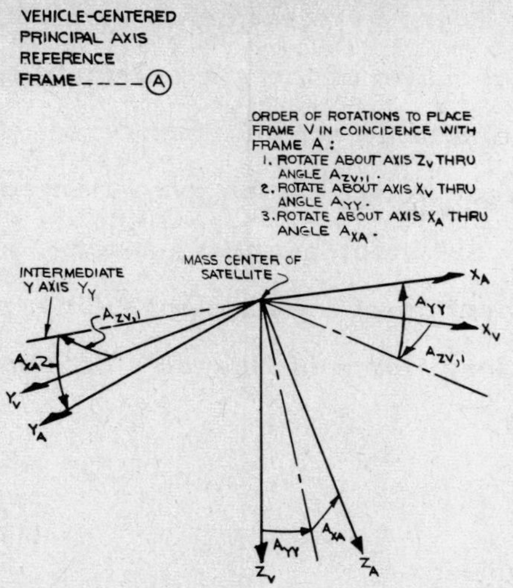

Fig. 5 Vehicle-Centered Principal Axis (A) Reference Coordinate Frame

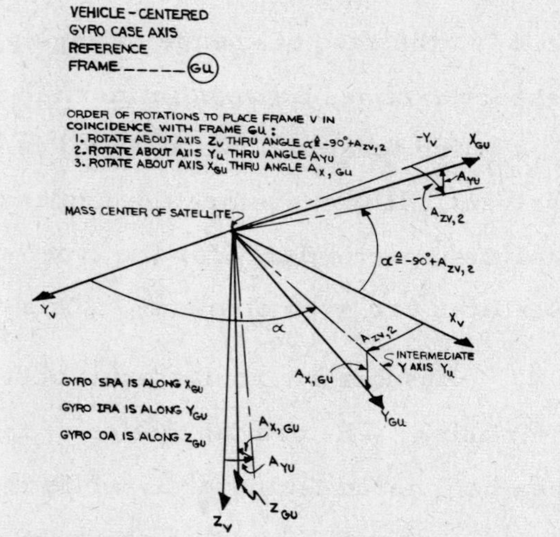

Fig. 6 Vehicle-Centered Gyro Case (GU) Reference Coordinate Frame

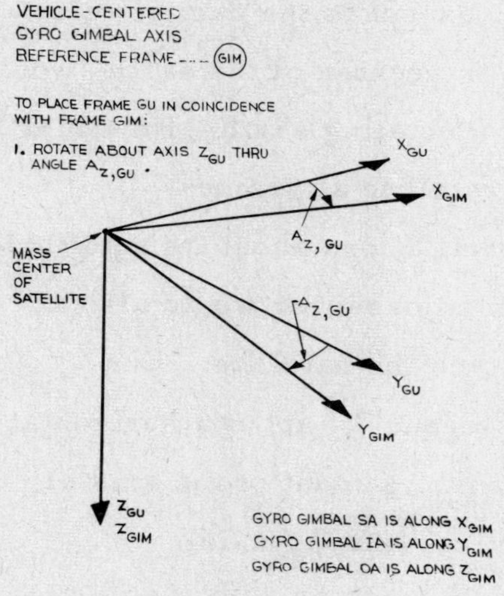

Fig. 7 Vehicle-Centered Gyro Gimbal Axis (GIM) Reference Coordinate Frame

frame, there is no unbalanced torque acting on the gyro element, then that same observer will see the total angular momentum of the gyro element as a vector that is invariant both in magnitude and in direction. Conversely, if the observer sees a vector unbalanced torque acting on the gyro element he will simultaneously observe, as an identical vector quantity, a non-zero time rate of change of the angular momentum of the element. These concepts are neatly condensed into the rotational form of Newton's Second Law of Motion:

$$\left.\frac{d\bar{H}}{dt}\right|_I = \Sigma\bar{M}_{\text{applied}} \tag{1}$$

That is, the rate of change of angular momentum of the rotating body with respect to time, as seen in inertial space, is equal to the vector sum of all applied torques. Equation (1) is the fundamental equation of gyroscopic analysis. It is therefore the fundamental equation of satellite motion analysis, particularly for the type of satellite studied here, since such satellites are gyro elements. This may be visualized as follows:

Assume a circular orbit about a spherically-symmetric earth for simplicity. The orbital motion of the mass center of the satellite may then be ignored temporarily while trying to achieve a clear understanding of the rotational dynamics about its mass center. An observer placed at the satellite mass center, non-rotating with respect to the orbital reference frame O, and looking downward toward the center of the earth, would consider himself to be inertially fixed above a rotating earth. He would also be aware of the libratory motion of the satellite as frame V - affixed to the vehicle structure - exhibited oscillations about its equilibrium orientation in which it tracks frame O. Such an observer would also anticipate certain of the external torques acting on the satellite. For example there is the gravitational gradient torque[2, 3] about a horizontal axis through the mass center, produced by misalignment of the axis of least inertia (the Z_A axis) with the local vertical (the Z_O axis).

The O frame observer might reason that since misalignment of the vehicle about only the Z_O axis (the local vertical) gives rise to no external gravitational gradient torque, the orientation of the vehicle about the local vertical is arbitrary. However, the vehicle, given such an initial misalignment in yaw and released, would exhibit not only oscillations about the local vertical but also coupled oscillations about the horizontal axis in the orbital plane (the X_O axis).

In order to justify the observed phenomena in the case of the undamped vehicle alone, the O frame observer might postulate a vector angular momentum associated with the vehicle itself and directed along the negative sense of the Y_V axis. Such an angular momentum vector is in fact associated with the vehicle due to its once-per-orbit revolution about its mass center in pointing nominally "downward" toward the earth's center at every point along its path. The vehicle's angular momentum $\overline{H}_B$ is illustrated in Fig. 8. The difficulty in presenting this point is in finding a single observer, associated with the satellite's position, that can be aware of both the rotational and libratory motions of the satellite. Perhaps a better choice would be an observer that is non-rotating with respect to frame K (see Fig. 2) but is translating along with the mass center of the satellite.

The simultaneous nature of roll and yaw motions may be clarified by plotting them as functions of each other, with time suppressed as a coordinate. If the roll error angle A_X is plotted as a function of the yaw error angle A_Z for an initial roll offset, the result for an undamped vehicle is ordinarily a complex curve that is described in a counterclockwise sense, within a smooth envelope that is nominally centered about the origin in absense of constant external torques. The shape of the curve is dependent upon the satellite vehicle configuration and the resultant angular momentum of any rotating components within the vehicle. Fig. 9 is a representative plot for a particular undamped vehicle with no internal rotating parts, while Fig. 10 shows the effect on the plot if the

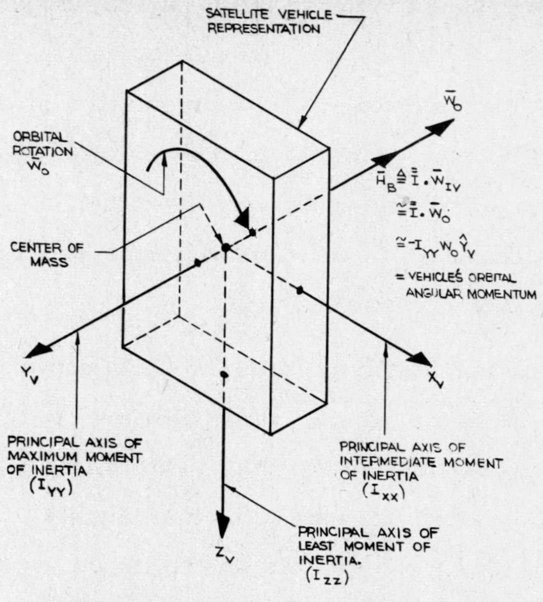

Fig. 8 Representation of the Satellite Vehicle As a Gyro Element

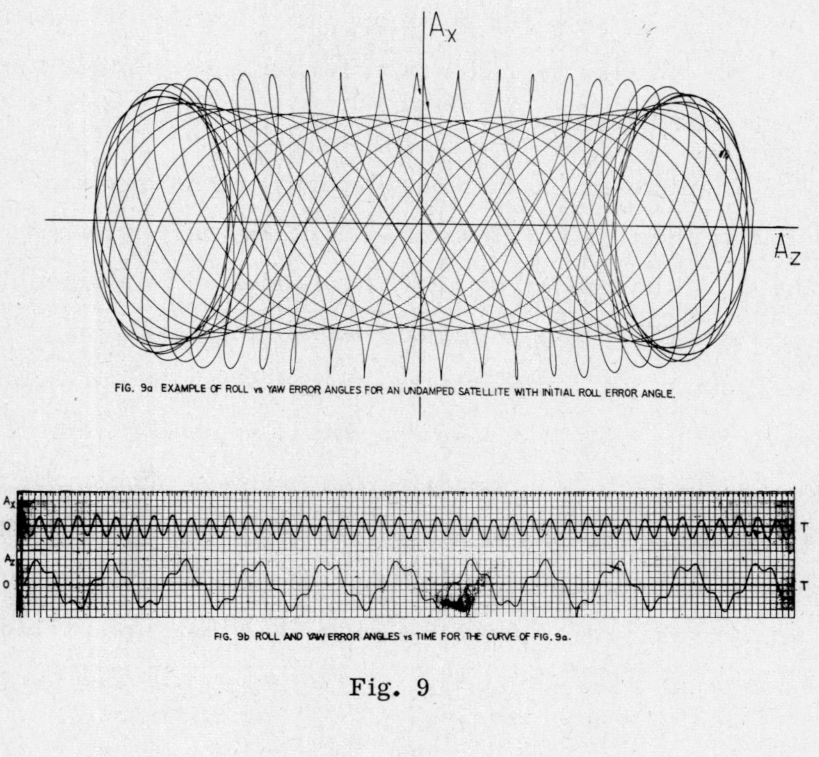

Fig. 9

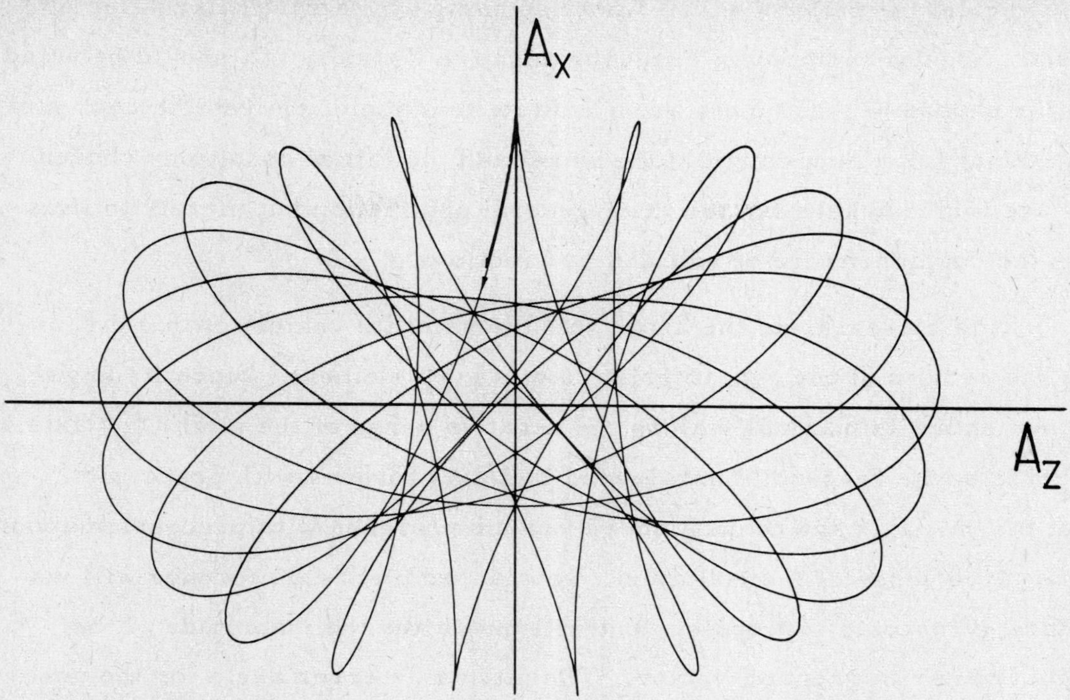

Fig. 10a Example of Roll vs Yaw Error Angles for an Undamped Satellite with Constant Speed Wheel Along Negative Pitch Axis, for Initial Roll Error Angle

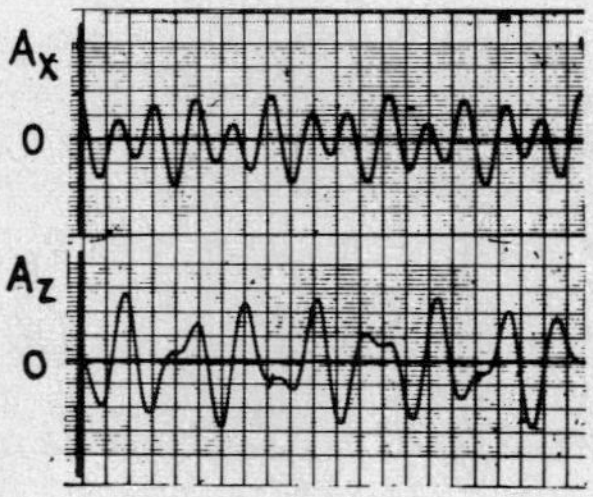

Fig. 10b Roll and Yaw Error Angles vs Time for the Curve of Fig. 10a

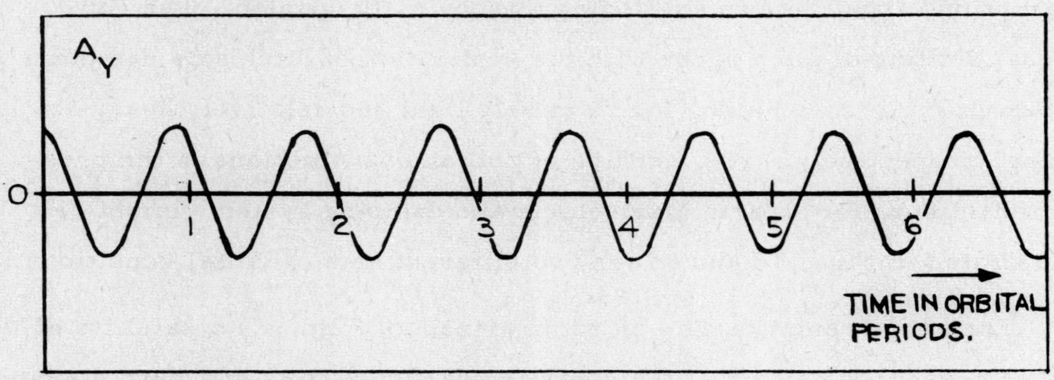

Fig. 11 Pitch Error Angle vs Time for an Undamped Satellite with Initial Pitch Error

same vehicle is equipped with a wheel spinning with a particular value of constant angular momentum about the negative Y_V axis. (It should be noted that the shapes of these plots are a definite function of the vehicle configuration, the angular momentum of the wheel and the initial conditions chosen; they are included here not for their general application but merely to illustrate the coupled nature of roll and yaw motions.)

It is evident from the above that the satellite vehicle of this discussion has certain of the characteristics of a gyro element. Since its angular momentum is nominally along the negative sense of the pitch (Y_V) axis, the vehicle will respond to unbalanced roll axis torques with precession about the yaw axis and to unbalanced yaw axis torques with precession about the negative sense of the roll axis. Unbalanced pitch axis torques will not produce gyroscopic precession, but will only alter the magnitude of the body's angular momentum vector. The pitch axis error angle for the undamped vehicle, plotted as a function of time after an initial pitch offset, is as shown in Fig. 11.

From a consideration of the error plots described above, it is evident that the vehicle axes may be caused to settle to an equilibrium condition in which they track the orbital reference axes, frame O, if damping torques are applied both about the pitch axis and in the roll-yaw plane. As in gyrocompass erection, the satellite roll-yaw damping may be accomplished by damping the motion about either the roll or the yaw axis or both, since motions about the two axes are gyroscopically coupled. The conventional settling of pitch error with the application of pitch axis damping is shown in Figs. 12a and 12b, for relatively light and relatively heavy damping, respectively. The settling of roll and yaw motions in the presence of damping, for a particular vehicle and damping system, might be as illustrated in Figs. 13 and 14 for two different sets of initial conditions.

The above remarks may be summarized as follows: A satellite of this type possesses certain of the dynamic characteristics of a gyro element;

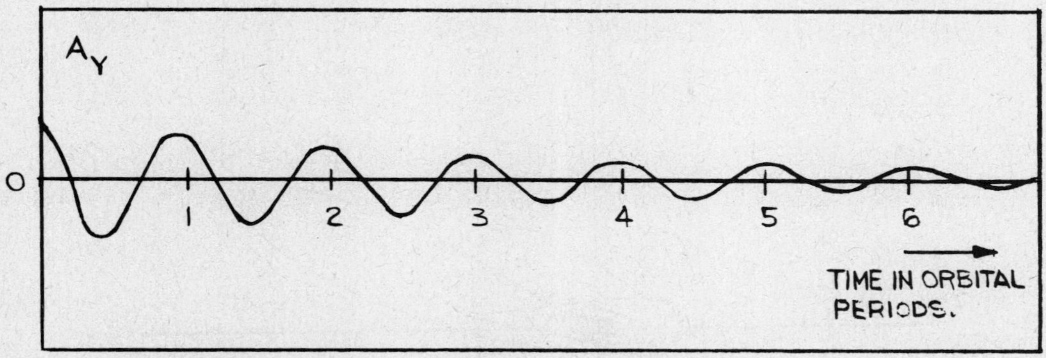

Fig. 12a Pitch Error Angle vs Time for Relatively Light Damping, with Initial Pitch Error

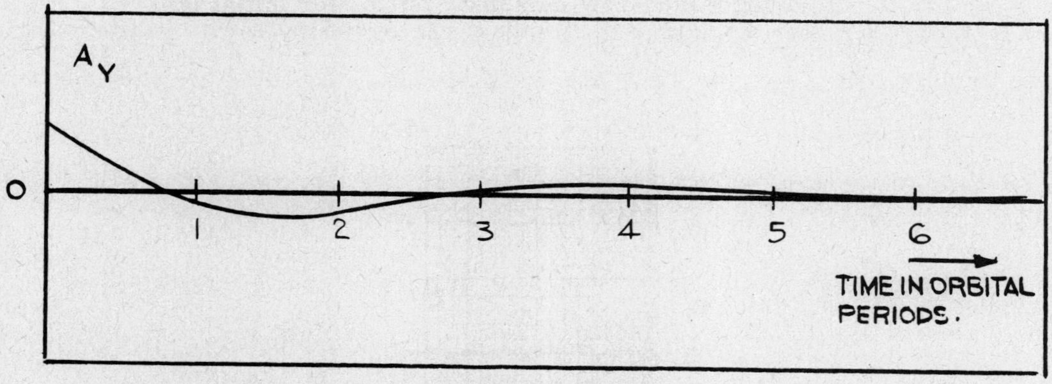

Fig. 12b Pitch Error Angle vs Time for Relatively Heavy Damping, with Initial Pitch Error

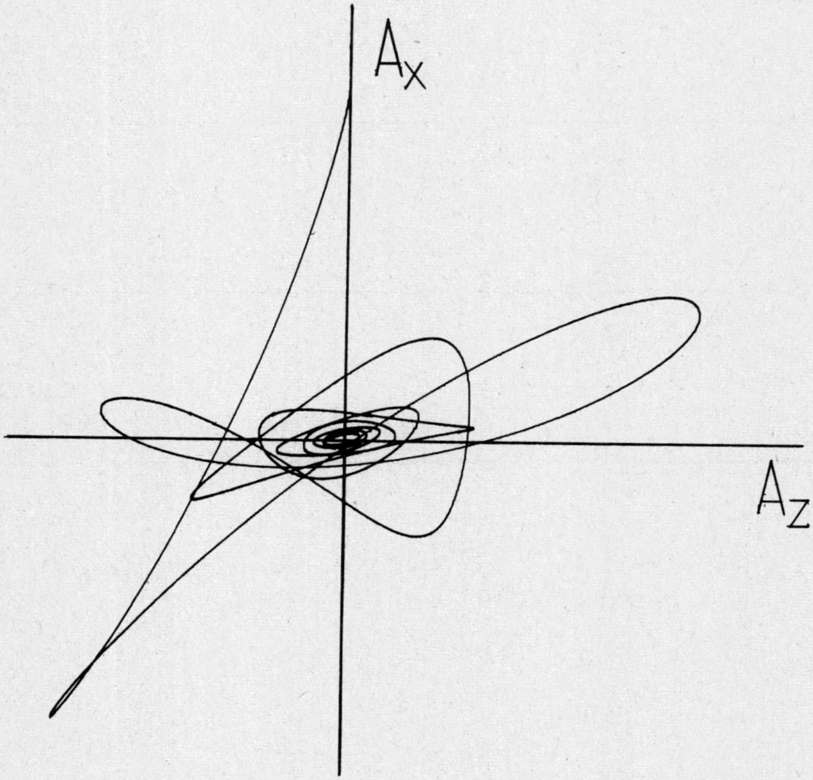

Fig. 13a Example of Roll vs Yaw Error Angles for a Satellite Having a Roll-Yaw Damping System, for Initial Roll Error Angle

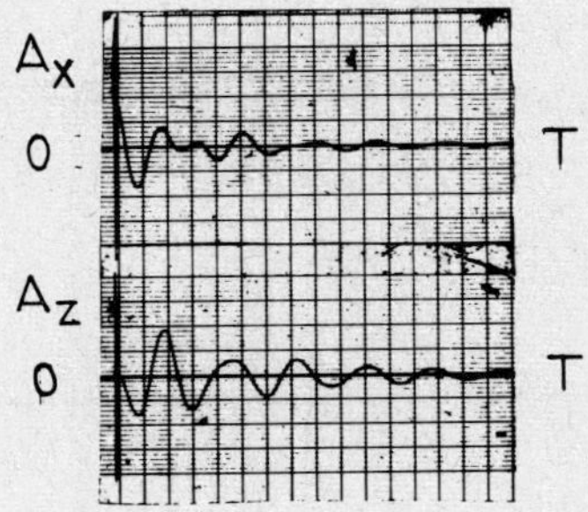

Fig. 13b Roll and Yaw Error Angles vs Time for the Curve of Fig. 13a

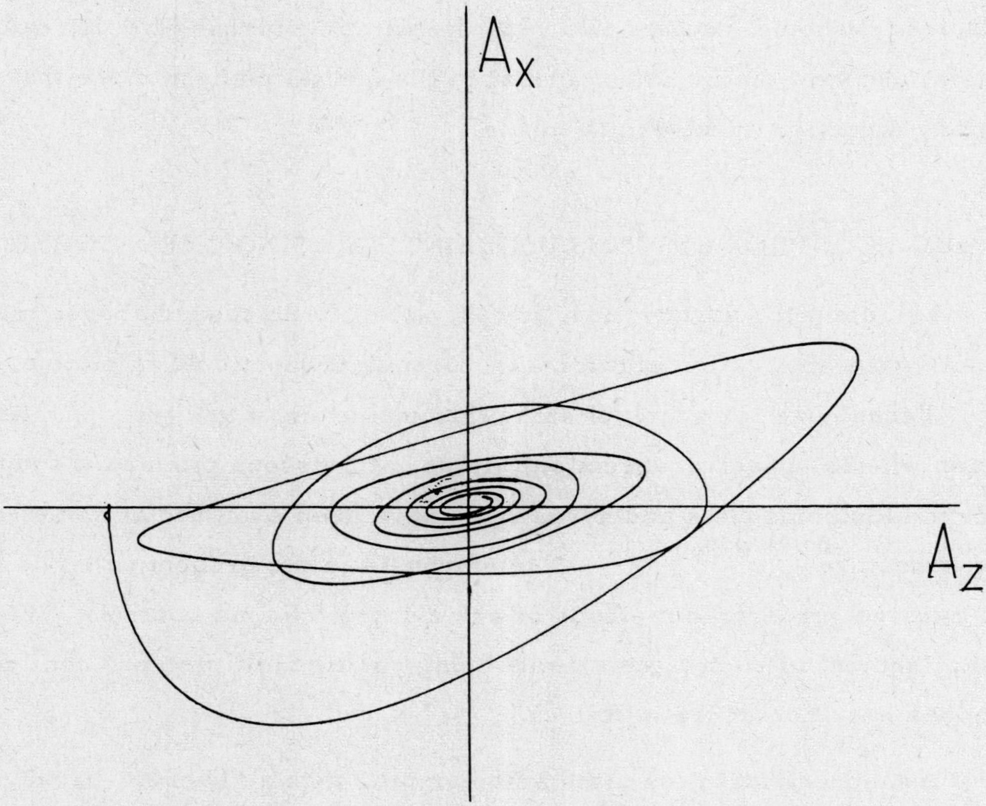

Fig. 14a Example of Roll vs Yaw Error Angles for a Satellite Having a Roll-Yaw Damping System for Negative Initial Steps in Yaw Error Angle and Roll Error Angle Rate

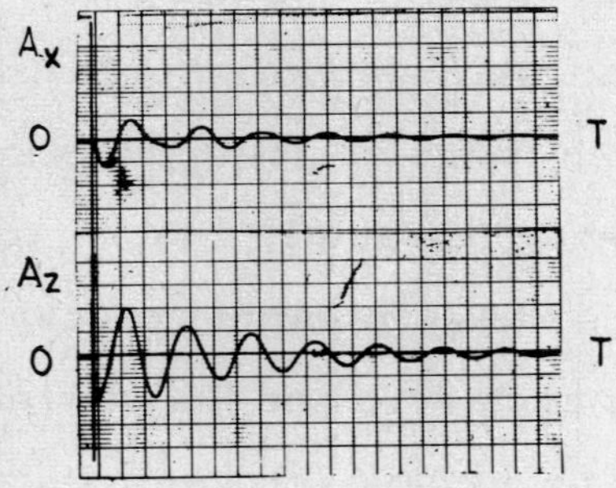

Fig. 14b Roll and Yaw Error Angles vs Time for the Curve of Fig. 14a

with suitable damping of its motions axes fixed in such a vehicle can be made to instrument a local orbital reference frame; the accomplishment of the desired damping involves two separate damping torque effects, namely, one along the axis that is to be normal to the orbital plane and one that is along any direction in the orbital plane.

ATTITUDE CONTROL SYSTEMS USING INTEGRATING GYROSCOPES

Vehicle pitch axis motion, for the satellite discussed above, may be damped by the application of pitch axis torques proportional to pitch error rates. Techniques for accomplishing this by means of gas jets, inertia reaction wheels, passive wheels, and fluids with viscous pitch axis coupling, on-board electromagnets and gyroscopes have been explored at some length in the literature[3,7 8,9,10,11] The emphasis in the present paper is on semi-passive gyroscopic methods of achieving pitch axis control. The accomplishment of such pitch axis damping, using multiple gyro configurations, has been described in Ref. 3.

A mechanization for pitch axis damping with a "back-to-back" pair of integrating gyros is illustrated in Fig. 15. The gyros are arranged to have no component of angular momentum along the roll axis, and for their components of angular momentum along the yaw axis to be continuously equal and opposite, in order to be self-cancelling. Thus, the system has no influence on the roll-yaw motion. As the gyros precess equally and oppositely in response to angular velocities occuring along their common input reference axis - which is the vehicle pitch axis - the vector sum of the two spin angular momenta $\bar{H}_1$ and $\bar{H}_2$, is a pitch axis angular momentum component, $\bar{H}_Y$. Since the gyro gimbal precession is restrained by the viscous coupling (C) in each gyro and the instruments hence are classed as integrating gyros, it follows that each gimbal's angular velocity - and hence the time rate of change of the pitch axis component of total gyro angular momentum, $\dot{H}_Y$ - is proportional to the sensed input axis angular velocity The vehicle's pitch axis angular velocity is composed of orbital rate (a constant for the circular orbit studied here) and pitch error rate. The constant component would produce a continuous gimbal precession if uncompensated. It must be cancelled in each gyro by the application of constant torque generator torques M_{TG1} and M_{TG2} leaving only the pitch error rate, $\dot{A}_Y$, influencing gimbal precession. Thus the rate of change of damping system angular momentum along the pitch axis (equivalent to an applied torque about the pitch axis) is directly proportional to pitch error angular velocity, and therefore constitutes a damping torque. Suitable selection of gyro angular momentum and coefficient of damping will yield any desired degree of vehicle pitch axis damping. Note that the system would perform in exactly the same manner if the roll and yaw axes in the figure were interchanged and phased properly to retain a right handed set.

It should also be noted that in the pitch axis damping system just described there is no requirement for separate sensing, computing and control devices. The integrating gyroscopes provide all three functions. It is only necessary to provide the correct orientation of the gyros in the vehicle, the power needed to spin their wheels at constant speed, and the steady gimbal torques needed to prevent gyroscopic precession in response to the body's orbital angular velocity about its mass center, and the pair of gyros then serve as a self-contained pitch damping system. Hence, the term "semi-passive" in describing these systems. This simplicity is characteristic of the gyroscopic control systems discussed in this paper. None of them require any external error sensing by means involving electromagnetic radiation, nor is any on-board computation beyond that inherent in the gyroscopic responses involved.

The use of two gyros in a back-to-back arrangement as described above is applicable to damping of motion about any axis, not just pitch. Special torques would have to be applied in each case via the torque generators (or similar means such as spring restraints) to keep the gyro angular momenta nominally zero in the plane normal to the controlled axis, but the damping function will be essentially the same along the roll or yaw axis, for example, as it is along pitch. The actual response in roll or yaw is, of course, complicated by the gyroscopic coupling of the two axes, but the resulting torques due to precession of the gyro gimbals are in fact damping torques, and two axis settling is obtained. Somewhat more damping is attainable by using a pair of gyros in this manner along each of the three axes due to the larger number of adjustable parameters, but the single pair in the roll-yaw plane ordinarily provides good settling if properly designed. It will be shown in the following that there are several simpler mechanizations offering roll/yaw as well as three axis damping.

Return now to the vehicle of Fig. 8 and recall that the body, in having an angular momentum normal to the orbital plane, has, in effect, an infinitude of gyroscopic input axes that form the orbital plane. We next consider simpler methods of achieving damping in roll and in yaw with integrating gyroscopes. Figures 16 and 17 show two roll-yaw damping systems each of which involves only a single gyro. In both cases, the single gyro's angular momentum $\bar{H}$ is nominally aligned with the body's angular velocity vector $\bar{W}_0$ in order not to exert unwanted body torques in the steady state due to gyroscopic coupling with $\bar{W}_0$. The input axis in Fig. 16 is in the yaw axis direction, while that in Fig. 17 is along the roll axis. The action of the gyro in either case is such that its precession in response to error angular velocities about the input axis produces damping torques about that axis. As shown previously in Fig. 13 and 14, this results in roll and yaw settling. Experience has shown that there is an upper bound to the combined roll and yaw axis damping that can be obtained with this sort of system. The optimum response generally occurs if the angular momentum of the single gyroscope is of the same order of magnitude as the orbital angular momentum $\bar{H}_B$ of the vehicle itself [3, 12]. This fact emphasizes the type of roll and yaw damping action that occurs in semi-passive systems utilizing integrating gyroscopes. It is this: The body's motion is damped by the relative precession of the gyro gimbal or, alternatively, the gyro's gimbal motion is damped by the relative precession of the body. The one "rubs against" the other, and, perhaps in some way analogous to the principle of maximum power transfer in electrical circuits, the maximum rate of energy dissipation in the gyro's damper seems to occur when the two rubbing parts (vehicle and gimbal) are about equal in angular momentum.

Since the gyro angular momentums of Figs. 16 and 17 are approximately determined by vehicle inertias and orbital angular velocity, the principal parameter for optimizing system response is the H/C ratio. (Actually, the torque constant of flex leads, other springs, or electronic feedback loops around the gyros give additional parameters, but in general the performance is degraded for positive spring constants, and the stability of the system is to some extent jeopardized in the case of negative spring constants.) By combining the concepts of Figs. 15 and 16, it is possible to achieve three axis control with only two gyros, and to have three adjustable parameters, within certain physical constraints. Such a system is illustrated in Fig. 18 and will be referred to here as a "Roll-Vee" configuration. The components of angular momentum of the two gyros along the yaw axis comprise the pitch damping system analogous to that of Fig. 15. The components of gyro angular momentum along the negative pitch axis provide roll-yaw damping in the same manner as the system of Fig. 16. Note that the gyros could have been arranged with their output axes along yaw rather than roll. This would lead to a system described here as a "Yaw-Vee" configuration. The equations of motion of the two Vee configurations as well as their responses to roll or yaw disturbances are quite similar. Their responses to pitch disturbances are identical.

The choice of skew angle α for a particular size of gyro angular momentum H in the two Vee configurations seems to follow the conclusions presented in connection with Figs. 16 and 17. That is, the sum of the projections of the two gyro angular momentum vectors on the negative pitch axis ($2H \cos \alpha$) should be of the order of magnitude of the vehicle's orbital

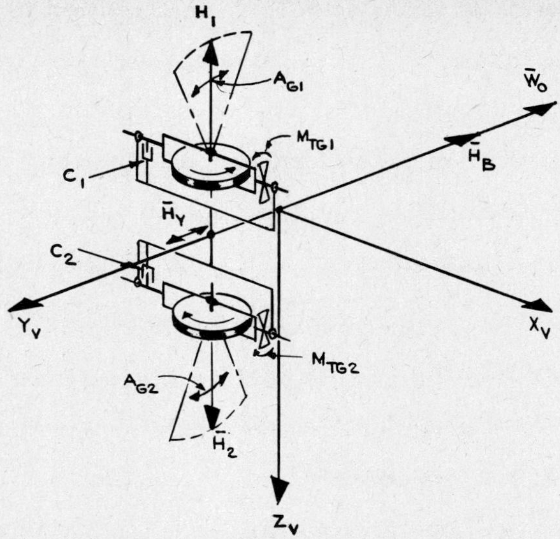

Fig. 15 Pitch Axis Damping System Using Two Integrating Gyroscopes in a "Back-to-Back" Configuration

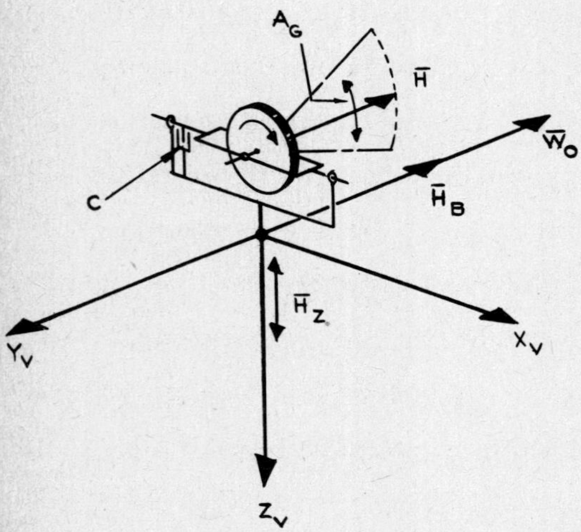

Fig. 16 Roll-Yaw Damping System Using a Single Integrating Gyroscope With Input Axis Along Yaw Axis

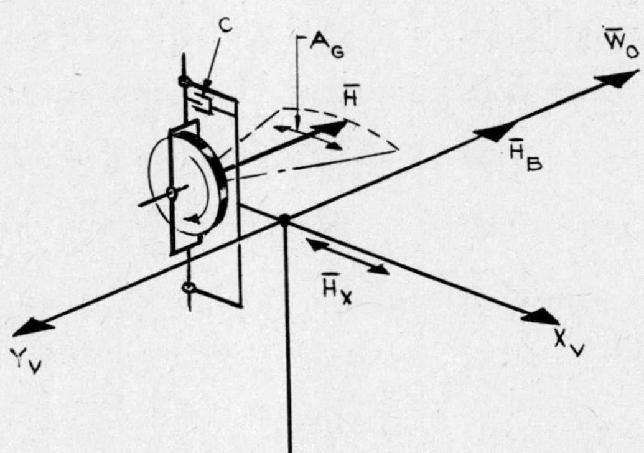

Fig. 17 Roll-Yaw Damping System Using a Single Integrating Gyroscope With Input Axis Along Roll Axis

angular momentum for optimum roll-yaw response. Obvously, as larger skew angles are employed, with attendant increase in gyro angular momentum (to keep $2H \cos \alpha$ near optimum), the pitch axis damping increases. Burt[12] has pointed out that in this sort of system, the optimum pitch transient response occurs with the system adjusted to provide three equal, real, negative roots in the pitch motion characteristic equation. On the other hand again referring to Burt's results, the pitch error response to the forcing effects of eccentricity may be greatly reduced by severely over-damping the pitch system, but this compromises its transient response, as expected.

Thus, the Vee configurations appear to offer possibilities of quite flexible three axis damping, combining the capabilities of the three-gyro systems represented by Figs. 15 and 16 or 15 and 17. As mentioned previously, there are bounds on the attainable roll and yaw axis damping, since the vehicle is called upon as a gyro element itself to work against the gyro angular momentum components that comprise the roll-yaw damping system. If negative spring constants are allowed between each gyro gimbal and its respective case, and a spring and dashpot arrangement (with negative spring constant allowed) linking the two gimbals directly is employed, Burt[12] has shown that remarkably good three axis response is attainable. Settling times (to 1/e of small initial steps in error angles) of the order of fractions of an orbital period are possible with certain vehicle and system configurations. The effects of flex lead or other elastic torques on the gimbals must be included in any analysis or simulations of these systems, as they may have a profound influence on the results as compared with those obtained when treating the gyros as pure integrators.

At this point it has been demonstrated that the satellite vehicle itself behaves as a gyro element (Fig. 8), and can be made to "work against" a conventional integrating gyro in damping roll and yaw motions, as in Figs. 16 and 17. Also, when two gyros are arranged in a Vee configuration as in Fig. 18, the combination will accomplish three axis vehicle damping. We might examine what would happen to the satellite equipped with such a Vee configuration of two gyros if one of the gyros was removed, leaving only a single gyro skewed about the vehicle's roll or yaw axis. The angular momentum of such a single skewed gyro can be resolved into a component along the negative pitch axis and (taking the yaw-skewed case as an example) the roll axis. Clearly, the component along the roll axis will couple with the orbital angular velocity $\overline{W}_0$ and, because of the presence of the torque generator, will result in the exertion of a gyroscopic torque in yaw on the vehicle. Such an unbalanced yaw torque will deflect the vehicle about the yaw axis. An equilibrium condition will result (assume for the moment that the response to the unbalanced torque has been damped, and that the torque generator torque has been reduced if required for maintenance of system stability), in which the body's own gyroscopic restoring torque in yaw is equal and opposite to the steady state torque exerted by the single gyro via its torque generator on the vehicle. What has happened here is that the body's orbital angular momentum vector has been deflected from its customary alignment with $\overline{W}_0$ and caused to be skewed about the yaw axis such that the cross product of body angular momentum with $\overline{W}_0$ is a vector torque equal in magnitude and opposite in sense to the cross product of gyro angular momentum and $\overline{W}_0$. (The magnitude of these torques is the new constant torque level that must be exerted by the gyro's torque generator. It will generally be a lower level than was needed in the Vee configuration since the body's yaw restoring torque is usually weaker than that of the removed gyro.) If the Vehicle Reference Coordinate Frame (V), associated with lines scribed in the structure of the satellite, is now properly relocated by rotation about the principal yaw axis, relative to the deflected Principal Axis Frame (A) described above, the V frame will again point nominally along the local vertical, the normal to the orbital plane, and the horizontal in the orbital plane. In other words, the Vehicle Reference Frame (V) will then instrument or indicate the Orbital Reference Frame (O) just as it and the Principal Axis Frame (A) did when using a full Vee configuration of two gyros.

By the above considerations we are now led to a damping system that is comprised of only one integrating gyroscope working against the vehicle which exhibits gyroscopic properties. The vehicle also provides the vital gravitational gradient torque that yields the customary bistable equilibrium orientation in pitch and, depending on the magnitude of the gyro's angular momentum, either a mono-stable or bistable equilibrium orientation in roll. The yaw equilibrium orientation is also either mono-stable or bistable, depending upon the gyro's angular momentum, but gravitational torque is of course, not involved. (Normal design will usually ensure that only a single stable roll and yaw orientation exist, but no device operating only on the gyroscopic principles discussed here is envisioned as capable of providing a single stable pitch orientation.)

By the same process as described above, the removal of one of the gyros of a Roll-Vee configuration will result in a roll skew of the vehicle principal axes. A single roll-skewed gyro system will result. Since a steady-state gravitational torque will exist in this case due to the roll axis deflection of the vehicle's principal axis frame, it is clear from Eq. (1) that the resultant angular momentum of the system (body orbital angular momentum plus gyro angular momentum) must be deflected from $\overline{W}_0$ such that its vector time rate of change, as seen from inertial space, is equal to the vector roll axis gravitational torque acting on the vehicle. That is, the total angular momentum of the system precesses conically in inertial space at orbital frequency about the normal to the orbital plane. It is therefore a nominally constant vector in the V frame (the V frame is appropriately skewed about roll to again instrument the O frame, as before). and is deflected from $\overline{W}_0$ in the vertical plane that is normal to the orbital plane. The level of torque required in the torque generator in this case is the magnitude of the body's combined gravitational and gyroscopic restoring torques in the equilibrium orientation.

The fact that the two single gyro systems just described actually provide three axis damping is clarified by the following arguments: As

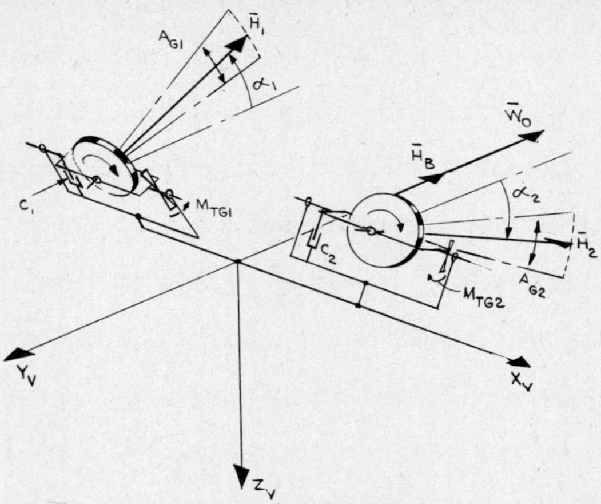

Fig. 18 Three-Axis Damping System Using a Roll-Vee Configuration of Two Integrating Gyroscopes

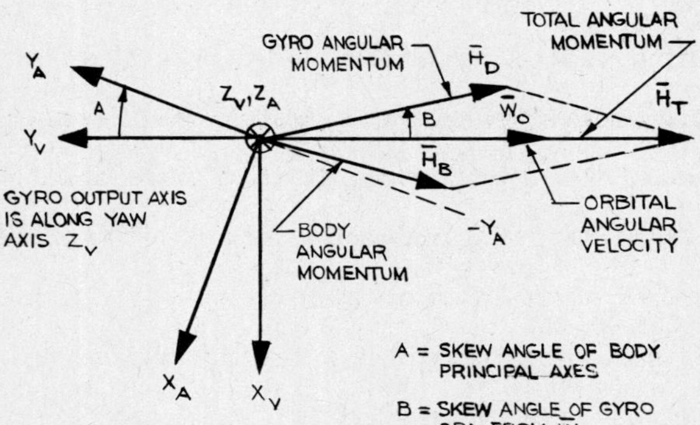

Fig. 19 Vector Diagram for Equilibrium Orientation of Single Yaw-Skewed Gyro Damping System

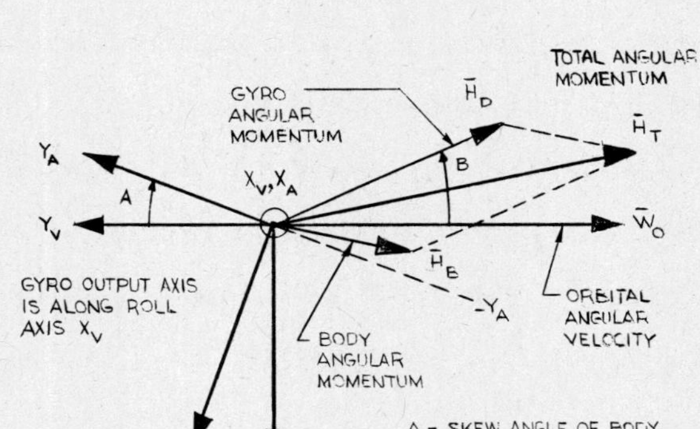

Fig. 20 Vector Diagram for Equilibrium Orientation of Single Roll-Skewed Gyro Damping System

long as the vehicle exhibits roll, pitch or yaw librations about its equilibrium orientation, either the body or the gyro gimbal - or both - will respond to such librations by relative precession. Any relative motion of the two elements must take place against the viscous restraint of the gyro's damper, and hence results in irreversible energy transfer through heat dissipation. Thus transient disturbances, which may be represented as discrete quanta of surplus mechanical energy stored initially in the system, must decay as the stored energy is dissipated in the gyro damper. Similarly, sinusoidal forcing functions will generally have reduced disturbing effect due to the dissipative system characteristics.

The two single gyro systems are illustrated by Figs. 19 and 20. Each figure shows the equilibrium orientation vector diagram in the plane normal to the skew axis. Figure 19 shows the relationships among the vector quantities for the single yaw-skewed gyro system, and Fig. 20 for the roll-skewed case. Note in Fig. 19 that the total angular momentum $\overline{H}_T$ is colinear with the orbital angular velocity $\overline{W}_0$ since the yaw deflection of the principal axes produces no gravitational torque. In the roll-skewed case, Fig. 20, the skew of the principal axes about roll gives rise to a roll-axis gravitational restoring torque which requires [by Eq. (1)] that $\overline{H}_T$ must have a non-zero time rate of change as seen in inertial space; that is, it cannot be colinear with $\overline{W}_0$ but must precess around it in a conical manner at orbital frequency as previously described. The gyro skew angle B in each figure is a readily calculable function of the principal axis skew angle A, as is the magnitude of the required torque generator torque to maintain the equilibrium orientation.

SUMMARY

In this paper, an attempt has been made to promote a more general awareness of the gyroscopic nature of a vertically-oriented satellite vehicle, and to show how gyroscopic attitude control systems utilizing this characteristic may be readily developed by intuitive processes. The mathematical analyses supporting these systems are not included here, but have been carried out, both by the present authors and by others. Analog and digital simulations have demonstrated the feasibility of the use of semi-passive integrating gyros as attitude control devices. Such systems can easily meet orientational requirements of the order of a few degrees. Because of the inherent simplicity of the systems they are well suited to applications in which longevity is a prime factor. The bearing problems of gyroscopes used in the acceleration environment of orbital motion have not as yet been fully defined, but it seems reasonable that, with the possible exception of self-acting gas-lubricated spin bearings (due to whirl resonance phenomena), the extreme reduction of g loadings in bearings in the satellite may significantly increase their useful life. The power requirements of the gyro systems are expected to be relatively low as compared with active systems offering similar damping. Analog simulations have shown that system responses to transient and sinusoidal disturbances are affected only slightly by relatively large (e.g., 50%) variations in gyro H/C ratio at orbital frequency; hence, it appears likely that temperature control may be a minor problem or no problem at all, even for the floated types of integrating gyros.

Of the several systems described, the Vee configurations and the skewed vehicle axis single gyro systems appear to be most attractive. The Vee configuration offers considerably improved damping capabilities, if widely skewed ordinary integrating gyros having the customary positive elastic restraint coefficient are to be employed. If the gyro spring constants are modified by positive feedback or similar methods, the single gyro system becomes a stronger contender due to its simplicity, but still has inherent limitations compared with the Vee configurations, particularly in regard to response to orbital eccentricity. For any proposed application, it is as necessary with these systems as with any other to weigh all factors in order to ensure selection of the proper system.

REFERENCES

1. W. Wrigley, P. B. Woodbury, and J. Hovorka, "Inertial Guidance," Sherman M. Fairchild Publication Fund Paper No. FF-16, Institute of Aeronautical Sciences, 2 East 64th St., New York 21, N. Y., 1957.

2. R. E. Roberson, "Gravitational Torques on a Satellite Vehicle," J. of Franklin Institute, Vol. 265, No. 1, Jan. 1958.

3. G. Ogletree, S. J. Sklar, and J. G. Mangan, "Satellite Attitude Control Study," Instrumentation Laboratory Report R-308, Part I, Massachusetts Institute of Technology. Cambridge 39, Massachusetts, July 1961.

4. L. Page, Introduction to Theoretical Physics, D. Van Nostrand Co., Inc. Princeton, N. J., 3rd. Ed., 1952.

5. H. Goldstein, Classical Mechanics, Addison-Wesley Publishing Company, Inc., Reading, Massachusetts, 1959.

6. C. S. Draper, W. Wrigley, and L. R. Grohe, "The Floating Integrating Gyro and its Application to Geometrical Stabilization Problems on Moving Bases," Sherman M. Fairchild Publication Fund Paper No. FF-13, Institute of Aeronautical Sciences, 2 East 64th St., New York 21, N. Y., 1955.

7. R. E. Roberson, Methods for the Control of Satellites and Space Vehicles, Vol. 2, Control System Mechanization and Analysis, WADC TR 60-643, Vol. 2, July, 1960, (ASTIA AD No. 250270).

8. R. E. Roberson, "Torques on a Satellite Vehicle from Internal Moving Parts," J. Appl. Mech., Vol. 25, pp 196-200, 1958.

9. R. E. Roberson, "Gyroscopic Control of Satellite Attitude," Proceedings of the First Symposium (International) on Rockets and Astronautics, Tokyo, Japan, 1959.

10. R. H. Cannon, Jr., "Gyroscopic Coupling in Space Vehicle Attitude Control Systems," ASME paper No. 61-JAC-8, presented at the 1961 Joint Automatic Control Conference, Boulder, Colorado, June, 1961.

11. A. G. Buckingham, "A New Method of Attitude Control Utilizing the Earth's Magnetic Field for Long Life Space Vehicles." presented at the National Conference on Guidance, Control and Navigation, American Rocket Society, at Stanford University, Palo Alto, California, August, 1961.

12. E. G. C. Burt, "On the Attitude Control of Earth Satellites," presented at the Eighth Anglo-American Aeronautical Conference, London, Sept. 1961, Royal Aeronautical Society, 4 Hamilton Place, London, W. 1, England.

PASSIVE MAGNETIC ATTITUDE CONTROL FOR EARTH SATELLITES

Robert E. Fischell*

This paper describes methods by which passive magnetic attitude control can be accomplished as well as experimental results from orbiting satellites. Furthermore, several applications of magnetic attitude control are discussed. These applications include methods for improving the operation of solar power systems, controlling the satellite's thermal balance and for performing scientific investigations concerning the earth's magnetic field and for determining the directional characteristics of corpuscular radiation in the earth's Van Allen belts. Furthermore, magnetic stabilization can be used to control initial conditions for capture into gravity gradient attitude stabilization.

INTRODUCTION

Passive attitude stabilization for earth satellites is particularly desirable for space missions that require a useful vehicle lifetime of several years. One system for attitude stabilization that has been used successfully on the Transit navigational satellite program is based on aligning a satellite axis along the direction of the Earth's magnetic field. The Transit 1B satellite, launched April 13, 1960, was the first space vehicle to be magnetically oriented. The Transit 2A satellite, launched approximately two months later, also had its symmetry axis aligned along the direction of the Earth's magnetic field. Transit 4A and Transit 4B were the third and fourth satellites to be magnetically oriented. The TRAAC satellite launched on November 15, 1961, has the unique capability of providing magnetic stabilization on command. Furthermore, the TRAAC satellite has the

---

* The Johns Hopkins University, Applied Physics Laboratory, Silver Spring, Md.

capability of demagnetizing itself so that it can be released by command from the condition of magnetic stabilization. Data received from each of the satellites, particularly Transit 4A, Transit 4B, and TRAAC, indicate that these space vehicles are magnetically stabilized.

Magnetic attitude control is defined as the alignment of one axis of the satellite along the local direction of the Earth's magnetic field. Furthermore, passive magnetic attitude control is defined as obtaining this orientation without the use of moving parts or active damping and without employing position sensors to maintain the stabilized condition. Controlled orientation of a satellite axis along the local direction of the Earth's magnetic field provides several capabilities that are not achievable with a randomly oriented space vehicle. These advantages are as follows:

1. Control of the electrical power generating capability of solar cells mounted on the satellite.
2. Control of the satellite's thermal balance.
3. Directional control of satellite receiving or transmitting antennas.
4. The capability of directing a narrow beam, brilliant, flashing light toward an observer on the ground for the purpose of optical tracking of the satellite.

5. The ability to perform scientific experiments to determine the directional properties of charged particles and other radiation which is present in the magnetic field of the earth.
6. The ability to use rockets fired in orbit along an accurately known direction to alter the orbit of the satellite.
7. The ability to control initial conditions of satellite orientation for proper capture into gravity gradient attitude stabilization.

THE MAGNETIC FIELD OF THE EARTH

In order to analyze the motions of satellites oriented in the earth's magnetic field it is necessary to understand its magnitude and direction at orbital altitudes. From a point of view of attitude control, the earth's magnetic field is quite complex both in magnitude and direction.

To a first approximation, the magnetic field of the earth is distributed as if there were a short (compared to the earth's diameter) bar magnet located at its center.

For a simple analysis of the motions of a magnetically oriented satellite, a dipole representation of the earth's magnetic field will first be considered. For this representation, the local horizontal, (H_x), and local vertical, (H_z), magnetic field components are given by:

$$H_x = \frac{M_e \cos \Phi}{r_e^3} \quad \text{(oersteds)},$$

and $$H_z = \frac{2M_e \sin \Phi}{r_e^3} \quad \text{(oersteds)}, \qquad (2)$$

where M_e = Strength of the earth's magnetic dipole (pole-cm),

Φ = Magnetic latitude (degrees),

and r_e = Distance from the center of the earth (cm).

It is also of importance to have the resultant field (H_t) and the angle (β) between H_t and the earth's spin axis. The relations between H_x, H_z, H_t, and β are given by:

$$H_t = \sqrt{H_x^2 + H_z^2} \quad \text{(oersteds)}, \qquad (3)$$

and $$\beta = \tan^{-1}\left[\frac{H_z}{H_x}\right] + \Phi \quad \text{(degrees)}. \qquad (4)$$

Figure 1 illustrates the total intensity of the earth's magnetic field as a function of altitude at the magnetic equator and at the magnetic pole. Figure 2 shows the total magnetic field intensity as a function of magnetic latitude at orbital altitudes of 400 and 600 nautical miles. From Figs. 1 and 2 it can be seen that the earth's magnetic field is greater in the polar region that it is at the equator. Even out to altitudes of 25,000 nautical miles there is sufficient field strength to obtain a significant torque when a strong magnetic dipole is contained in the satellite. Also, since modern magnetic alloys have some hysteresis loss even at field intensities as low as 0.001 oersted, it is possible to obtain sufficient magnetic damping effects at great altitudes to make passive magnetic stabilization a practical system of attitude control.

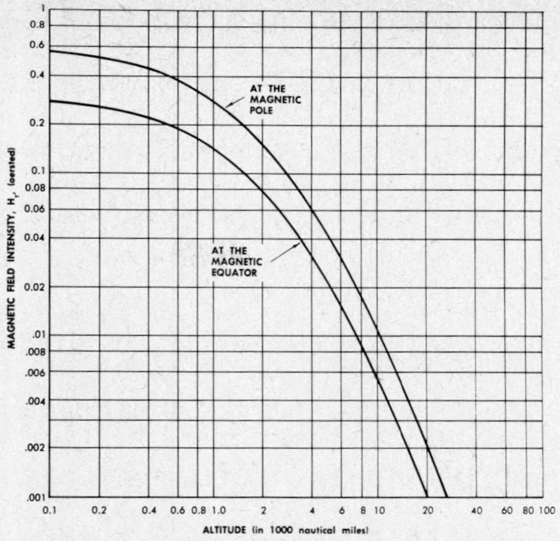

Fig. 1 The Earth's Total Magnetic Field Intensity as a Function of Altitude at the Magnetic Pole and at the Magnetic Equator

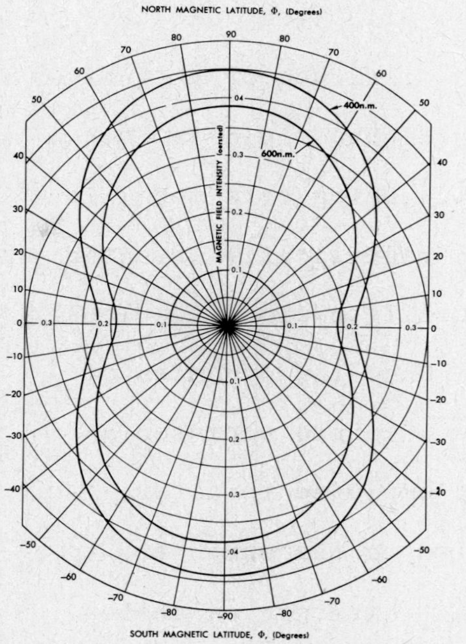

Fig. 2 The Earth's Total Magnetic Field Intensity as a Function of Latitude for Altitudes of 400 and 600 Nautical Miles

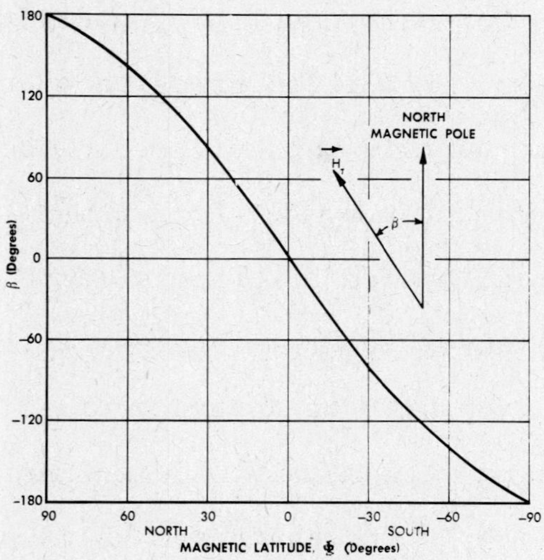

Fig. 3 Angle Between the Local Magnetic Field Vector and the Direction of the Magnetic Dipole as a Function of Magnetic Latitude

The angle β that the satellite makes with the earth's spin axis is shown in Figure 3 as a function of magnetic latitude. From this figure we see that there are large deviations of the angle β as the satellite moves from one magnetic pole to the other.

To analyze the angular motions of a magnetically oriented satellite in the earth's magnetic field we will first use the reasonably accurate assumption that at the altitude of an orbiting satellite, the center of the magnetic dipole field is coincident with the center of mass of the earth. Using this model it will be shown that particular designs for magnetically oriented satellites will result in aligning the satellite along the local direction of the earth's magnetic field. Once the dynamics of the satellite motion are analyzed, a more accurate representation of the earth's magnetic field can be employed to determine the position of the satellite relative to the earth and to the sun. For this purpose a 48 coefficient harmonic expansion of the earth's magnetic field has been programmed on the IBM 7090 computer by Dr. A. J. Zmuda and M. L. Neumann of the Applied Physics Laboratory (Ref. 1). It is believed that this model gives accurate angles of the magnetic field relative to the earth with an accuracy of ± 2 degrees. The results obtained in using this model to determine the attitude of the satellite relative to the earth and relative to the sun are presented in a later section of this paper.

DAMPING OF SATELLITE SPIN MOTION

Before a satellite can be oriented along either the earth's magnetic or gravitational field, the spin of the satellite that is usually imparted during the launching operation must be stopped. This is a result of the fact that a spinning satellite acts like a gyroscope and will therefore be spin stabilized.

The removal of the spin energy of the satellite can be accomplished by several means. One method that has been used successfully in the Transit satellite program is a set of weights attached to the satellite by means of long cables wrapped around the equator of the satellite (Ref. 2). After a prescribed time in orbit a wire connecting two weights which are located diametrically opposite is cut and the weights are allowed to fly off, unwinding the cables as they go. When the cable is completely unwound, the weights and cable fly off from the satellite. By proper selection of the mass of the weights and the cable length the spinning of the satellite can be decreased to a low rate (Fig. 4).

Another method for spin removal is based on the use of highly permeable rods of specially prepared magnetic materials (Ref. 3). This means of spin removal depends on the fact that a permeable rod when spinning in the earth's magnetic field will develop eddy current and hysteresis energy losses, and by this method remove the spin energy of the satellite. In Figure 5 is shown the spin rate of the Transit 1B satellite as a function of time after launch. For the first seven days the

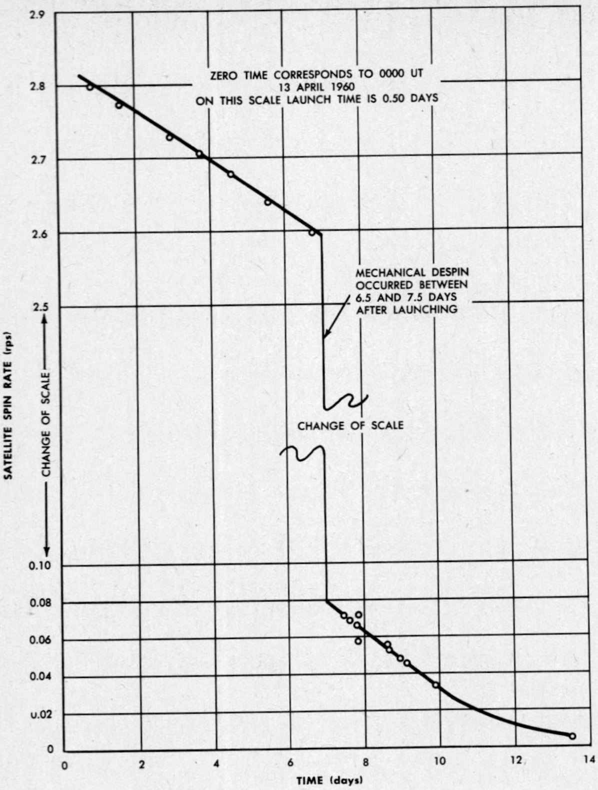

Fig. 4 Spin Rate as a Function of Time for the Transit 1B Satellite

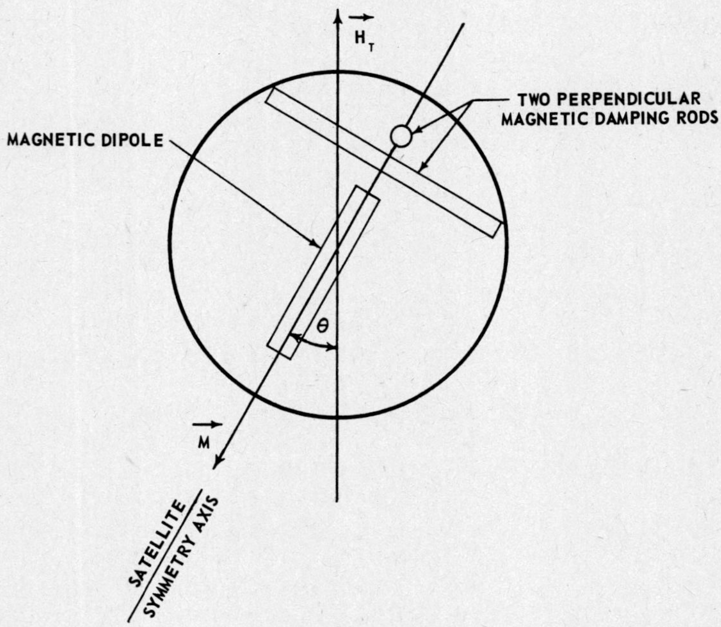

Fig. 5 Satellite With Magnetic Dipole Moment and Magnetic Damping Rods

satellite spin rate was reduced by means of the magnetic despin rods. On the seventh day a timer caused the despin weights to fly out from the satellite and thereby reduced the spin rate from 2.59 rps to 0.08 rps. The residual spin rate was then removed by the permeable rods.

After the satellite's spin has decreased to the point where gyroscopic torques are neglibible, magnetic attitude control can be put into effect. One remaining problem, however, is to damp out the oscillations which will surely be present. For magnetic attitude stabilization the damping of oscillatory motions can be accomplished by the same permeable rods that are used to remove the spin of the satellite. A description of this method of damping of oscillatory motions is discussed in the following section of this paper.

MAGNETIC ATTITUDE STABILIZATION

A satellite which has a large magnetic dipole moment will, like a simple compass, tend to align itself along the local direction of the earth's magnetic field. Throughout this and subsequent discussions we shall consider that the satellite has cylindrical symmetry about the axis of its magnetic dipole moment. For the discussion of magnetic attitude stabilization we will make the quite accurate assumption that, at the altitude of an orbiting satellite, the center of the earth's magnetic dipole field is coincident with the center of mass of the earth. Based on this assumption it follows that a satellite will orbit about the earth's magnetic dipole axis in a manner analogous to its motion about the earth's spin axis. The orbital plane will have a certain inclination with respect to the earth's magnetic equator, which is analogous to (but different from) the inclination with respect to the earth's geographical equator. The major difference between the orbit's inclination with respect to the magnetic equator and with respect to the geographic equator is that in the former case the inclination changes as the dipole axis revolves about the earth's spin axis, while in the latter case, the inclination angle is virtually constant over extended periods of time. This condition results from the fact that the inclination of the satellite's orbital plane and the inclination of the earth's equatorial plane remain constant with respect to the ecliptic plane; however, the inclination of the plane of the earth's magnetic equator with respect to the ecliptic plane varies with a 24 hour period. Since the orbital period for near earth satellites is short compared to a day (approximately 100 minutes for near-earth satellites as compared with 1440 minutes in a day) it is a good approximation to assume that the inclination of the satellite's orbit with respect to the plane of the earth's magnetic equator will remain essentially constant <u>during a single revolution</u>. The validity of this assumption has been verified in Reference 4, which shows that the maximum possible variation in the magnetic inclination of the orbit during a single 100 minute orbital period is approximately 3-1/2 degrees. This small variation in inclination during a single pass does not have an appreciable effect on the satellite's attitude relative to the magnetic field.

The assumptions that we shall use concerning the direction of the earth's magnetic field as observed at the altitude of an orbiting satellite are as follows:

1. The earth's field is that of a magnetic dipole,
2. The center of the dipole is at the center of mass of the earth, and
3. The inclination angle of the orbital plane with respect to the plane of the magnetic equator does not vary appreciably during a revolution.

Based on these assumptions it is possible to analyze the motion of a satellite having a strong magnetic dipole moment.

First consider the case of a satellite traversing an orbit inclined at 0 degrees with respect to the plane of the magnetic equator. i.e. The magnetic inclination, $i_m = 0$ degrees. The satellite dipole axis will tend to align itself parallel to the direction of the earth's magnetic dipole. If the satellite is initially displaced it will oscillate about this direction. If a mechanism for damping is present within the satellite this oscillatory motion can be stopped. The permeable magnetic rods which are used for spin removal also provide a mechanism for damping the oscillations of the satellite.

Figure 5 is an illustration of a satellite with magnetic dipole moment $\vec{M}$ displaced by the angle θ from the direction of the earth's magnetic field.

This satellite also contains two magnetic damping bars which are oriented perpendicular to each other and perpendicular to the direction of the magnetic dipole.

If the damping torque is negligible, the torque on the satellite is given by

$$\vec{\Upsilon} = \vec{M} \times \vec{H}_o , \quad (5)$$

and considering only the magnitude of the torque,

$$\Upsilon = MH_o \sin \theta \text{ (dyne-cm)}, \quad (6)$$

where

M = Magnitude of the satellite's magnetic dipole moment (unit-pole cm),

and

H_o = Magnetic field intensity (oersted).

Let

I_x = Moment of inertia of the satellite about an axis through the center of gravity of the satellite and perpendicular to $\vec{M}$ (gm-cm^2).

We then obtain the equation of motion

$$I_x \frac{d^2\theta}{dt^2} + MH_o \sin \theta = 0 \quad (7)$$

This equation can be solved for small values of θ and yields the familiar result

$$\theta_n = \theta_o \cos 2 \pi ft, \quad (8)$$

where the natural frequency f is given by

$$f = \frac{1}{2\pi} \sqrt{\frac{MH_o}{I_x}} \quad (\text{sec}^{-1}). \tag{9}$$

Even though the solution of Equation (7) will not hold exactly for large values of θ, Equation (9) is a good expression for the natural period. For example, an oscillation amplitude of 60 degrees results in an error of less than 8 per cent in the natural frequency.

The potential energy of the magnetic dipole when displaced by an angle θ from the local direction of the earth's magnetic field is found by integrating the torque over the angular displacement, i.e.,

$$E(\theta) = \int_o^{\theta} MH_o \sin \theta d\theta \quad (\text{ergs}), \tag{10}$$

which yields

$$E(\theta) = MH_o (1 - \cos \theta) \quad (\text{ergs}). \tag{11}$$

In the presence of damping, a solution for θ can be obtained in the form

$$\theta(t) = \theta_m(t) \, \theta_n(t) \tag{12}$$

where $\theta_m(t)$ is a damping function that approaches zero as $t \rightarrow \infty$, and $\theta_n(t)$ is a periodic function having a frequency defined by Equation (9) and which approaches $\sin 2\pi ft$ for small values of θ.

It can be easily shown that only those magnetic rods perpendicular to $\vec{M}$ and perpendicular to the axis about which the satellite is oscillating will be effective in damping the magnetic oscillation (Ref. 3).

For a single permeable rod perpendicular to $\vec{M}$, and perpendicular to the axis of oscillation the hysteresis loss per cycle is given by (Ref. 3)

$$E = \frac{V}{4\pi} \oint HdB \quad \text{(ergs/cycle)}, \tag{13}$$

where

$$V = \text{Volume of the rod (cm}^3\text{)},$$

and

$$\oint HdB = \text{Area of the hysteresis loop (gauss-oersted)}.$$

If there are N bars (each perpendicular to $\vec{M}$ and perpendicular to the axis about which the satellite is oscillating) then the energy loss per second is given by

$$\frac{dE}{dt} = \frac{-NV}{4\pi} f \oint HdB \quad \text{(ergs/sec)}, \tag{14}$$

where the negative sign is taken because the energy is decreasing. Substituting the value of f from Equation (9) gives

$$\frac{dE}{dt} = \frac{NV}{8\pi^2} \sqrt{\frac{MH_o}{I_x}} \oint HdB \quad \text{(ergs/sec)}. \tag{15}$$

Although there will be energy loss due to eddy currents in the rods, it can be shown that for the comparatively low oscillatory rates achievable with a magnetically oriented satellite, the eddy current loss can be neglected in comparison to the hysteresis loss (Ref. 3).

Experimental measurements have been made on the hysteresis loop area as a function of the peak magnetizing field H_m.

These results are shown in Figure 6 for a permalloy rod (47.5 per nickel, 52.5 per cent iron) having a length-to-diameter (L/D) ratio of 248. From the experimental curve it has been deduced that for this material and L/D ratio the value of $\oint$ HdB is very nearly proportional to the cube of the peak magnetizing field, that is,

$$\oint HdB = \alpha H_m^3 \text{ (gauss-oersted)}, \qquad (16)$$

where

α = constant of proportionally having the dimensions gauss/(oersted)^2.

The rod characteristics illustrated in Figure 6 can be closely approximated by setting $\alpha = 25.8 \times 10^3$ gauss/(oersted)^2.

We can write an expression for the peak magnetizing field along the permeable rods as

$$H_m = H_o \sin \theta_m \text{ (oersted)}, \qquad (17)$$

where

θ_m = the maximum angular displacement from the equilibrium position during a particular oscillation (degrees).

then

$$\oint HdB = \alpha H_o^3 \sin^3 \theta_m \text{ (gauss-oersted)}, \qquad (18)$$

and we can write the rate of energy loss as

$$\frac{dE}{dt} = \frac{-\alpha NV}{8\pi^2} \sqrt{\frac{M}{I_x}} H_o^{7/2} \sin^3\theta_m \text{ (ergs/secs)}, \qquad (19)$$

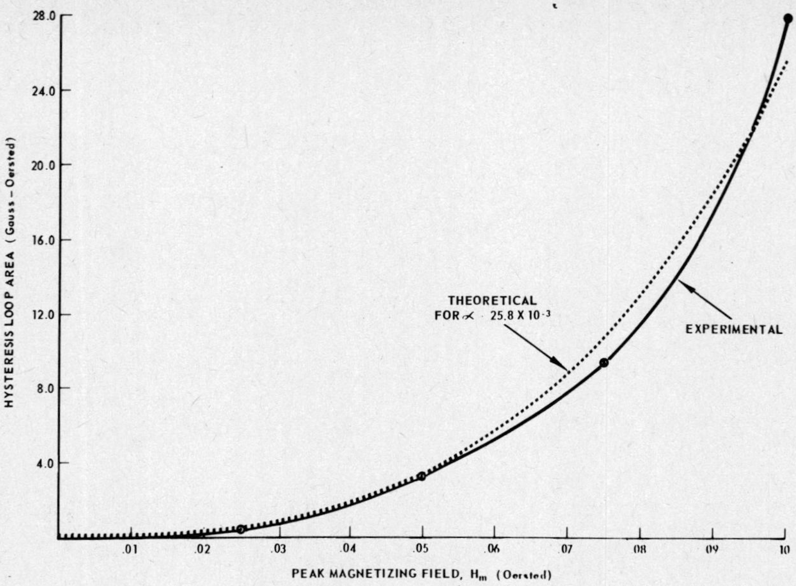

Fig. 6 Experimental and Theoretical Values for Hysteresis Loop Area as a Function of Peak Magnetizing Field

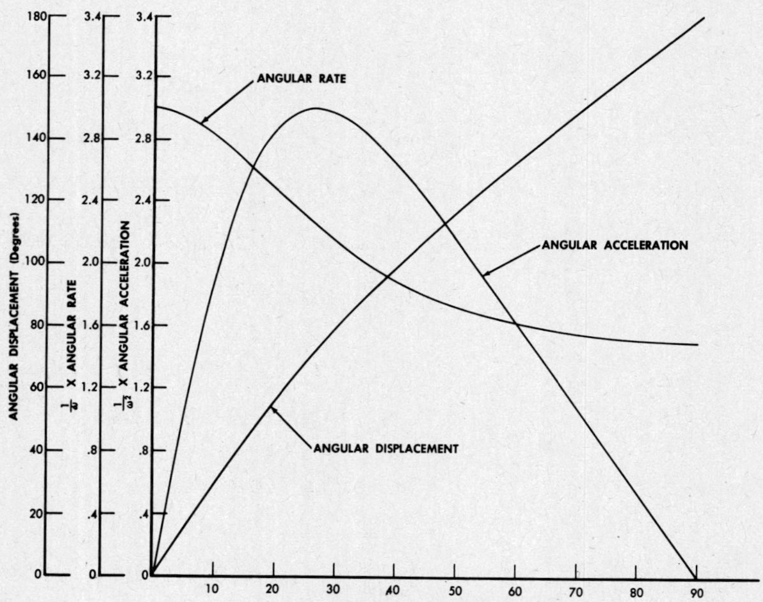

Fig. 7 Angular Displacement, Rate and Acceleration for a Magnetically Oriented Satellite

setting

$$k = \frac{\alpha\ NV}{8\ \pi^2} \sqrt{\frac{M}{I_x}}\ H_o^{7/2} \tag{20}$$

then

$$\frac{dE}{dt} = -k \sin^3 \theta_m \quad \text{(ergs/sec)}. \tag{21}$$

If we rewrite equation (11) in terms of the maximum angular displacement θ_m we obtain

$$E\ (\theta_m) = MH_o\ (1 - \cos \theta_m) \quad \text{(ergs)} \tag{22}$$

But

$$\frac{d}{dt} \left[E\ (\theta_m) \right] = MH_o \sin \theta_m \frac{d\theta_m}{dt} \quad \text{(ergs/sec)}. \tag{23}$$

Equating equations (21) and (23) yields

$$MH_o \sin \theta_m \frac{d\theta_m}{dt} = -k \sin^3 \theta_m, \tag{24}$$

which yields the solution

$$\cot \theta_m = \frac{k}{MH_o}\ (t) + c_1. \tag{25}$$

If we take the initial condition that the satellite is displaced an angle θ_o at time t = 0, then

$$c_1 = \cot \theta_o,$$

and

$$\theta_m(t) = \text{arc cot}\ \left(\frac{k}{MH_o}\ t + \cot \theta_o \right). \tag{26}$$

For reasonably small values of θ, it is possible to write the complete solution of θ (t) as

$$\theta(t) = \left[\text{arc cot}\ \left(\frac{k}{MH_o}\ t + \cot \theta_o\right) \right] \cos 2\ \pi\ ft. \tag{27}$$

The most significant portion of this expression is the damping term which is quite accurate even for large angular displacements of the satellite from its equilibrium position.

Although the result obtained is based on the assumption of a simple cubic relation of $\oint$ HdB with H_m, the differential equation can also be solved for a general polynomial of order n that best fits the experimental curve.

To find the time for a satellite to damp down from an initial peak angular displacement of θ_o to a final displacement θ_f, one solves equation (26) for t which yields

$$t = \frac{MH_o}{k} (\cot \theta_f - \cot \theta_o) \quad \text{(seconds)}. \tag{28}$$

This last result is based on the assumption that the satellite is not spinning about the axis of the magnetic dipole. If there is some spin rate of the satellite, this causes the oscillating satellite to precess, resulting in a considerably longer time to damp down to a final angular deviation θ_f.

For the Transit 3B satellite we had the following constants:

$N =$ 8 (effective bars for oscillatory damping)

$\alpha =$ 25.8×10^3 gauss/(oersted)^2

$V =$ 6.2 cm^3

$M =$ 7×10^4 pole-cm

$I_x =$ 1.1×10^8 gm-cm^2

Therefore, from Equation (20) taking H_o = 0.3 oersted, we get k = 6.09. We can then use Equation (28) (considering no satellite spin) to determine the time required for the satellite to damp down from an initial displacement angle of θ_o = 90 degrees to a final angle of θ_f = 2 degrees. This gives the result of t = 98,500 seconds, or roughly 1.1 days.

Therefore, for i_m = 0 degrees the satellite will align itself along the direction of the earth's magnetic dipole, and the oscillations will be damped out in a reasonably short period of time.

Next, consider the case of a satellite traversing a magnetically polar orbit, that is, i_m = 90 degrees. For this case the angular displacement, angular rate, and angular acceleration of a magnetically oriented satellite relative to the direction of the earth's magnetic dipole axis are given by

$$\beta = \arc \tan \left[\frac{3 \tan \omega t}{1-2 \tan^2 \omega t}\right] \quad \text{(degrees)}, \qquad (29)$$

$$\frac{d\beta}{dt} = \omega \left[1 + \frac{2}{1 + 3 \sin^2 \omega t}\right] \quad \text{(radians/sec)}, \qquad (30)$$

and

$$\frac{d^2\beta}{dt^2} = -\omega^2 \left[\frac{6 \sin 2\omega t}{1+3 \sin^2 \omega t}\right] \quad \text{(radians/sec}^2\text{)}, \qquad (31)$$

where for the polar orbit the magnetic latitude Φ is given by

$$\Phi = 57.3\, \omega t \quad \text{(degrees)}, \qquad (32)$$

where

ω= orbital angular rate of the satellite (radians/sec.).

These three equations are plotted in Figure 7 for 1/4 of an orbital revolution; i.e., for a satellite moving from the ascending node to the north magnetic pole. These results show that a magnetically oriented satellite in a polar orbit will undergo, on the average, two revolutions per orbit, as illustrated in Figure 8. These results also show that the instantaneous angular rate of the satellite is three (3) revolutions per orbit (rpo) at the equator and 1.5 revolutions per orbit (rpo) at the pole. However, the average rate over an entire orbital period is 2.0 rpo. Another result of this analysis is that the angular acceleration is 0 at the equator and at the poles, and has a maximum value of $3\omega^2$ at a magnetic latitude of 26 degrees.

For a maximum acceleration equal to $3\omega^2$ a rough estimate can be made of how well a magnetically oriented satellite will follow the dipole field for a polar orbit. To accomplish this it is necessary to equate the torque acting on the satellite due to an angular displacement from the local direction of the field to the product of the maximum acceleration that the satellite must be able to follow times the moment of inertia I_x; i.e.

$$I_x \left[\frac{d^2\beta}{dt^2}\right]_{max} = MH_o \sin \theta_{max} \text{ (dyne-cm)}, \qquad (33)$$

where θ_{max} = approximate angle by which the satellite will be displaced from the local direction of the magnetic field. Since $\left[\frac{d^2\beta}{dt^2}\right]_{max} = 3\omega^2$, we can solve for θ_{max} to give

$$\theta_{max} = \text{arc sin } \frac{3I_x\omega^2}{MH_o} \qquad (34)$$

For the magnetically stabilized INJUN satellite, launched with Transit 4A, I_x=24.4 x 10^6 gm-cm^2 and M = 2.5 x 10^3; this gives θ_{max} = 1.0 degree.

This simple analysis gives a fairly good maximum value of the angular deviation from the local direction of the earth's magnetic dipole field if the natural period of oscillation of the magnetically oriented satellite is short compared to half the orbital period and if the satellite has some magnetic damping. The natural period of the satellite is given by the reciprocal of f from Equation (9), i.e.,

$$T = 2\pi \sqrt{\frac{I_x}{MH_o}} \quad \text{(seconds).} \qquad (35)$$

For INJUN this yields a natural period of T = 480 seconds, which is quite short compared to a half orbital period of approximately 3000 seconds. Permeable rods were provided to give damping of the oscillatory motion.

F. F. Mobley, of the Applied Physics Laboratory, has devised a means for obtaining a computer solution for the attitude of a magnetically oriented satellite in a magnetic dipole field (Ref. 5). For this purpose a mathematical model of the hysteresis loop of the magnetic damping rods was employed. An IBM 7090 computer was used to obtain a steady state solution for the angle that the magnetically stabilized satellite deviates from the local direction of the magnetic dipole field of the earth. Figure 9 illustrates the steady state motion of the INJUN satellite for a magnetic inclination of 90 degrees. From this curve it can be seen that the maximum value for the angular deviation of the satellite axis from the direction of the magnetic field is in close agreement with the value of 1.0 degree obtained from the simplified analysis of Equation (34).

It has been shown here that a satellite whose natural period of oscillation is short compared to an orbital period will align itself along the local direction of the earth's magnetic field for orbits in which $i_m = 0$ degrees and $i_m = 90$ degrees. For magnetic inclinations below 90 degrees, the maximum deviation from alignment, θ_{max}, decreases as we approach $i_m = 0$ at which point $\theta_m = 0$ (Ref. 4).

If we consider the actual magnetic field of the earth instead of a simple dipole model we see that maximum angular accelerations of the satellite can be only slightly greater than the value $\left[\frac{d^2\theta}{dt^2}\right]_{max} = 3\omega^2$ derived from Equation (31). Therefore the satellite deviation from the local direction of the earth's magnetic field will not be appreciably greater than the value given by Equation (34) for θ_{max}.

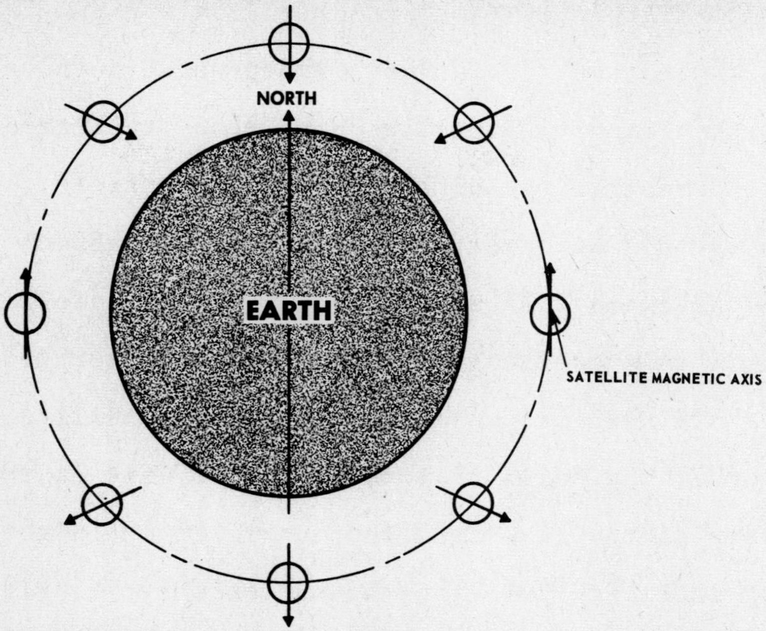

Fig. 8 Motion of a Magnetically Oriented Satellite in a Polar Orbit

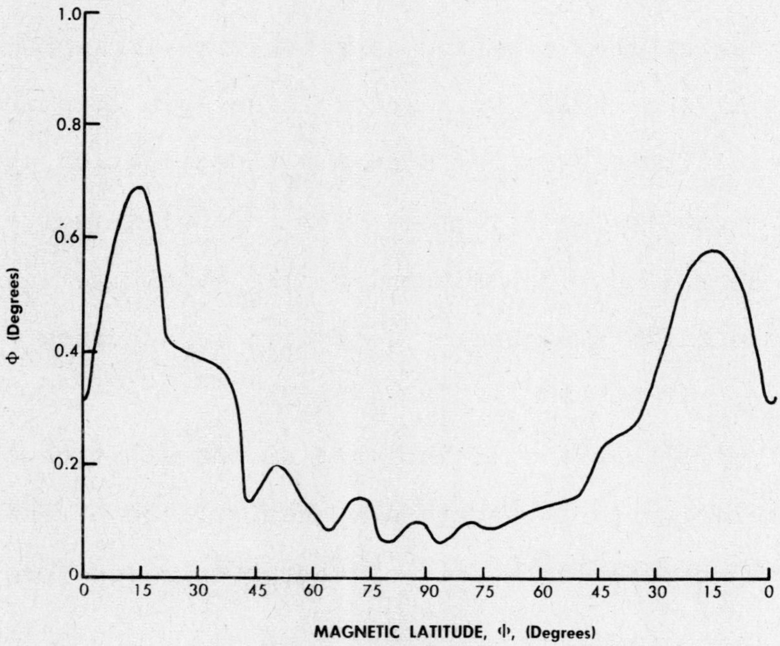

Fig. 9 Equilibrium Attitude State for Injun Satellite, Θ = Angle Between Satellite Magnet and Magnetic Field, Orbital Magnetic Inclination = 90 Degrees

EXPERIMENTAL RESULTS

There have been several independent methods for verifying that the Transit satellites have been stabilized along the local direction of the earth's magnetic field. One verification of this alignment is based on solar aspect data obtained from the Transit 4A satellite. For this satellite, presuming that it is magnetically oriented, a computer print-out has been obtained for the attitude of the satellite relative to the sun. The angle that the magnetic axis of the satellite makes with the line from the sun is given in the column marked "PSI" in Figure 10. The two curves shown in Figure 11 are a comparison of the measured and the computed attitude of the Transit 4A satellite relative to the sun. This data was taken 8 days after the launching, at which time it is believed there was still some small spin rate of the satellite about its magnetic dipole axis and some small oscillation about the local magnetic field vector. These results indicate that, at this short time after launching, the satellite was aligned with the earth's magnetic field with an accuracy of about 4 degrees.

A more accurate measurement of the attitude of the satellite relative to the earth's magnetic field direction can be obtained by placing a pair of vector magnetometers perpendicular to each other and perpendicular to the direction of the satellite's magnetic dipole axis. This was done on the TRAAC satellite. The TRAAC satellite also possessed the unique

SATELLITE SPIN-AXIS ORIENTATION TO SUN EARTH-LINE AND OBSERVER LINE OF SIGHT, MAGNETIC DIP AND SOLAR CELL OUTPUT

INPUTS- ASC NODE FROM GRNWH = 106.70 INCLINATION = 67.50 SM AXIS = 7303.2 ECC = 0.0001 ARG OF PER = 237.20
DAY NO. = 165 YEAR = 1961 OBSERVER COORDINATES- GEOC LAT = 39.16 GEOC LONG = -76.90

TIME (SEC UT)	LAT (DEG)	LONG (DEG)	ALT (NM)	DIP (DEG)	ALPHA (DEG)	AZIM (DEG)	ELEV (DEG)	PSI (DEG)	CURRENT (MA)	SUM MA DEL T
30840.00	-50.96	-42.59	504.52	55.83	SATELLITE	BELOW	HORIZON	77.72	955.25	955.25
30900.00	-53.68	-39.26	504.52	58.24	SATELLITE	BELOW	HORIZON	78.52	952.18	1907.43
30960.00	-56.28	-35.43	504.52	60.45	SATELLITE	BELOW	HORIZON	79.62	948.32	2855.75
31020.00	-58.70	-31.10	504.52	62.45	SATELLITE	BELOW	HORIZON	80.96	943.63	3799.38
31080.00	-60.97	-26.04	504.53	64.31	SATELLITE	BELOW	HORIZON	82.55	938.61	4737.99
31140.00	-62.96	-20.30	504.53	66.00	SATELLITE	BELOW	HORIZON	84.32	934.36	5672.35
31200.00	-64.69	-13.64	504.54	67.56	SATELLITE	BELOW	HORIZON	86.28	930.86	6603.21
31260.00	-66.04	-6.27	504.55	69.96	SATELLITE	BELOW	HORIZON	88.37	929.81	7533.02
31320.00	-66.99	1.95	504.56	70.24	SATELLITE	BELOW	HORIZON	89.38	929.31	8462.34
31380.00	-67.46	10.67	504.57	71.37	SATELLITE	BELOW	HORIZON	87.01	930.50	9392.83
31440.00	-67.42	19.42	504.59	72.34	SATELLITE	BELOW	HORIZON	84.58	933.85	10326.68
31500.00	-66.87	28.07	504.60	73.16	SATELLITE	BELOW	HORIZON	82.02	939.95	11266.63
31560.00	-65.87	36.07	504.62	73.80	SATELLITE	BELOW	HORIZON	79.43	949.01	12215.63
31620.00	-64.44	43.43	504.63	74.27	SATELLITE	BELOW	HORIZON	76.73	959.71	13175.35
31680.00	-62.69	49.86	504.65	74.53	SATELLITE	BELOW	HORIZON	74.02	976.87	14152.22
31740.00	-60.62	55.57	504.67	74.59	SATELLITE	BELOW	HORIZON	71.22	1001.76	15153.98
31800.00	-58.36	60.45	504.69	74.44	SATELLITE	BELOW	HORIZON	68.43	1033.58	16187.56
31860.00	-55.88	64.75	504.71	74.06	SATELLITE	BELOW	HORIZON	65.58	1067.68	17255.25
31920.00	-53.25	68.50	504.73	73.45	SATELLITE	BELOW	HORIZON	62.71	1106.39	18361.64
31980.00	-50.55	71.72	504.75	72.64	SATELLITE	BELOW	HORIZON	59.88	1141.44	19503.07
32040.00	-47.71	74.60	504.77	71.61	SATELLITE	BELOW	HORIZON	57.03	1176.11	20679.18
32100.00	-44.84	77.11	504.80	70.38	SATELLITE	BELOW	HORIZON	54.26	1204.55	21883.74
32160.00	-41.86	79.40	504.82	68.93	SATELLITE	BELOW	HORIZON	51.50	1229.74	23113.48
32220.00	-38.88	81.42	504.84	67.29	SATELLITE	BELOW	HORIZON	48.87	1253.01	24366.49
32280.00	-35.81	83.29	504.87	65.42	SATELLITE	BELOW	HORIZON	46.34	1271.62	25638.12
32340.00	-32.75	84.98	504.89	63.36	SATELLITE	BELOW	HORIZON	44.02	1288.83	26926.94
32400.00	-29.62	86.56	504.91	61.04	SATELLITE	BELOW	HORIZON	41.93	1300.41	28227.35
32460.00	-26.47	88.03	504.94	58.47	SATELLITE	BELOW	HORIZON	40.17	1310.08	29537.43
32520.00	-23.34	89.38	504.96	55.67	SATELLITE	BELOW	HORIZON	38.87	1316.10	30853.54
32580.00	-20.15	90.69	504.99	52.53	SATELLITE	BELOW	HORIZON	38.10	1319.56	32173.10
32640.00	-17.00	91.91	505.01	49.13	SATELLITE	BELOW	HORIZON	38.00	1320.00	33493.10
32700.00	-13.78	93.10	505.03	45.32	SATELLITE	BELOW	HORIZON	38.67	1316.99	34810.09
32760.00	-10.61	94.23	505.05	41.19	SATELLITE	BELOW	HORIZON	40.15	1310.17	36120.26
32820.00	-7.38	95.35	505.08	36.60	SATELLITE	BELOW	HORIZON	42.51	1297.17	37417.43
32880.00	-4.20	96.43	505.10	31.65	SATELLITE	BELOW	HORIZON	45.67	1276.49	38693.92
32940.00	-0.97	97.53	505.12	26.21	SATELLITE	BELOW	HORIZON	49.64	1246.92	39940.84
33000.00	2.27	98.61	505.14	20.36	SATELLITE	BELOW	HORIZON	54.27	1204.42	41145.26
33060.00	5.45	99.69	505.16	14.24	SATELLITE	BELOW	HORIZON	59.36	1147.95	42293.21
33120.00	8.68	100.80	505.18	7.77	SATELLITE	BELOW	HORIZON	64.90	1076.89	43370.10
33190.00	11.86	101.91	505.19	1.22	SATELLITE	BELOW	HORIZON	70.53	1008.68	44378.78
33240.00	15.08	103.08	505.21	-5.45	SATELLITE	BELOW	HORIZON	76.24	961.93	45340.71
33300.00	18.24	104.27	505.22	-11.94	SATELLITE	BELOW	HORIZON	81.67	941.17	46281.88
33360.00	21.44	105.53	505.24	-18.31	SATELLITE	BELOW	HORIZON	86.80	930.60	47212.48
33420.00	24.57	106.84	505.25	-24.28	SATELLITE	BELOW	HORIZON	88.62	929.69	48142.17
33480.00	27.74	108.25	505.26	-29.96	SATELLITE	BELOW	HORIZON	84.58	933.84	49076.01
33540.00	30.84	109.74	505.27	-35.17	SATELLITE	BELOW	HORIZON	81.24	942.67	50018.67
33600.00	33.96	111.37	505.28	-40.05	SATELLITE	BELOW	HORIZON	78.52	952.19	50970.87
33660.00	37.05	113.14	505.29	-44.54	SATELLITE	BELOW	HORIZON	76.46	960.93	51931.80
33720.00	40.05	115.05	505.29	-48.59	SATELLITE	BELOW	HORIZON	75.06	969.59	52901.39
33790.00	43.06	117.19	505.30	-52.35	SATELLITE	BELOW	HORIZON	74.23	975.42	53876.81
33640.00	45.97	119.54	505.30	-55.76	SATELLITE	BELOW	HORIZON	73.94	977.54	54854.35
33900.00	48.86	122.22	505.30	-58.93	SATELLITE	BELOW	HORIZON	74.13	976.12	55830.46
33960.00	50.27	123.56	505.30	-60.42	SATELLITE	BELOW	HORIZON	74.40	974.20	56804.67
34020.00	54.32	128.64	505.30	-64.50	SATELLITE	BELOW	HORIZON	75.73	964.91	57769.57
34030.00	56.86	132.53	505.30	-66.95	SATELLITE	BELOW	HORIZON	77.03	958.38	58727.96
34140.00	59.27	137.08	505.30	-69.23	SATELLITE	BELOW	HORIZON	78.62	951.83	59679.78
34200.00	61.48	142.32	505.29	-71.32	SATELLITE	BELOW	HORIZON	80.46	945.40	60625.18
34260.00	63.41	148.26	505.29	-73.20	SATELLITE	BELOW	HORIZON	82.47	938.83	61564.02
34320.00	65.06	155.11	505.28	-74.91	SATELLITE	BELOW	HORIZON	84.68	933.65	62497.67
34380.00	66.31	162.67	505.27	-76.39	SATELLITE	BELOW	HORIZON	87.00	930.50	63428.17
34440.00	67.15	171.03	505.26	-77.66	SATELLITE	BELOW	HORIZON	89.47	929.26	64357.43
34500.00	67.49	179.69	505.24	-78.65	SATELLITE	BELOW	HORIZON	87.99	930.01	65287.44
34560.00	67.33	-171.45	505.23	-79.34	SATELLITE	BELOW	HORIZON	85.33	932.34	66219.78
34620.00	66.68	-163.02	505.22	-79.68	SATELLITE	BELOW	HORIZON	82.64	938.40	67158.18
34680.00	65.56	-155.07	505.20	-79.67	SATELLITE	BELOW	HORIZON	79.86	947.49	68105.66
34740.00	64.07	-148.01	505.18	-79.30	SATELLITE	BELOW	HORIZON	77.08	958.14	69063.81
34800.00	62.22	-141.68	505.17	-78.62	SATELLITE	BELOW	HORIZON	74.23	975.40	70039.20
34860.00	60.09	-136.16	505.15	-77.67	SATELLITE	BELOW	HORIZON	71.36	1000.39	71039.59
34920.00	57.78	-131.44	505.13	-76.51	SATELLITE	BELOW	HORIZON	68.53	1032.38	72071.97
34980.00	55.26	-127.28	505.11	-75.15	SATELLITE	BELOW	HORIZON	65.66	1066.54	73138.51
35040.00	52.64	-123.71	505.09	-73.65	SATELLITE	BELOW	HORIZON	62.86	1104.39	74242.90
35100.00	49.88	-120.54	505.07	-71.99	SATELLITE	BELOW	HORIZON	60.05	1139.38	75382.28
35160.00	47.06	-117.78	505.04	-70.23	SATELLITE	BELOW	HORIZON	57.34	1172.61	76554.89
35220.00	44.13	-115.29	505.02	-68.30	60.85	292.50	0.31	54.66	1200.71	77755.60
35280.00	41.19	-113.10	505.00	-66.28	59.42	285.96	1.66	52.13	1223.84	78979.44
35340.00	38.15	-111.09	504.97	-64.08	57.33	278.77	2.71	49.70	1246.42	80225.85
35400.00	35.07	-109.26	504.95	-61.72	54.64	271.18	3.34	47.45	1263.84	81489.69
35460.00	32.01	-107.61	504.93	-59.23	51.45	263.50	3.50	45.47	1277.98	82767.67
35520.00	28.87	-106.06	504.90	-56.51	47.79	255.73	3.19	43.76	1290.33	84058.00
35580.00	25.76	-104.63	504.88	-53.62	43.86	248.35	2.44	42.44	1297.57	85355.56
35640.00	22.58	-103.28	504.85	-50.46	39.65	241.30	1.29	41.57	1302.38	86657.95
35700.00	19.44	-102.01	504.83	-47.08	SATELLITE	BELOW	HORIZON	41.24	1304.18	87962.13
35760.00	16.23	-100.78	504.81	-43.37	SATELLITE	BELOW	HORIZON	41.53	1302.59	89264.72
35820.00	13.06	-99.62	504.78	-39.40	SATELLITE	BELOW	HORIZON	42.49	1297.32	90562.03
35880.00	9.84	-98.48	504.76	-35.03	SATELLITE	BELOW	HORIZON	SATELLITE IN SHADOW OF EARTH		
35940.00	6.61	-97.36	504.74	-30.36	SATELLITE	BELOW	HORIZON	SATELLITE IN SHADOW OF EARTH		
36000.00	3.43	-96.28	504.72	-25.40	SATELLITE	BELOW	HORIZON	SATELLITE IN SHADOW OF EARTH		
36060.00	0.20	-95.19	504.70	-20.05	SATELLITE	BELOW	HORIZON	SATELLITE IN SHADOW OF EARTH		
36120.00	-2.99	-94.12	504.68	-14.50	SATELLITE	BELOW	HORIZON	SATELLITE IN SHADOW OF EARTH		
36180.00	-6.22	-93.02	504.66	-8.66	SATELLITE	BELOW	HORIZON	SATELLITE IN SHADOW OF EARTH		
36240.00	-9.40	-91.93	504.64	-2.78	SATELLITE	BELOW	HORIZON	SATELLITE IN SHADOW OF EARTH		
36300.00	-12.63	-90.79	504.62	3.21	SATELLITE	BELOW	HORIZON	SATELLITE IN SHADOW OF EARTH		
36360.00	-15.80	-89.63	504.61	9.05	SATELLITE	BELOW	HORIZON	SATELLITE IN SHADOW OF EARTH		
36420.00	-19.00	-88.41	504.59	14.79	SATELLITE	BELOW	HORIZON	SATELLITE IN SHADOW OF EARTH		
36480.00	-22.20	-87.13	504.58	20.27	SATELLITE	BELOW	HORIZON	SATELLITE IN SHADOW OF EARTH		
36540.00	-25.33	-85.81	504.57	25.37	SATELLITE	BELOW	HORIZON	SATELLITE IN SHADOW OF EARTH		
36600.00	-28.49	-84.37	504.56	30.20	SATELLITE	BELOW	HORIZON	89.44	929.28	91491.31
36660.00	-31.59	-82.86	504.55	34.60	SATELLITE	BELOW	HORIZON	86.96	930.52	92421.83
36720.00	-34.70	-81.20	504.54	38.70	SATELLITE	BELOW	HORIZON	84.98	933.04	93354.87
36790.00	-37.74	-79.42	504.53	42.40	SATELLITE	BELOW	HORIZON	83.54	936.16	94291.03
36840.00	-40.78	-77.44	504.53	45.83	SATELLITE	BELOW	HORIZON	82.56	938.59	95229.62
36900.00	-43.77	-75.24	504.52	48.95	SATELLITE	BELOW	HORIZON	82.06	939.86	96169.47
36960.00	-46.67	-72.82	504.52	51.75	SATELLITE	BELOW	HORIZON	81.97	940.10	97109.57
37020.00	-49.54	-70.07	504.52	54.33	SATELLITE	BELOW	HORIZON	82.27	939.33	98048.90
37080.00	-49.60	-70.25	504.52	54.39	SATELLITE	BELOW	HORIZON	82.32	939.21	98988.11

SATELLITE NUMBER AND POWER EQUATION IS 411
AVERAGE CURRENT IN MILLIAMPS FOR ONE REVOLUTION = 956.24

PERIOD = 103.518

Fig. 10 Computer Results for a Magnetically Oriented Satellite

capability of having a strong magnetic dipole which could be turned on or off by radio command. This was accomplished by means of iron-core electromagnets in which the current could be commanded to flow through the windings. When the current was commanded off, a capacitor connected across the electromagnet coil caused a damped oscillation of the magnetizing current which resulted in excellent demagnetization of the iron-core. After the TRAAC satellite was in orbit two days with the spin rate reduced to one revolution in 1.8 minutes, the electromagnet was commanded on. The resulting magnetic stabilization of the satellite is shown by the curves of Figure 12. After 6 days the electromagnet was commanded off. The reason that the satellite did not approach magnetic stabilization at a more rapid rate is due to spin about the dipole axis.

At the end of 5 days the satellite spin rate was 0.15 rpm. Approximately 20 days later when the satellite's spin rate was approximately .025 rpm, the electromagnet was again commanded on. After 3 days the satellite deviation from the local direction of the earth's magnetic field was less than 2 degrees. In this case additional damping was provided by a special damping spring that was to be used for a gravity gradient attitude stabilization experiment; also a part of the TRAAC satellite experiment.

The results of these experiments indicate that magnetic attitude control can be effectively accomplished.

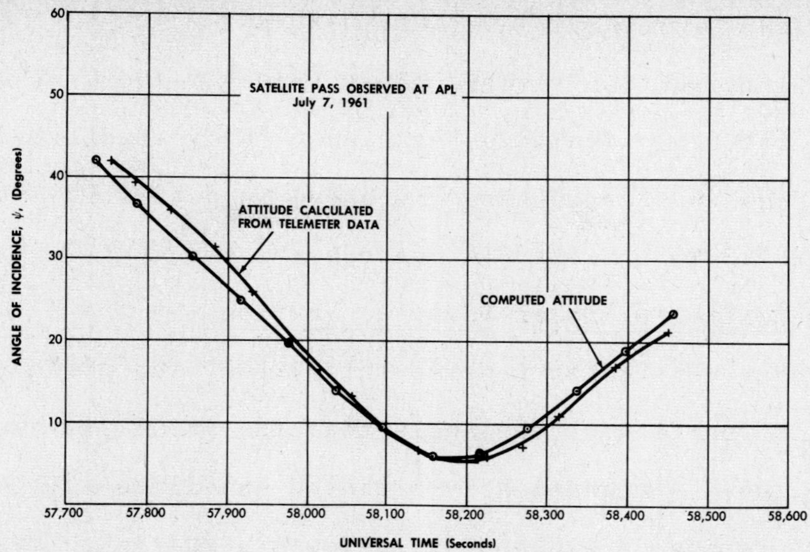

Fig. 11 Attitude of Magnetically Stabilized Transit 4A Satellite

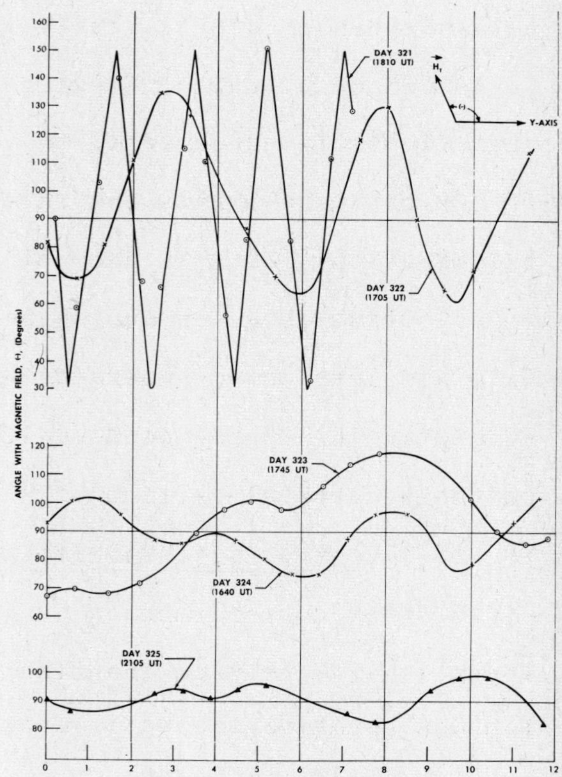

Fig. 12 Approach of TRAAC Satellite to Magnetic Stabilization

APPLICATIONS OF MAGNETIC ATTITUDE CONTROL

The introduction to this paper (pages 2 and 3) lists seven operational advantages of magnetically stabilized satellites. These advantages are discussed in detail in Reference 4 and will be only briefly discussed herein.

1. <u>Control of Solar Power Generation</u>

For most solar powered satellites, the power generated is a function of the angle that the satellite makes with the sun. A magnetically oriented satellite will undergo large excursions of angle of its symmetry axis with respect to the sun and will therefore guarantee that the satellite does not dwell for extended periods of time in a position of minimum power generation. For polar and near-polar orbits at altitudes of approximately 1,000 miles and below, solar cells can be placed on a satellite in such a way that, with magnetic attitude control, the same power will be generated in 100 per cent orbits and in the orbits of minimum solar illumination. This is accomplished by providing more solar power output when the sun shines on the poles of the satellite and less power generated when the sun shines on the equator. In 100 per cent sunlit orbits, the sun shines only on the equatorial regions of a magnetically oriented satellite.

2. <u>Control of Satellite's Thermal Balance</u>

By coating the equatorial region with a material having a lower α/ϵ ratio than that used on the polar regions the temperature of the satellite can be made quite independent of the per cent illumination. This can also be accomplished by controlling the exterior shape of the satellite.

3. Directional Control of Satellite Antennas

Magnetically stabilized satellites orbiting above certain portions of the earth will have a particular pole of the satellite directed downward toward a ground receiving station. Directional antennas can then be used to advantage in these regions. The column marked "APHA" in Figure 10 gives the angle between the satellite's magnetic dipole axis and the line of sight from a particular ground receiving station.

4. Directing of a Beam of Light

This can be accomplished in certain parts of the world in a manner analagous to a directional antenna.

5. Scientific Experiments

Experiments to study directional characteristics of corpuscular radiation in the earth's Van Allen belts are best accomplished on a magnetically stabilized satellite. Also, there are distinct advantages in making measurements of the intensity of the earth's magnetic field on a magnetically oriented vehicle.

6. Altering of Satellite Orbits by an Orbiting Rocket

A magnetically stabilized satellite has a controlled and constantly changing attitude with respect to its velocity vector. A particular time and place in the orbit can be selected, at which time firing a rocket could produce a desired change in the satellite's orbit.

7. Capture Into Gravity Gradient Attitude Stabilization

By initially magnetically orienting a satellite, one can get a particular pole of the satellite to face outward. Also the angular motions of the satellite can be controlled. These are desirable initial conditions for capture into gravity gradient attitude stabilization.

REFERENCES

1. Private Communication with A. J. Zmuda of The Johns Hopkins University, Applied Physics Laboratory.
2. J. F. Smola, "An Alternate Solution to the Problem of Despinning an Earth Satellite and an Extension of the Theory to Improve the Control over Residual Angular Velocity," SSD-5-6, The Johns Hopkins University, Applied Physics Laboratory Memorandum, May 20, 1960.
3. R. E. Fischell, "Magnetic Damping of the Angular Motions of Earth Satellites," ARS Paper No. 1502-60, presented at The American Rocket Society Annual Meeting, Washington, D. C., December 5-8, 1960.
4. R. E. Fischell, "Magnetic and Gravity Attitude Stabilization of Earth Satellites," Report CM-996, The Johns Hopkins University, Applied Physics Laboratory, May 1961.
5. Private Communication with F. F. Mobley of The Johns Hopkins University, Applied Physics Laboratory.

ANALYSIS OF GUIDANCE TECHNIQUES FOR ACHIEVING ORBITAL RENDEZVOUS

Philip G. Felleman

This paper examines several types of satellite rendezvous terminal guidance techniques by means of a phase plane analysis and numerical examples. The methods of rendezvous differ essentially in the type of propulsion used, for example on-off thrusting or continuous thrusting. The analysis demonstrates the relationship between propulsive characteristics, initial range and velocity and on board tracking equipment requirements. A mathematical derivation of a guidance philosophy commensurate with both propulsion capability and initial condition variations is presented and the relationship between throttling capability and launch timing accuracy is examined. A comparison between propellant requirements for the different types of rendezvous guidance systems is made and several numerical examples of each are presented.

---

* Staff Engineer, Instrumentation Laboratory, Massachusetts Institute of Technology, Cambridge, Mass.

INTRODUCTION

The satellite rendezvous problem is unique in that it requires a guidance and control system not only to match the space-time positions of two vehicles, but also to match the velocity vectors coincidentally with position and time. Several methods of velocity control have been proposed to accomplish a rendezvous mission. It will not be the purpose of this paper to indicate any "best" method, but to analytically examine features of several. While velocity control is primary in the rendezvous mission, the problem of path (or trajectory) control is also significant. This latter problem will be discussed later. The main emphasis will be placed upon methods of achieving the velocity boundary conditions peculiar to the rendezvous problem and the relationship to the other parameters of the mission, propulsion, tracking equipment and initial condition variations.

A convenient technique which can be used to analyze the rendezvous problem is that of the phase plane which is well known from the study of non-linear control systems. The phase plane as applied to rendezvous will examine the range-velocity relationship for several propulsion configurations.

In order to derive the applicable formulas, a one-dimensional rendezvous will be examined. In Fig. 1, the geometry is illustrated: two vehicles labelled A and B are depicted moving along a straight line with velocities at time t_0 of V_A and V_B, respectively. The initial separation range is R_0. It is assumed that V_A is constant while vehicle B is accelerating. Since vehicle B is a rocket and accelerates by expulsion of mass, the acceleration, a, is non-constant and can be expressed by means of the following equation:

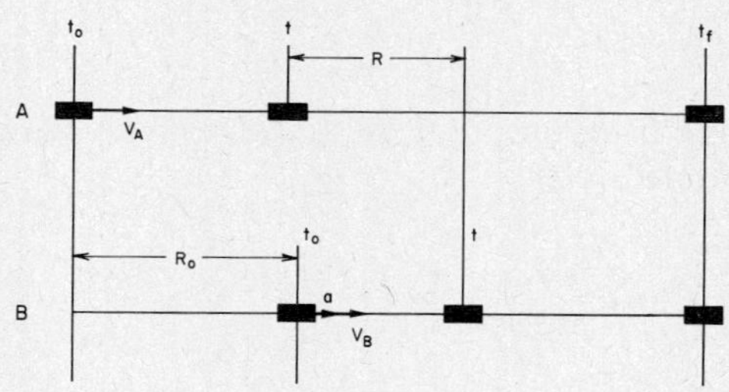

Fig. 1

$$a(t) = \frac{T}{m_0 + \dot{m}t} \tag{1}$$

where T = thrust (lb)

m_0 = initial mass (slugs)

$\dot{m}$ = mass flow rate (slugs/sec)

The mass flow rate in Eq.(1) is always negative.

The velocity, V_B, can be attained by integrating Eq.(1),

$$V_B(t) = V_{B_0} - \frac{T}{\dot{m}} \log_e \left(\frac{m_0}{m_0 + \dot{m}t}\right) \tag{2}$$

The relative velocity, R, can be expressed as the difference between the two velocities,

$$\dot{R}(t) = V_B(t) - V_A(t)$$

$$= V_{B_0} - \frac{T}{\dot{m}} \log_e \left(\frac{m_0}{m_0 + \dot{m}t}\right) - V_A \tag{3}$$

$$= \dot{R}_0 - \frac{T}{\dot{m}} \log_e \left(\frac{m_0}{m_0 + \dot{m}t}\right)$$

The range between A and B can be found by integrating Eq.(3),

$$R(t) = R_0 + \dot{R}_0 t$$

$$+ \frac{T m_0}{\dot{m}^2} \left\{ \left(\frac{m_0 + \dot{m}t}{m_0}\right) \left[\log_e \left(\frac{m_0 + \dot{m}t}{m_0}\right) - 1\right] + 1 \right\} \tag{4}$$

Since $m = m_0 + \dot{m}t$,
it follows that both range and range rate may be expressed as functions of mass.

In particular,

$$\dot{R}(m) = \dot{R}_0 - \frac{T}{\dot{m}} \log_e \left(\frac{m_0}{m}\right) \tag{5}$$

and

$$R(m) = R_0 + \dot{R}_0 \left(\frac{m - m_0}{\dot{m}}\right)$$

$$+ \frac{Tm_0}{\dot{m}^2}\left\{\frac{m}{m_0}\left[\log_e\left(\frac{m}{m_0}\right) - 1\right] + 1\right\} \qquad (6)$$

From Eq.(5),

$$m = m_0 \exp\left[\frac{\dot{m}(\dot{R} - \dot{R}_0)}{T}\right] \qquad (7)$$

Substituting Eq.(7) into Eq.(6) to eliminate mass from the expression,

$$R = R_0 + \left(\frac{Tm_0}{\dot{m}^2} - \frac{\dot{R}_0 m_0}{\dot{m}}\right)$$

$$- \left(\frac{Tm_0}{\dot{m}^2} - \frac{\dot{R}\dot{m}}{T}\,\frac{Tm_0}{\dot{m}^2}\right) \exp \frac{\dot{m}(\dot{R} - \dot{R}_0)}{T} \qquad (8)$$

By definition, $\frac{T}{\dot{m}} = -gI_{sp}$

where g = acceleration of gravity at sea level

I_{sp} = specific impulse of propellant

Then, since a boundary condition of rendezvous is that $R = 0$ when $\dot{R} = 0$, it follows that

$$R_0 = \left(\frac{gI_{sp} m_0}{\dot{m}} + \frac{\dot{R}_0 m_0}{\dot{m}}\right) - gI_{sp}\frac{m_0}{\dot{m}}\, e\,(\dot{R}_0/gI_{sp})$$

Thus Eq.(8) may now be written as a function of $\dot{R}$ as follows,

$$R(\dot{R}) = \frac{Tm_0}{\dot{m}^2}\, e^{(\dot{R}_0/gI_{sp})}\left[1 - \left(1 + \frac{\dot{R}}{gI_{sp}}\right) e^{(-\dot{R}/gI_{sp})}\right] \qquad (9)$$

Eq.(9) is the functional relationship between range and velocity which must be satisfied for a rendezvous to occur using the constant thrust engine in vehicle B.

RENDEZVOUS TECHNIQUES

A survey of proposed systems for accomplishing rendezvous between a satellite and another vehicle would indicate that most of these systems could be classified into two groups. The first and most frequently encountered is where the relative velocity of the two vehicles is small, on the order of 1000 fps or less. These conditions result from the rendezvous vehicle having been injected into a lower orbit and some "optimum" type transfer initiated or if the maneuver is of the "direct ascent" type where the rendezvous vehicle's ascent trajectory puts it near the altitude of the target, some acceleration has already been used to bring about the low relative velocity.

The second general type of system investigates the direct ascent maneuver, but allows the rendezvous guidance system to control the propulsion system in order to accomplish the orbital injection simultaneously with the rendezvous. These maneuvers in general result in much higher initial relative velocities and correspondingly larger initial ranges. While the pros and cons of these two approaches have been the subject of some controversy, no attempt will be made here to choose either one as being the more desirable. It is the purpose here to examine the characteristics of each of the above classes of maneuver and attempt to derive an analytic approach which could be used for either type.

For all the maneuvers discussed in this paper, one basic assumption is always made: the rendezvous vehicle velocity is less than the target velocity and the initial configuration is such that the rendezvous vehicle is ahead of the target with the target overtaking it. The advantage of this geometrical arrangement is obvious since the rendezvous vehicle must accelerate to orbital velocity in any case, any increase in velocity beyond this, which would be necessary if the target was in front of the rendezvous vehicle, would require acceleration to lower the velocity to the orbital velocity. Such maneuvers are, in general, inefficient.

RENDEZVOUS FOR LOW CLOSING VELOCITIES

In order to illustrate the utilization of the analytic approach, it is convenient to choose numerical values for an example and by varying some of the parameters of the problem see the interaction of guidance concept with propulsion, initial conditions, etc. For these purposes, let us choose the following:

W_0 = initial rendezvous vehicle weight = 10000 lb

gI_{sp} = vehicle exhaust velocity = 10000 fps

Two distinct propulsion systems will be analyzed:

1) An on-off or pulsed engine with fixed thrust of 2000 lb
2) A continuously throttleable or variable thrust engine with a thrust range of 2000 - 500 lb

Both types of propulsion, while perhaps not "off the shelf" items, are within state-of-the-art capability today.

If the above values are substituted into Eq (9), one obtains,

$$R(\dot{R}) = 15.54 \times 10^6 \, e^{(R_0 \times 10^{-4})} \left[1 - (1 + \dot{R} \times 10^{-4}) \, e^{-\dot{R} \times 10^{-4}}\right]$$

Now, if $\dot{R}_0 = -500$ fps, then

$$R(\dot{R}) = 14.78 \times 10^6 \left[1 - (1 + \dot{\rho}) \, e^{-\dot{\rho}}\right]$$

$$\text{where } \dot{\rho} = \dot{R} \times 10^{-4}$$

This relationship is plotted in Fig. 2. It will be noted that for this value of $\dot{R}_0$, the curve closely approximates a parabola (since the mass changes by only 5% and hence the acceleration is nearly constant).

A scheme for decelerating relative to the target is illustrated in Fig. 3. Two constant acceleration profiles are plotted for $a = 5 \text{ fps}^2$ and $a = 2.5 \text{ fps}^2$. These two levels are used as switching criteria in the following manner: the rendezvous vehicle, B, approaches the target, A, at 500 fps until the range has decreased to 25000 ft, (point 1). At this time, the vehicle thrusts until the lower acceleration profile is reached, point 2 of Fig. 3. At this time, the range has decreased to 9300 ft and the closing velocity is 215 fps. The first thrust period lasted 43.7 sec. Vehicle B then coasts until the higher acceleration profile is reached, point 3, and then repeats the procedure. The thrusting and coasting times are indicated on Fig. 3.

For an on-off thrusting system, the number of thrust pulses is a function of the two acceleration levels chosen as switching criteria, the thrust, and the initial velocity. It may be noted that if the two acceleration levels are nearly equal, a larger number of pulses would be required than when there is a large difference in the levels. In this system, the shut-down periods could be used for measuring range and range rate in order to compensate for any uncertainties in thrust levels, and to avoid the problem of measuring through the exhaust plume, if this is indeed a problem.

A second technique which has been proposed to accomplish satellite rendezvous is to use continuous thrust which is variable in magnitude. Several such engines have been built in recent years, the most widely known being the engine used in the X-15 rocket research aircraft. In order to illustrate the use of this technique, a phase plane analysis will again be used. For the example of this paper, the engine will be assumed to be continuously variable over a thrust range of 2000 lb to 500 lb. Fig. 4 shows the R-$\dot{R}$ plot for maximum and minimum thrust levels. It may be noted that any thrust level which lies between the maximum and minimum would also achieve a rendezvous from some set of initial conditions. If the range and range rate were accurately known, there exists a unique thrust level which would bring about the rendezvous. Since, in general, there exist inaccuracies in the knowledge of position and velocity, some guidance philosophy must be used in a closed-loop form to compensate for these uncertainties. One possibility is to define a reference thrust or

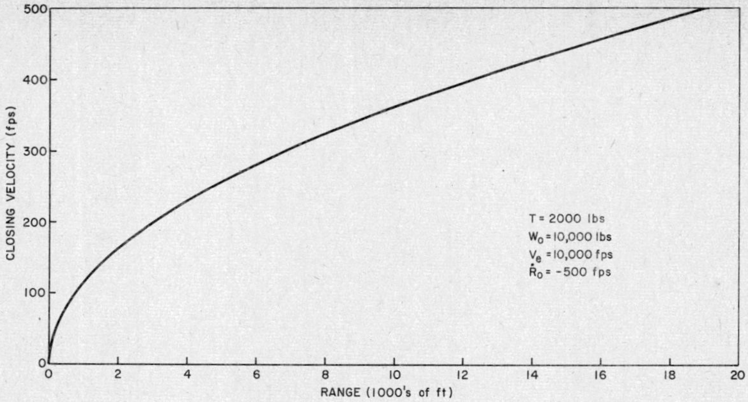

Fig. 2

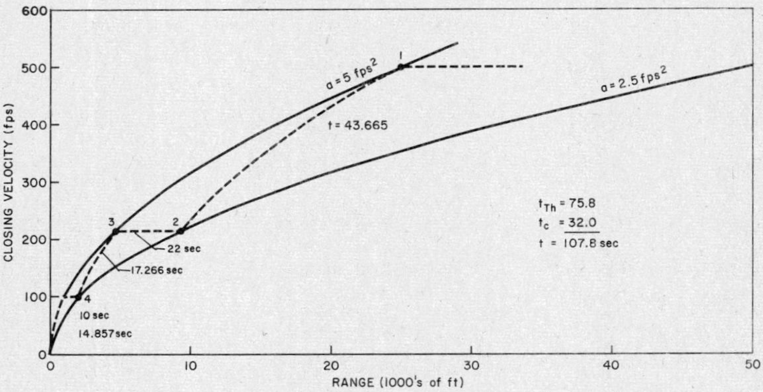

Fig. 3

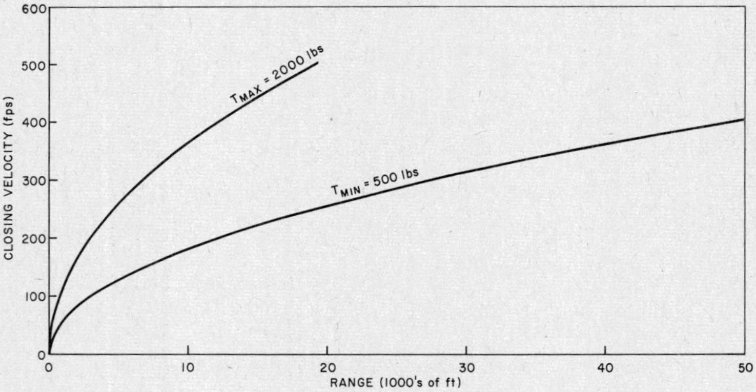

Fig. 4

acceleration level intermediate to the extremes and vary the thrust about this reference level to account for deviations from the reference trajectory conditions. For example, let T_{ref} be the intermediate thrust level. Then the following equation could be used to command a thrust, T_{com}, to bring about a rendezvous:

$$T_{com} = T_{ref} + S \left[R(\dot{R})_{T_{ref}} - R_{meas} \right]$$

where

S = System gain

$R(\dot{R})_{T_{ref}}$ = Desired range vs measured range rate for T_{ref}

R_{meas} = Measured value of range

From this equation, it will be seen that if $R_{meas} > R(\dot{R})_{T_{ref}}$, then the thrust is reduced to return the $R - \dot{R}$ conditions to the reference curve. If R_{meas} is below the desired range for the corresponding closing rate, then the thrust is increased.

In actual practice, the system described above could also be instrumented to command acceleration rather than thrust level, since devices for measuring acceleration are orders of magnitude more accurate than thrust measuring devices. A guidance equation could then be written:

$$a_{com} = a_{ref} + S \left[\dot{R}^2 - 2\, a_{ref} R \right] \quad (10)$$

In this case constant acceleration profiles are parabolas on the phase plane.

In Fig. 5, the maximum and minimum thrusts are plotted as well as the reference acceleration. In Fig. 6, the use of the guidance equation is illustrated. The initial range is below the desired range for the existing closing rate and hence the a_{com} will require the engine to thrust at its maximum level until the T_{max} curve intersects the a_{ref} curve at the throttle point. It

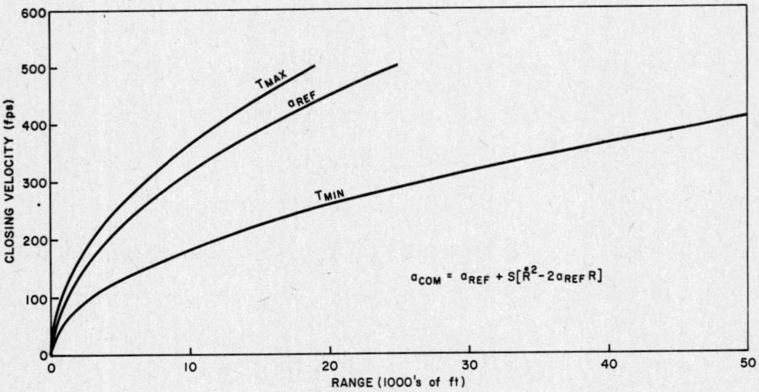

CLOSING VELOCITY (fps)
RANGE (1000's of ft)
T_{MAX}
a_{REF}
T_{MIN}
$a_{COM} = a_{REF} + S[\dot{R}^2 - 2a_{REF}R]$

Fig. 5

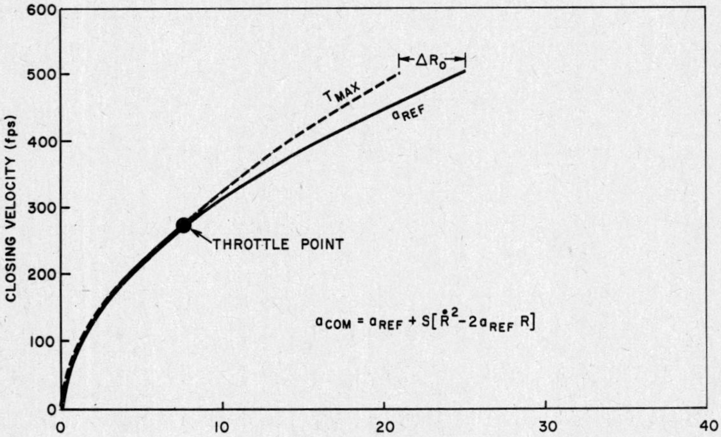

CLOSING VELOCITY (fps)
RANGE (1000's of ft)
ΔR_0
T_{MAX}
a_{REF}
THROTTLE POINT
$a_{COM} = a_{REF} + S[\dot{R}^2 - 2a_{REF}R]$

Fig. 6

then throttles down to whatever thrust will produce the desired acceleration.

Using Eq.(10) to guide the rendezvous vehicle, it has been found that the propellant requirement to accomplish the mission is essentially the same as that required to inject the vehicle into the satellite orbit without regard to the rendezvous constraint. The prime factors which influence the mission then are propulsion and initial conditions. The method of rendezvous outlined above required the rendezvous vehicle to remain in front of the satellite for the entire maneuver. For this constraint to be met, it is evident from the phase plane plots that for a given initial velocity, the initial range must be greater than some minimum level. This minimum range, in fact, can be determined from Eq.(9) by substituting R_0 for R. The equation then reduces to:

$$R_{0(min)} = \frac{Tm_0}{\dot{m}^2}\left[e^{\dot{R}_0/gI_{sp}} - \left(1 + \frac{\dot{R}_0}{gI_{sp}}\right)\right]$$

For the numerical example chosen here, this results in

$$R_{0(min)} = 20000 \text{ ft}$$

Since the above analysis considered only the simpler one-dimensional problem, this minimum range must actually be greater to account for the fact that the thrust is not applied impulsively and hence since the vehicle is moving in a gravity field, the thrust direction actually changes. Also, since the path control uses a small portion of the acceleration capability, some additional initial range must be included.

It may be noted that the minimum initial range is directly a function of thrust level, and hence if a throttleable engine is used to perform the rendezvous, the allowable variation in initial range is directly proportional to the ratio T_{max}/T_{min}. In particular, this variation may be expressed as:

$$\delta R_0 = \frac{m_0 (gI_{sp})^2}{T_{max}}\left[e^{\dot{R}_0/gI_{sp}} - \left(1 + \frac{\dot{R}_0}{gI_{sp}}\right)\right]\left(\frac{T_{max}}{T_{min}} - 1\right) \qquad (11)$$

Thus it can be seen that a 3 to 1 throttle range allows twice the initial range variation that a 2 to 1 throttle range permits. The main feature of having this initial range tolerance is to relax the requirement of launching the rendezvous vehicle at some exact instant of time in order that the vehicle arrive at the point of initiation of the maneuver exactly.

Rendezvous for High Initial Closing Velocities

The capability of extending the launch window by throttleable engines is more evident in direct ascent rendezvous maneuvers where the initial relative velocity is large (over 10000 fps). To cite an example, let

$$\text{Initial weight} = 10000 \text{ lb}$$

$$\text{Thrust max} = 10000 \text{ lb}$$

$$g I_{sp} = 10000 \text{ fps}$$

$$\dot{R}_0 = -10000 \text{ fps}$$

Then, Eq.(11) becomes

$$\delta R_0 = 188 \left[\frac{T_{max}}{T_{min}} - 1 \right]$$

where δR_0 is in nautical miles.

If the throttleability is 2 to 1, then the 188 n. mi. allowable variation corresponds to a launch window of 50 sec. This is for a 300 n. mi. orbit. The phase plane for this example is shown in Fig. 7.

When comparing thrust programs for velocity control in rendezvous maneuvers, it must be realized that, in general, the total velocity requirements are the same for all the various techniques. Hence, reducing the relative velocity by means of some open loop guidance system in order to have low closing velocities for the terminal phase may offer advantages in terms of seeker requirements but puts a considerable burden on the open loop guidance phase. There is, however, one technique where low closing velocities are very nearly inevitable. That is the use of a parking orbit as an intermediate to the ascent to the desired orbit. For reasonable altitudes, the orbit-to-orbit transfer results in low closing velocities. For direct ascent maneuvers, the closing velocity can be quite high and as shown earlier the rendezvous maneuver must be initiated at greater initial ranges. Here the trade-off between accurate launch guidance and terminal seeker capability becomes important.

TRAJECTORY CONTROL

The previous sections have considered only the one-dimensional rendezvous problem. In any real system, the problem of path control is equally as important as velocity control in providing for successful completion of the mission. The two propulsion systems discussed earlier (i. e., on-off or continuous, variable thrust) also could be used to provide path control. In the on-off system, pulses of thrust are applied at right angles to the line-of-sight to control the angular rate of the line-of-sight to some prescribed value (usually zero). In the continuous thrust system, the engine must be capable of swiveling in order to provide components of thrust to accomplish the line-of-sight control. The choice between the two systems would depend on the availability of multiple thrust units located about the vehicle to provide this lateral control versus the greater complexity of a throttleable engine. The required thrust levels as well as the expected variations in initial conditions will undoubtedly dictate which choice should be made.

COORDINATE SYSTEMS

Many authors in their study of satellite rendezvous problems have simulated the problem in a target centered coordinate system which has one axis along the local vertical, another in the horizontal plane and the plane of the orbit, the third normal to these two. This coordinate system is shown in Fig. 8. A linearized set of equations has been developed to express the unthrusted motion of the rendezvous vehicle relative to the target in this system:

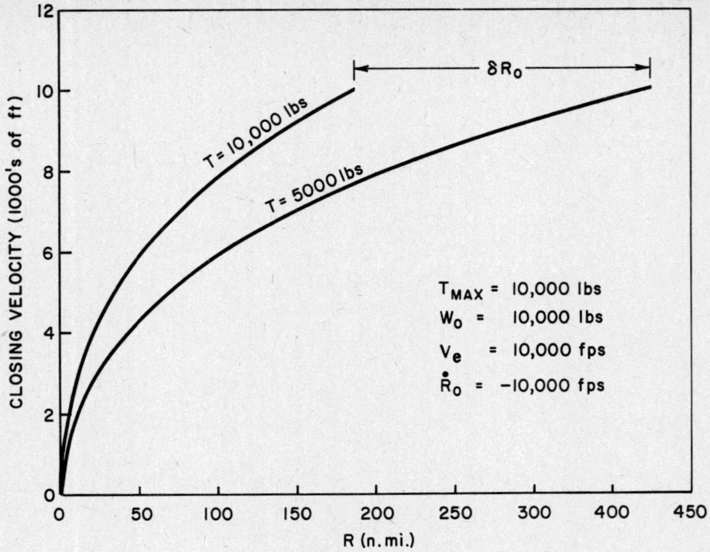
CLOSING VELOCITY (1000's of ft)
R (n. mi.)
δR₀
T= 10,000 lbs
T= 5000 lbs
T_{MAX} = 10,000 lbs
W_0 = 10,000 lbs
V_e = 10,000 fps
$\dot{R}_0$ = −10,000 fps
0
2
4
6
8
10
12
0
50
100
150
200
250
300
400
450

Fig. 7

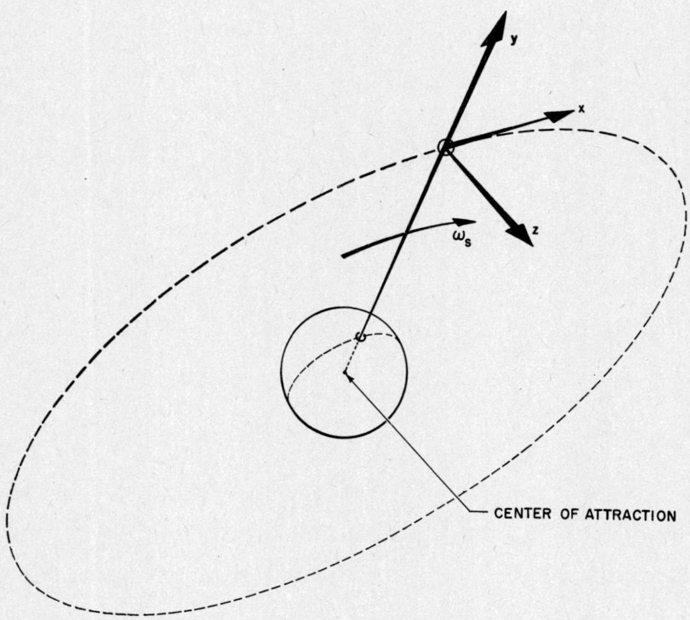
y
x
z
ω_s
CENTER OF ATTRACTION

Fig. 8

$$\ddot{x} = 2\omega_s \dot{y}$$

$$\ddot{y} = -2\omega_s \dot{x} + 3\omega_s^2 y$$

$$\ddot{z} = -\omega_s^2 z$$

These equations have proved adequate under certain restrictive conditions:

1) the times involved in the maneuver are small (less than one-fourth the orbital period of the target) and
2) the distances are relatively small when compared to the distance from the center of force

When these two conditions are not met, or if greater accuracy is required, a more sophisticated set of equations must be used. They may be derived in the same manner as the linearized equations, but more terms in the series expansion must be retained. These equations are:

$$\ddot{x} = 2\omega_s \dot{y} + \frac{3\omega_s^2 xy}{r_s} + \frac{3}{2}\frac{\omega_s^2 x^3}{r_s^2} - \frac{6\omega_s^2 xy^2}{r_s^2}$$

$$\ddot{y} = -2\omega_s \dot{x} + 3\omega_s^2 y + \frac{3}{2}\frac{\omega_s^2 x^2}{r_s} - \frac{3\omega_s^2 y^2}{r_s} - \frac{6\omega_s^2 y^3}{r_s^2} + \frac{3}{2}\frac{\omega_s^2 yx^2}{r_s^2}$$

$$\ddot{z} = -\omega_s^2 z + \frac{3\omega_s^2 yz}{r_s} + \frac{3}{2}\frac{\omega_s^2 x^2 z}{r_s^2} - \frac{6\omega_s^2 y^2 z}{r_s^2} + \frac{3}{2}\frac{\omega_s^2 z^3}{r_s^2}$$

These equations are not analytically solvable due to their non-linearity, but lend themselves to solution on analog computers where earth-centered coordinate systems are not feasible due to scaling problems. This author has had success in instrumenting this coordinate system to handle maneuvers taking an entire orbital period.

SUMMARY

This paper has demonstrated an analytic technique which is useful in determining mission parameters for satellite rendezvous problems. A comparison between methods of implementing the terminal guidance system as a function of the propulsion system and initial conditions is made. The problem of path control is discussed and an extension of a useful coordinate system for simulating relative motion is given.

AN AUTOMATIC FLIGHT PATH CONTROL SYSTEM FOR THE TERMINAL PHASE OF ORBITAL RENDEZVOUS

R. L. STAPLEFORD*

A proposed design of an automatic flight path control system for the terminal phase of orbital rendezvous is presented. The system approaches a minimum-complexity system, and as a result, should provide relatively high reliability and low cost, albeit at the expense of fuel consumption which is somewhat higher than the minimum. Simplicity is obtained by using nonmovable, fixed-thrust rockets which are operated in an on-off manner with a control logic simple enough to be mechanized by analog components alone. The system requires neither a digital computer nor an inertial platform.

INTRODUCTION

Within the last year or two, orbital rendezvous has become increasingly popular in the technical literature. Particular attention has been given to guidance and control during the terminal phase of the rendezvous operation+, and a number of automatic and manual flight path control systems have been proposed.[1-17] Some of these have emphasized system performance or fuel economy, while others stressed the reliability and cost advantages available with less complex systems. The designer is left with the familiar problem of having to evaluate the relative merits of reliability and cost vs performance. The objective of this paper is not to solve the designer's problem of selecting the best compromise for a given mission; instead, it is to add another automatic control system design, which may have application for some missions, to the assemblage of systems available to the designer.

In the proposed design, reliability and cost, rather than fuel consumption, are emphasized. This design approaches a minimum-complexity system. The interceptor's relative motion is controlled by nonmovable, fixed-thrust rockets which are operated in an on-off or "bang-bang" fashion. The control logic is simple enough to be easily mechanized with analog components alone, and the system requires neither a digital computer nor an inertial platform.

---

*Senior Staff Engineer, Systems Technology, Inc., Inglewood, California.

+The terminal phase of orbital rendezvous begins when the booster has placed the interceptor vehicle into an orbit nearly identical to that of the target satellite; this phase ends when the interceptor is in position for a parking or docking maneuver. (Parking is here defined as maintaining a position near the satellite without establishing physical contact.)

GENERAL DESCRIPTION

The proposed system is conceptually quite similar to a number of prior designs;[1,2,17] however, some of its details and some features of the analysis are new. The basic guidance and control concept is that proposed by Duke, Goldberg, and Pfeffer.[1] The interceptor's attitude control system maintains the vehicle longitudinal axis along the line of sight (LOS) between the interceptor and the target, while at the same time limiting the vehicle's roll rate to prevent excessive gyroscopic crosscoupling between the pitch and yaw motions. If the attitude system maintains the vehicle in the proper orientation within relatively broad limits, the attitude control system will have little effect on the flight path control system. Consequently, the attitude control system will not be considered here.

The flight path control system is composed of two major elements—a range controller and an LOS controller, Fig. 1. The tracker supplies range and range-rate data to the range controller which operates the single rocket aligned along the interceptor's longitudinal axis. The function of the range controller is to reduce the initial closing velocity (negative range-rate) as the interceptor nears the target. It is assumed that for fuel economy the initial range-rate will be negative.

The tracker also supplies range and LOS angular velocity data to the LOS controller which operates four transverse rockets mounted normal to the vehicle's longitudinal axis. These rockets are equally spaced around the vehicle, with their thrust vectors all directed through the nominal center of gravity, Fig. 2. The function of the LOS controller is to keep the vehicle on a collision course with the target by keeping small the LOS angular velocity, which is measured by rate gyros mounted directly on the tracker. The reason for the transverse integrating accelerometers shown in Fig. 1 will be explained later.

The following discussion will show that if certain small terms in the equations of motion are neglected, the range motion is independent of the LOS motion. Consequently, the range controller will be considered first, and then the resulting range motion will be used in the analysis of the LOS controller. The validity of the simplifying assumptions will be examined in a later portion of the paper.

EQUATIONS OF MOTION

Because of the guidance scheme being used, it is convenient to utilize the equations of relative motion in an LOS coordinate system. One such coordinate system is the axes set, RNT (Ref. 18). The R axis is along the LOS from the target to the interceptor. The N axis is in the direction of $\underline{\Omega}_{LOS} \times \underline{R}$, where $\underline{\Omega}_{LOS}$ is the angular velocity vector of the LOS, and $\underline{R}$ is the range vector from the target to the interceptor. The third axis, T, is normal to the other two, and together they form a right-handed coordinate system in the order RNT, Fig. 3. Since the angular velocity of the LOS about itself is not definable, $\underline{\Omega}_{LOS}$ lies along the T axis.

The components of the angular velocity of the coordinate system along the R, N, and T axes are defined as Ω_1, Ω_2, and Ω_3, respectively. To maintain the R axis along the instantaneous LOS, the angular velocity components normal to $\underline{R}$, for both the LOS and the coordinates, must be equal, or $\Omega_2 = 0$, $\Omega_3 = \Omega_{LOS}$. Thus, the coordinates rotate about T at the rate Ω_{LOS}, and about R at the rate necessary to keep $\underline{\Omega}_{LOS}$ along the T axis.

The complete equations of relative motion in these coordinates have been derived in Ref. 18. With the first-order approximation to the tidal acceleration included (the vector difference between the gravitational acceleration of the target and that of the interceptor), these equations are

$$\begin{aligned}
\ddot{R} - \Omega_3^2 R &= A_R - \omega^2 R\left[1 - 3(\underline{e}_r \cdot \underline{e}_R)^2\right] \\
R\dot{\Omega}_3 + 2\dot{R}\Omega_3 &= A_N + 3\omega^2 R(\underline{e}_r \cdot \underline{e}_R)(\underline{e}_r \cdot \underline{e}_N) \\
R\Omega_3\Omega_1 &= A_T + 3\omega^2 R(\underline{e}_r \cdot \underline{e}_R)(\underline{e}_r \cdot \underline{e}_T) \qquad (1)
\end{aligned}$$

where

$\underline{e}_R, \underline{e}_N, \underline{e}_T$ ≡ unit vectors in the R, N, T directions

A_R, A_N, A_T ≡ thrust acceleration components in the R, N, T directions

$\underline{e}_r$ ≡ unit vector along the geocentric radius from the earth to the target

ω ≡ angular rate of a circular orbit with radius equal to the instantaneous target geocentric radius

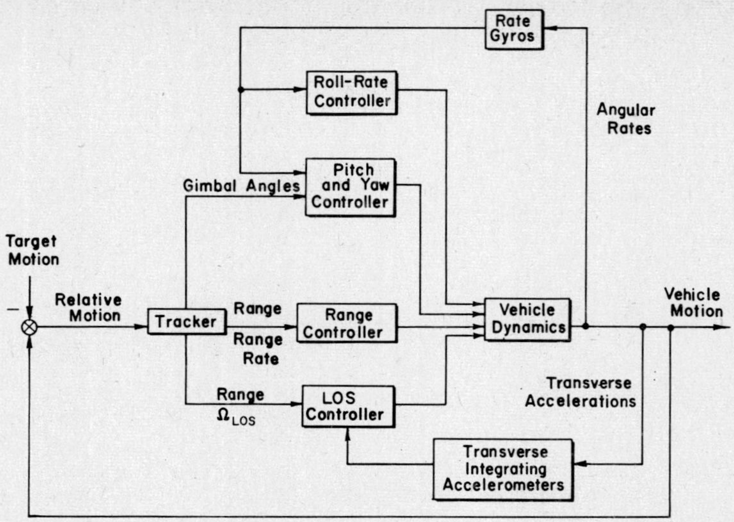

Fig. 1 Automatic Flight Control System

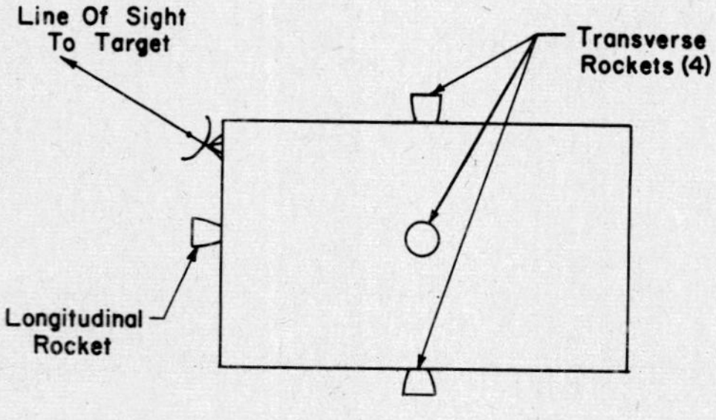

Fig. 2 Rendezvous Vehicle

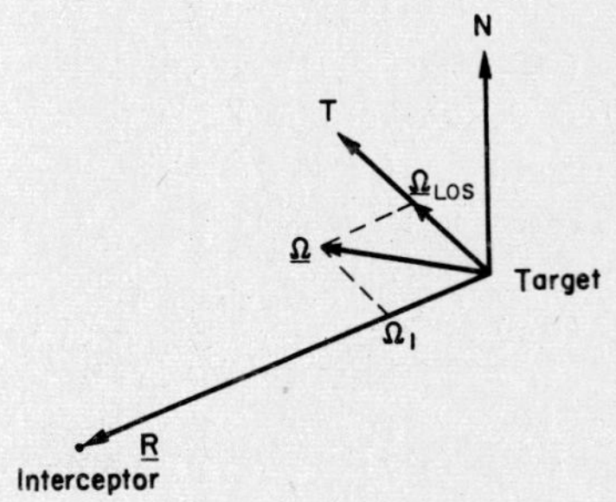

Fig. 3 Line-of-Sight Coordinates

These equations are both coupled and nonlinear, and are further complicated in that ω (for noncircular target orbits) and $\underline{e}_r$ are functions of time.

To reduce these equations to a more tractable form, two simplifying assumptions will be made. These are that the tidal acceleration terms (the $\omega^2 R$ terms) are negligible, and that the $\Omega_3^2 R$ term in the range equation is also negligible. The first assumption is justified on the grounds that during terminal rendezvous, the range is small compared to the geocentric radius, and the tidal accelerations are quite small. Since the function of the LOS controller is to keep Ω_3 near zero, the omission of the $\Omega_3^2 R$ term appears reasonable; however, the validity of both assumptions will be checked later.

With the tidal acceleration neglected, there is no need to apply thrust in the T direction, and it can also be assumed that $\Omega_1 = 0$; as a result, the problem is reduced to a two-dimensional one. This simplification is readily explained, since the relative motion will naturally remain in one plane as long as the relative acceleration normal to this plane is zero.

The simplified equations of motion are now

$$\ddot{R} = A_R$$

$$R\dot{\Omega}_3 + 2\dot{R}\Omega_3 = A_N \tag{2}$$

The range equation is now independent of Ω_3, and contains only an inertia term. For this reason, the range controller may now be considered independently of the LOS controller.

RANGE CONTROLLER

A block diagram of the proposed range controller is shown in Fig. 4. The tracker supplies a feedback signal to the controller; this signal is the sum of the range and a constant, T_E, times the range-rate. The forward path includes a one-sided relay and two transport lags. The first lag, τ_1, is to approximate the effects of any smoothing or filtering which is required on the range or range-rate data. The second, τ_2, is to approximate the thrust buildup and decay characteristics of the rocket.

From Fig. 4 it can be seen that the thrust will begin a time, $\tau = \tau_1 + \tau_2$, after

$$\epsilon = R_{COM} - (T_E\dot{R} + R) = a \tag{3}$$

and terminate a time, τ, after

$$\epsilon = R_{COM} - (T_E\dot{R} + R) = b \tag{4}$$

For convenience, the criteria of Eqs. (3) and (4) will be redefined so that thrust begins and terminates a time, τ, after

$$R + T_E\dot{R} = R_A \tag{5}$$

and

$$R + T_E\dot{R} = R_A + h_R \tag{6}$$

respectively.

The proposed range controller can best be analyzed by means of the phase plane technique. The Appendix shows that, in the notation of Eqs. (5) and (6), the thrust-on line is defined by

$$R + \dot{R}(T_E - \tau) = R_A \tag{7}$$

and the thrust-off line is defined by

$$R + \dot{R}(T_E - \tau) = R_A + h_R + \frac{A_R\tau}{2}(2T_E - \tau) \tag{8}$$

With the switching lines (thrust on or off) given by Eqs. (7) and (8), the phase plane trajectories can take either of the two forms shown in Fig. 5. In the first type, only one thrust correction or application is used, but the final value of range-rate may be quite large. For the second type, a number of corrections are necessary, but the final range-rate is limited.

To limit the final value of range-rate, the range-controller parameters will be selected so that the second type of trajectory is normally used. The switch to the parking or docking phase can then be made either after the range-rate becomes positive, or when its absolute value is very small.

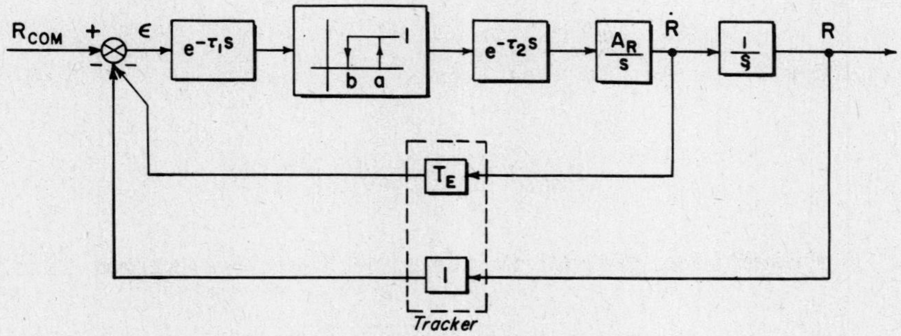

Fig. 4 Range Controller

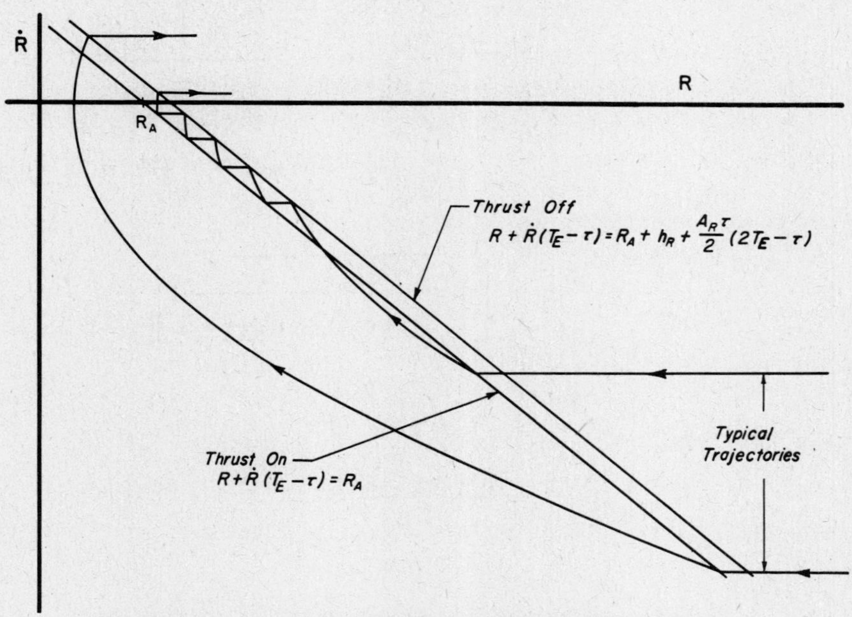

Fig. 5 Typical Range Phase Plane Trajectories

In setting the range-controller parameters, the possible variations in initial range-rate must be considered. Let the largest and smallest magnitudes of the initial range-rate be defined by

$$\dot{R}_{MAX} = -(1 + \lambda)V_{NOM}$$

$$\dot{R}_{MIN} = -(1 - \lambda)V_{NOM} \tag{9}$$

The parameters will be set so that

$$\dot{R}_{MAX} = -2A_R(T_E - \tau) \tag{10}$$

Then, from the equations derived in the Appendix, the maximum final value of range-rate is

$$\dot{R}_{f_{MAX}} = \sqrt{A_R^2T_E^2 + 2A_Rh_R} - A_R(T_E - \tau) \tag{11}$$

If $\lambda \leq 1/3$, the maximum number of thrust corrections which may be required is

$$N_{MAX} = \text{first integer} \geq \frac{-\dot{R}_{MIN}\left[2A_R(T_E - \tau) + \dot{R}_{MIN}\right]}{A_R^2\tau(2T_E - \tau) + 2A_Rh_R} \tag{12}$$

Combining Eqs. (9) through (12), and neglecting the fact that N_{MAX} must be an integer,

$$\frac{\dot{R}_{f_{MAX}}}{V_{NOM}} = \frac{1 + \lambda}{2}\left[\sqrt{1 + \frac{8\lambda(1 - \lambda)}{(1 + \lambda)^2N_{MAX}}} - 1\right] \tag{13}$$

and for

$$\frac{8\lambda(1 - \lambda)}{(1 + \lambda)^2N_{MAX}} \ll 1$$

$$\frac{\dot{R}_{f_{MAX}}}{V_{NOM}} \doteq \frac{2\lambda(1 - \lambda)}{(1 + \lambda)N_{MAX}}$$

Equation 13 is plotted in Fig. 6, which shows that $\dot{R}_{f_{MAX}}$ can be made quite small by using a large number of corrections. Very large values of N_{MAX} are undesirable, however, because of the reduction in reliability which must be associated with restarting the longitudinal rocket a great many times.

Some important features of the proposed range controller should be noted here before giving a numerical example of the controller:

1. When $\dot{R}_o = -2A_R(T_E - \tau)$, only one correction is used, the final range-rate is given by Eq. (11), and the minimum value of range is R_A.
2. When the maximum magnitude of $\dot{R}_o$ is less than or equal to $2A_R(T_E - \tau)$, the minimum possible range is R_A; thus R_A can be set at the desired closest approach distance.
3. From Eq. (11) and the fact that h_R must be greater than or equal to zero, it can be seen that $\dot{R}_{f_{MAX}}$ must be greater than or equal to $A_R\tau$.

As an example, assume:

$$V_{NOM} = 500 \text{ ft/sec}$$

$$\lambda = 0.1$$

$$\tau = 1 \text{ sec}$$

If it were necessary to restrict the final range-rate to 10 ft/sec or less, i.e.,

$$\dot{R}_{f_{MAX}} = 10 \text{ ft/sec}$$

then the range controller parameters might be set at:

$$A_R = 5 \text{ ft/sec}^2$$

$$T_E = 56 \text{ sec}$$

$$h_R = 282.5 \text{ ft}$$

$$N_{MAX} = \text{first integer} \geq 8.03$$

$$= 9$$

Phase plane trajectories for maximum, nominal, and minimum initial range rates are shown in Fig. 7. An enlarged view of the latter portions of these trajectories is shown in Fig. 8.

LINE-OF-SIGHT CONTROLLER

Having selected a range controller, the LOS controller will now be considered. The LOS controller will consist of two identical channels. Each channel will operate two diagonally opposite transverse rockets, and will control one component of the angular rate of the LOS. Each of these components could be measured directly by one of two rate gyros mounted on the tracker. Because the channels are identical, the following discussion applies to either channel when the proper component of $\Omega_{LOS} = \Omega_3$ is used.

Basically, the scheme is to apply a transverse correction whenever Ω_3 exceeds a threshold which may be a function of range. Because rather long smoothing times will probably be required for the Ω_3 data, it will not be desirable to use the measurement of Ω_3 to terminate the thrust. Instead, body-mounted integrating accelerometers will be used for thrust cutoff. The effects of the time lags due to the smoothing will be neglected for the present, but they will be considered later.

The accelerometers will measure $\int A_N dt$, and the thrust will be cut off when the integral equals the negative of the initial value of the transverse velocity, $V_N = \Omega_3 R$. Even without any errors in the thrust cutoff or in the measurements of V_{N_0} and A_N, this technique would not perfectly cancel the transverse velocity, but, as will be shown later, the errors are relatively small compared to the errors in the V_N measurements.

Normally, the main source of error in the transverse-velocity measurements will be tracking noise. For simplicity in the numerical example, the 3σ noise

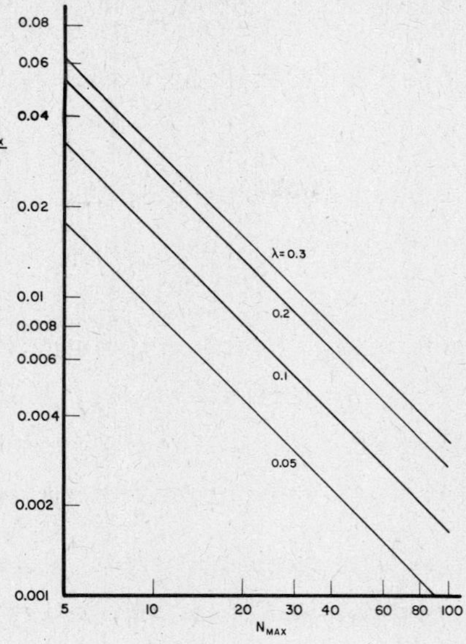

Fig. 6 Final Range Rate

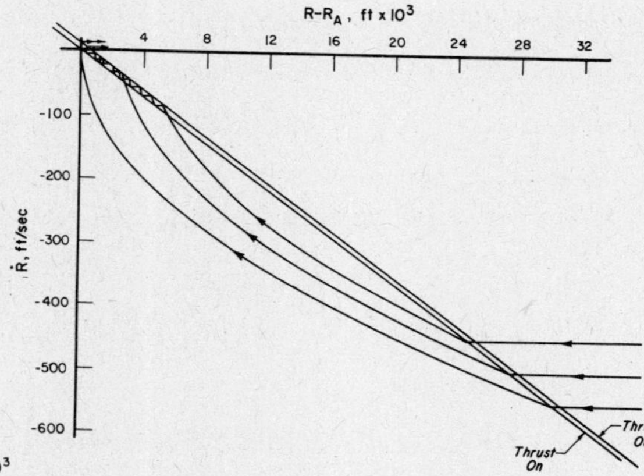

Fig. 7 Range Phase Plane Trajectories

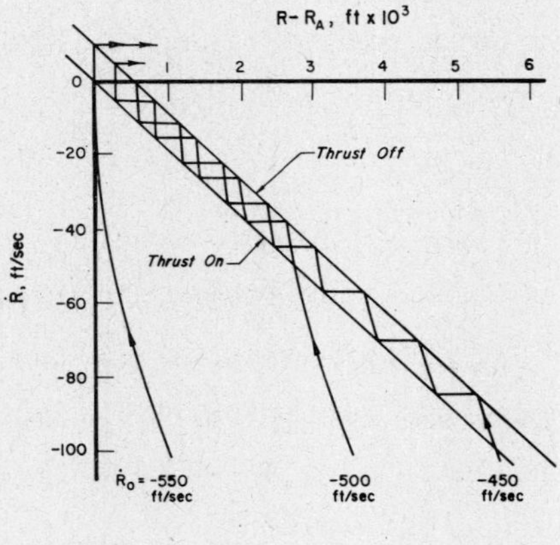

Fig. 8 Range Phase Plane Trajectories

envelope will be assumed to be a linear function of range, with a value of 0.5 ft/sec at zero range, and 18.5 ft/sec at 10^5 ft. These values appear reasonable for a radar which is tracking either a relatively small target or a large target which has a radar transponder.[1]

The transverse-velocity noise must be considered when selecting the correction threshold or switching line. If the switching line is too close to the noise envelope, a great many small corrections will be required, and there will not be sufficient smoothing time between them. Setting the line too high wastes fuel, because V_N will increase as the range decreases. For illustrative purposes, the switching line has arbitrarily been set at three times the 3σ noise envelope, and this appears to give quite satisfactory performance.

The behavior of V_N can be studied by a method similar to the phase plane technique;[19] the coordinates will not be a variable and its derivative, but will be two related variables, V_N and R. The LOS equation can be rewritten as

$$\dot{V}_N + \frac{\dot{R}}{R} V_N = A_N \tag{14}$$

This is a first-order differential equation, and has a solution

$$V_N = \frac{1}{R}\int_o^t A_N R dt + \frac{V_{N_o} R_o}{R} \tag{15}$$

The results of the range-controller design could be used to express range as a function of time, but it is more convenient to eliminate time from Eq. (15) by rewriting it as

$$V_N = \frac{1}{R}\int_{R_o}^{R} \frac{A_N R}{\dot{R}} dR + \frac{V_{N_o} R_o}{R} \tag{16}$$

With the on-off control, A_N will be constant between switching intervals, so that if $\dot{R}$ can be specified as a function of R, piecewise solutions can be constructed. In the following numerical example, the variation of $\dot{R}$ with R is based on a value of 500 ft for R_A and the nominal value of -500 ft/sec for $\dot{R}_o$.

From Eq. (16) the transverse velocity, as a function of range, can be determined. When the transverse acceleration is zero, the expression reduces to simply

$$V_N = \frac{V_{N_o} R_o}{R} \tag{17}$$

This elementary function can be evaluated very easily and quickly.

For a constant range-rate ($A_R = 0$), V_N is given by

$$V_N = V_{N_o} \frac{R_o}{R} + \frac{A_N}{2\dot{R}_o R} (R^2 - R_o^2) \tag{18}$$

Equation (18) can be cast into a more useful form by nondimensionalizing it to

$$\frac{V_N}{V_{N_o}} = \frac{R_o}{R} - \frac{A_N R_o}{2\dot{R}_o V_{N_o}} \left(\frac{R_o}{R} - \frac{R}{R_o} \right) \tag{19}$$

From Eq. (19), a nondimensional plot can be made of transverse velocity vs range for various values of the parameter $A_N R_o / \dot{R}_o V_{N_o}$, Fig. 9. Because the V_N switching line is a linear function of range, the magnitude of this parameter decreases as the range decreases, and reaches a minimum at the point where the range controller begins to make longitudinal corrections. For the example being used here, the minimum value is about 6.7 for a transverse acceleration of 2 ft/sec^2. The parameter is positive because $\dot{R}_o$ is negative, and A_N and V_{N_o} must be of opposite sign.

For the example, a transverse acceleration of 2 ft/sec^2 was arbitrarily selected as being a reasonable value. Too low an acceleration would result in excessive time required to make the transverse corrections, while too high a value would unnecessarily increase the weight of the rockets, and would increase their shutoff errors.

Even on the expanded scale used in Fig. 9, the trajectories are nearly straight lines. Thus, linear approximate solutions, which might be very easily calculated, could be used with little error. One approximation would be to assume

$$R = R_o + \dot{R}_o t$$

and

$$V_N = V_{N_o} + A_N t$$

or that

$$\frac{V_N}{V_{N_o}} = 1 - \frac{A_N R_o}{\dot{R}_o V_{N_o}}\left(1 - \frac{R}{R_o}\right)$$

Figure 9 shows that this approximate solution introduces only trivial errors.

In constructing an approximate transverse trajectory, the simplest method is to calculate the range at which V_N would be equal to zero by

$$(R)_{V_N=0} = R_o - \frac{\dot{R}_o V_{N_o}}{A_N} \tag{20}$$

The points $[R_o, V_{N_o}]$ and $[(R)_{V_N=0}, 0]$ can then be connected by a straight line to give a quite accurate trajectory.

When the range-rate is not constant, the expression for V_N becomes a great deal more complicated. Substituting

$$\dot{R} = \dot{R}_o \sqrt{1 + \frac{2A_R(R - R_o)}{\dot{R}_o^2}}$$

into Eq. (16), evaluating the integral, and rearranging it into a nondimensional form,

$$\frac{V_N}{V_{N_o}} = \frac{R_o}{R}\left\{1 + \frac{1}{3}\left(\frac{A_N R_o}{\dot{R}_o V_{N_o}}\right)\left(\frac{\dot{R}_o^2}{A_R R_o}\right)^2\left\{\left[\left(\frac{A_R R_o}{\dot{R}_o^2}\right)\left(2 + \frac{R}{R_o}\right) - 1\right]\left[\sqrt{1 + 2\left(\frac{A_R R_o}{\dot{R}_o^2}\right)\left(\frac{R}{R_o} - 1\right)}\right] - \left[3\left(\frac{A_R R_o}{\dot{R}_o^2}\right) - 1\right]\right\}\right\} \tag{21}$$

Three sample trajectories, using Eq. (21), have been calculated, and are shown in Fig. 10. To determine trajectory 3($\dot{R}_o = 0$), the limit of Eq. (21) as $\dot{R}_o$ approaches zero must be taken, or the integral of Eq. (16) must be reevaluated with the proper expression for $\dot{R}$. Either method gives

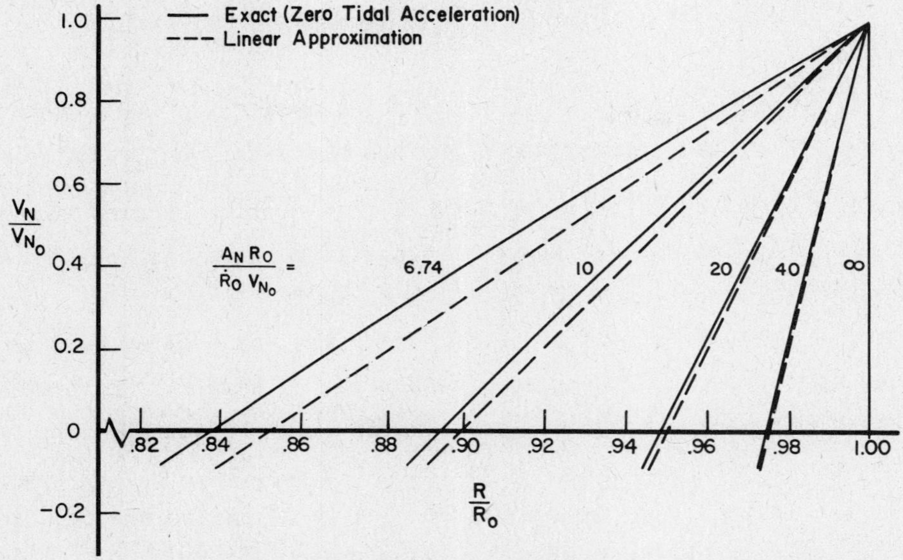

Fig. 9 Nondimensional Transverse Trajectories, $A_R = 0$

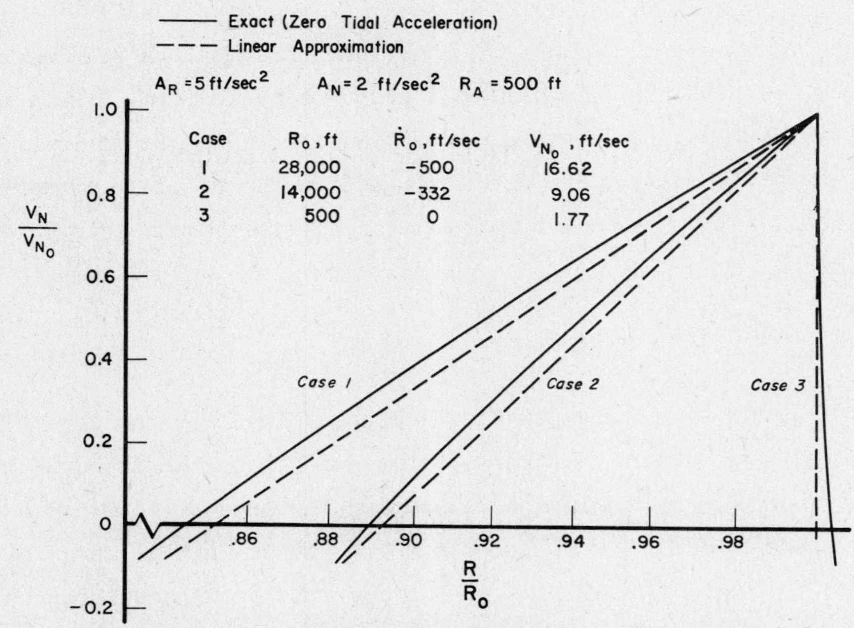

Fig. 10 Nondimensional Transverse Trajectories, $A_R \neq 0$

$$\left(\frac{V_N}{V_{N_o}}\right)_{\dot{R}_o=0} = \frac{R_o}{R}\left[1 + \frac{1}{3}\sqrt{\frac{2R_o}{A_R}}\left(\frac{A_N}{V_{N_o}}\right)\left(2 + \frac{R}{R_o}\right)\sqrt{\frac{R}{R_o} - 1}\right]$$

Figure 10 also shows the linear approximate solutions of Eq. (20); even with range acceleration, these approximate solutions are highly accurate.

As a result, the construction of V_N vs R trajectories becomes very simple. The approximate solution is used whenever transverse thrust is applied, and when the transverse thrust is zero, Eq. (17) is used. If the range is small enough so that range corrections are being made, the value of $\dot{R}_o$ must be taken from the appropriate range phase plane trajectory. A sample transverse trajectory, computed by this technique, is shown in Figs. 11 and 12. The conservative assumption was made that thrust cutoff always occurred at the 3σ noise envelope. This extreme case could be used to determine the fuel requirements for the LOS controller.

VALIDITY OF ASSUMPTIONS

In the preceding development, three important assumptions were made in order to reduce the particle dynamics to a more tractable form. These were that the effects of tidal acceleration, the $\Omega_3^2 R$ term in the range equation, and the lags in the LOS controller are negligible. The validity of each of these assumptions, and the errors due to the transverse thrust cutoff technique, will now be investigated.

Tidal Acceleration

From Eq. (1), the tidal acceleration component in the range direction is seen to vary from $-\omega^2 R$ to $2\omega^2 R$. At an altitude of 300 mi, ω^2 is equal to 1.2×10^{-6} rad^2/sec^2, and the tidal acceleration in range will be between $-1.2 \times 10^{-6} \times R$ sec^{-2} and $2.4 \times 10^{-6} \times R$ sec^{-2}. For an initial range of 200,000 ft, and a total approach time of 400 sec, the tidal acceleration would produce a change in range-rate of between -48 and 96 ft/sec.

The positive increment of 96 ft/sec is more significant than the negative increment since it reduces the closing velocity between the target and

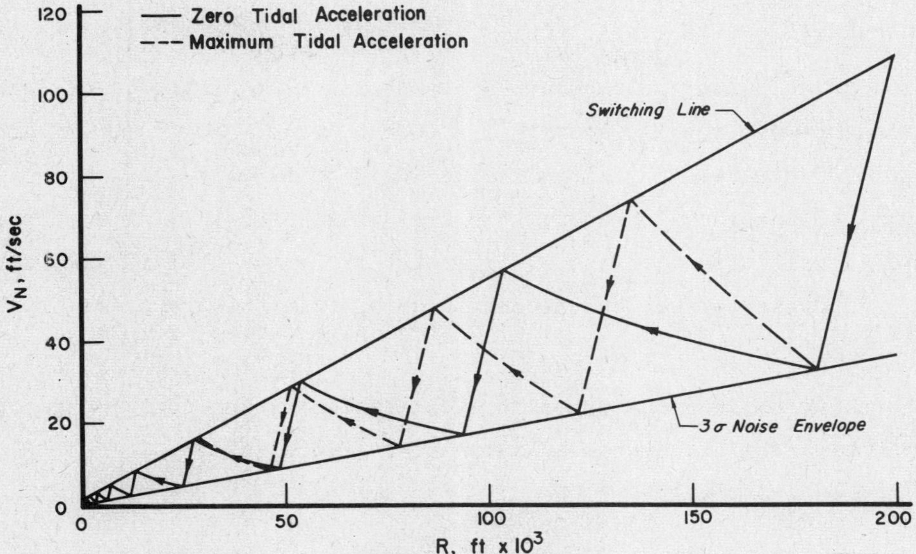

Fig. 11 Sample Transverse Trajectory

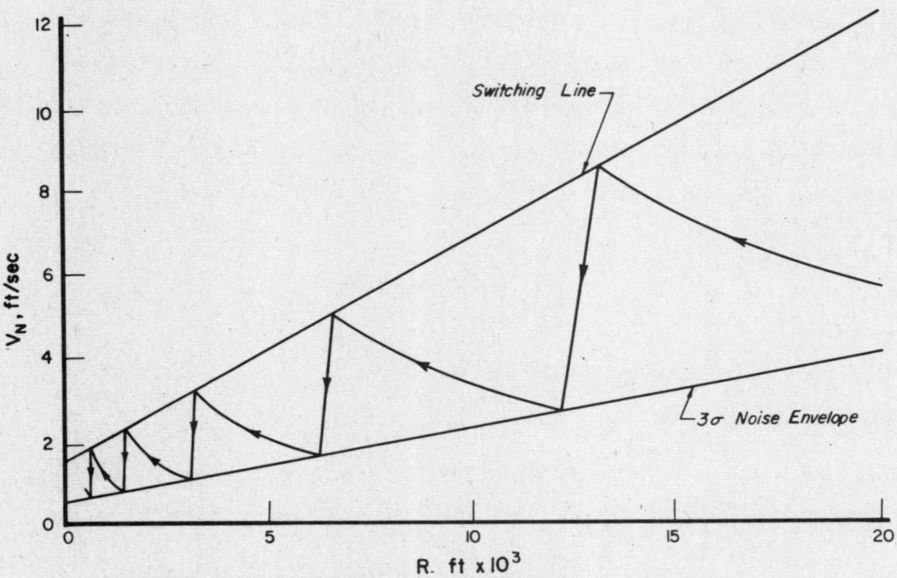

Fig. 12 Sample Transverse Trajectory

interceptor; however, this corresponds to the vehicle approaching the target from directly above or below. Approach from these directions is very unlikely, but even if it did occur, the reduction in closing velocity of 96 ft/sec would not be a severe effect. An increase in closing velocity of 48 ft/sec is of even less significance.

Equation (1) also shows that the transverse component of tidal acceleration varies between $\pm(3/2)\omega^2 R$ or $\pm 1.8 \times 10^{-6} \times R\ \mathrm{sec}^{-2}$ for a 300-mi orbit. At a range of 200,000 ft, the maximum value would be 0.36 ft/sec^2, or 18 percent of the nominal acceleration due to the transverse jets. Tidal acceleration will have a small effect on the transverse corrections, but it will have a much more important effect on the growth of V_N when the transverse rockets are off.

To examine the effects of tidal acceleration during coasting portions of the trajectory, the transverse equation is rewritten as

$$\dot{V}_N = -V_N \frac{\dot{R}}{R} + k_T R \qquad (22)$$

where k_T defines the magnitude of the transverse tidal acceleration ($|k_T| \leq 1.8 \times 10^{-6}\ \mathrm{sec}^{-2}$). Tidal acceleration effects will be important if the $k_T R$ term can be of the same order as the $-V_N \dot{R}/R$ term. At a range of 200,000 ft, the $k_T R$ term can be 0.36 ft/sec^2, while the $V_N \dot{R}/R$ term might be about 0.15 ft/sec^2. Obviously, tidal acceleration effects are not negligible. The importance of tidal acceleration will decrease as the range decreases, until such time as the range controller begins reducing the range-rate. As the range-rate becomes very small, Eq. (22) will reduce to

$$\dot{V}_N = k_T R \qquad (23)$$

At a range of 500 ft, the magnitude of $\dot{V}_N$ will be less than 10^{-3} ft/sec^2, which is negligible.

To examine the effects of tidal acceleration on the V_N vs R trajectories, refer again to Eq. (22). The parameter k_T will vary with time as the orientation of the LOS relative to the local vertical changes. The assumption that

k_T is constant and at its maximum value will somewhat overestimate the tidal acceleration effects, but will greatly simplify the solution of Eq. (22). With k_T held constant, the equation can be readily integrated to give

$$V_N = V_{N_o} \frac{R_o}{R} \left\{ 1 - \frac{k_T R_o^2}{3 \dot{R}_o V_{N_o}} \left[1 - \left(\frac{R}{R_o} \right)^3 \right] \right\} \tag{24}$$

In solving this equation, the effects of tidal acceleration on $\dot{R}$ are still neglected, i.e., $\dot{R}$ is assumed constant.

A sample trajectory was calculated from Eq. (24), and is included in Fig. 11 so that the tidal acceleration effects can readily be seen. An examination of Fig. 11 will show the pronounced effects of tidal acceleration at long range, and will indicate how the effects diminish as the range decreases. Apparently the primary result of including the effects of tidal acceleration will be to increase the amount of fuel expended by the transverse jets.

The $\Omega_3^2 R$ Term

The second assumption was that the $\Omega_3^2 R$ (or V_N^2/R) term in the range equation is negligible. The validity of this assumption can be shown by considering the change in $\dot{R}$ due to this term. For an average V_N of 60 ft/sec, an average R of 10^5 ft, and a time of 400 sec, $\Delta\dot{R}$ would be 14 ft/sec. This is certainly negligible.

Lags in the LOS Controller

The third assumption was to neglect the lags in the LOS controller. Since the thrust cutoff will be controlled by integrating accelerometers, there will be only very small lags in the thrust cutoff. The setting of the accelerometers might also be biased to include the effects of the thrust impulse which would occur after the cutoff signal was given.

There will be an appreciable lag between the actual and the measured values of V_N because of the required data-smoothing. As a result, the actual magnitude of V_N when the thrust is applied will be greater than the measured value. If the difference is a relatively large fraction of the measured value, the proposed scheme would have to be modified to compensate for the effects of the lag. The rate of change of V_N was given by Eq. (22):

$$\dot{V}_N = \frac{-V_N\dot{R}}{R} + k_T R$$

Dividing through by V_N,

$$\frac{\dot{V}_N}{V_N} = \frac{-\dot{R}}{R} + \frac{k_T R}{V_N}$$

The maximum value of the second term on the right side occurs at long range, and is 0.003 sec^{-1} or 0.3 percent/sec. The first term can be much larger than this. For $\dot{R}$ = -500 ft/sec at R = 28,000 ft, this first term is 0.018 sec^{-1} or 1.8 percent/sec. A 5-sec lag would cause a 9-percent error, which is approximately equal to the 1σ noise value. Therefore, the neglecting of lags in the LOS controller should not produce any significant effects.

<u>Transverse Thrust Cutoff Technique</u>

To examine the errors in V_N due to the proposed thrust cutoff scheme, refer to Eq. (14):

$$\dot{V}_N + \frac{\dot{R}}{R} V_N = A_N$$

The cutoff scheme neglects the $\dot{R}V_N/R$ term, and assumes that

$$\Delta V_N = \int_0^t A_N dt$$

Since the burning time will be $-V_{N_o}/A_N$, the contribution of the neglected term can be approximated by

$$\delta V_N \doteq \frac{\dot{R}_o}{R_o} \frac{V_{N_o}}{2} \frac{V_{N_o}}{A_N}$$

The factor of 2 arises because the average V_N over the burning time is approximately one-half V_{N_o}. Because of the short burning times, variations in R and $\dot{R}$ have been neglected. Converting to a percentage error,

$$\frac{\delta V_N}{V_{N_o}} \doteq \frac{V_{N_o}}{2A_N} \frac{\dot{R}_o}{R_o}$$

This quantity will have a maximum value for the minimum range before the range corrections are made, and this value is less than 8 percent. Compared with the errors due to noise, this value is unimportant.

SUMMARY

A proposed automatic flight path control system for the terminal phase of orbital rendezvous has been outlined. This design approaches a minimum-complexity system, and as a result, should have relatively high reliability and low cost. Simplicity has been obtained by using nonmovable, fixed-thrust rockets with on-off control and a control logic simple enough to be mechanized with analog components alone. This system requires neither a digital computer nor an inertial platform. Although the proposed design does not minimize fuel consumption, its reliability and cost advantages may be sufficient to justify its use for certain rendezvous missions.

ACKNOWLEDGEMENT

This paper is based on work carried out under Air Force Contract AF 33(616)-7094, monitored by the Directorate of Advanced Systems Planning, Aeronautical Systems Division, Air Force Systems Command.

REFERENCES

1. Duke, W. M., E. A. Goldberg, and I. Pfeffer, Error Analysis Considerations for a Satellite Rendezvous, ARS Preprint No. 1198-60, presented at the Semiannual ARS Meeting, Los Angeles, California, 9-12 May 1960.

2. Schroeder, W., A Terminal Guidance Scheme for Docking Satellites, ARS Paper No. 1952-61, presented at the ARS Guidance, Control, and Navigation Conference, Stanford University, Stanford, California, 7-9 August 1961.

3. Felleman, Philip G. and Norman E. Sears, Jr., "A Guidance Technique for Achieving Rendezvous," Proc. of the Manned Space Stations Symposium, Los Angeles, California, 20-22 April 1960, pp. 179-183.

4. Spradlin, Louis W., "The Long-Time Satellite Rendezvous Trajectory," Aero/Space Engineering, Vol. 19, No. 6, June 1960.

5. Swanson, R. S., N. V. Peterson, and L. R. Hoover, "An Astrovehicle Rendezvous-Guidance Concept," Western Aviation, Vol. 40, No. 2, February 1960.

6. Clohessy, W. H. and R. S. Wiltshire, "Terminal Guidance System for Satellite Rendezvous," Jour. of Aero/Space Sciences, Vol. 27, No. 9, September 1960.

7. Sears, Norman E., Jr., and Philip G. Felleman, Terminal Guidance for a Satellite Rendezvous, ARS Preprint No. 778-59, presented at the ARS Controllable Satellite Conference, Massachusetts Institute of Technology, Cambridge, Massachusetts, 30 April - 1 May 1960.

8. Levin, E. and J. Ward, "Manned Control of Orbital Rendezvous," Proc. of the Manned Space Stations Symposium, Los Angeles, California, 20-22 April 1960, pp. 184-187.

9. Kurbjun, Max C., Edwin C. Foudriat, Roy F. Brissenden, and Bert B. Burton, Pilot Control of Rendezvous, presented at the IAS 29th Annual Meeting, New York, New York, 23-25 January 1961.

10. Wolowicz, Chester H., Hubert M. Drake, and Edward N. Videan, Simulator Investigation of Controls and Display Required for Terminal Phase of Coplanar Orbital Rendezvous, NASA TN D-511, Flight Research Center, Edwards, California, October 1960.

11. Harrison, Eugene, "Some Considerations of Guidance and Control Techniques for Coplanar Orbital Rendezvous," Proc. of the National Specialists Meeting on Guidance of Aerospace Vehicles, Boston, Massachusetts, 25-27 May 1960, pp. 14-20.

12. Brissenden, Roy F., Bert B. Burton, Edwin C. Foudriat, and James B. Whitten, Analog Simulation of a Pilot-Controlled Rendezvous, NASA TN D-747, April 1961.

13. Stalony-Dobryanski, J. and O. Imai, "Attitude and Flight Path Control System for a Space Station Supply Vehicle," Proc. of the Manned Space Stations Symposium, Los Angeles, California, 20-22 April 1960. (Published by Institute of the Aeronautical Sciences, Inc., New York, 1960.)

14. Simon, E., Jr., A Proposed Control System to Facilitate the Terminal Stages of Manned Rendezvous, ARS Paper 1480-60, presented at the ARS 15th Annual Meeting, Washington, D. C., 5-8 December 1960.

15. Shapiro, M., "Attenuated Intercept Satellite Rendezvous System," ARS Journal, Vol. 31, No. 12, December 1961.

16. Cicolani, L. S., Trajectory Control in Rendezvous Problems Using Proportional Navigation, NASA TN D-772, April 1961.

17. Steffan, K. F., "Satellite Rendezvous Terminal Guidance System," ARS Journal, Vol. 31, No. 11, November 1961.

18. Thompson, H. B. and R. L. Stapleford, Basic Guidance and Control Information for the Terminal Phase of Orbital Rendezvous, ASD TR 61-344, August 1961.

19. Stapleford, R. L. and H. R. Jex, A Study of Manual and Automatic Control Systems for the Terminal Phase of Orbital Rendezvous (to be published as an ASD TR).

APPENDIX

The purpose of this Appendix is to derive the phase-plane equations for the range-controller. The longitudinal rocket thrust comes on a time, τ, after the quantity $R + \dot{R}T_E$ has decreased to a value R_A. The thrust is terminated a time, τ, after the parameter, $R + T_E\dot{R}$, has increased to a value $R_A + h_R$. If the subscript 1 is used to denote the values of R and $\dot{R}$ which satisfy:

$$R_1 + \dot{R}_1 T_E = R_A \tag{A-1}$$

Then, when the thrust is initiated,

$$R = R_1 + \dot{R}_1 \tau \tag{A-2}$$

$$\dot{R} = \dot{R}_1 \tag{A-3}$$

Combining Eqs. (A-1), (A-2), and (A-3), the thrust-on line is given by

$$R + \dot{R}(T_E - \tau) = R_A \tag{A-4}$$

Using the subscript 2 to denote the values of R and $\dot{R}$ which satisfy:

$$R_2 + \dot{R}_2 T_E = R_A + h_R \tag{A-5}$$

Then, when the thrust is terminated,

$$R = R_2 + \dot{R}_2 \tau + \frac{1}{2} A_R \tau^2 \tag{A-6}$$

$$\dot{R} = \dot{R}_2 + A_R \tau \tag{A-7}$$

Combining Eqs. (A-5), (A-6), and (A-7), the thrust-off line is given by

$$R + \dot{R}(T_E - \tau) = R_A + h_R + \frac{A_R \tau}{2} (2T_E - \tau) \tag{A-8}$$

An expression for the range-rate after n thrust corrections, $\dot{R}_n$, can be derived by considering the relationship between $\dot{R}_{n+1}$ and $\dot{R}_n$. If R_{n+1} is the range at the end of the (n + 1) correction, and R'_{n+1} is the range at the start of that same correction,

$$2A_R(R_{n+1} - R'_{n+1}) = \dot{R}^2_{n+1} - \dot{R}^2_n \tag{A-9}$$

Also from Eqs. (A-4) and (A-8),

$$R'_{n+1} + \dot{R}_n(T_E - \tau) = R_A \tag{A-10}$$

$$R_{n+1} + \dot{R}_{n+1}(T_E - \tau) = R_A + h_R + \frac{A_R\tau}{2}(2T_E - \tau) \tag{A-11}$$

Combining Eqs. (A-9), (A-10), and (A-11),

$$\left[\dot{R}_{n+1} + A_R(T_E - \tau)\right]^2 = \left[\dot{R}_n + A_R(T_E - \tau)\right]^2 + A_R^2\tau(2T_E - \tau) + 2A_Rh_R \tag{A-12}$$

Equation (A-12) is an elementary difference equation which has the solution

$$\left[\dot{R}_n + A_R(T_E - \tau)\right]^2 = \left[\dot{R}_o + A_R(T_E - \tau)\right]^2 + n\left[A_R^2\tau(2T_E - \tau) + 2A_Rh_R\right]$$

or

$$\dot{R}_n = -A_R(T_E - \tau) + \sqrt{\left[\dot{R}_o + A_R(T_E - \tau)\right]^2 + n\left[A_R^2\tau(2T_E - \tau) + 2A_Rh_R\right]} \tag{A-13}$$

Equation (A-13) is only valid for $n \leq N$ where N is defined by the inequalities:

$$\dot{R}_N \geq 0 \quad ; \quad \dot{R}_{N-1} < 0 \tag{A-14}$$

$\dot{R}_N$ is the final value of range-rate after all thrust corrections have been made. From Eqs. (A-13) and (A-14), it can be seen that if

$$-2A_R(T_E - \tau) \leq \dot{R}_o < 0 \tag{A-15}$$

then the maximum final value of range-rate, $\dot{R}_{f_{MAX}}$, is given by

$$\dot{R}_{f_{MAX}} = \sqrt{A_R^2 T_E^2 + 2A_R h_R} - A_R(T_E - \tau) \qquad \text{(A-16)}$$

Equations (A-13) and (A-14) can also be used to determine N. Letting $n = N$ and $\dot{R}_n = 0$ in Eq. (A-13) gives

$$N = \text{first integer} \geq \frac{-\dot{R}_o\left[\dot{R}_o + 2A_R(T_E - \tau)\right]}{A_R^2\tau(2T_E - \tau) + 2A_R h_R} \qquad \text{(A-17)}$$

Equation (A-17) has an absolute maximum for $\dot{R}_o = -A_R(T_E - \tau)$. Denoting the most positive (smallest magnitude) value of $\dot{R}_o$ by $\dot{R}_{MIN}$, the maximum value of N is

$$N_{MAX} = \text{first integer} \geq \begin{cases} \dfrac{-\dot{R}_{MIN}\left[\dot{R}_{MIN} + 2A_R(T_E - \tau)\right]}{A_R^2\tau(2T_E - \tau) + 2A_R h_R} & \text{for } -\dot{R}_{MIN} \geq A_R(T_E - \tau) \\[2ex] \dfrac{A_R(T_E - \tau)^2}{A_R\tau(2T_E - \tau) + 2h_R} & \text{for } -\dot{R}_{MIN} \leq A_R(T_E - \tau) \end{cases}$$

(A-18)

eighth
annual
meeting

AMERICAN ASTRONAUTICAL SOCIETY

INTRODUCTION

NEW CONCEPTS IN MONITORING OF ASTRONAUT BRAIN FUNCTION*

Roger M. Morrell, M.D. +

New findings in experimental and clinical neurophysiology, psychology, and instrumentation are enumerated as they might bear on techniques in monitoring of astronaut brain function. It is concluded that some form of representation and analysis of electrical activity of the brain is necessary but that monitoring of certain psycho-physiological or behavioral variables will also be useful. Methods are sought which allow prediction of significant changes in brain function, rather than waiting for them to occur. The following original proposals are made:

1. System for detecting and locating occlusion of active processes occurring in two or more areas of cerebral cortex or underlying neural pathways

2. Use of system in (1) to modify previous proposal for monitoring of alpha-rhythm and analysis of evoked potentials

3. Application of psychovisual techniques to brain monitoring

4. Analog simulation of harmonic oscillation of brain potentials as a test of attentiveness

5. Use of digital time-compression techniques in correlation analysis of the electroencephalogram (EEG)

6. Capacitance-sensing system for monitoring of movements of astronaut in vehicle

7. Experiment for testing a matrix theory of sensation (Oxley)

8. Modification of current methods of resolving scalp potentials from EEG

---

* A complete version of this paper (AAS 62-11) can be obtained from the American Astronautical Society or by contacting the author.

+ Jackson Memorial Hospital, Miami, Florida

9. Application of Wiener expansion theory and random process theory in predicting significant changes in the EEG treated as a signal space

As more data are gathered in physiology and psychology, the use of the traditional telemetered EEG will probably become outmoded and give way to radically new methods. The basis for successful monitoring will continue to be a unique representation of the information in abnormal brain function.

SUMMARY

In a previous report it was concluded that crosscorrelation and toposcopic techniques for EEG analysis would provide the greatest advantages in telemetered monitoring of astronaut brain function, in order to indicate impending or existing abnormalities in sensation, cognition, and behavior. Control commands were to be dependent upon computer data from EEG analysis, relayed as instructions to the astronaut, or as direct control over vehicular motion. This system based control commands on changes in the alpha brain rhythm, and the response to standardized visual or auditory stimuli.

The present communication describes recent findings in experimental and clinical neurophysiology which introduce unanticipated complexities into the problem of brain monitoring. The new perspectives forced by these findings also suggest innovations in the recording and interpretation of central nervous activity. Some of these will be facilitated by the new developments in instrumentation described in this paper.

It has been suggested that some of the lack of success in EEG analysis has been due to the fact that significant information has been discarded by performing statistical analysis rather than be determining a unique representation of the signal. For the determining of a unique number for each degree of freedom, those signals which are abnormal must be presented, and these may exist for brief periods only. If an average EEG is assumed to consist of eight channels for 30 minutes recording time and if 30 cps is the highest frequency, there are about 864, 000 possible degrees of freedom in the EEG, if each channel contains independent signals. In a paradox from entropy theory, we may be afforded a degree of freedom by removing uninformative variables. Unless there is prior knowledge of how information is coded into a signal like the EEG, a unique representation should be the goal of analysis. This approach should provide a useful guide to the development of practical systems for the monitoring of astronaut brain function.

BIBLIOGRAPHY

Adey, W. R. et al., EEG records from cortical and deep brain structures during centrifugal and vibrational acceleration in cats and monkeys, IRE Transactions on Biomedical Electronics, p. 182, July 1961

Anderson, V. C., Technical Memorandum No. 37 (NR -014-903) Harvard University Acoustics Research Laboratory, 5 Jan. 1956

Brooks, W. O., and Ervin, H. D., Detecting, locating, and mapping of neuron signal emissions within the human brain, Reprint 13, Ann. Conference on Electrical Techniques in Medicine and Biology. Digest of Sci. Papers, Nov. 1960

Brown, R. H., Weber ratio for visual discrimination of velocity, Science, 131, 1809, 17 June 1960

Cady, Lee., ECG wave amplitudes and areas for the IBM 650, private communication

Campbell, P. A. (Ed), Medical and Biological Aspects of the Energies of Space, Columbia Univ. Press, 1961

Clynes, M., Electronics, p. 6, 31 Mar. 1961

Clynes, M., Respiratory control of heart rate: laws derived from analog computer simulation, 1959 IRE National Convention Record, pt. 9, pp. 22-41

Drenick, R. F., Random processes in control and communications, Science, 132, 865, 30 Sept. 1960

Ecology of Space Flight, Physiological Journal, USSR, incomplete information

Fechner, G. T., Elemente der Psychophysik, 1860

Flaherty, B. E. (Ed.), Psychophysiological Aspects of Space Flight, Columbia University Press, 1961

Freeman, W. J., Harmonic oscillation as model for cortical excitability changes with attention in cats, Science, 133, 2058, 30 June 1961

Goertzel, G., Physiological Monitoring, What and How, New York Acad. Sci., Section of Physical Sciences, 18 Oct. 1961

Greenberg, I., Electroluminescent display and logic devices, Electronics, p. 31, 24 Mar. 1961

Haley, T. J., and H. Ganloff, Central nervous system effects following whole body x-irradiation of cats, Proc XII Internat'l Congress of Physiological Sciences, Buenos Aires, Argentina, August 1959

Holland, J. G., Human vigilance, Science, 128, 3315, 11 July 1958

Horton, G. A., and Koroncai, A. K., Radiotransmitter for remote heartbeat measurements, Electronics, p. 54, 23 Dec. 1960

Kiang, N. Y., et al., Evoked cortical activity from auditory cortex in anesthetized and unanesthetized cats, Science, 133, 1928, 16 June 1961

King, W. J., Continuous compensatory tracking by a Cebus monkey, Science, 134, 947, 20 Sept. 1961

Lilly, J. C., Sensorimotor cortex and patterns of activity in monkey (Macaque) with 600 implanted surface electrodes, private communication, 1957

Lindsley, D. B., and Emmons, W. H., Perception time and evoked potentials, Abstr. National Science Foundation, Science, 20 Sept. 1961

Lowenberg, E. C., Signal theory applied to the analysis of electroencephalograms, IRE Transactions on Biomedical Electronics, p. 7, Jan. 1961

Morrell, R. M., Autocorrelation and toposcopic technics in monitoring of astronaut brain function, Abstr. VII Annual Meeting, AAS, Dallas, Jan. 1961

Morrell, R. M., Monitoring of astronaut brain function by automatic analysis, Advances in the Astronautical Sciences, Vol. 8, Plenum Press, 1963

New York Academy of Sciences, Division of Mathematics and Division of Psychology, Conference on Human Decisions in Complex Systems, May 1960

Notterman, J. M., Cicala, G. A., Page, D. E., Demonstration of the influence of stimulus and response categories upon difference limens, Science, 131, 983, 1 April 1960

Peckham, R. H., Hart, W., Subthreshold retinal integration shown in low contrast flicker measurements, Science, 130, 1256, 6 Nov. 1959

Peterson, W. W., Birdsall, T. G., Fox W. C., IRE Transactions Profess. Group on Information Theory, PGIT-4 171, 1954

Purpura, D. P., Cohen, B., and Marini, G., Generalized neocortical responses and corticospinal neuron activity, Science, 134, 729-730, 15 Sept. 1961

Rosenbloom, M., Using time-compression techniques in digital correlation, Electronics, p. 191, 10 Mar. 1961

Section on Research and Development, Panel Displays Real-time or Programmed Data, Electronics, p. 68, 26 May 1961

Swets, J. A., Is there a sensory threshold?, Science, 134, 168, 21 July 1961

Swets, J. A., Tanner, W. P., Jr., and Bridsall, T. G., The evidence for a decision-making theory of visual detection, Electronic defense group, Univ. Michigan Tech. Rept. No. 40, 1955

Thomas, E. L., Howat, R., and Mackworth, N. H., TV tracker records eye focus movements, Electronics, p. 57, 22 April 1960

THE EFFECTS OF MAN ON THE WEIGHT OF A MANNED SPACE VEHICLE

Herbert C. Sanders*

The presence of man in space flight -- when compared to unmanned flight -- creates a series of new problems. One of the most important areas affected is that of the vehicle weight. This paper discusses the additional weight required by man's presence.

To accommodate man, vehicle design enters a new dimension. Systems to supply food, drink, and air will be required; a comfortable environment must be provided; pressure, temperature, and humidity must be controlled within definite limits. The psychological well-being of the occupants will influence the size and furnishings of the work and rest compartments. The larger vehicle surface areas will increase the amount of environmental gas leakage and also the probability of puncture by meteroroids. To accommodate humans, the vehicle system must be "man-rated" to provide an extra margin of reliability, safety, and escape.

To solve any one of these problems, the overall integrated vehicle must be considered since any system will affect other systems and structure. These interrelated problems are discussed from the aspect of added weight.

The weight data presented and discussed in this paper is associated with subjects such as:

1. An artificial gravity environment
2. The minimum crew living space as related to duration of space travel
3. Weights of food for space travel
4. Water requirements
5. Various components and assembled subsystems for four typical life-support concepts
6. Furnishings and instrumentation
7. Environment control.

A weight comparison is made between two vehicles, manned and unmanned, which perform similar missions in space. Weight statements and configuration drawings are presented for both vehicles.

Because the weight of the manned vehicle is approximately three times greater than the unmanned, it is necessary to design vehicle subsystems requiring a minimum crew and to provide (e.g., by stimulation) a means for the crew to live in the smallest possible space without inhibiting performance efficiency.

---

* The Boeing Company, Aero-Space Division, Seattle, Washington

INTRODUCTION

There has been a preponderance of technical papers and articles treating the problems of manned space travel. Some of these papers have dealt with the overall design problems that may be expected in a space vehicle while others have been devoted to more specialized subjects such as radiation effects, space feeding, meteoroid shielding, etc. The one question repeatedly asked is ''How much weight is added to a space vehicle because man is present?''

Man needs a unique environment to satisfy his mental and physical requirements. The various components of a space vehicle are so closely interrelated that the added weight of a suitable human environment will cause a chain of events which will have a large weight effect on many components throughout the vehicle. This effect is ultimately reflected in launch weight. The weight which must be added to a spacecraft to accommodate man will be discussed.

APPROACH

As presented in this paper, the subject matter is divided into three major sections as shown below:

Manned Space Vehicle Design Considerations: The required design considerations of a manned space vehicle are reviewed emphasizing those areas significantly affecting vehicle weight.

Vehicle Subsystem Description: A number of subsystems within the manned space vehicle needed to sustain the crew will be discussed together with their weights.

Manned and Unmanned Vehicles, a Comparison: Using similar missions, a weight comparison is made between two vehicles, one manned and the other unmanned.

I. MANNED SPACE VEHICLE DESIGN CONSIDERATIONS

Reliability and Safety

Reliability and crew safety are important design considerations for a manned space vehicle. Reliability may be defined as the probability of a vehicle performing its mission. Crew safety is the probability of crew survival and depends upon vehicle reliability; if flight failure occurs, crew safety is then dependent upon the probability of successful descent by escape means.

The overall vehicle reliability goals for both manned and unmanned vehicles could conceivably be the same.

Reliability may be maximized by using capable, highly trained crews performing maintenance and various other tasks while in flight. In addition to supplying spare parts, reliability can also be maximized by expert design and using redundant parts and systems. Living quarters may include more than one pressure compartment for crew safeguards in case of puncture or faulty seals.

Structural safety factors are generally higher when man is present. These higher factors are favored for the manned systems to ensure the highest possible crew safety.

Providing for high reliability would be greatly simplified if the designer had a comprehensive set of rules to govern each and every detail design. Lacking such a code, the designer must make his own design rules and thus tends to add unnecessary parts and structure when man is a design consideration. Only an expert analysis of all design problems can eliminate the excess weight of unneeded parts.

Artificial Gravity

Determining if an artificial gravity (by spin) should be provided is another important design consideration for a manned space vehicle.

If an artificial gravitational field could be economically and safely developed, the fears of many scientists and engineers on the problems of weightlessness and crew disorientation would diminish. Not only could urination, defecation, and wastes storage be handled in a conventional manner, but the danger of inhaling food particles and other liquids and solids that may be floating around the vehicle would be eliminated. Muscular degradation and decalcification of the bones would be lessened. The additional burdens of having to tie down everything after use would be eliminated. The control problems a crew member would experience in rapid movement from one place to another would not be so severe. Food preparation would be simplified.

In creating an artificial gravitational field, a number of physiological problems must be considered. To solve these problems would require a large amount of added weight. Consider, for example a space station 12-feet in diameter with sufficient spin to produce 1 "g" at the outer wall. A 6-foot man standing would feel 1 "g" at his feet and zero "g" at his head. Obviously, this would be a very poor design. Fig. 1 shows the gravity gradient from head to toe for a 6-foot man in percent of total local gravity as a function of vehicle diameter. It is not known what the maximum allowable gravity gradient should be; however, if 15 percent were selected (Ref. 1), then a 40-foot spin radius would be required.

Coriolis acceleration is another important consideration which will influence the size of a space vehicle and consequently its weight. Coriolis acceleration results from a body moving along a path that has rotation. The magnitude of Coriolis acceleration is expressed as $A_{cor} = 2\omega V$ where ω is the vehicle spin rate expressed in radians/sec and V is the velocity of a moving body with respect to the rotating vehicle expressed in feet/sec.

Fig. 1 is a plot of Coriolis force in percent of local "g" for various ω's as a function of vehicle spin radius. Local "g's" of .3, .5, 1, and 1.5 are shown. A crew walking velocity of 3.5 ft/sec. was assumed. If the maximum allowable Coriolis force is to be 25 percent (Ref. 2) of spin "g" and a spin "g" requirement of .5 is established, a spin radius of 50 feet would be required. When the disorienting and nauseating effects brought about by the movement of one's head while in a rotating system are considered, a much larger radius would be desirable (Ref. 3) in order to provide .5"g."

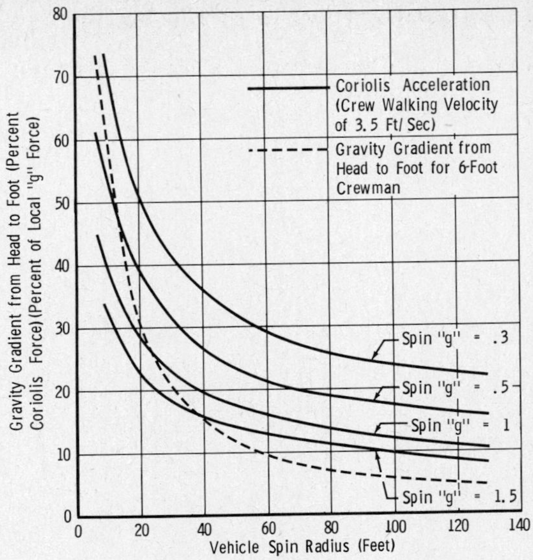

Fig. 1 Coriolis Acceleration and Gravity Gradient in Space Vehicle

There have been a number of methods presented throughout the past several years suggesting how vehicles could be constructed with a spin radius large enough to provide a suitable artificial gravitational field. The weights and sizes of many of the vehicles indicated were beyond the capabilities of a single booster system; two or more trips would be required to place the necessary parts in orbit, and the final vehicle was then to be assembled in space. High reliability and crew safety becomes even more difficult under these conditions. The problem of assembly is compounded by two things: (1) rendezvous in space; and (2) the ability of the crew to assemble the vehicle when hampered by working in a stress-inducing state, exposed to the physical hazards of radiation and meteoroids.

Because these vehicles were to have utility beyond just being in space, equipment such as telescopes, cameras, radar, etc., were placed in the hub of rotation to ensure an acceptable sensor resolution. This necessitates an additional section which must be provided with a suitable atmosphere. As the crew moves to and from the "weightless" hub in performing their normal duties, a possible degraded physiological state could result from a changing "g" loading.

The final vehicle design might well end up as a compromise between the heavier artificial gravity type and the smaller zero gravity vehicle. A part-time gravitational environment would be supplied only in case of an emergency to correct a severe case of crew disorientation. It has already been suggested by Clark and Hardy (Ref. 3) that the Mercury and Dyna-Soar vehicles be allowed to tumble with their last stages and the effects on the crew determined.

Gaseous Environment for Man

The gaseous environment for man is a compromise between the lowest pressure which can be effectively tolerated and the increasing fire hazards which accompany lower pressures when these pressures are accompanied by an oxygen partial pressure equivalent to that at sea level (158 mm Hg.). The extremes of low pressures to which man can adjust are not yet determined. While it is a fact that man can live quite well at both sea level and up to altitudes of 18,000 feet (oxygen partial pressure of 83 mm Hg.), it is felt by many researchers that the oxygen partial pressure should not go below 158 mm Hg. if a maximum of performance is to be maintained. There are limitations indicated as to the lowest pressures in which man can perform efficiently even with increased oxygen content. Experiments conducted by Space Research Laboratories of Litton Industries have shown that in an oxygen-rich atmosphere pressurized to approximately 5 psia (27,000 feet altitude) the subject experienced "excessive fatigue and increased insensible water loss amounting to several hundred cc's per hour." An observation by Sir John Hunt in his _Conquest of Everest_ pointed out that large amounts of body moisture were lost via the lungs in the rarefied atmosphere to which his expedition was subjected.

A probable compromise would be a cabin pressure of 10.3 psia (8,000 feet) with an allowable range of 7.3 to 11 psia. An oxygen partial pressure approximately equivalent to that of sea level should be provided (Ref. 5). The oxygen concentration at these conditions would be approximately 29 percent, an amount considered sufficiently low to avoid appreciable fire hazard. Other desired conditions are nitrogen 70 percent by volume, carbon dioxide .03 percent, relative humidity 40 percent and temperature 68 ^{o}F to 76°F.

Structural weight penalty caused by cabin pressure is of primary importance. Due to other design considerations, however, the structural integrity of the space craft could be such that a cabin pressure of 11 psia would not add additional weight. This point can only be determined for a particular design.

Food for Space Travel

The approximate average daily intake for one crewman is .9 pound of dry solid food. Since foods contain varying percentages of water depending on type, the average daily water requirements for man depend upon the type of food used.

Food is recognized as one of the most important single factors that contributes to the morale of an astronaut. Food for space travel must be nutritious and easily prepared. It is a morale builder when it is tasty and varied sufficiently to eliminate monotony. Hot foods and cold beverages give a lift to the crew. On the other hand, food will tend to lower morale if it is unsightly and doesn't contain a few extras such as candies, cookies, and various seasonings and spreads. Supplementary snacks can very well promote good fellowhip.

The types of foods which are now available to astronauts may be divided into three categories: low-, intermediate-, and high-levels of acceptability (Ref. 6). Foods of low acceptability level are dehydrated and instant foods. Monotony of this diet causes poor consumption and low morale. The second or intermediate category is composed of canned, dehydrated, and instant foods. The third or high-level category is the same as the second except that frozen foods are added, providing the crew with maximum variation. The food weights per man-day are approximately 1, 1.7, and 3 pounds respectively for the low-, intermediate-, and high-level categories. It should be stressed that as the mission time increases the importance of providing the crew with a larger variety also increases.

Crew Living Space Requirements

One of the most important design considerations affecting vehicle weight is ample living space for the crew; also ample crew space is one of the more important psychological considerations. Thus we must compromise between the greater weight required by larger living and working areas and the lesser weight of smaller crew areas.

Vehicle size affects the weight in a number of ways. For a given time period, the probability of sustaining a meteoroid puncture increases as the surface area of the vehicle increases. Additional seams and joints increase the possibility of environmental gas leakage. Greater loads are imposed on the attitude control system since the inertia of the space craft has increased. Of course the amount of these increases will depend on both missions and vehicle. These particular points will be covered in this paper for two vehicles.

Because of the many variables affecting the behavior of man, the existing literature has not satisfactorily presented the required minimum free-space requirements for crewmen. Experiments, using the space cabin simulator of approximately 50 cubic feet usable space and lasting as long as seven days for one man, have shown this arrangement to be unsatisfactory for extended periods of space travel. In attempting to establish the size (and hence the weight) of a manned space vehicle, it is imperative that some estimate of the required crew living space be available. Such an estimate is presented in Fig. 2 (Ref. 8). Some variations may result from individual crew and mission requirements.

Referring again to Fig. 2, note the minimum number of men for each time period. For a one-day mission, not more than three men would probably participate. For the one-week and one-month missions, a minimum of two and a maximum of eight men would participate. For six months in space travel, a minimum of three men would be used.

The minimum crew required will be determined by an analysis of the list of duties the crew must perform. It is highly probable that for missions of one to seven days, the crew duties will require a level of specialized training and mental ability exceeding the capacity of one man (Ref. 9).

<u>Acoustic Environment for Man</u>

It is expected that the noise of the life support systems and mission equipment within the space vehicle will eliminate the problem of the absolute silence of space. Nevertheless, sound equipment would be desirable to maintain a high crew morale. Taped music can serve as a valuable stimulus. However, it is important to vary the selections to avoid boring the occupants. Voice communication with Earth is highly desirable. Radio facilities, through a broad range of programing, can also create a feeling of closeness between the crew and Earth.

The sounds of meteoroids colliding with the outside wall could cause a degrading psychological effect, so it is desirable that these sounds be muffled from the crew (Ref. 10). However, this may be impractical due to large weight penalties. Using a meteor bumper a short distance from the major vehicle surface would lessen this sound.

Sanitation

There is a direct correlation between cleanliness and crew morale. Periodic bathing and shaving boosts an individual's personal feelings. Also, it is of prime importance to keep down body odors, especially when more than one man is present. Clothes will need washing if they are to be reworn. Human waste must be properly handled as odors and unsightliness can become quite a problem. Other waste articles, such as empty food containers, must also be disposed of.

Sanitation suggests the need for water. The daily water requirements for personal sanitation is approximately two lbs/man; and this amount may increase with mission time. Water estimates for clothes washings run as high as 6 lbs/man/day (Ref. 11) based on weekly washings. These water requirements will probably not be used daily in the amounts shown but will vary. Water storage tanks would be required to eliminate surge-type demands on the life support system.

II. VEHICLE SUBSYSTEM DESCRIPTION

This section discusses the systems and furnishings needed for maintaining man.

Life Support Systems

Life support systems as considered here are those systems which supply the crewman with air, food, and water. (Temperature control is covered later in this section.) Four types of life support systems will be discussed, but since the scope of this paper does not permit an optimization of the various components, typical systems will be presented.

System A: In this system, often called an open system, all of the necessary food, water, oxygen, and inert atmospheric gases are stored. Human waste, including liquids, solids, and exhaled gases, are expelled overboard. To control build up of waste gases, this system may employ vehicle leakage; none of the necessary life support essentials is reclaimed. A block diagram of this system is shown in Fig. 3 and the weights of stored components are presented in Fig. 4. The overall system weight is shown in Fig. 5. (Note, the weights for System B, which will be explained, are also included in Fig. 5.)

The N_2 weights (tanks included) are based on a consumption rate of 4.32 lbs/man/day plus a one-day reserve. The N_2 is stored in a liquid state at 60 psia. The spherical container is constructed of a steel inner shell, a layer of insulation, and an outer aluminum shell with a vacuum existing between the inner and outer shells. The insulation was sized assuming a cabin temperature of $70^{\circ}F$, a liquid nitrogen temperature of $-290^{\circ}F$, a boil-off rate equal to the daily requirement, and a heat of vaporization of 77 BTU/lb. Support and miscellaneous allowances have been included. A cabin pressure of 11 psia was assumed.

The O_2 consumption rates are estimated at 2 lbs/man/day. Like the N_2, the O_2 is stored as a liquid at 60 psia and tank construction is similar. The O_2 temperature is $-270^{\circ}F$ and the heat of vaporization is 86 BTU/lb.

The food weight shown on Fig. 4 for System A is based on using an intermediate level of acceptability menu as discussed in the previous section. The food weight is 1.5 lbs/man/day and the necessary food containers .2 lbs/man/day. No allowance has been included for food warming and cooling equipment.

The water weight assumes a daily consumption of 4.9 lbs/man (additional to the .6 lbs/man of water in the food). Water for sanitary purposes of 2 lbs/man/day is included except that no allowance is made for the first day and 1 pound is provided for the second day. Water for clothes washing and cabin and console cleaning

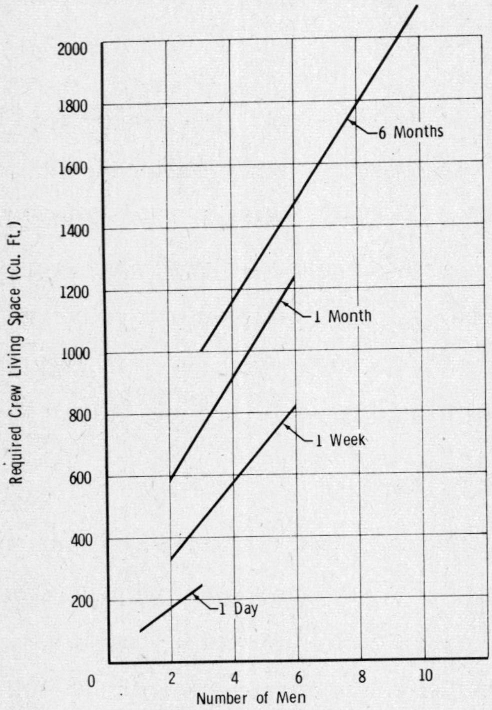

Fig. 2 Crew Space Requirements

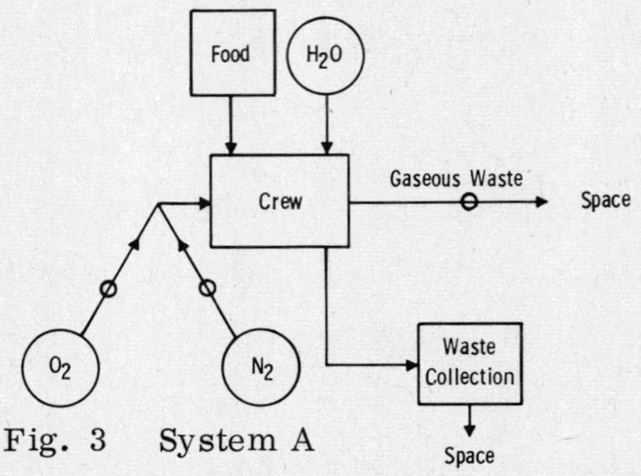

Fig. 3 System A

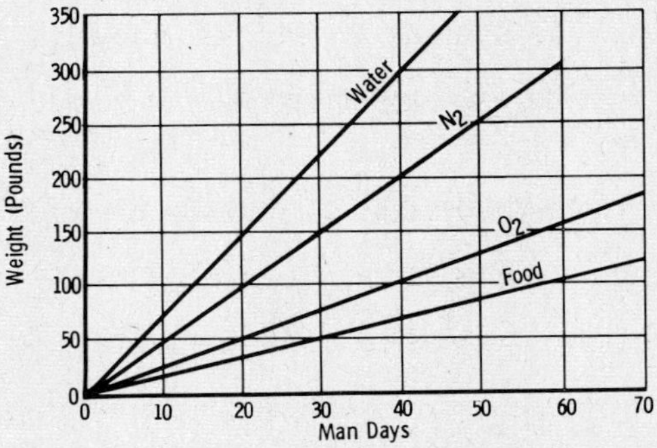

Fig. 4 Weights of Stored Supplies

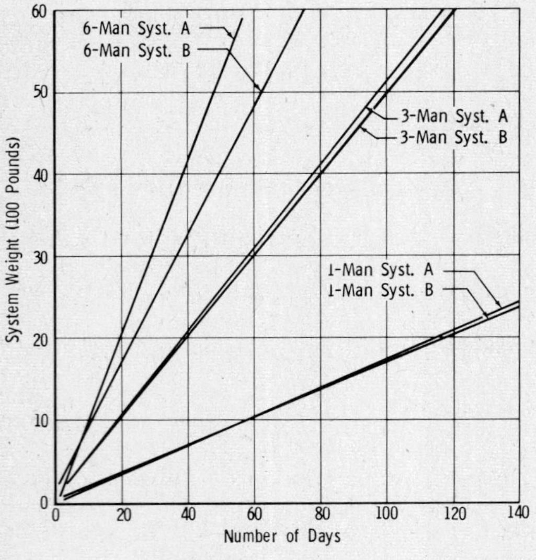

Fig. 5 Weights of Systems A and B

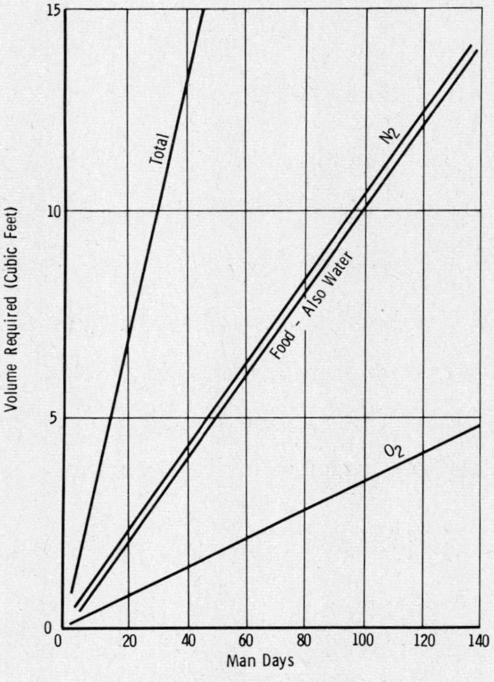

Fig. 6 Total Volume Required (System A)

is not included since this is a special problem dependent upon the particular mission and configuration. The water tankage weights are based on using an aluminum spherical tank with bladder. A separate He pressure tank supplying approximately 12 psia is included. Tank supports are approximately 5 percent of water weight.

Waste collection is not considered here as a part of the life support system. This subject will be treated later under crew furnishings.

The total weight of System A is the sum of the various components plus the weight of those items necessary to assemble the various parts into a working system. In this case, an addition of approximately 5.5 pounds of valves and 3 pounds of tubing is required for the N_2 tank and 7 pounds of valves and 4 pounds of tubing is required for the O_2 tank. For each man, approximately 4 pounds of valves are required to tie pressure suits into the O_2 tanks.

The total volume required within a space ship to accommodate this system is shown in Fig. 6.

System B: This system differs from System A in that the N_2 gas is reclaimed. A block diagram for this system is shown in Fig. 7. Exhaled gases are forced through a heat exchanger by means of a circulation blower. The temperature of the gases is lowered to approximately 40°F, thus causing the water vapor to condense. Next the gases and the condensed water are forced through a water separator. The gases then pass through a CO_2 absorber and the water is collected in small lightweight containers for disposal later. The remaining gases (chiefly N_2) pass through a toxic and odor remover, are combined with O_2 from a storage container, and are ready for use. Note that this system does not depend on cabin leakage to control buildup of waste gases. On the contrary, cabin leakage is to be avoided.

The number of circulation blowers required will vary with the shape and number of compartments in the space vehicle and with reliability considerations. The weights presented in Fig. 8 are based on a requirement for one blower. A cabin pressure of 11 psia and a pressure loss of 1.5 psia through the system is assumed. An allowance of approximately 5 percent for supports is included. The blower speed

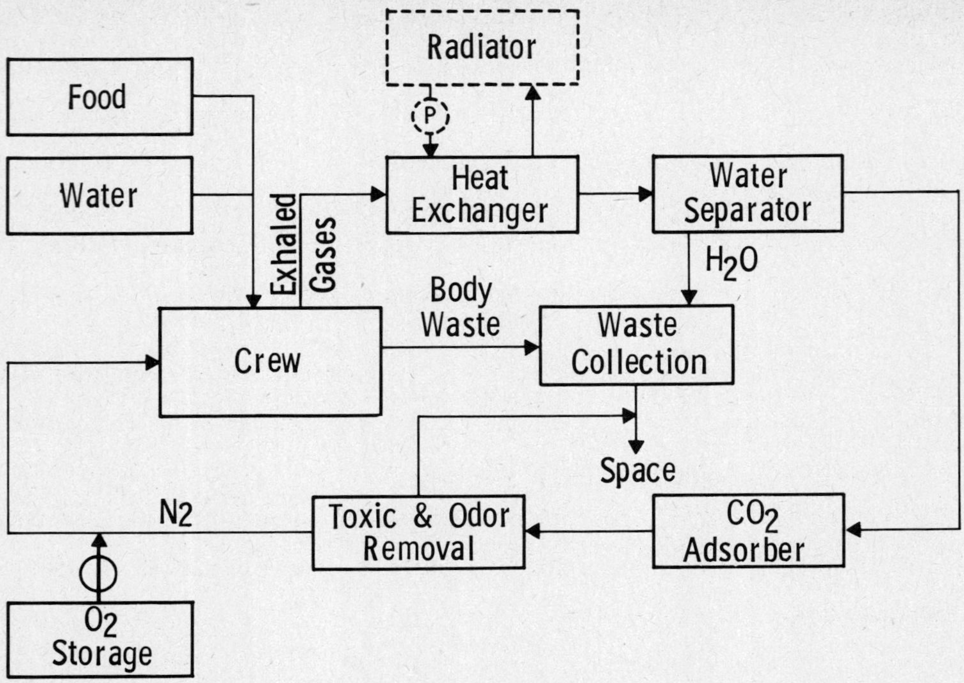

Fig. 7 System B

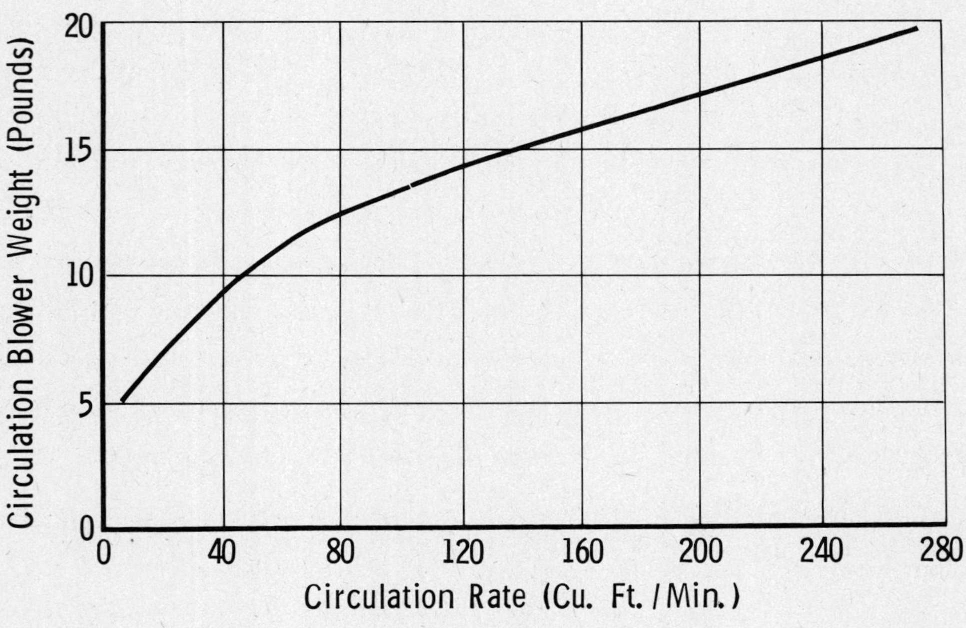

Fig. 8 Circulation Blower Weight

assumed is approximately 11,000 rpm. The blower weights for System B are based on a circulation rate of 30 cfm/man.

Heat exchanger loads are a function of individual design. In one case all of the electronic load may be taken by the crew environmental heat exchanger and in another only the crew heat load is taken. The heat exchanger weights for this estimate are based on a requirement of 820 BTU/hr/man. Counter flow is assumed with the coolant entering at $20^{o}F$ and leaving at $40^{o}F$. A cabin temperature of $70^{o}F$ is maintained by excess heat from other sources. Heat sink weights such as space radiators are not included in this system weight. Heat exchanger weights as a function of BTU/hr capacity are presented in Fig. 9.

The literature has shown a number of acceptable methods for separating water from gas under weightless conditions. A method which should prove acceptable for long flight periods employs a sponge-type absorbant and a plunger. After a given time interval, the plunger squeezes the water from the sponge through a tube into a lightweight collecting bag. The water is then taken to an ejection compartment or may be forced directly into a water storage tank to await purification. Such a unit would weigh approximately 6 pounds. Variations in capacity may be achieved by changing the rate at which the water is squeezed from the sponge.

Carbon dioxide may be adsorbed by exposing it to small particles of lithium hydroxide. The CO_2 reacts with the Li(OH) forming LI_2CO_3 and H_2O. In a system such as the one under discussion, the Li(OH) would be stored in containers having a small cross-sectional area. Approximately 1.35 pounds of Li(OH) is required to adsorb 1 pound of CO_2. The density of Li(OH) is 28 pounds/cubic foot. A plot presenting the CO_2 adsorber weight as a function of man days is presented in Fig. 10. The weights are based on using one unit, cylindrical in shape ranging from 4 by 15 inches for 1 man day to approximately 12 by 44 inches for 24 man days. An allowance of 10 percent of Li(OH) weight is included for screens at each end, supports, and connections for ducting. The cylinders are fabricated from fiberglass or aluminum.

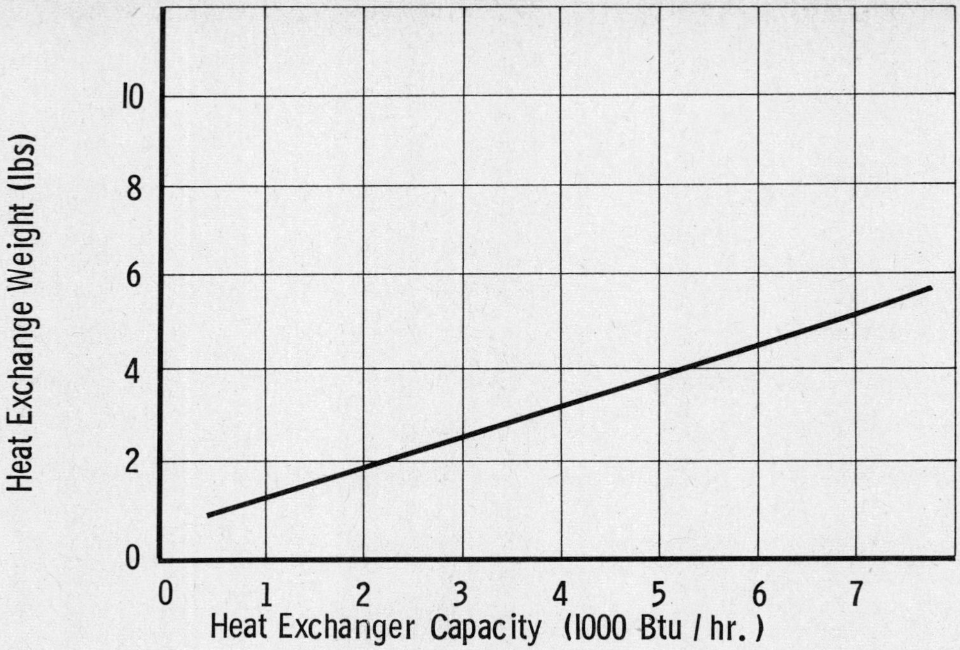

Fig. 9 Heat Exchanger Weight

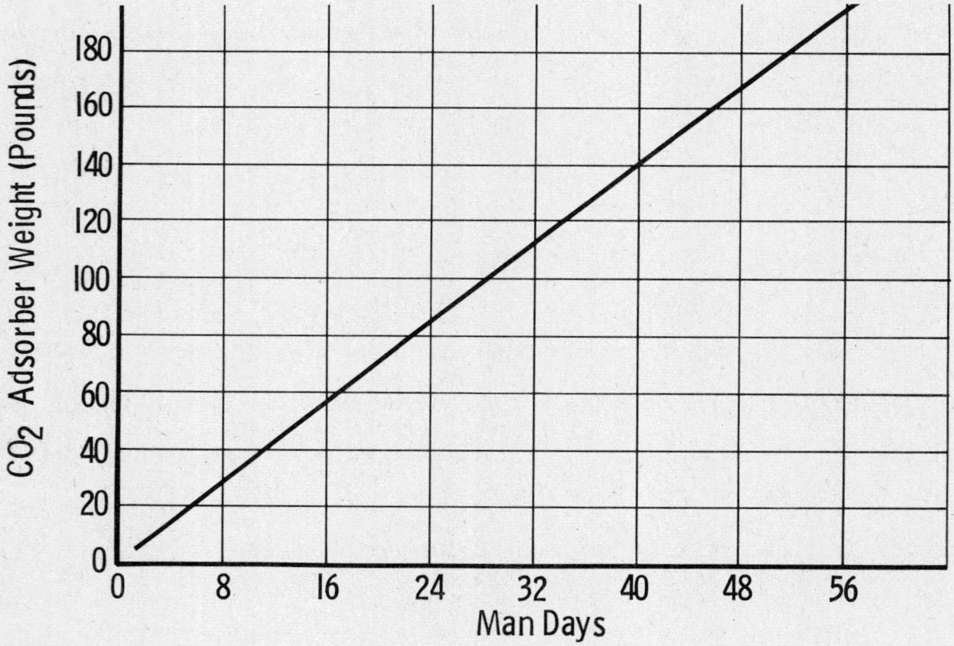

Fig. 10 CO_2 Adsorber Weight

For many years, activated charcoal has been important to sanitary engineering for removing tastes and odors from public water supplies. In the space age, activated charcoal may be used for removing toxic and odorous gases from the cabin atmosphere. Charcoal has a density of approximately 25 lbs/cu.ft. It is estimated that .33 lbs/man/day of activated charcoal will be required to satisfactorily remove the toxic and odorous gases. A flow rate of 30 cfm/man through the charcoal bed is required.

Charcoal may be regenerated by heating (from 570 to 700°F) and venting to space thereby enabling a single unit to be used over and over again. The odor removal units may be similar in shape to those of Li(OH) discussed above.

Fig. 11 presents the charcoal odor removal unit with and without regeneration. The units without regeneration capability are cylindrical in shape ranging in size from 3 by 3 inches for 1 man day up to 8.5 by 19.5 inches for 48 man days. The cases may be constructed of aluminum or fiberglass. The weights shown are based on aluminum containers. A factor of 10 percent of the charcoal weight has been added for fittings and supports.

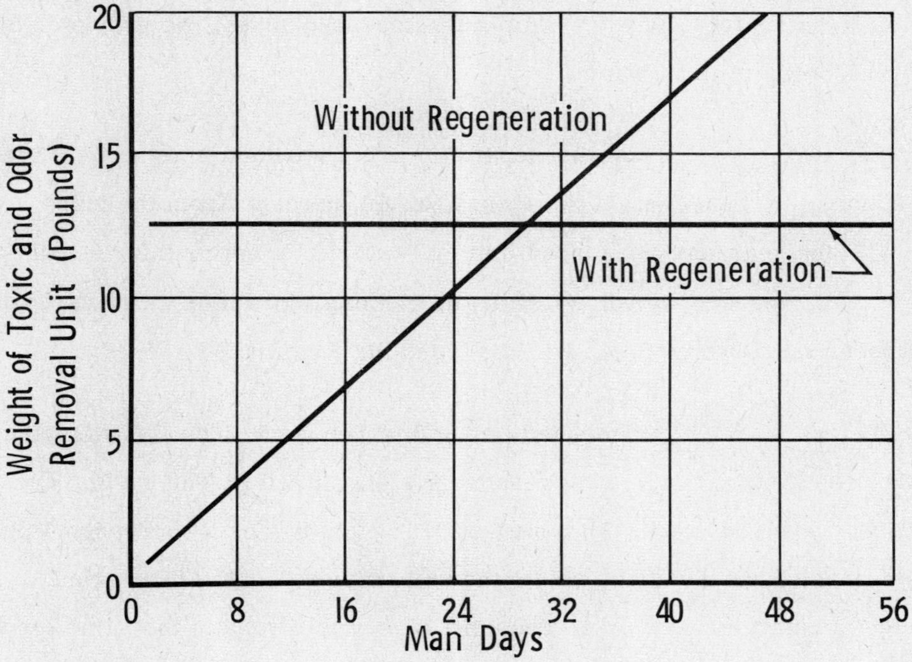

Fig. 11 Toxic and Odor Removal Weight

The weight shown for the odor removal unit with regeneration capability is based on placing two units parallel to one another, each being regenerated alternately. Each unit contains charcoal sufficient for 3 man days, has a steel (t = .02) inner shell and 1-inch-thick asbestos insulation. An allowance of 7.5 pounds has been included for valving, tubing and wiring. The total weight of these two units is approximately 12.5 pounds.

Ducting from the heat exchanger to water separator will be required. A duct will be required from the water separator to CO_2 remover then to the toxic and odor removal unit. The amount of ducting depends on cabin arrangement and size; however, 10 pounds has been included.

Estimated weights for System B are presented in Fig. 5. These weights include 1.9 lbs/man/day for N2 leakage and the valving and tubing weights as estimated for System A. The volume required to accommodate this system within the space vehicle is shown in Fig. 12.

System C: A block diagram for system C is shown in Fig. 13. This system differs from System B in that a molecular sieve is employed for CO_2 removal rather than Li(OH). A water recovery unit has also been added. No provision has been made for recovery of fecal water.

Crew urine and water from the separator are channeled into the water recovery unit. Assuming a reclamation efficiency of 80 percent from the urine, approximately 4.9 lbs/man/day is forced into a water storage tank for consumption at the crew's convenience. Wash water from clothing and other water used for sanitary purposes may also be piped into the water recovery unit.

Several methods are recommended for water recovery. The weights used in this study are based on a vapor compression system (Ref. 12). The system description will not be repeated here. The basic unit weight has been increased 4 pounds to allow for an electric motor to rotate the unit (rather than the hand crank as shown) plus three pounds for supports. The still size would remain the same for varying numbers of crewmen up to about 20 and would require approximately .6 cubic feet.

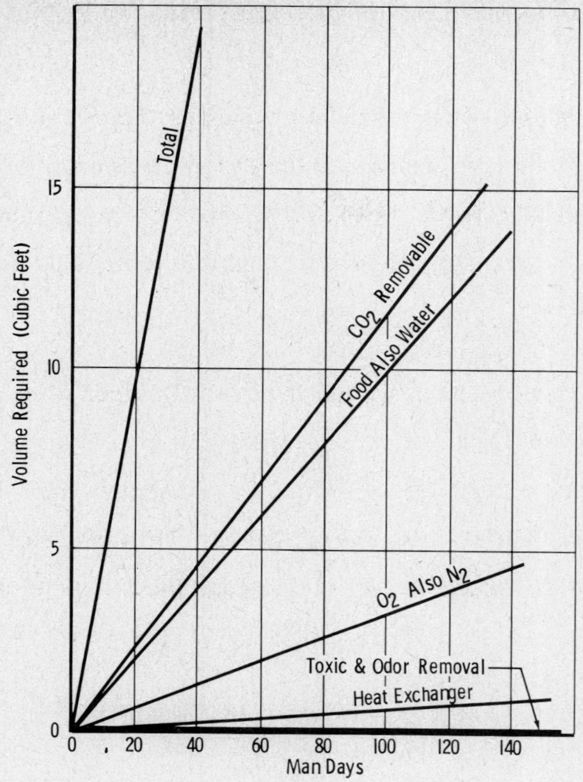

Fig. 12 Volume Required (System B)

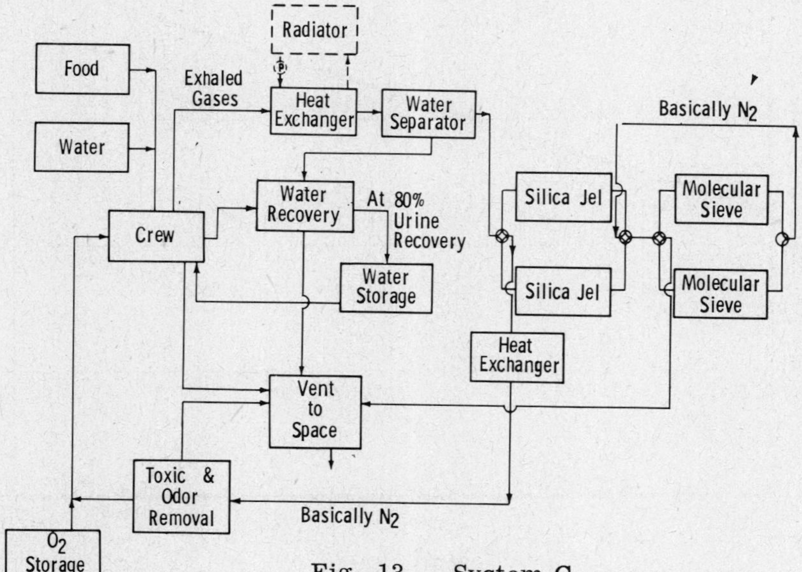

Fig. 13 System C

The total weight (without water) as a function of man days is shown in Fig. 14.

The nitrogen-rich air, after leaving the water separator, passes through a silica jel bed where any water which may have remained (due to inefficiencies of the water separator) is removed. (Molecule sieves favor water over CO_2. Therefore, a silica jel unit is necessary.) The gas then passes through the molecular sieve where the CO_2 is adsorbed.

Both the silica jel and molecular sieve are placed in parallel pairs. When one of each unit is adsorbing, the other is being regenerated. The molecular sieve is rejuvenated by heating to approximately $300^{\circ}F$ while in a vacuum. The silica jel unit is heated to approximately $340^{\circ}F$ then the N_2 gases from the adsorbing molecular sieve are passed in a reversed direction through the silica jel bed.

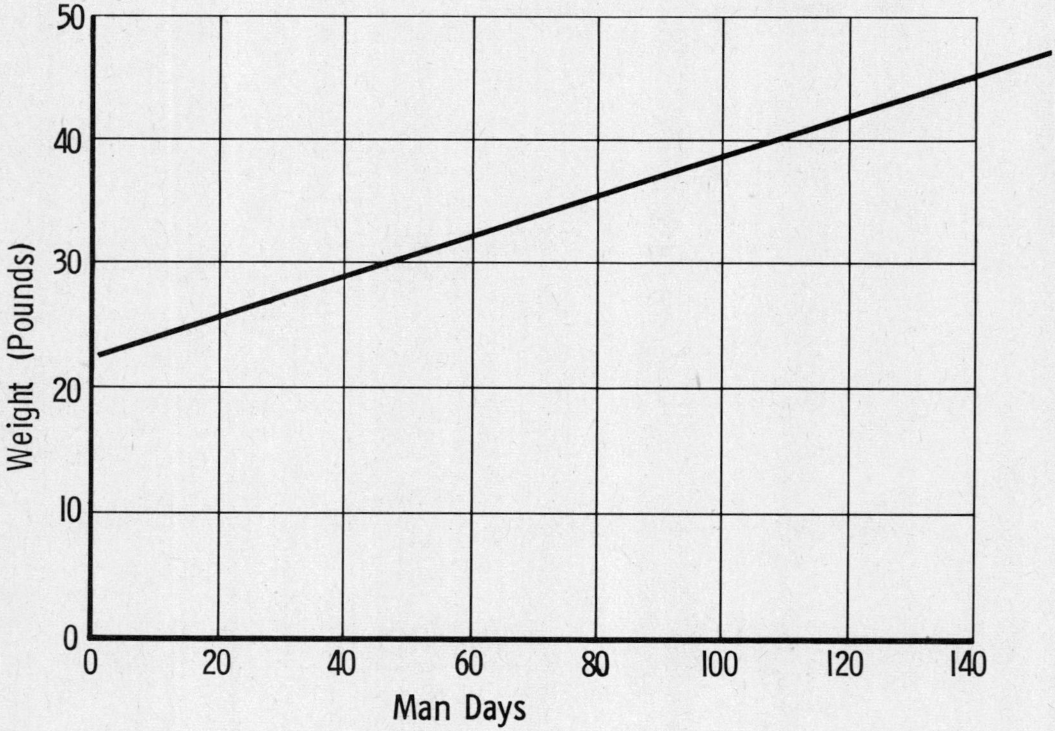

Fig. 14 Water Recovery Unit Weight

The hot N_2 gas, with some small amount of water vapor, is cooled, passed through the toxic and odor removal unit, and then to the cabin. This last heat exchanger may be a part of the environmental heat exchanger unit discussed previously. The frequency of regeneration determines the weight of both the molecular sieve and silica jel units.

Fig. 15 presents the weight of the combination of silica jel and molecular sieve as a function of number of men. The weight assumes a 4-hour cycle (2 hours adsorbing and 2 hours for regeneration). Two silica jel units, two molecular sieves, valves, ducts, and supports are included. The silica jel units are cylindrical in shape ranging in size from 2.2 by 30 inches for 1 man to 18 by 45 inches for 12 men. The molecular sieve units range in size from 2.8 by 30 inches for 1 man to 25 by 60 inches for 12 men.

Space cabin volumes required for System C are shown in Fig. 16. Volumes for System D are included and will be discussed shortly. Molecular sieve volumes (including silica jel) are .23, .75, and 1.5 cubic feet for the 1-, 3-, and 6-man crews.

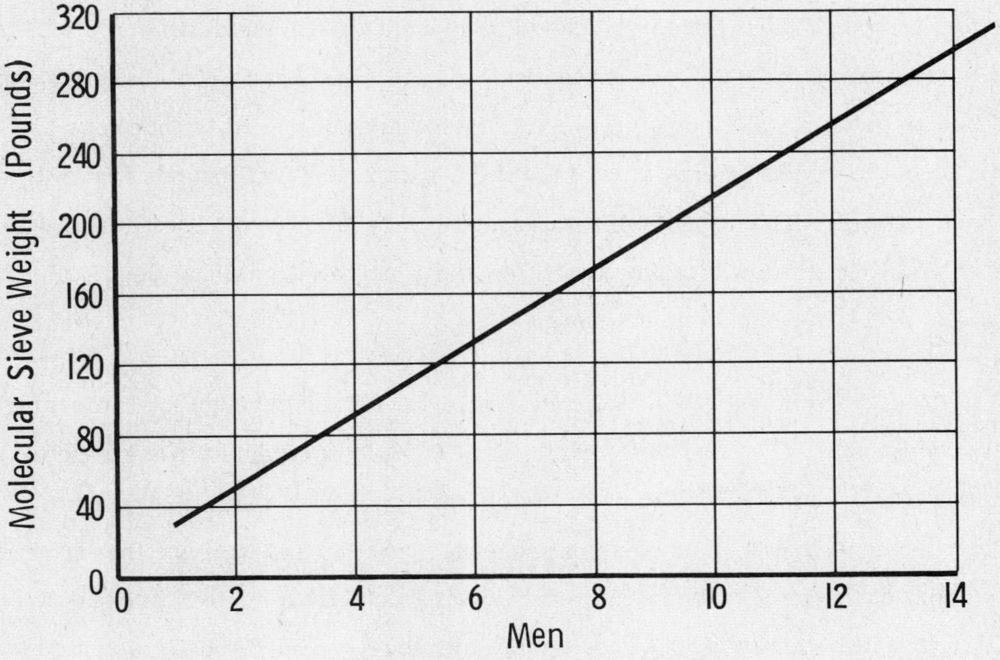

Fig. 15 Molecular Sieve Weight

Water requirements are 9.8 lbs/man for initial operation of water recovery system. The overall system weights are presented in Fig. 17.

System D: This system, often termed hydrogenation or Sabatier, is a progression from System C in that both oxygen and water are produced (see Fig. 18). Oxygen and hydrogen are produced by the electrolysis of water. The hydrogen is combined with CO_2 in the presence of a catalyst to produce water.

A possible type of electrolytic cell that may be used in a weightless state is depicted on Fig. 19. The weights for this system assume the use of such a unit. The hydrogen and oxygen molecules collect at the electrodes. Due to the rotation of the drum, the lighter bubbles are forced to the center section and a divider prevents the interchange of H_2 and O_2. The H_2 enters through holes into the center of the shaft and is forced out in one direction. The O_2 is channeled through the outside section of the shaft and in the other direction. Water is fed back into the drum at each end as shown. The size of this unit is a function of number of crew, ranging from approximately a drum dimension of 6 by 8 inches for one man up to 10 by 17 inches for 12 men. The weights are based on using a fiberglass drum and dividers, rubber insulation blocks at the base of electrodes, and a steel shaft. The electrode weights are based on using platinum. An electric motor and electrical connections are included. The initial requirements for water are 6.8, 21.2, and 40.5 pounds for 1, 3, and 6 men respectively. The weights of this assembly are shown in Fig. 20. The approximate volume required within the cabin for the electrolytic cell is .26, .7, and 1.27 cubic feet for 1-, 6-, and 12-man crews.

Hydrogen from the electrolytic cell may be forced through the molecular sieve as a purge gas. The mixture of $H_2 + CO_2$ is now directed to the reactor where H_2 and CO_2 combine in the presence of a catalyst and heat to form H_2O and CH_4 (some unused CO_2 may also be present). The CH_4 and H_2O are then passed through a heat exchanger, causing the water vapor to condense. Water is collected by means of a water separator as discussed previously, and the CH_4 is vented overboard.

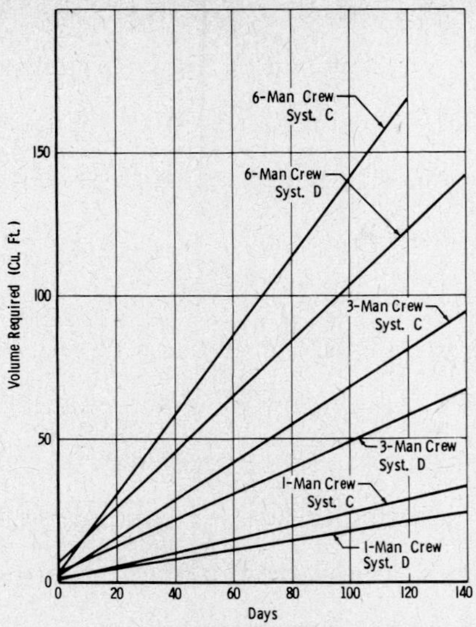

Fig. 16 Volumes Required (Systems C and D)

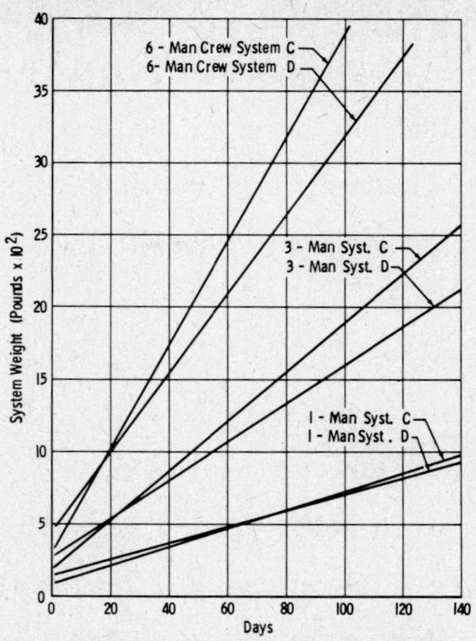

Fig. 17 Weights of Systems C and D

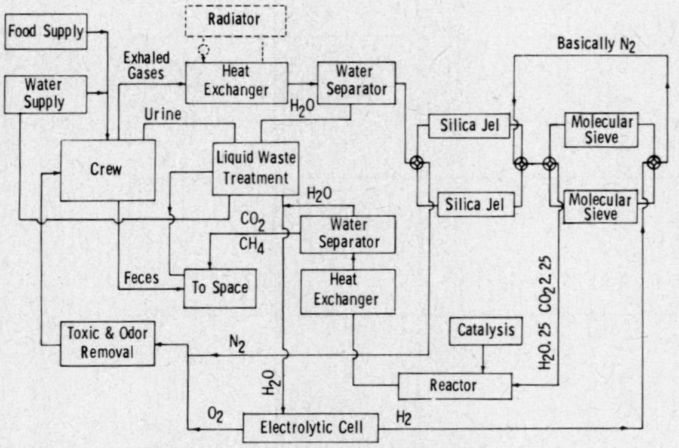

Fig. 18 System D

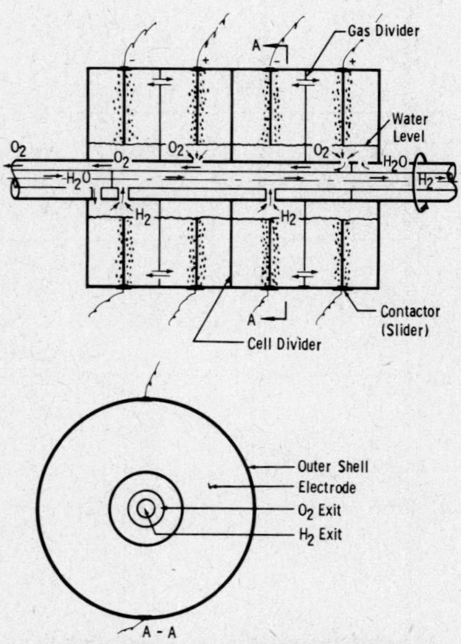

Fig. 19 Electrolytic Cell

The reactor unit is possibly cylindrical in shape with approximately 1.5 to 2 inch diameter. The catalyst required is approximately .1 pound of finely divided nickel on a geiselyhur base. Since heat must be added to ensure a reaction, an electrical coil is included. This unit, including wiring, housing, supports, etc., is estimated at 8 pounds for crews up to 12 men.

In this system there is a shortage of water due mainly to the addition of an electrolytic cell. A summary of water requirements is as follows: 9.8 lbs/man initially stored to allow for operation of water recovery unit, .67 lbs/man/day is make-up water due to shortages of system. This, in addition to water required for initial operation of 90 percent, was assumed for recovery of water from urine; the efficiency of the water recovery system has a large effect on total system weight especially for the longer periods in flight.

System D overall weights (excluding power) are presented in Fig. 16. The volumes required are shown on Fig. 15.

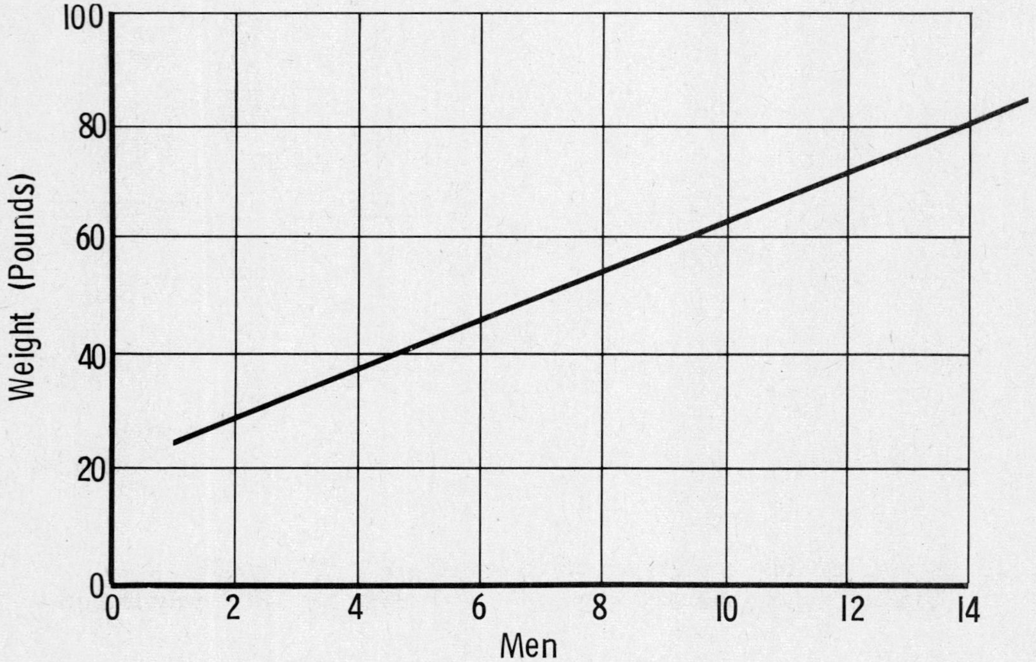

Fig. 20 Electrolytic Cell Weight

Temperature Control

Compared to electronic equipment, man is relatively inadaptable to changes in his environment. Thus his presence within a space craft creates additional burdens on the temperature control system. The allowable cabin temperature may range from 65 to 80°F when man is present; whereas, an environmental temperature ranging from approximately 170°F to -40°F may be satisfactory for electronic equipment. This reduction in allowable maximum cabin temperature adds considerable weight to the cooling system.

In pointing out the weight which is added to a vehicle cooling system to accommodate man, two types of temperature control systems will be discussed. These are: (1) active (involving a mechanical refrigeration vapor cycle and using space as a heat sink), and (2) passive (where no mechanical refrigeration cycle is involved and space is used as a heat sink).

The active (vapor cycle) cooling system (Fig. 21) uses Freon 114 as the refrigerant. Other pertinent data for this system is shown in Table I. It can be seen that the difference between the systems is in the operating temperatures. These operating temperatures were selected as typical and are not necessarily optimum for a space vehicle.

TABLE I

	Manned Vehicle	Unmanned Vehicle
Cabin Air into Evaporator	120°F	200°F
Condenser Temperature	150°F	150°F
Evaporator Temperature	40°F	100°F
Compressor Efficiency	70%	70%
Compressor Motor Efficiency	85%	85%
Radiator Emissivity	0.9	0.9
Radiator Absorptivity	0.2	0.2
Radiator Facing	Space	Space

A comparison of the vapor cycle cooling system weights for a manned and unmanned vehicle is shown in Fig. 22. The radiator weights used (see Fig. 23 for radiator area) are based on an average of 2.9 lbs/sq.ft. The radiators are constructed of 1-inch diameter .035 steel tubes spaced 1-inch apart as shown in Fig. 21. The material and gages are dependent upon the particular configuration being considered. Blower or cabin circulation fan weight is included. Volume requirements, including compressor evaporator, fan, and motor, are presented in Fig. 24. This radiator would constitute a portion of the space craft external wall.

Power requirements for the vapor cycle cooling system are given in Fig. 25. The specific weight for power (lbs/hp. hr.) varies with type of power system selected and length of time in space travel, probably ranging from 19 to 150 lbs/hp. hr.

A schematic for the passive cooling system is shown in Figure 26. In this system, air is assumed as the cooling medium. The radiator properties are the same as those shown in Table I. The air temperature into the radiator is 100°F for the manned vehicle and the exit temperature is 20°F. For the unmanned vehicle the radiator temperatures are 160°F and 70°F.

Radiative cooling system weights are presented in Fig. 27. The major weight difference between the manned and unmanned vehicle is in the radiator weight. The system weights shown on Fig. 27 consist of radiator, fan, fan motor, and an allowance for fittings. Curves showing radiator surface area, volume, and power requirements are shown on Figs. 23, 24, and 25.

Furnishings

Furnishings are additional items peculiar to a manned space vehicle. The type and duration of mission will dictate items required. A typical list follows including brief discussions and estimated weight.

Pressure Suit (1 per man): It is highly probable that pressure suits will be required for all missions. For some missions, pressure suits may be donned only during the boost, re-entry, and periods of emergency. (Weight = 25 lbs.)

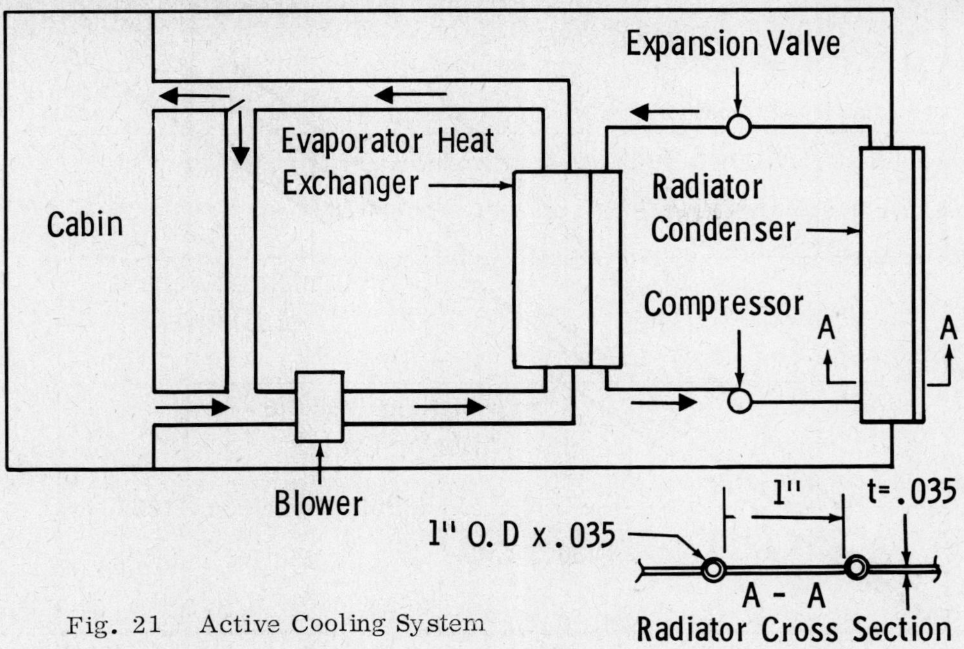

Fig. 21 Active Cooling System

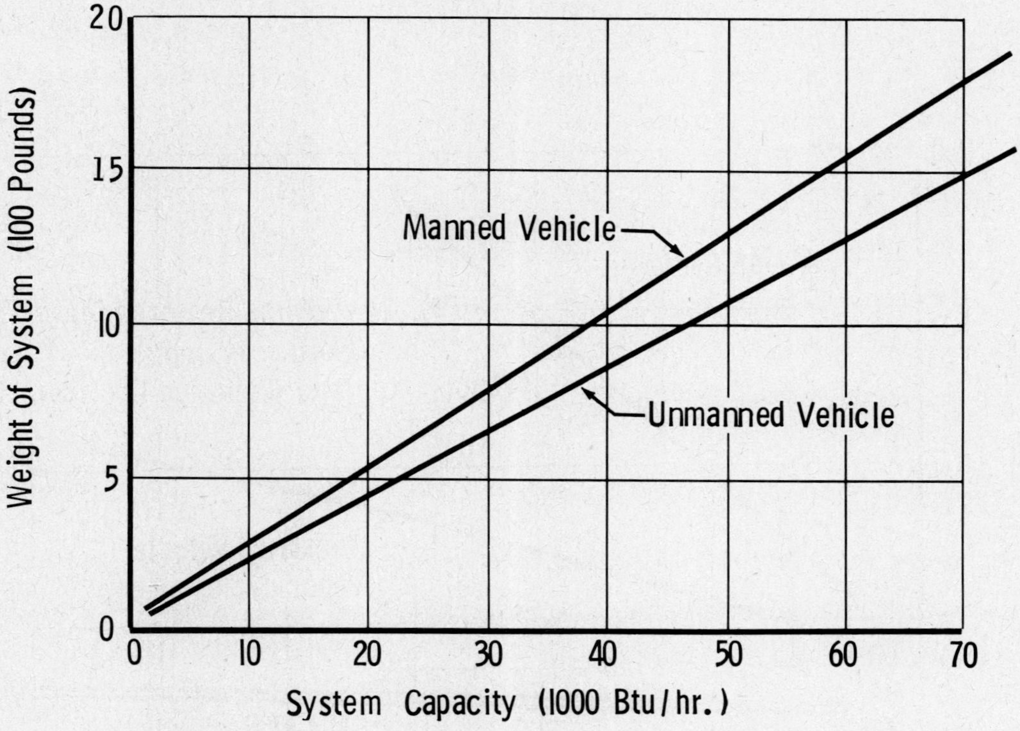

Fig. 22 Cooling System Weights (Vapor Cycle)

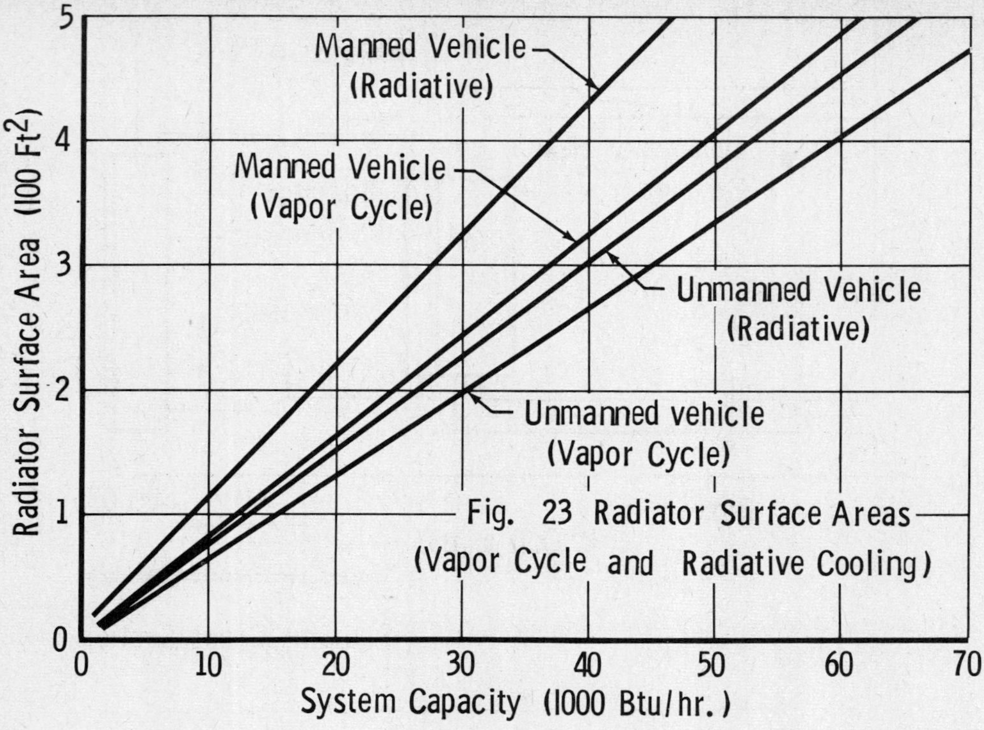
5
4
3
2
1
0
Radiator Surface Area (100 Ft2)
0
10
20
30
40
50
60
70
System Capacity (1000 Btu/hr.)
Manned Vehicle (Radiative)
Manned Vehicle (Vapor Cycle)
Unmanned Vehicle (Radiative)
Unmanned vehicle (Vapor Cycle)
Fig. 23 Radiator Surface Areas (Vapor Cycle and Radiative Cooling)

Fig. 23

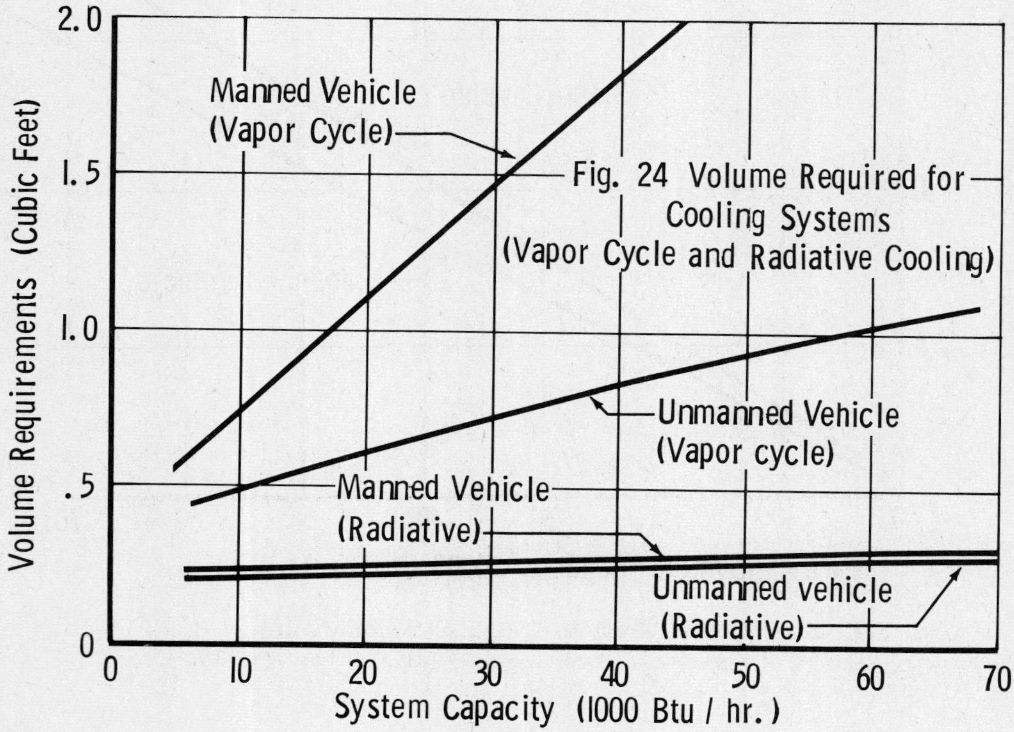
2.0
1.5
1.0
.5
0
Volume Requirements (Cubic Feet)
0
10
20
30
40
50
60
70
System Capacity (1000 Btu / hr.)
Manned Vehicle (Vapor Cycle)
Fig. 24 Volume Required for Cooling Systems (Vapor Cycle and Radiative Cooling)
Unmanned Vehicle (Vapor cycle)
Manned Vehicle (Radiative)
Unmanned vehicle (Radiative)

Fig. 24

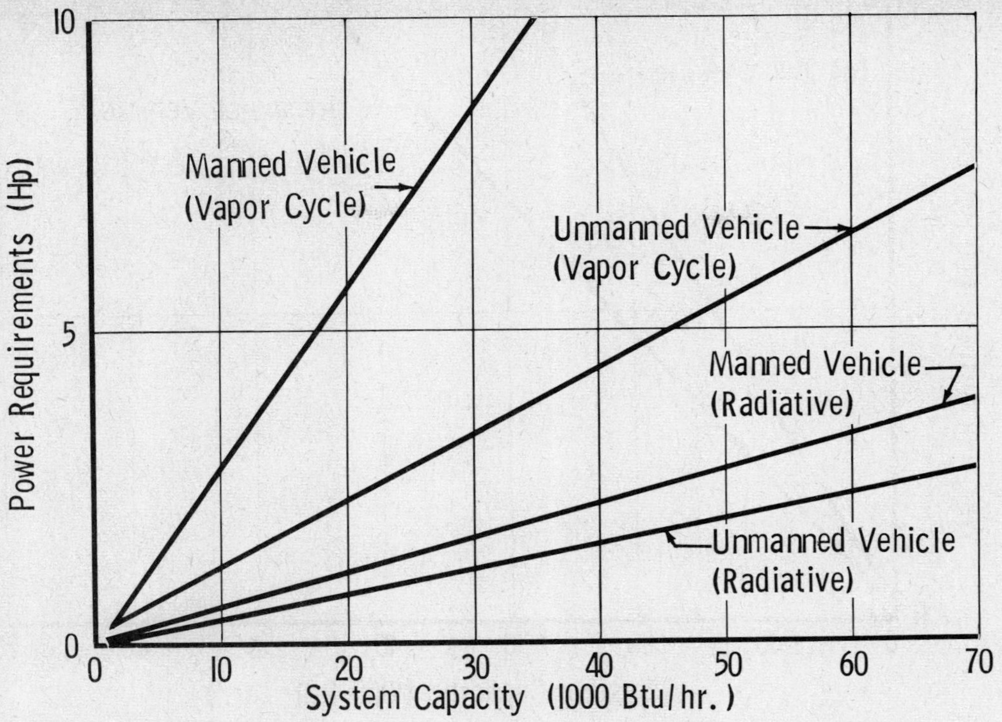

Fig. 25 Power Requirements for Cooling Systems (Vapor Cycle and Radiative Cooling)

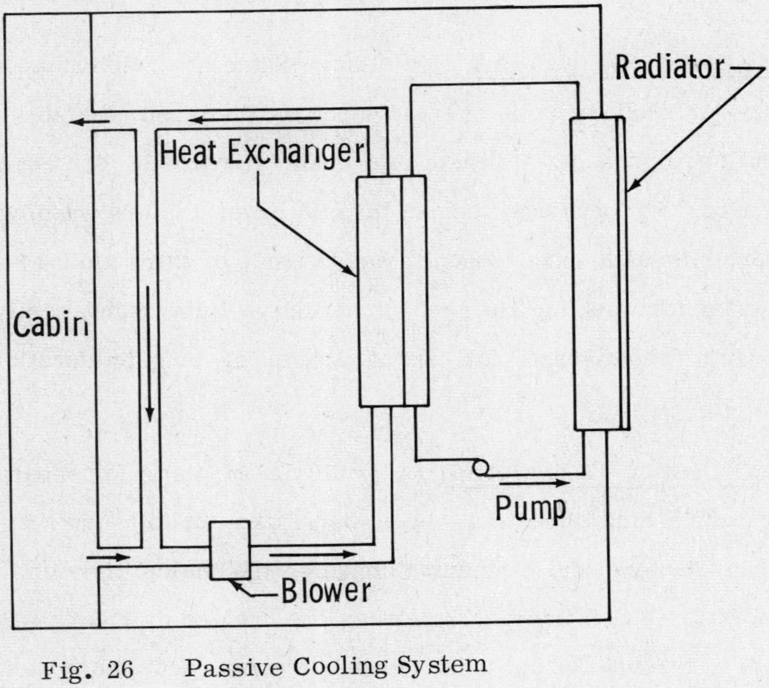

Fig. 26 Passive Cooling System

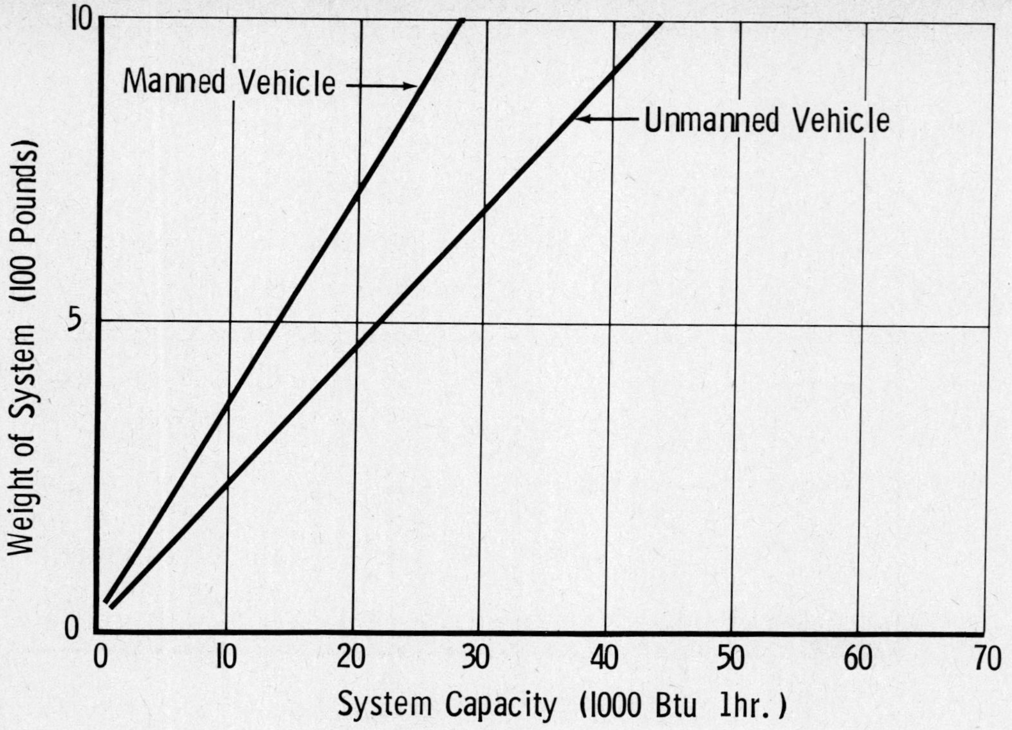

Fig. 27 Cooling System Weights (Radiative System)

Clothing for Crew (each set of clothing): Some provision should be made to permit a change in clothing both when it becomes dirty and when different clothing would help relieve boredom. A reasonable quantity would be one set of clothing for each three days. For periods longer than 9 days, clothes washing may be desirable. However, this will depend upon a weight trade of extra clothing as opposed to additional water for washing. The need for attractive, lightweight, easily washable clothing comes into prominence for space travel of lengthy duration. (Weight = 3 lbs.)

Waste Collection (1 per crew): A prototype of a unit for separately collecting and storing urine and feces has been developed under contract AF 33(616)-6132 by American Machine and Foundry Company, Mechanics Division, 7501 North Natchez Avenue, Niles 48, Illinois. Drag forces are created by bleeding cabin air around the anal area, through a personal feces container, and into a vacuum reservoir. The vacuum is created by a manually operated air evacuation pump. Flatulent gases are absorbed within the unit. Urine is collected in a portable bellow shaped container and can be directly transferred to a water recovery unit. It would be highly desirable to place this unit in an enclosed area—possibly surrounded by cloth curtains. The enclosed area would require 4.75 by 3 by 1.5 feet of cabin space. (Weight = 12 lbs.)

Bathing (1 per crew): The more frequently the crew bathes, the better its morale. Bathing becomes even more important when there are two or more crew members. Bathing may be accomplished in a weightless state by the use of a plastic bag. The bag is equipped with fittings for filling and removing water and is prevented from floating around by attaching the bottom and top to some firm structure. The crewman enters the bag, fastens it snugly about the neck, adds water, and washes himself. (Weight = 8 lbs.)

Crew Seat (1 per man): It would be highly desirable for the crew seat to also serve as the crew bunk if internal cabin arrangement could be worked out. Ejection seat rails, thrust structure, guides, etc., would add approximately 100 pounds; and the weight of encapsulated ejection seats would approach 300 pounds. (Weight = 12 lbs.)

Shaving and Haircutting (1 per crew): Since some crewmen may wish to remain clean shaven, an electric shaver-hair clipper may be provided with a vacuum attachment. (Weight = 2 lbs.)

Emergency First-Aid Kit (1 per man): The probability of injury by crewmen colliding with the vehicle's walls and equipment while in a weightless state should be considered. Provisions should be included for cuts and injured limbs. (Weight = 4 lbs.)

Reading Material (1 per crew): Books probably weigh too much, but reading material could be provided by using microfilm and a viewer. (Weight = 5 lbs.)

Recreational Equipment (1 per crew): Some purely recreational equipment may be taken, such as checkers and chess. In a weightless state, the use of magnetic pieces appears most practical. (Weight = 5 lbs.)

Crewman's Personal Items (1 per man): A weight allowance should be made for the crewman to bring personal items. These items would include pictures, diaries, special food items, religious books, etc. (Weight = 10 lbs.)

Exercise Equipment (for crew): Spring tension devices, resistance built into control levers, hand grips placed within the individual compartments, etc., could reduce muscular degradation and decalcification of the bones. (Weight = 8 lbs.)

Food Warming (1 per crew): WADD Technical Report 60-620 presents preliminary data on an electrical and solar powered food warming unit for space vehicles. (Weight = 3 lbs.)

Sound Stimulus: Radio and/or tape recorder. (Weight = 10 lbs.)

Clothes Washing (1 per crew): A plastic container with water inlet and outlet fitting may be sufficient for washing clothes. The clothing is placed in the container, as much air as possible is forced out, and water and soap are forced in. The clothing is washed by forcing the soapy water through and around the clothing. The dirty wash water is forced from the plastic container and rinse water added. (Weight = 3 lbs.)

Instrumentation Displays

During launch, boost, space travel, and re-entry, man will require certain intelligence of his flight conditions. This intelligence is provided by instrumentation displays which present data about vehicle velocity, location, attitude, internal environment, radiation, etc.

Intelligence derived from these displays may be divided into two major categories: (1) instruments for vehicle navigation, and (2) instruments for vehicle operation.

Display instruments for navigation should provide the following information: (Ref. 13)

(1) Orientation: This type of information enables crew to determine their location at a given time.

(2) Prediction: Having certain data, the crew may determine a course to the next destination.

(3) Control: This data enables the crew to maintain the required flight condition.

Instruments for vehicle operation are those that inform the pilot of the operation of the vehicle's internal and external systems.

Table II presents the estimated weight of display instrumentation for various functions and flight phases for a typical manned orbital vehicle. Instruments that may be required for particular missions are in addition to those shown.

III. MANNED AND UNMANNED VEHICLES—A COMPARISON

This section will compare weight effects of manned and unmanned vehicles. For illustrative purposes, this paper will compare manned and unmanned configurations of an astronomical observatory mission in which the vehicles are placed in a 500-nautical-mile orbit for six months. Each vehicle is equipped with a 36-inch telescope. The configurations are shown in Figs. 28 and 29.

The unmanned vehicle accomplishes its mission by storing data on film. At the end of the mission the film, along with the complete vehicle, will be returned. (It is probable that only the film and a small re-entry body would be returned from an unmanned vehicle; however, in an effort to use similar ground rules for weight comparisons, the complete vehicle will be returned.) A monitoring system will enable the ground observers to ascertain the proper functioning of the sensors. Although the vehicle will be preprogramed, it may be commanded from Earth.

The mission of the manned vehicle is to store astronomical data on film that will be returned to Earth with the three-man crew and the complete vehicle after six months.

Film developing facilities will permit the crew to analyze film and program astronomical areas to be investigated. Communication with Earth will be provided; however, mission decisions will remain with the crew. A supply of spare parts will enable proper maintenance of sensors and subsystems while in orbit.

TABLE II

DISPLAY INSTRUMENTATION

FOR TYPICAL MANNED ORBITAL VEHICLE

	Boost Phase	Spa.	Re-entry	Weight
Navigation				
Time Elapsed	X	X	X	2
Velocity	X	X	X	8
Acceleration	X	X		3
Attitude	X	X	X	10
Flight Path	X	X	X	10
Abort Flight	X			2
Time to Re-enter		X		3
Vehicle Location		X	X	35
Air Data			X	7
Thermal Monitor			X	4
Landing Information			X	3
Vehicle Operation				
Cabin Pressure	X	X	X	3
Decompression Alarm	X	X	X	3
Temperature	X	X	X	1
Fire Alarm	X	X	X	1
Stores, Quantity	X	X	X	4
Radiation Activity	X	X	X	4
Vehicle Systems APU, Hydraulic, etc.	X	X	X	10
Cabin Gaseous Composition	X	X	X	6
TOTAL				116

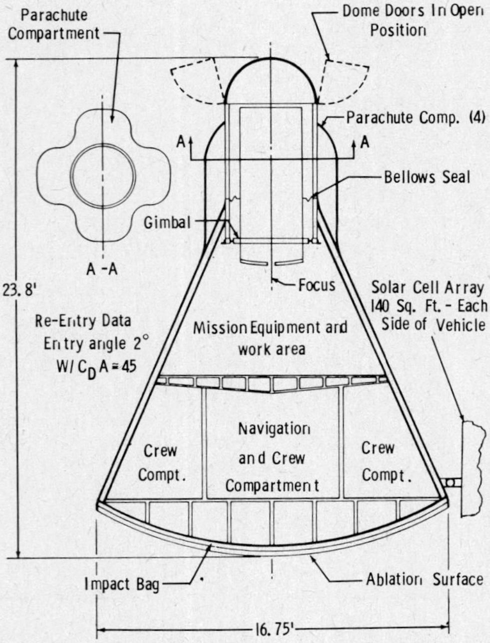

Fig. 28 Manned Space Vehicle

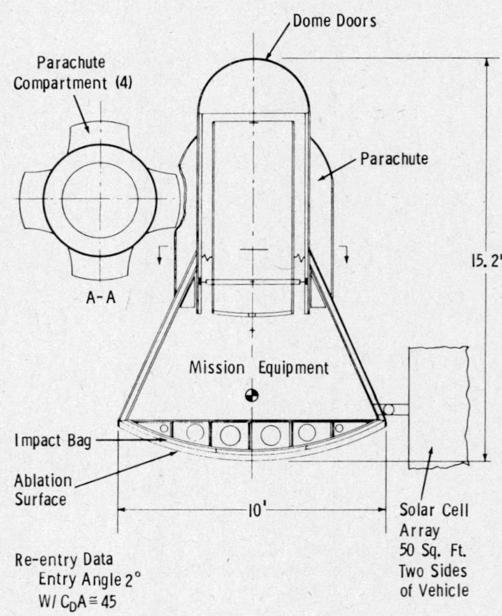

Fig. 29 Unmanned Space Vehicle

The interior volume requirements for the manned vehicle consists of 1000 cubic feet for crew (see Fig. 2) and an additional 150 cubic feet for sensors, subsystems, and furnishings. A 20-percent increase in volume requirements has been included in the latter for installation inefficiencies.

The unmanned vehicle interior volume requirement for sensors and subsystems is estimated at 20 cubic feet. Here again a 20-percent increase in volume due to installation inefficiencies has been included.

Meteoroid Shielding

It follows that a space vehicle requiring a greater internal volume will also require a larger external surface area, provided the shape characteristic is not changed. Larger surface areas are more likely to suffer damage and/or penetrations from meteoroids. Vehicle surface area affects both weight and skin thickness. The effect of surface area on skin thickness can best be shown by the plot in Fig. 30. For a given time in orbit,the skin thickness is a function of vehicle area exposed.

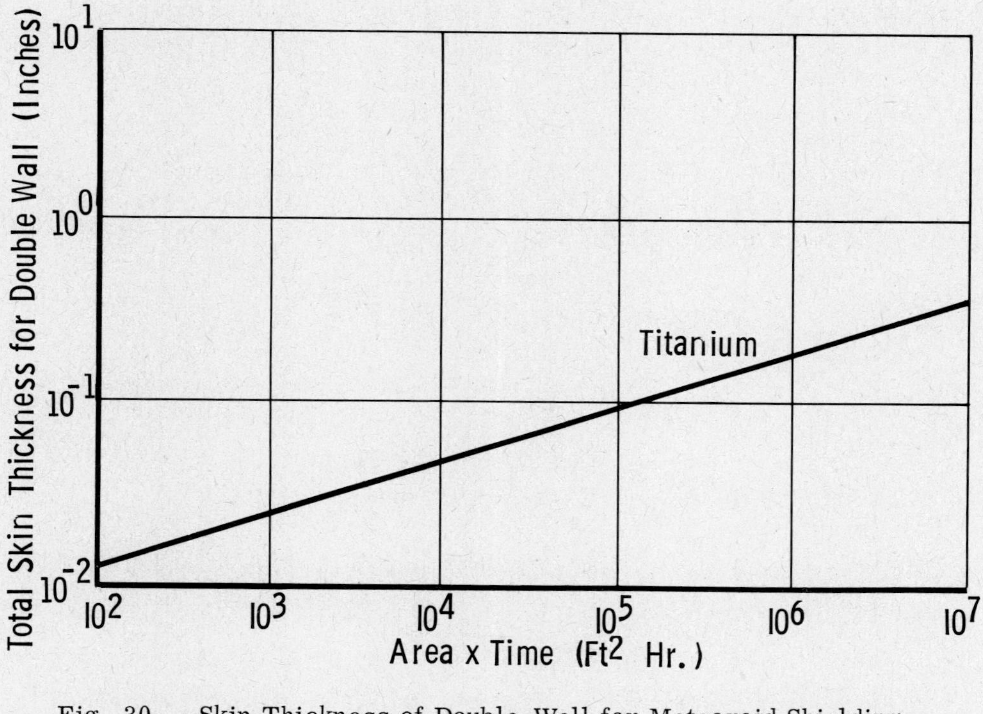

Fig. 30 Skin Thickness of Double-Wall for Meteoroid Shielding (P(o) = 0.99)

The plot in Fig. 30 is based on the following equations:

$$t_s = \frac{7.1356 \times 10^{-4} \left(\frac{V}{c}\right)^{1/3} \rho_T (A\tau)^{.281}}{\left[-\ln_e P(o)\right]^{.281}}$$

where

V = velocity of meteoroid ∽ ft/sec.

c = speed of sound in target material $\sqrt{\frac{E_T}{\rho'_T}}$ ∽ ft/sec.

ρ_T = density of target material ∽ lb/in^3

ρ'_T = density of target material ∽ $slugs/ft^3$

A = area of target ∽ ft^2

τ = time in orbit ∽ hours

t_s = thickness of target material ∽ inches

$P(o)$ = the probability of not receiving a puncture in the time period under consideration

The skin thickness equation above is based on the following:

(1) Flux equation[14]

$$F = 1.0 \times 10^{-12.6} m^{-1.186}$$

(2) Penetration equation[15]

$$t_s = 3d\left(\frac{V}{c}\right)^{1/3}\left(\frac{\rho_P}{\rho_T}\right)^{1/3}$$

where

ρ_P = density of meteoroid ∽ .101 lbs/in^3

m = weight of meteoroid ∽ gms

d = diameter of meteoroid ∽ inches

The surface area of the manned vehicle used in the example is 545 square feet, not including ablation material nor the section above the bellows seal. Assuming titanium is selected as the outer wall material and designing to a .99 probability of not sustaining a puncture, the skin thickness would be .62 inches. Based on experiments by Olshaher (Ref. 16), a double-wall construction may be used which results in a total thickness of 40 percent of the .62 inch thickness (i.e., .24 inches). This component for the manned vehicle would weigh 2260 pounds.

It is very likely that a higher penetration probability criterion would be used for the manned vehicle; however, for the purposes of this comparison .99 was used. The outside area, not including the ablation surface nor the section above the bellows seal, is 150 square feet. Applying the same rules as in the preceding paragraph, the titanium outside walls would weigh 585 pounds, based on a total skin thickness of .168 inch.

The difference in weight for this one item, resulting from the presence of the crewmen, is 1675 pounds. Had the probability of not receiving a puncture been .90 for the unmanned vehicle instead of .99 then this difference would have been increased to 2045 pounds.

Temperature Control

The temperature control system weights for the manned and unmanned vehicles are 800 and 145 pounds respectively. The major differences between the two systems are in the amount of heat to be dissipated and the vehicular surface areas requiring insulation.

The differences in radiative cooling systems for manned and unmanned vehicles have been discussed in the previous section. It is sufficient to point out the estimated heat loads for the two vehicles.

Manned Vehicle

Crew (3)	1800 BTU/hr
Life Support System	6500 BTU/hr

Electronics	3000 BTU/hr
Total	11,300 BTU/hr

Unmanned Vehicle

Electronics	3400 BTU/hr

Based on these heat loads, the radiative cooling system is 210 pounds for the manned, and 65 pounds for the unmanned vehicle.

The insulation weights for the manned and unmanned vehicles are affected by both the allowable operating temperature and by the vehicle surface area. The insulation weight for the manned vehicle is 590 pounds and is based on a 3-inch-thick material having a density of 1.08 lbs/sq. ft. The insulation weight for the unmanned vehicle is 80 pounds and is based on a 1.5-inch-thick material having a density of .54 lbs/sq. ft.

Radiation Shielding

At the altitude under consideration, 500 nautical miles, the major radiation hazard is likely to be proton energy originating with solar flares. The frequency of solar flares of sufficient magnitude to harm spacemen is very erratic. Considerations should be given radiation shielding for the crew on missions of considerable length, as is the case with the examples chosen for this study. However, the amount of shielding which should be provided is a matter of conjecture.

An examination of the manned vehicle skins, sized for meteoroid considerations, would indicate that proton energies of up to 50 Mev would be absorbed (based on studies by Barbieri and Lampert (Ref. 17)). For proton energies greater than 50 Mev, special shielding will have to be added (e.g., placing water or paraffin between the skins of the double wall). A small shielded compartment may be provided within the vehicle. No weight has been included to shield against proton energies greater than 50 Mev.

Preliminary investigations indicate that no shielding would be necessary for electronic equipment while in the orbit under consideration; therefore, no radiation considerations were given the unmanned vehicle.

Electrical System

Solar cells and batteries were chosen as the source of electrical energy for both the manned and unmanned vehicles. The weight difference (1015 pounds for the manned vehicle as compared to 575 pounds for the unmanned vehicle) is due to the differences in load. The estimated loads for the two vehicles are as follows:

Manned Vehicle	
Normal Load	1500 watts
Battery Load	920 watts
Total	2420 watts
Unmanned Vehicle	
Normal Load	530 watts
Battery Load	310 watts
Total	840 watts

The battery size is based on the vehicle being in the shadow of the Earth approximately 34 percent of the time.

Landing System

The landing systems for the two vehicles employ flat circular-type parachutes. At approximately 14,000 feet (and a velocity of 230 ft/sec) a cluster of four chutes is deployed. Should a failure occur with any one chute the three remaining would suffice. A torus impact bag, placed between ablation surface and main structure, is inflated at the time of parachute deployment. The design sink-rate at sea level is 30 ft/sec based on three chutes. The opening shock load on the manned system is 10 "g" which is within the allowable limits. The landing system weight for the manned vehicle is 720 pounds while the weight for the unmanned system is 350 pounds—again demonstrating the weight effects of the human accommodations.

Attitude Control

The presence of man and the resulting increases in volume and weight create greater demands upon the attitude control system. Larger and heavier vehicles are accompanied by increased inertias for a given shape characteristic. The equations that follow show the effect of inertia upon the propellant weight of a reaction-type attitude control system.

$$W_P = f F$$

and

$$F = \frac{\ddot{\theta} I}{R}$$

where

W_P = weight of propellant necessary for control in a single axis (yaw, pitch, or roll) for a particular case.

$\ddot{\theta}$ = angular acceleration

I = vehicle inertia about the axis in question — slug/ft^2

R = arm - distance from vehicle C.G. to reaction thrust ~ ft.

The above propellant weight equation is based on the following general equation:

$$W_P = \frac{(F \Delta t)^2 \quad R \; T}{I_{sp} (2 \theta_c I + FR\Delta t^2}$$

where

t = burn time for one pulse of the reaction motor

θ_c = maximum allowable deviation of vehicle

I_{sp} = specific impulse

T = time in space travel

Pertinent design data used for the attitude control system weights for this study are shown below:

For Re-entry

Maximum Deviation	1°
Angular Acceleration	$10^{o}/sec^2$
Re-entry Time	112 min
Burn Time (Δt), (pitch and yaw)	.006 sec
Burn Time, roll	.005 sec
Specific Impulse	300 sec

For Space Travel

Maximum Deviation	1 sec
Angular Acceleration	$.25^{o}/sec^2$
Burn Time, Pitch, Yaw, and Roll	.001 sec
Specific Impulse	300 sec

The inertias for manned and unmanned vehicles are as follows:

Manned Vehicle

Pitch and Yaw	15,000 slug ft^2
Roll	9,000 slug ft^2

Unmanned Vehicle

Pitch and Yaw	2,770 slug ft^2
Roll	1,400 slug ft^2

In solving for propellant weight and making allowances for the system components, we find that the attitude control system for the manned vehicle weighs 510 pounds while the system for the unmanned vehicle weighs 305 pounds. The system for the manned vehicle is 67 percent heavier.

The weight statements for the manned and unmanned vehicles are shown in Table III.

Table III

WEIGHT STATEMENT

Item	Manned	Unmanned
Structure	(5485)	(2075)
Ablation	2000	780
Ablation Support Structure	240	175
Trusses under Floor	125	50
Floor	145	70
Crew Compartment Walls	200	
Outside Shell	2260	585
Mission Area Flooring	60	
Mission Area Floor Trusses	20	
Telescope Support Structure	50	50
Structure Around Telescope	105	105
Dome Doors (Above Telescope)	200	200
Hatch (Egress and Ingress)	80	60
Electrical System	(1015)	(575)
Solar Cells	350	125
Battery	230	80
Converters	25	25
Voltage Regulator	10	10
Solar Cell Orientation	50	35
Electrical Wiring	350	300
Attitude Control	(510)	(305)
Fuel	355	230
Tanks, Plumbing, Valves	100	45
Fittings	10	5
Electrical Conversion	10	10
Nozzles	35	15
Life Support System	(2660)	
Using System D		
Mission Equipment	(145)	(385)
Eye Piece	2	
UV Sensor (2500 A and Down)	5	
IR Sensor	3	
Spectrometer	15	
Display	5	
Photo Multipliers	5	
Controls	30	
Photo Equipment Including Film	80	
Programer/controller		35
Clock		5
Map Matching		25
Attitude Control		80
Monitor		40
Antenna		20
Command Receiver-Transmitter		60
Camera		120
Telescope and Required Supports	(530)	(530)
Pilot's Display	(140)	
Instruments	116	
Instrument Board	24	
Landing System	(720)	(350)
Parachute	570	250
Automatic Ejection	20	20
Impact Bag	130	80
Guidance Electronics	(350)	(350)
Crew Furnishings	(430)	
Pressure Suit (3)	75	
Clothing (9 sets)	27	
Waste Collection	12	
Bathing Bags	8	
Seats (also bunks (3))	36	
Shaver	2	
Survival Kits	75	
First Aid	15	
Reading Material	5	
Recreational	5	
Personal Items	10	
Exercise Equipment	8	
Food Warming	3	
Food Cooling	10	
Radio	10	
Clothes Washing Bag	3	
Clothes Washing Water	90	
Emergency O_2	36	
Crew 3 at 165	(495)	
Temperature Control	(800)	(145)
Insulation	590	80
Radiator	180	40
Ducting, Blower, etc.	30	25
Retro (Escape Rockets)	(1740)	(710)
Propellant	1430	550
Inerts	230	100
Supports	80	60
Spares	(175)	
Electrical	45	
Attitude Control	35	
Mission Equipment	25	
Instruments	10	
Temperature Control	8	
Mission Equipment	23	
Total Space Vehicle	15,195	5,425

Booster System

The booster systems for the manned and unmanned vehicles are shown in Fig. 31. If man is present, more stringent stability and control requirements will be imposed upon the vehicle. Safety factors will probably be greater than for a manned system. To account for these two weight effects, the propellant loading fraction for the manned system has been made less than the unmanned system; .928 and .910 as compared to .936 and .915 for the first and second stages respectively.

The "g" loading during boost does not exceed 3.8 for the manned vehicle, which is well within the allowable limits.

The summary weight statements for the two vehicles are presented in Table IV.

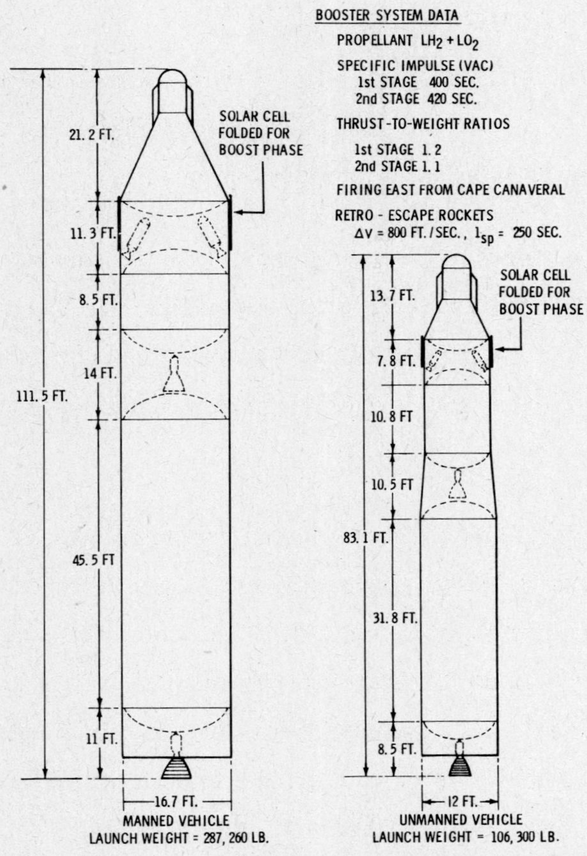

Fig. 31 Booster Systems (Manned and Unmanned Vehicles)

Table IV

SUMMARY WEIGHT STATEMENT

	Manned	Unmanned
Space Vehicle	(15,195)	(5,425)
Transition Section	(1,655)	(1,030)
2nd-Stage Propellant	52,700	19,995
2nd-Stage Inerts	5,200	1.900
Start Burn 2nd Stage	(74,750)	(28,350)
1st-Stage Propellant	197,210	67,950
1st-Stage Inerts	15,300	5,000
Start Burn 1st Stage	(287,260)	(106,300)

CONCLUSION

This paper does not attempt to justify nor discourage the use of man in space. The intent is to point out the many weight considerations which must be taken into account when man is integrated into a space vehicle.

An examination of the weight statements shows that the manned vehicle is approximately three times heavier than the unmanned. The weight difference for the example used may not be the case for two vehicles under a different set of circumstances. But the weights indicate the magnitude of the difference between manned and unmanned systems.

In today's highly specialized fields, a scientist or engineer is often so completely engrossed in his own research program that he fails to recognize the impact of his influence on total vehicle weight. By describing the weight effects resulting from man's presence, and by showing the overall weight added when all of the disciplines are integrated into a manned space vehicle, it is hoped that the specialist will better understand the interplay of the disciplines on system weights. It is further hoped that he will be stimulated into directing his research and design toward methods and criteria which result in manned space vehicles of minimum weight.

There are two criteria which account for the major weight difference: (1) the number of men required for a given mission, and (2) the vehicle internal volume provided for the crew. Of the 9770 lb difference between the manned and unmanned vehicle in the example above, 4950 lb were added because of the crewmen, and 4820 lb were added because of internal volume requirements. The criteria established for these two considerations are of prime importance. There is a fundamental need, therefore, to work toward the design of subsystems that (1) employ the least possible number of crewmen to accomplish a given mission, and (2) provide a means (for example, by stimulation) of permitting the crew to live in the smallest possible volumes and yet retain their ability to carry out a mission efficiently.

REFERENCES

1. F. A. Payne, "Work and Living Space Requirements for Manned Space Stations," Proceedings of the Manned Space Stations Symposium, IAS, April 1960, Los Angeles, Calif.

2. S. B. Kramer and R. A. Byers, "A Modular Concept for a Multimanned Space Station," Proceedings of the Manned Space Station Symposium, IAS, April 1960, Los Angeles, Calif.

3. C. C. Clark and J. D. Hardy, "Gravity Problems in Manned Space Stations," Proceedings of the Manned Space Station Symposium, IAS, April 1960, Los Angeles, Calif.

4. S. U. Miller and A. Stephenson, "Crew Stations of the Future," Proceedings of the Institute of Environmental Sciences, 1960, Los Angeles, Calif.

5. C. J. Lambertsen, "A Philosophy of Extremes for the Gaseous Environment of Manned Closed Ecological Systems," Proceedings of the Manned Space Station Symposium, IAS, April 1960, Los Angeles, Calif.

6. Lt. Col. A. A. Taylor, B. Finkelstein and R. E. Hayes, Food for Space Travel, ARDC TR 60-8, July 1960.

7. G. T. Hauty, "Psycho-Physiological Problems of Manned Space Vehicles," Lectures in Aerospace Medicine, 11-15 January 1960, School of Aviation Medicine.

8. P. H. Stern, Bioastronautic Section, The Boeing Company, through personal contact.

9. R. H. Lowry, "Crew Requirements for an Orbiting Space Station," Proceedings of the Manned Space Stations Symposium, IAS, April 1960, Los Angeles, Calif.

10. F. L. Whipple, "The Meteoritic Risk to Space Vehicles," Vistas in Astronautics, Pergamon Press, Los Angeles, Calif.

11. W. T. Ingram, B. Newman, G. Palevsky and L. Slote, "Exploratory Research on the Theoretical Considerations of Waste Water Cycles in a Closed Ecological System," Advances in the Astronautical Sciences, Vol. 4, 1958.

12. WADD Technical Report 60-243, Section VIII.

13. R. C. Kaehler and S. Romano, "Navigation and Energy Display Requirements for Pilot-Controlled Satelloid Flight," Advances in the Astronautical Sciences Vol. 6, 1960.

14. F. L. Whipple, "Particulate Contents of Space," paper presented at third Symposium on Medical and Biological Aspects of Energies of Space, U.S. School of Aviation Medicine, Brooks AFB, Texas, 1960.

15. R. L. Bjork, "Effects of a Meteoroid Impacting on Steel and Aluminum in Space," Rand Paper P-1662, presented at the Tenth International Astronautical Congress, London 1959, Springer-Verley, Vienna.

16. "An Experimental Investigation in Lead of the Whipple Meteor Bumper," A. E. Olshaker, MIT Thesis.

17. L.J. Barbieri and S. Lampert, "The Interdependence of Manned Spacecraft Design and Radiation Shielding," Aerospace Engineering, April 1961.

DISPLAY AND CONTROL
IN MANNED SPACE VEHICLES

A. Shulman*

This paper describes the results of a study aimed at formulating a display and control system particularly suited to the manual operation of manned space vehicles. A system plan has evolved that features closed-circuit television for presenting either symbolic displays derived from vehicle sensor data or television picture video from vehicle TV cameras. The system plan calls for the translation of sensor data and operator control commands to be carried out primarily by a digital control unit. The control unit is comparable to a special-purpose digital computer facility. The symbol-video generating equipment and the vehicle functions are required to be responsive to digital commands from the control unit. An important feature of the system is selective control over a large number of vehicle functions by a small assortment of operator controls. Provision is also made for the augmenting of manual controls by a limited repertoire of vocal commands through a speech-recognition system.

INTRODUCTION

A preliminary design for a display and control system in a manned space vehicle is summarized. The work was originally incorporated in a broad study for a lunar exploration project, but it will become clear as we proceed that these designs need

---

* Airborne Instruments Laboratory, a Division of Cutler-Hammer, Inc., Deer Park, Long Island, New York.

not be restricted to specific missions. Accordingly, the following description is oriented toward general applicability.

There is a considerable body of literature on proposed manned space projects including several excellent, authoritative reports. One example that can be cited (Ref. 1) describes a manned lunar mission; the plan involved a hard re-entry vehicle similar to that of Mercury, but large enough to hold two or three men. Other excellent studies of an earlier vintage are available (Refs. 2 and 3).

One important point that often does not receive sufficient attention in such proposals is that the center of interest in a manned space flight should be not so much the machine as the man. In this discussion, we have taken cognizance of the central role played by the operator in vehicle operations. We have sought to evaluate the restrictions that his presence imposes on the equipment and operations, and to most effectively adapt the system to these requirements. In short, the orientation is toward a systems engineering approach in the broad sense, which requires that the system be optimized within its constraints.

SYSTEM REQUIREMENTS

Many studies, for the most part of relatively narrow scope, have been made of the problems and needs that arise in manned space flight. Those results that are applicable to display and control problems have been drawn on freely to formulate the present requirements. The reviews are consolidated under two general headings: environmental restrictions and operational problems.

Environmental Restrictions

An operator must be able to function effectively in spite of a number of physical restrictions or burdens that may be imposed upon him by the vehicle environment. These restrictions fall into the following groups: (1) physiological limitations, (2) physical constraints, (3) radiation hazards, and (4) emergency escape.

Physiological Limitations. Physiological difficulties can be caused by such factors as severe acceleration, fatigue, vibration, noise, unusual maneuvers, or perhaps prolonged weightlessness. We ourselves have probably all suffered some experience in which excessive noise or vibration caused annoyance and perhaps reduced working efficiency. It is reasonable to suppose that operations must cease if such burdens are carried to excess.

Yet the ability to carry on under abnormal conditions is important because first, operator assistance during emergencies is a primary requirement, and second, it is precisely during emergencies that abnormal physical conditions are most likely to develop. There is, however, a disturbing paucity of published data that suitably quantify the effects of extreme physical factors. This is so much the case that determining the performance limitations of the operator under great stress can be described as one of the more serious problems involved in space flight.

There is evidence to suggest that the simpler manual operations can be carried on under surprisingly adverse conditions. Andrews and Holleman (Ref. 4) in describing centrifuge experiments, report that acceptable control was maintained at acceleration levels as high as 12 g. Furthermore, Useller and Algranti (Ref. 5) report successful pilot recovery from spins exceeding 25 rpm, but find vestibular nystagmus (an ocular disturbance) occurring briefly after exposure to the higher spin rates.

It has also been reported (Ref. 6) that extreme noise has little effect on higher mental processes, but some effect on lower physiological functions. Another item of interest is that moderate oxygen deprivation (hypoxia) may reportedly be accompanied by transitory deterioration in color perception ability (Ref. 7, p 101). There is not enough data to draw firm conclusions on this point, but until more is known it would seem wise to refrain from the use of color in our displays. An operator should not become accustomed to relying heavily on a feature that may prove valueless in certain emergencies.

Physical Constraints. Space suits, space suit connections, impact-protection equipment, and associated paraphernalia may all impose serious physical constraints on the space traveler. For example, Still (Ref. 8) in a recent study of internal environment, concludes that crew members will wear full pressure suits at least during the ascent and re-entry phases. Still furnishes interesting illustrations of such gear that show serious mobility restrictions. The Mercury capsule environment (Ref. 9) also furnishes a good example of the complexity involved in the pressure suit system. Couches and impact-protection gear comprise still another important type of constraint (Refs. 10 and 11).

Radiation Hazards. The requirements for shielding against radiation hazards are, on the other hand, largely undetermined and, in fact, quite controversial. The severity of the restrictions that will be imposed by the need for shielding were recently highlighted by Ganguly and Lence (Ref. 12), and also by Schaefer (Ref. 13) in studies of the radiation problem. These investigators thought that heavily shielded cubicles would be needed to protect astronauts, particularly against the proton streams of solar flares. It appears that an encounter with an especially persistent flare stream might require as many as several days of continuous residence under shielding.

As a matter of fact, Rosenbaum (Ref. 14) mentions the possibility that severe solar flares might require thousands of pounds of shield mass; a rather sensational comment and which, if true, may certainly create difficulties for man-in-space programs. Furthermore, little is known of the meteoroid-shielding requirements (Refs. 15 and 16), and account must be taken of this additional uncertainty.

At any rate, it would obviously be desirable to have display and control capability under protracted shield conditions, particularly in view of the emergency nature of the situation. Yet the available space in a shield cubicle, if such is employed, must clearly be at a premium.

Emergency Escape. The desirability of an emergency escape or abort capability is generally recognized. Emergency escape provisions in deep space are discussed, for example, by Bono (Ref. 17), and by Connor and Smith (Ref. 18) who refer to an escape vehicle as a "lifeboat."

Abort capability is most likely to be provided by a detachable capsule; an enclosure that would house the crew during critical periods, and that could be separated from the body of the vehicle should need arise. Since residence would have to be established in such a capsule, albeit for limited periods, it will evidently be necessary to establish operating stations therein.

This is a particularly difficult requirement because capsule space is sure to be even more restricted, and furthermore, cabling between such operator stations and the vehicle must be very limited to avoid compromising the abort capability.

Nevertheless, full display capability and facility for full control should probably be established at the operator stations inside the capsule.

Operational Problems

Normally, most of the processes in the operational control system will be fully automated. It is generally recognized that the most efficient guidance and energy management regimes involve problems that are too demanding for human control alone. We may cite, for example, studies (Refs. 19, 20, and 21), which indicate the desirability of automatic guidance in space, and also the results from the NASA multi-axis test facility (Ref. 5), which show that automation of the attitude control is desirable for good energy management. Furthermore, there is general agreement that control of the internal environment should normally be automatic (Refs. 8 and 9).

Nevertheless, fully automatic control cannot be depended upon in situations where the nature of the circumstances are not completely foreseeable. A prime example of such a case would occur in the making of a soft landing on alien terrain (Ref. 22). Another good example is furnished by the monitoring of the system for in-flight checkout, and manipulations for fault correction, since obviously fault occurrences are not readily predictable.

Emergencies comprise an important class of situations in which fully automatic control cannot be planned for. Here is where the human operator really comes into his own (Ref. 18). Morris and Kaehler (Ref. 23), in considering human versus automatic control, have concluded from the results of a series of simulation tests that about 25 percent of successful "flights" were saved at some critical control stage by action of the human control pilot. Of course, it is all very well to think, as most of us probably do, that human judgment and intervention are valuable in emergencies, but it is quite a different thing to quantify this value, in the way that these workers appear to have done. Their results are certainly most welcome.

The feasibility of human control over such a vehicle appears reasonably well established; see for example (Ref. 24). Clark (Ref. 25) also points to the many degrees of freedom that are practicable in human control over "Mobot" manipulators.

Nevertheless, there is general recognition of the need for computational or semiautomatic aids to human control. Such aid is certainly needed for astronavigation (Ref. 21). Another application of value is to furnish predictive information to the operator (Ref. 50). On the other hand, for life-support monitoring, Carlson (Ref. 26) recommends data processing, not just direct monitoring, of physiological information. When a computational or programmable facility is present there is also opportunity for man-computer cooperation in decision-making (Refs. 41, 28, and 29). Investigators such as Mayo (Ref. 31) have recommended that human control over spacecraft be mediated by a computer facility; Mayo, in fact, proposes a man-computer complex that is similar to that of Sells and Berry (Ref. 7, p 356). Also, Richardson (Ref. 32), in reference to an advanced aircraft design, describes a noteworthy crew-computer organization that forms the basis for an "electronic management system."

An important operational requirement, and one that militates against extensive console equipment, is the need for duplicate display and control facilities for each operator. These will enable any operator to act independently, and to support or replace another if necessary. In any case, one operator should be able to monitor every feature and to exercise control over every aspect of performance in the spacecraft. It should not be necessary to call on an engineer at another station for an in-flight checkout, for example; the checkout should be capable of being performed at the same station that controls the craft. Such a feature is desirable if stations are to back up one another in an emergency. Consequently, there is need for displays that cover every aspect--not just navigation, but also monitoring of life-support processes, engine performance, and in-flight checkouts. It is also interesting to note that the use of closed-circuit television, including remote-controlled television cameras, is recommended for observation and monitoring of both vehicle and crew (Refs. 30 and 33).

A display and control scheme for a manned orbital vehicle that is relevant to the present problem has been published by Hopkins et al. (Ref. 34). Their display formats conform in a general way to recommendations for "integrated display panels" (Ref. 35) and consequently may serve as valid guides.

The display and control problems outlined are considerably magnified when we take into account the hundreds of parameters that are involved. Hopkins (Ref. 29) gives a better idea of the magnitude of the task. He proposes detailed planning of the human function to fit the mission profile and furnishes illustrations of operational requirements for space-ship guidance.

To summarize the problems that we are faced with in considering crew stations for a space vehicle:

1. Displays and controls must be provided in an efficiently human-engineered design, and suited to operators who are expected to have very limited mobility for critical portions of the mission, as well as possible impairment of sensory capabilities.

2. Displays and controls must be provided in multiplicate for the interchangeable use of several operators.
3. Equipment packaging must be subject to difficult dimensional restrictions.
4. The system must permit an operator to obtain rapid and convenient access to any of a multitude of functions and parameters.
5. Provision should be made for almost fully automated operations, as well as those that can be almost wholly manually controlled.
6. The system operations must be capable of coping with a potentially great variety of unforeseen emergency situations.

SYSTEM DESIGN

A system design capable of meeting the requirements outlined demands the application of advanced display and control techniques. However, to be assured of remaining within the bounds of practicality the techniques that are considered here are restricted to electronics currently within the state of the art. Wherever necessary in the following description, attention is directed to the current usage of the specific method or device. It is apparent that to remain at the forefront of a rapidly advancing art, the system design would have to be in constant flux; therefore, let it be understood that the equipment plans are just for today, or perhaps we should say yesterday, and that there is no intention to anticipate what tomorrow may bring.

General Description

The system design depends on the use of multi-purpose displays and controls. It is intended to have just one television monitor at an operator's station, and to present all of the required indications and displays on its screen.

Furthermore, every aspect of vehicle performance is to be controllable by appropriate manipulation of two hand controls. No supplementary console controls are planned, since a small number of highly integrated manual controls is considered definitely preferable, from the human factors point of view, to a multitude of special-purpose buttons, knobs, and switches (Ref. 36).

The display system features closed-circuit television for displaying symbolic patterns and alpha-numerics from storage tubes and for picture video derived from television cameras. It is intended to synthesize displays on the television screen that will incorporate the best features of modern integrated display panels (Refs. 35 and 37).

The display system functions will be coordinated by a programmable control unit. This unit is to be a digital machine designed primarily for table-look-up from a fixed information store. Coordination of system functions will be accomplished by making the system equipment responsive to digital command programs obtained with the table-look-up procedure.

System Units

The planned display system may be described with reference to the semi-pictorial diagram (Fig. 1) and the block diagram (Fig. 2). The system is designed to display alpha-numeric information from a matrix character generating unit, patterns and symbols from a monoscope unit, photographs reduced to television pictures from rapid-access film transports, and picture video from television cameras. The information from these sources is to be displayed principally through a scan-conversion process under the coordinated control of the programmable control unit.

Scan Conversion. The scan-conversion process will make use of miniaturized scan-converter tubes in which information is recorded on a storage-tube target by the electron beam from a writing gun. At the same time, recorded information is read by the electron beam from a reading gun positioned on the opposite side of the target electrode. It is conventional practice to read the recorded information by television raster-scanning of the target. The resulting television video will furnish a picture of the recorded information that may be observed on a television monitor.

Scan-conversion permits writing and reading to proceed independently; thus, complex patterns will be written on the scan-converter tube targets at a relatively slow rate and may remain stored up to several minutes, while being observed from the reading side at conventional television rates. Half-tones are not needed for such patterns; therefore, all such information will be recorded in monotone. The absence of half-tones, incidentally, will greatly facilitate erasure.

This scan-conversion processing for stored patterns incorporates the basic elements of the "Pattern Read-Out and Storage Equipment" (Ref. 38) produced by Airborne Instruments Laboratory. This equipment uses a set of scan-conversion storage tubes that have digitally controlled writing beams, each beam being positioned and intensified by digital data from external equipment. The system is designed for the recording and temporary storage of complex patterns, and it incorporates a high-resolution closed-circuit television system for reading out recorded information.

Character Generation. The bulk of any text that is displayed will be derived from a matrix type of character-generating equipment. An example of a matrix character-generating technique is furnished by the "Vidiac" system

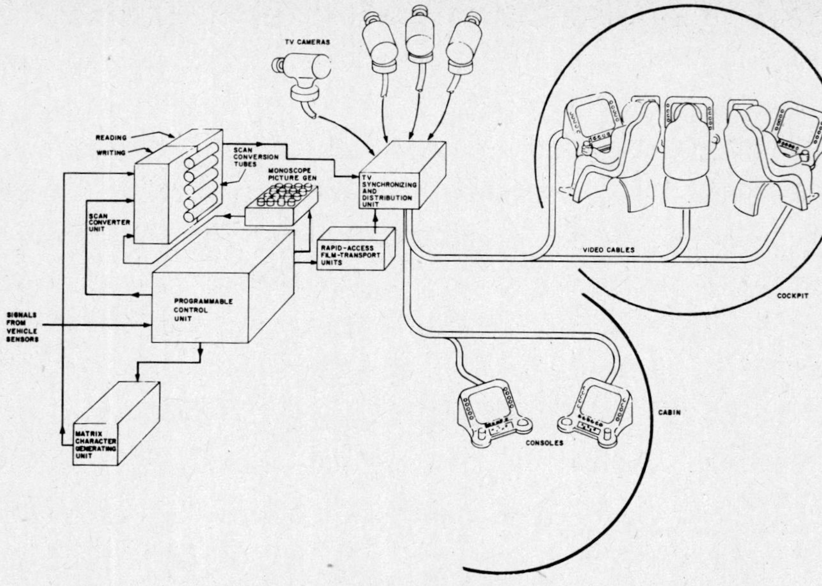

Fig. 1 Pictorial View of Display and Control System

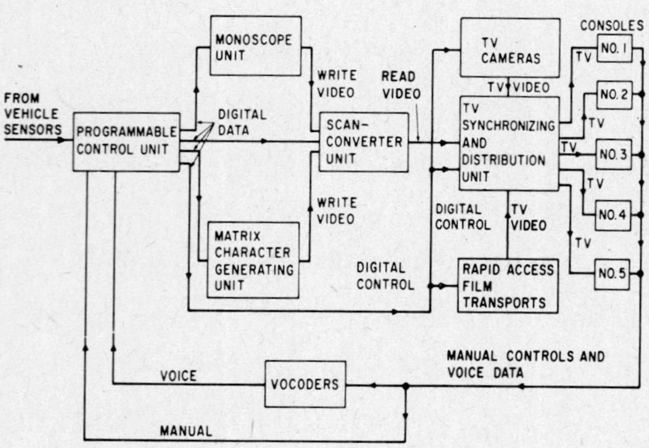

Fig. 2 Block Diagram of Display and Control System

Fig. 3 Simplified Functional Diagram of Pattern Synthesis

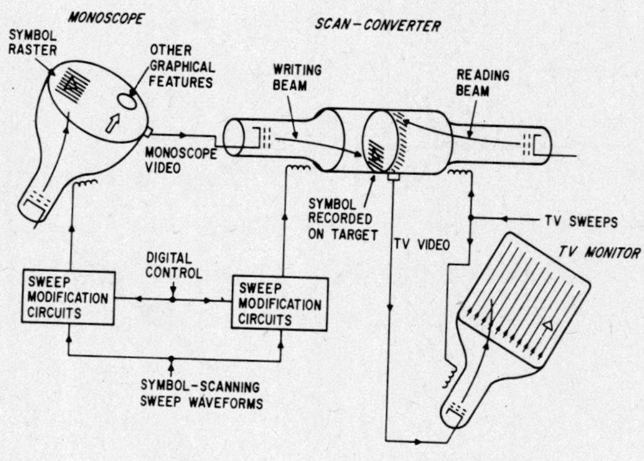

of CBS Laboratories (Ref. 39). In this system, the character is formed from video produced by the sampling of magnetic cores in a core switching matrix. The cores are positioned at appropriate junctions in the matrix so that video representative of the desired character results. This requires a separate matrix for each character, but has worthwhile potential from the viewpoint of compactness and reliability. The matrix equipment will be convenient for the readout of programmable or variable text, as in teletype, and matrix-generated characters will be added to pattern displays wherever needed.

Pattern Generation. The principal source of patterns for these displays will be miniature monoscopes from which patterns will be selected for recording in the scan-converters. Patterns will display, for example, bar charts, meter indications, and graphic representations of actual and predicted situations. The monoscopes will carry extensive pattern work as in the skeletal structure of bar charts, and also the rudimentary elements such as rectangles, arrows, and circles from which patterns may be constructed. Remote-controlled sweep modification circuitry will be provided in the monoscope system for symbol selection. Remote-controlled sweep modification circuitry will also be provided in the scan-conversion system for controlling the position, orientation, and affine distortions of monoscope patterns when they are being recorded.

The simplified functional diagram (Fig. 3) shows the basic operating scheme for pattern generation. A triangular symbol on a monoscope target is scanned by a diminutive raster derived from symbol-scanning waveforms. The raster is positioned on the monoscope target by digital commands from the coordinating equipment to select the desired symbol. Video from the monoscope then modulates the writing beam of a scan-converter tube, while the beam records a similar symbol scanning raster on the scan-converter target. The recorded symbol raster is also positioned by digital commands to properly integrate the symbol into the pattern that is being composed. The recorded symbols are converted to television video by a scan-converter reading beam that scans the entire target. In this way, patterns of any desired complexity may be built up from elements under program control.

Consequently, pattern generation is based on advance planning and programming so that the program that generates a pattern is always available in storage and may be quickly called up when desired. An example of the synthesis of video data extracted under program control from a monoscope is found in the "Videograph" system (Ref. 40). Those patterns that display variable features or moving indicators will have provision made in their programming for modifying the variable aspect during recording; this is done through the use of the sweep modification circuitry in the scan-conversion system.

When patterns require moving elements such as cursors or indicating arrows, provision is easily made for periodic erasure and regeneration. However, in order to conserve data-processing capacity the periodic recycling will be restricted primarily to just those features that move, or that change rapidly.

Programmable Control Unit. The organization and control of pattern generation and display is performed by a programmable control unit. This unit will be a digital machine designed primarily for table-look-up operations. The table that is to be "looked-up" consists of the complete list of pattern-generating and servo-controlling program sequences that have been prepared in advance.

Since the great majority of these sequences will not require modification during the course of a mission, the storage medium can be fixed. This is a great advantage since fixed stores of great capacity are available in relatively compact form. Licklider (Ref. 41) refers to such storage as "published memory"--a medium in which information is recorded in advance and which can be easily referred to by the machine, but which cannot be modified. The term "fixed storage" is due to Alonso (Ref. 27). Our plan is to provide 4 megabits of such storage capacity, which includes considerable reserve, in a 10-inch photoscopic disk, having about 30 milliseconds access time.

The photoscopic storage technique (Ref. 42) has been applied in the USAF Language Translator (Ref. 43). The medium provides digital coding in the form of photographically recorded marks on the surface of a glass disk. The disk rotates at a speed sufficient to provide the desired access time, in our case about 2000 rpm. The photographic marks are scanned by a CRT beam through an optical system; a phototube that senses the beam is used as the readout device.

For those portions of program sequences that are expected to require some modification, a scribable store of much smaller capacity will supplement the photoscopic disk. For this purpose it is planned to provide 100 kilobits of storage capacity in a magnetic-core type of digital memory. The scribable store will communicate directly with the address and information registers associated with the photoscopic disk so as to accommodate address modifications of table-look-up sequences.

The scribable storage unit will also be needed as an input-output buffer serving the fixed store. In addition, it will serve conveniently for maintaining periodically recycling data in buffer storage. An arithmetic unit will also be required in the programmable control unit, primarily for furnishing predictive data to the displays.

Remote Control System. The programmable control unit is designed to exercise control over other equipment by appropriate addressing of output data. It is convenient to transmit a digital control word from the programmable control unit indiscriminately to all equipment, but to "enable" the reception and storage of the digital word only in a selected equipment. Selective enabling can be accomplished by digital address data that accompanies the digital command and that affects only the register in the selected equipment.

The digital control data is either translated to analog form or else actuates gating devices and switch-gear. For example, when recording on a scan-converter tube, an electron beam is deflected by a digital command converted to analog form, after the digital registers controlling that tube have been enabled by an appropriate address.

Using the principle of digital control by selective enabling, every controllable device in the vehicle can be selected for control through appropriate addressing. Thus both the display system and the vehicle control system will be made subservient to digital commands emanating from the programmable control unit. On the other hand, the programmable control unit will itself be subject to operator control, and it is in this way that the operator attains complete command over the vehicle.

However, commands from the programmable control unit are intended to exercise parallel rather than serial control over most vehicle functions. In parallel control the commands insert occasional controlling or updating information into a process that is normally carried on automatically for a special function. For example, there will be a separate computer and autopilot for guidance (Ref. 21) which can be updated or overridden by the operator through the programmable control unit. Another example of an autonomous device subject to operator override would be a star-tracker (Ref. 44).

Centralized digital control is attractive for the expediting and coordinating of operations; however, it obviously demands potent measures for maintaining reliability at a tolerable level. For one thing such a project requires, of course, the application of rigorous quality control. In addition, system thinking is tending toward multiple-line redundancy combined with thorough in-flight checkout programs and facilities for in-flight maintenance (Refs. 45 and 46). At any rate the modern tendency is, of necessity, away from the idea that reliability necessitates crude or oversimplified apparatus.

Television Pictures. Television cameras are expected to provide the crew with convenient means for visually monitoring every important location both inside and outside the spacecraft. Miniature Vidicon cameras are favored for such applications. In many cases, servo-controlled lens turrets, and PAN and TILT mobility will be desirable. Camera servos are also expected to be under remote operator control by means of the digital control facility.

Bulk text such as manuals and handbooks, cartography, celestial maps, and drawings should be reserved in film storage. However, it will be necessary to furnish means for rapid remote readout from the film. For this purpose, it is planned to provide a number of rapid-access film transports in conjunction with flying-spot scanners. These units will be capable of storing up to 10,000 frames of high-resolution film in compact reels and of addressing and reading out the information in any frame under remote digital control. It will be appreciated that 10,000 frames can accommodate a library of very substantial size.

This method of presenting microfilmed library information on television screens is considered highly desirable for efficient vehicle operations. Otherwise, it would be necessary to carry along a considerable stock of literature that, quite apart from space and weight considerations, would be almost impossible for rapid access in emergencies.

The video pictures from the television cameras and film transports should be presented directly on the console screens without undergoing scan conversion since there is no need for synthesis or storage. Accordingly, this kind of video will be mixed directly within the television closed circuit. The mixing is to be performed in a TV distribution unit, where a separate remote-controlled mixer will be provided for each console station.

Manual Selection

A major problem in this design has been to devise an effective scheme for giving an operator access to the large number of available presentations with just one television screen, and to the many controllable processes with no more than the two hand controls. The solution that has been devised can best be described as "multi-level mode selection."

To select a specific display using the multi-level mode selection process, it will first be necessary to choose an operating mode. The mode is chosen from a mode selection list for a "first level" of selection, which may be called forth on the television screen. The following list illustrates a possible line-up of operating modes for "first level" selection:

MODE SELECTION

1. Environment
2. Medical

3. Vehicle Attitude
4. Guidance
5. Alarms
6. Vehicle TV Cameras
7. Library
8. Operation and Maintenance
9. Time-To-Go
10. Maps

Additional provision must, of course, be made for quickly bringing alarms to the attention of the operator. The alarm mode, listed above, may then serve to garner further details concerning the alarm situation.

The mode selection determines the type of display to be presented on the console. The selection of the mode is accomplished at the console by an electronic cursor, such as an arrow, that will be subject to manual control. When the selection is made a new listing for the designated mode will automatically pre-empt the screen. For example, if the "ENVIRONMENT" mode were selected from the MODE SELECTION list, the following new listing represents an illustrative example of what might appear:

ENVIRONMENT

1. Abort Capsule
2. Main Cabin
3. Air Lock No. 1
4. Air Lock No. 2
5. Air Lock No. 3
6. Pressurized Suit - No. 1
7. Pressurized Suit - No. 2
8. Pressurized Suit - No. 3

This listing would be considered a "second level" in the mode selection process. Provision will also be made for combined displays in which the new mode listing appears on another section of the same screen. The desired display will usually be selectable (also by means of an electronic cursor) from the second or third level of the mode selection listing.

Fig. 4 shows an illustrative display of internal environment that would appear after "Main Cabin" had been selected from the ENVIRONMENT listing. The main cabin environment display will be on a third level of selection and will furnish information on the important environmental life-support functions. Provision is made for either precise alphanumeric readout, or merely go/no-go indications. In any case, the aim at this selection level should be to permit an operator to survey an entire situation at a glance. More detailed information can be displayed, when needed, after further selection. In addition, provision can easily be made for accentuating and flashing out-of-tolerance parameters to alarm the operator.

Manual Control

The basic means for attaining control over a parameter or function will be by cursor designation at an appropriate mode selection level. For example, in Fig. 4 an electronic cursor consisting of an arrow and bracketing rectangle is shown designating the "Pressure" parameter. Such a designation will result in the specified parameter becoming subject to the manual controls. The operator will then be able to manually alter the pressure in the selected compartment of the vehicle while simultaneously observing on the screen the effects of the adjustment. In this way, not only may any aspect of performance be displayed, but also any parameter or process may be placed under individual control. Furthermore, effective visual feedback is provided to an operator, in that the effect of a controlling action is immediately visible on the display. Feedback is of considerable importance for effective control and it is therefore planned to supplement the visual feedback with kinesthetic feedback to the manual controls in a manner similar to that used in remote manipulators (Ref. 47).

In most cases, the commands that are initiated by an operator will call forth previously prepared controlling sequences from digital storage. These sequences may often be quite complex and may be used in conjunction with elaborate combination displays. Fig. 5 is an illustrative example of such a combination display as it might appear in a hypothetical soft lunar landing. The display shown combines a graphic representation of instantaneous vehicle position relative to a prescribed landing point, an attitude indicator, and a television camera picture of the terrain surrounding the landing site. The camera picture is supplemented with superimposed graphics. To such a combination there may be profitably added mode selection displays for sequencing rapidly between controlled functions.

Vocal Selection

In order to expedite the process of mode selection, it is considered highly desirable to augment the manual selection procedure with a facility for automatic recognition of vocal commands. For this purpose, it is planned to digitally encode the operator's voice signals in a "vocoder." Our plans for application of the vocoder to automatic recognition follows along the lines of work reported by Petrick and Willett (Ref. 48).

The operator will have at his disposal a relatively limited repertoire of vocal commands that he can deliver to the programmable control unit through the vocoder.

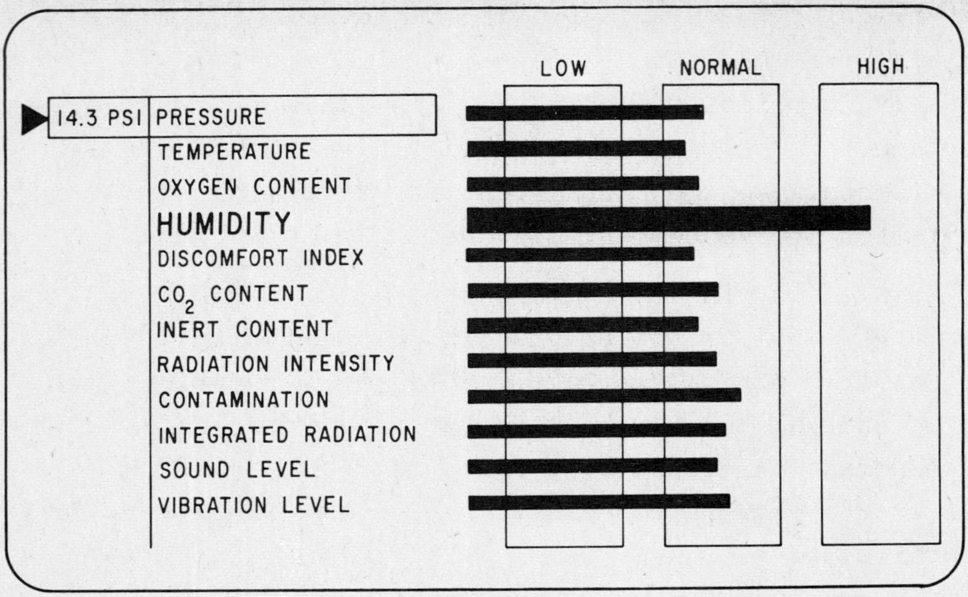

Fig. 4 Display of Internal Environment

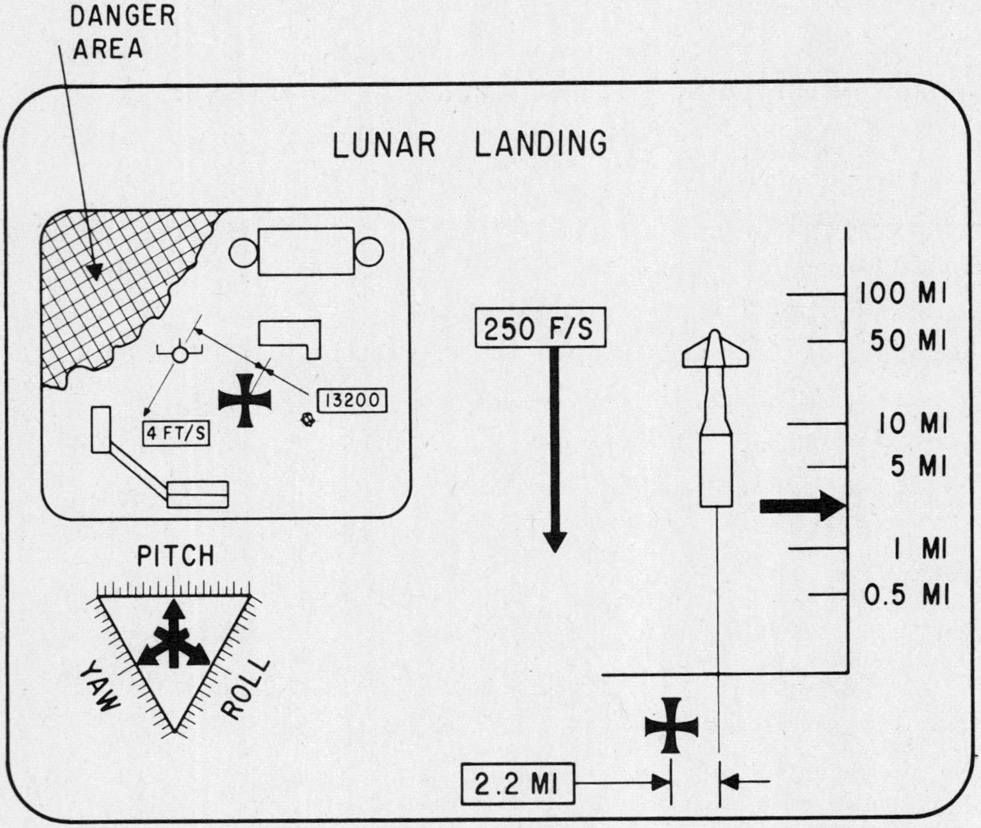

Fig. 5 Combination Display for Soft Lunar Landing

When the programmable control unit is altered by an actuating signal from the operator, digitally encoded vocal commands will be accepted and will be compared with information stored in a digital vocabulary that subsumes the command repertoire. The digitally encoded commands in the vocabulary are referred to as "masks."

Each operator will have prepared his own masks before the mission, and the mask data will be stored in the fixed storage of the programmable control unit. The storage capacity required for a typical set of masks is not expected to exceed 50 kilobits, which represents an insignificant proportion of the available fixed storage capacity. When a vocal command has been delivered and identified with an operator's mask, the command will be registered and executed at the conclusion of the command sequence. Suitable initiating and concluding vocal codes will be chosen for the command sequence so that manual control over the vocal command system will not be required.

The vocoder equipment that is planned for this system is based on the original design of H. Dudley (Ref. 49). It consists of a set of band-pass audio filters that divide the voice spectrum into channels. The spectral-energy density in each channel is sampled rapidly and the samples are digitally encoded. The arrangement chosen uses 18 channels that are each sampled sequentially at a rate of 50 cps. Each sample is encoded into a 3-bit digital character. Normalization in both amplitude and time has been found to be very desirable and is also to be provided.

The principal advantage of the vocal command system will be the speed with which modes can be selected for display and parameters selected for control. The vocal command feature is expected to be of maximum benefit during emergency situations when it supplements or substitutes for manual control. This benefit far outweighs the relatively small penalty in vocoder equipment.

Petrick and Willett have amply demonstrated that very effective speech recognition can be attained when the operator has prepared an individual mask for the digital comparison process. In such cases, the reliability of machine recognition can be comparable to human recognition. Recognition is known to be much more difficult when masks cannot be individually prepared in advance, but fortunately this problem need not concern us here.

It is important to realize, however, that visual confirmation concerning mode or parameter selection will be available on the display immediately after delivery of a vocal command. Furthermore, it is not intended that the vital systems, such as the propulsive or guidance systems, be under direct vocal command; the vocal control feature is to be restricted to mode or parameter selection only. For these reasons, perfect recognition reliability is quite unnecessary.

CONCLUSIONS

It is believed that the system described will effectively meet the requirements imposed by space travel. The system architecture is adaptable in the sense that operational capabilities can be reduced or increased in many ways without necessitating functional redesign. Furthermore, it is "general-purpose" in view of the fact that specific displays and control processes are variable just by reprogramming and by the changing of monoscope patterns.

With such a system a great number of displays and controls can be efficiently handled by an operator with relatively simple equipment, and the design is therefore believed to propose a significant advance in man-machine symbiosis.

ACKNOWLEDGMENTS

Acknowledgment is made for the kind assistance of Messrs. L. Penner, B. Levy, and W. Butler of Airborne Instruments Laboratory.

BIBLIOGRAPHY

1. M. W. Rosen and F. C. Schwenk, "A Rocket for Manned Lunar Exploration," IRE Trans. on Space Electronics and Telemetry, December 1959.

2. C. Ryan, Ed., "Across the Space Frontier," Viking Press, 1952.

3. C. Ryan, Ed., "Conquest of the Moon," Viking Press, 1953.

4. W. H. Andrews and E. C. Holleman, "Experience with a Three-Axis Side-Located Controller During a Static and Centrifuge Simulation of the Piloted Launch of a Manned Multistage Vehicle," NASA TN D-546, November 1960.

5. J. W. Useller and J. S. Algranti, "Pilot Control of Space Vehicle Tumbling," Proceedings of National Specialists Meeting on Guidance of Aerospace Vehicles, Boston, May 1960.

6. J. F. Corso, "The Effects of Noise on Human Behavior," WADC TN 53-81, December 1952.

7. S. B. Sells and C. A. Berry, "Human Factors in Jet and Space Travel," Ronald Press, 1961.

8. E. W. Still, "High-Altitude Chambers and Pressure Suits and Their Part in Manned Flight to the Moon," J. Brit. Interplanetary Soc., March-April 1960.

9. R. S. Johnston, "The Control and Measurement of the Mercury Capsule," Proceedings of Fifth National Symposium on Space Electronics and Telemetry, 1960.

10. G. A. Holcomb, "Human Experiments to Determine Human Tolerance to Landing Impact in Capsule Systems," Ballistic Missile and Space Technology, Vol. 1, Academic Press, 1960.

11. J. W. Brinkley, "Man Protection During Impact Landing of Aerospace Vehicles," Ballistic Missile and Space Technology, Vol. 1, Academic Press, 1960.

12. N. K. Ganguly and J. T. Lence, "Shielding Manned Space Vehicles from Space Radiations," J. Brit Interplanetary Soc., Vol. 18, May-June 1961.

13. H. J. Schaefer, "Proton Radiation Hazards in Space," Astronautics, February 1961.

14. M. Rosenbaum, "Elements of Typical Space Systems," Proceedings of National Aeronautical Electronics Conference, 1960.

15. M. W. Ovenden, "On the Nature and Distribution of Meteoric Matter," J. Brit Interplanetary Soc., Vol. 6, 1947.

16. D. Rodriguez, "Meteoroid Shielding for Space Vehicles," Aerospace Engineering, December 1960.

17. P. Bono, "Conceptual Design for a Manned Space Vehicle," Proceedings of Eleventh International Astronaut. Congress, Vol. 1, 1960.

18. R. W. Connor and E. A. Smith, "Emphasis on Survival in the Design of Interplanetary Vehicles," Proceedings of Fifth Symposium on Ballistic Missile and Space Technology, 1960.

19. C. A. Cross, "A Practical Investigation of Spaceship Control Problems," Proceedings of Tenth International Astronaut. Congress, 1959.

20. R. E. Roberson, "Astronavigation--Guidance and Control in Space," J. Astronaut. Sci., Vol. 7, No. 4, 1960.

21. H. B. Haake and J. D. Welch, "Mechanizing a Space Navigation System," Proceedings of Fifth National Symposium on Space Electronics and Telemetry, 1960.

22. K. R. Stehling, "Lunar Landing Problems," Proceedings of Eleventh International Astronaut. Congress, 1960.

23. W. L. Morris and R. C. Kaehler, "Synthesis of Human-Automatic Control Systems for High-Performance Vehicles," J. Astronaut. Sci., Vol. 5, No. 2, 1958.

24. J. A. O'Malley, Jr., "Control Development for Jet VTOL," Proceedings of National Specialists Meeting on Guidance of Aerospace Vehicles, May 1960.

25. J. W. Clark, "Mobotry--The New Art of Remote Handling," IRE Trans. on Vehicular Communications, August 1961.

26. L. D. Carlson, "Requirements for Monitoring Physiological Function in Space," Proceedings of Fifth National Symposium on Space Electronics and Telemetry, 1960.

27. R. L. Alonso, "Design Principles for a Control Computer for Space Guidance," Proceedings of National Specialists Meeting on Guidance of Aerospace Vehicles, May 1960.

28. D. B. Yutema and W. S. Torgerson, "Man-Computer Cooperation in Decisions Requiring Common Sense," IRE Trans. on Human Factors in Electronics, March 1961.

29. C. O. Hopkins, "Determination of Human Operator Functions in a Manned Space Vehicle," IRE Trans. on Human Factors in Electronics, September 1960.

30. T. E. Garrigan, "Lunar Flight Control Display," Astronautics, June 1961.

31. A. M. Mayo, "Space Cabin Design," J. Astronaut. Sci., Vol. 5, Spring 1958.

32. D. W. Richardson, "The Integrated Crew-Computer Team," Aerospace Engineering, December 1960.

33. A. W. Hetherington, "Performance Assessing Human Capability in Space," IRE Wescon Conv. Rec., 1958.

34. C. O. Hopkins, D. K. Bauerschmidt, and M. J. Anderson, "Display and Control Requirements of Manned Space Flight," WADD TR 60-197, April 1960.

35. S. N. Roscoe, "The Development of Integrated Display Panels at Hughes Aircraft Company," Psychological Aspects of Cockpit Design: A Symposium Report, WADC TR 57-117, 1957.

36. F. Ludwig, D. Ellis, and L. Steinman, "The Tunnel Display Concept," Proceedings of Sixth Annual Meeting of American Astronautical Society, Vol. 6, 1957.

37. L. J. Fogel, "A New Concept--Integrated Anti-Vertigo Cockpit Display and Control," Proceedings of Second International Congress on Cybernetics, 1958.

38. "Pattern Readout and Storage Equipment," Product Bulletin, Airborne Instruments Laboratory, 1961.

39. "New High-Resolution Symbol Generator," Automatic Control, June 1960.

40. "Model 980 Character Generator: Videograph," Applications Bulletin, A. B. Dick Co., 1960.

41. J. C. Licklider, "Man-Computer Symbiosis," IRE Trans. on Human Factors in Electronics, March 1960.

42. G. W. King, G. W. Brown, and L. N. Ridenour, "Photographic Techniques for Information Storage," Proc. IRE, October 1953.

43. G. Shiver, "The USAF Automatic Language Translator Mark I," IRE Conv. Rec., 1958.

44. J. E. Abate, "Star Tracking and Scanning Systems--Their Performance and Parametric Design," IRE Int. Conv. Rec., Part 5, 1961.

45. T. B. Lewis, "Digital Computer Equipment for an Advanced Bombing, Navigation and Missile Guidance Subsystem for the B-70 Air Vehicle," Proc. IRE, January 1961.

46. W. C. Mann, "Systematically Introduced Redundancy in Logical Systems," IRE Int. Conv. Rec., 1961.

47. B. M. Crawford and D. F. Baker, "Human Factors in Remote Handling--Survey and Bibliography," WADD TR 60-476, July 1960.

48. S. R. Petrick and H. M. Willett, "A Method of Voice Communication with a Digital Computer," Proceedings of the Eastern Joint Computer Conference, December 1960.

49. H. Dudley, "Remaking Speech," J. Acoust. Soc. Amer., October 1939.

50. L. J. Fogel and M. Dwonczyk, "Anticipating Display Design Through the Use of an Analog Computer," IRE Wescon Conv. Rec., Part 4, 1958.

PRACTICALITY AND THE SELECTION OF A SPACE CABIN ENVIRONMENT

Donald L. Foster[+] and Malcolm D. Ross[+]

Current emphasis placed upon the "Space Program" and, in particular that portion involving the sustenance of human life in a space environment has resulted in numerous conceptual solutions for the provision of an adequate life support environment.

It is proposed that, insofar as possible, a normal sea-level environment be utilized for our early manned space operations. The practicality of modifying this natural environment for reasons of engineering expediency, or because of potential hazards associated with loss of system integrity, is questioned. It is suggested that the advantages of a sea-level environment are far more important than inadequate resolution of problems arising from anticipatory situations.

It is not intended to discourage the continuation of efforts which may lead to the use of modified environments at such a time when the results of such modification are better understood.

[+] Biological Sciences and Systems Dept., Defense Research Laboratories, General Motors Corporation, Warren, Michigan

INTRODUCTION

We are all aware of the importance and urgency associated with our space program and the emphasis currently being placed on the man-in-space missions. If we are to meet the challenge which has been placed before us, our scientists and engineers must mold together propulsion, guidance, and life support systems to accomplish each mission objective. There are many additional facets, such as communications, tracking, recovery techniques, remote information acquisition and display, to specify a few, which must be interrelated into the operation of a manned extra-terrestrial vehicle. The critical areas, however, are the three stated above.

The technologies of propulsion and guidance are based on well described physical phenomena, and apparently progress in these areas is encouraging; however, greater advancements are required. On the other hand, man is a very complex organism and our knowledge is comparatively limited, as pointed out by Biedler (4). Because of our meager understanding of man, we are somewhat concerned about some of the current thinking devoted to the development of life support systems for manned space explorations. We believe two basic approaches must be made to this problem. These may be identified simply as short-range efforts (for missions to be carried out prior to 1970) and long-range efforts (missions to be realized after 1969). These identifications, although arbitrary, are germane because our present concern is with the short-range efforts.

To define this more clearly, short-range objectives are related to "hardened" systems which must have proven components and subsystems and be operational between now and 1970 to meet the basic national goals of the manned space flight program outlined by President Kennedy in his message of May 25, 1961. One further assumption is that the duration of any single mission will be greater than 24 hours, but will not exceed two weeks, regardless of whether it is manned by a crew of one, two, or three astronauts.

REQUIREMENTS OF A LIFE SUPPORT SYSTEM

Serious thought must be given to the practical requirements (11) of a short-range life support system. Proper consideration must also be given to the fact that our knowledge of man is relatively limited. Mission profiles will be limited in duration by the status of our booster technology, our guidance capability, and the state of development of advanced operational techniques, such as orbital rendezvous. The basic requirements of the life support system are, to some degree, dictated by these limitations. It appears to us that a large percentage of the current efforts are directed to the development of highly conceptual techniques with too little thought directed to the well-being of the man. His status has been considered by Lambertson (15), and by one of the present authors in an earlier paper (24). We are not attempting to discourage conceptual efforts, but to emphasize the critical need for realistic approaches now to obtain practical solutions. In order to meet the short-range objectives within a reasonable period of time, it is necessary that the requirements be examined in terms of conservative practicality.

Let us look at the basic requirements of a life support system for manned space explorations. It must provide:

1. An Atmosphere of Appropriate Chemical Composition,

2. A Suitable Atmospheric Pressure,

3. Environmental Temperatures and Humidities Within Normal Comfort Limits,

4. A Means to Permit Acceptance of Necessary Variations of Accelerative Forces from One G,

5. Adequate Nutrition and Fluids; also a Means for Waste Product Disposal, and

6. A Means to Maintain Radiation Exposures Within Acceptable Limits.

We believe that it is primarily within the areas of the first three of these requirements that clear objective thinking has not always prevailed (2). Although the remaining three requirements are also important, there are many excellent papers which discuss these areas in detail. We shall, therefore, confine our discussion to the first three requirements and consider them in greater detail.

DISCUSSION OF SELECTED LIFE SUPPORT SYSTEM REQUIREMENTS

Because of the interrelationship between the atmospheric composition and the pressure within a sealed cabin system, it is difficult to discuss these two aspects separately. We shall, therefore, treat them together.

1. Consideration of Chemical Composition and Pressure of Cabin Atmosphere

When variations are contemplated for either of these parameters, compensating changes in the other become necessary to maintain an apparently acceptable life support environment. We wish to emphasize the two words: "apparently acceptable". There are relatively little data or experience which suggest that man's natural environment should be modified. On the other hand, there are well-substantiated indications that such modifications may have extremely detrimental effects, and may well result in complete mission failure.

Langdon and Reynolds (16) have reported that breathing 100% oxygen at reduced pressure and under moderate accelerative loads has resulted in mild cases of atelectasis in approximately 20% of the subjects investigated. The effects of longer periods of breathing a 100% oxygen atmosphere at reduced pressure, and with more severe accelerative loads, are not known. Yet, 100% oxygen at a total pressure of 1/3 atmosphere is being used on our current manned space programs (22).

In the literature, there are various reasons which attempt to justify the use of a modified atmospheric composition and pressure. For purposes of review, we shall present some of them here.

a. Arguments Used by Others to Justify a Modified Atmospheric Composition and Reduced Pressure

(1) Weight Saving

Substantial weight reductions in the system may be achieved by decreasing the pressure stresses imposed upon the pressure shell. Such reduction of pressure stresses will permit the use of lighter materials without a corresponding decrease in the factor of safety for the system (12,29).

(2) Minimize Cabin Atmosphere Gas Loss

The loss of cabin atmosphere gases through inherent leakage may be considerably reduced (3,12).

(3) Prevent Aeroembolism

In the event of loss of pressure shell integrity with a resultant rapid decompression, aeroembolism within the body tissues may be reduced or prevented (26).

(4) Allow Inflight Cabin Repair

If a relatively slow leakage from the sealed cabin system were to develop, the rate of loss of atmospheric gases would be considerably reduced if the total pressure is less than one atmosphere. Conceivably, this would permit repairs and continuation of the mission (33).

(5) Increase Reliability

A single gas system, being less complex than a multi-gas system, will be more reliable. Therefore, because of the problems of oxygen toxicity at high partial pressures, a reduction in total pressure is mandatory to achieve the desired reliability (10,15).

(6) Reduce Fire Hazards

There are other reasons given for reducing the total pressure, including the reduction of fire hazards in the presence of 100% oxygen as reported by various investigators (10,26).

These arguments for selecting a modified sealed cabin atmosphere environment are based on engineering expediency, or to minimize the potential hazards associated with a loss of system integrity. Let us re-examine these arguments, one by one, to determine their validity when considering all aspects of a space mission from the standpoint of practicality.

b. Analysis of Arguments Justifying a Modified Atmosphere

(1) Weight Saving

Actual experience in the design and fabrication of sealed cabin systems for employment of sea level pressures during stratospheric balloon flights has indicated that the pressure shell weight is less than 10% of the total weight of the flight system. Based on this figure, it would appear that the total weight of the flight system might be reduced by as much as 7%, merely by reduction of the internal pressure to 1/3 atmosphere. This is not necessarily the case. For instance, the U. S. Navy's Strato-Lab Program utilized a sea level pressure environment within a spherical gondola 7' 2" in diameter. The shell of this gondola, originally fabricated in 1947, was constructed from 3003-0 aluminum, one-tenth inch in thickness, and weighed approximately 230 pounds. Using materials available today, this gondola might be fabricated from 6061-T6 aluminum, .030 inches in thickness, and would weigh approximately 70 pounds. Thus, a saving of 160 pounds could be achieved by proper material selection without any sacrifice in shell strength.

Theoretically, one could maintain the same factor of safety, from the standpoint of pressure stress, by reducing the shell to .010 inches in thickness and operating with an internal pressure of 1/3 atmosphere. This would result in an additional reduction of shell weight of about 46 pounds. For a one-ton system, this would merely provide a 2.3% savings in weight. However, because of practical fabrication limitations and the probability of accidental shell penetration, one probably would not consider seriously the reduction of the shell thickness below the .030 inch figure even for the 1/3 atmosphere design. We have purposely disregarded consideration of the requirements for withstanding the dynamic forces imposed during boost or re-entry phases of a space flight profile, or for radiation and thermal shielding. Materials utilized for these purposes would add to the strength of the basic structure. Therefore, we feel that reduction in the pressure within a sealed cabin system, for purposes of reducing the system weight slightly, or to provide a higher factor of safety, is not justified because the physiological aspects of such modifications are not well defined and may well actually result in a significant decrease in the reliability of the man/machine system.

(2) Minimize Cabin Atmosphere Gas Loss

The second reason given for reducing total pressure within the cabin is to minimize loss of cabin atmosphere gases due to inherent leakage. This is such an insignificant contribution that it is worthy of very little consideration and is certainly rejected by us as a sufficient reason for using a reduced total pressure. By proper engineering and fabrication techniques, the pressure integrity of the sealed cabin system can be optimized with gas leakage very low.

Once again, using the Strato-Lab system as an example, it is of interest that this sphere, after four flights, each of which resulted in considerable damage and required many modifications, was tested by hydrostatic and pneumatic techniques to check its pressure integrity. During one of these tests the unit was pressurized to 20 psi gauge (34.7 psi absolute) and allowed to sit for a period of 5 days. At the end of this period, the total pressure had dropped 1/2 pound psi to approximately 19.5 psi gauge. This would indicate that less than one pound (weight) of make-up gas would be required during a ten-day mission to compensate for this leakage. From an engineering point of view, this would certainly be an acceptable requirement for a short-range manned space mission.

There is little that can be done to prevent serious or fatal consequences in the event of a rapid loss of cabin pressure, unless there is a pressure suit or cabin redundancy. Loss of pressure shell integrity would make it mandatory that a mission be completed by the use of personnel pressure suits. Although personnel of the U. S. Navy have worn the Mark IV full pressure suit for a period of five days in the unpressurized condition (20), we believe it is not practical to wear a complete pressure suit, pressurized or not, throughout the duration of the space missions presently contemplated. This has been pointed out by Stubbs (29), as well as by one of us in an earlier paper (24).

In the event of a rapid decompression, it would be physically impossible to don the suit before death results. At most, it appears that the complete suits would be worn only during the boost and re-entry phases of the space mission. The use of pressure suits will be discussed further below.

(3) Prevent Aeroembolism

In the event of a rapid decompression of a system in which a pressure suit provides emergency pressurization for flight personnel, there are substantial data by many investigators which indicate that aeroembolism may be prevented by use of reduced total pressures and 100% oxygen within the sealed cabin system; but what have we actually accomplished by this modification?

In order that this system be fully effective, the personnel must wear their complete pressure suits throughout the duration of the mission in anticipation of a system failure. We have previously indicated that this is unacceptable because of the physio-psychological aspects and performance decrement imposed upon the crew. In accordance with our definition of short-range missions, we are referring in our discussions to flights exceeding 24 hours which is clearly beyond the present U. S. manned space flight experience of Project Mercury. Let us face reality. It seems much more practical to allow the man to exist within a more normal physiologically and psychologically suitable environment and to reduce the probability of loss of pressure integrity by proper vehicle design. A finite probability of failure will always exist, however, but this must be accepted as a calculated risk to the mission success.

(4) Allow Inflight Cabin Repair

If a leak occurs and the loss of cabin pressure is sufficiently slow, there would be time, presumably, for the flight personnel to put on their pressure suits. It might even be possible to locate the leak and to effect adequate repairs; however, this is not considered a very practical possibility. The most probable occurrence would be the development of a leak or leaks of sufficient size which would result in an unacceptable depletion rate of the atmosphere supplies, but which could not be repaired because of physical size or location. Even in the event of a slow leak, it is likely that the crew would find it necessary to put on their pressure suits to reduce depletion of their oxygen supply. The physiologic and psychologic performance decrement which results from long periods in the present pressure suit can only have extremely adverse effect upon the outcome of the mission. The authors believe that full-time use of pressure suits cannot be tolerated for our astronauts and astroscientists participating in even the short-range missions. We believe they are an unnecessary redundancy and, indeed, are a liability for missions which require fully operant, comfortable astronauts and space scientists for periods of up to two weeks.

We do believe, however, that pressure suits will serve a useful purpose if utilized in a sensible manner. The procedure which we recommend is very simple. Each member of the flight crew will be fully suited at lift-off. Whether the suit is actually pressurized or not, it is immediately ready and available for use if the cabin loses pressure early in the mission.

After arriving in space, whether it be on an earth orbital mission, or the long free flight trip to orbit the moon and return, the pressure suit should merely become a flight suit by removing the helmet, gloves, and loosening it from the body. A certain degree of desired system redundancy would be exchanged for some other important features, such as mobility, comfort, and maximum operant utilization of man. These features, however, are essential if man is ever to be much more than a mere captive of the space system with severe limitations to his capabilities.

With minimum effort the flight suit would become a pressure suit again to meet various eventualities, such as slow leaks or failure of the gas system. Unfortunately, it would not be available immediately if that should be necessary. However, participation in space flight is not a guarantee of long life; some hazards will simply have to be accepted as calculated risks. Whether we like it or not, some of the men pioneering in space, as in almost any prior type of exploration, will contribute their lives as a partial cost of the effort.

The pressure suit philosophy expressed above has been presented before (24,25). Once again, the Navy Strato-Lab balloon program flight experiences are cited because a sea level atmosphere and the pressure suit-flight suit technique suggested above have been utilized on two of the four sealed cabin

Strato-Lab flights. (Other flights in the program were made with "open" systems, therefore, are not pertinent here.) The actual atmospheric composition, cabin pressure used, subjective crew reactions, and pressure suit techniques have been reported (5, 6, 23,24,25, 32) for the first three of the four Strato-Lab High sealed cabin balloon flights. Although a report (34) of the fourth flight has been issued, it is inadequate for this discussion.

The authors, before our present association, were closely connected with all Strato-Lab flights, so can report on all four which comprise a total time of more than 3 days aloft or, approximately 150 man hours. One of us (MDR) participated as pilot on all four flights; the other (DLF) was project engineer for developing the system. The sea level atmosphere and pressure suit technique pioneered on flights #3 and #4 in the series were heavily favored by comparison. On those flights, the pressure suits were fully intact, but not used (face plates were open) until reaching ceiling (floating) altitude, then became flight suits by removing gloves,helmets, and loosening the suit by relaxing all zippers. For descent, the suits were fully rigged again as pressure suits. Subjective flight crew evaluation concluded that the philosophy and techniques may be as useful to space missions as they are to balloon flights.

(5) Increase Reliability

If a full sea-level pressure is used, it is well known that a 100% oxygen atmosphere is unacceptable for more than relatively short durations (8, 27), and many writers have pointed out that a multi-gas system is undesirable because of increased complexity with a resultant decrease in reliability (33). We are not willing to accept these "engineering excuses" as valid reasons for avoiding simple solutions to practical problems which are of utmost importance to total success for the short-range U. S. manned space missions.* It is of interest to note that Yazdovsky, according to Konecci (14), has pointed out that the Russian manned orbital flights have employed a sea-level pressure as an integral part of the life support system. Apparently, they also employed a simple multi-gas supply and maintained an extremely low loss rate due to leakage. We have previously indicated that more consideration must be given to the physiologic and psychologic effects upon the human organism before modifying the pressure and oxygen environments within a life support system. We have also suggested that there is little cause to modify the total pressure for reasons of engineering expediency or to provide inadequate solutions to anticipatory problems.

---

* Data by Thompson and Gell (30) related to the effects of meteoroid penetrations into a sealed cabin system strongly support our recommendation for use of an atmosphere of normal sea-level pressure and composition.

(6) Reduce Fire Hazards

It has been pointed out by Simons and Archibald ([26]), also Heldenbrand ([10]), that a fire hazard exists in a sea-level pressure atmosphere if the composition is 100% oxygen. There is no quarrel, of course, with this statement. This has been carried a step farther in design, however, as an argument for a pressure considerably less than sea level. We think this is extrapolating limited test data and generalizing without a sound physiological basis for short-range space missions. We also think this is avoiding the simple solution of employing pressures and chemical composition of the cabin atmosphere so that it very nearly approximates man's normal sea-level environment. Based on lack of evidence, we prefer to remain conservative on this point until physiologists have acquired enough information to be certain that man's environment can, and must, be changed for space flight missions.

2. Consideration of Environmental Temperatures and Humidities

Temperature and humidity control within a sealed cabin system during manned space missions will be of extreme importance. Jannsen (13) recognizes that there is no universal temperature or humidity at which human comfort is assured. One reliable source (9) prescribes the comfort zone for individuals on a seasonal basis. In winter, the range would be 72 $\pm 3^\circ$ F; in summer, it would be 76 $\pm 5^\circ$ F. In all seasons, 50% relative humidity is prescribed for the comfort zone. Therefore, we might consider 75 $\pm 5^\circ$ F as a reasonable temperature when the relative humidity is maintained at 50%. These values are valid, of course, for sea level pressures. Variations between individuals, variations as a function of time for a single individual, relative activity level, atmosphere composition and pressure, are modifying factors in the comfort level temperature-humidity relationship. Because of this variation in subjective responses to temperatures, it is important that means be provided for controlling the thermal environment within the life support system. The degree of control necessary, will, to some extent, depend upon the duration of the mission. Inasmuch as precise control probably will not be possible within the limits of current technological capabilities and weight allowances, the effects of variations from the desired temperature level must be considered.

We are also realistic enough to suggest that temperatures lower than the comfort zone for hours, or perhaps even days, would be acceptable if the physiological discomfort can be solved, much as we do here on earth, by the use of protective clothing. By effecting compromises such as this, it is considered quite feasible to provide a comfortable temperature, within practical limitations, without the severe penalty for a complete engineering solution in terms of fully automatic active or passive control.

It has been rather well established that major temperature variations for short exposure periods do not result in appreciable detrimental effects to the human body, while long-term exposure to comparatively minor changes in the thermal environment may have serious consequences (19). Operations within a physiologically suitable temperature range may not be psychologically acceptable for long-term exposures. In addition, cyclic temperature variations within the life support system or stratification of thermal layers may result in either physiological or psychological performance decrement.

There are three distinct phases of a man-in-space mission during which the thermal environment is affected by variations in the heat input to the system. They are:

a. Boost and exit from earth's atmosphere,

b. Coast along the mission path,

c. Re-entry into the earth's atmosphere.

During the boost and exit from the earth's atmosphere, the life support system will be affected by conduction, convection, and radiation heat transfer; however, the relatively brief exposure time and the thermal lag of the system will undoubtedly negate, to a large extent, significant changes in the environmental temperature. Excellent data are certainly available to describe the changes during this phase.

The temperature within the life support cabin during the coast phase of any space mission will be primarily a function of the radiation characteristics of the vehicle and the mission profile. During earth equatorial orbital missions, the short duration of the sunlight/darkness cycles will result in relatively small fluctuations in the temperature of the life support system (17). Heat gained while the vehicle is exposed to incident solar radiation will be re-radiated to the heat sink of space while in the earth's shadow. A more severe temperature control problem will be those missions in which the durations of exposure to either incident solar radiation or the darkness of space become very much longer. Lunar orbital or polar earth orbital missions will result in relatively long periods during which the vehicle will be subjected to incident solar radiation. It is possible that during the initial stages of a lunar orbiting mission, and surely during that portion of such a mission that the vehicle was in the lunar shadow, the vehicle would be subjected to long periods of radiation losses to space without the benefit of solar radiation heat input. The possible extreme temperature variations within the life support cabin will challenge the capability of the engineer to provide a suitable temperature control system.

The third phase, also of considerable concern, is re-entry into the atmosphere. During this phase, the kinetic energy of the system must be dissipated. This energy dissipation is accomplished by transformation of the kinetic energy to heat. Because the re-entry phase is accomplished within an extremely short time interval, the vehicle cannot rid itself of this heat sufficiently fast to prevent an appreciable temperature increase. This, in turn, may be reflected in higher temperatures within the life support environment. Since the time duration is short, however, any departure above the comfort zone must be considered with the duration in mind to determine whether or not it will be acceptable.

We have looked at the problem, and are once again faced with the necessity of providing a practical solution. Because of the stringent requirement to reduce the weight of the life support system to that of an absolute practical minimum, it will be necessary to accept temperature variations in excess of the ideal, although they must be well within the limits of human tolerance. This probably means that a resulting performance decrement must be accepted as a portion of the total risk of the mission.

PRACTICAL SPECIFICATIONS TO MEET REQUIREMENTS

In the specifications listed below, emphasis should be placed on the word "practical" because we fully realize that our knowledge is too limited to know what ideal specifications might include Philosophically, however, the practical specifications we propose will meet the general requirements, they are indeed within engineering feasibility, and, finally, primary emphasis is given to the mobility and comfort of the crew. Even for the short-range missions under discussion, we believe (and hope) that our national space program goals contemplate the use of man's scientific capabilities for exploratory research, as well as his abilities in astronautics and his courage. We believe, therefore, that radical departures from man's normal environment cannot be considered acceptable until substantial data are available so that physiologists and medical scientists would agree on a modified environment for man.

Before listing our specifications, there are some additional comments which we believe are appropriate observations to our major theme. There is apparently a problem in communications and a grave misunderstanding between workers in the biological sciences and engineers, although both are trying to merge their knowledge, experience, and abilities to develop our manned space systems and assure mission success. There is another factor, management, which is responsible for a successful marriage of biological knowledge and engineering application to achieve success.

No one is apparently at fault, but it appears that the many engineering problems have become so well emphasized that the biological scientists are extremely sympathetic. So much so, in fact, that material by many outstanding individuals (7,18,28,31) stresses the "ragged" limits of man's physiological tolerance, rather than his known basic environmental requirements. The engineers seem pleased that man can apparently be squeezed and environmentally "engineered" into a system in this manner and are using the information as design criteria. A lack of adequate understanding obviously exists between the two groups.

Management is also apparently not sufficiently cognizant of the problem in depth to halt the general poor philosophy of designing a system which treats man as a "black box". If we are to succeed in our manned space missions in the near future, however, the philosophy of designing a system for man requires more scrutiny. The responsibility for this, and the important subsequent decisions, belongs to management.

Our specifications, which we believe are practical, even if somewhat general, are submitted below:

1. Appropriate Chemical Composition

There should be an arrangement of oxygen and nitrogen so that the percentages are approximately the same as found at sea level in the earth's atmosphere. In quantitative terms, we will specify 160 ± 5 mm Hg as the O_2 partial pressure and 600 ± 50 mm Hg for the partial pressure of N_2. Carbon dioxide and trace elements should also approximate the partial pressures found at sea level in the earth's atmosphere.

2. Suitable Pressure

Using 760 mm Hg as the total internal pressure, acceptable variations should not exceed $\pm 7\%$.

3. Temperatures and Humidities

Temperatures inside the space cabin should be maintained at 75 $\pm 5^{\circ}$ F with relative humidity at 50 $\pm 10\%$. Means must be made available for maintaining a reasonable degree of comfort even though the internal temperatures should become lower than those specified.

4. A Means to Permit Acceptance of Necessary Variations of Accelerative Forces from One G,

5. Adequate Nutrition and Fluids; also a Means for Waste Product Disposal, and

6. A Means to Maintain Radiation Exposures Within Acceptable Limits.

Although the requirements 4, 5, and 6 are not under discussion here, the engineering philosophy should be to establish practical specifications for each of these items so that the departure from "normal" is minimized. It is recognized that the problem of gravity is important and great variations, ranging from near zero to many times gravity, may have to be accepted for the short range missions until man's relationship to the effect is better explored and understood.

PRACTICAL ENGINEERING SOLUTIONS TO MEET SPECIFICATIONS

Once again, the chemical composition and pressure will be treated together in proposing a solution to meet the simple specifications.

1. Appropriate Chemical Composition

2. Suitable Pressure

It is recommended that a two-gas system be employed to maintain a full sea-level pressure and provide the appropriate chemical composition. Although oxygen may be combined with either helium or nitrogen, or possibly both, it is recognized that our statistical data are ponderously in favor of an oxygen-nitrogen combination. These are the basic constituents of man's normal terrestrial environment. For this reason, the oxygen-nitrogen combination is recommended It is proposed that the sealed cabin environment be maintained at a partial pressure of approximately 160 mm O_2 and 600 mm N_2. Again, these values closely approximate man's normal environment. It is interesting to note that the Soviets have apparently followed the above reasoning and have even indicated that nitrogen may not actually be an inert gas when considered physiologically (14).

Although a make-up supply of N_2 will be required to compensate for the loss of this gas through inherent leakage from the cabin, it is believed that system complexity and reliability need not suffer. A suitable O_2 partial pressure sensor and a total pressure regulator can be integrated into an automatic control system. If necessary, the system could be monitored and controlled manually by the flight personnel. Most likely, the system should be automatically controlled with flight personnel provided with a proper display of information and a manual override capability. A simple schematic of such a two-gas system is presented in Figure 1.

The technique employed in this system envisions primary regulation of the oxygen partial pressure, with secondary control of the total cabin pressure. Make-up gas is provided to the cabin from either oxygen or nitrogen supply sources. When an under-pressure of oxygen is sensed, the oxygen partial pressure sensor, in addition to providing a demand on the oxygen supply, will cause the nitrogen flow control valve to close, preventing the flow of nitrogen. If there is a deficiency in the total cabin pressure while the partial pressure of oxygen is within control limits (160 ± 5 mm Hg), nitrogen will be fed into the cabin to re-establish the desired total pressure and balance of oxygen and nitrogen in the atmosphere. Such a system provides a selective replenishment of the atmospheric gases because oxygen will be returned to the control level prior to the replenishment of the inert gas to control the total cabin pressure. This is as it should be.

This system would be used until such time that proven regenerative-type systems become available and would utilize stored supplies to provide the necessary oxygen and nitrogen. The manner in which these items are to be stored requires a careful evaluation of the methods available to accomplish this vital function. There are basically four:

1. Gaseous

2. Liquid

3. Chemical

4. Cryogenic

The storage of oxygen and nitrogen in the gaseous state, while necessitating the use of comparatively large, heavy containers, has two desirable features which are noted below:

1. Storage is in its usable form which reduces system complexity and improves reliability.

2. It has a relatively long storage life without loss.

While the use of liquid or cryogenic storage techniques offers substantial volume reductions, the added complexity results in a net increase in the weight for the shorter duration missions. In addition, these systems do not have the proven operational reliability which is associated with the gaseous storage systems.

Chemical storage of oxygen in the form of superoxides, and similar forms, is of extreme interest because of the dual capability for liberating oxygen while absorbing CO_2 and H_2O. Presently, this technique does not provide a balanced oxygen output. An additional capability is thus required for removing carbon dioxide and water in conjunction with a make-up supply of nitrogen to maintain the desired balance of partial pressures. The authors believe that the storage of oxygen and nitrogen in gaseous form offers the most satisfactory solution to our short-range man-in-space program requirements.

The control of carbon dioxide, water vapor, and trace contaminants within a sealed cabin system has been the subject of extensive study and developmental effort. These studies and developments have been directed, to a large degree, to the provision of a suitable system for use on our long-range programs, and, in this context, are most worthwhile. However, the truth is, we do not have a single suitable proven system for our imminent short-range requirements. The problems inherent in the development of fully regenerative techniques also indicate a considerable time lapse before the evolution of hardened systems. Our astronauts cannot fly with theoretical systems. It seems timely to survey the techniques which promise a proven capability within a reasonable period of time, and then to concentrate our major efforts within this sphere of development to meet the requirements of our short-range goals.

Probably the greatest problem associated with the development of a carbon dioxide, water, and trace contaminant scrubbing system is that of reduced gravity, or weightlessness. It is possible, through the use of a minimum amount of electrical energy, to remove carbon dioxide, water and trace contaminants by chemical absorption techniques. However, these processes normally produce by-products in a liquid phase. Separation of the gas and liquid phases is a major problem. The phase separation techniques which are easiest to apply normally require a large amount of electrical power -- power which will not be available during our short-range efforts. The technique for separating water vapor from air utilized on "Project Mercury" (22) is a rather crude, inefficient method. Although it has met the relatively simple Mercury requirement, considerable improvement is essential to meet the more stringent requirements of future systems which are included here as short-range goals. This is easier said than done, and immediate work is required.

We believe that it may be fruitful to consider the reactivation of the water and carbon dioxide scrubbing systems by vacuum desorption alone, without any attempt to reclaim these metabolic products. Arnoldi (1) has reported on a study considering vacuum desorption. The system which seems to be most practical for this purpose would, in addition to removing CO_2 and H_2O, provide some degree of

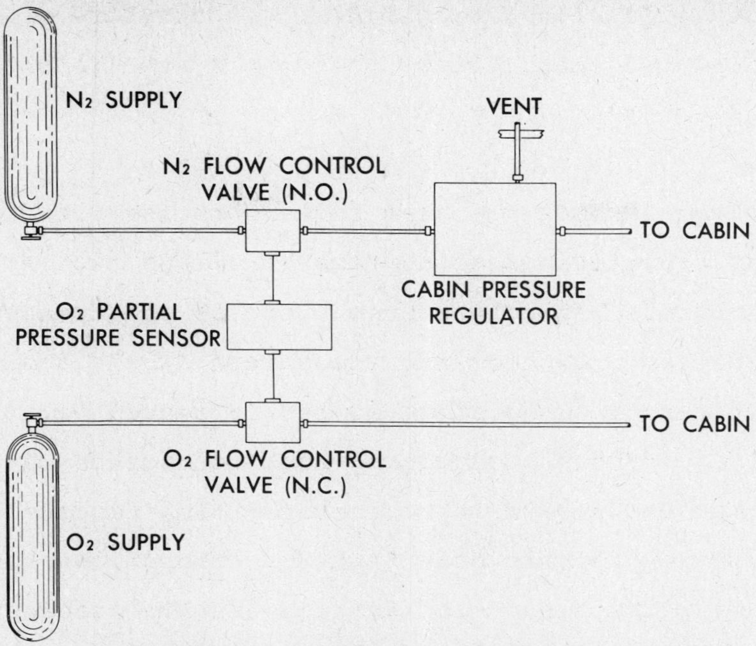

Fig. 1 Two-Gas Sealed Cabin Atmosphere Control System

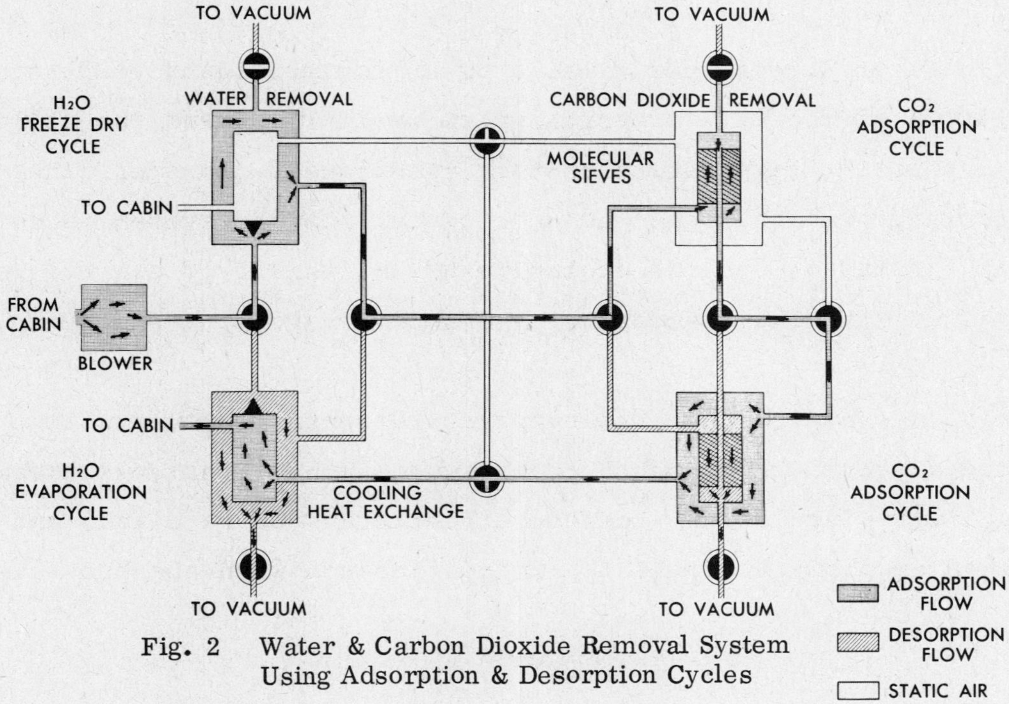

Fig. 2 Water & Carbon Dioxide Removal System Using Adsorption & Desorption Cycles

temperature stabilization with a minimum expenditure of electrical energy. A simple schematic diagram of the system is shown in Figure 2.

This system utilizes two heat exchangers for freezing the water from the circulating cabin atmosphere, and two carbon dioxide scrubbing chambers. During operation, one of the freeze-out heat exchangers removes water from the atmosphere, while the second is regenerated by vacuum evaporation of the collected water. Only one of the CO_2 adsorbers, containing zeolite, scrubs carbon dioxide at any one time; the second is regenerated simultaneously by vacuum desorption. The exothermic heat from the adsorption process will be used to assist the desorption and evaporation processes, providing some degree of cooling of the purified air as it is returned to the cabin.

3. Temperatures and Humidities

The limitations in available power for providing temperature regulation of the life support system would recommend the utilization of passive temperature control techniques. However, the variability of heat inputs to and losses from the system is sufficiently great to preclude exclusive use of this type control in accordance with the discussion by Pinc (21).

It is expected that the temperature control system will utilize a hybrid-type control technique, such as variable thermal conductivity insulation (to reduce heat transfer rate into and from the life support environment), in conjunction with the provision of variable absorption/emission characteristics of the external surfaces of the vehicle and a minimum of stored thermal energy. Attitude stabilization and control would be utilized during the incident solar radiation phase of the flight to present, by preferential orientation, contrasting absorption/emission characteristics to the sources and sinks of thermal radiation. Variations in the thermal conductivity across the walls of the vehicle may be achieved by inflation or deflation of the insulating barrier.

Extremely long periods of operation in the earth or lunar shadow may require supplemental heating from stored thermal energy sources and, as presented earlier, the simple use of protective clothing is considered quite acceptable as a partial practical solution to the problem of temperatures well below the comfort zone. It is expected that the techniques discussed above can provide reasonably satisfactory control of the system temperatures with minimum payload penalties.

Use of a sea level pressure for a manned space cabin is probably also helpful in regulating the interior environmental temperatures. This is because the cabin atmosphere itself is the means by which heat is lost to the heat sink of space and gained from the major heat sources, such as the sun and earth. This heat transfer process is dependent on the density, thus would change in direct relation to a change in pressure.

During the re-entry phase of the mission, the kinetic energy of the system must be dissipated over a very limited time interval. This results in extremely high loads and thermal stresses being imparted to the system. If the energy dissipation period is increased, some of these factors will be mitigated. Modulated drag techniques may be the most satisfactory solution to this problem. These techniques would be used in conjunction with suitable ablating materials for minimizing temperature increases.

SUMMARY

In conclusion, let us summarize. In order to meet the short-range goals (during this decade) of a national manned space flight program outlined by President Kennedy, we will be forced to accept risks to the flight personnel and our national prestige. These risks might be less if it were not for the time element. Ultimately, we will develop the capabilities required to optimize the probability of success. However, let us be practical and utilize all available data for an appropriate evaluation before determining what calculated risks we must accept. It is possible to fail in our efforts, but let us not invite failure by making unnecessary biological compromises for reasons of engineering expediency or to provide inadequate solutions to low probability situations. Let us also not concentrate our efforts too far in the future, because we also need solutions to the immediate needs of our programs. It is within our current technological capability to provide the necessary hardware. Let us concentrate on practical technology for our short-range goals while continuing parallel research which will contribute to long-range system requirements, for goals in a program not yet specified.

ACKNOWLEDGEMENT

Appreciation is expressed to Dr. Norman Lee Barr and Dr. Min-Shuey Lin, who reviewed this paper with helpful comments and suggestions.

REFERENCES

1. W. E. Arnoldi, "Regenerable Solid Adsorbent Systems for Carbon Dioxide Control," WADD Technical Report 60-574. Closed Circuit Respiratory Systems Symposium. August, 1960.

2. R. G. Bartlett, Jr., "Problems of Nitrogen-Free and Carbon Dioxide Rich Extraterrestrial Atmospheres," WADD Technical Report 60-574. Closed Circuit Respiratory Systems Symposium. August, 1960.

3. E. E. Beson, "Design Considerations of a Balloon Borne Pressurized Capsule for High Altitude Bailout Study," Journal of Aviation Medicine, July, 1958.

4. J. L. Biedler, "Furthering Basic Biological Knowledge Through Space Research," Journal of Astronautical Sciences, Vol. 8, No. 3, Fall 1961.

5. D. L. Foster, M. L. Lewis, R. M. Enderson, W. L. Murray and E. F. Lewis, "Strato-Lab High #3, Post Flight Report," Report No. 1246-R, Winzen Research Inc., 25 May 1959.

6. D. L. Foster, "Strato-Lab High #2, Post Flight Report," Report No. 1266-R, Winzen Research Inc., 19 January 1961.

7. C. F. Gell, "Breathing Oxygen," Chapter 10, Aerospace Medicine, Editor, H. G. Armstrong. The Williams & Wilkins Company, Baltimore, 1961.

8. A. L. Hall and R. J. Martin, "Prolonged Exposure in the Navy Full Pressure Suit at 'Space Equivalent' Altitudes," Aerospace Medicine, February, 1960.

9. Heating, Ventilating, Air Conditioning Guide. American Society of Heating and Air-Conditioning Engineers, Inc., New York, 1958.

10. J. L. Heldenbrand, "Selection of the Closed Atmosphere," WADD Technical Report 60-574. Closed Circuit Respiratory Systems Symposium. August, 1960.

11. G. W. Hoover, "The Man-Machine System in Space Vehicles," Advances in Astronautical Sciences, Vol. 4. Plenum Press, Inc., New York, 1959.

12. S. L. Jacobson, "Engineering of the Sealed Cabin Atmosphere Control System," Aerospace Medicine, May, 1960.

13. J. E. Jannsen, "Thermal Comfort in Space Vehicles," ASME Paper 59-A-207, 1959.

14. E. B. Konecci, "Human Factors and Bioastronautics," Astronautics, December, 1961.

15. C. J. Lambertson, "A Philosophy of Extremes for the Gaseous Environment of Manned Closed Ecological Systems," Proceedings of the Manned Space Stations Symposium, 1960.

16. D. E. Langdon and G. E. Reynolds, "Post Flight Respiratory Symptoms Associated with 100 Percent Oxygen and G-Forces," Aerospace Medicine, August, 1961.

17. H. H. Lettau, "A Theoretical Model of Temperature Variations at the Surface of an Orbiting Satellite," Journal of Geophysical Research, November, 1961.

18. U. C. Luft, "Altitude Sickness," Chapter 9, Aerospace Medicine, Ed., H. G. Armstrong. The Williams & Wilkins Co., Baltimore, 1961.

19. Man-High III, USAF Manned Balloon Flight into the Stratosphere, AFMDC-TR-60-16, Holloman Air Force Base, New Mexico, April, 1961.

20. Naval Aviation News, August, 1961.

21. B. W. Pinc, "Thermal Management Techniques in Bioastronautic Packages: A Review," ARS Paper 2109-61, October, 1961.

22. W. S. Reed, "System Controls Capsule Environment," Aviation Week and Space Technology, July 4, 1960.

23. M. D. Ross and M. L. Lewis, "The Role of Manned Balloons in the Exploration of Space," Aero/Space Engineering, August, 1958.

24. M. D. Ross, "Reactions of a Balloon Crew in a Controlled Environment," Journal of Aviation Medicine, May, 1959.

25. M. D. Ross, "A Comparison of Artificial Environments Used in Sealed Cabins During Flights into the Stratosphere," Advances in Astronautical Sciences, Vol. 4. Plenum Press, Inc., New York, 1959.

26. D. G. Simons and E. R. Archibald, "Selection of a Sealed Cabin Atmosphere," Journal of Aviation Medicine, May, 1958.

27. F. E. Snapp and H. F. Adler, "Oxygen Toxicity," Report J U2505, School of Aviation Medicine, AF, Randolph Field, Texas, 1948.

28. H. Strughold, "Space Medicine," Chapter 31, Aerospace Medicine, Ed., H. G. Armstrong. The Williams & Wilkins Co., Baltimore, 1961.

29. R. A. Stubbs, "Some Engineering Considerations for the Manned Orbiting Vehicle," Canadian Aeronautical Journal, November, 1960.

30. A. B. Thompson and C. F. Gell, "Meteoroids as a Hazard in Space Flight, A Survey of Present Information," Space Flight Report to the Nation, American Rocket Society, October 1961.

31. J. N. Waggoner and J. L. Mason, "Goal: Safety and Less Duplication," Missiles and Rockets, December 11, 1961.

32. B. Welinski and D. L. Foster, "Strato-Lab High #1, Post Flight Report of System Development," Report No. 1248-R, Winzen Research Inc., 1 February 1959.

33. G. B. Whisenhunt, Jr., "Living in Space," Industrial Research, June-July, 1961.

34. O. C. Winzen, "Strato-Lab High #4, Post Flight Report," Report No. 1262-R, Winzen Research Inc., 23 August 1961.

A CURRENT SURVEY OF SPACE RADIATION HAZARDS TO ASTRONAUTS

Richard Madey*

The most penetrating radiations in space consist of the energetic particles in the cosmic radiations of solar and non-solar origin and in the geomagnetically trapped radiation. This survey reviews the composition, location, intensity, angular and energy distributions, temporal variations, and interrelationships of energetic particles in space. The description in this paper of the radiation environment emphasizes the background necessary to estimate and to protect man against the natural radiation hazards in space.

Estimates of natural radiation hazards to manned space travel are given on the basis of the best measurements currently available. The uncertainty arising from extrapolation of spectral data is avoided by restriction of the calculations of the absorbed dose rates to the observed energy regions. Similarly, a restriction of calculations of the absorbed dose arising from solar flare protons to observed time intervals avoids the uncertainty resulting from extrapolation of the observed time dependence to unknown regions.

INTRODUCTION

Although data on space radiation are still relatively meager, our present knowledge of the penetrating radiations in space indicates that natural radiation hazards will impose limitations on manned space travel. The most penetrating radiations in space consist of the energetic particles in the geomagnetically trapped radiation and in the cosmic radiations of solar and non-solar origin. These energetic particles constitute the important natural radiation hazards to astronauts. The objectives of this paper are twofold: first, to review the properties of energetic particles in space, as currently known; and then to estimate natural radiation hazards to astronauts on the basis of the best measurements currently available.

---

* Chief Staff Scientist, Scientific Research Staff, Republic Aviation Corporation, Farmingdale, L. I., New York.

The discussion of space radiation in this paper treats briefly the nature, the spatial and energy distributions, the intensity, the temporal variations, and the interrelations of energetic particles in space. The most energetic particles in space are the charged particles that constitute the cosmic radiation. Cosmic rays consist primarily of hydrogen nuclei plus a small percentage of nuclei of heavier atoms. Cosmic ray energies from about 10^8 electron volts[(i)] up to 10^{19} electron volts have been observed at the earth. These extremely energetic particles originate outside the solar system. Similar energetic particles, but of much lower energy, are emitted from the neighborhood of the sun. Energetic particles of solar origin are called "solar" cosmic rays. Those of non-solar origin are called "galactic" cosmic rays to indicate that they originate in some distant part of our galaxy, or perhaps outside our galaxy. The energies of protons from the sun have been observed in the neighborhood of the earth from about 10^7 electron volts to 10^{10} electron volts. Protons in about this same energy range have been found also to be temporarily "trapped" in the magnetic field of the earth. In addition to these trapped protons, the geomagnetically trapped radiation consists of lower energy electrons. Measurements thus far have shown electron energies from about 10^4 electron volts to about 10^7 electron volts. The geomagnetically trapped radiation is often referred to as Van Allen radiation in honor of James Van Allen, who led the research team that first detected the trapped radiation.

Still lower energy protons and electrons of solar origin constitute a plasma stream or "solar wind." The solar wind consists of protons and electrons moving through space at the same velocity. The kinetic energy of the protons in this stream varies between a few hundred and a few thousand electron volts.

As we move lower on the energy scale, we find that the radiation environment in space comprises x-rays, broad-band radio noise, and ultraviolet, optical, and thermal radiation. In this paper we shall not discuss these or other low energy radiations such as solar wind. We shall focus our attention on radiations having energies above approximately 1 Mev. Thus we shall discuss only the most penetrating radiations in space, viz., the galactic cosmic rays, the solar cosmic rays, and the Van Allen radiation. These energetic particles constitute the important natural radiation hazards to man in space. In our description of this radiation environment, we shall endeavor to develop the background necessary to estimate these natural radiation hazards in space and to protect man against them.

(i) An energy of one electron volt is equal to the energy acquired by an electron that is accelerated from rest through a potential difference of one volt.

GALACTIC COSMIC RAYS

Nature

Historically, our ideas about the nature of the primary cosmic radiation have undergone large changes. Originally, the electromagnetic effects occurring in the atmosphere were attributed to very penetrating extraterrestrial gamma rays. In 1928, J. Clay demonstrated that the intensity of cosmic rays on the earth's surface decreases with decreasing geomagnetic latitude. The change in intensity from the geomagnetic pole to the geomagnetic equator is about 10 to 14 percent at sea level. Thus, since the incoming radiation is affected by the earth's magnetic field, the discovery of the latitude effect by Clay proved that the primary radiation must consist of particles carrying an electric charge. In 1938, T. H. Johnson(1) observed that more particles reach the earth from the west than from the east. After the motion of charged particles in the magnetic dipole field of the earth had been studied, mainly by Stoermer, (2) Lemaitre, and Vallarta, it was concluded from the observed east-west asymmetry that most if not all primary particles are positively charged. To determine whether the positively charged cosmic ray primaries consisted of protons or positrons, Schein et al. (3) in 1941 showed that most of the particles near the top of the atmosphere do not initiate the cascade showers typical of high energy electronic radiation. A recent experiment(4) showed that the primary electron flux above 500 Mev is 3 $\pm$1 percent of the proton flux on May 12, 1960 at a geomagnetic latitude of 55° N; the reported electron flux above 500 Mev is 32 $\pm$10 electrons/meter^2-sec-steradian. In 1948, nuclear emulsion and cloud chamber measurements by Freier et al. (5) indicated the presence of primary helium nuclei and heavier nuclei. Many balloon and rocket measurements have since been made of the composition and energy spectrum of the various components of the primary radiation.

Composition

Primary cosmic rays in the vicinity of the earth consist mainly of protons and alpha particles. Nuclei of all other elements together make up less than 2 percent of the cosmic radiation in the vicinity of the earth. Table I compares the relative abundance of nuclei in the primary cosmic radiation with the relative abundance of the elements in the universe, as obtained from meteorites and from spectroscopic analysis of starlight. It appears that the galactic cosmic rays represent a distribution of the elements in the universe. Although protons make up about 85 percent of the incoming primary particles, the number of nucleons carried by the alpha particles and the heavy nuclei is almost equal to the number of primary protons. The particle fractions and the nucleon fractions are listed in Table II.

The data in Table I on the composition of the primary cosmic radiation is valid for energies up to about 10^{13} ev/nucleon. The question of the composition of cosmic radiation

TABLE I

ABUNDANCES DETERMINED FROM COSMIC RAY RATIOS COMPARED WITH "UNIVERSAL ABUNDANCES"

Group	H	He	Light Nuclei Li, Be, B	Medium Nuclei C, N, O, F	Heavy Nuclei	"Fe" Group	Very Heavy Nuclei
	$Z=1$	$Z=2$	$3 \leq Z \leq 5$	$6 \leq Z \leq 9$	$Z \geq 10$	$23 \leq Z \leq 30$	$Z \geq 30$
Relative abundance in cosmic rays (E.P. Ney)	100	15.5	0.24	1.20	0.4	0.1	$\lesssim 10^{-4}$
"Universal" abundance (Suess and Urey)	100	7.7	$\approx 10^{-6}$	0.20	0.03	0.003	$\approx 10^{-6}$

TABLE II

COMPOSITION OF PRIMARY COSMIC RADIATION

Nuclear Particle Radiation

Z	Particle	Particle Fraction	Nucleon Fraction
0	neutron	0	0
1	proton	0.85	$\gtrsim 1/2$
2	alpha	0.13	$\lesssim 1/2$ (alpha and heavy nuclei together)
>2	heavy nuclei	0.016	

Electromagnetic Radiation

Particle	Particle Fraction	Energy Flux
electron (>500 Mev)	0.03	
photon		<0.001

at energies greater than 10^{13} ev/nucleon and up to the highest observed energies of about 10^{19} ev remains open. It remains for experiments to clarify whether the particles with highest energies are mainly protons or heavy nuclei.

A small flux of primary cosmic ray electrons has been found[4, 6] recently in the vicinity of the earth. Earl has reported that the electron flux above 500 Mev is 3 $\pm$1 percent of the primary proton flux on May 12, 1960.

Another essential feature of the primary cosmic radiation that reaches the earth is that photons and neutrons have not been found. Rocket experiments by Perlow and Kissinger[7] have set an upper limit of 0.1 percent to the energy flux of photons above the atmosphere. Jones[8] placed an upper limit of 0.000 $\pm$0.034 photons/cm^2-sec-sterad for the vertical flux of gamma rays in the energy region about 0.5 Mev at altitudes between 5.4 and 6.0 gm/cm^2 atmospheric depth.

Spectra

The energy spectrum of primary protons and of other nuclei may be represented over limited energy intervals by a power law of the form

$$n(E) = CE^{-\gamma} \tag{1}$$

where $n(E)\,dE$ is the differential intensity between E and $E + dE$, C and γ are constants, and E is the total energy per nucleon. For protons,

$$E = E_k + Mc^2 = E_k + 0.938 \text{ Bev} \tag{2}$$

where E_k is the proton kinetic energy.

The integral intensity, which gives the number of particles with energy greater than E, is

$$N(>E) = \int_E^{\infty} n(E)\, dE = \frac{C}{\gamma-1} E^{-(\gamma-1)} . \tag{3}$$

For protons with kinetic energy in the range $0.5 \leq E_k \leq 10$ Bev, Peters[9] gives

$$N(>E_k \text{ in Bev}) = \frac{0.38}{1 + E_k} \quad (\text{protons/cm}^2\text{-sec-steradian}). \tag{4}$$

Slightly different expressions are to be found in the literature.[10-13] Based upon Eq. (4), the values of the constants C and γ for protons with kinetic energy in the range $0.5 \leq E_k \leq 10$ Bev are $\gamma = 2$ and $C = 0.38$ p/cm^2-sec-steradian-Bev$^{-(\gamma-1)}$.

For heavy nuclei in the range $0.35 \leq E_k \leq 10\text{-}20$ Bev/nucleon, Peters[9] gives

$$N\,(>E_k \text{ in Bev/nucleon}) = \frac{K}{(1 + E_k)^{1.2}} \text{ (particles/m}^2\text{-sec-steradian)} \tag{5}$$

where

$$K_\alpha = 380; \; K_{CNO} = 20; \; K_{Z\geq 9} = 6.$$

These examples illustrate the fact that the spectra of protons and other nuclei in this energy interval ($0.35 \leq E_k \leq 10\text{-}20$ Bev/nucleon) have approximately the same shape; that is, for protons $\gamma = 2$, and for heavier nuclei $\gamma = 2.2$. The parameter γ increases slowly with increasing energy; for $5 \times 10^{10} < E < 10^{12}$ ev/nucleon, $\gamma = 2.5$, and for $10^{13} < E < 10^{18}$ ev/nucleon, $\gamma = 2.7 - 3.0$. (As indicated earlier, the composition of the primary radiation is not known for $E > 10^{13}$ ev/nucleon.) Figure 1 plots the integral energy spectra of the primary cosmic rays versus kinetic energy in Bev per nucleon for the kinetic energy range $0.35 \leq E_k \leq 20$ Bev/nucleon. The high energy proton spectrum is shown in Fig. 2.

At low energies ($E_k < 0.35 - .50$ Bev/nucleon), the primary spectrum is very poorly known. The low energy limit to the galactic cosmic rays is perhaps less than 100 Mev. It is still an open question whether these low energy particles are absent in the source of cosmic rays, or whether these particles are screened out by a magnetic field or by some absorption process acting within the limits of the solar system.

We can estimate the concentration of particles in space from the integral spectrum: The omnidirectional flux[ii] of protons with kinetic energies greater than 1 Bev is obtained from Eq. (4):

$$J_o\,(E_k > 1 \text{ Bev}) = 4\pi N\,(E_k > 1 \text{ Bev}) = 2.4 \text{ protons/cm}^2\text{-sec} \tag{6}$$

The omnidirectional intensity of all particles in the vicinity of the earth is 3.7 particles/cm^2-sec[14, 15]. Thus, in the vicinity of the earth, the density of cosmic ray primaries with energies greater than 1 Bev is approximately

$$\rho_n\,(E_k > 1 \text{ Bev}) \approx J_o/c \approx 1.2 \times 10^{-10} \text{ protons/cc} \tag{7}$$

(ii) If $n(E, \Omega)dE$ represents the number of protons in the incident spectrum with energies between E and $E + dE$, crossing a unit area per unit time per unit solid angle in a direction Ω normal to the unit area, then $n(E, \Omega)$ is the unidirectional flux and has units of protons/sec-cm^2-steradian-Mev. The omnidirectional differential energy flux $j_o(E)$ is defined as

$$j_o \equiv 4\pi\,\bar{n}(E)$$

where $\bar{n}(E)$ is the unidirectional flux averaged over a sphere.

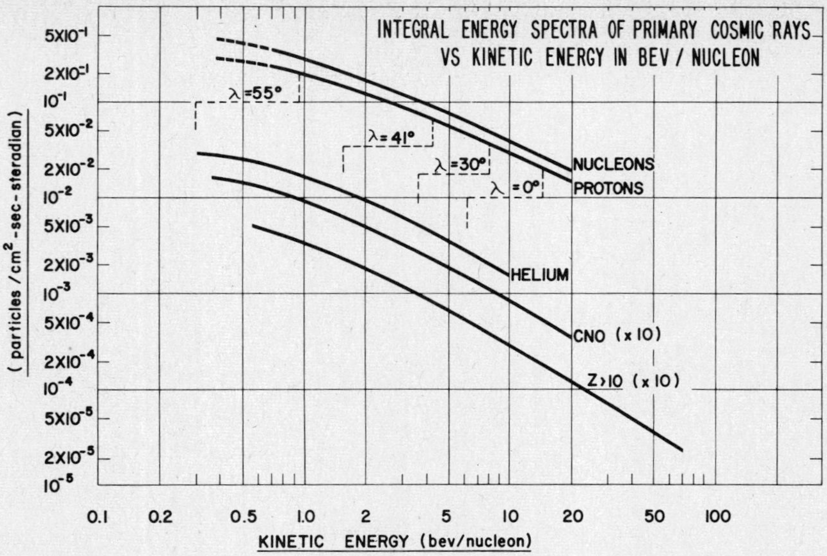

Fig. 1 Integral Energy Spectra of Primary Cosmic Rays vs Kinetic Energy in Bev/Nucleon

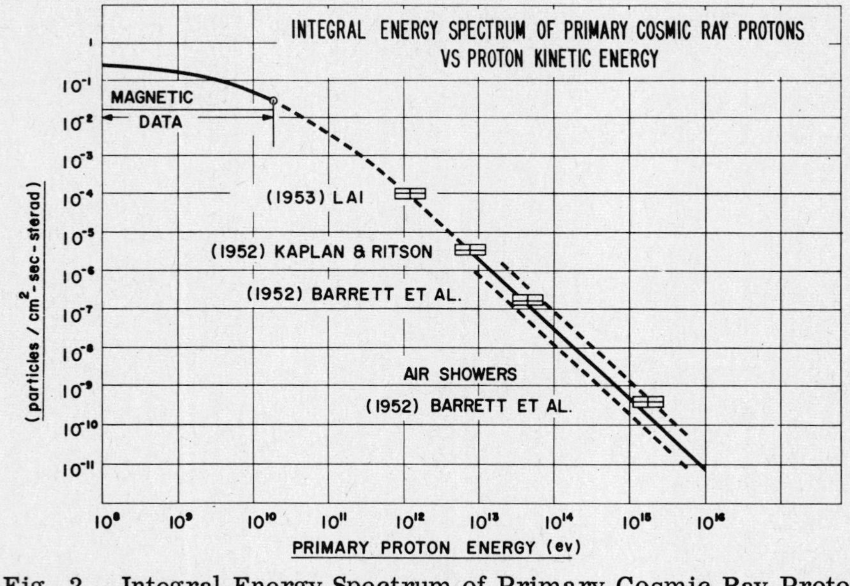

Fig. 2 Integral Energy Spectrum of Primary Cosmic Ray Protons vs Proton Kinetic Energy

where we have assumed that the particle velocity $v = c = 3 \times 10^{10}$ cm/sec for all primaries.

It is interesting to estimate the energy flux and the energy density of cosmic radiation:

$$W \text{ (energy flux)} = \int_0^\infty E\, n(E)\, dE \tag{8}$$

Using Eqs. (1)-(4), we may estimate W as follows:

$$W\ (E_k > 0.5 \text{ Bev}) = 0.38 \int_{1.5}^{11} \frac{dE}{E} = 0.75 \text{ Bev/cm}^2\text{-sec-steradian} \tag{9}$$

This estimate for the energy flux can be compared with measurements discussed by Neher.[16] For 60° N geomagnetic latitude, where the minimum kinetic energy at the vertical for a proton is 0.4 Bev, the measured energy flux is 0.774 Bev/cm^2-sec-steradian at the vertical.

The energy density can be estimated as follows:

$$U = \frac{4\pi}{c} W \approx 0.3 \text{ ev/cc} \tag{10}$$

The energy density of cosmic radiation is of the same order as the density of other forms of energy within the galaxy, such as the energy density of starlight, the density of kinetic energy of turbulence in interstellar space, the density of magnetic energy, and the thermal energy density in ionized regions of interstellar space.[iii]

The average energy of the primary cosmic radiation in the vicinity of the earth is about 4 Bev/nucleon.

Origin

Whether the origin of cosmic rays is galactic or extragalactic is still unsettled. A satisfactory theory of the acceleration of cosmic rays should explain the observed composition and energy spectra of cosmic ray nuclei.

(iii) Estimates of the energy density of various forms of energy within galaxy:

$U_{turbulence} = \frac{1}{2} \rho u^2 \approx 1$ ev/cc ($\rho \approx 10^{-24}$ g/cc, $u \approx 10^6$ cm/sec)

$U_{magnetic} = B^2/8\pi \approx 1$ ev/cc ($B \approx 3 \times 10^{-6} - 10^{-5}$ gauss)

$U_{thermal} = n\,k\,T \approx 1$ ev/cc ($n \approx 1$ ion/cc, $T \approx 10^4\ {}^\circ$k)

The great energies of at least 10^{19} ev are thought to arise from a slow acceleration in the galaxy.[17] In the process suggested by Fermi,[18] the particles derive their energy from the kinetic energy of interstellar matter by colliding with magnetic fields assumed to be attached to highly ionized gaseous clouds traveling in our galaxy. Particles are lost by nuclear collisions with the interstellar material. Based upon the assumption that the cross section or probability for nuclear collisions in interstellar matter is proportional to the geometric area of the nucleus and independent of its energy, the large nuclei would spend less time in the accelerating process and fewer should reach very high energies. Thus, elimination of particles by nuclear collisions will reduce the high energy intensity of a heavy nucleus relative to a light nucleus.

An electromagnetic accelerating process, which produces cosmic ray particles in a time short compared to the time between nuclear collisions in interstellar matter, would produce the same velocity spectrum for all heavy nuclei, provided that the acceleration takes place after the heavy nuclei have been completely stripped of electrons. In this case, we note that the charge-to-mass ratio of all nuclei from helium to iron is very nearly the same once they are completely stripped of electrons.

Similarity of spectra for different nuclear species suggests that cosmic rays are accelerated in large bundles of plasma or in shock waves, rather than as individual particles in a synchrotron or cyclotron. A particle accelerator type mechanism would separate the various values of e/m.

Time Dependence

The intensity of the galactic cosmic radiation in the vicinity of the earth is essentially constant in time. The intensity for particles with energies greater than 10 Bev is constant to within a few tenths of 1 percent. The intensity for particles of all energies varies by a few percent per year.

Small periodic variations in the cosmic ray intensity occur with periods of 1 day, 27 days, and 11 years. The solar diurnal variation is a fraction of 1 percent. The 27-day recurrence of the order of several percent is associated with the solar rotational period of 27 days. The 11-year variation is a result of magnetic activity associated with the 11-year sun spot cycle. This 11-year variation is only about 2 percent for observatory measurements at an atmospheric depth greater than 700 gm/cm^2 (10, 000-foot altitude), but it is a factor of approximately two near the top of the atmosphere above the north geomagnetic pole. The intensity decreases with increasing solar activity; that is, the intensity at a period of maximum solar activity is about one-half the intensity during a period of minimum solar activity.

In addition to these small periodic variations in intensity, there are world-wide decreases which usually accompany magnetic storms. These intensity decreases, first described by Forbush,[19] are quite striking and in some cases as large as 30 percent. On the basis of sea level measurements, Forbush concluded that the cosmic ray intensity decreases as solar activity increases. The Forbush-type decreases are associated with large solar eruptions. Presumably, a solar plasma carries a magnetic field which deflects away the galactic cosmic rays. The Pioneer V space probe observed[20] relatively large magnetic fields of the order of 50 γ (1 $\gamma = 10^{-5}$ gauss) during a Forbush decrease.

Angular Distribution

At large distances from the earth, the galactic cosmic radiation is isotropic (i.e., the intensity is equal from all directions). Near the earth, a slight degree of anistropy has been observed. If the degree of anisotropy δ is defined as

$$\delta \equiv \Delta F/F \tag{11}$$

where F is the cosmic ray flux in any chosen direction, and $\Delta F = F_{max} - F_{min}$ is the difference between the maximum and minimum fluxes in any direction, then, according to data given by Cocconi,[21] $\delta < 10^{-3}$ in the energy range up to 10^{15} ev and $\delta < 0.1$ for an average energy of 10^{17} ev.

SOLAR COSMIC RAYS

Phenomenology

The term "solar flare" refers to eruptions on the sun that result in the release of ionized gas clouds (plasma), energetic particles (mostly protons), and other radiations (e.g., optical and ultraviolet radiations, x-rays, and broad-band radio noise).[iv] Solar flares are separated into classes according to the brightness and the size of the optical flare observed on the sun's disk. In a flare of class 1 or importance 1, the flare region occupies an area less than 1×10^{-4} of the solar disk; in a class 2 flare, 2×10^{-4}; in a class 3 flare, 3×10^{-4}; and in a class 3^+, the area occupied is greater than 3×10^{-4} of the solar disk. Historically, from 1942 to 1956, the effects of energetic particles from the sun were detected at the surface of the earth on a world-wide bases in five events.[v] The largest and best known of these five events is the one of February 23, 1956. The characteristics of the

---

(iv) Type IV radio noise extends from the ionospheric cutoff of about 18 megacycles up into the 2000-megacycle region.

(v) The dates of these five high energy cases are February 28 and March 7, 1942; July 25, 1946; November 19, 1949; and February 23, 1956.

primary particles are consistent with protons. Since the cosmic ray intensity at sea level increased even at the equator, the proton spectrum extended beyond 15 Bev, which is the magnetic cutoff at the equator. These very energetic events are termed "giant" or relativistic cosmic ray flares.

The giant flare of February 23, 1956, is also the first known case of cosmic rays producing ionospheric absorption of radio waves.[22] The solar cosmic rays ionize the atmosphere thereby increasing the electron density of the D-layer of the ionosphere, which in turn increases the absorption of radio waves. This disruption of radio communications at earth occurs most frequently in the polar regions which are more accessible to low energy solar cosmic rays. These events are called "polar blackouts" or "polar cap absorptions" because receivers measuring the level of radio noise from the galaxy suddenly become quiet when the ionizing solar cosmic rays abruptly close the atmospheric "window" through which the galactic radio waves reach the earth. A world-wide network of sensitive radio receivers was set up during the International Geophysical Year to keep a continuous record of the level of cosmic radio noise. By measuring the varying opacity of the atmospheric radio window, these receivers furnish a reliable index to the ionization of the atmosphere, which is related to the flux of protons. These radio receivers are called "Relative Ionospheric Opacity Meters" or riometers.

Since the International Geophysical Year in 1957, a much more numerous class of cosmic ray flares was detected at high altitudes by balloons, rockets, satellites, and space probes. Energetic protons were definitely identified from the solar flare of August 22, 1958, both by balloon[23] and satellite[24] observations. Since the kinetic energies of most of the protons in these cosmic ray events are substantially less than the proton rest energy of 938 Mev, these events are termed nonrelativistic cosmic ray flares to distinguish them from the relativistic "giant" flares observed at sea level. This distinction is probably an artificial one; it represents a historical separation of the early work which was limited to solar cosmic rays detectable at sea level from those which are now easily detected at high altitudes or outside of the atmosphere. During the recent period of high solar activity, these low energy events have occurred about once per month on the average, in contrast to the high energy flares that occur about once every four years. The nonrelativistic cosmic ray events are observed to be incident over the north polar region of the earth north of about 60° geomagnitude latitude, as allowed by theory for charged particles of the observed energy. If the solar flare is accompanied by solar plasma that causes a magnetic storm and perturbs the magnetic field of the earth when the particles are incident, these particles may enter the earth's atmosphere at normally forbidden latitudes below the polar region. In this case, the nonrelativistic protons may be incident as far south as 50° N geomagnetic latitude.

Production[25-30]

The processes of injection, acceleration, and propagation of protons from the sun are not established; however, observed solar flare cosmic ray effects are in agreement with a simple model for the magnetic fields in the inner solar system. Experimental observations have led to the concept of the storage of cosmic ray flare particles in large scale, expanding magnetic loops attached to the sun.

The model has been illustrated schematically in the paper by Steljes, Carmichael, and McCracken[29] for the sequence of solar flares that occurred on November 10, 12, and 15, 1960. Figure 3 is a schematic diagram of the magnetic fields in the inner solar system assumed to exist at the time of visible flare on November 12, 1960. The magnetic loops, which constitute a "magnetic bottle" attached to the sun, are assumed to be produced by a gas cloud from a previous solar outburst. In this case, the region of ordered magnetic fields behind the leading edge of the gas cloud did not envelop the earth. The earth was situated in fairly weak magnetic fields that were the result of the general outflow of matter from the previously quiet sun. The solar cosmic rays injected into the magnetic bottle by the new solar outburst of November 12, 1960, travelled along the magnetic lines of force and reached points within the bottle very quickly. To have reached the earth, the solar cosmic rays must have diffused across the lines of force in the bottle, through the disordered fields at the leading edge of the cloud, and then through the weak magnetic field in which the earth was situated. This diffusion process increased the sun-to-earth time of flight, thereby slowing the rise time of the signal observed by sea level neutron monitors. Thus, while the earth is still outside the advancing gas cloud, this model predicts:

(1) The rise time of a solar cosmic ray increase should be relatively slow. (This is the case for the November 12th cosmic ray increase which resulted from a flare 10° west of the central solar meridian.

(2) The solar cosmic rays should arrive at the earth from a large solid angle (i.e., the radiation is relatively isotropic).

Figure 4 is a schematic diagram of the magnetic fields in the inner solar system assumed to exist at the time of the visible flare on November 15, 1960. In this case, the earth was inside a magnetic bottle formed by loops of force attached to the sun. Cosmic rays produced in the flare of November 15 gained quick access to the earth by spiraling down the lines of force in this ordered magnetic field. Thus, when the earth is within the magnetic bottle, this model predicts:

(1) The rise time of a solar cosmic ray increase should be relatively fast. (This is the case for the November 15th cosmic ray increase which resulted from a flare about 43° west of the central solar meridian.)

(2) The solar cosmic rays should arrive at the earth from a relatively small solid angle (i. e., the radiation should exhibit marked anisotropy).

In general, the rise time of a cosmic ray increase at the earth has been correlated with the location of the visible flare on the sun. A cosmic ray increase observed at the earth after a flare that is well to the west of the central solar meridian exhibits a rapid rise to maximum intensity, whereas the rise to maximum is slow if the flare occurs near or to the east of the central meridian.[10]

It has also been conjectured that the periodic fluctuations observed in the neutron monitors in this event are the result of magnetic mirror action; that is, an injected pulse of particles is first observed at the earth, then continues to travel along the loops back toward the sun, and returns to the earth after this particle bunch has mirrored near the sun.

Spectra

The first good spectral measurements with nuclear emulsions were obtained for the solar flare of May 10, 1959.[31] The omnidirectional differential energy spectrum measured 33 hours after the peak of the optical flare can be represented by the power law:

$$j_o = 4\pi CE^{-\gamma} \text{ p/cm}^2\text{-sec-Mev} \qquad (12)$$

where E is the kinetic energy. For protons with kinetic energies in the range from 110 Mev < E < 220 Mev, $\gamma = 4.8$ and $C = 2.86 \times 10^{10}$ protons/cm^2-sec-steradian-Mev$^{-(\gamma-1)}$.[vi]

Good measurements were also obtained[33] for the intense flare of July 15, 1959, at 28.8 hours after the peak of the optical flare. In this case, for 85 Mev < E < 300 Mev, $\gamma = 5.0$ and $C = 3.18 \times 10^{11}$ protons/cm^2-sec-steradian-Mev$^{-\gamma}$.[vi]

Note that the spectral measurements of solar protons did not extend below 110 Mev for the May 1959 flare and below 85 Mev for the July 1959 event. The steep energy dependence of the proton flux in the measured energy intervals demands an accurate knowledge of the spectrum at lower proton energies in order to make reliable calculations of the absorbed dose rate. Several shielding calculations[34-41] have already been made which extrapolate the flare spectra at the low energy end. The spread in the calculated shield requirements for protection against solar protons fluctuates wildly as a result of different extrapolations of the flare spectra.

(vi) In another paper[32] we have evaluated the coefficient C from the reported measurements.

Davis et al.[42] used sounding rockets to measure the proton energy spectrum down to 13 Mev at a maximum altitude of 130 kilometers and at 13.5 hours after the peak of the optical flare of September 3, 1960. The spectrum for this event flattened at the lower energies. Ogilvie, Bryant, and Davis[43] reported similar rocket measurements of proton spectra from the solar flare of November 12, 1960. From the reported values (viz., $\gamma - 1 = 1.7 \pm 0.2$ and $N(>2 \text{ Mev}) = 3 \times 10^4$ protons/cm^2-sec-steradian), the constants are $\gamma = 2.7$ and $C = 1.66 \times 10^5$ in the energy interval from 2 Mev to 100 Mev and at 5 hours and 27 minutes after the peak of the optical flare. Additional measurements pertaining to the November 1960 series of flares have not yet been fully reported.

Time Dependence

Solar cosmic rays reach the vicinity of the earth within approximately one hour after a major flare appears on the sun. The particles are first noted in the polar region as soon as 30 minutes after some prominent feature of a flare, by the observation of such phenomena as a burst of radio noise or a maximum in the optical radiation. The particle intensity perhaps reaches a peak within 10 to 15 minutes to a few hours. As mentioned earlier, the rise time of the increase has been correlated with the location of the visible flare on the sun's disk: Rapid rise times are associated with western flares, and slow rise times with eastern and central flares. After reaching peak intensity, the particle intensity decays rapidly in the environment of the earth in accordance with a time dependent power law, t^{-n} with $3/2 \leq n \leq 3$. For the solar flare of July 15, 1959, the time dependence of the particle intensity was observed[33] to follow a t^{-3} law for times $t \geq t_2 = 1.2$ days after the peak of the optical flare. Thus, the omnidirectional differential proton spectrum for this flare of July 15, 1959, was found[33] to be

$$j_o = \frac{4 \times 10^{12}}{E^5} \left(\frac{t_2}{t} \right)^3 \text{ protons/cm}^2\text{-sec-Mev} \tag{13}$$

where $t \geq t_2 = 1.2$ days.

Although many hours or days may elapse before the cosmic ray intensity returns to galactic background level, the steep decay of the particle flux with time shows that the main integrated dose occurs early in the event. Again, good measurements of the time dependence of the particle flux near the peak flux are necessary in order to make reliable calculations of the time integrated absorbed dose. The assumptions regarding the time dependence of the proton flux strongly affect the results of shielding calculations that have been reported to date.[34-41]

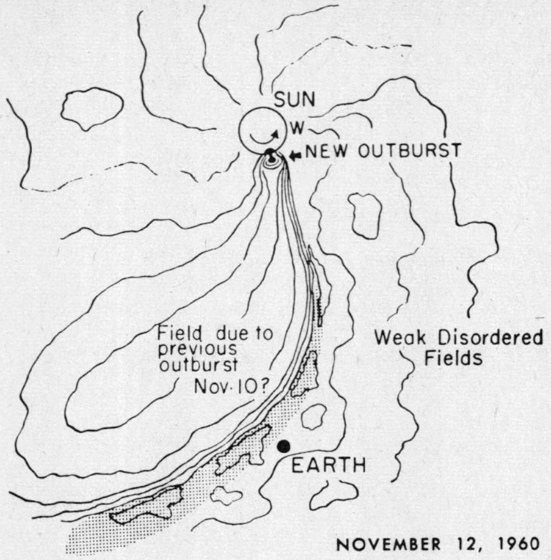

Fig. 3 Model for Magnetic Fields in the Inner Solar System at the Time of the Visible Flare on November 12, 1960

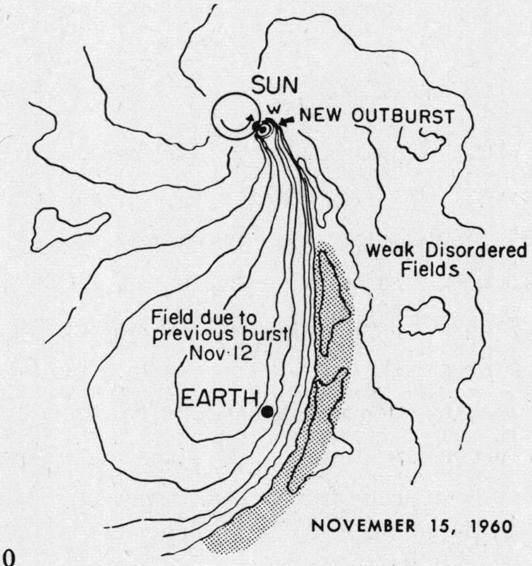

Fig. 4 Model for Magnetic Fields in the Inner Solar System at the Time of the Visible Flare on November 15, 1960

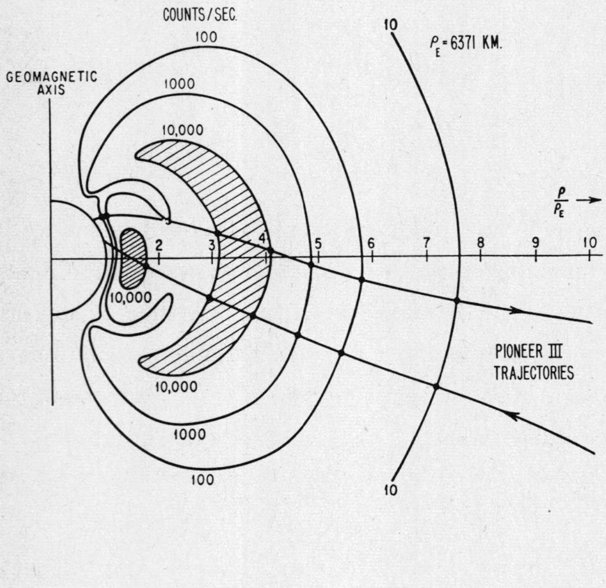

Fig. 5 Intensity Structure of the Terrestrial Radiation Belts

Composition

In addition to the solar protons, recent balloon, rocket and satellite observations[44-47] have revealed the presence of small flux of alpha particles and nuclei with charge $Z \geq 6$. The abundances are characteristic of solar origin. Although the intensities are higher, the energies are much lower than those of the galactic heavy primaries. The solar heavy particles have residual ranges in nuclear emulsions of less than one centimeter.

Angular Distribution

During the rise time of the solar cosmic ray particle flux, the angular distribution in the vicinity of the earth may exhibit marked anisotropy or may be relatively isotropic, depending upon the heliographic position of the flare, as discussed earlier. Observations of the particle flux during the decay phase are consistent with isotropy.

VAN ALLEN RADIATION

Location

The geomagnetically trapped radiation is located within two principal zones. Figure 5 has become a classical diagram which shows the intensity structure of the terrestrial radiation belts, as derived by Van Allen[48, 49] from Geiger counter observations in satellites Explorer IV and Pioneer III. Russian results[50] from Sputnik III and Mechta are in general agreement.

Composition and Tentative Fluxes[51-54]

Heart of Inner Zone: At an altitude of about 2200 miles on the magnetic equator, the composition and tentative fluxes are as follows:

Electrons:

Maximum unidirectional[vii] intensity

$$N\ (>20\ \mathrm{Kev})^{(viii)} \simeq 2 \times 10^{9}\ \beta/\mathrm{cm}^{2}\text{-sec-sterad}$$

$$N\ (>600\ \mathrm{Kev})^{(ix)} \simeq 1 \times 10^{7}\ \beta/\mathrm{cm}^{2}\text{-sec-sterad}$$

(vii) The omnidirectional intensity is about equal to the unidirectional intensity since the angular distribution is spread over a span of approximately 1 steradian.

(viii) Trustworthy to factor 10

(ix) Trustworthy to factor 2

Protons:

Omnidirectional intensity

$$J_o(>40\ \text{Mev})^{(ix)} \simeq 2 \times 10^4\ \text{p/cm}^2\text{-sec}$$

Heart of Outer Zone: At an altitude of about 9930 miles on the magnetic equator, the composition and tentative fluxes are as follows:

Electrons:

$$J_o(>40\ \text{Kev}) \approx 1 \times 10^8\ \beta/\text{cm}^2\text{-sec}$$

$$J_o(>2\ \text{Mev}) \approx 1 \times 10^5\ \beta/\text{cm}^2\text{-sec}$$

$$J_o(>5\ \text{Mev}) < 10^2\ \beta/\text{cm}^2\text{-sec}$$

Protons:

$$J_o(>60\ \text{Mev}) < 10^2\ \text{p/cm}^2\text{-sec}$$

$$J_o(<30\ \text{Mev})^{(x)} = ?$$

Results quoted by various investigators for the outer zone electron flux have differed markedly. The figures quoted here were reported by Van Allen in an oral paper presented in October 1961 at the New York meeting of the American Rocket Society. These electron flux figures are about 1000 times smaller than earlier numbers reported by Van Allen(51). The new flux values represent the results of later measurements with Explorer XII on September 5, 1961 by a more direct measurement technique. The earlier results obtained with space probes Pioneer III and Pioneer IV are subject to reinterpretation. Results quoted by other investigators are:

Arnoldy et al.(55) -- $J_o(>30\ \text{Kev}) \simeq 1.2 \times 10^{11}\ \beta/\text{cm}^2\text{-sec}$

Vernov and Chudakov(56) -- $J_o(>20\ \text{Kev}) \simeq 10^{10}\ \beta/\text{cm}^2\text{-sec}$

Lockheed Report NR-104(57) -- $J_o(>20\ \text{Kev}) \simeq 10^8\ \beta/\text{cm}^2\text{-sec}$

The electron flux value given by Vernov and Chudakov(56) was deduced from the energy flux measured by Mechta (Lunik I), on the assumption that the integral spectrum of electrons

---

(x) No significant information.

follows an E^{-5} dependence down to 20 Kev.

The Lockheed value of the electron flux is also based on the Russian measurement of the energy flux of electrons of 2×10^{11} ev/cm^2-sec-steradian, with electron energies greater than 50 Kev, but the Lockheed result(57) is based upon an electron spectrum, measured by Lockheed near the bottom of the outer zone,(58, 59) that appears to flatten out very sharply at energies below about 100 Kev.

The flux value given by Arnoldy et al.(55) represents a quiet-day figure obtained by Explorer III. The Lockheed authors(57) argue that this figure is too high because they deduce that the quiet-day electron energy density is too great to be contained in the geomagnetic field.

Relative Composition vs Position: The relative composition of the radiation is a function of position. The radiation becomes rapidly and progressively softer (i.e., less penetrating) as one moves outward from the earth.

Origin

Van Allen(51, 60) originally suggested that the origin of the trapped radiation was the injection of solar plasma into the geomagnetic field with an increase in the observed energies by some geomagnetic acceleration process. Freden and White(61) observed that the proton content of the inner zone was not increased by the solar flare of May 10, 1959, and concluded that this is evidence against solar injection.

Various other workers(62-66) have discussed the neutron component of the cosmic ray albedo as a source of the trapped particles. In the cosmic ray neutron albedo hypothesis,(xi) neutrons from nuclear disintegrations induced in the atmosphere by cosmic rays move outward through the geomagnetic field until they undergo radioactive decay:

$$n \xrightarrow[(\tau_{1/2} = 11.7 \text{ m})]{} P + e + \nu$$

The kinetic energy of the proton is comparable to that of the parent neutron. The electron has a well known beta decay spectrum with an upper limit of 782 Kev for a neutron at rest.

(xi) Cosmic ray protons enter into (p, n) nuclear reactions with oxygen and nitrogen of atmosphere. Since the atmosphere can be considered to be a subcritical reactor, neutrons diffuse through this homogeneous medium and leak out of the infinite slab of finite thickness. These leakage neutrons decay into protons which are trapped by the earth's magnetic field.

The observed spectra and composition of the radiation in the inner zone resemble those expected from the neutron decay hypothesis of origin. In the inner zone, the major component consists of electrons with an energy spectrum resembling that of neutron decay electrons; the minor (but penetrating) component consists of protons of the order of 100 Mev with an energy spectrum that has the same shape as that expected from the neutron decay hypothesis. Although present knowledge favors the neutron albedo hypothesis, some authors, including Van Allen, are somewhat reluctant to accept certain quantitative features[xii] necessary for the success of this hypothesis. Van Allen plans to seek illuminating evidence on temporal changes in both inner and outer zones during the approaching period of minimum solar activity.

The origin of the particles in the outer zone has been attributed directly to solar gas injection. The great temporal fluctuations observed with Pioneer IV following strong corpuscular emission from the sun constitutes the most direct evidence for the solar origin of the outer zone. The acceleration mechanism and its location (whether near the sun or near the earth) constitute major, unresolved problems.

Time Variations

Inner Zone: On the basis of the early observations, the intensity and structure of the inner zone appear relatively stable[xiii]; however, if the trapped particles originate from cosmic ray albedo neutron decays, then an increase in the flux of electrons and protons in the inner zone should be observed during the oncoming period of minimum solar activity, because the primary cosmic ray flux and hence the induced neutron intensity increase by a factor of approximately two during such periods of minimum solar activity.

Outer Zone: Large fluctuations in intensity and spatial structure are associated with geomagnetic storms. The observed time variations[51, 55, 67, 68] do not fit into any simple pattern. Several conflicting points of view have been reported in the literature.[51, 55, 62, 69-71]

Spectra

Inner Zone Protons: Freden and White[61] identified the proton component of the inner zone and measured an omnidirectional energy distribution at an altitude of about 1200 km.

(xii) The source strength may be inadequate. Gold has suggested that some particles may be "convected" inward from the outer zone. The required trapping lifetimes of several years may be long.

(xiii) On the basis of observations with Explorer VII, fluctuations have been found in the inner zone.

Measurements on April 7, 1959, included proton energies from 75 Mev to 700 Mev;[xiv] later measurements on October 13, 1959, extended these results down to a proton energy of about 18 Mev. The results of both measurements[72] are plotted in Fig. 6. In Fig. 6 the solid curve is a theoretical curve calculated from 10 to 700 Mev by Freden and White[61] on the basis of the albedo neutron decay source. The curve is normalized to the data at 100 Mev.

Armstrong et al.[73] observed inner zone protons from 42 Mev to 570 Mev at an altitude of 1125 km a few days after the intense solar flares of July 10-16, 1959. The total proton flux is about one-third that reported by Freden and White; this difference may be due to the lower peak altitude of 1125 km compared to 1230 km. The data for the proton spectrum above 100 Mev are consistent with the neutron albedo hypothesis; however, below 100 Mev there is indication of structure with a maximum at about 80 Mev and a minimum at 60 Mev. This structure may be attributable to solar protons. Although Freden and White found no evidence for a solar contribution to the cosmic ray albedo spectrum for protons with energies between 60 and 100 Mev nine days after the solar flare of May 12, 1959, the proton flux from the May 12, 1959 flare was about one-tenth as large as that from the July 14, 1959 flare.

Naugle and Kniffen[74] measured the energy spectrum of the trapped protons and the unidirectional flux as a function of position in the northern edge of the inner zone. The nose cone was spin stabilized nearly parallel to the magnetic field lines and the plane of the emulsions was normal to the spin axis of the rocket so that a unidirectional flux in the emulsion was measured at large angles with respect to the field lines. Their results are shown in Fig. 7. Their data show that (1) the shape of the spectrum and the flux change with latitude, (2) at a comparable position in the belt, the flux and shape of the spectrum agree with previous data, and (3) at the higher latitudes, the slope of the spectrum below 40 Mev is very steep compared to predictions from galactic cosmic ray neutron albedo theory and

---

(xiv) The data can be represented by a differential energy spectrum of the form

$$n(E) = C\,E^{-\gamma}$$

for 75 Mev < E < 700 Mev, with $\gamma = 1.84 \pm 0.08$ and $C = (2.1 \, {}^{+1.0}_{-0.7}) \times 10^3$ protons/cm^2-sec-$Mev^{-\gamma+1}$-steradian. In the measured energy interval, the data can also fit an exponential

$$n(E) = n_o e^{-E/E_o}$$

with $E_o = 120 \pm 5$ Mev and $n_o = 0.84 \pm 0.07$ proton/cm^2-sec-Mev-steradian. The omnidirectional integral flux for protons above 75 Mev is

$$J_o\,(E > 75\ \text{Mev}) = 800 \pm 200\ \text{protons/cm}^2\text{-sec.}$$

to measurements at lower geomagnetic latitudes[54]. The reported unidirectional differential spectra are listed below[xv]:

Naugle and Kniffen Point "3" (1600 km north): 1600 km, 28.4^{o} N lat., 237.7^{o}W long., $B = 0.231$ gauss, $I = 1.67\ r_e$, $L = 1.793\ r_e$

For $10\ \text{Mev} \leq E \leq 50\ \text{Mev}$,

$n(E) = (3.2 \pm 0.6) \times 10^6/E^{4.5 \pm 0.5}$ p/cm^2-sec-Mev-steradian

Naugle and Kniffen Point "4" (apogee): 1884 km, 25.2^{o}N lat., 236.5^{o}W long., $B = 0.198$ gauss, $I = 1.34\ r_e$, $L = 1.722\ r_e$

For $10\ \text{Mev} \leq E \leq 50\ \text{Mev}$,

$n(E) = (6.8 \pm 0.3) \times 10^6/E^{4.5 \pm 0.2}$ p/cm^2-sec-Mev-steradian

Naugle and Kniffen Point "5" (1600 km south): 1600 km, 21.5^{o}N lat., 235.4^{o}W long., $B = 0.209$ gauss, $I = 0.946\ r_e$, $L = 1.543\ r_e$

For $40\ \text{Mev} \leq E \leq 100\ \text{Mev}$,

$n(E) = (1 \pm 0.5) \times 10^3/E^{1.7 \pm 0.3}$ p/cm^2-sec-Mev-steradian

The spectra obtained at these three points are shown in Fig. 7.

Preferential trapping of low energy protons at the higher latitudes is expected (if solar protons constitute the source) either by the mechanism of direct trapping in the geomagnetic field or by trapping of decay protons from albedo neutrons produced in nuclear interactions of solar protons over the polar cap.

Outer Zone Electrons: A direct measurement of the electron spectrum at the foot of the outer zone was made with a magnetic spectrometer[58, 59]. The rocket that carried the spectrometer was launched from Wallops Island, Virginia, and reached an altitude of 1045 km. The omnidirectional differential electron spectrum at apogee is shown in Fig. 8, At apogee the omnidirectional integral flux of electrons with energies greater than 50 Kev was found to be $(4.2 \pm 0.8) \times 10^6$ electrons/cm^2-sec. The spectrum at an altitude of 920 km is shown in Fig. 9. In Fig. 9, the solid line gives the best fit of the data to the mathematical form

$$n\,(E \text{ in Kev}) \sim E^p\, e^{-E/E_o} .$$

The best fit was obtained[59] for $p = 1.60$ and $E_o = 45$ Kev.

Inner Zone Electrons: Holly and Johnson[53, 76] used counters shielded with absorbers of

(xv) The parameter L was introduced by McIlwain[75] to specify the magnetic shell on which measurements are made.

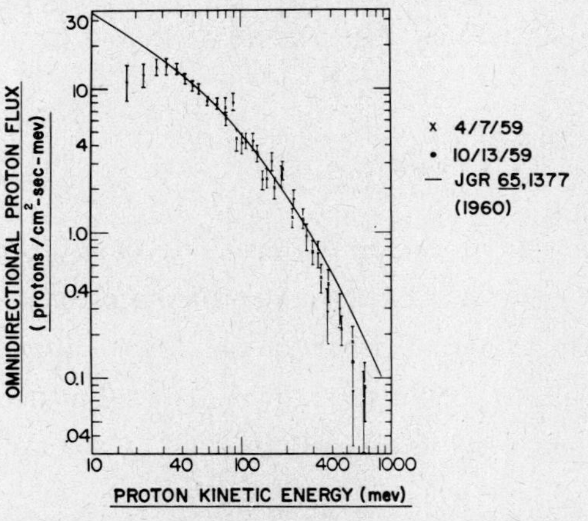

Fig. 6 Onmidirectional Differential Energy Spectrum of Protons in the Inner Radiation Belt

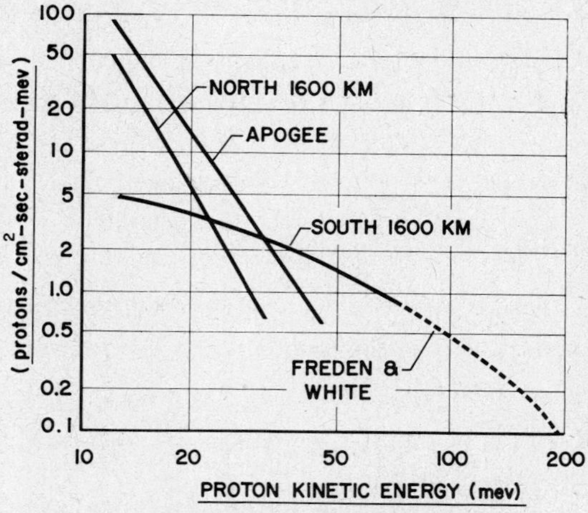

Fig. 7 Unidirectional Differential Energy Spectrum of Protons in the Inner Radiation Belt for Different Latitudes

Fig. 8 Omnidirectional Differential Electron Spectrum at Foot of Outer Zone (Altitude = 1040 km)

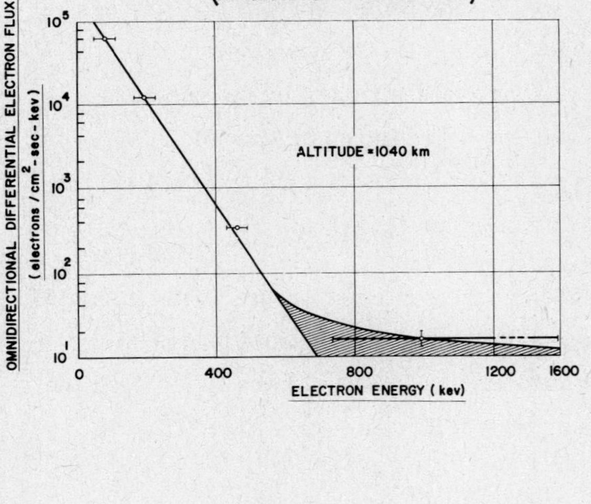

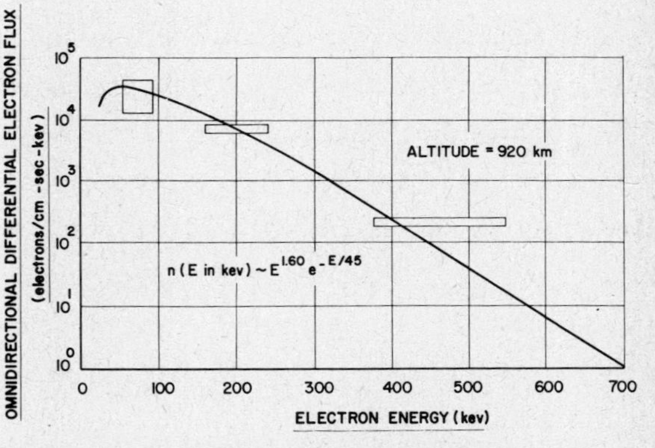

Fig. 9 Omnidirectional Differential Electron Spectrum at Foot of Outer Zone (Altitude = 920 km)

various thicknesses to obtain an idea of the differential spectrum of the observed radiation at a latitude representative of the inner zone. These data, which involve broad energy intervals, indicate a steep energy dependence similar to that observed at the foot of the outer zone by Walt et al.[58] and Cladis et al.[59].

Angular Distribution

The angular distribution of the electron flux with respect to the magnetic field direction has been measured[58, 59] at the foot of the outer zone. As expected, the flux is strongly peaked normal to the magnetic field direction. The angular distribution found at an altitude of about 1000 km is shown in Fig. 10; at lower altitudes, the distributions are still narrower. The widths of these distributions at half-height were found to be 41.6^{o} at 920 ± 50 km and 31.2^{o} at 750 ± 100 km.

ESTIMATES OF SPACE RADIATION HAZARDS

Proton Dose Rates Behind Material Shields in Space

On the basis of present knowledge, the dominant natural radiation hazards to astronauts arise from solar and Van Allen protons. For exposures of a few weeks, the galactic cosmic radiation is not intense enough to deliver a serious whole-body radiation dose. Recent experimental work at the Brookhaven National Laboratory[77] indicates that the heavy nuclei in the cosmic radiation may not present an unusual hazard. In this paper, estimates of the absorbed dose rates from solar and Van Allen protons are based upon measured spectra. Figure 11 plots the omnidirectional differential intensity of protons versus the proton kinetic energy corresponding to the observed energy regions. In Fig. 11, the proton spectra that are designated by the letters (A), (B), and (C) were measured at different locations in the inner zone of the Van Allen belt. The solar proton spectra that are designated by the numbers (1), (2), and (3) were measured at specific times after the solar flares of May 12, 1959, July 15, 1959 and November 12, 1960. The prompt spectrum (as estimated by Winckler) for the giant flare of February 23, 1956 is plotted as curve number (4) in Fig. 11.

In Fig. 12, absorbed dose rates in rads per hour at the center of spherical shell shields are plotted for both solar and Van Allen protons as a function of the threshold proton energy required to penetrate the shields. The proton dose rate curves in Fig. 12 are based upon the spectra in Fig. 11 and are calculated from a formula derived in a previous paper[32]. A useful approximate form of the formula is

$$W \text{ (ergs/gram-hour)} \approx E_t^{0.2}\, j_o (E_t)$$

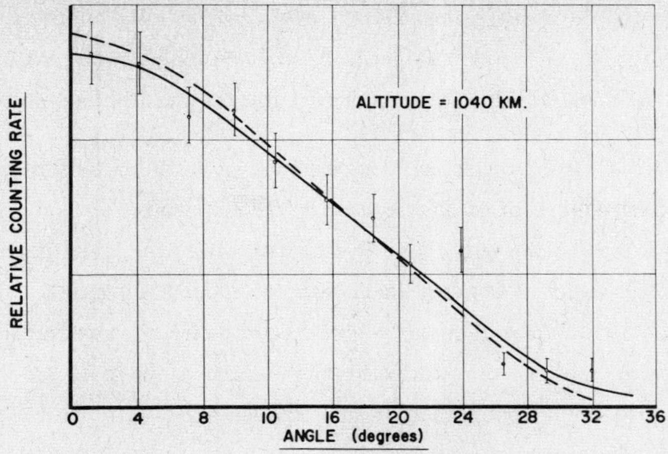

Fig. 10 Angular Distribution of the Electron Flux With Respect to Magnetic Field Direction at Foot of Outer Zone

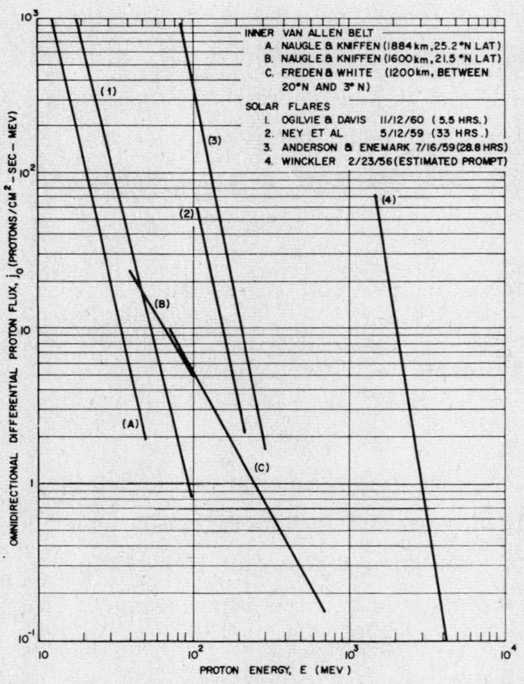

Fig. 11 Omnidirectional Differential Flux for Solar Flare and Van Allen Protons vs Proton Energy

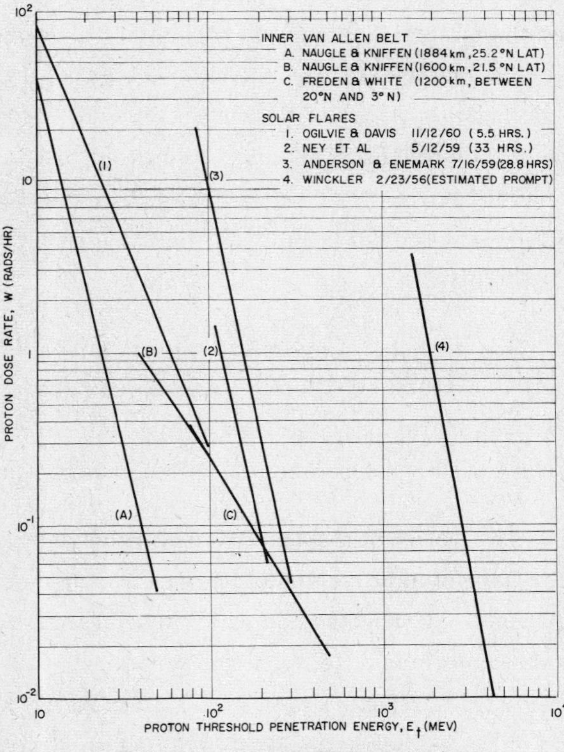

Fig. 12 Proton Dose Rates at Center of Spherical Shell Shields in Space vs Proton Threshold Penetration Energy

where E_t is the proton threshold penetration energy, and $j_o (E_t)$ is the incident omnidirectional differential spectrum. This simple form can be used to make rapid estimates of proton dose rates behind spherical shell shields of arbitrary material and thickness.

Proton dose rates from spectra measured in the inner Van Allen belt range from 40 rads per hour for a stopping energy of 10 Mev to about 250 millirads per hour for a stopping energy of 100 Mev. It is worth noting that these dose rates would increase perhaps ten or twenty times in the heart of the inner zone at an altitude of about 3500 miles. This rough estimate is based upon the assumption that the spectra do not differ appreciably in the two locations. Then, a comparison of the omnidirectional intensities above 40 Mev, as measured by Van Allen[(51)] at the heart of the inner zone and by Freden and White[(54, 72)] at an altitude of 1200 kilometers, gives a factor of about 20.

For a given proton threshold energy, the dose rate is practically independent of the material; however, the weight of a shield to stop protons of a specified energy will depend upon the material. Figure 13 shows the range-energy relationship for protons in various shielding materials. The maximum weight of the spherical shell shield depends upon the ratio of the thickness of the material required to stop protons of given energy, E_t, to the radius of the spherical cavity to be shielded. Figures 14 and 15 plot the weights of spherical shell shields for shielding 6- and 12-foot-diameter cavities, respectively, as functions of the proton threshold penetration energy for liquid hydrogen, polyethylene (CH_2), and aluminum shields.

The weight of a spherical shell required to reduce the dose rate at the center of the cavity to a specified value is illustrated in Fig. 16. Figure 16 plots the weight of polyethylene spherical shells for shielding a 6-foot-diameter cavity from solar flare and Van Allen spectra versus the proton dose rate at the center of the cavity.

The weight of a spherical shell required to reduce the proton dose (in rads) from the solar flare of July 16, 1959 to a specified value is illustrated in Fig. 17. Figure 17 plots the weight of a ployethylene (CH_2) spherical shell for shielding a 6-foot-diameter cavity versus the proton dose in rads absorbed from the solar flare proton spectrum measured by Anderson and Enemark at 1.2 days after the peak of the optical flare. Since the time dependence of the omnidirectional intensity for this flare was measured only after 1.2 days, it is not possible to obtain a reliable estimate of the total absorbed dose. The dashed curve in Fig. 17 is shown for orientation purposes; it is the weight-dose relationship calculated on the basis of a dose rate that is constant in time during the first 1.2 days of the flare.

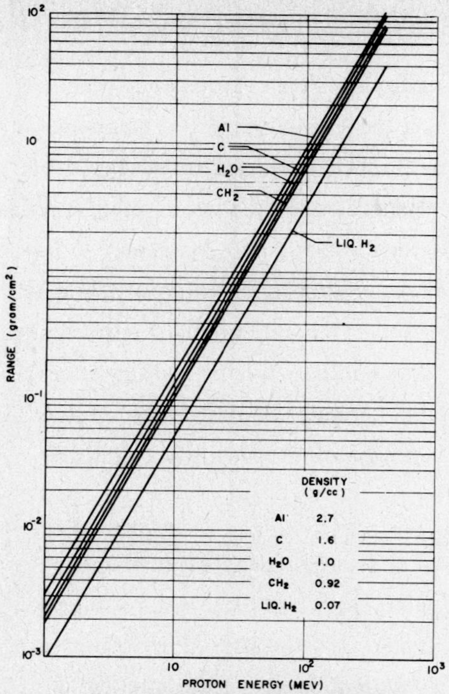

Fig. 13 **Range vs Energy for Protons in Various Shielding Materials**

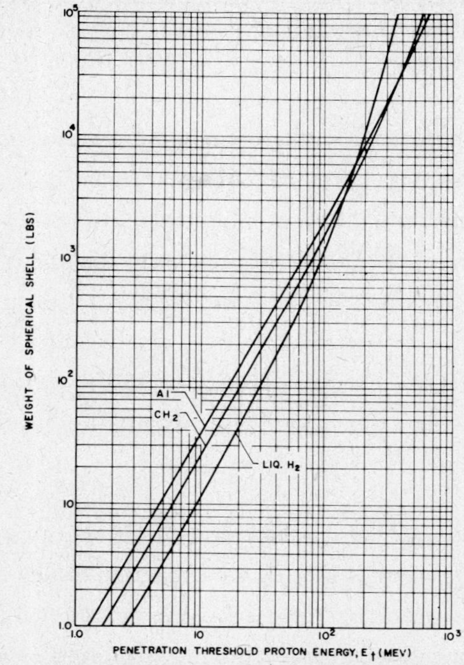

Fig. 14 **Weight of a Spherical Shell for Shielding a 6-Foot-Diameter Cavity vs Penetration Threshold Proton Energy**

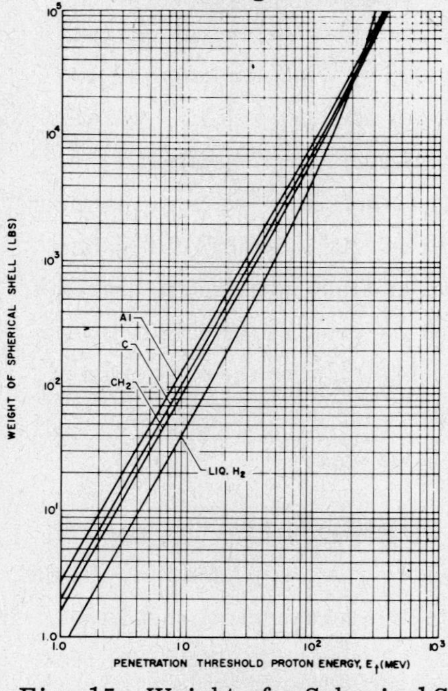

Fig. 15 **Weight of a Spherical Shell for Shielding a 12-Foot-Diameter Cavity vs Penetration Threshold Proton Energy**

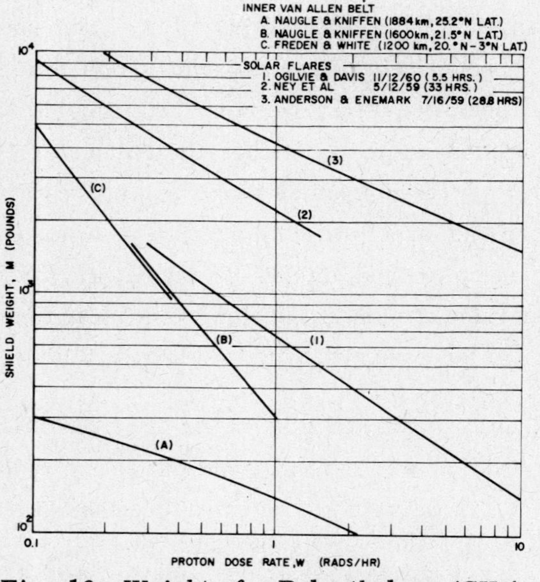

Fig. 16 **Weight of a Polyethylene (CH_2) Spherical Shell for Shielding a 6-Foot-Diameter Cavity From Solar and Van Allen Proton Spectra vs Proton Dose Rate at the Center of the Cavity**

On the basis of present evidence, it appears that a radiation safe region for astronauts exists in the equatorial region of the earth at altitudes below the Van Allen belt. Measurements[78] indicate that the intensity of the geomagnetically trapped radiation is negligible below about 400 kilometers. The geomagnetic field of the earth offers protection against solar flare protons, and this protection would extend over a wide latitude region of perhaps 45 degrees above and below the geomagnetic equator provided that the proton energies which are repelled or "cut off" by the geomagnetic field are not changed very much during a solar flare. A study by Ray[79] showed that the geomagnetic cutoffs are essentially unchanged for energetic protons (in the range from 2 to 15 Bev corresponding to geomagnetic latitudes up to about 45°), and are changed only slightly for lower energy protons that normally reach the earth at geomagnetic latitudes above 45°.

In Fig. 18, the geomagnetic cutoff energy for protons approaching the earth from various directions is plotted as a function of geomagnetic latitude. Cutoff energies for protons in the region of a few Bev are indicated in Fig. 18. In Fig. 19, the proton cutoff energy in the geomagnetic equatorial plane is plotted as a function of altitude above the surface of the earth for a simple dipole field. The protection offered by the geomagnetic field may be illustrated for the giant flare of February 23, 1956. Figure 20 plots the unshielded proton dose rate in rads per hour as a function of the proton cutoff energy in the energy interval from 1.5 Bev to 10 Bev for the prompt spectrum, as estimated by Winckler from the data for this flare. Indicated dose rates range from about 1 rad per hour down to 10^{-4} rad per hour for proton cutoff energies from 1.7 Bev to 9 Bev. These considerations neglect the effect of nuclear interactions in material shields which may be present. Sea level cosmic ray neutron monitors at geomagnetic latitude 54°N indicated that the intensity of this flare was reduced tenfold from its peak value in about four hours, and then continued to decrease to background level for a total duration of about 20 hours. If the proton spectrum is indeed cut off magnetically at the low-energy end at about 2 Bev or higher, then the unshielded proton dose rates are small. In view of the steep time dependence and the relatively short duration of the flare spectrum, the total unshielded dose is also small.

By way of contrast, the weight of a material shield required to furnish this same radiation protection to astronauts outside of this radiation safe corridor is prohibitive. To the best of our knowledge, no one has offered satisfactory protective measures against an infrequent but "giant" relativistic solar flare such as the one of February 23, 1956. To protect against the spectra observed from other "giant" flares, shield weights for interplanetary travel are substantial in terms of launch-vehicle lifting capacity; however, missions in the "safe corridor" can be performed without the need for radiation shielding.

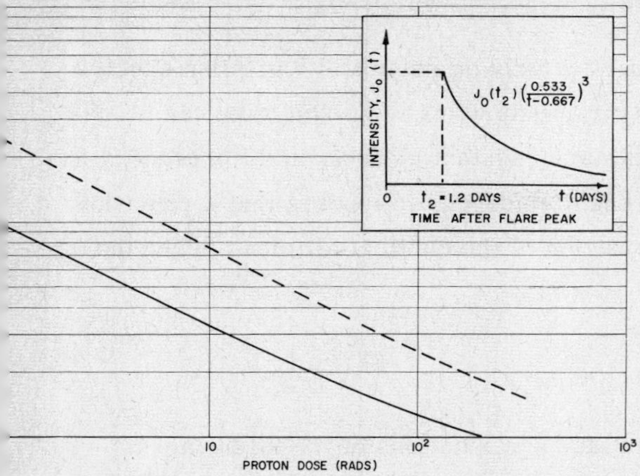

Fig. 17 Weight of a Polyethylene (CH_2) Spherical Shell for Shielding a 6-Foot-Diameter Cavity vs Proton Dose Absorbed From the Solar Flare of July 16, 1959

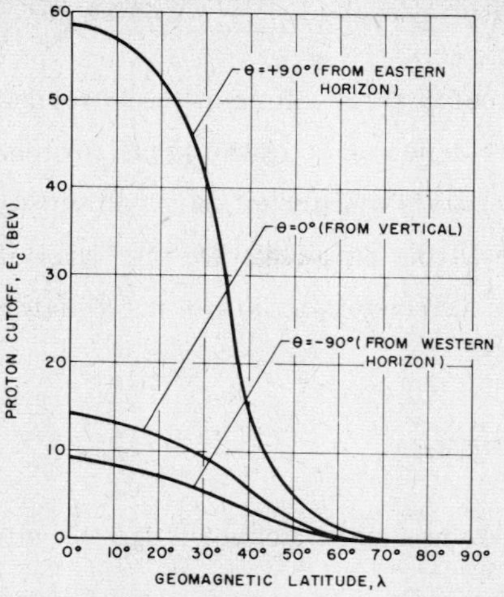

Fig. 18 Cutoff Energies for Protons Approaching the Earth from Various Directions vs Geomagnetic Latitude

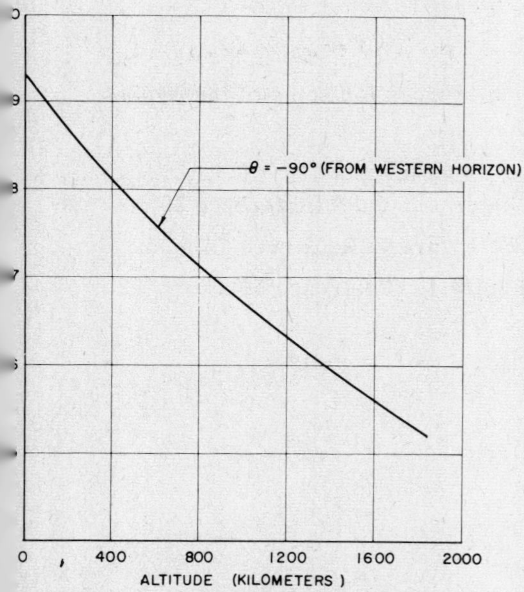

Fig. 19 Proton Cutoff Energy in the Geomagnetic Equatorial Plane vs Altitude Above Surface of Earth

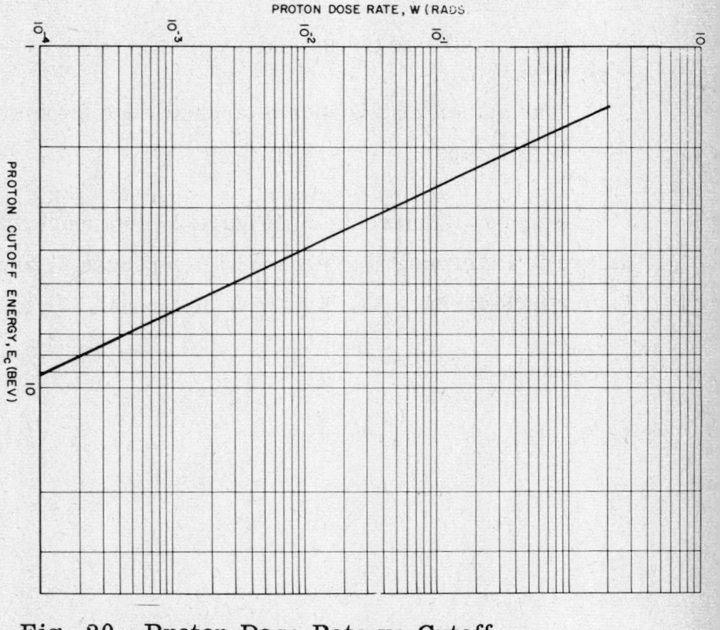

Fig. 20 Proton Dose Rate vs Cutoff Energy for Winckler's Estimated Prompt Spectrum for Giant Flare of February 23, 1956

CONCLUSIONS

More experimental information is needed, particularly on solar and Van Allen spectra and on the time dependence of the solar proton spectra, before optimum shielding can be designed. On the basis of current data, shield weights are substantial in terms of present launch-vehicle lifting capacity. On the basis of present evidence, it appears that a radiation safe region for astronauts exists in the equatorial region of the earth at altitudes below the Van Allen belt.

ACKNOWLEDGEMENT

The author wishes to acknowledge the interest and contributions of the following personnel:

Dr. William Helvey, who suggested that a lecture series on this subject be given at Republic Aviation Corporation; Drs. A. Duneer and T. Krieger, who collaborated with the author in the preparation of a paper from which much of the material in the section "Proton Dose Rates Behind Material Shields in Space" is taken; Mr. E. Kuhn, who provided the information for Figure 19; Mr. R. Fiore, who assisted in the preparation of Figures 1-10 and 13-15, and who read and commented on the completed notes; and Dr. G. Albright, who read and commented on the manuscript.

The author also wishes to acknowledge the following sources for some of the figures used in this paper:

Figure 1 (Reference 9); Figure 2 (Reference 10); Figures 3 and 4 (Reference 29); Figure 5 (Reference 51); Figure 6 (References 61 and 72); Figure 7 (Reference 74); Figure 8 (Reference 58); Figure 9 (Reference 59); and Figure 10 (Reference 58).

REFERENCES

1. Johnson, T. H., Revs. Modern Phys. 10, 193, 219 (1938).

2. Stoermer, C., C. Zeit. f. Astrophys. 1, 237 (1930).

3. Schein, M., Jesse, W. P., and Woolan, E. O., Phys. Rev. 59, 615 (1941).

4. Earl, James A., "Cloud Chamber Observations of Primary Cosmic Ray Electrons," Phys. Rev. Letters 6, 125 (1961).

5. Freier, P., Lofgren, E. J., Ney, E. P., and Oppenheimer, F., Phys. Rev. 74, 1818 (1948).

6. Meyer, P. and Vogt, R., "Electrons in the Primary Cosmic Radiation," Phys. Rev. Letters 6, 193 (1961).

7. Perlow, G. J., and Kissinger, C. W., Phys. Rev. 81, 552 (1952); Phys. Rev. 84, 572 (1952).

8. Jones, Frank C., "Cosmic Ray Production of Low Energy Gamma Rays," J. Geophys. Research 66, 2029 (1961).

9. Peters, B., "Progress in Cosmic Ray Physics," I, 226, North-Holland Publishing Company, Amsterdam (1952).

10. Singer, S. F., "Progress in Elementary Particle and Cosmic Ray Physics," IV, 282, North-Holland Publishing Co., Amsterdam (1958).

11. Winckler, J. R., Rad. Research 14 (5), 521 (1951).

12. Scarsi, L., Am. J. Phys. 28, 213 (see Figure 5) (1960).

13. Hillberry, N., Phys. Rev. 60, 1 (1941).

14. Meredith, L. H., Van Allen, J. A., and Gottlieb, M. B., Phys. Rev. 99, 198 (1955).

15. Neher, H. V., Phys. Rev. 103, 228 (1956).

16. Neher, H. V., "Progress in Cosmic Ray Physics," I, 298, North-Holland Publishing Company, Amsterdam (1952).

17. Linsley, J., Scarsi, L., and Rossi, B., Phys. Rev. Letters 6, 485 (1961).

18. Fermi, E., Astrophysical Jour. 119, 1 (January 1954).

19. Forbush, S. E., Terr, Mag. 43, 207 (1938).

20. Coleman, P. J., Judge, D. L., Smith, E. J., and Sonett, C. P., "Measurements of the Geomagnetic and Interplanetary Magnetic Fields, Pioneer V," Paper presented at Washington AGU meeting, Planetary Sciences Program, Fifth Session (April 29, 1960).

21. Cocconi, G., Nuovo Cimento 3, 1433 (1956).

22. Bailey, D. K., Proc. Inst. Radio Engrs. 47, 255 (1959).

23. Anderson, K. A., Arnoldy, R., Hoffman, R., Peterson, L., and Winckler, J. R., J. Geophys. Research 64, 1113 (1959).

24. Rothwell, P., and McIlwain, C., Nature 184, 138 (1959).

25. McCracken, K. G., and Palmeira, R. A. R., J. Geophys. Research 65, 2673 (1960).

26. Obayaski, T., and Hakura, Y., J. Geophys. Research 65, 3143 (1960).

27. Cocconi, G., et al., Nuovo Cimento 8, Suppl. 161 (1958).

28. Gold, T., J. Geophys. Research 64, 1665 (1959).

29. Steljes, J. F., Carmichael, H., and McCracken, K. G., J. Geophys. Research 66, 1363 (1961).

30. McCracken, K. G., Bull. Am. Phys. Soc. II 6, 264A (April 24, 1961).

31. Ney, E. P., Winckler, J. R., and Freier, P. S., Phys. Rev. Letters 3, 183 (1959).

32. Madey, R., Duneer Jr., A. G., and Krieger, T. J., "Proton Dose Rates in Manned Space Vehicles," Republic Aviation Corporation Report RAC 433 (1 September 1961).

33. Anderson, K. A., and Enemark, D. C., J. Geophys. Research 65, 2657 (1960).

34. Robey, D. H., Astronautica Acta VI (4), 206 (1960).

35. Dew, N. F., Aero/Space Engineering 19 (5), 46 (1960).

36. Keller, J. W., and Schaeffer, N. M., "Radiation Shielding for Space Vehicles," General Dynamics Corp. FZM-1987 (June 15, 1960).

37. Wallner, E., and Kaufman, R., "Radiation Shielding for Manned Space Flight," NASA TN-D-681 (July 1961).

38. Unpublished NRL data reported in NASA TN-D-681.

39. Schaeffer, H. J., Jour. Av. Med. 28, 387 (1957).

40. Foelsche, T., "Protection Against Solar Flare Protons," AAS Preprint 61-34 (January, 1961).

41. Barnes, T. G., Finkelman, E. M., and Barazotti, A. L., "Radiation Shielding of Lunar Spacecraft," presented at AAS Lunar Flight Symposium, New York (December 27, 1960).

42. Davis, L. R., Fichtel, C. E., Guss, D. E., and Ogilvie, K. W., "Rocket Observations of Solar Protons on September 3, 1960," Phys. Rev. Letters 6, 492 (1961); NASA Goddard Space Flight Center Contributions to 1961 Kyoto Conference on Cosmic Rays and the Earth Storm.

43. Ogilvie, K. W., Bryant, D. A., and Davis, L. R., "Rocket Observations of Solar Protons During the November 12, 1960 Event," NASA Goddard Space Flight Center Contributions to 1961 Kyoto Conference on Cosmic Rays and the Earth Storm.

44. Yagoda, H., Filz, R., and Fukui, K., Phys. Rev. Letters 6 (11), 626 (June 1, 1961).

45. Ney, E. P., and Stein, W., J. Geophys. Research 66 (8), 2550A (1961).

46. Freier, P., Biswas, S., and Stein, W., J. Geophys. Research 66 (8), 2529A (1961).

47. Fichtel, C. E., and Guss, D. E., Phys. Rev. Letters 6 (9), 495 (1961).

48. Van Allen, J. A., McIlwain, C. E., and Ludwig, G. H., J. Geophys. Research 64, 271 (1959).

49. Van Allen, J. A., and Frank L. A., Nature 183, 430 (1959).

50. Vernov, S. N., Chudakov, A. Y., Vakulev, P. W., and Logachev, Y. I., Doklady Akad. Nauk S. S. S. R. 125, 304 (1959); Planetary and Space Science 1, 86 (1959).

51. Van Allen, J. A., J. Geophys. Research 64, 1683 (1959).

52. Van Allen, J. A., and Frank, L. A., Nature 184, 219 (1959).

53. Holly, F. E., and Johnson, R. G., J. Geophys. Research 65, 771 (1960).

54. Freden, S. C., and White, R. S., Phys. Rev. Letters 3, 9 (1959); J. Geophys. Research 65, 1377 (1960).

55. Arnoldy, R., Hoffman R., and Winckler, J. R., "Observations of Van Allen Radiation Regions During Geomagnetic Storms," Space Research, Proceedings of the

First International Space Science Symposium, Nice, January 11-16, 1960. Edited by Hilde Kallman Bijl, North-Holland Publishing Company, Amsterdam (1960).

56. Vernov, S. N., and Chudakov, A. E., "Investigating Cosmic Rays and the Terrestrial Corpuscular Radiation by Rockets and Artificial Satellites," Space Research, Proceedings of the First International Space Science Symposium, Nice, January 11-16, 1960. Edited by Hilde Kallman Bijl, North-Holland Publishing Company, Amsterdam (1960).

57. Allen, R. I., Dessler, A. J., Perkins, J. F., and Price, H. C., "Shielding Problems in Manned Space Vehicles," Lockheed Report NR-104 (July 1960).

58. Walt, J., Chase, L. F. Jr., Cladis, J., Imhof, W. L., and Knecht, D. J., "Energy Spectra and Altitude Dependence of Electrons Trapped in the Earth's Magnetic Field," Space Research, Proceedings of the First International Space Science Symposium, Nice, January 11-16, 1960. Edited by Hilde Kallman Bijl, North-Holland Publishing Company, Amsterdam (1960). (See also Figure 2 of Reference 11.)

59. Cladis, J. B., Chase, L. F. Jr., Imhof, W. L., and Knecht, D. J., J. Geophys. Research 66, 2297 (1961).

60. Van Allen, J. A., Paper presented at joint meeting of National Academy of Sciences and American Physical Society on May 1, 1958.

61. Freden, S. C., and White, R. S., J. Geophys. Research 65 (5), 1377 (1960).

62. Kellogg, P. J., Nuovo Cimento 11, 48 (1959).

63. Singer, S. F., Phys. Rev. Letters 1, 173, 181 (1958).

64. Hess, W. N., J. Geophys. Research 65 (10), 3107 (1960); Phys. Rev. Letters 3, 11 (1959).

65. Hess, W. N., Canfield, E. H., and Lingenfelter, R. E., J. Geophys. Research 65 (3), 665 (March 1961).

66. Vernov, S. N., Special Lecture, Fifth General Assembly of C. S. A. G. I. in Moscow, July 30 - August 9, 1958.

67. Fan, C. Y., Meyer, P., and Simpson, J. A., "Trapped and Cosmic Radiation Measurements from Explorer VI," Space Research, Proceedings of the First International Space Science Symposium, Nice, January 11-16, 1960. Edited by Hilde Kallman Bijl, North-Holland Publishing Company, Amsterdam (1960).

68. Rosen, A., Farley, T. A., and Sonett, C. P., "Soft Radiation Measurements on Explorer VI Earth Satellite," Space Research, Proceedings of the First International Space Science Symposium, Nice, January 11-16, 1960. Edited by Hilde Kallman Bijl, North-Holland Publishing Company, Amsterdam (1960).

69. Dessler, A. J., and Karplus, R., Phys. Rev. Letters 4, 271 (1960).

70. Singer, S. F., "Artificial Modification of the Earth's Radiation Belt," Advances in Astronautical Sciences, IV, 335-354, Plenum Press, New York (1959).

71. Vernov, S. N., Chudakov, A. E., Vakulov, P. V., and Logachev, Y. I., Soviet Physics "Doklady" 4, 338 (1959).

72. White, R. S., Private Communication (April 1961).

73. Armstrong, A. H., Harrison, F. B., Heckman, H. H., and Rosen, L., J. Geophys. Research 66 (2), 351 (1961).

74. Naugle, J. E., and Kniffen, D. A., Phys. Rev. Letters 1, 3 (1961).

75. McIlwain, Carl E., "Coordinates for Mapping the Distribution of Magnetically Trapped Particles," J. Geophys. Research 66, 3681 (1961).

76. Holly, F. E., "Radiation Measurements to 1500 Kilometers with Atlas Pads," Air Force Special Weapons Center Report AFSWC-TR-60-9 (May, 1960).

77. Curtis, Howard J., "Limitations on Space Flight Due to Cosmic Radiations," Science 133, 312 (3 February 1961).

78. Yoshida, S., Ludwig, G. H., and Van Allen, J. A., "Distribution of the Trapped Radiation in the Geomagnetic Field," J. Geophys. Research 65, 807 (1960).

79. Ray, Ernest C., "Effects of a Ring Current on Cosmic Radiation," Phys. Rev. 101, 1142 (1956).

eighth annual meeting

AMERICAN ASTRONAUTICAL SOCIETY

INTRODUCTION

PART I — BASIC RESEARCH
Chairman: Robert Jastrow

PART II — GUIDANCE AND CONTROL
Chairman: Robert E. Roberson

PART III — BIOASTRONAUTICS
Chairman: William M. Helvey M.D.

PART IV — COMMUNICATIONS
Chairman: Ray A. Glaser

PART V — LUNAR AND PLANETARY EXPLORATION
Chairman: Eliot T. Benedikt

PART VI — ASTRODYNAMICS
Chairman: Hans Lieske

PART VII — APPLICATIONS OF ASTRONAUTICAL SYSTEMS
Chairman: G. L. Hollingsworth

PART VIII — SPACE VEHICLE DESIGN
Chairman: Saunders B. Kramer

SPACESCAN - A RADIO BROADCAST SYSTEM OF SPACE NAVIGATION

A. Tatz*

SPACESCAN is a radio-broadcast system of navigation guidance with total-volume coverage for on-board space-position measurement of geographical position, velocity and path angle for orbit, re-entry and recovery of manned vehicles.

SPACESCAN provides a system of radio navigation aids that broadcast position signals useful in forming a space-position, three-dimensional grid with total-volume radio coverage. The vehicle equipment determines its position in the grid by receiving and processing signals that are radiated from earth-based sites over volume-scanning plane beams.

The beams are essentially thin, flat fans of K_u-band energy, modulated with elevation-angle data, and scanning in elevation angle from horizon to horizon over the site of an earth-based transmitter.

The vehicle-borne receiver measures elevation angle by decoding the beam contents as the beam scans through the vehicle. The elevation angle defines a plane of position; three intersecting plans of position are thus measured from two sites as far apart as permitted by line-of-sight coverage; the intersection of the three planes of position determines the vehicle position in rectangular coordinates relative to the center of the earth.

---

* Airborne Instruments Laboratory, A Division of Cutler-Hammer, Inc.

I. INTRODUCTION

For manned space vehicles that are either in earth orbit or in earth-atmosphere re-entry, it is desired to make continuously available in the vehicle a set of outputs representing the vehicle's geographic position in a recognizable set of coordinates.

A coordinate set that is readily recognizable is, of course, the altitude, latitude, and longitude of the vertical reference line down from the vehicle.

Another set of useful outputs consists of the velocity relative to the center of the earth and the rate of change of altitude, from which is computed the relative path angle (gamma).

Various ways to accomplish the on-board navigation functions outlined above include the use of optical sensors, inertial platform, and radio communication with earth-based tracking nets.

It is the purpose of this paper to present a new method of geographic position fixing, using radio beams that are transmitted from earth sites in such a manner that the received signals can be processed into the desired outputs aboard the vehicle directly from the data demodulated from the radio beams, without using tracking radars at the earth sites.

The system under discussion can be characterized as a radio-navigational aid, with the same connotation that is evoked when speaking of the network of VOR stations as a radio-navigational aid for aircraft. The correct implication is that navigation service is provided to any user who is equipped with compatible radio receiving equipment, provided that the vehicle is within radio coverage from two ground stations, each of which broadcasts modulated energy in a hemisphere centered over the zenith of each station site.

The usual way to visualize radio-navigational position fixing is to regard a datum of measurement as the basis for defining a surface of position. It is then axiomatic that three surfaces of position, intersecting in a point, are sufficient to define the vehicle's geographic position.

For convenience in visualizing the point of intersection, and for mathematical simplicity in solving for position coordinates, it is advantageous to use flat planes for the surfaces of position. Then a triplet of equations of the form $Ax + By + Cz = D$ expresses the three planes of position, and the solution for x, y, and z is straightforward.

The set of rectangular coordinates is taken with respect to the origin at the center of the earth, with the x axis passing through the point of zero latitude and zero longitude. Then the geographic coordinates are obtained by a transformation to spherical coordinates.

The radio-navigational aid to be described provides geographic position fixing directly computed aboard the vehicle without dependence on computing look information, such as is needed for radar tracking techniques, and without dependence on initial-condition values, such as are needed for inertial devices. For these reasons, it is worth-while to consider this new system as a useful complement to the other prevailing systems, especially where emphasis is placed on maximum reliability and safety through redundancy of independent techniques.

II. TECHNIQUES

A predesign analysis of a radio-navigational aid for geographic position fixing has resulted in the formulation of a set of techniques embodied in a project called SPACESCAN. As the name implies, a beam-scanning principle is involved in the broadcasting of surface-of-position data from a ground site into the hemisphere of radio coverage over the site.

The beams are essentially thin, flat fans of K_u-band energy, modulated with elevation-angle data, and scanning in elevation angle from horizon to horizon over the site of an earth-based transmitter (see Fig. 1).

The vehicle-borne receiver measures elevation angle by decoding the beam contents as the beam scans through the vehicle. The elevation angle defines a plane of position;

three intersecting planes of position are thus measured from two sites as far apart as permitted by line-of-sight coverage; the intersection of the three planes of position determines the vehicle position in rectangular coordinates relative to the center of the earth (see Fig. 2).

A SPACESCAN radiation pattern is very narrow (0.5 degree) in elevation angle and very wide (170 degrees) in azimuth angle. The beam can be approximately visualized as a flat sheet, like the page of a book. This analogy is useful in describing the scanning action. Imagine an open book lying horizontally on the table. As each page is turned, it represents the scanning of a beam, the center of the book binding representing the site of the station. In this context, the binding represents the axis of the scanning motion, and the line of scan is taken as the horizontal perpendicular bisector of the axis of scan. As each page is turned, the beam position is defined by the elevation angle between the plane of the beam and the horizontal plane of the book covers.

The peak-amplitude point on the radiation pattern is a point on the scanning plane of the page, directly over the horizontal line of scan. The peak-amplitude point is made to scan in a semicircle through 180 degrees of elevation angle, from one horizontal direction of the line of scan, through the station zenith, and over to the opposite direction of the line of scan (see Fig. 1).

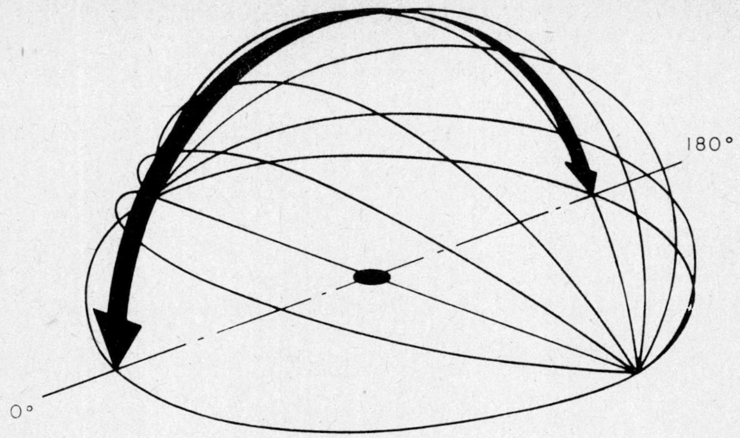

Fig. 1 One-Plane Scanning

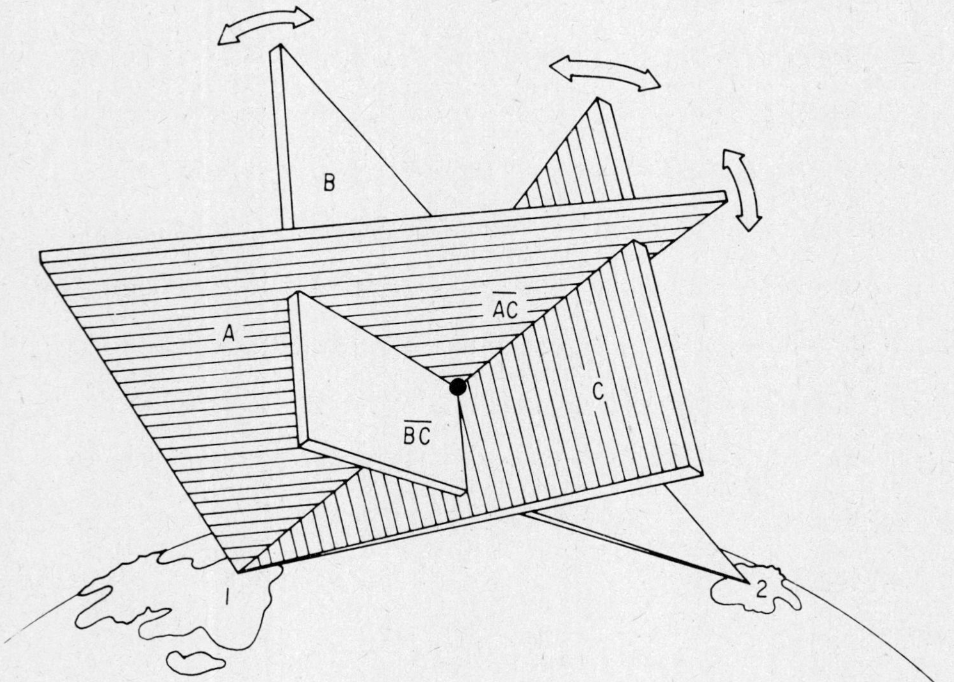

Fig. 2 Intersecting Planes of Position

The scan semicircle lies in the vertical plane (perpendicular to the earth-tangent plane of the station site). Thus, the vertical semicircle is directly over the line of scan, and an elevation angle is formed between the line of scan and the semicircle's rotating radius vector. This elevation angle defines the plane of position of the vehicle when the center of the scanning beam is detected in the vehicle receiver.

The microwave energy in the scanning beam is pulse-modulated at the transmitter site, so that the time between pulses accurately represents the elevation angle of the peak of the beam. This beam is only 0.5 degree wide in the direction in which it scans 180 degrees in 1 second; therefore, the beam scans through the vehicle in a period of about 3000 μsec, the exact duration depending on the level of receiver detection below the peak amplitude of the beam.

The beam modulation consists of an interpulse spacing of between 30 and 150 μsec, where the interpulse spacing is linearly related to the elevation angle of scan at a scale factor of 4 μsec for degree of scan* (see Fig. 3). Taking into account the period of scan through the vehicle

---

* The modulation range over 30 degrees is repeated six times to cover 180 degrees of scan.

(3000 μsec) and the interpulse spacing, it is seen that between 20 and 100 samples of elevation angle are measured by the vehicle receiver during each passage of a beam.

As each pulse is detected in the receiver, the interpulse spacing is measured and compared with the previously tracked and stored memory analog of the elevation angle. Throughout the entire train of received pulses, a tracking error voltage is generated, so that after the beam contents have been sampled in each scan, the corrected memory represents the plane of position with an accuracy of 0.25 milliradian in terms of elevation angle.

An elevation angle defining a plane of position is measured at the rate of three times per second, since three beams of the same type sweep across the vehicle each second. The tracking circuit in the receiver contains angle-velocity memory as well as angle memory; the angle-velocity data are used to update the angle data, so that the angle readout is current at the time it is sampled by the coordinate computer.

III. GEOMETRY

In this discussion, a plane of position has thus far been defined by an elevation-angle value relative to a known line of scan in the earth-tangent plane containing the transmitter site.

Ultimately, the position of the vehicle must be computed from the intersection of three planes of position, and it is desired to express the computed position in a universal geographic coordinate system.

The universal geographic coordinate system is a rectangular coordinate system with the origin at the center of the earth. The positive X axis passes through the zero meridian at the equator; the position Y axis is oriented 90 degrees east in the earth's equatorial plane; the positive Z axis is oriented northward, perpendicular to the equatorial plane.

A given scan line is specified in terms of its three direction cosines relative to the XYZ axes; the station zenith is specified by three other direction cosines relative to the XYZ axes; the seventh constant required for coordinate transformation is the radius of the earth at the station site.

The following steps are used in deriving the mathematical expressions for a plane of position in geographic coordinates (XYZ).

1. State the plane of position in terms of the sine and cosine of the elevation angle relative to the earth-tangent-plane origin.
2. Translate the origin along the earth's radius to the center of the earth.
3. Rotate the axes of the scan line and the station zenith to the XYZ axes, using the six direction-cosine constants for the rotational transformation.

After these steps are performed, the plane of position is expressed in the following form:

$$Ax + By + Cz = D \tag{1}$$

where the coefficients A, B, C, and D are arithmetic combinations of the seven constants and the sine and cosine of the elevation angle.

Three equations of the form of equation 1 make up a system of linear equations from which the XYZ coordinates are computed, provided that the three equations represent three planes of position that intersect in a point fix.

In the foregoing discussion, consideration was given to one line of scan and one axis of scan, both lying in the earth-tangent plane, and one scanning beam whose peak-amplitude point moved in a semicircle over the line of scan.

If the roles of the line of scan and the axis of scan were interchanged, the result would be a scanning beam whose semicircle lay in a vertical plane rotated 90 degrees around the station-zenith axis.

A complete SPACESCAN station uses two mutually perpendicular lines of scan, and thus two simultaneous systems of scanning beams. The line of scan for one scan serves as the axis of scan for the other scan. For example, if one scan line runs north and south from the transmitter site, the second line of scan runs east and west from the same transmitter site (see Fig. 4).

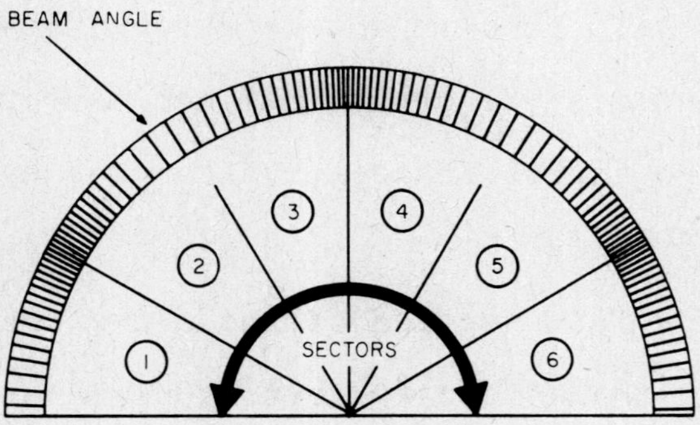

Fig. 3 Angle Data Transmission and Reception

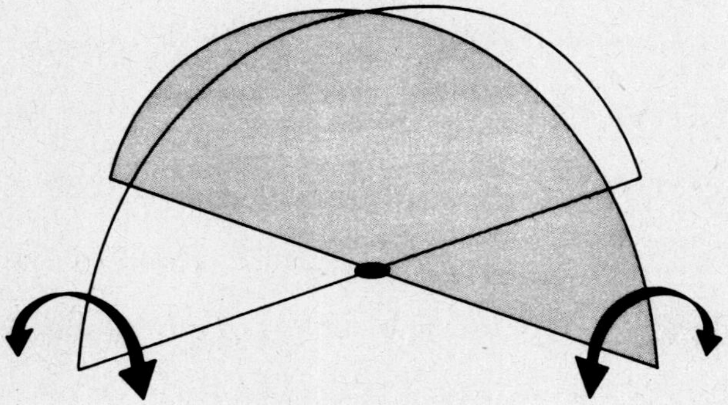

Fig. 4 Scanning Beams from Each Ground Site

A plane of position is measured from each of the two scans, and the result is that two intersecting planes of position are measured from a single transmitter site.

In a typical situation, one line of scan is oriented so that it lies nearly in the plane of the expected trajectory of the vehicle. This arrangement signifies an along-trajectory scanning system. Its orthogonal counterpart, the second scan, signifies the across-trajectory scanning system (see Fig. 2, planes A and C).

The intersection of the along-trajectory plane of position and the across-trajectory plane of position is a line of position from the transmitter site to the vehicle (see Fig. 2, $\overline{AC}$).

The third plane of position is measured advantageously from an along-trajectory scan that originates from another transmitter site. Preferably, one of the sites would be behind the vehicle, and its second site ahead of the vehicle in its trajectory. The intersection of the two along-trajectory planes of position is a line of position approximately perpendicular to the trajectory (see Fig. 2, line $\overline{BC}$).

For best results in terms of minimum geometric dilution of position fix and maximum utilization of site configurations, the sites should be as far apart as line-of-sight ratio coverage permits, taking into account the curvature of the earth.

In low-altitude orbits and for a re-entry trajectory, the expected altitude of the trajectory over the approaching site will be the principal limiting factor in siting for optimum coverage.

A second limiting factor is the change in the index of refraction for radio propagation at low elevation angles. The region of most useful coverage is limited to elevation angles above 6 degrees in order to minimize the angular errors introduced by errors in predicting the change in the index of refraction.

Taking into account the earth curvature and the 6-degree lower limit, the maximum site spacing is approximately 500 miles for 100-mile altitudes, 300 miles for 50-mile altitudes, and 200 miles for 30-mile altitudes.

A chain of about 12 stations is required to completely cover a typical re-entry trajectory of about 3000 miles and to provide continuous re-entry guidance data by SPACESCAN, from re-entry to the recovery area. (Trajectory data for this estimate were supplied by the Navigation and Control Section, Space Task Group, NASA, Langley Field, Virginia, during the study program.)

Fig. 2 shows three planes of position intersecting at the circled point. Planes A and C are the planes of position for the across-trajectory and along-trajectory scans from site 1. Plane B is the plane of position for the along-trajectory scan from site 2. The circled position fix is visualized as the intersection of two lines of position, one formed by planes A and C (line $\overline{AC}$) and the other formed by planes B and C (line $\overline{BC}$).

The elevation angles that define the planes of position are each measured three times per second. This rate is estimated to be sufficiently rapid so that rates of change of X, Y, and Z are computable with about 1-second smoothing. Thus, the directly computed outputs are the rectangular coordinates of geographic position (XYZ) and their rates of change ($\dot{X}$, $\dot{Y}$, and $\dot{Z}$).

GUIDANCE OUTPUTS

The following positional data outputs are computed from the rectangular coordinate data:

1. Altitude (H) and its rate ($\dot{H}$), where H is the root sum square of X, Y, and Z (range R_o) minus the earth radius.
2. Relative velocity (V), which is the root sum square of $\dot{X}$, $\dot{Y}$, and $\dot{Z}$.
3. Relative flight path angle (gamma), which is the angle whose sine is $\dot{H}/V$.
4. Geocentric longitude, which is the angle whose tangent is Y/X.
5. Geocentric latitude, which is the angle whose sine is Z/R_o.

As an incidental feature, if it is desired to align the vehicle trajectory with a particular vertical plane containing a landing site, it is feasible to obtain lateral-alignment guidance directly from the across-trajectory scanning beam. The station site for this localizer-type operation would be in the desired vertical alignment plane, located on the earth's surface at a point to be passed over before reaching the landing area. The 90-degree value on the across-trajectory scan is taken as the reference plane, and left-right guidance is obtained as the deviation from 90 degrees of the elevation-angle output of the across-trajectory plane of position.

IV. BLOCK DIAGRAMS

Fig. 5 is a block diagram of the major functional features of a SPACESCAN station. At each station, there are two systems--one for an along-trajectory scan and one for an across-trajectory scan. As the antenna rotates in a vertical circle, the scan of the peak of the beam is picked off the rotating mechanism at a rate of one pulse per 0.01 degree of movement. The accumulated pulse train is encoded to a time-delay scale (4 μsec per degree) by a digital-counting encoder. The time delay establishes the interpulse spacing between successive pulses of the modulator output and thus between successive pulses of the transmitter output. The K_u-band (16-kMc, 0.35-μsec) pulses are radiated over the scanning beam. The result is that the pulse contents of the scanning beam contain the data representing the elevation angle of the beam throughout 180 degrees of scan.

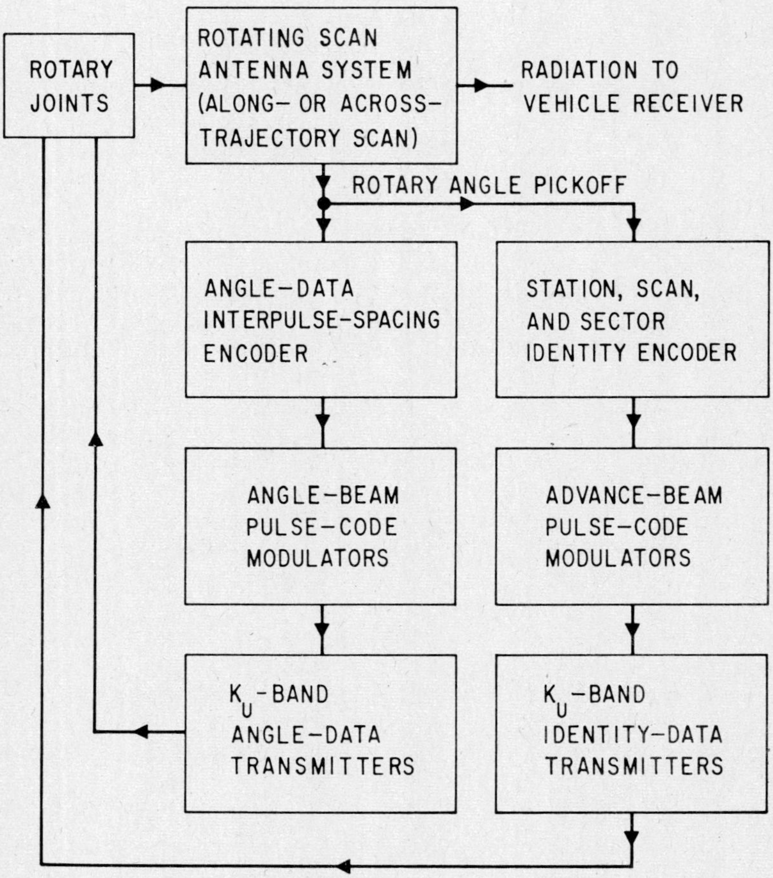

Fig. 5 Functional Block Diagram of Spacescan Ground Station

The modulation code is repeated over six 30-degree sectors. The sector number (one to six) is encoded on a beam that precedes the angle-data beam by 1 degree in the scan (see Fig. 6). The advance beam also contains pulse-pair coding that identifies the station site and the scan direction (along- or across-trajectory). The identity code is used in setting up the vehicle receiver circuits for unambiguous angle tracking, and in the vehicle coordinate computer for selecting the station constants involved in computing geographical coordinates of position.

Fig. 7 is a block diagram of the major functional features of the vehicle equipment. The RF and IF sections of the receiver supply detected pulses to the identity decoder and the angle-data decoder. The decoder supplies a voltage analog of the interpulse spacing to one of three angle-tracking memory circuits, correcting the angle memory to define the plane of position of the vehicle. The automatic calibration operates just before the angle beam is received, establishing the decoder levels for highly accurate tracking outputs.

The computer contains stored constants relating to the identity of the station and the scan; it operates on the three angle-data inputs to provide a position fix in earth-centered rectangular coordinates. These coordinates are then processed into the outputs shown in Fig. 7. The outputs are used as back-up for the inertial guidance system, and are displayed for monitoring and for manual-control back-up.

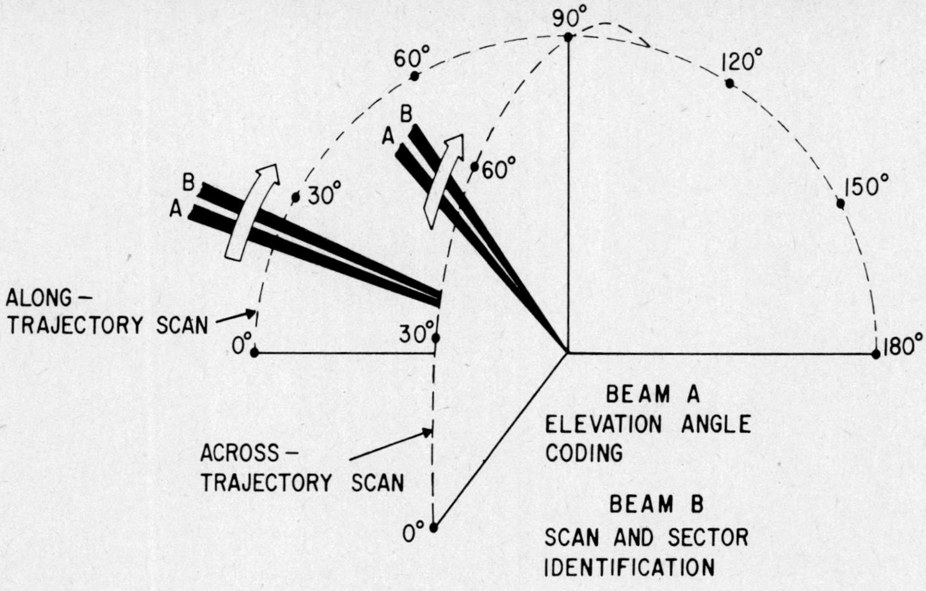

Fig. 6 Relationship Between Advance (Identity) Beam and Angle-Data Beam

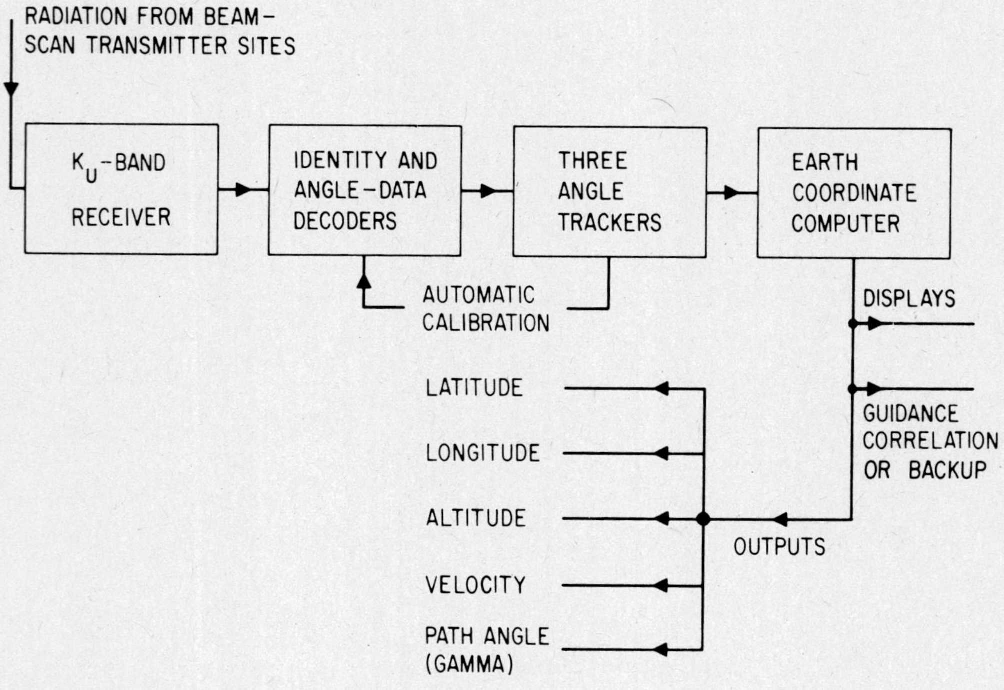

Fig. 7 Functional Block Diagram of Spacescan Vehicle Receiver

V. ANTENNA

The 0.5-degree beamwidth is formed by an 8-foot antenna aperture in the vertical plane of the semicircular scan. This aperture is energized as soon as the beam rises above the plus axis, and the energy is cut off when the beam drops below the minus axis. Another antenna aperture, the back-to-back mate of its forerunner, will now rise above the plus axis, and the transmitted energy will be switched to the newly rising beam, as shown in Fig. 8.

The antenna system for each line of scan consists of six linear apertures, arranged in the shape of two overlapping equilateral triangles (Fig. 9). While their beams are above the tangent plane, three of the six apertures are energized. Since their beams would be directed below the tangent plane, the remaining three apertures are de-energized.

As the beam of an active aperture scans the semicircle over the tangent plane, the elevation angle of the pointing line of the beam is picked off the antenna structure at the rate of one pulse per 0.01 degree of antenna motion. The pulse count is run up on a register during a 30-degree sector of scan, run down on the next 30-degree sector of scan, run up again, and so forth. The pulse count, therefore, represents the angle of the pointing line in a 30-degree sector.

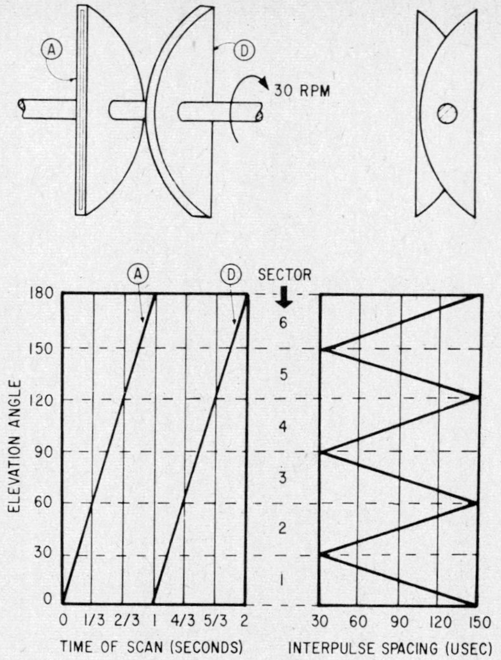

Fig. 8 Back-to-Back Scanner

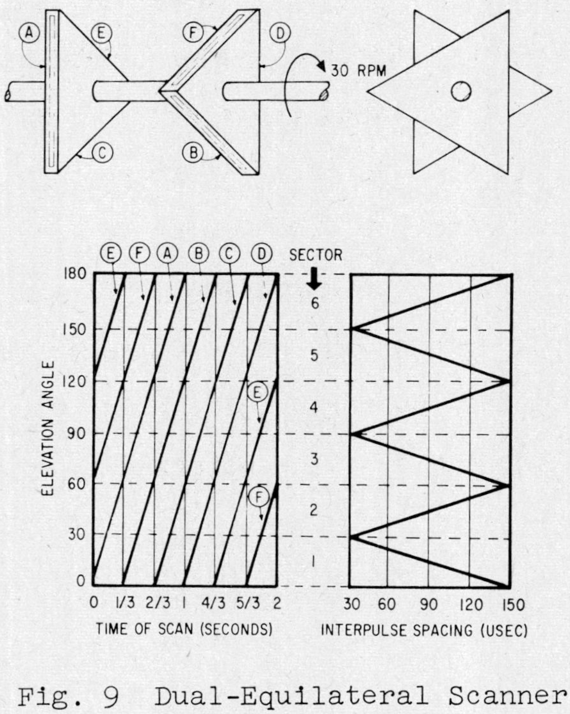

Fig. 9 Dual-Equilateral Scanner

In a parellel register, the angle pulse count is run down at a standard rate (25 counts per microsecond). The time required to run down the count is taken as the measure of the interpulse spacing for modulating the transmitter.

The interpulse spacing is arranged to vary from 30 to 150 μsec, corresponding to a scanned sector of 30 degrees.

It is convenient to use a code sector that is one-half the angular spacing between beams in each of the semicircular scans. As shown in Fig. 9, the angle-data beams for the three active antenna apertures are separated by 60 degrees of scan angle. The range of interpulse spacing (30 to 150 μsec) encodes a 30-degree sector of scan. Under these conditions, the three beams that are on the air carry the same angle-data code (except for sector number). This permits some simplification in the techniques of picking off the rotational angle from the antenna system and for encoding the modulators of the RF transmitters.

The scanning speed is 180 degrees per second, so that 1 degree of scan takes 5555 μsec. With an interpulse spacing of 150 μsec, a receiver will detect a train of about 40 pulses within 1 degree of scan. The half-power beamwidth is 0.5 degree, and the system can be designed so that only pulses more than 16 db below the strongest ones in a beam fail to contribute to the angle-data measurement. Consequently, the effective measurement beamwidth is practically 1 degree, and at least 35 to 40 pulses are contained in the effective beam envelope. Based on experience with angle measurements in AIL Landing System equipment, it is expected that the angular accuracy will not be significantly affected by variations in interpulse spacing for the system of SPACESCAN techniques described here and that an angular accuracy of 0.25 milliradian is technically feasible.

VI. IDENTITY CODING

To interpret the angle readout operationally, the vehicle equipment must have information relative to the identity of the station (one of about twelve stations), the identity of the scan (one of two possible directions for each station), and the identity of the 30-degree encoded sector (one of six possible sectors for each scan).

The information relating to station identity, scan identity, and sector identity is contained on a microwave beam that scans over the vehicle just before the reception of the angle-data beam. An advance beam is associated, on a one-for-one basis, with each of the six angle-data beams from a station. Using a separate transmitter and a microwave antenna feed slightly displaced from the focal point of the angle-data feed, the advance beam is formed by the same antenna aperture that forms the associated angle-data beam. The small displacement from the aperture focus gives the advance beam pattern a 1-degree spatial displacement toward the direction of the final end of the semicircle of scan. Consequently, the vehicle equipment will first receive the advance beam and then the angle-data beam, all within a period of about 1/60 second (see Fig. 6).

The principal function of the advance beam is to establish the identity of the station, the scan, and the sector. Secondary functions of the advance beam are to set the automatic gain control of the receiver for the associated angle-data beam and to gate on the receiver circuits to anticipate the advent of a train of angle-data pulses.

Taking advantage of the fact that the angle-data pulses are single pulses spaced not less than 30 μsec apart, the identity messages in the advance beam are distinguished by pulse pairs whose interpulse pair spacing is 20 μsec for the station and scan identity and 24 μsec for the sector identity.

Each scan will have a constant address message for its site number (one of about ten) and its scan direction (one of two). Accordingly, the 20-μsec address pairs of pulses are spaced 41 to 65 μsec apart (leading pulse to leading pulse), thus allowing 24 possible addresses at increments of 1 μsec. A typical address message thus consists of two 20-μsec pulse pairs within an interval of 85 μsec maximum.

The address messages are interlaced, on a one-for-one basis, with sector identity messages. These are 24-μsec pulse pairs whose interpair spacing (leading pulse to leading pulse) is between 40 and 46 μsec. This allows six possible sector identities at increments of 1 μsec per sector. A typical sector identity message thus consists of two 24-μsec pulse pairs within an interval of 70 μsec.

A total set of messages (station, scan, and sector) is thus completed within less than 200 μsec. Since a 1/2-degree beam takes about 2800 μsec to scan through the vehicle equipment, the advance beam will deliver at least 14 complete scan address and sector identity messages before the pulses of the angle-data beam are received by the vehicle equipment.

Since the three message-data beams for one scanner carry the same information relating to site identity and scan identity, the only difference among the transmitted data for a given scanner is the sector identity number; the difference between sector identity numbers is precisely two at any instant. As previously discussed, the angle-data beams for one scanner carry the same angle-data code. These universal features among the data are useful in simplifying the monitoring equipment required at the station sites.

To obtain more accurate updating of the angle data during the condition of angular acceleration, it is advantageous to obtain several beam measurements per second. Accordingly, the system is designed to provide three beam measurements per second per scan, using the dual-equilateral-shape antenna structure that radiates three beams 60 degrees apart (see Fig. 9).

VII. PERFORMANCE ANALYSIS

Fig. 10 shows the predicted acquisition ranges for the achievable transmission and reception characteristics and for a relatively severe weather model. It is seen that the K_u-band system is good for maximum ranges of about 1000 to 3000 miles, depending on weather and angle of elevation. This range capability is more than adequate for re-entry applications, where the curvature of the earth limits radio coverage to ranges less than 1000 miles.

Accuracy of position fix is determined by the angular accuracy of the system and by the geometric dilution inherent in the intersections of the three planes of position. The angular system accuracy is 0.25 milliradian, so that very roughly the position accuracy is 0.025 percent of the distance to the more remote station--for example, 0.1 mile for a site spacing of 400 miles.

A more precise error analysis requires that the partial errors due to angular error in each plane of position be combined by taking into account the geometric relations between the trajectory and the site locations. A precisely computed error analysis is shown in Fig. 11. A possible trajectory is shown for a 2000-mile re-entry toward a landing area.

The station sites are shown on the bottom scale of the figure. Stations 1 through 7 are spaced 240 miles apart,

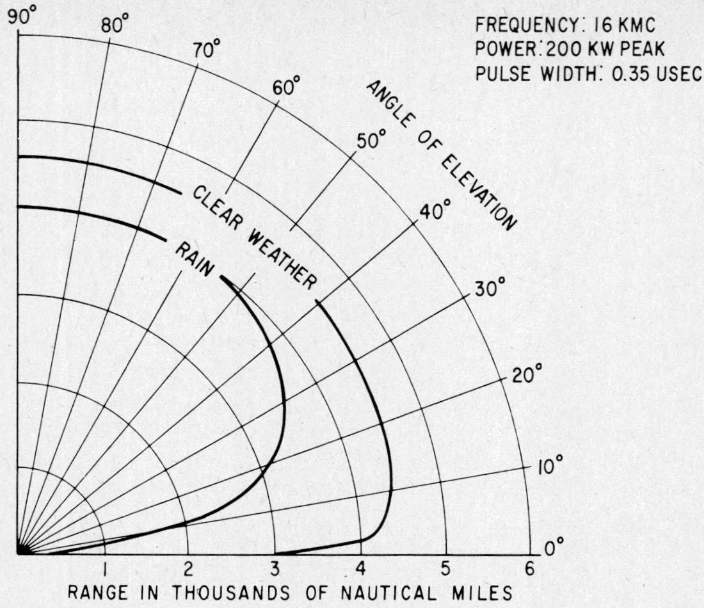

Fig. 10 Predicted Acquisition Ranges for Spacescan

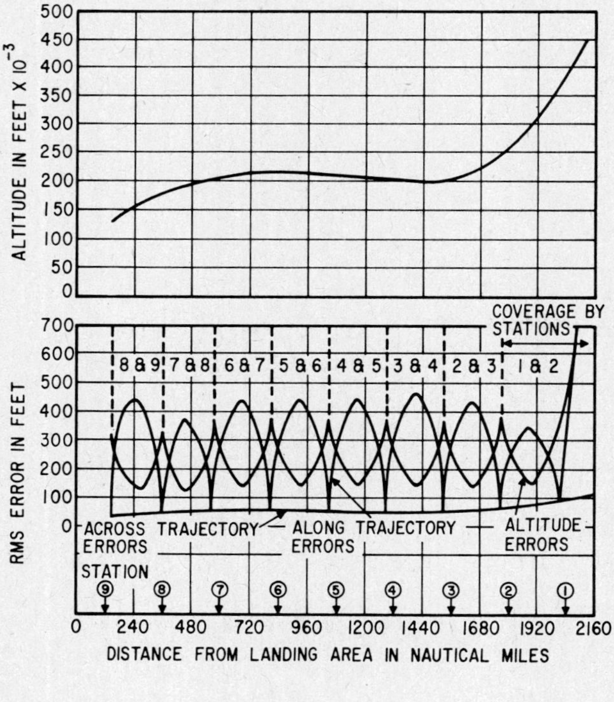

Fig. 11 Re-entry Trajectory and Error Plot

stations 7 through 9 are spaced 210 miles apart, and station 10 is located 90 miles from station 9. For the re-entry trajectory shown at the top of the figure, the chain of stations provides solid coverage for three planes of position for re-entry trajectories that are within about 100 miles to either side of the line of stations.

Altitude error, along-course error, and across-course error were computed for a trajectory directly over the station chain. The errors were computed as the square root of the total variance resulting from a standard deviation of 0.25 milliradian for each of the three angle values that define the three planes of position used in determining a position fix.

From the figure it is seen that the maximum altitude errors (about 350 feet) occur directly above the stations; the minimum altitude errors (about 120 to 150 feet) occur midway between stations.

The maximum along-trajectory errors (about 450 feet) occur midway between stations and the minimum along-course errors (about 50 to 100 feet) occur directly over a station.

The across-trajectory errors (about 50 to 100 feet) are the smallest of the three errors, mainly because the trajectory is over the chain of stations.

The differences among the distributions for the three types of errors are due to the differences in geometric dilution at different positions with respect to the pairs of sites from which the planes of position originate. The transients in the error plots are due to changes in the particular pairs of sites used along the trajectory.

Throughout the entire trajectory over the chain of stations, the position error in any direction is uniformly less than 500 feet.

ELECTROMAGNETIC RADIATIONS FROM SATELLITES AND SPACE VEHICLES - PART I*

K. N. Satyendra **

The literature is full of classical and contemporary work, both theoretical and observational, on the radiations from celestial bodies. Satellites and space probes have brought new objects into the celestial environment and give out electromagnetic radiations which are of interest to the astronomer, the physicist and the astronaut. Liberated from the absorptions and attenuations associated with atmospheric transmission, the radiations from space vehicles far above the atmosphere deserve detailed theoretical and experimental investigation.

These radiations, spanning the entire electromagnetic spectrum, may be generated internally due to active sources or externally due to reflections from the vehicle surfaces due to irradiation from the sun, the stars and nebulae, the planets and man-made radiative sources.

The first part of this paper presents a qualitative and quantitative examination of space vehicle radiations and the factors influencing them. Using the Echo satellite as an example, the detection theory of satellites including photometric considerations, space background problems and methods of identification of the satellite against a celestial background and theoretical calculations of detection ranges due to reflected sunlight are presented. A brief analysis is made of the nature and magnitude of the variation of the radiation as affected

---

* The complete paper may be made available by contacting the author.

** Director of Research, Nortronics, Northrop Corporation, Hawthorne, California.

by their motion relative to the earth and the sun. Conclusions are drawn that for space-to-space detection, the visible part of the spectrum offers many advantages over other regions in terms of size, weight, and power restrictions of spaceborne apparatus. An aggressive research effort is, however, required in the sensor technology.

Some interesting applications of the detection theory for navigational purposes are presented and a review of experimental and other considerations pertaining to observational physics, which will follow in Part II of this paper, are indicated.

PROPAGATION OF ELECTROMAGNETIC WAVES THROUGH REENTRY-INDUCED PLASMAS

P.J.Friel* and B. Rosenbaum*

The absorption and reflection of UHF telemetry signals in the ion layer are examined. The dependence of absorption and reflection upon frequency are further examined to determine whether there is a practical frequency that can be used for transmission throughout the reentry phase of the trajectory.

INTRODUCTION

When a space vehicle impinges on the atmosphere at hypersonic velocities, UHF signals abruptly black out, the precise interval of blackout and the degree of signal fade being dependent on the vehicle configuration and trajectory. The consensus is that the disappearance of the signal is associated with the formation of a shock-ionized flow about the vehicle. To explain signal fade, investigators have considered: (a) attenuation from absorption and reflection, (b) antenna detuning resulting from impedance mismatch between antenna and ion layer, and (c) voltage breakdown generated by the electric field in the antenna region. The results obtained are sensitive to the calculated plasma properties in the shock layer, and these properties, in turn, are sensitive to complex physicochemical processes within the shock layer (Section 1). Subsequently described in this report are the methods of calculating the flow field in the shock layer, the thermodynamic properties and plasma properties, the interaction of electromagnetic signals with the plasma, and signal attenuation.

---

* General Electric, Missile and Space Vehicle Dept., Philadelphia, Pa.

1. FORMATION OF THE PLASMA SHEATH

In the hypersonic flight of a re-entry vehicle, the quiescent atmosphere experiences a strong thermal excitation through the bow shock. The ionized gas formed in the process is swept by the streaming gas from the stagnation region along the body wall into the wake. Electromagnetic signals emitted from the vehicle and seeking to traverse the streaming plasma may thereby be strongly absorbed or reflected. The plasma thus constitutes a barrier for signals and is termed a plasma sheath. The properties of the plasma are characterized by two parameters: (a) electron concentration and (b) collision frequency of electrons with neutral atoms and ions. The determination of these parameters is important in ascertaining the optimum location of an antenna and in establishing the transmission properties of the plasma. The calculations necessary for evaluating these parameters with any degree of accuracy are highly complex. They are governed by the hypersonic flow in the shock layer, to which is coupled a complicated set of relations on the kinetics of various reacting chemical species.

The salient parameter in the plasma is the electron concentration. For a gas in thermodynamic and chemical equilibrium, the methods of statistical mechanics provide, in principle at least, a procedure for calculating this concentration. In the shock layer, the thermodynamic properties are the quantities least amenable to calculation. Not only are the thermodynamic properties of real gases at high temperatures complicated and incompletely understood, but in the flow there prevail wide departures from chemical equilibrium caused by the hypersonic speed.

The energy imparted by the shock wave to the atmosphere is initially manifest in the translational degrees of molecular motion and then in the internal molecular energy states. The redistribution of the energy as the system tends toward equilibrium is a sequential process determined by the relaxation time associated with each mode of freedom and by the reaction rates for chemical combination and dissociation. At altitudes of 150,000 ft or less, the internal states of molecular motion are fully excited in a few collisions immediately after crossing the normal bow shock wave. The gas then goes through the stagnation region where the enthalpy has a maximum value. In this region, the high temperature and density form a favorable environment for dissociation, ionization, and recombination of the atomic and molecular species that are subsequently swept downstream. Theoretical and experimental studies of the complex milieu prevailing in the stagnation region have not yet been able to treat all the reactions in a consistent and satisfactory manner. In the present state-of-the-art, the best we can do to describe the gas is to say that it is in quasi-equilibrium with the various atomic and molecular constituents having their own "temperature."

The flow of the gas downstream into the supersonic region is a similar complex process. The most reasonable description is that the flow is quasi-adiabatic, involving both temperature and pressure drop. The nonequilibrium state of the downstream flow is measured by a relaxation length. Thus, the nonequilibrium parameter for the flow is a ratio, Λ, of relaxation length to some characteristic body dimension. For $\Lambda = 0$, the flow is in a chemical equilibrium; for $\Lambda = \infty$, the flow is said to be frozen. The recombination rate for chemical reactions which is a three-body collision, hence, slow, will be further reduced by the low temperature and pressure. Dissociative equilibrium is thus impeded. Once freezing of the dissociation state is initiated, it will probably persist downstream to the wake region.

The flow described above is an inviscid flow prevailing between the shock wave and a boundary layer adjacent to the body wall. The boundary layer has properties quite distinct from the inviscid flow. In this layer, there is a large velocity gradient, and viscous and transport phenomena are important. In an analysis of the inviscid flow field, the presence of the boundary layer often can be ignored. It cannot be ignored, however, in the rarefied high-altitude regime of 250,000 ft, where the boundary layer becomes thick, comparable in thickness to the shock layer itself. The analysis of this flow is much more complicated than that at the denser atmosphere at 150,000 ft and has consequently received limited investigation. Efforts to determine the plasma properties in the rarefied high altitude are currently based on bold extrapolation from calculations of those at the 150,000-foot level.

2. THE INVISCID FLOW FIELD AROUND A 9-DEGREE SPHERE-CONE DURING RE-ENTRY

The properties of the gas surrounding a re-entry vehicle can be determined by the analysis of the hypersonic flow around axisymmetric bodies for air in chemical equilibrium developed by Gravalos, Edelfelt, and Emmons (Ref. 3). The

IBM 704 program developed from this analysis has been used by Langelo and Stamm (Ref. 4) to determine the inviscid flow around a 9-degree sphere-cone at Mach numbers of 20, 21, and 21.2 for altitudes of 60,000, 75,000, and 160,000 ft, respectively. These results can be used to predict the flow field around a body of the sphere-cone type (Fig. 1).

The vehicle trajectory considered is shown in Fig. 2. The temperature and density of the gas through the shock layer surrounding the body have been determined by Langelo and Stamm (Ref. 4) at several locations along the body, indicated in Fig. 1 and normals 1, 2, 3, and 4. The temperature and density along two of these normals (1 and 3) are shown in Figs. 3, 4, 5, and 6, where both quantities have been normalized to the difference in the temperature and density between the body and the shock, i.e., to $T_B - T_S$ and ρ_S and ρ_B, respectively.

We will be concerned with the state-of-the-gas along normals 1 and 3; however, the state-of-the-gas along normals 2 and 4 is not drastically different from that along normals 1 and 3, respectively. Consequently, the properties of the plasma along normals 1 and 3 should be reasonably representative of most of the plasma surrounding the body for a side aspect.

Since the analysis concerns the inviscid flow field, the conditions at the body actually refer to the conditions at the edge of the boundary layer, which was neglected in the present work. It is interesting to note that the profiles for the two low-altitude cases are essentially equal and that they, in turn, are quite similar to the profiles for the high-altitude case. Consequently, to utilize the available

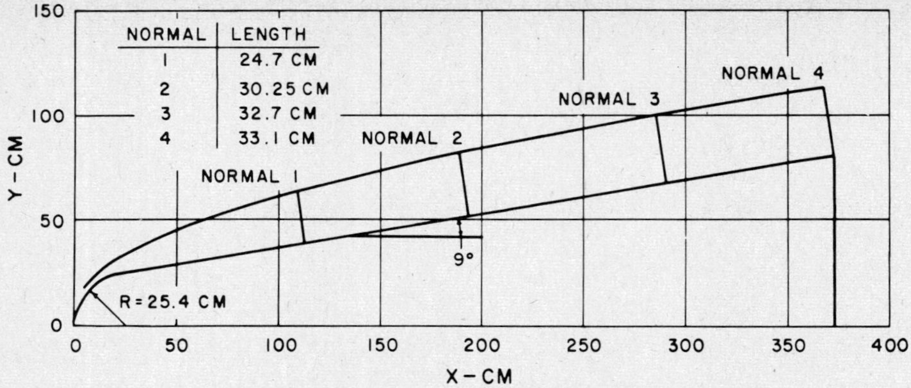

Fig. 1 Vehicle Configuration

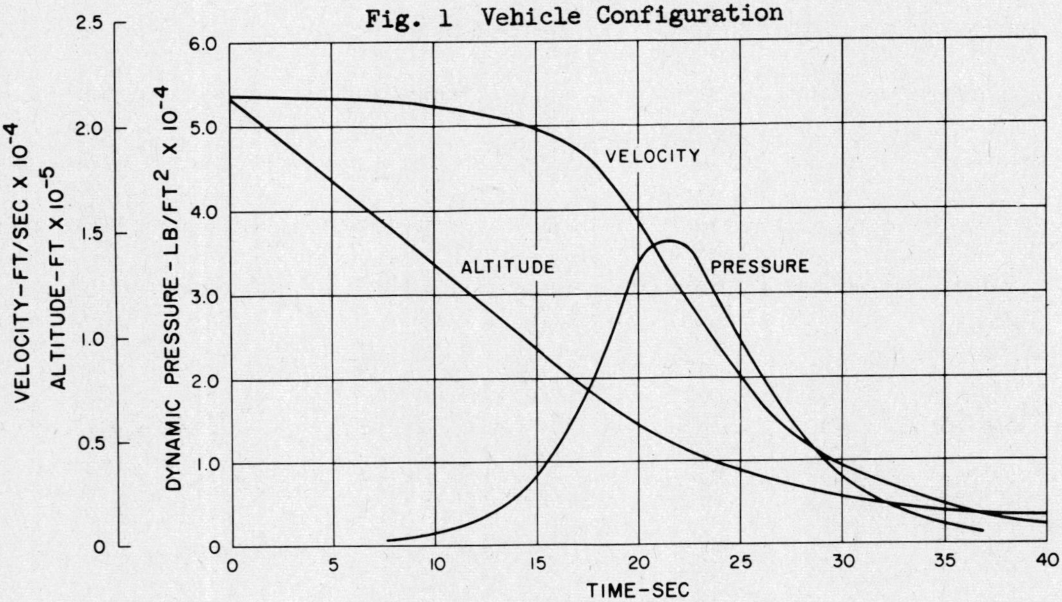

Fig. 2 Vehicle Trajectory

Fig. 3
Temperature Correlation
Normal 1

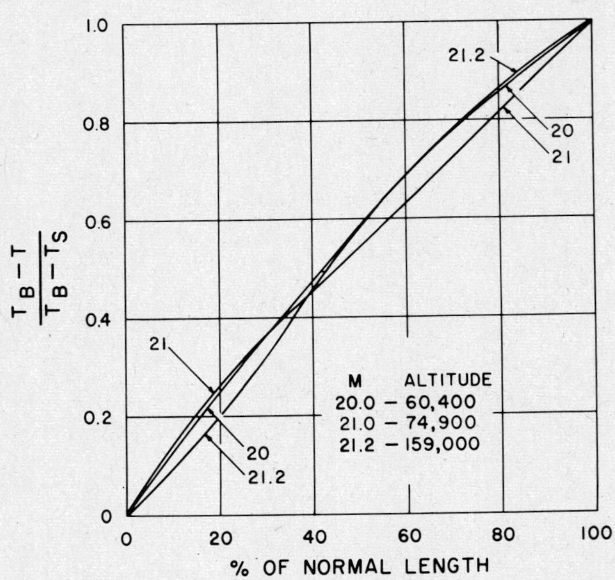

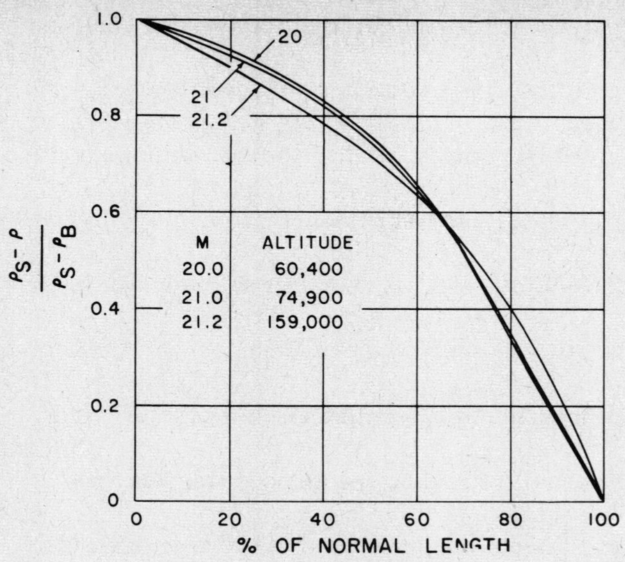

Fig. 4
Density Correlation
Normal 1

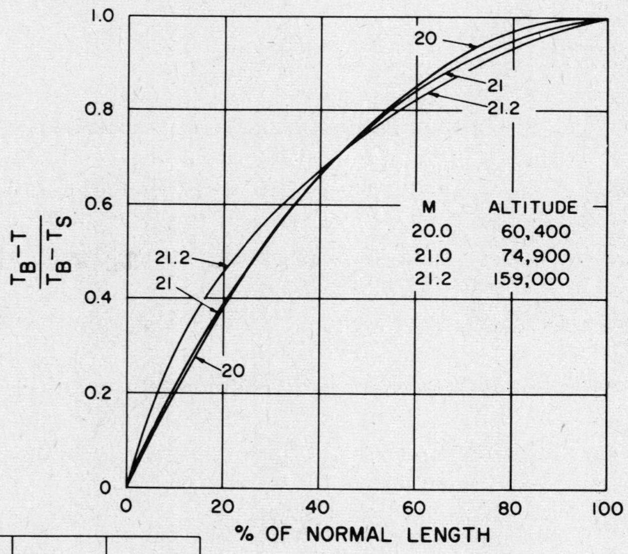

Fig. 5
Temperature Correlation
Normal 3

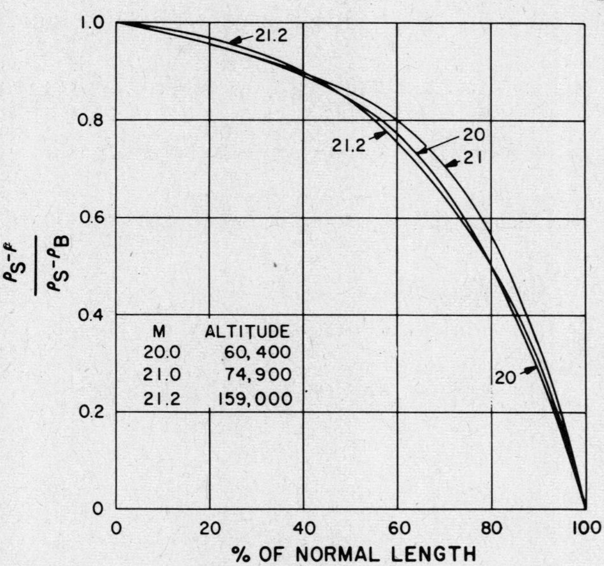

Fig. 6
Density Correlation
Normal 3

inviscid flow fields to determine the plasma properties around the body at various altitudes, it was assumed that these profiles would represent the profiles of temperature and density that would be obtained at other altitudes for the same body at velocities around 20,000 ft per second. Thus, the temperature and density profiles through the high-altitude flow field (160,000 ft) were assumed to be the shock layer profiles above 150,000 ft, and the low-altitude profiles were assumed to represent the flow field below 150,000 ft. Consequently, to determine the temperature and density through the shock layer along normals 1 and 3, it was necessary to know only the temperature and density behind the oblique shock and at the body for the two normals.

The conditions behind the oblique shock were determined by assuming that the shock angle at all altitudes was that predicted by the exact inviscid flow fields, which was very close to 10 degrees. Since the oblique shock is very weak, the temperature behind the shock is low enough that no significant amount of dissociation takes place and the conditions behind the shock can be determined by means of the shock tables for an ideal diatomic gas (Ref. 5).

The temperature and density of the gas at the body were determined by assuming that the gas at the body had originally passed through the stagnation region. The stagnation conditions were determined by first assuming that the stagnation pressure is not significantly affected by dissociation, so that the stagnation pressure can be determined from Ref. 5. The stagnation enthalpy is

$$h_T = h + \frac{V^2}{2}. \qquad (1)\ *$$

* A table of nomenclature is appended.

The temperature and entropy of the gas in the stagnation region can then be determined from the charts prepared by Moeckel and Weston (Ref. 6). The gas conditions at the body for the various normals were determined by isentropically expanding the gas from the stagnation pressure to the local body pressure. The local body pressures, in turn, were calculated from the local pressure coefficient C_p, which was determined by the exact inviscid flow field program. Values of the pressure coefficient at the body for the various normals are shown in Table I for the velocities and altitudes considered. It can be seen that the pressure coefficient was reasonably constant at each normal. Again with the aid of the charts in Ref. 6, the temperature, density, and compressibility of the gas at the body for the various normals can be determined after expansion from the stagnation pressure to the local body pressure. Thus, the temperature and density of the gas behind the shock and at the body can be determined for the various normals at any altitude and velocity of interest. Then with the aid of the correlations of the temperature and density through the shock layer shown in Figs. 3, 4, 5, and 6, the temperature and density profiles through the shock layer at normals 1 and 3 can also be determined for any altitude and velocity.

TABLE I. PRESSURE COEFFICIENTS AT VARIOUS BODY STATIONS OF THE VEHICLE

Altitude (ft)	Mach	Normal	C_p
60,430	20.0	1	0.0401
74,900	21.0	1	0.0493
159,000	21.2	1	0.0423
60,430	20.0	2	0.0363
74,900	21.0	2	0.0373
159,000	21.2	2	0.0360
60,430	20.0	3	0.0320
74,900	21.0	3	0.0324
159,000	21.2	3	0.0336
60,430	20.0	4	0.0317
74,900	21.0	4	0.0323
159,000	21.2	4	0.0325

TABLE II. ELECTRON DENSITY AND COLLISION FREQUENCY IN STAGNATION REGION

Altitude (ft)	N_e	ν_c
285,000	3.683×10^{11}	5.444×10^{9}
200,000	1.035×10^{13}	2.886×10^{10}
175,000	2.321×10^{13}	4.322×10^{10}
150,000	5.527×10^{13}	6.668×10^{10}
100,000	1.146×10^{14}	9.607×10^{10}

3. THE ELECTRON DENSITY AND THE COLLISION FREQUENCY IN THE PLASMA SURROUNDING THE BODY

The temperature and density of the plasma through the shock layer at normals 1 and 3 were determined as described above for the vehicle at six different altitudes in its re-entry trajectory (285,000, 200,000, 175,000, 150,000, 100,000, and 50,000 ft). The electron density at each point along a normal can then be determined from the tables of the composition of high-temperature air developed by Hilsenrath and Beckett (Ref. 7). The electron density is calculated from the ideal gas law and can be shown to be given by

$$N_e = \frac{X_e PA}{RT} , \qquad (2)$$

where X_e is mole fraction of electrons. The electron collision frequency is defined by Menzel (Ref. 8) as

$$\nu_c = \frac{C_{Ae} + 2C_{ee}}{N_e} , \qquad (3)$$

where C_{Ae} is the number of collisions between electrons and neutral or other charged particles and C_{ee} is the number of electron-electron collisions. The number of electron-electron collisions is very small and can be neglected. The quantity C_{Ae} can be calculated from the kinetic theory of gases (Ref. 8). Assuming that the effective electron collision cross section in high-temperature air is that determined by Lin et al (Ref. 9) (about 7×10^{-16} cm^2), it can be shown that the collision frequency is given by

$$\nu_c = 1.0 \times 10^{13} \frac{P}{\sqrt{T}} . \qquad (4)$$

Thus, the electron density and collision frequency along normals 1 and 3 were determined at six altitudes in the re-entry trajectory. The results are shown in Figs. 7, 8, 9, and 10. It is important to note that the electron density is, to a reasonable approximation, a simple logarithmic function of the distance from the body in the high electron density region near the surface and that the collision frequency through the plasma for a particular altitude is essentially constant. The electron density and collision frequency in the stagnation region are shown in Table II.

In all the calculations discussed above, it was assumed that the gas in the plasma around the body is in chemical equilibrium. However, there is considerable evidence that the flow beyond the sonic point is frozen above 150,000 ft, at least with respect to the atom recombination reactions of oxygen and nitrogen (Ref. 10).

4. THE IMPORTANCE OF NONEQUILIBRIUM EFFECTS AT HIGH ALTITUDES

In the calculation of the electron density and collision frequency through the plasma, we have assumed, as mentioned above, that the gas is in chemical equilibrium even at high altitudes. Above 150,000 ft, however, the flow is probably frozen beyond the sonic point of the body at least with respect to the atom reactions of oxygen and nitrogen (Ref. 10). Hence, as the gas expands around the body beyond the sonic point, the atoms of oxygen and nitrogen will not recombine fast enough and some of the enthalpy of the gas will be lost in the heat of formation of these atoms and thus the stream temperature in the frozen flow case will be lower than that in the equilibrium case. The reactions forming ions and electrons are not important in determining the temperature and density of the gas, since their

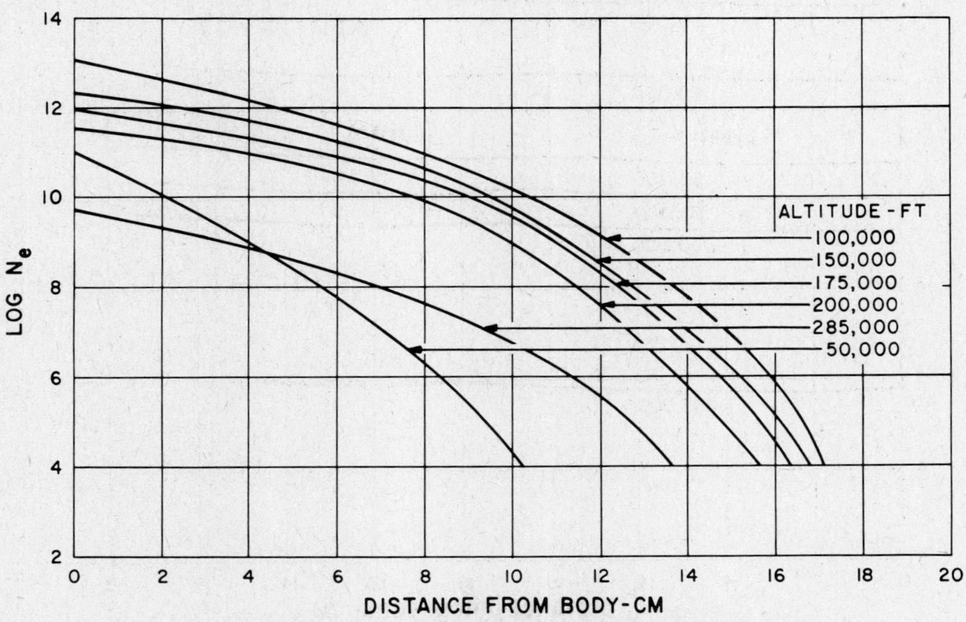

Fig. 7 Electron Density, Normal 1

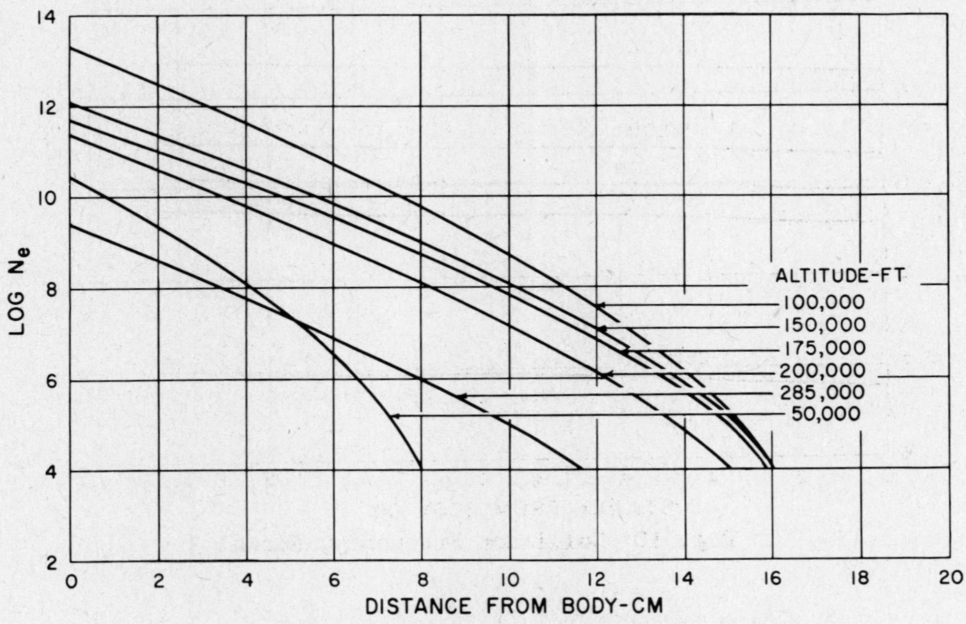

Fig. 8 Electron Density, Normal 3

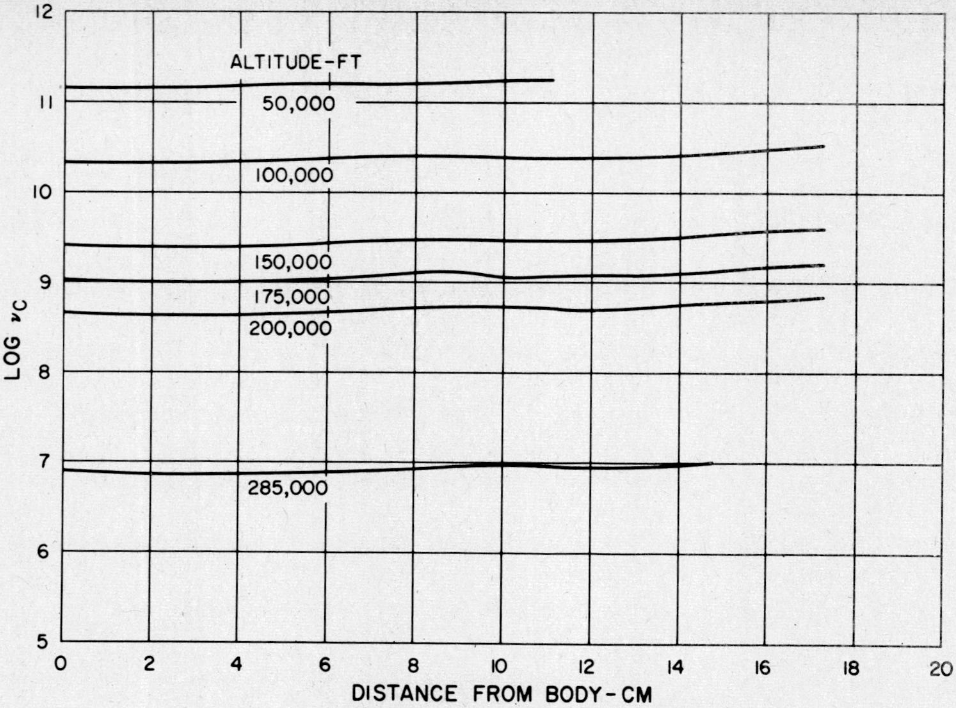

Fig. 9 Collision Frequency, Normal 1

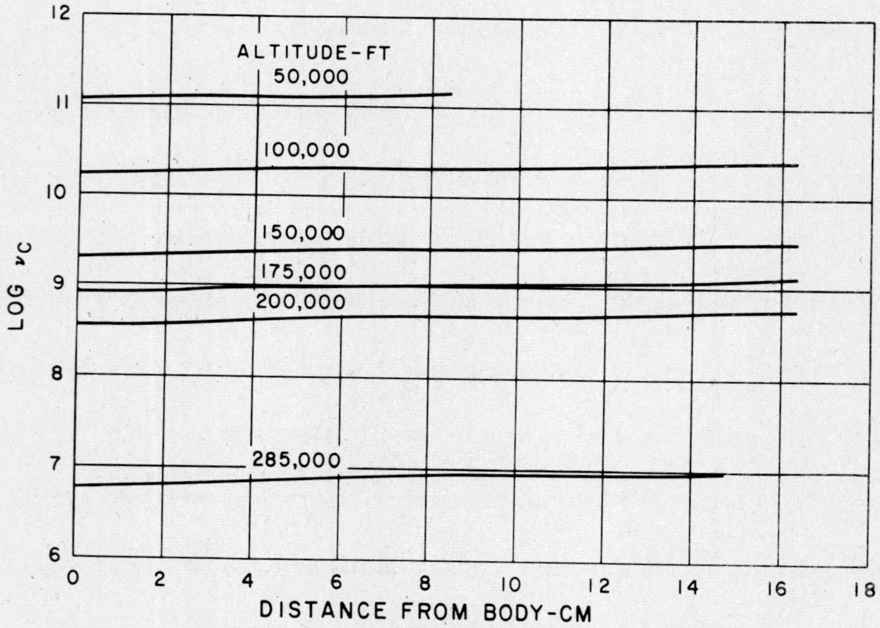

Fig. 10 Collision Frequency, Normal 3

concentration is always small compared with the concentration of atoms and molecules. A temperature decrease in the nonequilibrium case will be accompanied by a density increase, so that the local pressure will probably be the same in nonequilibrium and in equilibrium flow. The electron collision frequency is directly proportional to the pressure and inversely proportional to the square root of the temperature, so that the temperature decrease in nonequilibrium flow will increase the electron collision frequency. But a temperature decrease by a factor of 2 will increase the collision frequency by only 40 percent, assuming the pressure remains unchanged in the two types of flow. Therefore, it is reasonable to assume that nonequilibrium effects in the supersonic part of the flow field will not drastically alter the electron collision frequency. Also, it was shown above that the plasma surrounding the body at high altitudes is essentially collisionless (i.e., nonabsorptive), and nonequilibrium effects would have to increase the collision frequency by one to three orders of magnitude to have a significant effect on the attenuation of the telemetry signal.

The electron density, and possibly the electron density gradients, would, of course, be significantly different in nonequilibrium flow compared with equilibrium flow. In the case of the collisionless high-altitude plasma, if the plasma frequency at some point in the plasma is near the incident frequency, a relatively small change in the electron density will drastically affect the reflection coefficient. Therefore, nonequilibrium effects with respect to the electron-producing reactions could determine whether the plasma is reflective or transparent. It is difficult to say at this time whether the electron density in the supersonic region will be increased or decreased in nonequilibrium flow compared with equilibrium flow. If there is a point in the plasma at which the plasma and signal frequencies are equal in equilibrium flow, a small decrease in the electron density will render the plasma transparent, so that the signal will pass through the gas. On the other hand, an increase in the electron density in a transparent, collisionless plasma will introduce large reflection losses (see below). There is also the intermediate case in which nonequilibrium effects would simply change the point in the plasma at which the reflection occurs.

5. THE PROPAGATION OF AN ELECTROMAGNETIC WAVE IN A PLASMA

A plasma is an ionized gas, electrically neutral, whose volume elements have equal numbers of electrons and positive ions. An external electromagnetic wave propagating through the plasma will interact with the medium via the electrons. The role of the relatively immobile positive ions is to maintain the gas electrically neutral. Under the impressed field, the electrons oscillate and experience collisions with the neutral atoms, thereby transforming the electromagnetic energy of the wave to random thermal motion of the gas. For purposes of determining the propagation characteristics of the plasma, it is adequate to represent the equation of motion of the electrons by

$$m \dot{\vec{V}} + \nu_c m \vec{V} = e \vec{E}, \tag{5}$$

where v_c is the collision frequency of electrons with neutral atoms, m is mass of electron, $\vec{V}$ is the mean velocity, and $\vec{E}$ is the impressed external electric field.

If the collision frequency is taken to be independent of velocity (the Lorentz approximation), the current density is

$$\vec{J} = \frac{N_e e^2 \vec{E}}{m\,(\nu_c - i\omega)}, \qquad (6)$$

where w is the angular frequency of the impressed field and N_e is the electron density. The constitutive parameters--the complex conductivity σ and the complex dielectric constant ϵ--are conveniently represented in terms of the plasma frequency w_p given by

$$\omega_p^2 = \frac{4\pi N_e e^2}{m} \qquad (7)$$

Thus, from Eqs. (6) and (7),

$$\sigma = \frac{\omega_p^2}{4\pi(\nu_c - i\omega)}, \qquad (8)$$

and from Equation (A-10) in the Appendix,

$$\epsilon = 1 + \frac{4\pi i\sigma}{w}. \qquad (9)$$

For wave propagation in a nonuniform medium where the variation in electron concentration is in the z direction, the wave equation for propagation in this direction (see the Appendix) is

$$\nabla^2 \vec{E} + k^2 \vec{E} = 0, \qquad (10)$$

since E . $\nabla \epsilon = 0$ and where the propagator is

$$k = \frac{\omega}{c}\sqrt{\epsilon}. \qquad (11)$$

As an approximation, we shall take as the plane wave solution of Eq. (10).

$$E_x = E_x^{(o)} e^{i \int k dz} = E_x^{(o)} e^{-(\omega/c) \int \beta \, dz} e^{i(\omega/c) \int (\alpha - 1) \, dz} e^{i(\omega/c) z} \tag{12}$$

This approximation is good if k varies slowly over a wavelength but may not be so where $w \approx w_p$, but we shall consider Eq. (12) an adequate approximation for the accuracy required in this report.

6. ATTENUATION OF A TELEMETRY SIGNAL FROM A SPACE VEHICLE OF THE SPHERE-CONE TYPE

The transmission characteristics of a plasma will be dependent upon signal frequency. The transmission characteristics of typical plasma properties over a broad frequency spectrum (20 mc to 35 kmc) were studied to ascertain a frequency with which satisfactory transmission could be obtained over the entire re-entry phase. The plasma properties of the vehicle as discussed above were chosen for study of the db attenuation. The plasma properties at normal 3 were considered in the analysis. It is adequate to represent the radiation pattern emitted at the antenna cavity as a simple plane wave. The signal while propagating normally through the plasma sheath suffers both partial reflection and absorption. The reflected part will contribute to antenna detuning, but this effect is disregarded since the interest here is in the interaction of the signal and plasma only. The db absorptive loss is given (see Appendix) by

$$8.7 \frac{\omega}{c} \int \beta dz. \tag{13}$$

The integration takes account of the nonuniformity of the plasma absorption. The path of integration is through the entire plasma sheath. The methods of calculation in this section are similar to those outlined by Taylor (Ref. 11) for estimating attenuation. Plots of $(\omega/c)\beta$ as a function of distance through the plasma are shown in Figs. 11, 12, 13, 14 and 15 for various altitudes. The integration indicated in Eq. (13) was performed graphically.

Although the plasma is nonuniform, the reflection coefficient can be calculated in a simple way if one observes that the signal wavelength is much larger than the plasma thickness. We used the rough procedure of approximating the nonuniform by a uniform slab whose electron concentration is near the maximum of the nonuniform plasma. We therefore write the reflective db loss as approximately

$$2.10 \log_{10} \frac{1}{T}. \tag{14}$$

Here $T = (1 - R)$ = the transmission coefficient, where R is the reflection coefficient. The factor 2 accounts for reflection from the two surfaces of the model sheath. The effect of multiple reflection has been ignored in Eq. (14). This is a good approximation when absorption is high. When the effect of multiple reflection is significant, Eq. (14) can be modified in a simple manner. The reflection coefficient R is given by Equation (A-33) in the Appendix as

$$R = \frac{(\alpha - 1)^2 + \beta^2}{(\alpha + 1)^2 + \beta^2}, \tag{15}$$

where α and β are calculated from the properties of the uniform model. The total attenuation (in db) is obtained by combining Eqs. (14) and (15). Thus,

$$A = 2.10 \log_{10} \frac{1}{T} + 8.7 \frac{\omega}{c} \int \beta_{dz}, \tag{16}$$

where A is total attenuation and T is the transmission coefficient.

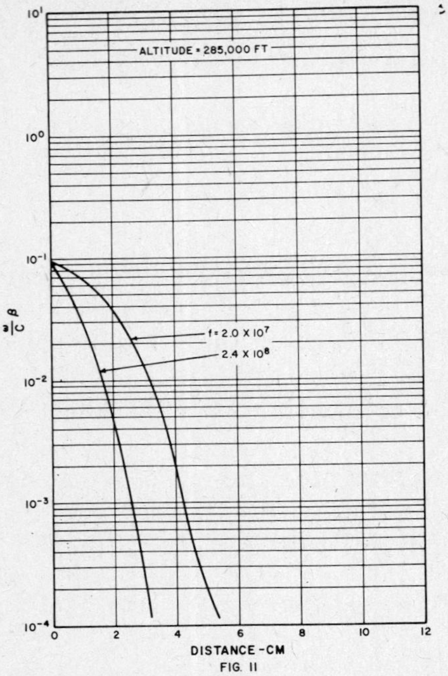

Fig. 11 Signal Attenuation vs Distance Through the Plasma, Altitude 285,000 Ft

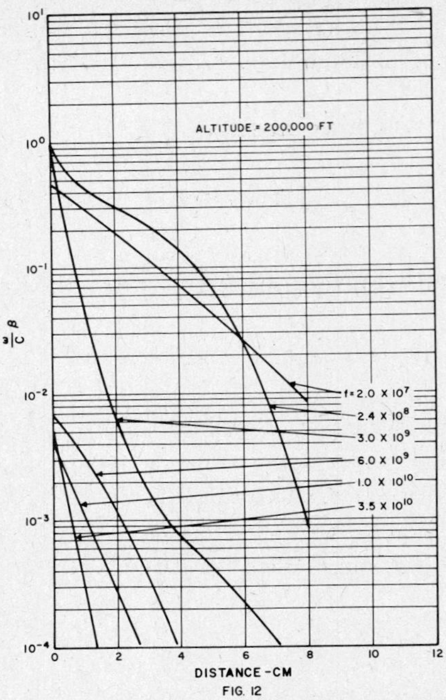

Fig. 12 Signal Attenuation vs Distance Through the Plasma, Altitude = 200,000 Ft

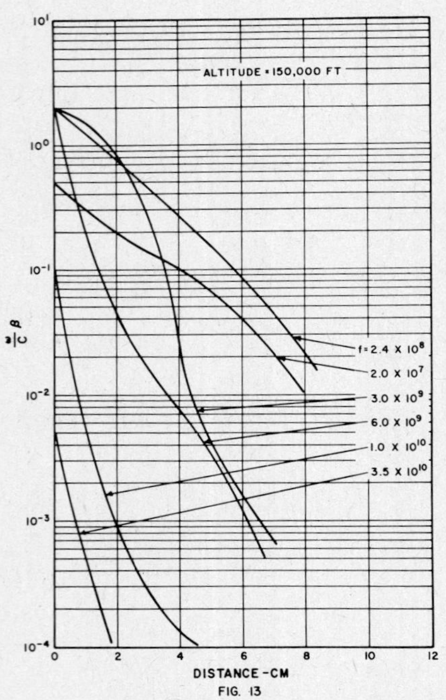

Fig. 13 Signal Attenuation vs Distance Through the Plasma, Altitude = 150,000 Ft

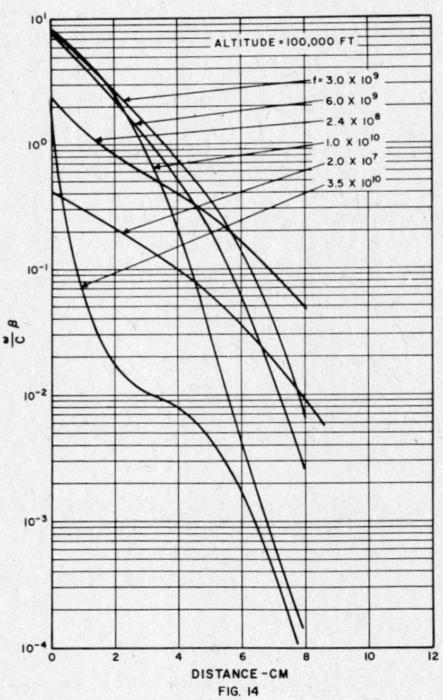

Fig. 14 Signal Attenuation vs Distance Through the Plasma, Altitude = 100,000 Ft

RESULTS

The attenuation factor (db) has been calculated for various frequencies in the spectrum from 20 mc to 35 kmc. We note first on examining the results as given in Table III and in Figs 16, 17, and 18 that the reflectivity is high at the lower frequencies. This is notably so in the high-altitude regime where the plasma is collisionless. There the plasma frequency exceeds the signal frequency, making the real part of the dielectric constant negative, and the plasma is consequently highly reflective. At lower altitudes, the medium behaves much like a lossy dielectric due principally to the high electron collision frequency. The frequency dependence of the reflection coefficient is evident in the systematic decrease of the attenuation as the signal frequency increases.

We observe further that absorption is very low at high altitudes due to the low electron concentration and collision frequency. Also, at 50,000 feet the absorptive attenuation is low, due here only to low electron concentration. In the intermediate-altitude regime (100,000 ft), both electron concentration and collision frequency are large. The absorption is found to be, as a result, most intense (approximately 100 db) for all frequencies except 35 kmc, where the attenuation drops precipitously--over 100 db. This occurred when the signal frequency exceeded the plasma frequency.

The calculations predict that a 240--mc signal will black out at all altitudes except 50,000 ft. Our results also show that a low-frequency (20-mc) telemetry signal will suffer high reflective losses. At 35 kmc, the plasma is found to be uniformly transparent at all altitudes.

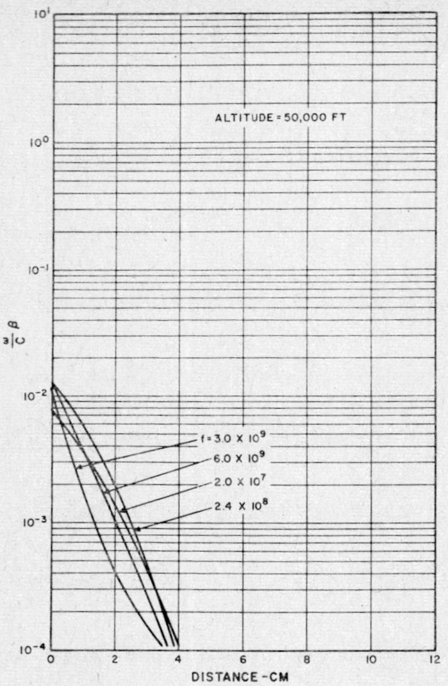

FIG. 15

Fig. 15 Signal Attenuation vs Distance Through the Plasma, Altitude = 50,000 Ft

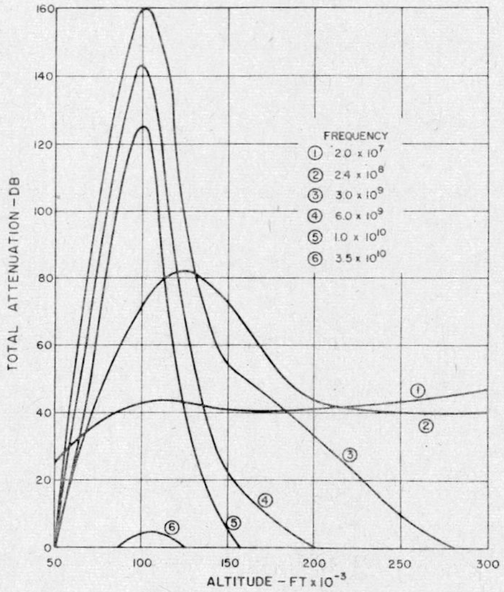

FIGURE 16

Fig. 16 Absorption vs Altitude

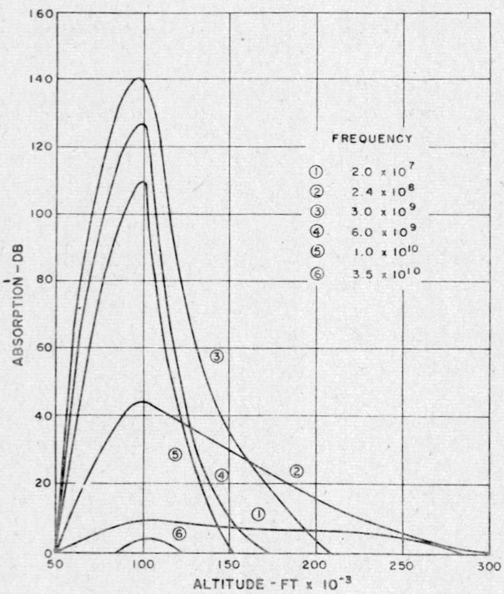

FIGURE 17

Fig. 17 Total Attenuation vs Altitude

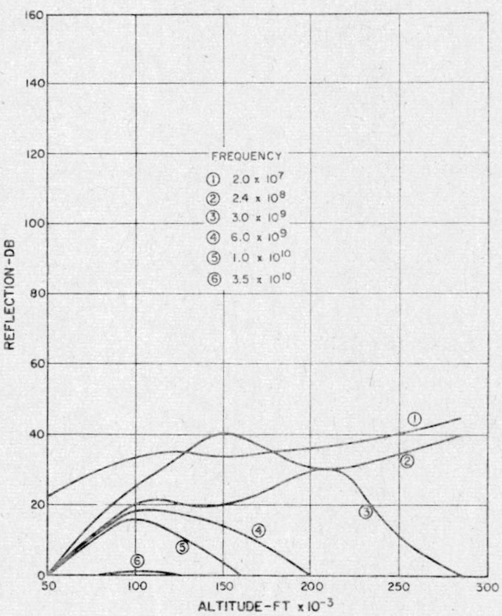

FIGURE 18

Fig. 18 Reflection vs Altitude

CONCLUSIONS

The predicted attenuation at 240 mc is 40 db at re-entry and rises to a maximum of 80 db at about 125, 000 ft. For a low-frequency (20-mc) signal, the reflection losses are high through most of the re-entry period but the absorptive losses are never greater than 5 to 10 db so that the total attenuation of a 20-mc signal is about 40 db through the blackout interval. For the S (3, 000 mc), C (6, 000 mc), and X (10, 000 mc) bands, significant attenuation (greater than 20 db) initiates at 225, 000 ft, 150, 000 ft, and 125, 000 ft, respectively; the maximum attenuation occurs at about 100, 000 ft (where the electron concentration is maximum), reaching values between 120 and 140 db in these frequency ranges. For the 35 kmc telemetry signal, the maximum attenuation is not greater than 5 db at 100, 000 ft. The calculations in this region are extremely sensitive to our knowledge of the plasma properties, however. For example, an error of a factor of 3 in the calculation of the plasma frequency would increase the attenuation of the 35-kmc signal to about 120 db. Consequently, the effect of ablation products or uncertainties in the calculations of the plasma properties could very well result in a severe attenuation of the 35-kmc signal which would initiate at about 125, 000 ft.

NOMENCLATURE

A	Avogadro's number = 6.02×10^{23}
$\vec{B}$	Magnetic field vector
c	Velocity of light
C_{Ae}	Number of electron-neutral molecule collisions
C_{ee}	Number of electron-electron collisions
C_p	Pressure coefficient
e	Electronic charge
$\vec{E}$	Electric field vector
$E_x^{(o)}$	Amplitude of incident wave
f_p	Plasma frequency
h	Enthalpy
h_T	Stagnation enthalpy
$\vec{J}$	Current density
k	Propagator, $k = (\omega/c)\sqrt{\epsilon}$
m	Mass of electron
N_e	Electron concentration
P	Pressure
$\vec{r}$	Position vector
R	Gas constant
Re	Real part of
$\vec{S}$	Poynting's vector
t	Time

T	Temperature
$\vec{V}$	Velocity
V	Velocity vector
X_e	Mole fraction of electrons
α	Real part of index of refraction
β	Imaginary part of index of refraction
ϵ	Dielectric constant
Λ	Relaxation length
ν_c	Electron collision frequency
ρ	Density
σ	Conductivity
ω	Angular frequency of radiation
ω_p	Plasma frequency

Subscripts

B	Body
S	Shock

REFERENCES

1. F. G. Gravalos, I. H. Edelfelt, and H. Emmons, "The Supersonic Flow About a Blunt Body of Revolution for Gases in Chemical Equilibrium," Ninth Annual Congress of the International Astrounautical Federation (Amsterdam, August 1958)

2. V. A. Langelo and S. L. Stamm, "Real Gas Flow Field Data," General Electric Company, Missile and Space Vehicle Department, Program Information Release 538-631

3. Ames Research Staff, "Equations, Tables, and Charts for Compressible Flow," National Advisory Committee on Aeronautics Report 1135

4. W. E. Moeckel and K. C. Weston, "Composition and Thermal Properties of Air in Chemical Equilibrium," National Advisory Committee on Aeronautics Technical Note 4265

5. J. Hilsenrath and C. Beckett, "Tables of the Thermodynamic Properties of Argon-Free Air to 15,000°K," Arnold Engineering Development Center Technical Note 56-12 (and addendum).

6. D. H. Menzel, "Formulas of Physics," Prentice-Hall, 1955, p 298

7. S. C. Lin, B. Kivel, and P. Hammerling, Bulletin of the American Physical Society, II:4 (1959), p 61

8. M. Bloom and M. Steiger, Institute of Aeronautical Sciences Paper 60-26 (January 1960)

9. W. C. Taylor, "Analysis and Prediction of Radio Signal Interference Effects Due to Ionized Layer Around a Re-entry Vehicle," Lockheed Missiles and Space Division Report LMSD-49754

APPENDIX

ELECTROMAGNETIC WAVE PROPAGATION AND ATTENUATION IN PLASMAS

WAVE PROPAGATION

Under a weak external electromagnetic field, the free electrons in the plasma will oscillate. This motion generates a polarization current $\vec{J}$ density given by

$$\vec{J} = N_e e \vec{V}, \tag{A-1}$$

where N_e is the electron density and V is the mean velocity. If the plasma is inhomogeneous, there will be local variation in the charge density related to the current by the equation of continuity,

$$\dot{\rho} + \nabla \cdot \vec{J} = 0. \tag{A-2}$$

The positive ions are relatively immobile and serve only to render the plasma electrically neutral. The polarization current is calculated from the equation of motion of the electron,

$$m \dot{\vec{V}} + \nu_c m \vec{V} = e \vec{E}, \tag{A-3}$$

where ν_c is the collision frequency of electrons with neutral atoms and $\vec{E}$ is the impressed external electric field. Taking the time dependence of the field as $e^{-i\omega t}$, the current is readily found to be

$$\vec{J} = \frac{N_e e^2 \vec{V}}{m(\nu_c - i\omega)}. \tag{A-4}$$

In terms of the plasma frequency

$$\omega_p^{\;2} = \frac{4\pi N_e e^2}{m}, \tag{A-5}$$

the current $\vec{J}$ can be written

$$\vec{J} = \frac{\omega_p^{\;2}\vec{E}}{4\pi(\nu_c - i\omega)}. \tag{A-6}$$

Maxwell's equations for the field are

$$\nabla \cdot \vec{E} = 4\pi\rho, \tag{A-7a}$$

$$\nabla \cdot \vec{B} = 0, \tag{A-7b}$$

$$\nabla \times \vec{E} = -\frac{1}{C}\dot{\vec{B}} \tag{A-7c}$$

$$\nabla \times \vec{B} = +\frac{1}{C}\dot{\vec{E}} + \frac{4\pi\vec{J}}{c}. \tag{A-7d}$$

For the time dependence $e^{-i\omega t}$, these equations (A-7) can be rewritten as

$$\nabla \cdot \vec{E} = \frac{4\pi}{i\omega}\nabla \cdot \vec{J}, \tag{A-8a}$$

$$\nabla \times \vec{E} = \frac{i\omega}{c}\vec{B}, \tag{A-8b}$$

$$\nabla \times \vec{B} = -\frac{i\omega}{c}\left(1 + \frac{4\pi i\sigma}{\omega}\right)\vec{E} = -\frac{i\omega}{c}\epsilon\vec{E}, \tag{A-8c}$$

where we used Equation (A-2) and introduced the constitutive complex conductive and dielectric parameters defined by

$$\vec{J} = \sigma\vec{E}, \tag{A-9}$$

$$\epsilon = 1 + \frac{4\pi i\sigma}{\omega}. \tag{A-10}$$

Now, inserting Equation (A-9) into Equation (A-8a) and using Equation (A-10) yields

$$\nabla \cdot \vec{E} = - \vec{E} \cdot \frac{\nabla \epsilon}{\epsilon}. \tag{A-11}$$

Then Equation (A-11) combined with Equations (A-8b) and (A-8c) gives the wave equations for $\vec{E}$ and $\vec{B}$,

$$\nabla^2 \vec{E} + k^2 \vec{E} = -\nabla \left\{ \vec{E} \cdot \frac{\nabla \epsilon}{\epsilon} \right\}, \tag{A-12}$$

$$\nabla^2 \vec{B} + k^2 \vec{B} = (\nabla \times \vec{B}) \times \frac{\nabla \epsilon}{\epsilon}, \tag{A-13}$$

If the solution of Equation (A-12) is known, $\vec{B}$ is most simply found from Equation (A-8c). The propagator k is given by

$$k^2 = \frac{\omega^2}{c^2} \epsilon . \tag{A-14}$$

The complex dielectric is found from Equations (A-10), (A-9), and (A-6) to be

$$\epsilon = 1 - \frac{(\omega_p / \omega)^2}{1 + (\nu_c / \omega)^2} + i \frac{\nu_c}{\omega} \frac{(\omega_p / \omega)^2}{1 + (\nu_c / \omega)^2} \tag{A-15}$$

An imaginary part for ϵ signifies that the medium is absorptive.

ATTENUATION IN UNIFORM MEDIA

Absorption

The wave equation for a uniform medium is from Equation (A-12)

$$\nabla^2 \vec{E} + k^2 \vec{E} = 0. \tag{A-16}$$

The plane wave solutions for $\vec{E}$ and $\vec{B}$ are

$$\vec{E} = \vec{E}^{(o)} e^{i(\vec{k} \cdot \vec{r} - \omega t)}, \tag{A-17a}$$

$$\vec{B} = \vec{B}^{(o)} e^{i(\vec{k} \cdot \vec{r} - \omega t)}, \tag{A-17b}$$

where, from Equation (A-8c),

$$\vec{B}^{(o)} = \frac{c}{\omega} \vec{k} \times \vec{E}^{(o)}, \tag{A-18}$$

or, explicitly in terms of the plasma properties,

$$\vec{B}^{(o)} = (\alpha^2 + \beta^2)^{1/2} e^{i\gamma} \vec{E}^{(o)}, \tag{A-19}$$

since

$$k = \frac{\omega}{c} (\alpha^2 + \beta^2)^{1/2} e^{i\gamma} \tag{A-20}$$

and where γ is defined by the relation

$$\tan \gamma \equiv \frac{\beta}{\alpha}. \tag{A-21}$$

The parameters α and β can be expressed in terms of the real and imaginary parts of ϵ. Thus.

$$\alpha = \frac{1}{\sqrt{2}} \left\{ (\epsilon_r^2 + \epsilon_i^2)^{1/2} + \epsilon_r \right\}^{1/2}, \tag{A-22a}$$

$$\beta = \frac{1}{\sqrt{2}} \left\{ (\epsilon_r^2 + \epsilon_i^2)^{1/2} - \epsilon_r \right\}^{1/2}. \tag{A-22b}$$

The real part time average of the complex Poynting's vector gives the energy flow

$$\bar{\vec{S}} = \frac{1}{2} \mathrm{Re} \frac{c}{4\pi} \vec{E} \times \vec{B}^*, \tag{A-23}$$

where $\vec{B}^*$ is the complex conjugate of $\vec{B}$. Substituting Equation (A-17) into Equation (A-23) and using Equation (A-19), we obtain

$$\bar{\vec{S}} = \frac{c}{8\pi} \frac{(\alpha^2 + \beta^2)^{1/2}}{\omega} \left| E_o \right|^2 e^{-2\beta(\omega/c) z} \cos \gamma \, \hat{z}, \tag{A-24}$$

where γ is given by Equation (A-21). The attenuation of the energy flow is

$$A\ (\mathrm{db}) = 8.7 \frac{\omega}{c} \beta \frac{(\mathrm{db})}{\mathrm{cm}}. \tag{A-25}$$

Reflection

Consider two semi-infinite media, with propagation constants k_0 and k_1 separated by a plane surface

Let a plane wave $\vec{E}_i$ in I be normally incident upon the interface, and let $\vec{E}_r$ be the reflected wave and $\vec{E}_t$ the transmitted wave. The boundary conditions on $\vec{E}$ and $\vec{B}$ require that

$$\vec{E}_i + \vec{E}_r = \vec{E}_t, \tag{A-26a}$$

$$k_0 (\vec{E}_i - \vec{E}_r) = k_1 \vec{E}_t \tag{A-26b}$$

We readily find that

$$\left| \frac{\vec{E}_r}{\vec{E}_t} \right| = \frac{|k_0 - k_1|}{|k_0 + k_1|} \tag{A-27a}$$

and

$$\frac{|\vec{E}_t|}{|\vec{E}_i|} = \frac{2\,|k_0|}{|k_0 + k_1|} \, . \tag{A-27b}$$

The reflection and transmission coefficients are given by the ratio of the Poynting's vectors. Thus,

$$R = \frac{S_r}{S_i} = \left| \frac{E_r}{E_i} \right|^2 = \frac{|k_0 - k_1|^2}{|k_0 + k_1|^2}, \tag{A-28a}$$

$$T = \frac{S_t}{S_i} \, . \tag{A-28b}$$

Multiplying Equation (A-26a) and complex conjugate of Equation (A-26v) together yields

$$k_0^* \; (E_i^2 - E_r^2) = k_1^* E_t^2 . \tag{A-29}$$

Hence,

$$S_i - S_r = S_t, \tag{A-30}$$

verifying the conservation of energy flow. Now dividing Equation (A-30) by S_i, we get the transmission coefficient,

$$1 - R = T$$

The attenuation (db) in transmission is

$$A\ (\text{db}) = 10 \log_{10} \frac{1}{T}, \tag{A-32}$$

and, from Equation (A-28a), the reflection coefficient is

$$R = \frac{(\alpha_0 - \alpha_1)^2 + (\beta_0 - \beta_1)^2}{(\alpha_0 + \alpha_1)^2 + (\beta_0 + \beta_1)^2} . \tag{A-32}$$

A TRULY GLOBAL SATELLITE COMMUNICATION SYSTEM

Herbert Trotter Jr.*

This paper describes a versatile global communications concept and a system which, by achieving the most efficient and flexible use of available channels, is most conservative of bandwidth. The synchronous 24-hour stationary satellite is an essential element in the realization of this concept. Its use affords an opportunity for a completely new concept of channel sharing between the various ground stations concerned. The philosophy of the system visualized differs radically from that of the submarine cable in that a high degree of flexibility of channel utilization is obtained among all the ground stations in widely diverse geographical locations.

The concept allows multiple access to the satellite, with the ground stations in the sector of a particular stationary satellite simultaneously sharing the total input band of the satellite. Channels are allocated for connection only as there is a demand for service with channel selection automatically made through a special control channel. On a demand for service, an idle channel is selected and automatically routed from the calling to the called ground station.

This operation concept has the following important features: (a) Each satellite circuit can be used by each ground station and remains unassigned to a specific route except when in use. (b) New ground stations and new routes can easily and economically be integrated into the proposed global system. (c) The system facilitates participation by nations which have small or modest traffic demands with a minimum investment.

---

* President, General Telephone and Electronics Laboratories, Inc.

INTRODUCTION

Any satellite communication system which is to provide the basis of a global communication system must meet certain basic requirements:

1. Access to the system must be available at all population centers of the world.
2. The system must serve the needs of small countries as well as large and of developing countries as well as developed.
3. The system must directly link the nations of the world.
4. The system must make the most efficient use of the already crowded frequency spectrum.

If such a system is to become a reality without becoming a burden on the national and world economies, it must be carefully planned. System objectives must be defined; technology must be reviewed for temporary compromises that can be made without jeopardizing eventual system capabilities.

This paper describes the characteristics and system concepts of a world-wide communication system developed by the General Telephone & Electronics Laboratories. In approaching this study, we have first established the characteristics that a global satellite communication system should have, and then described a satellite system, available within the limits of present-day technology, which satisfies these requirements and which is capable of growth with technological advances.

CHARACTERISTICS OF A GLOBAL SATELLITE COMMUNICATIONS SYSTEM

A world-wide satellite communication system must embody the following features:

Multiple Access

The satellite communications transponder must be capable of being used simultaneously by a large number of ground stations. The system must be capable of absorbing new ground stations without major redesign and these ground stations when added must be able to communicate directly with any other ground stations.

Flexibility of Use

Communications satellite channels must be capable of transmitting signals originating from any device that is permitted connection to common-carrier networks. (These include telegraph, voice, data, facsimile, and TV.)

Flexibility of Service

Channels should be available on either a "demand" or lease basis. Demand service will be used by customers whose requirements are small. Leased services will reduce costs to large users of particular links.

Flexibility of Channel Assignments

Channels should be assigned only in response to a user's demand for service; thus, channel assignment would automatically adjust to traffic variations among ground stations in different time zones. Leased channels would remain assigned for the duration of the lease. Flexibility of channel assignment could be extended to channel bandwidth so that the customer could not only demand a channel, but a channel of specific bandwidth (selected from a list of available channel bandwidths).

Capacity for Service Growth

A system which can make multiple use of R. F. channels is essential if the forecasted demand for service is realized. Of almost equal importance is the bandwidth efficiency of the R. F. carrier modulation. For example, one proposed system would provide 600 voice channels in 200 mc of spectrum. A modulation method with greater bandwidth-utilization efficiency would permit more than 10,000 voice channels in the same spectrum.

Early system will require excess bandwith on the satellite-to-Earth transmission link to obtain adequate signal-to-noise performance for television transmission. No such limitation exists on the ground-to-satellite link. Since a major cost of ground equipment will be the high-power transmitter, it is desirable that it not be rendered prematurely obsolete by early technological advances. Satellite repeaters can incorporate new developments every time one is replaced, and ground receivers can be modified when satellite transmitter power reaches the point where bandwidth-S/N trade is no longer necessary.

The system should be easily expandable to meet future demand without significant loss of original capital investment.

DESCRIPTION OF THE GT&E SYSTEM CONCEPT

As a part of Federal Communications Commission Docket 13522, General Telephone & Electronics proposed a system which embodies the features outlined above. This communications system can be implemented with available technology and will satisfy the estimated future global traffic requirements. Only limited modifications will be necessary to permit ultimate growth to millions of voice channels (or equivalent bandwidth in other services) with modest R.F. spectrum requirements.

The GT&E communications satellite system combines the so-called "synchronous stationary satellite" with certain known techniques of transmission and new techniques of channel switching.

The stationary satellite has two distinct advantages: 1) positioned at approximately 22,300 miles above the equator, the period of the satellite is 24 hours and thus remains fixed with respect to the Earth, enabling stationary ground antennas to be used, and 2) the stationary satellite is visible to nearly one-half of the Earth's surface (see Fig. 1). For example, a single satellite over the equator at longitude 25°W is visible from approximately 100 countries on the continents of North America, Central and South America, Africa, Europe and Asia. These areas contain approximately 91.8% of the telephones of the world. Three such satellites (Fig. 2) can provide a world-wide system covering all the Earth's surface except for the polar regions.

Single-sideband modulation on the Earth-to-satellite link permits access to any unassigned channel from any ground station in the service area of a repeater. SSB provides maximum bandwidth utilization efficiency with "state of the art" techniques (high-power klystron amplifiers). SSB is easily converted to phase modulation in the satellite for wideband

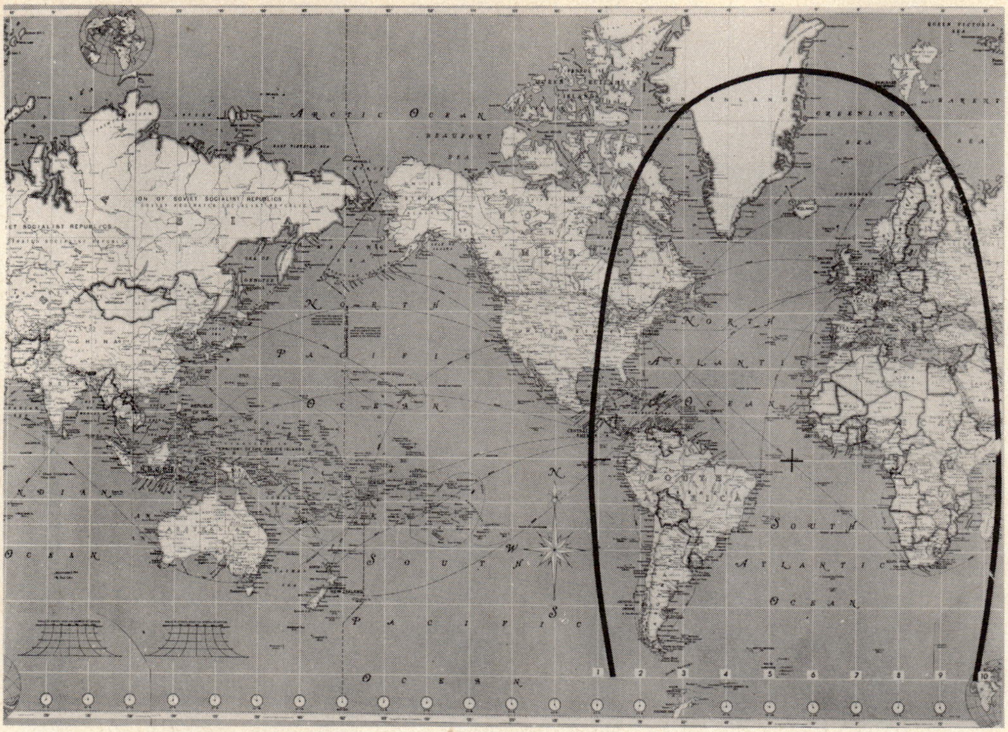

Fig. 1 Area Covered by One Satellite at 22° West Longitude

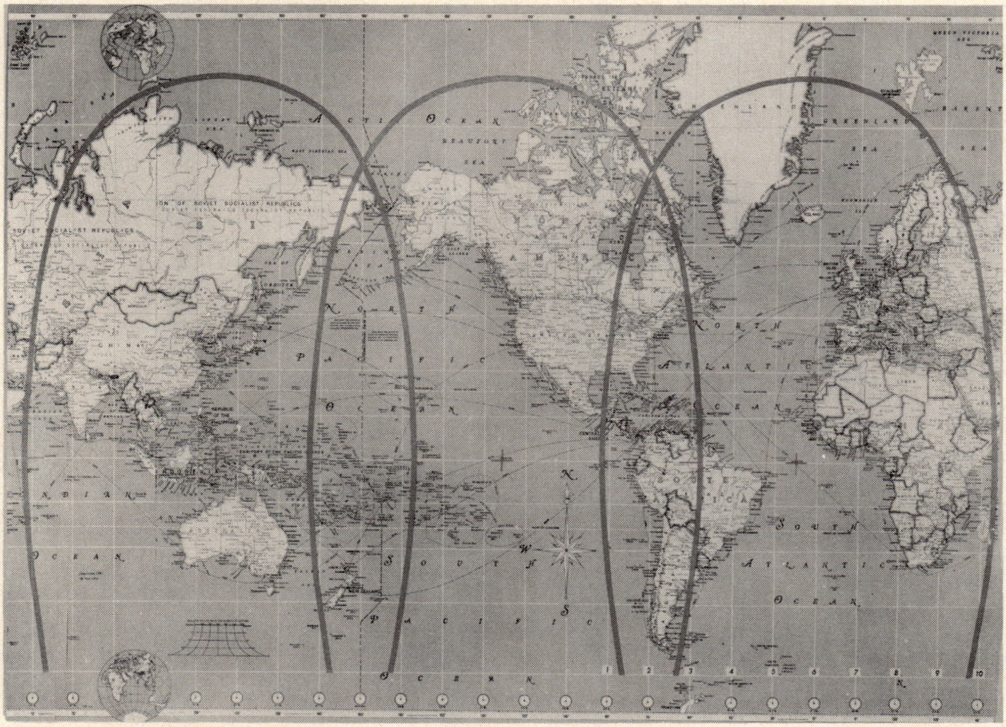

Fig. 2 Coverage of Three Satellites

transmission back to Earth with presently available satellite transmitter power. When higher powers become available, simple modification to the ground receivers will adapt them to receive SSB.

The system operates as follows. A call originating in Rome, Italy and destined for New York is routed to the ground station nearest Rome over ordinary telephone facilities. At the ground station (Fig. 3), a channel request is initiated and routed over a satellite control channel to a central control computer. The computer checks the circuits and selects an available channel (747). It then notifies New York and Rome (Fig. 4) of the identity of the channel, and the circuit between Rome and New York is established (Fig. 5). The process of establishing the circuit between the New York and Rome ground stations is automatic and is accomplished in a matter of seconds. The call is then connected by the New York ground station to the called party over ordinary telephone facilities. When the call is terminated (Fig. 6), the computer notes that the satellite channel is again available and computes the duration of the call. The latter information is used to bill the ground stations according to their use of this system.

By the same process, each ground station can simultaneously transmit calls to many other ground stations. The capacity of a ground station is determined by the needs of the area which it serves. Small countries may require only five channels. However, switching equipment is designed so that each of the five channels will have access to any of a large number of satellite channels. Hence, a small ground station has the opportunity to call or be called if there exists at least one unused satellite channel. Since having fewer channels will require less switching equipment and less powerful transmitters, the cost of a small ground station will be less than the cost of a large one.

A significant savings in ground equipment costs will result from integration of switching and multiplexing functions in the channelizing equipment. The savings will be particularly significant for ground stations requiring only a few connecting trunks. It is anticipated that total single ground station costs (for small channel requirements) will be well under one million dollars. Therefore, for this modest

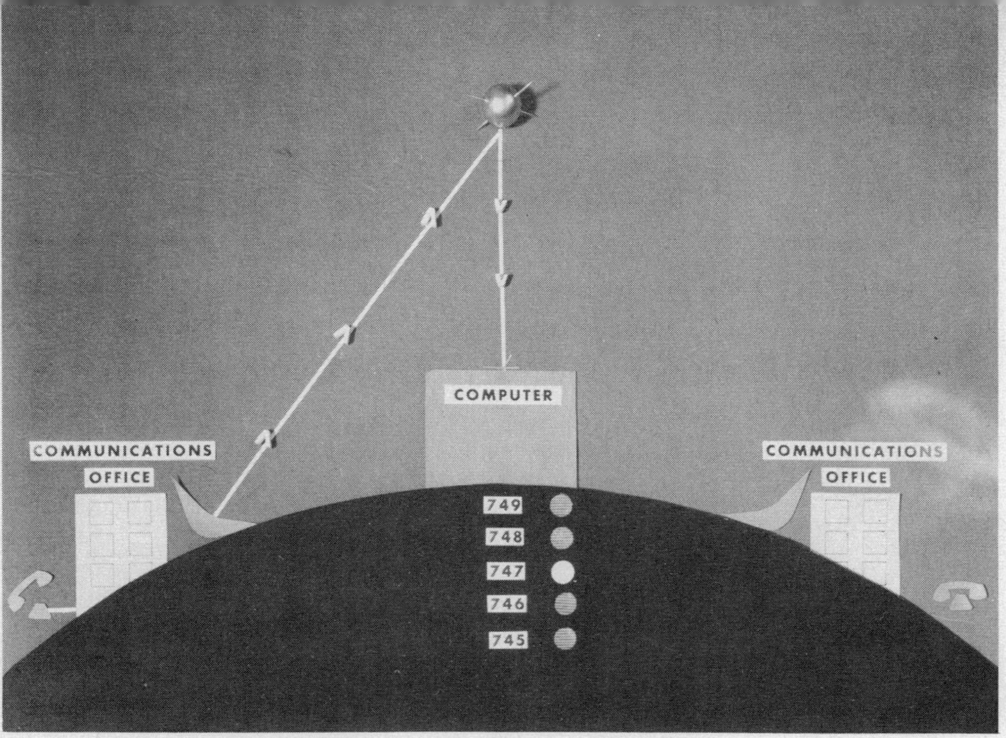

Fig. 3 Start of Call from Rome to New York

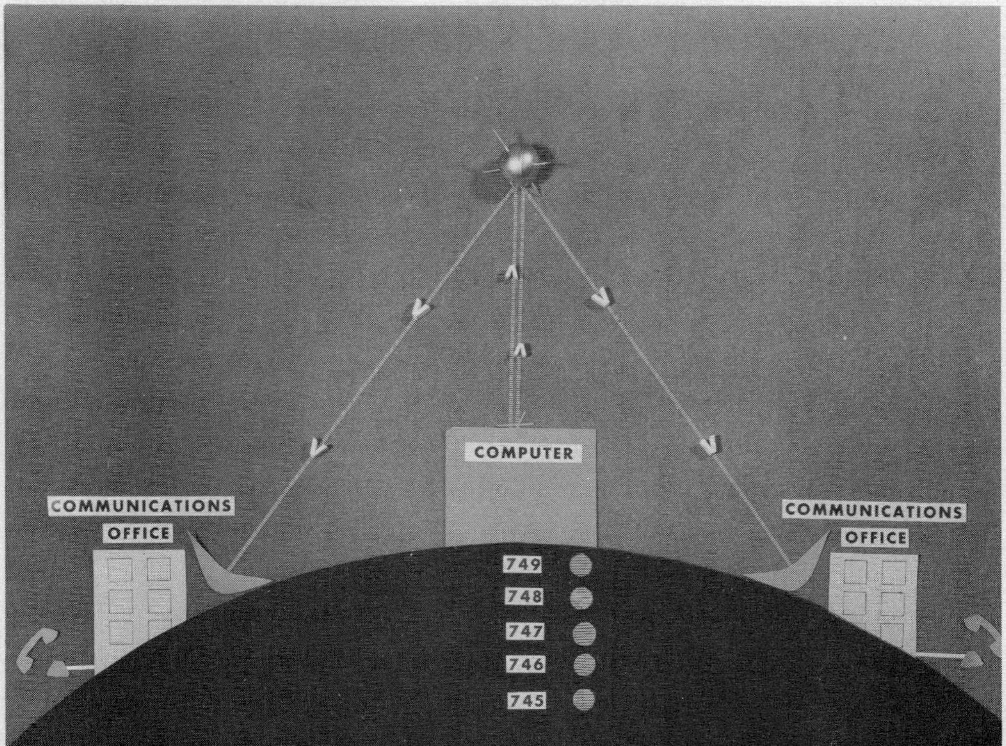

Fig. 4 Circuit Being Assigned by Computer. Channel 747 is Now Shown in Use

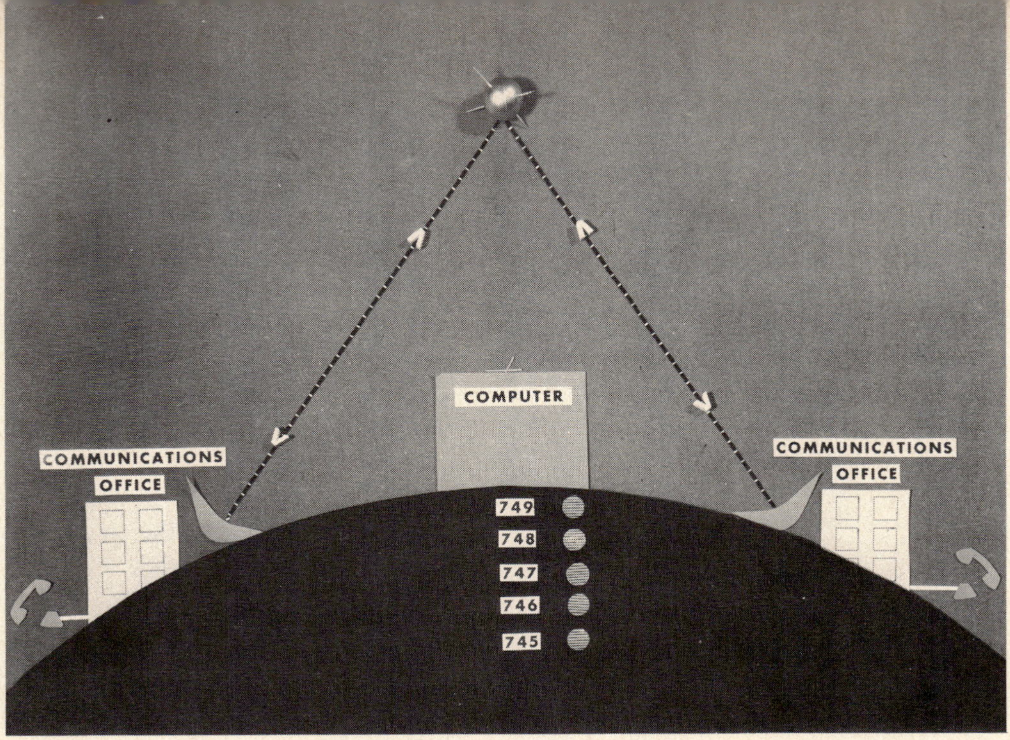

Fig. 5 Circuit in Use

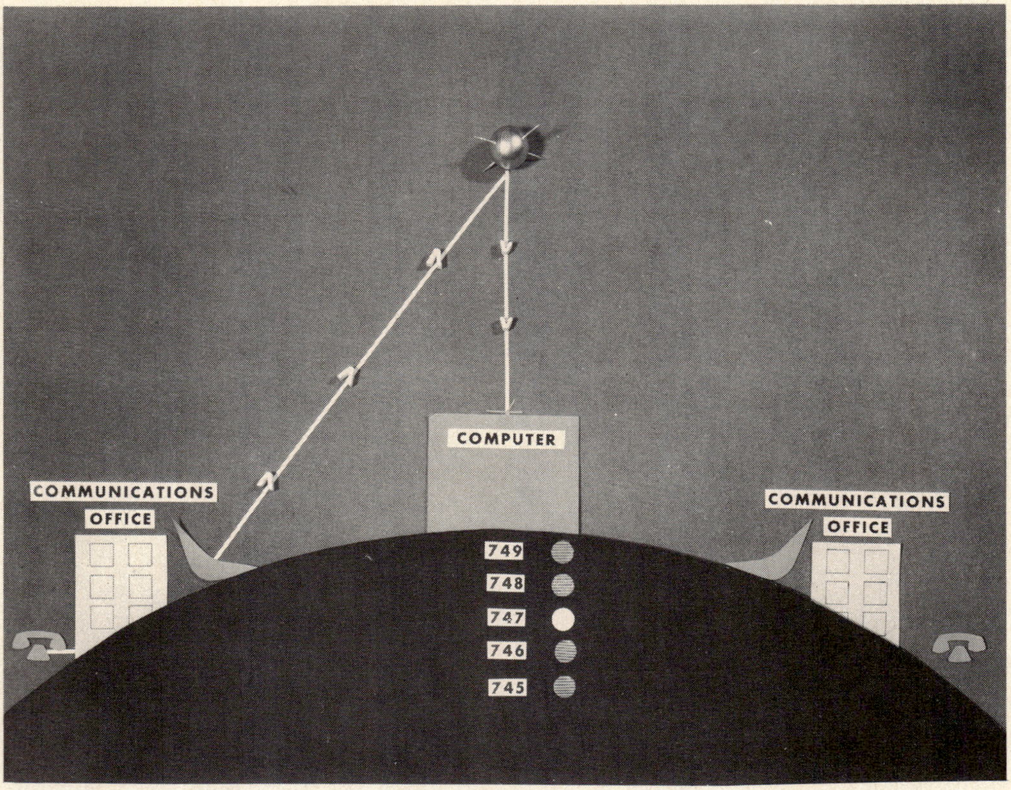

Fig. 6 Computer Being Notified of Completion of Call and Thereby Channel 747 is Being Made Available for Calls Between Any Two Ground Stations

investment a small country or city could have direct communications to all of the telephones of the world.

Because of the relatively low cost of proving-in small ground stations, new areas of communications can be opened up in regions where hitherto geographic or economic considerations prevented achievement of adequate communications. Each ground station can be designed to meet only the anticipated traffic load requirements for the region or area it serves. Traffic loads can be calculated through application of well-established statistical methods, and the total load through the spacecraft can take advantage of time-zone differences applicable to the ground stations. This is a unique advantage of the spaceborne relay system.

It is obvious that the peak demand for telephone communication moves around the world with the daylight hours. In the GT&E system, each channel is assigned only as actually needed; therefore, the system is able to accommodate the peak load as it moves westward.

The system consists of seven major subsystems:

1. Ground transmitter
2. Antennas
3. Satellite repeater
4. Frequency and level control
5. Channel control
6. Switching and multiplex
7. Echo suppressor
8. Ground receiver

Ground Transmitter

The ground transmitter consists of a single sideband exciter followed by a klystron power amplifier. Multi-channel SSB transmitters have been built for tropospheric scatter service. Transmitter power requirements are equivalent. Although most troposcatter applications have used frequencies between 400 and 2000 mc, no particular difficulty is anticipated in extending designs to 8 or 10 kmc.

The output power requirement of a particular ground station depends on the maximum number of voice channels to be transmitted and whether or not television is to be transmitted. Small ground stations in emerging nations might transmit less than 20 voice channels. If 30-ft antennas are used, then these stations need peak-power output capabilities of only 1 or 2 kw. This compares to 10 kw and a 60-ft antenna required for large, full capacity stations.

Antennas

Two classes of antenna are required: ground antennas and vehicle antennas. A single antenna can serve both transmitter and receiver on the ground, but separate antennas are desirable for the satellite.

Because of limited transmitter power in the satellite, the satellite transmitting antenna should be designed to just cover the Earth (18db of gain).

The complexity of the ground antenna is dictated by the accuracy with which the spacecraft is kept in orbit. For initial programs, some limited degree of elevation and azimuth correction will be desirable. Subsequently, however, ground antennas can be of fixed design with the small tracking requirement met by variation of the antenna feed position. Novel methods are being investigated for fabrication of large, inexpensive, low-noise, fixed parabolic reflectors. These fixed reflectors can be located in such terrain that interference to other services will be minimized.

Satellite Repeater

For the single-sideband wideband-FM transmission system being proposed, the satellite repeater must change the composite received SSB signal into a wideband-FM or phase-modulated signal. A simple method of achieving this transformation is injecting a high-level carrier and limiting the combined signal.[1]

---

1 An SSB-PM satellite repeater has been built and demonstrated by Hughes Aircraft Company.

For the linearity required, only a very low modulation index can be obtained directly, but frequency multiplication increases the index by the multiplication factor. In the repeater, the modulation conversion is carried out at a low intermediate frequency and frequency multiplication is used to translate the spectrum into the desired microwave region. Solid-state multipliers have been developed at the Sylvania-Amherst Engineering Laboratories which could deliver almost 1 watt at 4 to 6 kmc with bandwidths up to 100 mc. In the near future, further development will make possible all-solid-state satellite repeaters of suitable transmitter power. Currently, traveling-wave-tube amplifiers can be used to obtain transmitter output power of 2 to 10 watts.

Frequency and Level Control

The use of single-sideband modulation on the Earth-to-satellite link accentuates two transmission problems; frequency control and level control. Either one of these problems would be formidable if the satellite were moving appreciably with respect to the ground stations, because of the large transport delay in the control loop. With an almost stationary satellite, adequate control can be easily achieved. Important sources of frequency error are:

1. Doppler translation on the single-sideband link
2. Repeater translation-oscillator drift
3. Repeater carrier-injection-oscillator drift
4. Difference in frequency of the ground-station clocks

By transmitting a pilot from the master ground station and using it in the satellite repeater to derive the translation carrier and the injection carrier, the Doppler frequency error is reduced to the difference in Doppler on a particular ground station-to-satellite path and that on the master ground station-to-satellite path. If atomic clocks are used at all ground stations, the voice-channel frequency translation without further correction would be about 6 cps for 6000 mc on the SSB link (assuming Doppler frequency shift approximately 1 part in 10^7 and difference in Doppler between master and other ground stations = 1%). This can be reduced or eliminated by means of a control loop for each slave ground station.

Transmission path loss variations in the SSB link are translated directly into level variations in the communication channels. Fortunately, the transmission path loss for the stationary satellite system is expected to be extremely stable with maximum variations of the order of ± 1 to 3 db. Equivalent variations can be expected through equipment aging. Thus the necessary control range is very small in comparison with that required on earthbound long-distance single-sideband systems. The control can make use of the same pilot that is used for frequency control. Again, the main design problem is the loop transport delay.

Channel Control

The channel-control subsystem consists of a central computer, a high-speed signaling channel for broadcasting computer instructions to all ground stations, a low-speed signaling channel from each ground station to the central computer, control circuitry for channel switching at each ground station, and circuitry necessary for communicating signaling instructions with the telephone central office.

A request for one of the available classes of service connections to a specific party is received by a ground-station control unit from the local telephone central office. Identification of the called party and the class of service is transmitted on the slow-speed channel to the central computer. The computer requests, via the high-speed channel, that a connection be established between the applicable ground station and the called subscriber for the specified service. On receipt of information that this connection is complete (over the low-speed channel associated with the particular ground station), the computer transmits instructions to each ground station for establishing the connection between them. If a busy signal or a "no answer" signal is returned to the computer, it forwards this information to the calling ground station which in turn informs the calling subscriber. The "end of the call" is signaled to the computer by the calling ground station, and the computer then issues "disconnect" instructions to the two ground stations, clears the call from its memory, and records the charge information.

Switching and Multiplex

The switching multiplex subsystems make possible the routing of a call between two ground stations via a portion of the transmitted frequency band not already occupied. The specific frequency allocation is selected by the central computer, which keeps track of occupied spectrum. Switching control is a function of the channel-control subsystem.

By integrating the switching and multiplexing functions, the number of switch contacts and the number of multiplex channels is reduced when the number of trunks required at the ground station is less than the system channel capacity. The saving in equipment is most significant when the traffic warrants only a few trunks.

Echo Suppressors

The long transmission delay associated with satellite communications affects the operation of echo suppressors of conventional design. A new type of echo suppressor (with a new philosophy and method of operation) has been developed by the General Telephone and Electronics Laboratories, Inc. especially to overcome the effects of the actual duration of transmission delays. In this new suppressor, the echo is detected by means of a reference signal that is in proportion to the amplitude of the reflected signal so that no time factor is involved. In addition, these suppressors allow each party to interrupt the conversation at will with no "speech chopping" or "speech clipping" of either party. As a result, these suppressors are completely undetectable in operation.

Ground Receiver

The ground receivers consist of low-noise preamplifiers followed by FM receivers. Phase-modulation reception can be achieved by use of a suitable de-emphasis network. For receiver bandwidths of 20 mc or less, threshold requirements can be met without bandwidth compression with 60-foot receiving antennas and 80° Kelvin effective noise temperature. Threshold requirements for wider bandwidths will require compression, lower noise temperatures, larger antennas, or a combination of these.

MODIFICATIONS FOR FUTURE GROWTH

Present technology allows for continuous powered operation of more than one RF channel and for the possibility of narrow-beam satellite antennas permitting one RF channel to be allocated to high usage multiple access trunks having greater bandwidth efficiency, a second for general multiple access use, a third for TV, etc. Satellite repeaters can be identical, so that additional channels provide standby protection for the most important services. After a fixed number of the channels fail, a new satellite will be orbited. It is important that there not be any modification necessary to existing ground stations to absorb growth. Stage two in system growth would occur when satellite transmitter power can be increased by an order of magnitude, making possible large-capacity SSB transmission between satellite and ground. For this transition, ground receivers and frequency control subsystems would need modifications.

What about ultimate growth? Any forecast of future capacity must be based on some assumed technological status. We believe that the following assumptions are conservative:

1. Co-channel interference for SSB transmission can be controlled by antenna design to permit the use of the same frequency spectrum every 10° in space. Consequently, a minimum of three and a maximum of 14 satellites are mutually visible to two ground stations in a "one-hop" transmission area (120° on earth's surface).
2. Spectrum usage can be 100 two-way voice channels per megacycle.
3. Spectrum allocation for satellite communications is 1000 mc.

With these assumptions, a minimum of 300,000 and a maximum of 1,400,000 voice channels or equivalent bandwidth can be made available between any two ground stations, depending on the east-west distance between ground stations on the longest hop used. For ground stations in the same time zone, a minimum of 1,200,000 channels can be available. This compares to a projected total demand for about 5000 voice channels in 1980 for international communications.

CONCLUSIONS

I have described a global communications system capable of implementation with available technology and capable of expansion to accommodate global communications requirements in the foreseeable future. In summary, the GT&E global satellite communication system satisfies the following requirements:

1. It provides efficient spectrum usage through bandwidth-efficient modulation and through time-sharing of all channels by all ground stations.
2. It provides transmission bandwidth and service flexibility available to ground based common carrier facilities.
3. It provides efficient equipment usage through the multiple access concept and through integration of the switching and multiplex functions.
4. It provides the basis for a technological evolution toward "the ultimate system."

INTERNATIONAL PLANNING FOR COMMUNICATIONS SATELLITES

Philip J. Farley*

It is indeed an honor to be invited to speak at the Eighth Annual National Meeting of the American Astronautical Society and to discuss international planning for communications satellites.

What I say at this time will have to be of the nature of an interim report. You probably recall that the President said in his State of the Union message last week that he will "soon send to the Congress a measure to govern the financing and operation of an international communications satellite system in a manner consistent with the public interest and our foreign policy." This forthcoming legislation will propose an organizational basis for an active planning and operational program. Until an appropriate organ has been established and gone to work, our planning will inevitably be preliminary, to a degree, in the international field as well as domestically.

Nevertheless, we feel a keen sense of urgency and much is being done while the organizational and legislative enabling action is underway. Furthermore, it can never be forgotten that by the very nature of these satellites, they are best adapted to intercontinental communications and are thus inherently international in their application. So I shall tell you about three areas in which things are being done -- frequency allocation planning in the ITU; UN General Assembly discussions of the broader UN interest; and foreign participation in this year's experimentation. I shall then say a few words about how the arrangements between the United States and other countries for establishment and operation of a system of satellite relays might take shape.

---

* Special Assistant to the Secretary of State

Background

Our basic policy objectives, and the impetus for much of our planning, are the principles included in the statement of the President of July 24, 1961 on communications satellite policy. Of particular interest to this forum, the President in his policy pronouncement stated "I again invite all nations to participate in a communication satellite system, in the interest of world peace and brotherhood among peoples throughout the world." More specifically in delineating his policy of ownership and operation, he stated that the system must be global in coverage so as to provide satellite communications services throughout the whole world as soon as technically feasible, including service where individual portions of the coverage are not profitable, and must provide opportunities for foreign participation, through ownership or otherwise, in the communications satellite system. Furthermore, he directed that the Executive agencies of the government examine with other countries the most constructive role for the United Nations, including the International Telecommunications Union, in international space communications. His policy pronouncement established the guidelines within which the numerous interested agencies of this Government could proceed. The National Aeronautics and Space Council, chaired by the Vice President and operating under the stewardship of your speaker at the banquet tonight, Dr. Edward S. Welsh, has proceeded vigorously with the implementation of the President's policy pronouncement.

Congress as well as the Executive have been vitally interested and active in planning for communications via space satellites. Hearings have been held by the several Congressional committees to determine whether or not the United States is prepared to negotiate internationally for an adequate allocation of the international radio frequency spectrum for communications by space satellites and also to determine whether adequate machinery existed within the Executive Branch of the Government to formulate and implement a sound long-range communications satellite policy.

Frequency Allocations

Fortunately, with regard to frequency allocations, there has been in existence for many years the Inter-Departmental Radio Advisory Committee, known as IRAC, which is composed of interested agencies of the Government, which are claimants of government radio frequencies. The Federal Communications Commission, which is charged with the assignment of radio frequencies to non-governmental users, has worked very closely with the IRAC over the years, and I can say with some measure of confidence that these two bodies working together have done a good job in anticipating and accommodating United States radio frequency requirements.

As you can see, much has been done which shows real foresight and can be safely inscribed on the credit side of the ledger. At the ordinary Administrative Radio Conference, held in Geneva in 1959 under the aegis of the International Telecommunications Union, certain bands in the radio spectrum were, at the urging of the U.S. Delegation led by FCC Commissioner Craven, set aside for experimental purposes for space communications. Immediately following that conference, the IRAC and the FCC, which were fully aware of the fact that the band allocated would be insufficient to accommodate communications to and from satellites, took the question of longer term needs under study. Realizing that the matter of the use of the radio spectrum has been under study by members of the ITU for many years and that there is no foreseeable reduction in the use of frequencies for global communications, the U.S. planners sought to determine how communications with satellites could be sandwiched into the already congested radio spectrum without causing undue interference to existing facilities such as microwave, radar and other services. After many months of unremitting effort, a plan has been produced which has been circulated to all of the 115 members and associate members of the International Telecommunication Union for their study and, hopefully, approval. The U.S. plan will be the object of considerable scrutiny by two international study groups of the Consultative Committee on International Radio (CCIR) of the International Telecommunication Union which will meet in Washington in March of this year.

There are other questions, initiated by the United States, which will also be studied by the CCIR in addition to the sharing of the radio spectrum by earth-space links. Some of these problems are the technical characteristics of earth-space communications links; re-entry communications; study of preferred modulation characteristics by communications satellite space service; feasibility of spectrum sharing between communications satellite space systems; feasibility of direct broadcasting from earth satellites; space service antennas; noise limitation in the space service; and passive radio astronomy. These U.S. proposed topics are but a few of the questions which will be studied, as undoubtedly other countries will have questions of their own. I have given you the foregoing to illustrate the initiative that the United States has taken in planning for communications satellites.

To look into the future, the answers and findings that will evolve from the work of the CCIR meeting will be circulated to all members of the International Telecommunications Union and will be considered at the next Plenary Assembly of the CCIR which will be convened in New Delhi in January 1963. In May of this year, the Administrative Council or governing body of the ITU will hold its annual meeting in Geneva. Hopefully, the Administrative Coucil will designate a time and place for an Extraordinary Administrative Radio Conference, preferably in 1963, which will be charged with the responsibility, based on the CCIR studies, of making firm allocations within the radio spectrum for communications with communications satellites.

United Nations Action

Last December 20 the General Assembly of the United Nations passed unanimously a resolution introduced by the United States dealing with international cooperation in the peaceful uses of outer space. In its resolution, the United Nations noted with satisfaction that the ITU planned to call a special conference in 1963 to make allocations of radio frequency bands for outer space activities and recommended that the International Telecommunication Union go one step further and consider at that conference those aspects of space communications in which international cooperation will be required. It also invited the expanded program of technical assistance and the United Nations special fund in consultation with the ITU to give sympathetic consideration to requests from member states for technical and other assistance for the survey of the communications needs and further development of domestic communications facilities so they can make more effective use of space communications. As a continuing mechanism, the United Nations has in being a Committee on the Peaceful Uses of Outer Space which will convene in March of this year for the purpose of giving further impetus to international cooperation and technical assistance.

In passing its space resolution, the United Nations endorsed the principle of broad international cooperation first enunciated by President Kennedy last July. This action should encourage a feeling by many countries that they have a stake in the success of programs for use of communications satellites, and thus improve prospects for agreement on frequencies and on arrangements for installation and operation of a satellite system. It is also useful to have the UN call attention to the fact that satellites and associated facilities do not constitute in themselves a communications system, and that internal communications must be developed if advantage is to be taken of the new relay techniques.

International Participation in Experimental Program

With much effort and expenditure of funds, private industry and NASA have planned a program of experimental communication satellite launches for this year which should be productive of much valuable scientific information. For example, in May NASA will launch a wide-band communications satellite developed by the A T & T which will orbit the earth at a relatively low altitude. This launch will if necessary be followed by a back-up launch of a similar satellite in July. In August a different type of satellite developed under NASA contract will be launched in low altitude and will provide facilities for one-way transmission. In October a similar launch will be made with another A T & T satellite. If present plans hold, a third type of satellite will be placed in synchronous orbit which will provide two-way communications facilities. The data derived therefrom will be useful in determining the feasibility of a particular type of system or systems to be used.

All of these launches will require cooperation by one or more foreign countries. Ground stations are now under construction in France and the United Kingdom, and plans for participation are being considered in a number of other countires in Latin America, Europe, and Asia.

International Participation in Operational System

I believe that an operational communications satellite system must be a cooperative international venture, on practical as well as broad foreign policy grounds. I have cited already the principles of the President's policy statement in this regard. Similar conclusions are reached in the thoughtful report of the Ad Hoc Carrier Committee established by the FCC last year. This committee, representative of the extensive experience of private U.S. communications interests, pointed out the basic fact that the active interest of foreign communications administration is essential to a global communications satellite system -- since these administrations operate the international communications facilities emanating from other countries. Such administrations will wish to own their own ground stations, and many have expressed interest in the satellites themselves and will want substantial ownership interests in the satellites.

The Carrier Committee suggests, on the basis of experience in the shared ownership of overseas telephone cables, that foreign ownership participation might take the form of undivided interests in the satellites owned by overseas administrations. This is a useful suggestion. On the other hand, as the Committee also recognizes, "it will be necessary to agree upon and establish machinery to operate and maintain the system, administer and promote its use, handle accounting matters, and plan for growth." Machinery for these essential purposes might well take the form of an entity which would also own the system. Specifically, the other alternative which merits careful consideration is an international joint venture, in which shares would be held by interested private or public communications administrations from participating countries, and which would own and manage the system. The United States would participate in such an overall joint venture through the envisaged U.S. Communications Satellite Corporation, rather than having foreign interests participate in the U.S. corporation. I believe there are powerful practical forces pointing toward such a solution, which leaves ample scope for U.S. technical, industrial, and operating leadership. Decision as to the form of international organization obviously must, however, await the outcome of extensive exploratory discussions here and abroad.

Conclusion

As I have outlined, planning for international communications by satellites is being pursued concurrently in many areas. In looking at the past we can take satisfaction in the productive effort to date, including:

1. The technical development and experimentation by private industry and government.

2. The enabling international technical arrangements, such as the construction and operation of tracking stations and other ground facilities and the international planning for frequency allocations.

3. The policy pronouncement of the President.

4. Preparations for creation of a U.S. entity to own, create, operate and direct the U.S. interest.

5. The broad basis for international impetus, preparation and participation established in the United Nations.

6. And the keen Congressional interest and participation.

In viewing the future, it seems clear that 1962 will be a critical year, not only for technical experiments and the creation of the U.S. agent, but also for the emergence of the patterns of international cooperation.

eighth
annual
meeting

AMERICAN ASTRONAUTICAL SOCIETY

INTRODUCTION

PART I — BASIC RESEARCH
Chairman: Robert Jastrow

PART II — GUIDANCE AND CONTROL
Chairman: Robert E. Roberson

PART III — BIOASTRONAUTICS
Chairman: William M. Helvey M.D.

PART IV — COMMUNICATIONS
Chairman: Ray A. Glaser

PART V — LUNAR AND PLANETARY EXPLORATION
Chairman: Eliot T. Benedikt

PART VI — ASTRODYNAMICS
Chairman: Hans Lieske

PART VII — APPLICATIONS OF ASTRONAUTICAL SYSTEMS
Chairman: G. L. Hollingsworth

PART VIII — SPACE VEHICLE DESIGN
Chairman: Saunders B. Kramer

THE GEOLOGY OF THE LUNAR BASE

Jack Green*

A greater sensitivity to volcanic phenomena might exist on the moon because of tidal and gravity effects. Earth-induced tides would tend to fracture the lunar crust and promote the generation of magma. Bubbles would tend to nucleate deeper on the moon and expand faster than they rise relative to terrestrial rates. A more explosive type of volcanism might therefore be or have been produced. Details of lunar surface features support a volcanic origin. Base site selection may have to compromise the requirements of many scientific and engineering disciplines. Early base operations will probably be controlled by trajectory considerations and terrain and mineralization features of the base site. For this reason, a nonpolar caldera-type lunar crater will probably be superior to an impact crater or a basaltic flood plain. In later base operations, more independence may permit base location, perhaps, in a polar location adjacent to eternally shadowed zones which may be mineralized; here, continuous sunlight is available for ecological functions. At least seven technologies may be required to maintain the lunar base, all of which involve the geoscientist : (1) tunneling and base construction, (2) extraction of gases from rock froths, (3) extraction of water from volcanics, (4) the processing of sulfur, (5) mineral dressing using existing raw materials, (6) the metallurgy of basalt, and (7) the selective volatilization of rocks using plasma techniques. The significance of the topographic law in lunar craters is discussed from the standpoint of localizing mineralization, including water.

INTRODUCTION

The selection of a site and the construction and maintenance of the lunar base itself will involve many new scientific and engineering disciplines. On the moon, initial requirements of scientists and engineers for vital materials, facilities, and space may have to be compromised. The services of the geosciences will be basic to most of these new requirements in both the early and late phases of base operations. In the early phases, the terrain and mineralization may be the controlling factors, and in later phases, the technological processing of existing raw materials may assume the dominant role.

---

* North American Aviation, Inc., Space and Information Systems Division, Downey, Calif.

I. LUNAR VOLCANIC MECHANISMS

Tidal And Gravity Effects

Of the two major theories on the origin of the moon's surface, the volcanic theory offers more advantages in basing than does the impact theory[1]. This, however, is not to be taken as justification for study of lunar volcanic processes. Justification for study arises from consideration of the writings and observations of many students of lunar phenomena among whom have been many eminent volcanologists.

It is of interest to inquire if conditions for the evolution of volcanic forms may be different on the moon than on the earth. If most of the major lunar surface features are of an internal origin, what might be suggested as mechanisms intensifying lunar volcanism?

Two forces operative on the moon to a degree not present on earth are tides and gravity as shown in Table 1. These forces may trigger regional lunar volcanism. Earth-induced tides would be created because the lunar orbit is not a circle; its mean eccentricity is 0.0549. Therefore at apogee it is 50 thousand kilometers further from the earth than at perigee.

Tides of the magnitude suggested in Table 2 while not sufficient to disrupt the moon might induce pressure release in the lunar crust in two ways: (1) production of deep seated fractures and (2) movement of the base of one lunar crustal block relative to the other. As possible evidence for this tidal activity Bülow[2] has shown the lunar crust to be highly fractured (Figure 1). Both the effect of crustal fracturing and displacement of blocks are closely related in time and place. Since the lunar crust would tend to fail by fracture, and the sub-crust would tend to fail by flow, a pressure release at the base of the crustal plate would be uncompensated for until the subcrustal material flowed into the low pressure zone. It is important to point out that the crustal plate need not move outward relative to adjacent blocks but need only pivot.

Rocks at depth, on release of pressure, would possess a temperature in excess of that required for initial melting at the reduced pressure, and magma would be generated under adiabatic conditions. Production of magma at depth by pressure release would not necessarily produce volcanism at the surface unless the magma actively stoped its chamber roof or unless a fissure system developed along which magma could ascend.

Table 1. Factors Possibly Intensifying Lunar Volcanism

Tidal Effects in Lunar Crust	
Primary Effect	**Secondary Effect**
Greater tendency to fracture crust than on earth	
Periodic pressure release	Generation of magma at depth
Periodic temperature drop	Onset of magmatic crystallization
Intensification of the second boiling point effect	Explosive pressure rise under crustal plates
Gravity Effects in Lunar Crust	
Primary Effect	**Secondary Effect**
Deeper nucleation of bubbles in magma column than on earth	Slower rise of bubbles
	Greater expansion and coalescence of bubbles during rise
Greater viscosity of magma column top	Enhanced throwout effect during eruption
	Rock froths
Greater degree of subsidence in magma column after degassing	

Table 2. Ratio of Tidal Forces

$f = 2g \frac{M}{r^3}$	Present	Geological Past	
		Half present perigee radius	
	Apogee-perigee distance of 50,000 km	Apogee-perigee distance of 25,000 km	Apogee-perigee distance of 50,000 km
1 $f_{\text{perigee moon on earth}}$	1	8	8
2 $f_{\text{apogee moon on earth}}$	0.7	5.5	3.8
3 $f_{\text{earth on perigee moon}}$	13.5	108	108
4 $f_{\text{earth on apogee moon}}$	9.1	73	51.5
Difference 3-4	4.4	35	56.5
Resulting difference in tide height on moon between apogee and perigee based on a 30 cm body tide at present on the earth	1.3 meters	10.5 meters	> 17 meters

FROM RUSSEL ET AL (1952), WHERE f IS THE TIDAL FORCE, g IS ACCELERATION IN TERMS OF EARTH'S GRAVITY, M IS MASS IN LUNAR UNITS, AND r IS MINIMUM EARTH-MOON DISTANCE.

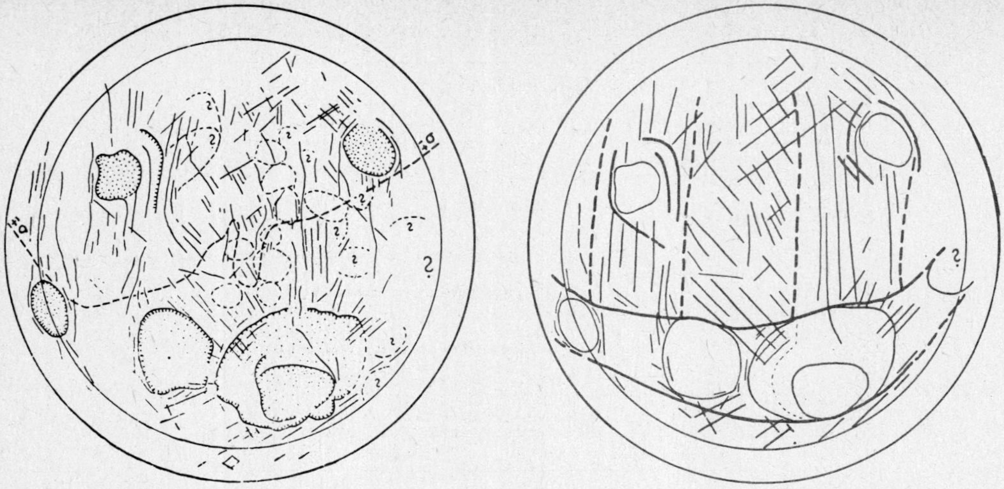

SIMPLIFIED FRACTURE PATTERN SCHEMATIC FRACTURE PATTERN

(After Bülow, 1957, p. 595)

Figure 1. Lunar Tectonics

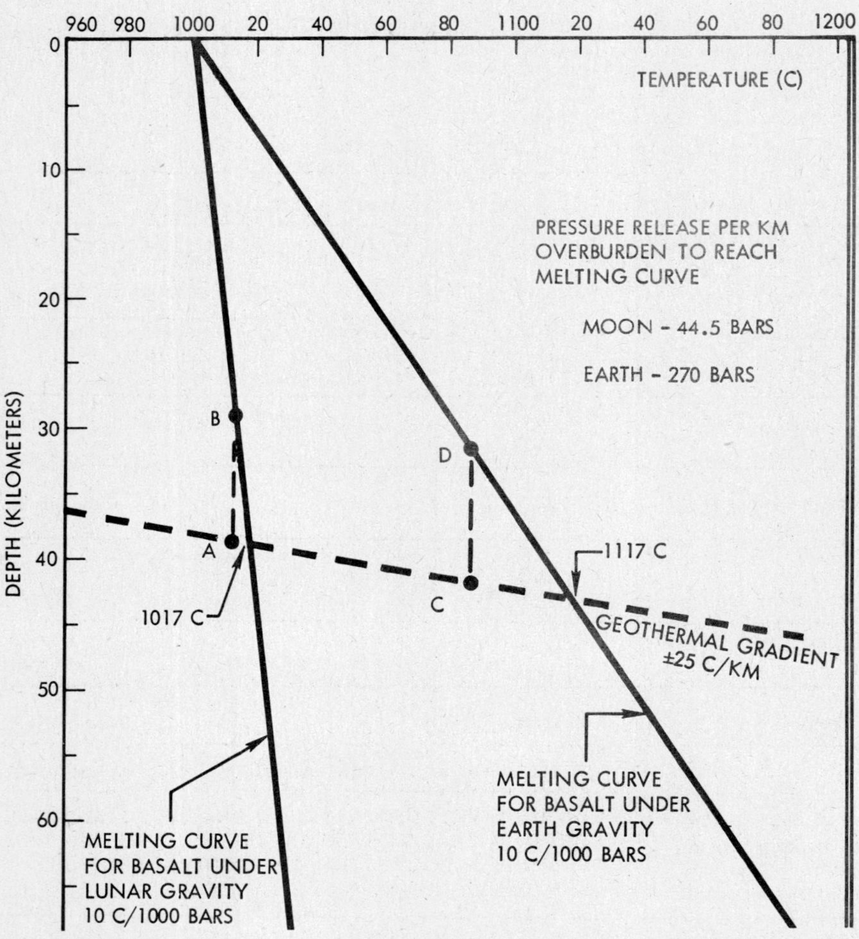

Figure 2. Relationship of Lunar and Terrestrial Melting Curves of Basalt to Thermal Gradient

Figure 2 is essentially a lunar application of a study by Yoder.[3] On earth we find that the melting curve of basalt has a lower slope than on the moon because of the difference in gravity. If the geothermal gradient on the moon approximates that in the earth, basalt will melt at a shallower depth and at a lower temperature in the moon than in the earth. Because of the steeper lunar melting curve, it will intersect the thermal gradient at a temperature of 1017°C at 30 kilometers of depth in contrast to 1117°C at 43 kilometers of depth in the earth. However, the assumption of a thermal gradient on the moon equal to that in the earth cannot be defended without a better estimate of the lunar heat balance. What is undeniable is the fact that a pressure release per kilometer of overburden to reach the melting curve of basalt would be less on the moon than the earth. As Figure 2 indicates, for the same rock column thickness (10 kilometers for A-B), a pressure release of 445 bars is required on the moon for a rock on any assumed thermal gradient to reach the melting curve. The pressure release in the earth for a rock on the geothermal gradient to reach the melting curve over the same rock column thickness of 10 kilometers (C-D) is 2700 bars.

Thus because of the relationship of the melting curves of rocks to the geothermal gradient on the moon, one would need a smaller pressure release per kilometer of overburden to produce a rock melt.

When and if a magma occurs by pressure release at the base of the lunar crust, a phenomenon known as the second boiling point may occur. This effect is produced when a fluid such as a magma cools and crystallizes. The volatiles which are evolved during this crystallization may reach a critical concentration at a temperature sufficient to cause a rapid rise in pressure. If this occurs, the pressure rise produces a second or retrograde boiling. The amount of volatiles (water in the case of a magma) and temperature control the amount of pressure generated. At the beginning of crystallization of a magma there is not enough volatile material yet differentiated to produce high pressure. At an intermediate stage the concentration of volatiles, might be sufficient at some point i.e., the second boiling point in the cooling history, to result in a large pressure buildup. At lower temperatures, even though large quantities of volatiles have been fractionated off, the pressures decrease rapidly. The pressure rise created by second boiling would tend to further fracture the crustal plates thereby inducing surface volcanism.

Miyamoto[4] first invoked the second boiling point effect to account for the evolution of gases from the lunar surface by cooling of magma within the moon. The gases are assumed by Miyamoto to occupy or to have occupied large underground caverns. While the writer subscribes to the second boiling point concept with the reservations noted above, he does not endorse the underground cavern concept nor the entrapment of large quantities of gases in such chambers. The depth at which the caverns might occur would be at the depth at which magma would cool. Unless the magma were formed extremely close to the surface, the writer believes that large openings within the lunar crust would be closed by overburden loading. Furthermore, the second boiling point would tend to promote the generation of fractures that would (1) enhance collapse of such caverns even if they were formed and (2) permit escape of gases that might collect there. Normal collapse features observable in the volcanic terrains on earth resulting from magma withdrawal at depth are therefore thought to have been operative on the moon.

The writer suggests that tidal influence on the moon by the earth is and has been more intense than the trivial effects noted in promoting activity in terrestrial volcanoes. Many of the effects noted on earth result in recurrence of activity in a volcano already in existence or along fractures of specific orientation. This is possibly in contrast to conditions imposed in the lunar crust. Conditions for the production of lunar magma may occur much more frequently in the moon than in the earth over geological time and therefore favorable conditions for initiating the second boiling point may be enhanced. If the second boiling point does or did occur more frequently in the moon the attendant pressure rise would have given rise to both passive or explosive types of volcanism depending on the depth and disposition of the magma chamber, the type of magma, volatile content, and fracture system associated with it. As implied previously, however, the occurrence of the second boiling point is not a requirement for surface volcanicity; the most important triggering agent would be periodic tidal deformation which would tend to fracture the lunar crust.

Presuming the moon to have been closer to the earth in the geological past[5, 6] then crustal fracturing and attendant phenomena may have been intensified conceivably on a weekly basis.

Another mechanism independent of possible periodic generation of magma at depth would be maar-type lunar cratering. From considerations of this type of cratering discussed by Green[7], one might predict that lunar volcanic craters exceeding 15 km in diameter would be predominantly of the caldera type, particularly if they contain central mountains. A large proportion of the smaller craters, perhaps those under 15 km in diameter, may be of the maar type. It is further suggested that maars within volcanic craters on the moon may be confined to topographic lows in these craters.

Combined with the suggested tidal effects are the gravity effects (Table 1). Under the 1/6 gravity that prevails on the moon, there would be a deeper nucleation and slower rise of bubbles in a lunar magma column than in a terrestrial magma column. In general, the upper portion of this column would tend to become viscous because of the abstraction of heat from the melt by bubble expansion. On the moon, the lesser gravity causing slower rise of gas bubbles than on earth would permit the growth of larger bubbles over a longer time period, permitting a more effective heat extraction in the upper portions of the melt chamber or column.

Another factor promoting a greater degree of lunar volcanic explosivity is the greater rate of bubble expansion relative to rise which would induce coalescence of bubbles in the magma or lava conduits or flows. Interpolating from Usiskin and Siegel's data[53], we note that bubbles on the moon will rise half as fast as on earth and will be 1.7 times larger because of the reduced gravity. Thus, larger bubbles will coalesce to a greater degree and at a greater depth in magma in the moon than in the earth; production of larger lunar calderas and maars would be favored. In addition, the effect of vacuum at the moon's surface would promote a more intense vesiculation of molten surface flows. A molten flow or lava lake some 20 meters thick would be under one atmosphere pressure on the moon. Extrapolating from Niskikawa and Urakawa's graphs[54], we can see that water under one atmosphere pressure will produce bubbles about 4 mm in diameter, whereas water under 0.01 atmosphere will produce bubbles 50 mm in diameter. The frequency of bubble production under these two pressures is about 30 and 1 per second, respectively. Thus, approximately 1/4 meter below a molten lava surface on the moon, bubbles will be some 20 times larger than bubbles nucleating the same distance below the same type of lava on earth when the lunar effects of reduced gravity and pressure are considered. The rate of bubble production would possibly be at least as low as 1/30 under these conditions.

The resulting higher viscosity of this lunar magma "cap" may tend to produce a more violent degree of volcanicity if a rapid pressure buildup were to occur below the more viscous portions of the magma column. The lack of atmosphere would promote a more extensive throwout of material because of the absence of air resistance to ejectamenta. The lesser gravity would also add to the increased throwout effects. Finally, crater for crater, the degassing of a longer bubble-entrained magma column on the moon would tend to produce a greater degree of crater floor subsidence and wall collapse than on earth.

It is the writer's opinion that tidal heat in the moon produced by the earth may be important in calculating the heat budget of the moon, assuming the moon once rotated as it revolved. Earth tides slowing the lunar period of rotation to a month could have resulted in the dissipation of large amounts of heat which may have contributed to volcanicity in the geological past. However, this assumption may not be necessary to invoke if it can be established that radioactive heating is sufficient to induce volcanicity.

Internal heat produced by the radioactive decay of uranium, thorium, potassium, and rubidium would be at least an order of magnitude greater from volcanic rocks than from stony meteorites. One can calculate that a thermally insulated silicic rock will heat up some 40° C in one million years; a basalt will heat up $6^{\circ}C$. During the same time period, an insulated stony meteorite would heat up less than $0.5^{\circ}C$. The heat rise per million years in these insulated rock types three billion years ago would be much greater because potassium was much more abundant. However, chondrites would heat up a degree or so, in contrast to the volcanic rock types which could generate as much as 80 °C per million years, three billion years ago. Now extinct radionuclides, for example I^{129}, may have been present when the moon was formed, which could have provided additional heat sources for intense defluidization and volcanism during the moon's early history.*

The heat balance of the moon has been discussed by Urey[8], MacDonald[9], and Levin[10]. On the basis of the heat evolved by the rock types noted above, one can say that widely different values for the rate of cooling of the moon can be proposed, depending upon the lithology assumed for the lunar crust and mantle. For example, a silicic crust could evolve one hundred times more heat than a chondritic crust.

The average heat production of a chondritic moon is 2.3×10^{-7} cal/cm/sec.[11]. Using this value and assuming a conductivity of 2×10^{-4} cal/cm/sec/°C, the subsurface temperature is calculated to be 3° C. Fremlin[12], however, points out that this calculation must be in error inasmuch as the eruption observed by Kozyrev[13], if volcanic, could not then have occurred. Thus, on this basis, the assumption that the moon's surface is chondritic is thought to be incorrect, because chondrites cannot supply the amount of heat necessary to induce volcanism.

---

* I am indebted to Dr. E. Anders of the Enrico Fermi Institute for Nuclear Studies for calling this possibility to my attention.

II. THE MICRORELIEF OF LUNAR TERRAIN

Microrelief And Radar Data

The term lunar microrelief requires definition. Does the term apply to the rugosity of steeply inclined walls of fractures or to the local fine structure of the rock surfaces themselves? If the term microrelief should apply only to the surface of a rock, then there is no ambiguity. However, does it take into consideration rock fragments or fractures within rock masses? Also, how much of the moon's surface has a "microrelief" of between 0.1 millimeter and 0.1 meter, the limits suggested by Sytinskaya[14]?

Sytinskaya's thesis is that (1) the moon is bright out to the edges which is optical proof of an extremely dissected microrelief and that (2) radar data evidence points to a smooth moon where smooth is less than the wavelength of the radar beam considered (1.7 meters in this case). Microrelief of the moon according to Sytinskaya is between one micron and 1.7 meters where one micron satisfies the conditions of a zero q factor. The factor q is a smoothness parameter defined by

$$B = B_o \cos q_i \text{ (i is less than } 90°\text{)}$$

where B is brightness, and i is angle of incidence of rays on a spherical surface. The parameter q expresses the distribution of brightness along the radius of a disc. However, these limits may be narrowed. By assuming that the minimum particle diameter of rock dust approximates 0.1 millimeter, the lower limit may be raised, and realizing that the radar beam wavelength is much greater than the optically derived q value, the upper limit may be reduced. An arbitrary 1/20 of the approximated maximum two meters was selected by Sytinskaya to yield the revised upper limit of 0.1 meter and the value is reasonable.

The interpretation of low slopes from radar data also appears reasonable providing one accepts the relatively small percentage of steep slopes that would border faults, rift systems, mountain scarps, terrace fronts and caldera or impact-crater walls. Gradients on lava flows and radial slopes from calderas on earth are often around three degrees which is the mean gradient measured on the lunar surface by Hey and Hughes[15] using radar techniques. Kopal and his coworkers at the University of Manchester also report low slopes of a few degrees from time lapse studies of lunar shadows[16].

For the purposes of this paper the writer has categorized the lunar surface into micro-, Mega-, and macro-surfaces in an attempt to relegate relief features to common frames of reference (Figure 3). One may adopt detailed terrain analysis techniques for this[17, 18]. However, Figure 3 approaches the problem from the geological standpoint of lunar mobility (disregarding traction aspects). The concept of rough and smooth are presented by reference to a bicycle wheel. The word rough in radar equivalent cross section parlance[19] is reserved for isotropically scattering material (such as frosted glass in optics) and depends on the wavelength of the radar beam employed. From a geological standpoint, the word is related to a geologist's mobility which is within more restricted bounds. Thus, points one and three on the ordinate of Figure 3 refer to 10- and 170-centimeter radar wavelengths respectively. Interpretation of the response of radar surveys at these wavelengths would indicate for the 10 centimeters a somewhat smooth moon and for the 170 centimeters a fairly rough one. The word rough should not be defined by or be dependent upon a radar wavelength. Perhaps the geologically accepted grain sizes for volcanic ejectamenta may provide a worthwhile scale for roughness indices that would be acceptable to both geologists and radar specialists. The range selected by Sytinskaya from 0.1 millimeter to 0.1 meter includes a wide spectrum of geological features coinciding with lava vesicle sizes, and an appreciable portion of fractures and faults. Sytinskaya's limits would include pahoehoe but would concede only a small amount of aa lava. The surface materials could include dust, ash, lapilli and cinders, and some small blocks; on the Wentworth scale, the materials could include all varieties except boulders.

Through the telescope one can observe large scale dislocations and unevennesses on the lunar surface just as a high altitude photograph shows such irregularities on the earth's surface. A geologist can often predict where local faulting and fracturing will occur by examining a high altitude photograph even if most of these faults and fractures are below the limit of resolution of the photograph. Microfeatures are often duplicated by macro-features. In other words, a large terrace fault in a terrestrial volcanic crater visible at high altitude will more than likely have a multitude of smaller subparallel fractures associated with it that may be invisible from

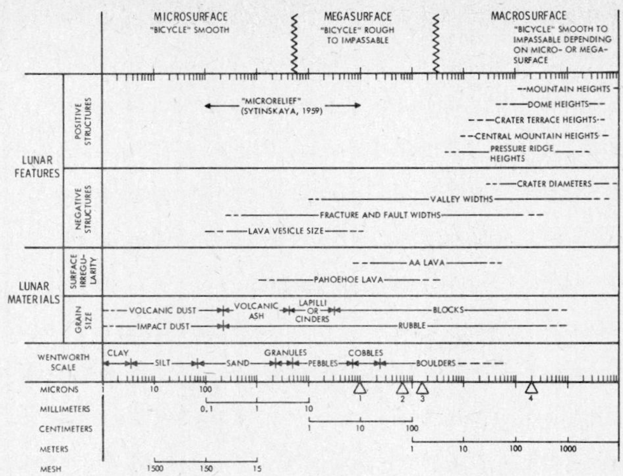

Note:

1. Traction effects are not considered.

2. Steep slopes are assumed to be present in the immediate vicinity of inside crater walls, crater terraces, rift walls, faults, fissures, and some pressure ridges.

3. Grain size of consolidated materials is not considered. For example, a coarse-grained tuff breccia can break across the large breccia fragments to yield a relatively smooth surface.

4. The combination of features listed can yield a wide spectrum of terrains. However, the associations must respect geological reality!

Triangles on abscissa refer to:

1. radar wave length used by Hey and Hughes (1959)
2. radar wave length used by Pettengill (1960)
3. radar wave length used by Evans (1957)
4. limit of telescopic resolution

Figure 3. Classification of Lunar Surface Relief

Figure 4. Snow-Covered Lava Plain, Craters of the Moon National Monument, Idaho

these high altitudes. Likewise, from visible alignments of spatter cones, one can often predict the occurrence of fractures that have localized these cones, as well as their directions and sometimes extent. Therefore, the writer believes that the lunar megafeatures - the resolution cut-off of which is indicated at four in Figure 3 - imply irregularities of specific types and intensities. From terrestrial analogy one might predict that extremely rough lunar areas should be smaller than less rough areas.

Surface-Shaping Mechanisms

In understanding the factors that would affect the lunar microrelief, we must recognize that lunar lava flows would not all be evolved at the same time and a recent aa lava flow would be more jagged than an old one accepting spallation, micrometeorite impact and sputtering as the erosional mechanisms. The older formations should be smoother than the younger ones. The same reasoning would apply if one wished to appeal to impact to create lava flows, a mechanism which the writer considers geologically improbable.

Which of the three mechanisms mentioned has produced the lunar microrelief? The spallation mechanism is self limiting in depth and dust produced would be thin. Micrometeorite impact would also tend to erode although, in contrast to the spallation mechanism, some authors assume a fresh surface would continuously be exposed with the ejecta being thrown into space. Other authors assume a fusion crust would develop on the lunar surface by micrometeorite impact.

The effect of solar winds in producing a lunar microrelief has been discussed by Wehner, et al.[20] On the basis of experimental data forty meters of lunar surface material is assumed to have been eroded away by solar winds in a billion years. In this process, the light fraction of the material sputtered escapes into space leaving a metal rich residue. Their experimental data support the assumption that loose material on the surface of the moon has been fused to form a solid but porous crust. This process according to Wehner, et al, explains the 10-centimeter smoothness of the lunar surface as interpreted from radar data.

The writer believes the interpretation given by Wehner, et al, to be inconsistent with certain observational facts. The estimate of 40 meters of erosion on the lunar surface per billion years is obtained by Wehner, et al, by summing up the sputtering yield for silver under a normal ion incidence of 5-kev protons and 20-kev He ions as determined by Grønlund and

More[5]. However, silver is a substance with one of the highest sputtering rates. Quantitative estimates on solar wind erosion on a metallic moon model may be risky. A more realistic model would have been radiation, weighted 90-percent protons and 10-percent He ions, incident on silicates, which is the most likely lunar surface material. Until this is done, it may be misleading to appeal to the erosion rate suggested.

Forty meters of erosion of the lunar globe would tend to homogenize the lun surface if all lunar surface features are older than one billion years. However wide range in time is apparently represented in the sequence: highland craters mare, postmare craters, rays, on-ray craters, central mountains in on-ray craters, and mare obscuration of ray sectors.* On this basis, the white highlands would have been more severely radiation-damaged and sputtered than the mare. A darker surface enriched in the residual metallics advocated by Wehne et al., would thus be expected on the highland areas. However, the reverse appears to be true.

Erosion of the highlands should invariably be more severe than the mare if a severe sputtering effect has been operative, but this is not true either. Assum a lunar magnetic field of at least a few gammas, solar winds would not be perfe ly omnidirectional. In this case, the reflecting qualities of east and west walls highland craters should be markedly different from the north and south walls if meters of material have been preferentially sputtered away. The east and west walls of highland craters under vertical illumination would reflect more highly than the north and south walls. The difference would also tend to be much less the younger postmare craters. However, no such effects are observed. The p mare craters may or may not be brighter than the highland craters, and highlar craters are not preferentially reflective on any portion of their perimeters. Sputtering effects are though to modify only the uppermost few centimeters, at most, of the lunar surface.

Infall And Volcanic Dust

In addition to spallation, micrometeorites and sputtering effects, dust produ from particulate infall debris, volcanism, and large impacts would tend to mute the ruggedness of lunar surface rocks. Gold[21], Gilvarry[22], and Wesselink[23] favor a very thick dust layer. Gold explains the smooth mare surface by an ero sion process that creates dust in the highlands and is coupled with a transport mechanism depositing it in the lower mare. Because no real homogenization of lunar surface details can be demonstrated, however, the concept of a thin dust layer appears more reasonable.

* I am indebted to Mrs. W. Cameron of NASA Goddard Space Flight Center for pointing out these ray observation features to me, particularly in the area north of Schröter.

Proponents of lunar volcanism contend that varying quantities of dust would be produced by volcanoes. This is because the sustained evolution of dust during the eruption cycle often varies in duration. The amount of dust resulting from impact would probably be much less by comparison.

In the absence of a lunar atmosphere, layering of any type of dust would be more poorly defined than on earth because coarse and fine material would settle equally as fast. Marked heterogeneity would be common in accumulations of both pyroclastics or meteoritic rubble. However, if volcanism were the dominant source of lunar ejectamenta, intercalations of fine ash and dust would be formed during those periods when only this ash or dust was being expelled from volcanic vents. Therefore, from the recurrence and periodicity of volcanism within a given caldera, volcanic strata might be distinguished from impact rubble strata. Because of the absence of erosional processes in the vicinity of lunar craters, a higher degree of continuity of either impact or volcanic strata should be in evidence.

According to the volcanic theory, local accumulation of thick dust layers is possible in the nonmare areas. Such accumulations are common on earth, for example at Parícutin, Mexico. Thin dust layers, thickening in topographic basins near postmare craters, are predicted for mare areas. The impact mechanism of crater origin would produce a thicker dust layer on the crater floor than on the crater flank, which is quite the opposite effect in volcanism. Again, the maria would have a thinner dust cover if basaltic rocks were formed by melting of impacted material. However, creation of basalt by impact melting cannot be demonstrated on earth.

The shear strength of rock dust under vacuum is of direct concern to the civil engineer in designing base facilities. The writer cannot accept Brunschwig's, _et al_,[24] suggestion that men and equipment would be exposed to foundering hazards in lunar dust until shear strength tests are made in vibration-compacted volcanic and chondritic dusts under vacuums exceeding 10^{-5} millimeters.

Perhaps a realistic picture of a lunar mare terrain would be simulated by a partially snow-covered lava plain with the snow considered equivalent to spallation, volcanic, cosmic infall, and micrometeorite impact dust. Figure 4 illustrates broken pahoehoe lava terrain at Craters of the Moon

National Monument, Idaho, in winter which might approximate a large portion of the mare surface. The lava surface is scoriaceous with vesicle pitting ranging from about one-half millimeter to as much as five centimeters with an average vesicle diameter of 0.3 centimeter. On the moon the vesicle size may be larger[25]. The maximum surface relief of the snow covered lava terrain in Idaho is estimated to be about two meters with an average of about one-fourth meter.

As yet we do not know to what degree dust covers the lunar surface nor its depth. We do not know if the exposed rocks are vesicular or surface-sintered. We do not know what kind of rocks exist there. It is the writer's opinion that volcanic extrusives exist, some vesicular, some not, of varying types covered at places with variable thicknesses of predominantly volcanic dust in different degrees of induration.

In conclusion, the term "microrelief" and the boundaries of one-tenth millimeter to one-tenth meter are useful as applying only to a given rock surface on the moon but not necessarily to the terrain in which this rock may occur. Sytinskaya's use of the word microrelief corresponds to geological reality in terms of rock surface textures. A unit rock surface can vary within one-tenth millimeter and one-tenth meter but a unit terrain area may not. However, whether or not impact and associated vaporization effects would produce a lava-like surface covered with deep depressions and sharp edges[26, 27, 28, 14] is a worthy problem for laboratory simulation. Normal volcanic processes, however, yield scoriaceous surfaces which are often pitted over an entire flow. It seems geologically difficult to accept a micrometeorite impact today and one adjacent to it a thousand or a million years hence, and thus build up a vaporized surface that is deeply but uniformly pitted over a large area. Yet, as in all arguments relating to a subject as black and white as the moon, we must realize that grays exist and how much and to what degree are the important questions to resolve for intelligent discussion.

III. MEGA-AND MACRO RELIEF TERRAIN FEATURES

Lunar And Terrestrial Analogs

There are many similarities between earth volcanic features and visible lunar features which support the volcanic hypothesis. Structures observed in and around these lunar features which are suggestive of a volcanic origin are given in Table 3 and may be divided into two categories — nonmare structures and mare structures.

On the nonmare area, many craters contain central mountains. Baldwin[29] states that these central mountains do not look like volcanoes within calderas. However, such is not the case.

The shapes of the mountains within lunar craters range from the crisp centrally pitted volcano-like mountains in Regiomontanus A to the irregular nondescript accumulations in craters such as Petavius. This range in form is duplicated in volcanic terrain on earth (Figure 5). Moreover, the lunar "central" mountains are not always central, as illustrated by Regiomontanus A in Figure 5 nor are terrestrial volcanic cones in calderas, as illustrated by Mokuaweoweo. With the variety of rock types found in calderas, the complexity of fracture patterns in them, and the variables involved in volcanic activity, this range in size, shape and disposition of volcanoes on crater floors is quite normal.

Features observed on earth which bear resemblance to mare features include surface irregularities of lava flows. The maria exhibit what appear to be pressure ridges, some of which are hundreds of kilometers long and hundreds of meters high. Commonly they are split by a fracture running along the summit of the ridge. This morphology is duplicated on a smaller scale on earth. Other pressure ridges are caused by sublaval drag which may have caused the structure present in the Sierra Negra caldera in the Galapagos Islands. Serpentine Ridge on the moon resembles it (Figure 6). Pressure ridges on the moon would tend to control the distribution of volcanic ash as first pointed out by Spurr[30].

Aligned groups of cones are common on the earth and moon[31]. Figure 7 illustrates the relationship of lunar and terrestrial crater alignments tangential to a parent crater. In many places on earth, this relationship is quite normal inasmuch as calderas are localized by fractures which are

Table 3. Caldera Features Found in Lunar Craters

Feature	Number observed
Polygonal shape	25
Central mountain	234
Central crater	43
Median ridge	26
Pressure ridge	48
Dark floor	32
Rim 'dikes' and radiating structures	20
Internal faulting (incl. terracing)	250
Breaching	40
Lava infilling	19
Changes (e.g., mist, lava flow, eruption)	15
Alignments	180*

*Each entry includes alignment of many craters.

Table 4. Wall or Peak Elevations in Possible Polar Base Sites

FORMATION	WALL OR PEAK ELEVATION		DIAMETER	
South Pole	Meters	Feet	Kilometers	Miles
†Leibnitz Mountains	4,900 to 10,000	16,000 to 33,000		
Malapert (Overshadowed by Leibnitz Beta)	4,600	15,000	56	35
Cartes	*	*	*	*
Cabaeus (North wall)	6,100	20,000	97	60
Scott	*	*	106	66
Amundsen	*	*	106	66
North Pole				
†Shackleton	2,700	9,000		
Delta	2,400	8,000	*	*
Beta, "lofty"	*	*		
Perry			24	15
Hansen			113	70
Beta	3,400	11,000		
Gamma	2,400	8,000		
Main	*	*	48	30
Vaisala	*	*	81	50
Gioja				
(1)	2,700	9,000		
(2)	2,400	8,000		
(3) and (4)	1,100	3,500		
Gamma	2,100	7,000		
Alpha	2,900	9,500		
Beta (1)	2,700	8,800		
Beta (unspecified)	2,400	8,000		

† Considered most favorable for a polar base site

*No data

Altitude and diameter data from Wilkins and Moore [51]

Off-center 1940 cinder cone in Mokuaweoweo caldera, Hawaii. Major diameter of caldera is 5.6 km (3.5 mi); minor diameter, 3.2 km (2 mi). Photo, courtesy of National Geographic.

Irregular central volcano in Awoe caldera, Indonesia. Major diameter of caldera is 1.5 km (0.9 mi); minor diameter, 1.2 km (0.7 mi). Photo, courtesy of Dr. R.W. van Bemmelen.

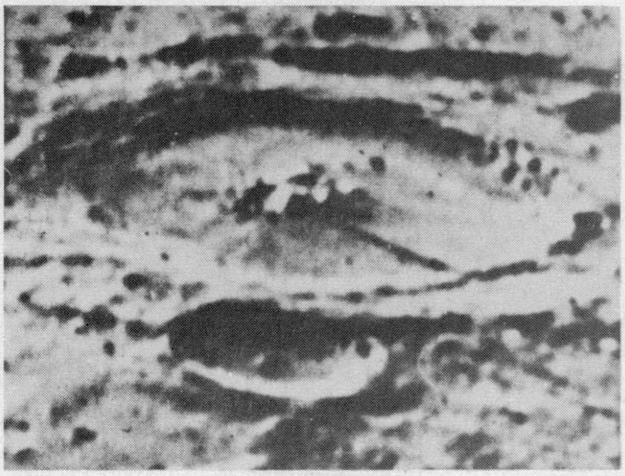

Irregular central mountain in Petavius. Diameter of crater is 160 km (100 mi). Photo, courtesy of Dr. A. Dollfus.

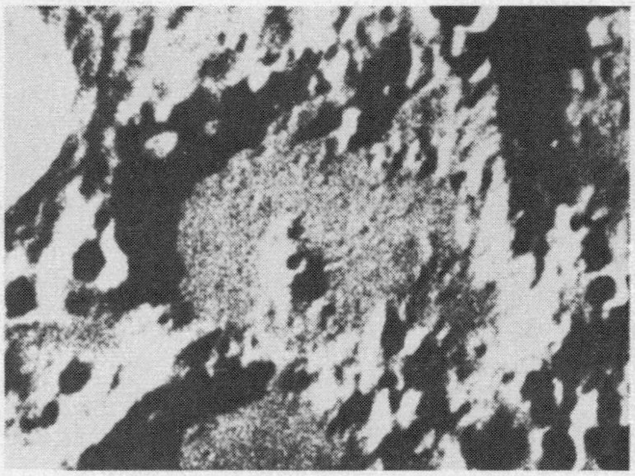

Off-center mountain in Regiomontanus A. Major diameter of crater is 128 km (80 mi); minor diameter, 104 km (65 mi). Photo, courtesy of Dr. Z. Kopal.

Fig. 5. Morphological Features of Mountains in Lunar Craters Compared with Volcanoes in Terrestrial Calderas

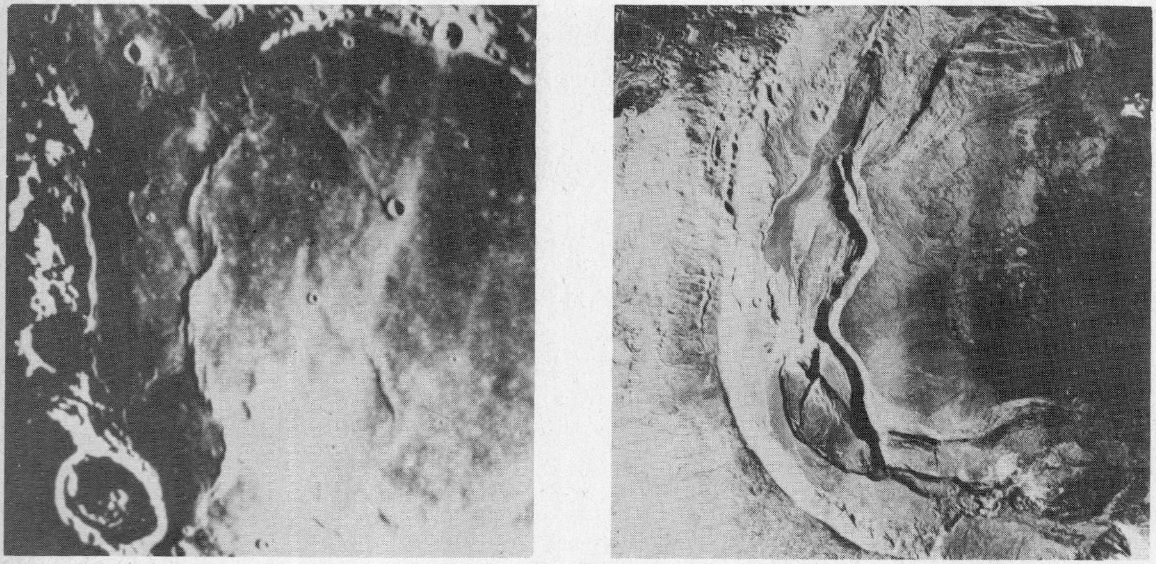

Ridge in Mare Serenitatis is over 260 km long (160 mi) and is about 240 m (800 ft) high. Ridge of basalt in the caldera is about 6 km (3-1/2 mi) long. Terrestrial photograph taken from 6100 m (20, 000 ft) by the U.S. Air Force in cooperation with the Government of Ecuador. Lunar photograph, courtesy of Dr. G.P. Kuiper.

Fig. 6 Comparison of Ridging Features in Mare Serenitatis (left) and Sierra Negra Caldera (right)

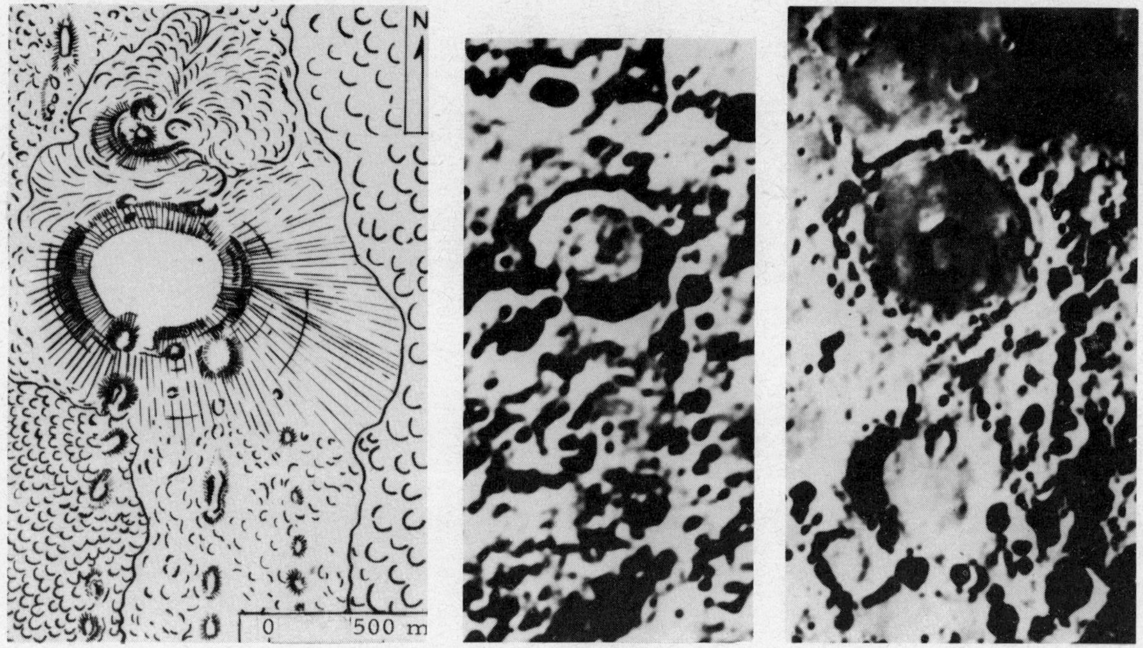

a. Iceland (volcanic crater) b. Abulfeda c. Pitatus

Courtesy of Dr. S. Thorarinsson

Figure 7. Comparison of Tangential Crater Alignments on Earth and on the Moon

often tangential to the caldera outline. In other words, fractures often serve to define a polygonal plate which is subsequently engulfed or otherwise fragmented by volcanic activity. These boundary fractures may subsequently provide zones of weakness along which extrusives may issue. Thus, a fracture that formed the boundary of a large crater could also provide a locus for an alignment of smaller craters or domes which would be tangential to the parent. An excellent example on the moon is the crater row forming the boundary of Müller and extending into the northern segment of the walled plain, Ptolemaeus. The tangential alignment of craters on the moon is in strong support of the volcanic origin of the lunar surface.

Lunar Television Surveys

Present-day astronomical resolution is inadequate to settle the question of the impact or volcanic origin of the lunar surface. Many features most characteristic of volcanism are beyond the reach of terrestrial telescopes. Future low-altitude or surface-television surveys will probably throw much light on this question.

Three techniques that can be displayed on televised pictures are differentiation, outlining, and contouring. Differentiation permits enhancement of slight differences in shade or color. Figure 8 exemplifies this high resolution television simulation technique as developed by the RCA Astro-Electronics Laboratory. The lava flows shown in the original photograph although conspicuously "stretched" in the televised equivalent, have been differentiated (in a nongeological sense) so that the fresh and hence blacker lava flows stand out against the older and slightly grayer lava flows. The ability to distinguish among different age groups of lunar lavas would be useful inasmuch as fresh lava flows might be more likely to contain intact lava tubes and caves. The fresh flow sources also would be more likely to harbor more mineralization than the older vents. The televised pattern in spite of the differentiation and stretching distortions clearly shows that the flows are onlapping and have moved from the lower right hand to the upper left hand corner. If this pattern is present it would support a volcanic origin of lunar surface features inasmuch as this pattern is absent around meteor craters on earth.

As an example of outlining, another photograph of the craters in the Galapagos Islands was televised using the RCA equipment. In Figure 9 the crater outlines are conspicuously defined as are the central volcanic hills in the Beagle crater (middle crater). This crater has been invaded by basaltic lava. The central crater within the Tagus crater (top) is also

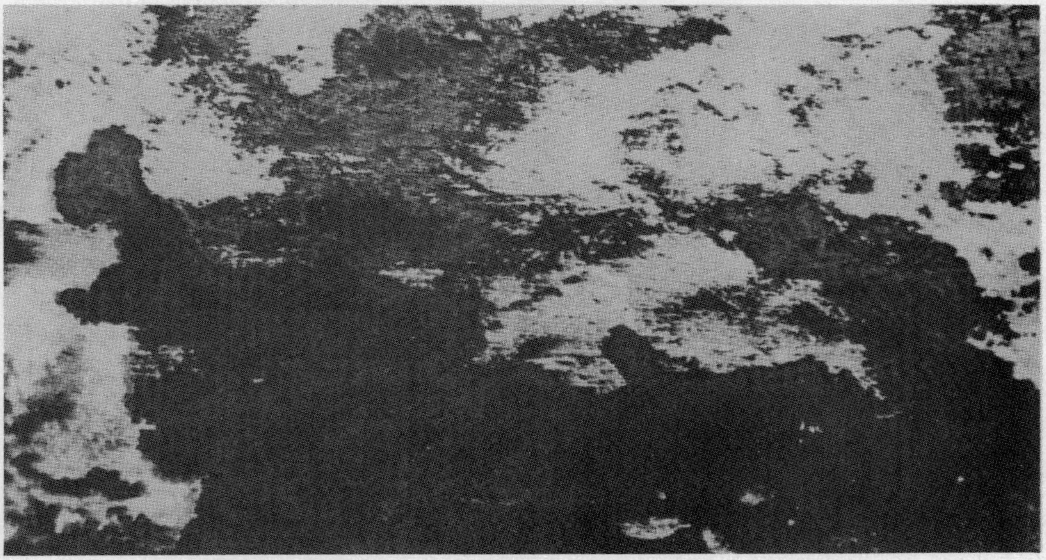

Original and television differentiation and "stretching" of photograph of lava flows of Isabela Island, Galapagos. Note flow directions of lava. Also, the fresher the lava, the blacker it appears. Photograph taken at 6000 meters (20,000 feet) by U.S. Air Force in cooperation with Ecuadorian government. Simulation techniques courtesy of RCA Astro-Electronics Laboratory. View encompasses about 3 miles (lengthwise direction of photograph).

Figure 8. Lava Flows of Isabela Island, Galapagos

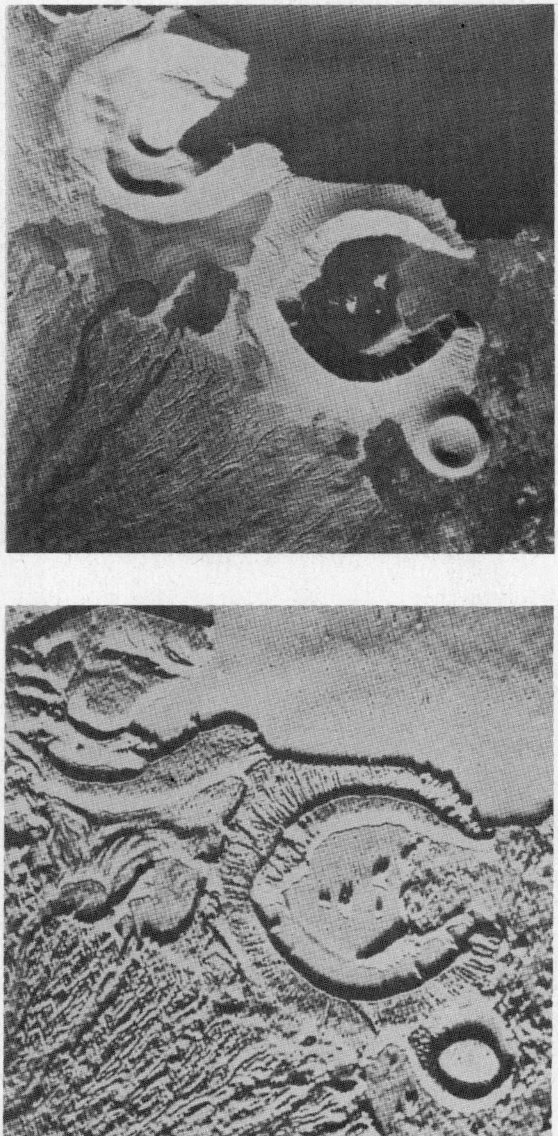

Original and television outlining of photograph of invaded Tagus and Beagle craters on Isabela Island, Galapagos. Radial grooves caused by tumbling lava blocks and subsequent erosion are enhanced. Photograph taken at 6000 meters (20,000 feet) by U.S. Air Force in cooperation with Ecuadorian government. Simulation techniques courtesy of RCA Astro-Electronics Laboratory. Diameter of largest crater approximately 1 mile.

Figure 9. Invaded Tagus and Beagle Craters on Isabela Island, Galapagos

outlined by this television simulation process. Even the flow pattern in the lava abutting the walls of Tagus crater can be distinguished.

Contouring by television techniques provides enhancement of the outlines of slight changes in light intensity. The gases escaping from the 1940 cinder cone in the crater of Mokuaweoweo are clearly outlined at the particular setting of the television simulator. Different "contours" on the light intensity of the gas cloud configuration can easily be obtained by this technique. The lip of the cone is also profiled. The likelihood of a subsequent impact on a central hill in a lunar crater to produce a similar morphology to that shown in Figure 10 is remote.

Television surveying of the lunar surface employing the techniques exemplified in Figures 8, 9, and 10 would greatly aid in the interpretation of the processes that formed the moon's surface.

Lunar Installation Sites

If the lunar surface is predominantly or even partially volcanic, the characteristics of volcanic terrain may prove valuable in manned lunar exploration and establishment of the lunar base. Judging by the known characteristics of terrestrial analogs, lunar volcanic terrain, where it exists, would provide many beneficial features.

In the craters of the nonmare area, natural protection against radiation and temperature extremes would be provided by overhangs and caves associated with at least four structures: (1) central cavities within the volcanoes on the caldera floors, (2) fractures bisecting pressure ridges on caldera floors, (3) peripheral caldera terracing and (4) major and minor faults and fractures attendant the localization of calderas.

Figure 11 illustrates fractures which commonly develop on the lips of caldera walls. Here openings many tens of meters long and deep and a few meters wide are accessible from the relatively gentle outer slopes of the San Fernandina caldera in the Galapagos Islands. If this situation were to exist on the moon, such features would make ideal primitive installation sites. The collapse hazard of these fractures is neglible in terms of human habitation time. However, the risk would be greater in those fractures immediately adjacent to the cliff face. If the San Fernandina caldera were a lunar crater selected for manned landing, optimum installation sites might be those which are encircled in Figure 11; these fractures are relatively remote from the dropoff and occur in rather flat terrain.

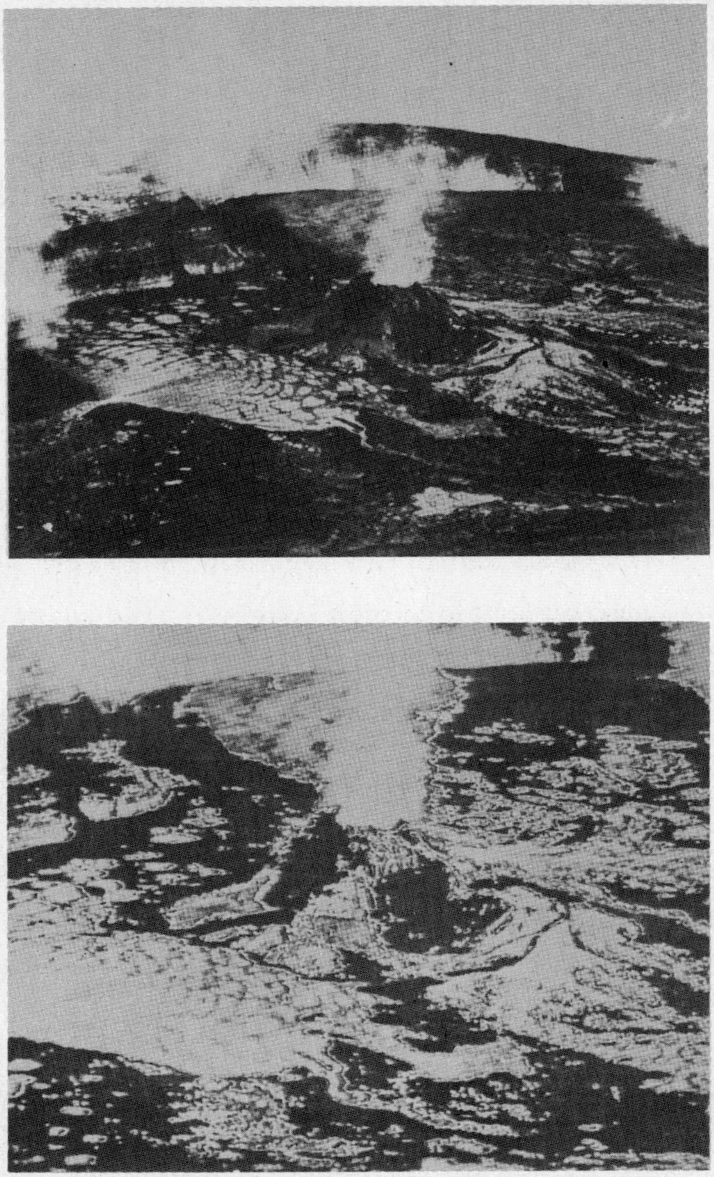

Original and television contouring of photograph of 1940 cinder cone in crater of Mokuaweoweo, Hawaii. Contours set to outline cone lip and steam area. Photograph courtesy of the National Geographic. Simulation techniques courtesy of RCA Astro-Electronics Laboratory. Diameter of cone base approximately 0.1 mile. View looking northwestward.

Figure 10. Cinder Cone of 1940 in Mokuaweoweo Caldera, Hawaii

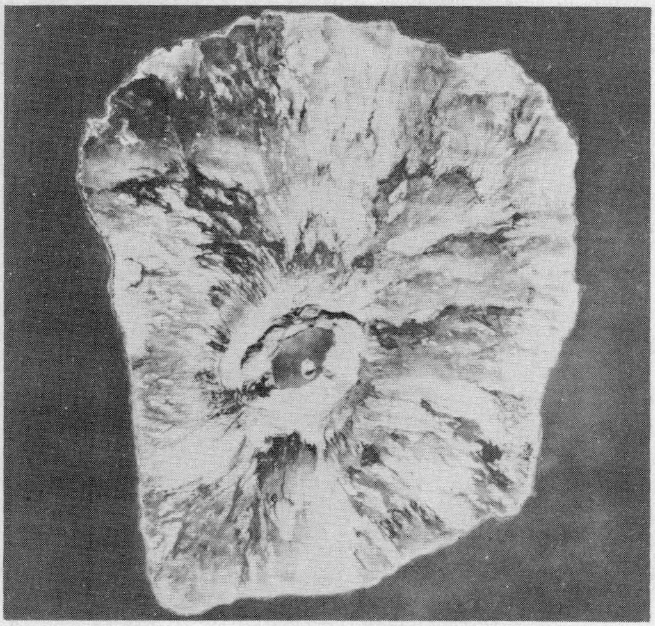

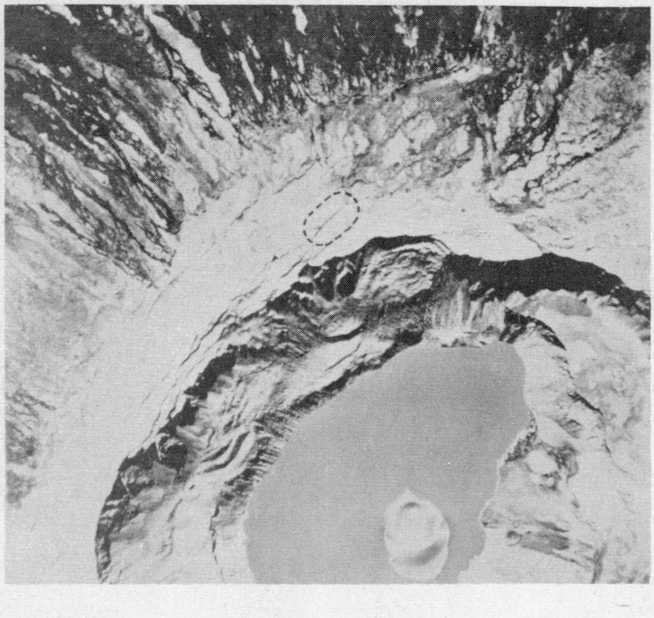

Lower photograph taken at 6000 meters (20,000 feet) by U. S. Air Force in cooperation with Equatorian government. Upper photograph is a mosaic.

Figure 11. Lava Flow Pattern on Flanks (Top) and Fracture Pattern on Rim (Bottom) of San Fernandina Caldera, Galapagos Islands

In the nonmare areas devoid of craters one might find suitable installation sites associated with the rift systems such as the Hyginus, the Sirsalis or the Triesnecker. Impact features on earth do not possess such a variety of shelters. However, steep dips and overturning of strata bordering those impact craters present in the nonmare province would possibly provide limited protection.

The most important groups of surface structures that might serve as natural installation sites on the maria are the lava tubes, caves, and pressure ridge fractures. Figure 12 illustrates a classic example of a cave in basaltic lava in the Dimmuborgir region of Iceland. The cave diameter is five meters. The roof of this cave is not excessively fractured and could be easily sealed. One such sealant would be a natural waterless cement, sulfur. If a generous supply of electrical power were available, a plasmatron might also be used in vacuum to spray molten rock to fill these fractures. Caves and lava tubes could thus be used as pressurized chambers for early basing activity. Fractures in pressure ridges are more common than lava tubes on earth and possibly the first human to explore the lunar surface may take refuge in such a fracture.

Lava tubes which are caused by drainage of subsurface channels of lava are numerous in terrestrial basaltic provinces. Some lava tubes are large enough to drive vehicles into. Many in Lava Beds National Monument and Lassen National Park in California are over three meters high and wide and tens of meters long. The one illustrated in Figure 13 is in Lava Caves State Park, Oregon, and supports a road built over it. Lava tubes by the nature of their origin and geometry are extremely strong; the collapse hazard of these structures is practically nil in terms of human habitation times.

Thus, if volcanic terrain exists on the moon, as telescopically-visible features lead us to believe, such terrain can be very useful in the establishment of manned bases. The potential benefits warrant the early employment of high resolution television surveys to resolve the important question of the origin of lunar terrain features.

White Line Across Cave Entrance Is Metric Tape

Figure 12. Cave in Basalt, Dimmuborgir, Iceland

Figure 13. Lava Tube at Lava Caves State Park, Oregon
(Courtesy of Oregon Highway Department)

IV. MINERALIZATION

Volcanic Sublimates

Volcanic mineralization on the moon will possibly involve the elements concentrated in sea water on earth. The reasoning followed is based on the defluidization process endorsed by Rubey[32], Vinogradov[33], and Ronov[34]. Defluidization is the evolution of fluids from a cosmic body at any given rate and at any given time. The defluidization hypothesis would appeal to submarine volcanism on earth to have enriched sea water in certain elements and to have even supplied part of the water. On the moon, any water furnished from the interior by an analogous defluidization process has obviously escaped. However, what has become of the elements that could have likewise been enriched in volcanic vapors contributing to a possible pre-existing lunar hydrosphere? What minerals might persist on the lunar surface?

An approach to the problem of mineral retentivity is to calculate the linear rates of decomposition according to formulae given in Figure 14. In this figure sulfur, sal ammoniac and chalcanthite would all persist in eternally shadowed zones (-150°C) on the lunar surface over geological time - some 4.5×10^9 years. In fact, a temperature approximating -100°C would permit these minerals to persist for this time period.

Meteoritic Minerals

Aside from nickel-iron, meteorites are almost devoid of useful minerals. There are some speculations as to the nature of cometary impact materials but there are no data as to their mineralogy. Meteorites do contain a large number of unusual minerals of technological importance but, with the exception of nickel-iron, they are volumetrically insignificant compared with the common rock-forming minerals. On the moon there may be more reason to appeal to larger relative volumes of troilite (iron sulfide) existing as surface debris.

Material	Years Required to Vaporize Layer One Centimeter Thick in Vacuum		
	-150°C. (lunar shadow)	0°C.	+100°C. (lunar sunlit surface)
Sulfur (S, rhombic)	10^{25}	10^{2}	10^{-3}
Sal Ammoniac (NH_4Cl)	10^{15}	10^{2}	10^{-1}
Chalcanthite ($CuSO_4 \cdot 5H_2O$)	10^{18}	10^{1}	10^{-3}
($CuSO_4 \cdot 3H_2O$)	10^{19}	10^{3}	10^{0}

Age of the Earth-Moon crust ± 4.5 x 10^9 years. Maximum temperature required for these compounds to exist over geological time is -70°C. for S; -114°C. for NH_4Cl; -105°C. for $CuSO_4 \cdot 5H_2O$; and -90°C. for $CuSO_4 \cdot 3H_2O$

Calculated from the equations:

1. $B = 1.1 \times 10^{10}\ e^{\frac{-23,890}{RT}}$ for rhombic sulfur
2. $B = 1.2 \times 10^{2}\ e^{\frac{-13,500}{RT}}$ for sal ammoniac (NH_4Cl)
3. $B = 3.3 \times 10^{7}\ e^{\frac{-18,300}{RT}}$ for chalcanthite ($CuSO_4 \cdot 5H_2O$)
4. $B = 2.8 \times 10^{2}\ e^{\frac{-15,600}{RT}}$ for ($CuSO_4 \cdot 3H_2O$)

where B = rate of sublimation in cm/sec; R = 1.99 cal/K°/mole; T = degrees K
Equations 1 and 2 from experimental data by Schultz and Dekker (1955, p. 2136 and 1956, p. 1099) and 3 and 4 from experimental data in the literature recalculated by Schultz and Dekker (1956, p. 1098).

Figure 14. Rates of Decomposition of Certain Volcanic Products in Vacuum

Figure 15. Ice Under 1885 Pumice Fall, Askja Caldera, Central Iceland

V. THERMAL SOURCES

If certain lunar craters are volcanic in origin, these areas should be considered in the selection of a base because of the heat and power potential of calderas relative to impact craters. The subsurface of volcanics on earth is warmer than the subsurface in non-volcanic regions and presumably the same might be true for the moon.

Numerous changes have been observed on the lunar surface. At least three craters have had a long history of reported changes, Linné, Plato, and Alphonsus. Unfortunately these changes are difficult to document although the spectrogram of the eruption in Alphonsus by Kozyrev[35] is now indisputable.

Bülow has made an extended study of the fracture patterns of the lunar surface from the tectonic point of view and mapped the major fracture systems, exclusive of ray patterns. Hundreds of fracture intersections are shown in Bülow's maps, and close inspection will undoubtedly reveal thousands more. Fracture intersections and rejuvenated fractures may define localized heat sources convenient in maintaining a lunar base. Rejuvenated fractures are also potential loci for hot-spring activity. Infrared photography and bolometry, while amenable to the mapping of rejuvenated fracture trends and their intersections, may be an indirect means for seeking out potentially hydrous areas.

To summarize, availability of subsurface and surface heat may be an important factor in locating the lunar base. Heat release to the lunar surface may be along volcanic conduits or fracture systems. If central mountains in craters represent volcanoes in calderas then these mountains would be potential heat sources. Concentric terrace faults in postmare craters, such as Copernicus or Bullialdus, if of recent and internal origin, may be warmer than faults in older nonmare craters. The same may be true for fracture intersections both within craters or on mare surfaces. Finally, if the ray systems represent fractures or fracture zones they would be among the warmest of surface features as recent bolometric work tends to indicate.

VI. BASE SITE CONSIDERATIONS

Evaluation

Presumably, the general site for lunar basing will be established by surveys of Surveyor- and Prospector-type vehicles with final appraisal by Apollo spacecraft. The data obtained from these early surveys will define the topographic nature of the base site and the presence or absence of mineralization. The lunar base, once established will go through many stages of evolution as detailed in Appendix I; and whereas the final development of the bases may be more or less independent of the terrain, the early phases will not. It is obvious that lunar basing operations should be preceded with carefully conceived simulated basing operations in cratered areas on the earth. A possible testing program of this type is outlined in Appendix II.

There are three considerations that may cautiously be entertained with regard to base selection. These are base location with respect to impact craters versus volcanic craters, mare areas versus nonmare areas, and polar areas versus nonpolar areas.

Primitive basing operations may find naturally occurring rifts and fractures valuable for protection against radiation, high temperatures, and micrometeorite bombardment. Both the mare and nonmare polar and nonpolar areas evidence such fracture systems and, therefore, no preferred location can be based on this criterion. However, volcanic craters generally offer more protective areas than impact craters. These sites are in the form of cupola-like spatter cones, terrace rifts, and lava caves. From the standpoint of protection, then, a volcanic crater is superior to an impact crater. However, as a base site a volcanic crater may not necessarily be superior to a basaltic flow terrain relative to its abundance of caves and fractures.

From the standpoint of mineralization and availability of water, a volcanic crater is superior to an impact crater. Because more craters occur in the nonmare areas, there would be a greater probability of finding a caldera in these areas. The nonmare polar areas are not necessarily superior to the nonmare nonpolar areas inasmuch as centers for volcanic mineralization on the nonmare areas would probably have random distribution. However, the occurrence of eternally shadowed zones in the polar regions would make them favorable prospecting sites.

From the standpoint of technology, the desirability of specific basalt types for cast basalt might prescribe a mare base site and the desirability of sulfur mineralization might prescribe a caldera base site. Therefore, calderas or volcanoes in the maria or calderas bordering the maria should be regarded favorably. In such a location, basalt for metallurgical processing and sulfur in mineral dressing operations might both be associated. Many calderas on earth composed of silicic rocks have basaltic flows and sulfur mineralization on their floors. This variety of substances of technological importance is not present in craters formed by impact. Impact craters offer only iron-nickel as useful products in a lunar technology. Presumably, iron-nickel would be available in any lunar terrain because of cosmic infall dust. A polar region would not necessarily offer more technologically important advantages than a nonpolar region.

However, from both an ecological and power standpoint, the polar lunar base does have benefits over those of the nonpolar base and a site in this location is discussed in detail below.

To summarize, if both terrain and mineralization are important considerations in early lunar basing operations, then a volcanic crater adjacent to a mare area or a volcanic crater in any location but containing a variety of rock types appears to be the ideal base site. It may or may not be in a polar location. If terrain, mineralization, light, and power are of prime concern then a volcanic crater adjacent to a mare in a polar location may be desirable. The rocket fuel penalty for polar landings versus nonpolar landings is slight. It would appear reasonable that a nonpolar base might be established first where background data has become available from early probes. After this base is established, other bases could be located possibly in a polar location. The early base in the nonpolar region would be suitable as an astronomical observatory which may not be the case for a polar base. We do not know where precisely to locate a lunar base. Only future investigations can provide the detailed control data that will define the base site.

The following discussion attempts to trace the basing operations that might take place first in a nonpolar probably volcanic crater adjacent to a mare, and subsequently in a polar crater in the Leibnitz or Shackleton areas.

<u>Nonpolar Base - Alphonsus Type</u>

The crater Alphonsus has been selected as an example of a base site in a probable volcanic crater adjacent to a mare. It is located in a favorable region from the standpoint of rocket trajectories, evidences obvious peripheral fracturing, has known internal heat possibilities, and contains possible mineralization in the dark spots on the crater floor. Alphonsus is about 115 kilometers (70 miles) from rim to rim with maximum wall heights in the north approximating 2100 meters (6900 feet) above the crater floor. The numerous objects on this floor have been described by Alter[36], Moore[37], and Both[38]. The association of the dark spots on the floor of Alphonsus with rift systems and crater alignments is strongly suggestive of volcanic activity, which is supported by the Kozyrev observation of eruptive activity on the central mountain.

In many ways Alphonsus resembles the Askja caldera in central Iceland. Askja is a circular crater of the Kilauean type approximately seven kilometers (4.3 miles) in diameter with walls up to 400 meters (1300 feet) in height. After collapse of the crustal plate defining the Askja caldera, a ring of cones developed along the rim and subsequently some maar-type activity took place on the floor adjacent to the north crater wall (the crater Viti). Flows emanating from the rim fractures filled the crater floor, and a sector graben developed which has resulted in a break in the northern section of the Askja caldera. A suggestion of rim cones and a graben leading into the southern end of Ptolemaeus is present in Alphonsus.

In 1875 a new vent opened on the crater floor of Askja, violently erupting great masses of rhyolitic pumice concurrently with basaltic flows far on the outside flanks of the crater. In 1927 a basaltic flow, starkly black in contrast to the pumiceous crater floor, flowed into the Knebel Sea, a lake filling the crater floor. This basaltic flow, vented from a fracture on the north crater wall, is similar in pattern to the dark lobate mass emanating from a point on the inside east wall of Alphonsus.

The next exploration objective significant in basing operations in a lunar crater may be the topographic low points. In the eruption of volcanoes on lunar crater floors, considerable amounts of fluids (liquids and/or gases) may have collected in the topographic lows and conceivably have formed surface accumulations either temporarily as water, semi-permanently as evaporates, or permanently as water-deposited ash and dust beds. The subsurface in areas of the topographic lows may therefore contain mineral deposits including water, as ice, and possibly elements which we find enriched in sea water on earth such as chlorine or boron.

The hydrous environment which may have existed for a short time in the topographic low areas in lunar volcanic craters may have localized maar-type craters in contrast to cinder cones and other positive volcanic structures. On earth, maars are formed in topographically low areas in volcanic provinces. This is strikingly shown in the Tarawera volcanic area in New Zealand. In 1886, an explosive eruption opened a nearly continuous trench about 14 kilometers (9 miles) long and in places, 150 meters (500 feet) wide and from 90 to 430 meters (300 feet to 1400 feet) deep. No lava was emitted. The ejected material consisted solely of fragments and fine powder derived in part from the scattering of the previously existing volcanic rock destroyed in the formation of the rift. Maar craters and tuff rings are found on topographic lows on the rift floor such as at Rotomahana in contrast to the volcanic cinder cones produced at higher elevations. Thus, superficial factors such as elevation and height of water table may control the formation of maars.

It is interesting to suggest (1) that topographic lows may contain more varied mineralization than higher portions of lunar crater floors, (2) that the topographic lows of lunar craters may be loci for and conversely defined by maar-type cratering, and (3) that the fine dust produced in maar-type eruptions or the fine grained pyroclastics of eruptive centers may produce smooth flat expanses in the topographic low areas in the volcanic craters on the moon.

Polar Base - Leibnitz Mountain Type

Although several critical factors are to be considered in the selection of a lunar base site, Brown, *et al*,[39] believe that the duration of the lunar photoperiod is one of the most important. Outside of the polar regions, the surface of the moon is bathed in sunlight for about 13.5 days in each 27-day period. However, since the moon's equator is inclined one degree 35 minutes to the ecliptic, the incidence of the sun's rays will always lie in the angle bounded by the horizon one degree 35 minutes at the poles. Thus, small areas of continuous light and nearly continuous light may be found in the polar regions, and certainly large areas of continuous darkness. The Mountains of Eternal Light at the north lunar pole have been described by Moore[40].

Theoretically, continuous full intensity sunlight would be available at an elevation of 660 meters (2170 feet) or more from the mean surface at either pole, providing no obstruction exists in any direction from a particular site in the polar regions. Since the sun's disk subtends an angle of 0 degrees 30 minutes, the sun is not completely blocked by obstructions depressed by more than one degree 20 minutes from the polar horizon. The minimum full-sunlight elevation increases to 3050 meters (10, 000 feet) 58 kilometers (36 miles) in any direction from either pole. A mountain of this altitude and at this distance would be intermittently shadowed by a polar mountain 660 meters (2170 feet) high.

Inspection of Table 4 indicates that of all the formations within a 48 kilometer (30 mile) radius of the poles, the Shackleton Mountains on the North Pole, or selected ridges, spires, or plateaus on the Leibnitz Mountains on the South Pole may serve as optimum polar base sites.

Associated with the eternally sunbathed peaks of the Leibnitz in the south and Shackleton in the north are the eternally shadowed portions of the floors of Newton and Peary, respectively. If defluidization has occurred on the moon, volcanic minerals of low vapor pressure may persist in eternally shadowed zones on the moon. The close association of mineralization of possible technological importance with eternal sunlight argues for consideration of a polar lunar base for three reasons:

1. Use of minerals such as sulfur in base technology
2. Use of sunlight for ecological facilities
3. Use of sunlight for the generation of power

In conclusion, the "best" place to locate a lunar base may be in or near a caldera adjacent to a mare. For the non-polar base — Alphonsus for example — the optimum base site would be in a caldera containing fractures on the inside rim floor and an extinct volcano somewhere on the floor. Sulfur mineralization should be present as well as some basalt flows and pumice. For the polar base, such as on the Leibnitz Mountains, the optimum base site would be on a caldera flank with fractures paralleling the crater lip, ideally with an extinct volcano on the rim. The crater floor should be eternal darkness and should contain mineralization, possibly sulfur, boron, or the halogens. Basalt flows and pumice should be present on the crater flank.

VII. BASE TECHNOLOGY

Prospecting Tools

Base technology will depend on the correct use of proper tools to seek out the raw materials necessary for survival and maintenance. Among these raw materials may be sulfur, basalt, and pumice. The economics of material transfer from the earth to the moon will probably require redesign of equipment for processing of what can be possibly useful. The choice of prospecting tools will be dictated by these vital needs. Of first priority will be water, then working materials for mineral processing and construction, as discussed in detail by Green[41]. Table 5 summarizes the possible logging devices that could be used either vertically or horizontally to find primarily water and other survival materials.

Extraction of Gases From Rock Froths

Inasmuch as the extrusion of a silicate in vacuum would tend to produce surficial high-gas activity promoting flow turbulence, a lava may possibly represent the more dominant form of surface flow on the moon. This is because molten silicates intumesce and become frothy when extruded into a vacuum.

Active gas content and viscosity of the lava at the temperature of chilling control vesiculation both on the earth and moon. But two factors which promote increased vesiculation of lunar lavas are extrusion into vacuum, as noted by Spurr, and lesser gravity. A much greater abundance of rock froths and a larger vesicle size is predicted for lunar surface extrusives.

Rock froths could be utilized as insulation and as a possible source of atmospheric gases.* Because of their cellular structure, they may be easily worked to provide building or cap materials for installation sites. Their extremely high porosity, which can exceed 25 percent, may supply a source of nitrogen or carbon dioxide by melting or simple mechanical crushing of these rocks. Vesicular rocks are easily crushed. Therefore, gas extraction from rock froths for the manufacture of an artificial atmosphere is qualitatively

---

* Entrained gases in rocks are primarily CO_2, HCl, Cl_2, and H_2 with minor amounts of CO, CH_4, N_2, and O_2, and range from 0.1 to 6 cc/g. Mafic rocks appear to be enriched in SO_2 and CO_2 with respect to silicic rocks. Recent obsidians and some recent basalts, e.g., those of Krakatoa, are enriched in Cl relative to older ones. Some pitchstones may have appreciable quantities of nitrogen and entrained methane, particularly those of Arran.

Table 5. Possible Logging Devices for Space Exploration

Parameter Measured	Conventional Well Logs				New Well Logs					
	Dry-Hole Resistivity	Neutron	Neutron-Neutron	Gamma	Discriminated Gamma	Density	Sonic Velocity	Accelerator Types	Magnetic Susceptibility	Nuclear Magnetism
SOLIDS										
Lithology				+	++			++	+	
Mineralization and Cementation				*	++			++	*	*
Provenance, Assemblages, and Facies				*	+			++	+	
Alteration				*	+	*	*	++	*	*
Physical Properties	*	*	+			++	++		++	
LIQUIDS										
Hydrogen Content	+	*	++					++		+
GASES										
Composition								+		

LEGEND

Blank: Probably not applicable in non-hydrous extraterrestrial environments
* Limited or specific isolated applications
\+ Useful or applicable
++ Highest potential in space exploration problems

NOTE: The conventional self-potential, resistivity, induction, and mud logs require a hydrous environment to function. They are omitted from this table for that reason.

Table 6. Lunar Base Applications of Processed Basalt

Cast Basalt	Sintered Basalt	Spun Basalt (Fibers)
Furnace material for water extraction operations	Nozzles	Cloth and Bedding
Crusher jaws	Tubing	Resilient shock absorbing pads
Pipes and conduits	Wire-drawing dies	Acoustic insulation
Conveyor material (pneumatic, hydraulic, sliding)	Ball bearings	Thermal insulation
Linings for ball, tube or pug mills and for flue ducts, ventilators, cyclers, drains, mixers, tanks, electrolyzers, mineral dressing equipment	Wheels	Filler in sulfur cement
Tiles and bricks	Low torque fasteners	Fine springs
Sidings	Studs	Packing material
Nose cones (?)	Furniture and tableware	Strainers or filters for industrial or agricultural use
Track rails	Low load axles	
Ties	Scientific equipment frames and yokes	
Pylons	Light tools	
Heavy duty containers for "garden and orchard" use	Light duty containers and flasks for laboratory use	
Radar dish frames	Pump housings	
Mirror bases		
Thermal rods		

simpler than the dissociation of the bulk rock into cations and anions using high energy and fluid-consuming techniques such as the plasma flame requires.

A thorough study of the gas content in rocks particularly the rock froths similar in scope to that of Chaigneau, et al,[42] is needed to determine the efficiency of recovery of gases needed for lunar basing and technology.

Water Sources

In the search for critical raw materials on the moon, no single substance is as important as water. From all standpoints - supply of oxygen and fuels, maintenance of equipment, support of life - water rates first as an exploration objective.

In order of probability of occurrence, five water sources are conceivably present on the lunar surface or near-surface.

1. Combined water present in volcanic extrusives
2. Ice in permanently shadowed zones in volcanic craters
3. Water of crystallization in volcanic sublimates
4. Permafrost in dust basins
5. Water in carbonaceous chondrites and nonvolcanic rock types

In discussing item two we note that the vapor pressure of ice at -150°C., i.e. the shadow temperature on the moon, is 10^{-11} millimeters of mercury. This means that if the value for the lunar atmosphere is above 10^{-11} millimeters of mercury, then it is possible for ice to exist and conceivably accrue in eternally shadowed zones on the lunar surface.

If the lunar atmosphere is at a pressure exceeding 10^{-11} atmospheres in lunar caves, the origin and age of ice in terrestrial lava caves is of interest even though the water is derived from surface runoff. More significant perhaps is the existence of ice under an 1885 pumice fall at Askja caldera in Iceland (Figure 15). Here ice has been insulated since that date. It would be extremely optimistic, however, to predict such occurrences of ice on the moon derived from meteoritic infall[43]. However, it is not impossible for defluidization and surface volcanism to have supplied this substance to the lunar surface.

More realistic is the use of rocks to supply water if defluidization (and attendant volcanism) has been the generally operative process on the lunar surface. Three guesses concerning the nature of the rocks that make up the surface of the moon are (1) meteoritic materials, (2) volcanic extrusives, and (3) ultrabasic rocks, including serpentines. The water content of these rock groups differs considerably, and even within a single group there is considerable range of water content. Although this discussion is in terms of weight percent, it should be noted that five percent by weight of water in the rock types considered is equivalent to 125 liters of water per cubic meter of rock (one gallon per cubic foot).

Figure 16 illustrates the water content in the three classes of rocks. The water percentages are for $H_2O^{+110°C}$ and not for $H_2O^{-110°C}$. $H_2O^{+110°C}$ is the water given off from 110° to 1000°C. In general, the water content of stony meteorites is on the order of a 0.1 percent, with the exception of carbonaceous chondrites which have as much as 10 percent water. More data are available on the water content of non-carbonaceous chondrites. Superior analyses were selected where the $H_2O^{+110°C}$ was tabulated separately from the $H_2O^{-110°C}$.

Together with consideration of the H_2O^+ water content of individual volcanic samples, the data averaged from rock groups shown in Figure 17 indicate that the water content of volcanic extrusives, approximating one percent, will probably be ten times greater than meteoritic rocks on the moon even if the volumetrically small amounts of "dry" obsidians are included. Many of these relatively anhydrous obsidians represent the final stages of eruptive activity.

<u>The Use And Processing Of Sulfur</u>

We will be obliged to use what we find on the moon for survival, for comfort, for sustenance, and for fuel. Among the possible substances that could be exploited in volcanic terrain on the moon is sulfur. Sulfur is the most abundant volcanic mineral and, on the basis of either thermal decomposition or vapor pressure characteristics, this mineral, if produced in appreciable quantities, will persist over geological time in eternally shadowed zones on the lunar surface. Sulfur has a fairly high thermal neutron capture cross section of a half a barn and could easily be sought out by nuclear prospecting techniques even if covered by a dust layer as thick as ten centimeters.

Occasionally, entire crater floors on earth become filled with pure sulfur during volcanic eruptions as in Mt. Etna in Italy and Mt. Adams in Oregon. In the latter crater, many tons of sulfur exist under a perennial ice cap. If central mountains in lunar craters are volcanoes, they should be prospected for this mineral. From terrestrial analogy, the inside walls of the craters themselves may also be sites for sulfur mineralization on the moon. In the

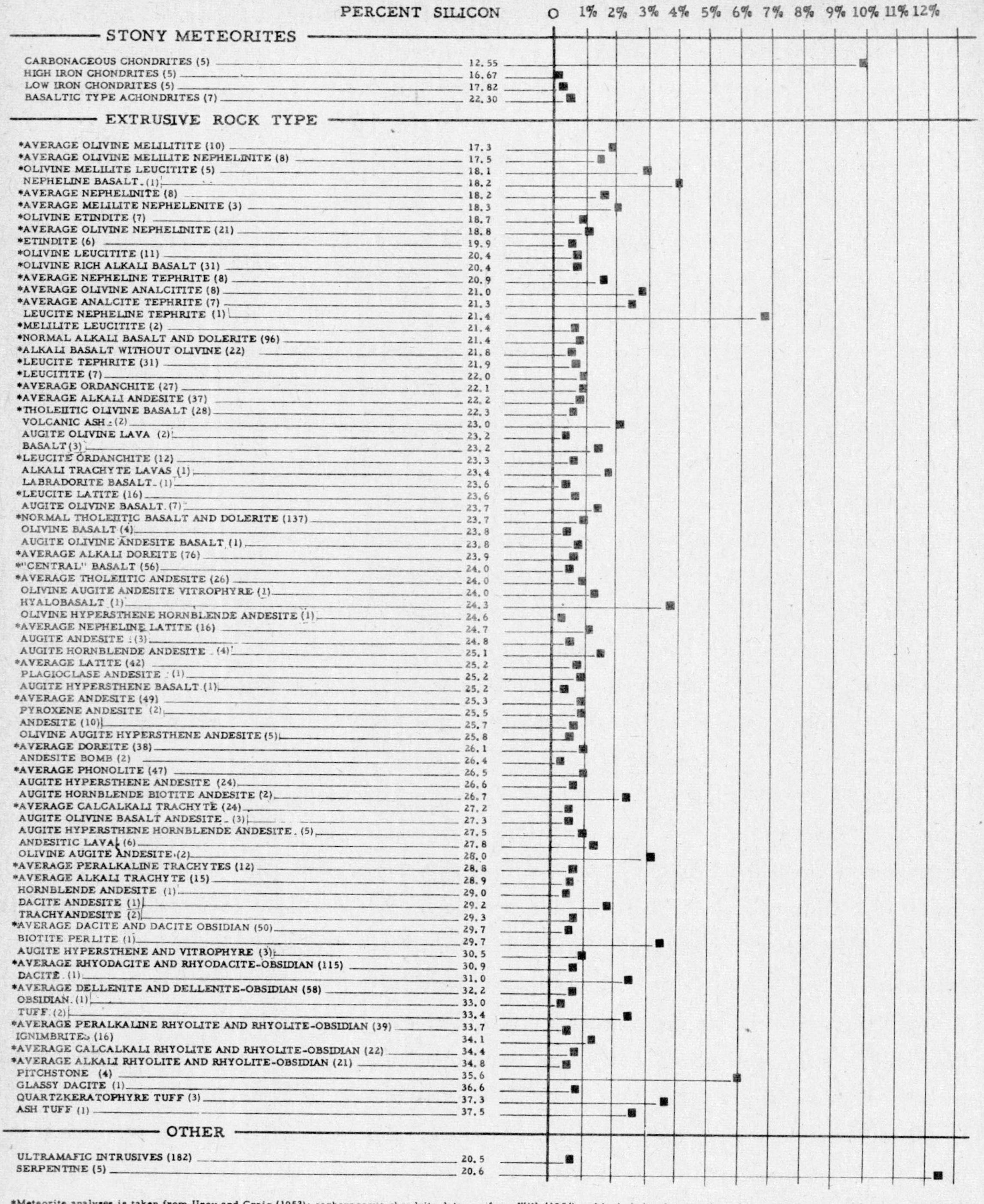

*Meteorite analyses is taken from Urey and Craig (1953); carbonaceous chondrite data are from Wiik (1956) and ignimbrite data are from Steiner (1958, pp. 330-331). Asterisked volcanic extrusive data are from Nockolds (1954); all others, except pitchstone average from Tilley (1957), are from an undated compilation by Van Bemmelen. Serpentine average is from Nagy and Faust (1956).

Numbers in parentheses indicate the number of individual analyses involved in determining the average.

Figure 16 Water Content in Rocks

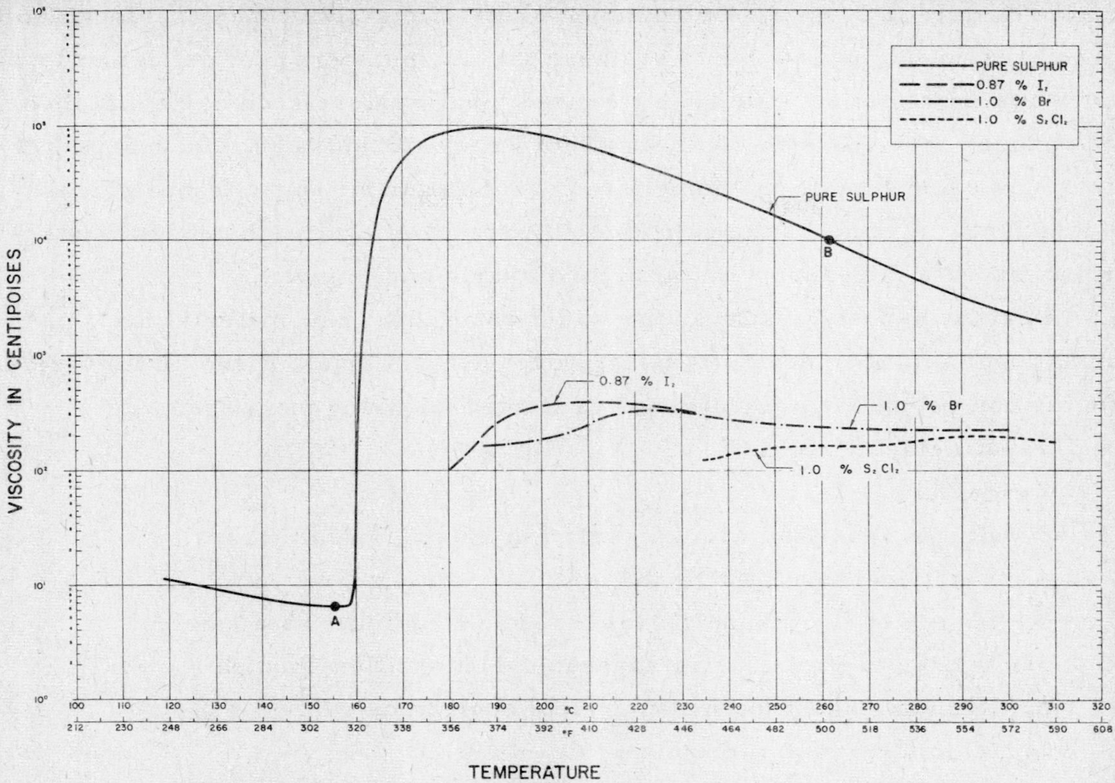

Figure 17. Effect of Viscosity Inhibitors on the Viscosity of Molten Sulfur

(After Bixson, et al 1960, p. 82-84)

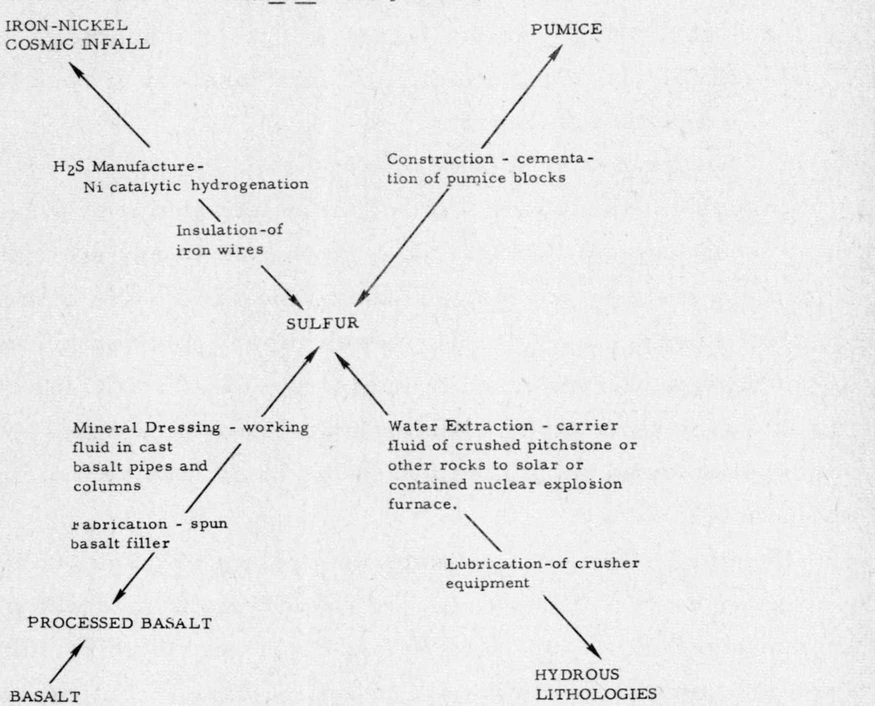

Figure 18. Relationship of Sulfur to a Possible Primitive Lunar Technology

Aksja caldera in Iceland, for example, fumarolic activity along caldera floor fractures bordering the inside southern and southeastern rim defines abundant sulfur mineralization. Multiple terraced lunar craters such as Bullialdis or Posidonius or visibly yellowed ones like Proclus deserve careful mapping from the standpoint of providing a source of sulfur for lunar technological operations. A study of the selective absorption of light in the Aristarchus area supports the concept of sulfur compounds on the moon.[44]

In many aspects, lunar sulfur would be an ideal raw material for metallurgical and mineral dressing processes. In brief, sulfur in its various forms can be used to varying degrees for the following purposes:

1. Waterless cement
2. Sealant
3. Bricks (admixed with volcanic ash or spun basalt fibers)
4. Electrical insulation
5. Source of H_2S using Ni from cosmic infall dust as a catalyst

5a. Source of H_2SO_4 using anaerobic Thiobacillus species

6. Reducing agent for metallurgical operations
7. Inhibitor of polymerization
8. Polished surfaces and mirrors
9. Low viscosity working fluid for mineral separation
10. Lubricating fluid for pumps and other machines
11. High viscosity carrier fluid for transport of rock fragments to water extraction chamber
12. Working fluid in power plants

Figure 18 illustrates some of the relationships of a sulfur technology with other lunar base operations. It is important to realize that sulfur can be stored and transported in cast basalt pipes which are more corrosion resistant to sulfur than are metal pipes. Preliminary thin section examination of cast basalt fragments immersed in molten sulfur at 150°C for over a month indicate almost no surface alteration or loss of compressive strength. A combination of basalt and sulfur on the moon may be a fortunate happenstance for base technology.

Pumice is another raw material present in large quantities in volcanic terrain on earth. It can be easily shaped into blocks which sulfur could easily cement together. Pumice walls and roofs cemented by sulfur would provide adequate thermal insulation for primitive base facilities. Sulfur can also be used to seal fractures in lava tubes and cave roofs.

Volcanic ash may be one of the most abundant comminuted raw materials on the lunar surface. When mixed with sulfur, bricks and structural components may be made which have rather high compressive strengths of up to 350 kilograms per square centimeter (5000 pounds per square inch) for an 85 percent volcanic ash, 15 percent sulfur mix. The volcanic material used was crushed basaltic lapilli and ash from Sunset Crater, Arizona.

Just as basalt, pumice, and volcanic ash can be used in conjunction with sulfur, so may cosmic infall material. The latter, if similar to the spherules dredged from the deep sea floor, will probably be iron and nickel rich. Such material may serve as a catalyst for the hydrogenation of sulfur to produce H_2S. Hydrogen required in this process could be derived from the electrolytic decomposition of water extracted from lunar rocks.

But the fine particle size of cosmic infall dust as well as its high nickel content[45] works to our advantage in the sintering of infall dust into wire and structural components. Again, a fortuitous combination of substances may be present on the moon for effecting a lunar base technology. Iron powder can be more effectively sintered when nickel-bearing.[46] Nickel apparently induces sintering in iron powders by creating surface defect structures on the iron particles. Electrical conductors can thus be fabricated and coated with sulfur as insulation material (Figure 18).

Another major use of sulfur may be in the mineral dressing field. Heavy liquids used to separate, say, lunar pyrite from volcanic ash and dust will be prohibitively expensive to import from the earth. Molten sulfur may be a useful natural heavy liquid for this purpose. Its melting point is low (123°C) and the specific gravity is 1.8. Admixing sulfur with naturally occurring selenium can increase the density of the melt up to three, which is that of selenium sulfide (SeS). Interestingly enough, selenium is an impurity characteristic of volcanic sulfur; the S:Se ratio varies from 600:1 to 140,000:1.

Whereas the highly viscous sulfur is not a predominantly thixotropic fluid it has a greater tendency to behave as a non-Newtonian fluid than the seven centipoise sulfur. For certain mineral separation techniques discussed in the next section it is desirable to use Newtonian fluids in mineral separation operations where one wishes to apply Stoke's equations. Therefore, additives such as I, Br, or S_2Cl_2 can be employed to reduce the viscosity of sulfur at higher temperatures and render it less thixotropic (Figure 17). Thus, the properties of low to high temperature sulfur will more closely approach each other except for a marked increase in viscosity.

The effect of I, Br, or S_2Cl_2 is to break the long S polymers which develop at higher temperatures. There are two reasons for attempting to reduce these high viscosities. The first is that for certain lunar base operations such as in power plants, high temperatures are necessary for optimum turbine operations and the use of sulfur as a working fluid at these temperatures requires efficient pumping of low viscosity fluids.

Realizing that addition of I, Br, or S_2Cl_2 can effectively control the viscosity of sulfur at temperatures above 200°C, then the use of sulfur as a lubricant in pumps and other types of machinery required to operate at high temperatures is optimized. The use of sulfur as a lubricant at low temperatures (ca. 150°C) is even more desirable because of the seven centipoise viscosity of the fluid. The viscosity of sulfur is capable of careful control by either temperature adjustment of a few degrees or by additives by a few tenths of a percent. In passing, the additives that control the viscosity of sulfur are all substances predicted by the defluidization hypothesis. The halogens as a group are enriched in sea water on earth and, if a similar process of enrichment has occurred on the moon, then iodine, bromine, and chlorine compounds should be locally enriched in shadow zones of crater areas on the moon. All of these elements have high neutron capture cross sections which make them eligible for rapid and definitive detection by nuclear prospecting techniques even under a dust layer.

Pure sulfur at 200°C may be a suitable carrier fluid because of its high viscosity. For example, rock fragments may be carried in suspension in a highly viscous medium such as 200°C sulfur. The rock fragments could be transported in large-diameter cast basalt pipes into contained nuclear explosion heat exchangers or to solar furnaces for extraction of water from hydrous rock types. Some of the combined water would be released at 200°C during the transport but most would be extracted in the 200° to 1000°C temperature range.

Finally, the use of sulfur as a working fluid in power plants is not only a distinct possibility but has been quantitatively evaluated from a thermodynamic, chemical, and engineering standpoint. In a report by the Sundstrand Turbo Corporation[47], the properties of sulfur as a working fluid in closed cycle, chemically fueled five-kilowatt turbogenerator were studied for application in a space vehicle. Hydrogen and oxygen burning stoichiometrically represent the heat source.

Other substances such as aluminum bromide have proven to be superior to sulfur as a working fluid albeit with problems, but if sulfur is already on the moon in large quantities, then more effort should be supported at this time to realize a lunar turbogenerator power plant using sulfur as a working fluid and cast basalt as corrosion proof conduit material.

Mineral Dressing Techniques

Gaudin[48] has discussed mineral dressing operations under lunar conditions. In essence there would be a general slowdown in most separation processes because of the lesser gravity. According to Gaudin, processes that depend in some way on the resistance of an enveloping fluid to the propulsion of an immersed solid are not usable on the moon under vacuum conditions. These processes include classification, jigging, tabling, heavy media separation, flotation, leaching, and amalgamation. More study might show to what degree these operations are impractical on the moon and to what degree they might not be useful if the systems were enclosed but not pressurized.

If the lunar base were in a volcanic area, one of the more practical mineral dressing working fluids to use would be sulfur because sulfur is the most abundant mineral in volcanics. In addition to the uses discussed under sulfur technology, sulfur might be quite suitable in lunar mineral dressing operations. For example, in the comparison of a pyrite settling in water on earth compared with pyrite settling in molten sulfur on the moon, several differences are apparent. Using Stoke's equation, for the same terminal velocity a net increase in particle size of pyrite is effected in a lunar settling column because of differences of densities, viscosities, and gravity.

By increasing the viscosity seven times that of water by using 155°C sulfur*, the particle size of pyrite that would fall at the same terminal velocity on the moon as it would on earth in water is six times that of the terrestrial equivalent. By increasing the viscosity ten thousand fold over that of water – sulfur at 262°C** – the particle radius of pyrite on the moon would be 212 greater than terrestrial pyrite settling at the same velocity in water. Sulfur at this viscosity, although not notably thixotropic, is difficult to work with and lower viscosities approximating a one thousand fold increase over that of water can be achieved by adding less than two percent of Br, I, or S_2Cl_2. This would still give a particle size increase of 67 for lunar pyrite particles over that of terrestrial pyrite particles settling in water at the same terminal velocity

---

* Point A, Fig. 17

** Point B, Fig. 17

The Metallurgy Of Basalt

If there is one point of agreement among most students of lunar petrology it is the probability of the occurrence of basaltic rocks on the dark portions of the moon. Whether originating from impact or volcanism, basalt may turn out to be the most abundant and uniform raw material to be found on the moon. However, impact does not produce basalt on earth and only negligible amounts of melting occur in terrestrial meteor craters. Furthermore, it can be demonstrated that some lunar craters, including Clavius, if caused by impact, apparently do not induce secondary volcanism.

Regardless of theories, the group that produces structural components and fasteners from lunar basalt may be the first to maintain a lunar base.

The results of the studies by Kopecký and Voldán made in the gradient furnace were applied to the manufacture of cast basalt (Figure 19). The most apparent feature in cast basalt is the zonality of the product which is a function of the cooling rate. The internal zone of a casting, usually two to eight millimeters thick, is always somewhat glassy and pigmented with submicroscopic magnetite, corresponding to a temperature of 1150° C to 1000°C from the gradient furnace results.

The central zone is thickest and is marked by the maximum amount of crystallized products with crystals always more skeletal than the border phases. However, some residual glass phase is also present.

The external zone is usually two millimeters thick with a prevailing glass phase. The crystalline phases predominate over the glass in the area bordering on the internal zone. The texture is spherulitic with solidification temperatures approximating 1000°C.

The Technology Of Processing Basalt. The gradient melting data obtained in the laboratory were applied to manufacturing operations. At present, the raw material, 8 to 15 mesh basalt, is smelted at 1300° to 1350°C in shaft furnaces similar to open hearth processing of steel. The molten material then is conducted into a homogenizer drum where, at carefully controlled temperatures and slightly reduced temperatures, the melt begins to crystallize. The casting process following this is similar to conventional metallurgical processing except for differences imposed by the greater viscosity and cooling rates. Lehr or chamber furnaces are used to cool the castings from 800°C to room temperature over a 24-hour period.

The sintering process is similar to that used in power metallurgy. The basalt frit made from molten material is finely ground (1600 mesh). impregnated with a plasticizer, shaped under a pressure of one thousand kilograms per square centimeter (14, 200 pounds per square inch) and sintered in electric furnaces at 1120°C to 1140°C (2048° to 2084°F). Sintered basalt is valuable in the manufacture of small articles as nozzles, wire-drawing dies, spheres, and special fabrications.

Fibers or basalt wool are produced by processes similar to those used for producing certain kinds of glass fibers. The thickness of the fibers range from 15 to 25 microns. Artificial basalt wool is used as a filler in asbestos-cement manufacture, and for acoustic and thermal insulation[49].

Lunar Application Of Processed Basalt. The use of basalt is important because of its high probability of occurrence on the moon. It, together with sulfur and iron and pumice, constitutes the four major raw materials that may co-exist on the moon.

Considering what importance has been attached to the use of sulfur as a working fluid on the lunar surface, what better way to store and move liquid sulfur than in basalt tanks and pipes? Preliminary data on the alteration effects on cast basalt by molten sulfur at 150°C show no adverse corrosion effects.

On the moon, the reduced gravity would tend to slightly improve the casting process by reducing the onset of turbulence of a given crystal size. Stoke's equation would apply to a higher value for the terminal velocity of particles. Laminar flow on the moon would persist at higher terminal velocities than on earth. Thus the effect of gravitation separation of mineral phases often noted during melting, and inhomogeneities produced in casting, would also tend to be less apparent lunar cast basalt. Applications of cast, sintered and spun basalt (fibers) to the maintenance of the lunar base are summarized in Table 6.

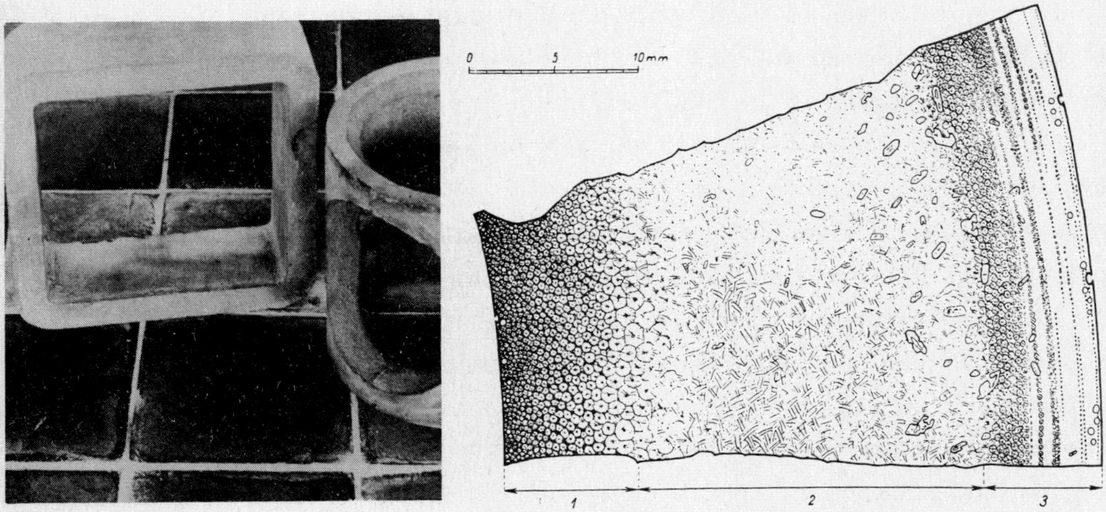

Figure 19. Pipes and Cross Section of Cast Basalt

Figure 20. Lunar Base Chronology

VIII. CONCLUSIONS

From a consideration of tidal and gravity effects operative or once operative on the moon, one might expect a greater sensitivity to volcanism on the moon than on the earth. Details of lunar surface features support widespread volcanic activity. If most lunar craters are calderas and most maria are basaltic flows, then base selection should be guided by a different set of objectives than if the lunar surface were dominantly impacted.

Volcanic terrain advantages offered the lunar base include lava tubes, caves, and fissures all with a negligible collapse hazard and all offering protection against radiation and temperature extremes. In addition, mineralization, such as ice or sulfur, can conceivably exist in volcanic environments but not in impacted ones. Thermal sources including volcanoes in lunar calderas, and fracture and ray intersections should be considered in selecting the lunar base.

This selection may have to compromise the requirements of many scientific disciplines. The first base will probably be determined by terrain and mineralization advantages. For this reason a caldera will be superior to either an impact crater or a basaltic flood plain. Therefore, when considering early trajectory restrictions, a nonpolar nonmare caldera would constitute a favorable lunar-base site. From the standpoint of a possible cast basalt technology, the caldera should either contain basaltic flows or be located near a mare. The lunar crater Alphonsus may possibly meet these requirements and is used as an example.

In later phases of lunar activity, it may be desirable to have continuous sunlight both for power and ecological requirements. Continuous sunlight is available at the lunar poles. The polar base site should be adjacent to eternally shadowed zones where mineralization needed for base technologies may be present. The examples cited involve the Leibnitz and Shackleton Mountains on the South and North Pole, respectively. After the base is selected, it will have to be established and then maintained. Specific lunar technologies will require the use of tools specific to needed raw materials. The neutron-neutron log or variation thereof is ideal for the detection of water; the density or velocity log for rock froths; and the neutron-gamma accelerator log for the identification of elements of high neutron capture cross section (sulfur, boron, chlorine, and iodine). At least seven technologies may be developed on the moon which will involve the earth scientist:

(1) Tunneling and base construction
(2) Extraction of gases from rock froths
(3) Extraction of water from volcanics
(4) The processing of sulfur
(5) Mineral dressing using existing raw materials
(6) The metallurgy of basalt
(7) The selective vaporization of rocks using plasmatron techniques

The topographic low or lows in some lunar craters may have significance in localizing maars and/or accumulations of subsurface water and mineralization, and in defining large flat smooth areas.

Appendix I. Lunar Base Chronology

Within the last few years, the division between science and fiction has become so fine that the reader should be less shocked at what follows than, say one living at the time of Jules Verne. Also, this chronology is but one of an infinite number.

The first phase of lunar basing may possibly begin in the mid-1960's, when a soft-landed vehicle of the Surveyor type would land at a predetermined site (1) and define a beacon possibly at a site defined by number 1 in Figure 20. Using Surveyor as a guide, an unmanned traverse would be made in an area of interest using a Prospector-type mobile vehicle. The traverse (2) might be used to investigate preselected fracture intersections and possible mineralized areas appearing on the floor of the lunar crater.

Several Prospector or Surveyor vehicles might be located in a crater with potential for basing and perhaps the placement of geophones at (3) to obtain some velocity data concerning the geophysical make-up of the crater floor would be achieved perhaps in 1967. These data would bear directly on future lunar base construction.

The next phase would probably be a touchdown and almost immediate return of man in an area of highest interest (4) early in the 1970's. A second expedition might place a booster tank in a crevasse (5) followed by installation of a beacon (6) at a prominent landmark. The booster tank would serve as a temporary housing site and as a center for further activity. With sophisticated vehicles of the Apollo type, a temporary power station (7), possibly a SNAP II reactor, would be assembled near a subsequently established primitive water extraction operation (8). In the middle of the 1970's it is hoped that enough sophistication in basing operation would permit construction of a cupola over the original fissure system as well as a primitive radar dish (9). The rock to be processed for water and other needs would be mined from a vertical shaft (10) directed perhaps to intersect critical mineralized fissure systems (11).

Realizing that underground installations would be the bonus obtained from the processing of rock for water and fuel, the tunneling operation (12) is assumed to continue to where a semi-permanent base storage and power station is excavated (13). This area also would serve as the first temporary communications center. Following this, areas on the lunar surface could be cleared (14) and simplified launch centers (15) constructed in the late 1970's. One of these centers would possibly function as a temporary geological and astronomical observatory and ecological installation. During this period, surface geological excursions (16), some of which locate additional beacons, would take place concurrently with horizontal tunneling to the crater rim summit (17) and the launch area (18). A permanent radar dish on top of the central mountain (19), completed in the early 1980's, may be timed with construction of a sulfur technology and solar energy laboratory with ecological facilities in a mineralized area (20). Likewise timed is the construction of a surface transportation tube to the summit of the central mountain with the completion of the horizontal tunnel below the central mountain peak (21 and 22).

In the late 1980's, the scope of operations may reach a point where a permanent power station and base supply depot can be constructed (23). Additional surface transportation routes (24) and more permanent cupolas would be required at this time. The master ecological center, at the top of the central mountain (25), would be connected by a vertical shaft (26) to the base supply depot and power station.

Final construction on a base astronomical observatory (27) may be completed in 1990 as well as additional surface transportation to a permanent geological cupola (28). Routine earth-return missions from a completed launch site would coincide with completion of surface transportation routes to physics and solar energy laboratories (29), which are subsequently completed.

Refinements in the master ecological center will undoubtedly continue through the mid-1990's. During this time, technology centers for processing basalt and other raw materials will be built at geologically favorable sites on the crater floor (30). A permanent communications center with supplementary lasar radar equipment (31) will probably guide large geological expeditions in hovering traverses (32). The final surface transportation network (33), powered by a master reactor power station (34) may be realized at the turn of the century.

Figure 21. Basaltic Flow Terrain, Lava Butte Area, Oregon

Appendix II. - Simulated Lunar Base Sites

Before going to the moon, landing and base construction should be simulated in terrain resembling the moon. If a vehicle can land in a rugged terrain on earth, move a tank into a crevasse, and cover it with surface material under normal gravity in a pressurized system that does not leak, then we may feel more hopeful of success of a similar operation at one-sixth gravity under vacuum.

Two simulated lunar-landing and base-construction phases are recommended: the first with highway, railroad, power, and water facilities; the second under extremely rugged conditions with no access routes or power facilities.

The first area suggested is in the Bend, Oregon, area. A basaltic flow issuing from Lava Butte a few miles south of town is adjacent to the Deschutes River. The flow itself is bounded by a U.S. Highway 97 and traversed by the Great Northern railroad. Adequate power, subsidiary access roads suitable for heavy duty trucks, and housing are available for operation of master control stations and slave vehicles. The basaltic flows and associated cinder cones in the Bend, Oregon, area are ideal for test traverses and missions[49]. Aerial photographs of the flow showing access routes and facilities are shown in Figure 21.

The basalt of Lava Butte occurs as pahoehoe and aa flows. On the pahoehoe, some tumuli about ten meters across often cracked on the periphery and tops could serve as simulated installation sites. The aa lava is rugged with maximum vertical irregularities of two meters and locally in the vent area on the south side of Lava Butte reaching four meters. The lava flow terrain provides a realistic environment for both vehicle and manned traverses.

To the west 24 miles on the paved Highway 46, access to silicic lava flows would provide another type of terrain and lithology. Dacitic lava flows issuing from the southeast side of South Sister present a wavelike topography to the flow surface which is visible from aerial surveys in contrast to the terrain seen by the ground observer.

Area two is in the Galapagos Islands. No services or facilities are available. For all practical purposes, conditions would be lunar except for vacuum and radiation extremes. Specific calderas of interest are the Sierra Negra, Darwin, Wolf and Fernandina Calderas on Isla Isabela and Isla Fernandina. The floors and flanks of the calderas are composed of bare basalt.

Ground surveys complemented by aerial traverses are intended to precede vehicular operations. Ground mapping will permit evaluation of density, magnetic susceptibility, and radioactivity with the air-borne gravimetric, magnetic, and scintillometric surveys, respectively. The specific directions of traverse across the Sierra Negra, Wolf and Darwin calderas would be determined by the aerial photographs determined in the early exploration phases.

By conducting aerial geophysical surveys with ground control, a large amount of terrain specific to lunar base operations can be covered. Such terrain would include large flat areas for landing and launch centers, thermal highs at fumaroles and fracture intersections for primitive installation and technology sites, and mineralized centers at solfataras for simulated lunar base technology experiments.

The Bend, Oregon, site would permit evaluation of reliability of components and systems with optimum conditions for change and evaluation of re-designed or modified equipment. The Galapagos site would constitute the final terrestrial test both with respect to landing mobility and operation of the scientific mission. If it fails on earth there is no reason to expect success on the moon.

BIBLIOGRAPHY

1. J. Green, "Geochemical Implications of Lunar Degassing," Proc. Lunar and Planetary Exploration Colloquium, vol. 1, No. 4, 1959, pp. 1-18
2. K. von Bulow, "Tektonische Analyse der Mondrinde," (Tectonic Analysis of the Moon Crust), Geologie, vol. 6, No. 6-7, 1957, pp. 565-609
3. H.S. Yoder, "Change of Melting Point of Diopside with Pressure," J. Geol., vol. 60, pp. 364-374
4. S. Miyamoto, "Magmatic Boiling and Underground Structure of the Moon," Contrib. Inst. Astrophys. and Kwasan Observatory of Kyoto, vol. 96, 1960, p. 6
5. F. Gronlund and W.V. More, J. Chem. Phys., 1960, vol. 32, p. 1540
6. H. Jeffreys, The Earth, Cambridge University Press, 1959
7. J. Green, "Geology for the Man on the Moon," New Scientist, vol. 9 No. 223, 1961, pp. 465-469
8. H.C. Urey and H. Craig, "The Composition of the Stony Meteorites and the Origin of the Meteorites," Geochem. Cosmochem. Acta, vol. 4, 1953, pp. 36-82
9. G.F. MacDonald, "A Critical Review of Geologically Important Thermochemical Data," Ph.D. dissertation, Harvard University, 1954
10. B.V. Levin, Voprosi Kozmogonij, vol. 6, 1958, p. 56
11. J.C. Jaeger, "Subsurface Temperatures on the Moon," Nature, vol. 183, No. 4571, 1959, pp. 1316-1317
12. J.H. Fremlin, "Volcanoes on the Moon," Nature, vol. 183, No. 4656, 1959, p. 239
13. N.A. Kozyrev, 1959, "Observation of a Volcanic Process on the Moon," Sky and Telescope, vol. 18, No. 4, 1959, p. 184
14. N. N. Sytinskaya, "New Data on Meteor-Slag Theory of the Formation of Outer Layers of the Lunar Surface," Akad. Nauk SSSR, Astron. Zhurnal, vol. 36, No. 2, 1959, pp. 315-321 (in Russian)
15. J.S. Hey and V.A. Hughes, "Radar Observations of the Moon at 10 cm Wavelength," Paris Symp. on Radio Astronomy, Stanford University Press, 1959, pp. 13-18
16. Z Kopal, Astronomical Contributions from the University of Manchester, vol. 3, 1959, pp. 63, 66, 71

17. M. A. Melton, "List of Sample Parameters of Quantitative Properties of Landforms - Their Use in Determining the Size of Geomorphic Experiments," Columbia Univ. Dept. of Geology, Tech. Rept. No. 16, 1958

18. J. R. van Lopik and C. R. Kolb, Handbook, A Technique for Preparing Desert Terrain Analogs, 1959. U. S. Army Eng. Waterways Exper. Sta. Vicksburg, Miss. Tech. Rept. 3-506

19. R. L. Leadabrand, "Radar Astronomy Symposium Report," J. Geophys. Res., vol. 65, 1960, pp. 1103-1118

20. G. K. Wehner, N. Laegreid and R. V. Stuart, "Study of Sputtering of Materials," General Mills, Inc., Rept. No. 2133 (Project 80087), 1960 (Sponsored by Geophys. Res. Directorate, AFCRC, ARDC, Bedford, Mass., Rept. AFCRL-TR-60-418)

21. T. Gold, "Dust on the Moon," Vistas in Astronautics, 2nd Ann. Astron. Sym., vol. II, 1960, Pergamon Press, N.Y., pp. 261-266

22. J. J. Gilvarry, "The Nature of the Lunar Maria," Nature, vol. 181, 1958, pp. 751-762

23. A. M. Wesselink, "Heat Conductivity And the Nature of the Lunar Surface," Bull. of the Astrom. Inst. of the Netherlands, vol. 10, 1948, p. 351

24. M. Brunschwig, et al, "Estimation of Physical Constants of the Lunar Surface." Univ. of Michigan, Rept 3544-1-F, 1960, (Prepared for Autometric Corp., Contract DA 49-018 eng 2133 (E), Army Map Service)

25. J. E. Spurr, "Geology as Applied to Selenology," The Imbrium Plain Region of the Moon, The Science Press, Lancaster, Penn., 1945a

26. N. Orlova, "A Study of the Slope Angle of Some Dry Substances in Connection with the Dust Layer Hypothesis of the Lunar Surface," Presented at Internat. Astron. Union, Leningrad, Dec. 1960

27. V. Sharanov, "The Microrelief of the Lunar Surface and Probable Ways of its Formation," presented at Internat. Astron. Union, Leningrad, Dec. 1960

28. N. P. Barabashev and V. Ezersky, "On the Photometric Uniformity of the Lunar Surface," presented at Internat. Astron. Union, Leningrad, Dec. 1960

29. R. B. Baldwin, The Face of the Moon, Chicago University Press, 1949

30. J. E. Spurr, "Geology as Applied to Selenology," The Features of the Moon, vol. II, The Science Press, Lancaster, Penn., 1945b

31. N. Boneff, "Les Probabilitiés des Causes et l'Origine des Cratères Lunaires," Astr. Nachr, vol. 260, No. 6239, 1936, pp. 422-426

32. W. W. Rubey, 1955. "Development of the Hydrosphere and Atmosphere with Special Reference to Probable Composition of the Early Atmosphere," Geol. Soc. Am., vol. 62, 1955, pp. 631-650

33. A. P. Vinogradov, 1956. The Regularity of Distribution of Chemical Elements in the Earth's Crust, Geokhimiya, vol. 1, 1956, pp. 6-52 (In Russian)

34. A. B. Ronov, "On the Post-Precambrian Geo-chemical History of the Atmosphere and Hydrosphere," Geokhimiya, No. 5, pp. 493-506, (pagination that of a translation of Geokhimiya published by the Geochemical Society)

35. N. A. Kozyrev, 1959. "Volcanic Activity on the Moon," Priroda, vol. 48, No. 3, 1959, pp. 84-88 (In Russian) (English translation in Internat. Geol. Rev., vol. 1, No. 10, pp. 40-44

36. D. Alter, "A Suspected Partial Obscuration of the Floor of Alphonsus," Astron. So. Pacific Pub., vol. 69, 1956, pp. 158-161

37. P. A. Moore, Guide to the Moon, Comet Books, Collins Paper Book, Ltd., London, 1953

38. E. E. Both, "The Floor of Alphonsus." J. Internat. Lunar Soc., vol. 1, 1960, pp. 140-144

39. O.D.R. Brown, J.C. Finn, Jr., and J. Green, "Certain Biological and Geological Advantages of a Polar Lunar Base," Submitted to Astronautics, 1961

40. P. A. Moore, "The History of Alphonsus," The Strolling Astronomer, vol. 13, No. 5-8, 1959, pp. 83-85

41. J. Green, "Geophysics as Applied to Lunar Exploration," Geophys. Res. Directorate, AFCRL, Bedford, Mass., 1960, Contract AF 19(604)-5886

42. M. Chaigneau, R. Fabre, and H. Tazieff, "Sur l'Extraction et l'Analyse des Gaz Occlus dans la Lave du Volcan Nyiragongo," Annales de Geophysique, vol. 16, No. 4, pp. 561-564

43. K. Watson, B. Murray, and H. Brown, "On the Possible Presence of Ice on the Moon," J. Geophys. Res., vol. 66, 1961, pp. 1958-1600

44. R.W. Wood , "Selective Absorption of Light on the Moon's Surface and Lunar Petrography," Astrophys. J., vol. 36, 1912, pp. 75-84

45. R. Castaing and K. Fredriksson, "Analyses of Cosmic Spherules with an X-ray Microanalyzer," Geochim. Cosmochim.Acta., vol. 14, 1958, pp. 114-117

46. R.A. Andrievskiy and I.M.Fedorchenko, "Effect of Small Additions of Nickel and Cobalt on the Sintering of Iron Powder," Izvest. Akad. Nauk SSSR, Otdel. Tekh. Nauk, Met. i Toplivo, vol. 3, 1961, pp. 50-54 (in Russian)

47. L. Bixon, et al., "Design Study for a Long Duration, Low Power, Chemically Fueled Power Supply for Space Vehicles, Phase I, Final Report," Sundstrand Turbo, Rept. S/TD 1812, Wright Air Develop. Div., Wright-Patterson AFB, AF 33(616)-6421, 1960

48. A.M. Gaudin, "Separation Engineering on the Moon," presented at Am. Inst. Mining Metallurgical and Petroleum Engrs., 26 Feb to 2 Mar, 1961

49. R.E. Corcoran, "Preliminary Report on Possible Sites in Oregon for Volcanic Terrain Studies on Lunar or Space Research," 1961 (unpublished manuscript)

50. J. Lehner, Cedicova vlakna, vyroba a fysikalni vlastnosti cedicovych vlaken vyrobenych obstredivou metodou (Basalt Fibers), Veda a Vyzkum v Prumyslu, Rada IV, Praha Statni Nakladalelstvi Technicke Literatury, pp. 149-171 (in Czechoclovakian); p. 173 (English abstract), 1958

51. H.P. Wilkins and P. Moore, The Moon, Faber and Faber, Ltd., London, 1958

52. H.N. Russell, R.S. Dugan, and V.Q. Stewart, Astronomy, vol. 1, Ginn and Co., New York, 1945

53. C.M. Usiskin and R. Siegel, "An Experimental Study of Boiling in Reduced and Zero Gravity Fields," J. Heat Transfer, 1961, pp. 243-253

54. K. Nishikawa and K Urakawa, "Nucleate Boiling under Reduced Pressure," Mem. Fac. Eng. Kyushu Univ., vol. 19, 1960, pp. 63-71

PROTON DOSE RATES IN MANNED SPACE VEHICLES*

Richard Madey, Arthur G. Duneer Jr., and Theodore J. Krieger**

An analytic expression is obtained for the surface dose rate at the center of a spherical shell shield of arbitrary material and thickness resulting from omnidirectionally incident protons with a differential flux spectrum representable by a power law. The dose rate formula is derived on the basis that protons are slowed down by electronic collisions only.

Surface dose rates are calculated for measured proton spectra in both solar flares and Van Allen belts. By restricting threshold penetration energies to values greater than the minimum observed proton energies, the uncertainty arising from extrapolation of spectral data to lower energies is avoided. The shield materials considered are air, Al, C, H_2O, CH_2, and liquid H_2. The calculated surface dose rates for a given spectrum and threshold energy E_t are practically independent of shield material in this group. Thus, for example, in the case of the solar flare of May 12, 1959, at 33 hours after the optical peak, the tissue dose rates for the various shield materials at fixed E_t are within 5 percent of one another for all values of E_t between 110 and 220 Mev. The calculated dose rate for this flare is approximately 1.4 rads/hr at E_t = 110 Mev and decreases with increasing E_t to 0.06 rad/hr at E_t = 220 Mev.

A similar analytic expression is given also for the absorbed dose rate arising from a power-law spectrum of protons incident on a low energy cutoff (magnetic) type shield. The expression is used to obtain the cutoff energy dependence of the dose rate that would result from the prompt spectrum, as given by Winckler, for the giant flare of February 23, 1956. It is estimated that cutoff energies near the geomagnetic equator would reduce the dose rate from this flare to tolerable levels. This result indicates that a radiation-safe corridor for manned space flight exists in the equatorial region of the earth at altitudes below the Van Allen belt.

A comparison of doses obtained with the two shield types and for the same threshold energy shows that the magnetic shield permits a dose of only one-half that of the material shield. This result is independent of the threshold energy, provided nuclear interactions can be neglected.

Preliminary calculations of the dose rate arising from gamma rays produced by primary protons in a material shield show that the gamma ray dose contribution may be significant for high intensity flares.

---

* A complete version of this paper (AAS 62-22) may be obtained from the American Astronautical Society or by contacting the authors.

** Chief Staff Scientist, Principal Scientist, and Senior Scientist, respectively, Republic Aviation Corporation, Farmingdale, L.I., New York.

REFERENCES

1. Ney, E. P., Winckler, J. R., and Freier, P. S., "Protons from the Sun on May 12, 1959," Phys. Rev. Letters 3, 183 (1959).

2. Anderson, K. A., and Enemark, D. C., "Observations of Solar Cosmic Rays Near the North Magnetic Pole," J. Geophys. Research 65, 2657 (1960).

3. Anderson, K. A., and Enemark, D. C., "Observations of Auroral Zone X-rays and Solar Cosmic Rays," Space Research, Proceedings of the First International Space Science Symposium, Edited by Hilde Kallmann Bijl, 702-714 (1960).

4. Ogilvie, K. W., and Davis, L. R., "Rocket Observations of Solar Protons," presented at Forty-Second Annual Meeting, American Geophysical Union, Washington, D. C., April 18-21, 1961. Also, Ogilvie, K. W., Bryant, D. A., and Davis, L. R., NASA Goddard Space Flight Center Contributions to the 1961 Kyoto Conference on Cosmic Rays and the Earth Storm.

5. Freden, S. C., and White, R. S., "Protons in the Earth's Magnetic Field," Phys. Rev. Letters 3, 9 (1959).

6. Freden, S. C., and White, R. S., "Particle Fluxes in the Inner Radiation Belt," J. Geophys. Research 65, 1377 (1960) and private communication.

7. Winckler, J. R., "Primary Cosmic Rays," Radiation Research Society Symposium on Radiation in Space, May 10, 1960; also in "Proceedings of Conference on Radiation Problems in Manned Space Flight," NASA TN D-588 (December 1960).

8. Meyer, P., Parker, E. N., and Simpson, J. A., "Solar Cosmic Rays of February 1956 and Their Propagation Through Interplanetary Space," Phys. Rev. 104, 768 (1956).

9. Naugle, J. E., and Kniffen, D. A., "Flux and Energy Spectra of the Protons in the Inner Van Allen Belt," Phys. Rev. Letters 7, 3 (1961).

10. Ray, E. C., "Effects of a Ring Current on Cosmic Radiation," Phys. Rev. 101, 1142 (1956).

11. Wakatsuki, T., et al., "Gamma Rays from Several Elements Bombarded by 10 and 14 Mev Protons," J. Phys. Soc. Japan 15, 1141 (1960).

12. Krieger, T. J., Duneer Jr., A. G., and Madey, R., "Production of Gamma Rays by Solar Flare Protons Incident on Material Shields in Space," Republic Aviation Corporation Technical Memorandum SRS-TM-61-122 (25 August 1961).

13. Varshni, Y. P., "Range-Energy Relation for Protons in Various Substances," Indian J. Phys. 32, 373 (1958).

14. Rich, M., and Madey, R., "Range-Energy Tables," UCRL-2301 (1954).

15. Steinheimer, R. M., "Range-Energy Relations for Protons in Be, C, Al, Cu, Pb, and Air," Phys. Rev. 115, 137 (1959).

THE CRATER CONTRIBUTION TO SURFACE ROUGHNESS OF THE MOON

M. Kornhauser *

By plotting numbers of craters versus size and extrapolating below the size limit of telescopic resolution, it is found that less than ten per cent of the Moon's surface is covered by craters. In local areas, however, the surface may be almost entirely covered by craters. Slopes inside the craters are an inverse function of size, the 10,000 ft diameter crater having a maximum slope of 28° while the 1,000 ft diameter crater has a maximum slope of 46°. It is concluded that some areas of the Moon should be easy to traverse, while others must be impassable.

The crater size distribution on the Moon does not correlate well with the size distribution of sporadic meteoroids currently existing in Earth-Moon space, indicating that this may be a function of geologic time.

INTRODUCTION

The surface of the Moon is the subject of a considerable amount of theorizing and speculation in regard to the physical properties of the surface material (chemical composition, consistency, depth of dust, if any, thermal properties, etc.) and in regard to the gross surface roughness of the Moon. On this latter surface feature, the popular opinion has changed recently. Artists' conceptions of the Moon used to show jagged peaks and precipitous terrain current thinking would eliminate all this in favor of very gentle slopes, not exceeding about 15°, covered by dust. The basis for this change of concept is the recent work on photographic analysis of lunar features, which indicates that the larger surface formations do indeed have gentle slopes, despite

* Missile and Space Vehicle Department, General Electric Company, Philadelphia, Pa.

their rough appearance through the telescope. Unfortunately, this information has been applied to all lunar surface features, resulting in very unconservative specifications for lunar surface vehicles. It is the purpose of the following discussion to indicate that the smaller (and more frequently encountered) surface features of the Moon have relatively steep slopes, and the the lunar surface vehicle may have to negotiate slopes much greater than 15°. The argument will be based on lunar craters only, since these form a continuous family with features which may be extrapolated below the limit of telescopic resolution. Whether or not other surface features contribute to small-scale roughness is not resolved in this discussion; as a result, the total roughness of the lunar surface may exceed that contributed by craters alone.

SHAPES OF CRATERS

Baldwin [1] has plotted crater depth versus crater diameter for lunar craters as well as terrestrial craters and explosion pits, obtaining a continuous curve with the following equation:

$$\text{Log } D = 0.1083 (\log d)^2 + 0.6917 \log d + 0.75 \qquad (1)$$

with D = crater diameter in feet and d = crater depth in feet. This expression is presented in Fig. 1 as a shape factor; d/D plotted versus D. The curve is shown dotted for D less than about 5, 000 feet, which is the limit of telescopic resolution, even though one might presume terrestrial craters to belong to the same family as lunar craters. It is considered reasonable to extrapolate the curve for lunar craters to the region of $D < 5,000$ ft, however, even though the craters are not visible. This extrapolation is the basis for the argument that steep slopes do indeed exist on the surface of the Moon, as evidenced by the plots on Fig. 2 of average slope (line from the crater bottom to crater rim) and of maximum slopes for reasonable crater profiles.

Also shown on Fig. 1 and 2 are depth/diameter ratios and slopes taken from sources which disagree with Baldwin's data. Note, however, that

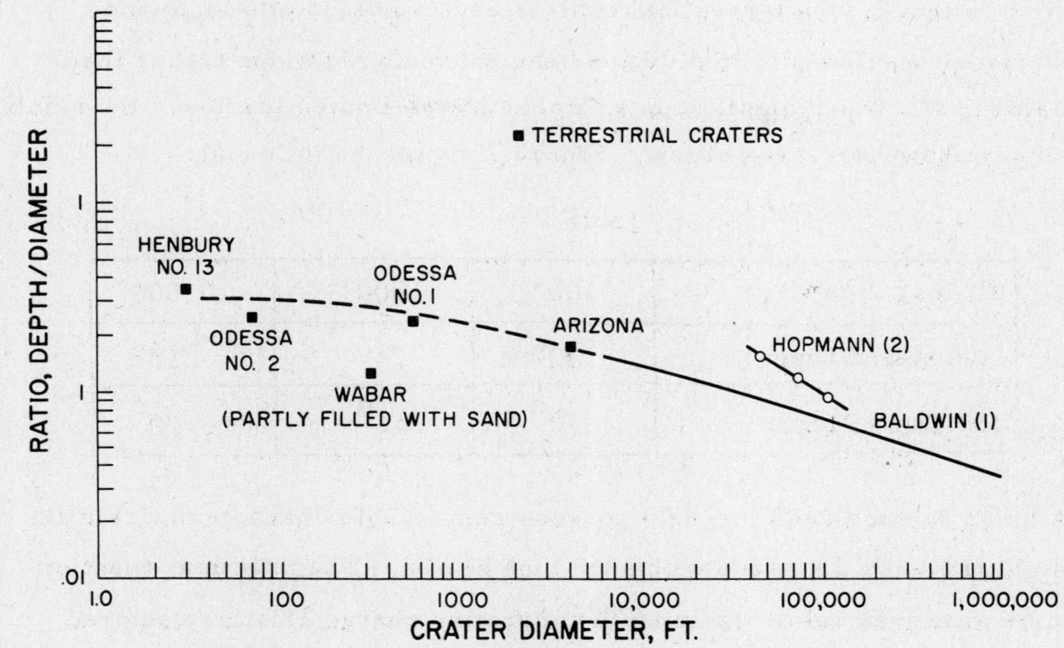

Fig. 1. Overall Shapes of Lunar Craters

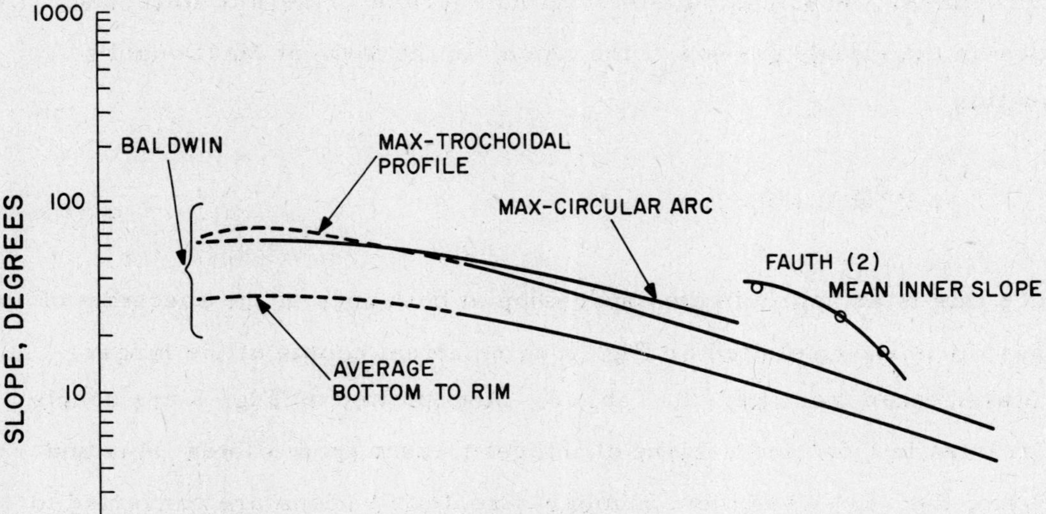

Fig. 2. Slopes in the Lunar Craters

these estimates result in slopes higher than Baldwin's. Also note that, had the curves been extrapolated back to the region of small crater sizes without using terrestrial craters and explosion pits as a guide, the result would undoubtedly have been estimates of slope higher than Baldwin's. Thus, on all counts, Baldwin's estimates are low. Baldwin's slopes, however, are not low. Figure 2 shows the following:

Table I

Crater Diameter	100'	1000'	10,000'
Average Slope	30°	24°	16°
Maximum Slope	70°	46°	28°

A lunar surface vehicle would possess remarkable characteristics if it could traverse a surface with such high slopes. The practical question to be answered before defining the operating characteristics required for lunar surface vehicles is "Can the steep-slope lunar craters be avoided easily?"

AREA COVERED BY CRATERS

Bobrovnikoff [2] presents data on total numbers of craters of different sizes on the visible portion of the Moon, in the form of MacDonald's equation

$$n\,D^2 \cong 7,150$$

Since this is an approximate expression at both ends of the spectrum of sizes, it is shown plotted on Fig. 3 using actual counts of the larger craters, taken from Ref. 1, Table 4. Also plotted on Fig. 3 are counts of craters in localized regions of interest, taken from Plates D4-a and C6-a of Ref. 3. When these number-size distributions are converted to percentage of lunar surface covered by craters, the results appear as in Fig. 4.

Figure 4 shows the interesting fact that less than 10% of the Moon's surface is covered by craters. Note, however, that one of the local* areas studied has about 30% of its area covered by craters; and this is visibly not the roughest area of the Moon. Thus, one must conclude that some areas appear easy to negotiate by surface vehicle, while others must be literally covered by rough surface formations. In particular, when one adds the roughness contributed by the exterior slopes of the crater rims, the roughness contributed by cracks, rills, escarpments, hills, boulders, etc., it is clear that some portions of the Moon's surface must be impassable using surface vehicles. The question, then, of specifying the operating characteristics of a surface vehicle becomes unanswerable in general. Only when specific lunar regions are considered can one make a reasonable estimate of vehicular requirements.

METEOROIDS AND CRATERS

It is of general interest to attempt to correlate our information on the frequency of meteoroid impacts with the crater distribution actually existing on the Moon. Figure 5, taken from Ref. 4, presents the population of sporadic meteoroids entering earth's atmosphere per unit time. Although Shoemaker and Hackman[5] estimate that impact frequencies or Earth are 2.15 times the impact frequencies on the Moon, the approximate nature of the data of Fig. 5 (three orders-of-magnitude spread) does not warrant such refinements. Thus, for purpose of discussion, the upper solid curve of Fig. 5 will be used. This curve may be converted to an analytic expression as follows:

$$N = 10^{-12}/M \tag{3}$$

For purposes of calculating cratering effects, Eq (3) is expressed in terms of kinetic energy, U, of the meteoroid:

---

* Areas on the plates of Reference 3 are about 165,000 sq. mi.

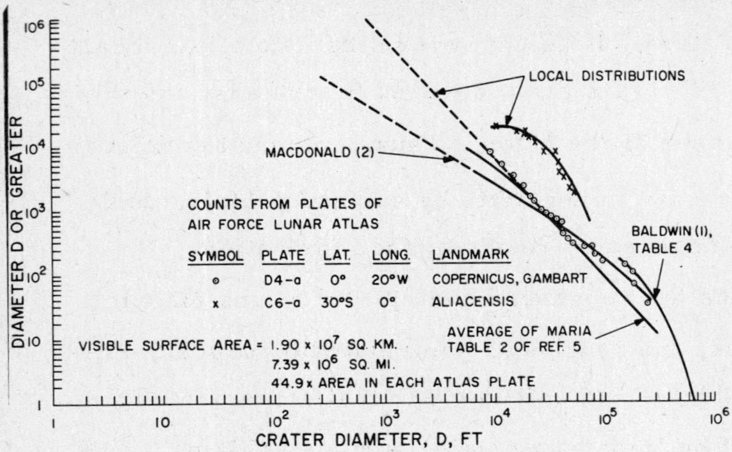

Fig 3. Distribution of Crater Sizes on the Moon.

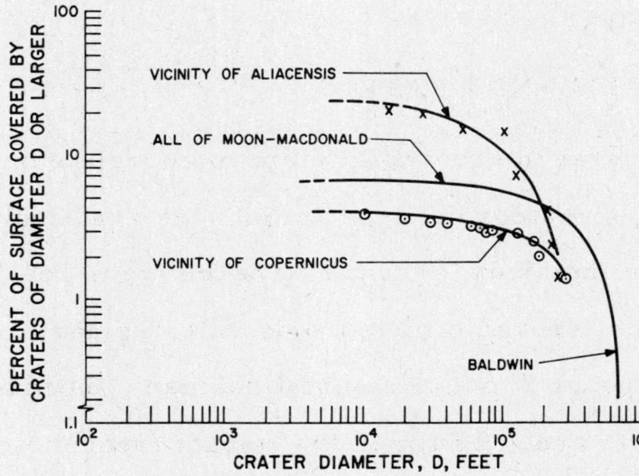

Fig. 4. Percent of Lunar Surface Covered by Craters.

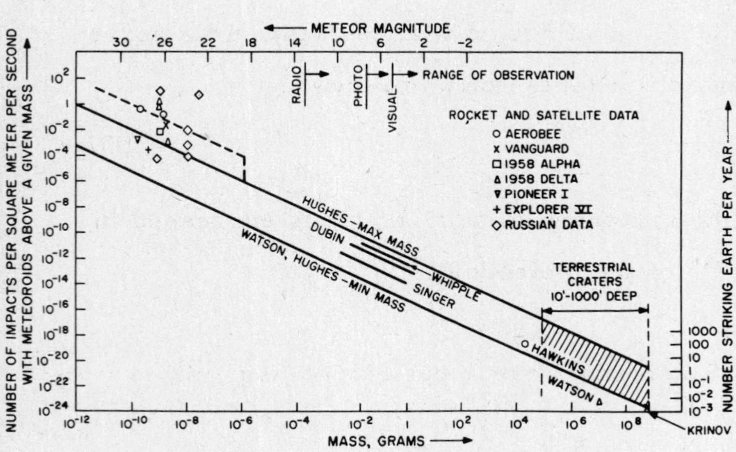

Fig. 5. Population of Sporadic Meteoroids.

$$U = 1/2 \left(\frac{m}{454}\right) (10.2 \times 3281)^2 \frac{(12)}{(32.2)} =$$

$$= 4.60\ m \times 10^5 \text{ in-lb} \qquad (4)$$

with 10.2 km/sec used as modal impact velocity on the Moon [5] as follows:

*N (per year on visible portion of Moon) =

$$\frac{2.76 \times 10^{14}}{U, \text{ in-lbs}} \qquad (5)$$

In order to predict crater diameters, it will be assumed that crater volume, V, is proportional to the kinetic energy, U, of the impacting meteoroid as follows:[6]

$$V/U = 0.00146 \text{ cu. in./in.-lb.} \qquad (6)$$

and that the ratio d/D shown in Fig. 1 (Baldwin's curve) may be expressed analytically as follows (for the visible range of crater sizes):

$$\frac{d}{D} = 2.24\ D^{-0.30} \qquad (7)$$

for diameter, D, in feet. Assuming the crater profile to be the arc of a circle:

$$V = \frac{\pi D^3}{6}\left(\frac{3}{4}\frac{d}{D} + \frac{d^3}{D^3}\right) \cong \frac{\pi}{8} D^3 \left(\frac{d}{D}\right) \qquad (8)$$

for shallow craters.

---

*Moon's visible area $\cong 1.90 \times 10^{13}$ sq. meters; number of seconds per year = 3.16×10^7

Rewriting Eq. (5) in terms of Eq. (6), (7) and (8):

$$N \cong \frac{2.65 \times 10^8}{D^{2.7}} \qquad (9)$$

for a total bombardment period of 10^9 years

Table II

D, ft.	10^4	4×10^4	10^5	4×10^5
N	4.21×10^6	1.00×10^5	8430	200

The absolute values of Table II are from 1 to 3 orders of magnitude higher than those shown in Fig. 3. This disagreement is understandable in view of the 3-order-of-magnitude spread in the impact frequency figures of Fig. 5, and also the uncertainty that meteoroid bombardment has been constant in the last 1 billion years. Likewise, the distributions of sizes expressed by Eq. (9) are not very close to the averages existing on the Moon.

For example:

Eq. (9)	$N \sim D^{-2.7}$	} *
Shoemaker and Hackman[5]	$N \sim D^{-2.4}$	} *
Also	$N \sim D^{-2.72}$	} *
Average of Maria, Fig. 3	$N \sim D^{-1.83}$	

It appears, therefore, much more satisfactory to work with the observed distributions and extrapolate to the smaller sizes than to rely on theoretical predictions.

*These are theoretical predictions.

REFERENCES

1. Baldwin, R. B., "The Face of the Moon" - U. of Chicago Press, 1949, pp. 131-133.

2. Bobrovnikoff, N. T., "Natural Environment of the Moon" Ohio State Univ. Research Foundation, WADC Phase Technical Note 3, June 1959, AD242177 pp. 43, 48, 50.

3. United States Air Force, "Lunar Atlas' - 1st Edition, February 1960, Prepared under direction of G. P. Kuiper at the Yerkes Observatory.

4. Kornhauser, M., "The Meteorite Danger on Earth" - Fourth Western Regional Meeting, AAS, Preprint 61-80, San Francisco, August 1-3 1961.

5. Shoemaker, E. M. and Hackman, R. J., "Interplanetary Correlation of Geologic Time" - American Astronautical Soc. Preprint 61-7, 7th Annual Meeting, January 16-18 1961.

6. Kornhauser, M., "Prediction of Cratering Caused by Meteoroid Impacts" - J. Astronaut. Sci., Vol. V, No. 3-4, Autumn-Winter 1958.

ORBITS FOR PLANETARY SATELLITES FROM DOPPLER DATA ALONE*

Armin J. Deutsch**

When an artificial satellite is put in orbit around another planet, it will be difficult or impossible from terrestrial observations to obtain its direction with sufficient precision to define its orbit. However, if the satellite carries a suitable transmitter, Doppler observations are capable of giving precise values for the line-of-sight component of the velocity vector. From such data alone, it is possible to obtain all the orbital elements and the mass of the planet. The analytical apparatus used by astronomers for the discussion of spectroscopic binary stars is here adapted to the interpretation of Doppler observations of a planetary or lunar satellite. An example is given of the orbit determination for a satellite of Venus. The orbit is assumed to be Keplerian, i.e., unperturbed. An examination of the propagation of errors in the determination from radio Doppler data alone of the orbital elements of this satellite shows that with the assumption of reasonable values for the frequency stability of the signal and for the rate of acquisition of Doppler data, one can establish the orbital elements with the precision of the order of one percent.

---

* A complete version of this paper (AAS 62-24) may be obtainable by contacting the author. This paper is not available from the American Astronautical Society.

** Missile and Space Systems Division, Douglas Aircraft Company, Inc., Santa Monica, Calif.

STAR PRODUCTION ALONG POLAR SATELLITE ORBITS *

Herman Yagoda**

---

* This paper (AAS 62-25) may be obtainable by contacting the author; it is not available through the American Astronautical Society.

** Air Force Cambridge Research Laboratories, Bedford, Mass.

eighth
annual
meeting

AMERICAN ASTRONAUTICAL SOCIETY

INTRODUCTION

A HIGHER ORDER THEORY OF BALLISTIC ENTRY

W. H. T. Loh *

The fundamental equations of motion for ballistic entry is unsolvable analytically. However, at large initial angles of inclination where the angle of inclination may be treated as a constant along the trajectory, approximate solutions were obtained in the last four years by Gazley[1], Allen and Eggers[2], Chapman[3], and Loh[4]. (These solutions are valid only for initial angle of inclination greater than 5 or 10 degrees . Large initial angles of inclination result in large missile deceleration and large aerodynamic heating. Both large deceleration and large aerodynamic heating at large initial angles of inclination are usually intolerable for ballistic entry; therefore, it is always desirable to have ballistic entry at small initial angles of inclination. The solution for small initial angles of inclination has been solved numerically by IBM machines.

Only very recently, a second order theory of entry mechanics[5] was developed. This theory covers both gliding entry and ballistic en ry at small and large angles of inclination. However, it suffers slight inaccuracies at the terminal phase of flight. It is the purpose of this paper to present a higher order approximation obtained recently for ballistic entry into atmosphere. When higher order effects are neglected, the present solution reduces precisely to the second order solution for ballistic entry into atmosphere[7]. When both higher order and second order effects are disregarded, the present solution yields exactly the previous solution valid only for ballistic entry into atmospheres at large initial angles of inclination[1, 2, 3, 4].

*Vought Research Center, Ling-Temco Vought, Inc., Dallas, Texas.

SYMBOLS

A	reference area for drag expressions, sq ft
C_D	drag coefficient
C_F'	equivalent skin-friction coefficient

$$C_F' = \frac{1}{S} \int_S \sqrt{P_r}\; C_{F_\ell} \left(\frac{\rho_\ell}{\rho}\right) \left(\frac{V_\ell}{V}\right) dS$$

C_F	skin friction coefficient
D	drag, lb
g	acceleration due to force of gravity, ft/sec^2
m	mass of the vehicle, slugs
H	convective heat transferred per unit area, ft-lb/ft^2
K	constant = 6.8×10^{-6} (Ref. 2)
R	distance of vehicle measured from center of planet or Earth, ft. When altitude of the vehicle is small in comparison with planet or Earth radius, R may be taken approximately as R_o, the radius of Earth or planet
Q	convective heat transferred, ft-lb
s	distance along flight path, ft
S	surface area, ft^2
t	time, sec
V	velocity, ft/sec
ρ	atmospheric density, slugs/ft^3
θ	angle of inclination or angle of flight path to local planet or Earth horizontal, degrees, positive for descent
σ	nose or leading edge radius of body or wing, ft
β	constant in planet or Earth's density-altitude relation $\rho = \rho_o e^{-\beta y}$. Here ρ_o = reference density slugs/ft^3, β = constant and y = altitude in feet.

$\rho_o = 0.0027,\ 1/\beta = 23,500$ for Earth

$$x = \frac{V^2}{gR_o}$$

$$C_1 = \left[\left(\frac{\cos\theta_m}{\rho_m R_o}\right)\left(\frac{m}{C_D A}\right)\left(\frac{1}{x_m} - 1\right)\right]^{-1}$$

Here subscript "m" indicates a reference point which may be taken anywhere except $x = 1$. The reference point is arbitrarily chosen in this paper at the point where maximum deceleration occurs. For reference point chosen at maximum deceleration (where $\frac{C_D A}{m}\rho \cong \sin\theta$),

$$C_1 = \left[\left(\frac{1}{\beta R_o}\right)\cot\theta_m \left(\frac{1}{x_m} - 1\right)\right]^{-1}$$

and θ_m and x_m may be obtained by solving the following two algebraic equations simultaneously (obtained from second order theory 7):

$$\cos\theta_m = \frac{\cos\theta_f}{1 + \left(\frac{1}{\beta R_o}\right)\left(\frac{1}{x_m} - 1\right)}$$

$$\sin\theta_m = \frac{\left(\frac{1}{\beta R_o}\right)\left(\frac{1}{x_m} - 1\right)\cos\theta_m}{(\theta_m - \theta_f)/\ln(x_f/x_m)}$$

After the constant C_1 is known, the following constants C_{15}, C_{16}, C_{17}, may be evaluated:

$$C_{15} = 1/2\left[\frac{\frac{1}{\cos\theta_f}\left(1 + \frac{1}{\cos^2\theta_f}\right)(C_1^2 + 3^2) - 6\left(\frac{1}{\cos^3\theta_f}\right)}{(C_1^2 + 1)(C_1^2 + 3^2)}\right]$$

$$C_{16} = 1/2\left[\cos^2\theta_f\,(C_1^2 + 2^2)\right]^{-1}$$

$$C_{17} = 1/2\left[\cos^3\theta_f\,(C_1^2 + 3^2)\right]^{-1}$$

$$F_1(\theta) \cong \left[\ln\left(\frac{\theta}{2}\right) + (C_1\theta) + \frac{1}{4}(C_1\theta)^2 + \frac{1}{18}(C_1\theta)^3\right]e^{-C_1\theta} - \left[\frac{(C_1\theta)^3}{720\,C_1^4} - \frac{(C_1\theta)^2}{240\,C_1^4} + \left(\frac{1}{120\,C_1^4} + \frac{1}{12\,C_1^2}\right)C_1\theta - \left(\frac{1}{120\,C_1^4} + \frac{1}{12\,C_1^2}\right)\right]$$

These are preferable to use when Θ is small or medium.

$$F_1(\Theta) = C_{15}(C_1 \sin\theta - \cos\theta) + C_{16}(\tfrac{1}{2}C_1 \sin 2\theta - \cos 2\theta) - C_{17}\sin^2\theta(C_1 \sin\theta - 3\cos\theta)$$

This is preferable to use when Θ is large or very large.

Subscripts

f = condition at beginning of ballistic entry

s = stagnation condition

av = average value

max = maximum value

INTRODUCTION

The fundamental equations of ballistic entry are

$$mV^2 \frac{d\theta}{ds} + mV^2 \frac{\cos\theta}{R} = mg\cos\theta \tag{1}$$

$$\frac{m}{2}\frac{dV^2}{ds} - mg\sin\theta + D = 0 \tag{2}$$

These two equations are not solvable analytically. However, for ballistic entry at large angles of inclination, Gazley[1], Allen and Eggers[2], Chapman[3], and Loh[4], have obtained analytical approximate solutions by neglecting the gravity force term and the centrifugal force term in the fundamental equations (1) and (2). However, using an approach similar to the approach used in the second order theory[5], Eqs. (1) and (2) may now be solved in an approximate manner without neglecting, in a sense, those terms which were neglected previously. Analytical approximate solutions obtained here are then compared with exact numerical solutions, and they show excellent agreement at both small and large angles of inclination.

ANALYSIS

In a nonrotating two-dimensional inertial coordinate system with its origin at the center of the planet or Earth and the gravitational field "G" during the entry portion assumed as constant, the equations of motion in the directions normal and tangential to the ballistic trajectory are

$$m v^2 \frac{d\theta}{ds} + m v^2 \frac{\cos\theta}{R} = mg \cos\theta \tag{1}$$

$$\frac{m}{2}\frac{dv^2}{ds} - mg \sin\theta + D = 0 \tag{2}$$

From the usual approximate exponential planetary atmosphere*, one obtains

$$\rho = \rho_0 e^{-\beta y} \tag{3}$$

In order to study the aerodynamic heating aspects of the ballistic re-entry, the heating parameters developed originally by Allen and Eggers[2] are the best, and therefore they are borrowed for use here. They are

(1) The total aerodynamic heat input to the vehicle, Q

(2) The time rate and maximum time rate of average heat input per unit area, $\left(\frac{dH_{av}}{dt}\right)$ and $\left(\frac{dH_{av}}{dt}\right)_{max}$

(3) The time rate and maximum time rate of local stagnation heat input per unit area, $\left(\frac{dH_g}{dt}\right)$ and $\left(\frac{dH_g}{dt}\right)_{max}$

The analytical expressions for these parameters are given by Refs. 2 and 6 as

*This density relationship is based on the assumption of an isothermal atmosphere in a uniform gravitational field (Ref. 3).

$$Q = \frac{1}{4} c_F' S \int \rho V^2 ds = \frac{1}{4} c_F' S \int \frac{V^2}{\beta \sin\theta} d\rho \tag{4}$$

$$\frac{dH_{av}}{dt} = \frac{1}{4} c_F' S \rho V^3 \tag{5}$$

$$\frac{dH_s}{dt} = K \sqrt{\frac{\rho}{\theta}} V^3 \tag{6}$$

Now putting

$$D = \frac{1}{2} C_D \rho V^2 A$$

Letting

$$x = V^2 / g R_0$$

and using the kinematic relation,

$$- \sin\theta \, ds = dy$$

into Eqs. (1) and (2) and rearranging terms, one obtains

$$\frac{d \cos\theta}{d(\beta y)} - \frac{1}{\beta R_0} \cos\theta \left(\frac{1}{x} - 1\right) = 0 \tag{7}$$

$$\frac{dx}{d(\beta y)} - \frac{C_D A}{m\beta} \rho \frac{x}{\sin\theta} = - \frac{2}{\beta R_0} \tag{8}$$

Substituting Eq. (3) into Eqs. (7) and (8) and simplifying, one obtains

$$\frac{d \cos\theta}{d\rho} + \frac{1}{\beta R_0} \frac{\cos\theta}{\rho} \left(\frac{1}{x} - 1\right) = 0 \tag{9}$$

$$\frac{dx}{d\rho} + \frac{C_D A}{m\beta} \frac{x}{\sin\theta} = \frac{x}{\beta R_0} \frac{1}{\rho} \tag{10}$$

Equations (9) and (10) are the exact equations of motion of ballistic entry into atmosphere, but they are written into a somewhat different form. The reason for this 5 may be briefly stated here as follows: Although neither the parameters "θ", "ρ", "V" are constants nor the term "$\frac{1}{\beta R_0} \frac{\cos\theta}{\rho} \left(\frac{1}{x} - 1\right)$" is constant, the later term is found to be very much insensitive to the "ρ" or "θ" integration; therefore, it may be taken out of the integral sign in Eq. (9) as a constant without introducing serious errors in the basic equation.

This reasoning is further verified by comparisons with exact numerical machine solutions.

With the assumption as mentioned and as explained, Eq. (9) may now be integrated out as

$$\cos\theta = \cos\theta_f - \left(\frac{1}{\beta R_0}\right)\frac{\cos\theta}{\rho}\left(\frac{1}{x}-1\right)(\rho-\rho_f) \tag{11}$$

Equation (11) may also be rearranged into

$$\cos\theta = \frac{\cos\theta_f}{1 + \frac{1}{\beta R_0}\left(\frac{1}{x}-1\right)\left(1-\frac{\rho_f}{\rho}\right)} \tag{11a}$$

Substituting Eq. (9) with the approximate assumption as explained in mind and Eq. (11) into Eq. (10), and after simplification, one obtains:

$$\frac{dx}{d\theta} + \left[\left(\frac{1}{R_0}\right)\frac{\cos\theta}{\rho}\left(\frac{m}{C_D A}\right)\left(\frac{1}{x}-1\right)\right]^{-1} x = \frac{2}{\beta R_0}\left(\frac{\sin\theta}{\cos\theta_f - \cos\theta}\right) \tag{10a}$$

$$\frac{dx}{d\theta} + \left[\left(\frac{1}{R_0}\right)\frac{\cos\theta}{\rho}\left(\frac{m}{C_D A}\right)\left(\frac{1}{x}-1\right)\right]^{-1} x \cong \frac{2}{\beta R_0}\cot\frac{\theta}{2} \tag{10b}$$

Expanding $(\text{Cos}\,\Theta_1 - \text{Cos}\,\Theta)^{-1}$ by binominal series for large Θ or expanding Cot ( ) by trigonometric series for small Θ, Eq. (10a) or (10b) may be solved by the conventional method of ordinary differential equations and its solution is:

$$\frac{x}{x_f} = \frac{4}{\beta R_0}\left(\frac{1}{x_f}\right)F_1(\theta) + \left[1 - \frac{4}{\beta R_0}\left(\frac{1}{x_f}\right)F_1(\theta_f)\right] e^{\frac{-R_0 \frac{C_D A}{m}(\theta-\theta_f)\rho}{\cos\theta\left(\frac{1}{x}-1\right)}} \tag{12}$$

$$F_1(\theta) \cong C_{15}(C_1\sin\theta - \cos\theta) + C_{16}\left(\tfrac{1}{2}C_1\sin 2\theta - \cos 2\theta\right) - C_{17}\sin^2\theta\,(C_1\sin\theta - 3\cos\theta) \tag{12a}$$

$$F_1(\theta) \cong \left[\ln\left(\frac{\theta}{2}\right) + (C_1\theta) + \frac{1}{4}(C_1\theta)^2 + \frac{1}{18}(C_1\theta)^3\right] e^{-C_1\theta} - \left[\frac{(C_1\theta)^3}{720\,C_1^4} - \frac{(C_1\theta)^2}{240\,C_1^4} + \left(\frac{1}{120\,C_1^4} + \frac{1}{12\,C_1^4}\right)(C_1\theta) - \left(\frac{1}{120\,C_1^4} + \frac{1}{12\,C_1^4}\right)\right] \tag{12b}$$

With Eqs (11 or 11a) and (12a or 12b), any two of the three parameters Θ, ρ, V, may be solved in terms of the other one, and the complete trajectory is therefore defined. In order to expedite calculations, the following approximations in Eq. (11a) may be made in almost all cases for $\frac{\rho_f}{\rho} < 0.1$:

$$\cos\theta = \frac{\cos\theta_f}{1+\left(\frac{1}{\beta R_o}\right)\left(\frac{1}{x}-1\right)} = \frac{\cos\theta_f}{1+\left(\frac{1}{\beta R_o}\right)\left(\frac{g R_o}{V^2}-1\right)} \qquad (11b)$$

Equation (11b) then solves out "θ" in terms of "V" only. Substituting corresponding values of "θ" and "V" into Eqs. (12a) or (12b), one determines the value of ρ, and consequently one determines completely the trajectory.

SPECIAL CASES

I. Reduce to Second Order Theory of Ballistic Reentry

For large values of (βR_o), say $\beta R_o > 200$, and small and medium values of θ, say $\theta < 45°$, the terms in the square bracket in Eq. (12) becomes very small. In other words,

$$\left(\frac{4}{\beta R_o}\right)\left(\frac{1}{x_f}\right) F_1(\theta) \cong 0 \qquad \left[1-\left(\frac{4}{\beta R_o}\right)\left(\frac{1}{x_f}\right) F_1(\theta_f)\right] \cong 1$$

Only when these higher order effects are disregarded, Eq. (12) reduces to

$$\left(\frac{x}{x_f}\right) = e^{-\left(\frac{C_D A}{m}\right) R_o \left(\frac{\theta-\theta_f}{\cos\theta}\right) \rho \left(\frac{1}{x}-1\right)^{-1}} \qquad (13a)$$

$$\left(\frac{x}{x_f}\right) = e^{-\left(\frac{C_D A}{m\beta}\right) \rho \left(1-\frac{\rho_f}{\rho}\right) \frac{\theta-\theta_f}{\cos\theta_f-\cos\theta}} \qquad (13b)$$

Equation (13) is precisely the most recent ballistic entry solution obtained in the second Order Theory of Entry Mechanics[7] (Since $\beta R_o \cong$ 900, 900, 200, 3600 for Earth, Venus, Mars and Jupiter respectively, the second order theory[7] holds valid for these planets, but it does suffer slight inaccuracies for these planets at the terminal phase of flight where "θ" becomes very large.) Again, since the terms in square brackets in Eq. (12) are small, in order to expedite calculations, the coefficients C_1, C_{15}, C_{16}, C_{17} of $F_1(\theta)$ function may be treated as constants. This was done in the "Symbols Section" where the constants C_1, C_{15}, C_{16}, C_{17} were being evaluated at a reference point arbitrarily taken in this paper at the maximum deceleration point.

II. Reduce to Previous Constant Angle of Inclination Ballistic Reentry

From Eq. (11a) it is to be particularly noted that when "ρ" is small or close to ρ_f or when V is large or close to (gR_o), then Θ becomes a constant. Even when V is not large or ρ is not small; since $\left(\frac{1}{\beta R_o}\right) \cong \frac{1}{900}$ is very small for Earth, the term

$$\frac{1}{\beta R_o}\left(\frac{gR_o}{V^2} - 1\right)\left(1 - \frac{\rho_f}{\rho}\right) << 1$$

so the Θ is nearly a constant for most of the ballistic trajectories with initial $\Theta_f > 5$ or $10°$. (In the theory of Gazley[1], Allen and Eggers[2], Θ was assumed to be a constant. The present approximate solution shows it is nearly a constant, but it is a function of V and only very slightly affected by ρ).

If the second order effects are further being disregarded, Eqs. (11a) and (13a or b) reduce respectively to

$$\cos\Theta = \cos\Theta_f$$

$$\frac{\theta - \theta_f}{\cos\theta_f - \cos\theta} = \frac{\theta_f - \theta}{2\sin\frac{1}{2}(\theta_f - \theta)\sin\frac{1}{2}(\theta_f + \theta)} \qquad (14)$$

$$\cong \frac{\theta_f - \theta}{2 \times \frac{1}{2}(\theta_f - \theta)\sin\frac{1}{2}(\theta_f + \theta)} \cong \frac{1}{\sin\theta}$$

$$\frac{x}{x_f} = e^{-\frac{C_D A}{m\beta}(\rho - \rho_f)\frac{1}{\sin\theta}} \qquad (15)$$

Equations (14) and (15) are the previous approximate solutions for ballistic entry at large angles of inclination given in Refs. 1, 2, 3, 4, and 6.

DECELERATION AND AERODYNAMIC HEATING

I. Deceleration

With the trajectory variables "ρ", "Θ" and "V" known at various points along the trajectory, other important parameters such as deceleration, and aerodynamic heating may now be easily determined along the trajectory in terms of these known values of "ρ" "V" and "Θ". For example, from Eq. (2) the deceleration becomes

$$\frac{dV}{dt} = -\frac{1}{2}\left(\frac{C_D A}{m}\right)\rho V^2 + g \sin\theta \cong -\frac{1}{2}\left(\frac{C_D A}{m}\right)\rho V^2$$

Since the corresponding set of "ρ", "V" and "Θ" at any point along the trajectory is known, the deceleration at any point becomes known also simply by substituting these known values of ρ, V and Θ into the above deceleration equation. So far as the maximum deceleration is concerned, it occurs (from Equation of deceleration shown above) at $\frac{d}{dV}\left(\frac{dV}{dt}\right) = 0$ or approximately $\frac{d}{dV}(\rho V^2) = 0$ this gives $\frac{dV^2}{d\rho} = -\frac{V^2}{\rho}$ From Eq. (10), it is then clear that the maximum deceleration occurs at

$$\frac{C_D A}{m\beta}\rho \cong \sin\theta$$

Substituting this relation into Eqs. (12a or 12b) or more easily into Eq. (13a), one obtains from (13a)

$$\sin\theta = \frac{1}{\beta R_o}\left(\frac{gR_o}{V^2} - 1\right)\frac{\cos\theta}{\theta - \theta_f}\ln\left(\frac{V_f^2/gR_o}{V^2/gR_o}\right) \qquad (13a')$$

Solving Eq. (13a') with Equation (11b) one obtains the numerical values of Θ and V at which the maximum deceleration occurs. Substituting these values back into Equation (13a), one then obtains the altitude at which maximum deceleration occurs.

II. Aerodynamic Heating

In the previous section under Eqs. (4), (5), (6) for heating[2], a number of assumptions which were originally used[2] for small angles of inclinations were also used here for large angles of inclination. Certainly, these assumptions in certain cases may not be completely justified in the case of large angles of inclination; however, for simplicity and consistency they were simply assumed to be also valid for large angles of inclination in the present analysis[5]. A detailed analysis of heating at large angles of inclination is to be given in a separate paper later. With this approximation in mind, Eqs. (4), (5), and (6), which are in terms of "ρ" and "V", may now be used to obtain various values of aerodynamic heating along the trajectory by simply substituting the known values of "ρ" and "V" at various points from the trajectory analysis into the appropriate Eqs (4), (5), and (6). (For Eq. (4), a simple graphitical integration is needed here.

The above Eqs (5), and (6) give, respectively, the time rate of average heat input per unit area and the time rate of stagnation region heat input per unit area. So far as the maximum time rate of average heat input per unit area is concerned, it occurs at

$$\frac{d}{dV}\left(\frac{dH_{av}}{dt}\right) \cong \frac{d}{dV}(\rho V^3) = 0 \qquad \text{(from Eq. (5)}$$

or

$$\frac{dV^2}{d\rho} = -\frac{2}{3}\left(\frac{V^2}{\rho}\right)$$

From Equation (10), one obtains, approximately[5], the condition at which the $\left(\frac{dHav}{dt}\right)_{max}$ occurs

$$\left(\frac{C_D A}{m\beta}\right)\rho = \frac{2}{3}\sin\theta$$

Substituting this relationship into Eq. (13a), one obtains

$$\sin\theta = \frac{3}{2}\left(\frac{1}{\beta R_0}\right)\left(\frac{gR_0}{V^2} - 1\right)\left(\frac{\cos\theta}{\theta - \theta_f}\right)\ln\left(\frac{V_f^2/gR_0}{V^2/gR_0}\right) \qquad (16)$$

Solving Eq. (16) with Eq. (11b), one obtains the numerical values of Θ and V at which $\left(\frac{dHav}{dt}\right)_{max}$ occurs. Substituting these values of Θ and V into Eq. (13a), one obtains the value of "ρ", the altitude density at which $\left(\frac{dHav}{dt}\right)_{max}$ occurs.

In a similar manner, one obtains the following approximate equation for solving the conditions at which $\left(\frac{dHs}{dt}\right)_{max}$ occurs.

$$\sin\theta = 3\left(\frac{1}{\beta R_0}\right)\left(\frac{gR_0}{V^2} - 1\right)\left(\frac{\cos\theta}{\theta - \theta_f}\right)\ln\left(\frac{V_f^2/gR_0}{V^2/gR_0}\right) \qquad (17)$$

Substituting the numerical values obtained from similtaneous solutions of Eq. (17) and (11b) into Eq. (13a) one obtains Θ, V and ρ at which

$(\frac{dHs}{dt})_{max}$ occurs.

REFERENCES

1. Carl Gazley, "Deceleration and Heating of a Body Entering a Planetary Atmosphere from Space", RAND Corporation Report P-955, Feb. 18, 1957

2. H. J. Allen, and A. J. Eggers, Jr., "A Study of the Motion and Aerodynamic Heating of Missiles Entering the Earth's Atmosphere at High Supersonic Speeds", NACA TN 4047, October 1957

3. D. R. Chapman, An Approximate Analytical Method for Studying Entry into Planetary Atmospheres", NACA TN 4276, May 1958

4. W. H. T. Loh, "Dynamics and Thermodynamics of ReEntry", J. Aerospace Sci. Vol. 27, No. 10, Oct. 1960

5. W. H. T. Loh, "A Second Order Theory of Entry Mechanics Into a Planetary Atmosphere", AAS preprint No. 61-116-1810, presented at the Joint National Meeting of IAS and ARS at Los Angeles, Calif., June 13 1961

6. A. J. Eggers Jr., H. J. Allen and S. E. Neice "A Comparative Analysis of the Performance of Long Range Hypervelocity Vehicles" NACA TN 4046, Oct. 1957

7. W. H. T. Loh, "Ballistic ReEntry at Small Angles of Inclination," presented at the American Astronautical Society Meeting in San Francisco, Calif., Aug. 1-3, 1961

SATURN DYNA-SOAR OPTIMUM TRAJECTORY CONSIDERATIONS

Ramon L. Chase* and L. E. Philips*

This paper considers the Saturn Dyna-Soar trajectory and performance optimization. The mating of the manned Dyna-Soar vehicle with the Saturn C-1 launch vehicle results in a very restricted ascent corridor. The characteristics of the Saturn C-1 combined with the abort and capture limitations of Dyna-Soar prescribe the set of constraints which bound the trajectory optimization problem. Currently the Saturn C-1 second-stage thrust level limits the vehicle's gross weight to a thrust-to-weight ratio of 1.5 at launch. Modifications to the second-stage thrust level are discussed in relation to the capabilities of the Saturn C-1 and its effects on the ascent trajectory.

The appendix describes the combination analog-digital computational technique employed to determine the Saturn Dyna-Soar trajectories. This technique proved effective in carrying out the trial and error procedure, currently required for bounded mixed end value trajectory optimization problems, with a certain degree of assurance as to the uniqueness of the answer.

INTRODUCTION

Two basic types of problems arise in trajectory optimization. The first type requires the determination of a steering program for a prescribed vehicle and a stated mission. This steering program optimizes the flight path with respect to a preselected parameter such as range, apogee altitude, and payload.. The second type of problem adds a considerable degree of complexity as only a mission and a payload or gross weight is specified. It then remains to determine simultaneously the optimum vehicle (minimum gross weight, maximum payload or minimum cost)

---

*Chrysler Corporation, Missile Division, Detroit, Michigan

and steering program to accomplish the required mission.

The Saturn Dyna-Soar problem presented in this paper is a problem which required only the determination of a steering program since the launch vehicle is specified for all practical purposes. Some modification to the Saturn C-1 vehicle is considered, extending the capability of the vehicle to deliver a larger payload. These modifications center on the thrust level of the second stage and an increase in the gross weight capability of the Saturn C-1 launch vehicle.

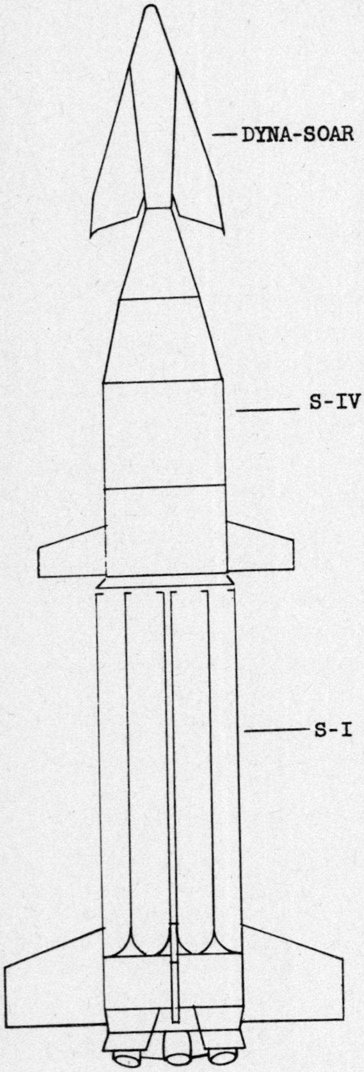

Fig. 1 Saturn C-1 Dyna-Soar Vehicle

SATURN DYNA-SOAR VEHICLE DESCRIPTION AND MISSION

The Saturn Dyna-Soar is composed of the Saturn C-1 launch vehicle and the manned Dyna-Soar spacecraft (Fig. 1). The Saturn C-1 is a two-stage launch vehicle of approximately one million pounds gross weight. The first stage (S-I) uses a combination liquid oxygen and hydrocarbon fuel to develop a sea-level thrust just over one and one half million pounds of thrust from a cluster of eight engines. The second stage (S-IV) uses liquid oxygen and liquid hydrogen as the propellants. The S-IV has a cluster of four engines which develop a total vacuum thrust of 70, 000 lb.

The initial thrust-to-weight ratio of the Saturn Dyna-Soar vehicle is limited to 1.5 because of the second-stage thrust level and propellant loading capability of the first stage. The thrust-to-weight ratio of the second stage, corresponding to a launch thrust to weight ratio of 1.5, is approximately 0.5. Additional second-stage thrust levels of 90,000 and 200,000 lb thrust are considered for purposes of trajectory comparison and performance analysis.

The Dyna-Soar vehicle can be described as a class of manned winged spacecraft having a significant maneuverability capability over a Mercury ballistic-type space capsule.

The mission of the Saturn C-1 launch vehicle is to boost the Dyna-Soar spacecraft into a low-altitude earth orbit. Because of the combined characteristics of the Saturn C-1 and the mission, thrust is employed continually from launch to orbit injection, except for a brief period during the separation of the first and second stages, to maximize performance.

ASCENT TRAJECTORY

The operational flight corridor is defined by the performance characteristics of the Saturn C-1 and the Dyna-Soar spacecraft. Figure 2 presents a plot of the permissible flight envelope in terms of altitude and velocity. The upper left-hand restriction curve is established by the abort acceleration limit on the Dyna-Soar vehicle. The launch vehicle cannot exceed this restriction during ascent and still assure safe recovery in case of an abort operation. The lower curve is set by a combination aerodynamic heating and structural limitation boundary. The upper right-hand curve represents the capture capability of the Dyna-Soar vehicle. The capture capability is determined by the negative lift capability of the reentering vehicle. This capability allows capture within one orbital period. The payload requirements, together with the permissible flight envelope, make it necessary to consider a class of minimum fuel optimum trajectories.

The Saturn second-stage thrust level of 70,000 lb thrust resulted in a second-stage thrust-to-weight ratio of approximately one half for maximum payload. The nature of the optimum trajectories associated with

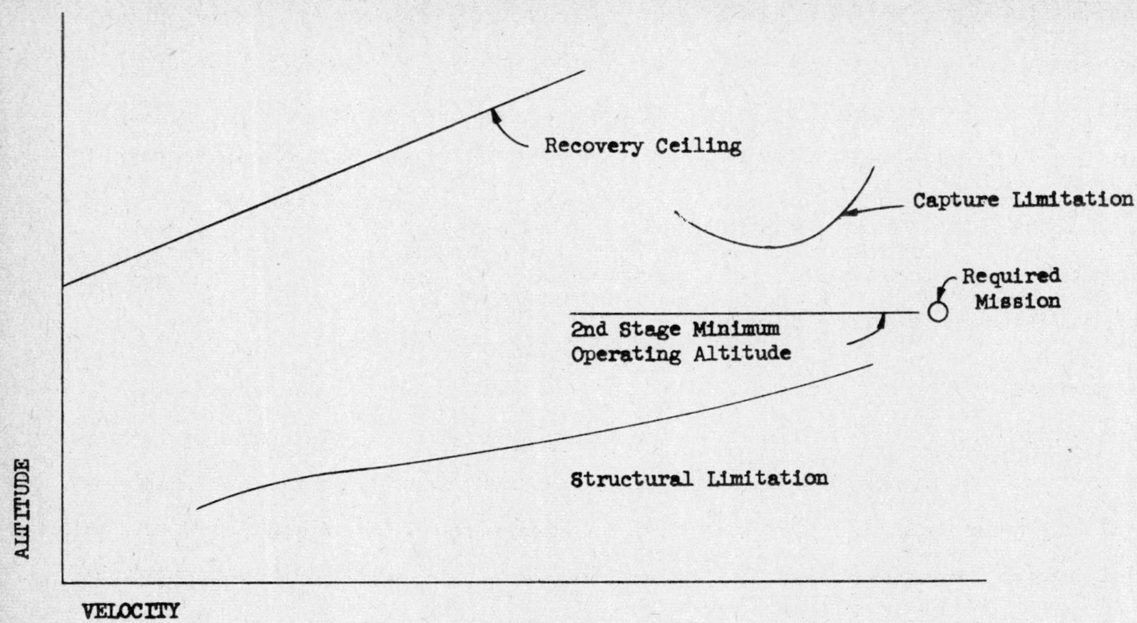

Fig. 2 Saturn Dyna-Soar Flight Envelope

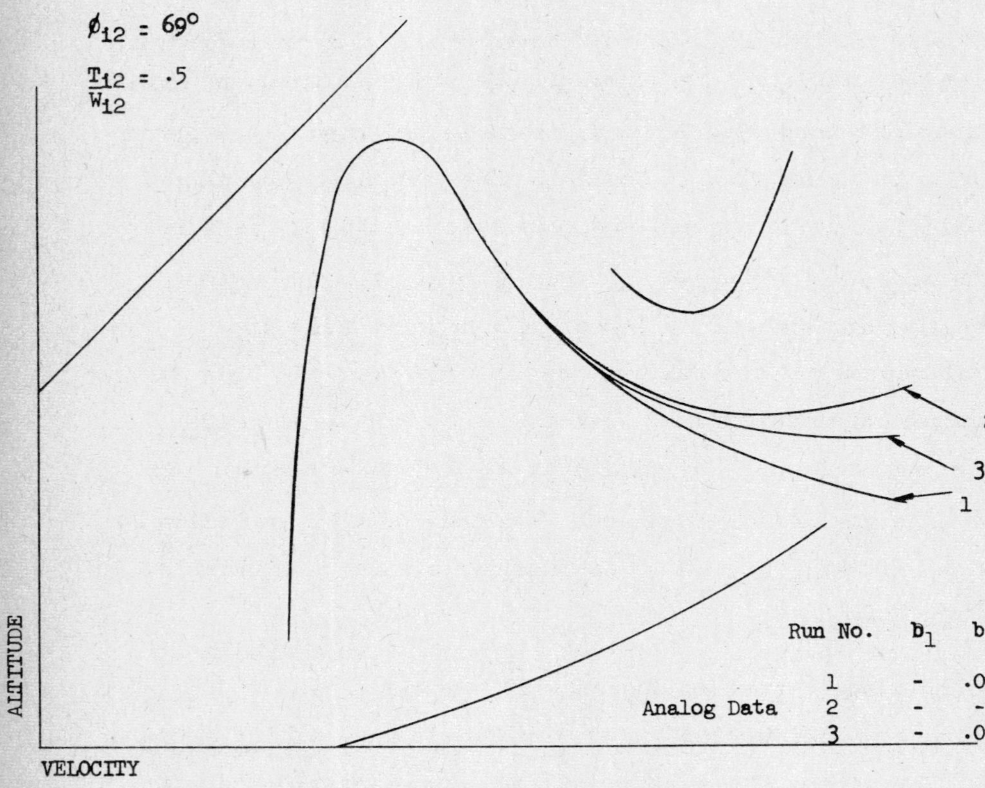

	Run No.	b_1	b_2	b_3	$\phi_{B.O.}$
	1	-	.001	.0040	90.6
Analog Data	2	-	-	.0035	88.8
	3	-	.001	.0037	89.5

Fig. 3 Trajectory Data Saturn Dyna-Soar Second Stage

this condition are rather peculiar. Because the initial thrust-to-weight ratio plus centrifugal acceleration is insufficient to balance gravitational attraction, the first stage is required to provide a unique set of second-stage initial conditions. The kinetic energy in the radial direction and, in some cases, potential energy are exchanged for tangential velocity. Such trajectories are therefore characterized by a rather large reduction in altitude during upper stage operation until sufficient balance between radial acceleration and gravitational acceleration is obtained.

RESULTS AND DISCUSSION

The final analog trajectory data used to determine the optimum flight path for the Saturn Dyna-Soar vehicle having a 70, 000 lb second-stage thrust level are shown on Fig. 3. In order to fly within the prescribed flight envelope, it was necessary to use three separate tilt programs during second-stage operation. The first tilt program pitched the vehicle over until the vehicle flew under the upper left hand restriction curve shown on Fig. 2. It was found that the closer the trajectory approached this boundary, the more the payload increased. Once the flight path was within the upper left hand boundary, a second tilt program was used to recover the dive resulting from the vehicle's inability to maintain altitude during this portion of second-stage operation. The optimum trajectory was such that complete recovery was obtained at orbital injection. A third tilt program was employed as soon as recovery capability was assured, to meet attitude conditions for orbit injection. The second and third tilting programs were determined by trial and error to minimize the required maneuverability of the second stage. Minimizing the required maneuvering maximized the payload capability.

It should be emphasized that the mission could not be completed at all within the specified flight envelope without using multiple tilt programs. A trial and error analysis indicated that three programs were optimum with respect to maximizing payload for Saturn Dyna-Soar. A restricted minimum fuel tilt program was used in each case during second stage operation (Appendix). This is mentioned at this time, as a question concerning whether the trajectories are really optimum or not can certainly be raised. Lacking complete analytical solution to bounded mixed end value optimum trajectory problems, these trajectories are offered as the best definition of the optimum trajectories presently available.

Additional trajectory data for the 90, 000 and 200, 000 lb second-stage thrust levels are shown on Fig. 4. No attempt was made to determine the optimum first-stage burnout conditions corresponding to the increased thrust levels because the burnout conditions for the 70, 000 lb second-stage thrust level were sufficiently close to the maximum velocity capability of the S-I to have little effect on payload for the close earth orbit mission under consideration.

The 90, 000 lb thrust level was obtained from a cluster of six engines. The performance results showed that the performance increase obtained from this thrust package was sufficient to offset the structural weight penalty of the two added engines. Also, an important feature of this thrust package is an engine out capability which is unobtainable with the four engine thrust package. The loss of an engine would result in an abort condition.

The 200, 000 lb thrust level was obtained from a single engine. The performance results show that a small increase in payload was obtained. However, the significant factor in using a second-stage thrust level of 200, 000 lb is the capability of increasing the gross weight of the Saturn C-1 to a thrust-to-weight ratio of 1. 25 at liftoff. This does reduce the second-stage thrust-to-weight ratio to approximately 0. 6 which is right back to the original second-stage thrust-to-weight ratio considered in this analysis. However, the increase in gross weight enables the payload to be increased two and one-half times. From the previous analysis, adding a second engine to the thrust package would not decrease the payload, and further, would provide the advantage of engine out capability.

The resulting Saturn C-1 launch vehicle has significantly extended capability over the original vehicle presented in this analysis.

Figure 5 presents the angle-of-attack history for the second-stage thrust levels considered. This data is presented since the Saturn Dyna-Soar trajectories require angles of attack which are considerably beyond that which are usually encountered. Despite the large angles of attack, these trajectories for second-stage thrust levels of less than one do not indicate a reduction in payload capability under Saturn Dyna-Soar conditions.

The Saturn Dyna-Soar trajectory and performance optimization analysis brings to light certain characteristics pertaining to the optimum delivery of payload into orbit using a two-stage launch vehicle. The optimum vehicle for a close earth orbit would have an initial thrust-to-gross weight ratio of 1. 25 (NASA minimum value) and a second-stage thrust-to-weight ratio less than one (approximately 0. 5). The vehicle would use thrust continually from launch to orbit injection except for a brief period during stage separation. The initial orbit would be elliptical with injection occuring just above the perceptible atmosphere (approximately 100KM). The optimum ascent into this "transfer orbit" would be characterized by an overshooting of the injection altitude during second-stage operation. The reduction in altitude would terminate simultaneously with second-stage burnout and orbit injection. Final orbit stabilization would then take place, with an impulse delivered by either a re-light capability of the second stage or a third stage, for the initial perigee would not be sufficiently high to maintain an extended orbital lifetime.

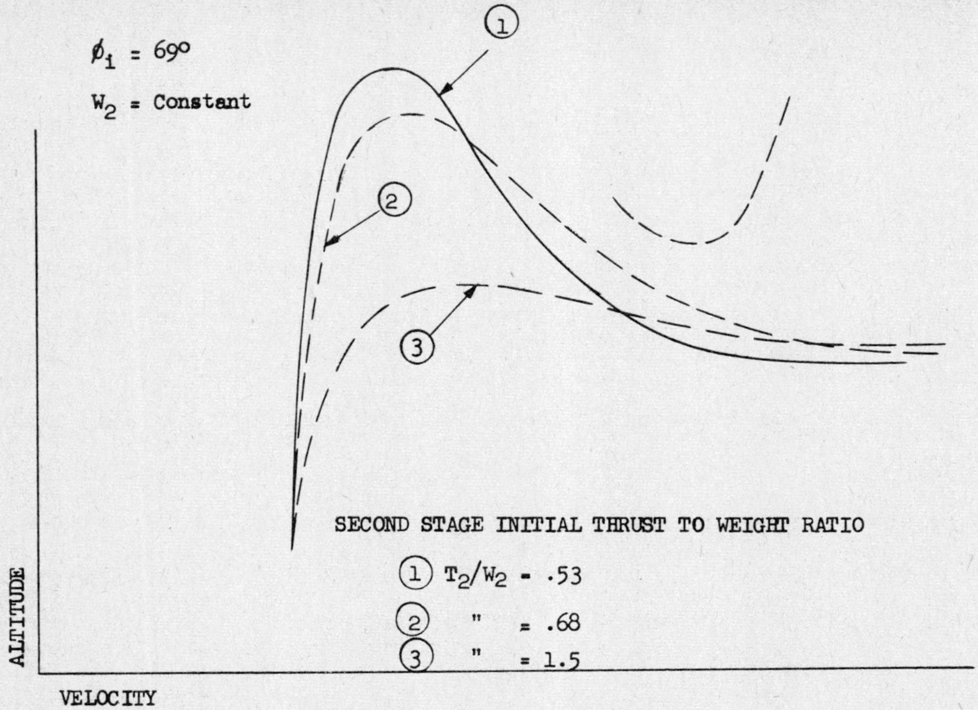

Fig. 4 Saturn Dyna-Soar Trajectory Data

Second Stage Thrust Level Comparison

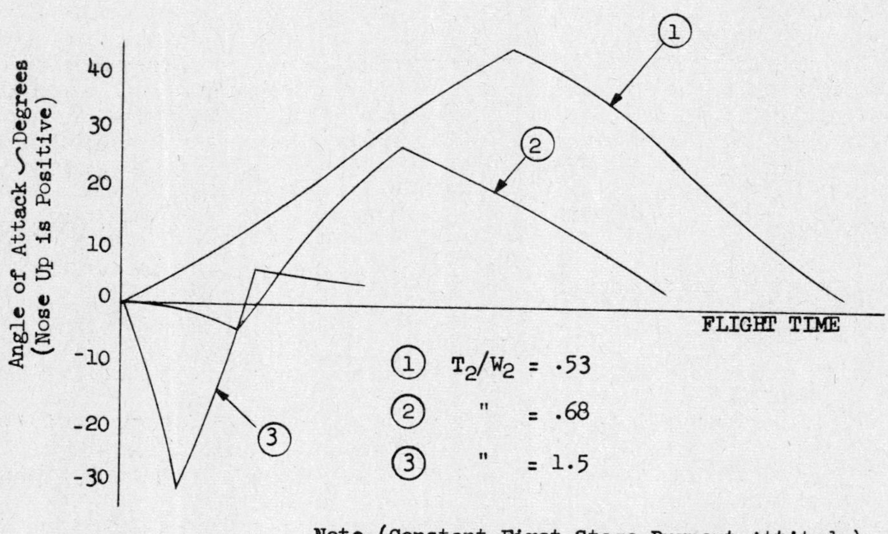

Fig. 5 Saturn Dyna-Soar Trajectory Data

Second Stage Angle of Attack Data

APPENDIX

INTRODUCTION

The current analytical investigation of optimum bounded mixed end value trajectory problems is centered around applications of the calculus of variations to determine general solutions and successive approximation techniques such as steepest descent to prescribe the particular solution. Such techniques for determining the particular solution are in reality mechanization of the trial and error approach and, as such, have the inherent disadvantages of convergence, time-consumption, and uniqueness.

The common characteristic of digital computer programs employed to determine optimum trajectories is the inclusion of successive approximation sub-routines, along with the equations of motion and the general steering equations. The computer solution is begun by a set of initial and terminal conditions. In actual problems, a given set of vehicle parameters and a mission are explicitly expressed. Initial conditions remain to be prescribed in some manner, usually predicated on previous personnel experience. Unfortunately, this implies an approximate foreknowledge of the answer. The selection of the initial conditions is extremely important because the computer time required to solve the problem depends on how close the initial conditions are to the final answer. The possibility exists that initial conditions could be prescribed which are outside the limits of convergence. Attempts to widen the convergence limits introduce the problem of local optimums which are a characteristic inherent in the degree of precision in numerically integrating equations of motion along with auxiliary equations. Increasing precision introduces additional local optimums, whereas decreasing the precision results in answers which do not meet the stated end conditions within acceptable error. One is then faced with selecting the degree of precision which assures acceptable mission error and concentrating on methods of determining satisfactory initial conditions.

COMBINATION ANALOG-DIGITAL TRAJECTORY PROCEDURE

Confronted with the problem of determining optimum trajectories for the Saturn Dyna-Soar boost vehicle under very restricted conditions, the present combination analog-digital computer trajectory optimization procedure was developed. Basically, the analog computer was used to effect a rapid search for regions of acceptable solutions and to determine the trade-off relationships between variables. Once the optimum trajectory was roughed out on the analog computer the data was used to supply initial conditions for a more precise digital computer simulation.

The present analog program is formulated around two dimensional equations of motion for a spherical nonrotating earth model. The equations programmed on the analog computer are

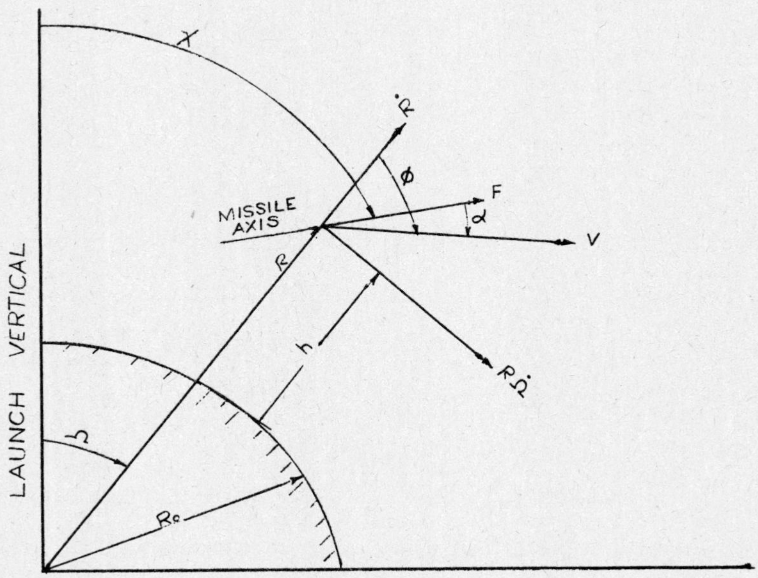

COORDINATE SYSTEM

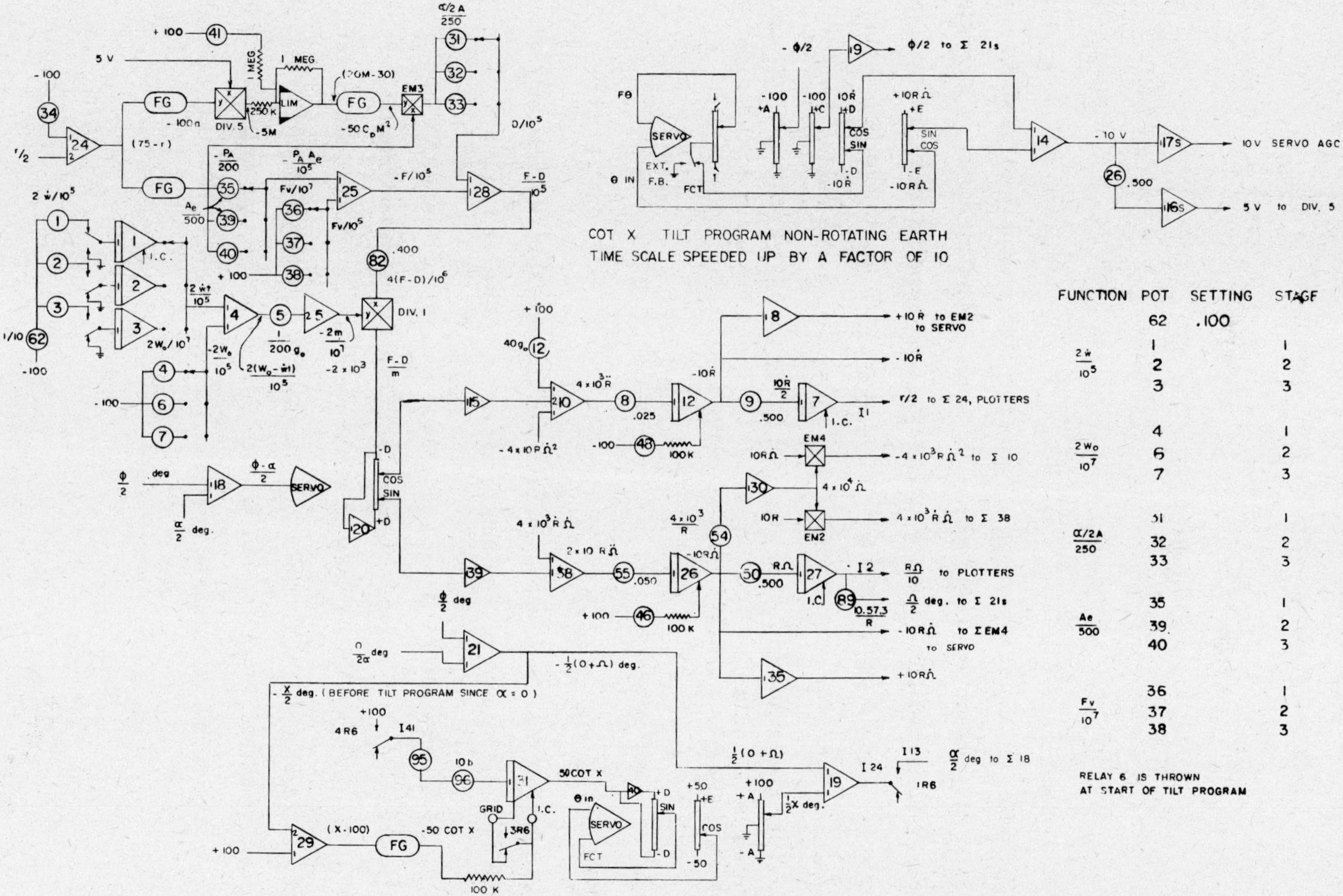

Fig 6 Two Dimensional Analog Mechanization Diagram

FUNCTION	POT	SETTING	STAGE
	62	.100	
$\frac{2\dot{w}}{10^5}$	1		1
	2		2
	3		3
$\frac{2W_o}{10^7}$	4		1
	6		2
	7		3
$\frac{\alpha/2A}{250}$	31		1
	32		2
	33		3
$\frac{A_e}{500}$	35		1
	39		2
	40		3
$\frac{F_v}{10^7}$	36		1
	37		2
	38		3

RELAY 6 IS THROWN AT START OF TILT PROGRAM

$$\ddot{R} = \frac{F_v g_0}{W} \cos(\phi - \alpha) + R\dot{\Omega}^2 - g_0 \left(\frac{R_0}{R}\right)^2 \quad (1)$$

$$R\ddot{\Omega} = \frac{F_v g_0}{W} \sin(\phi - \alpha) - 2\dot{R}\dot{\Omega} \quad (2)$$

where

$$R = R_0 + h \quad (3)$$

$$W = W_0 - \dot{w}t \quad (4)$$

$$\chi = \Omega + \phi - \alpha \quad (5)$$

and

$\left(\frac{R_0}{R}\right)^2$ is an average value.

The deletion of aerodynamic forces from these equations restricts their application to upper-stage simulation only. The inclusion of aerodynamic forces would increase the complexity and sensitivity of the program to errors while contributing no significant addition to the implementation of the program in this presentation. Included in the analog program is a restricted minimum fuel steering program based on the equations presented by Dr. Stanley Ross in the November 1959 issue of the American Rocket Society Journal. The steering equation is

$$\cot \chi = b - at \quad (6)$$

and

$$b = \cot \chi_0 \quad (7)$$

The analog mechanization diagram for the equations of motion including the linear contangent steering program is shown on Figure 6.

The analog trajectory program is used in conjunction with a two-dimensional digital trajectory program which includes aerodynamic forces. The digital program is used to generate a family of first stage data as shown on Fig. 7. This data accounts approximately for the effects of earth rotation, including launch site latitude and firing azimuth. The digital program assumes a vertical launch, an angle-of-attack program

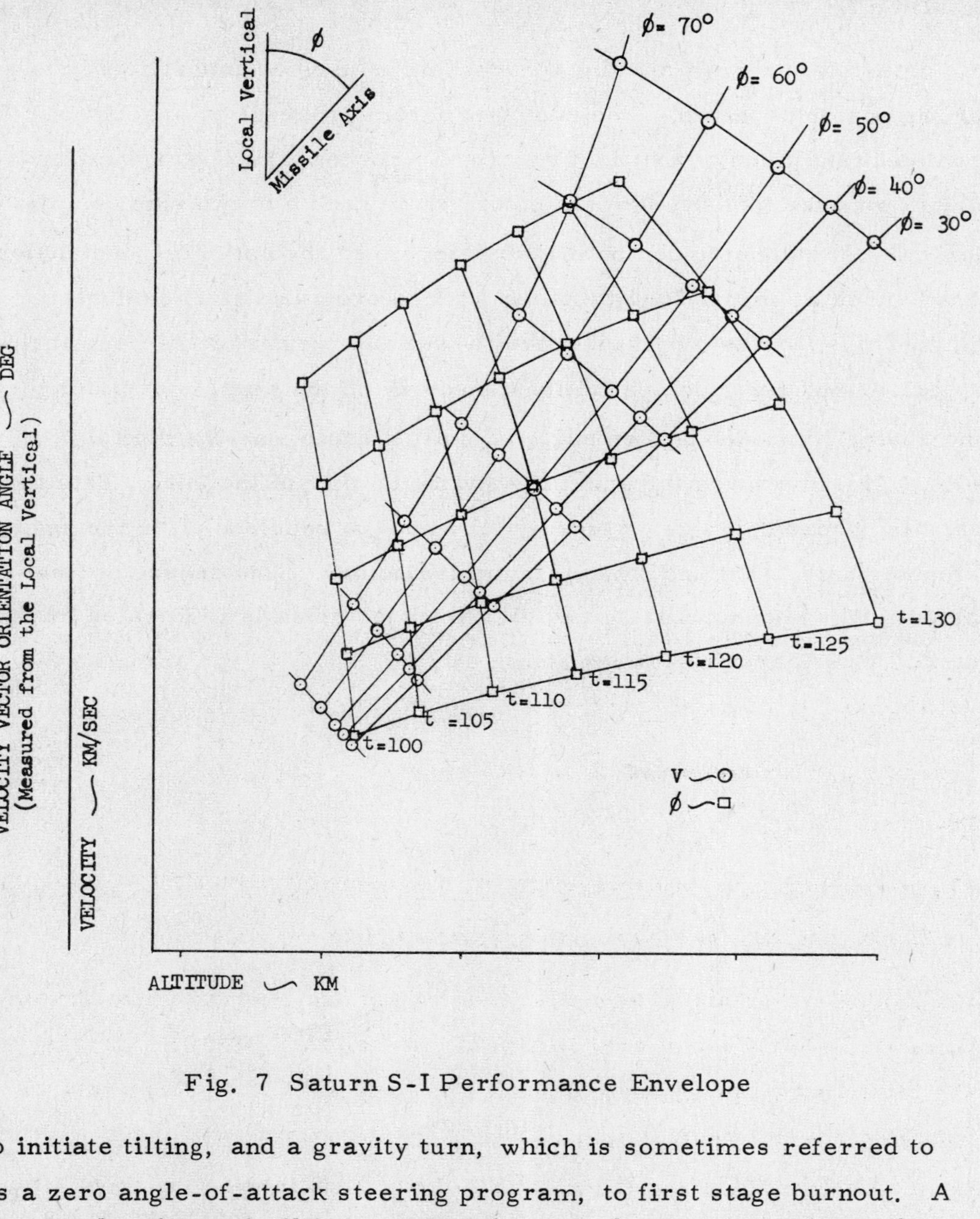

Fig. 7 Saturn S-I Performance Envelope

to initiate tilting, and a gravity turn, which is sometimes referred to as a zero angle-of-attack steering program, to first stage burnout. A zero angle-of-attack tilting program is very close to the optimum for large vehicles during atmospheric penetration and very easy to simulate. The family of first-stage trajectory data is obtained by varying the amplitude or duration of the angle-of-attack program which initiates the gravity turn.

The data obtained from the digital program is used as initial conditions for the analog program. The analog program proceeds to search the terminal conditions of the first stage and matches a first-stage trajectory to produce the optimum trajectory from launch to mission completion. Once the analog computer has determined the optimum, the analog data is used as initial conditions for a more precise digital computer simulation. It is usually necessary to take the results of the first analog trajectory and run a compatibility trajectory on the digital computer. The analog computer is then adjusted until it duplicates the digital results. This improves the accuracy and reliability of the analog data to the point where only one or two digital runs are required after the analog computer has determined the optimum trajectory. The degree of final compatibility between analog and digital trajectories is illustrated by the Saturn Dyna-Soar trajectories shown on Fig. 8.

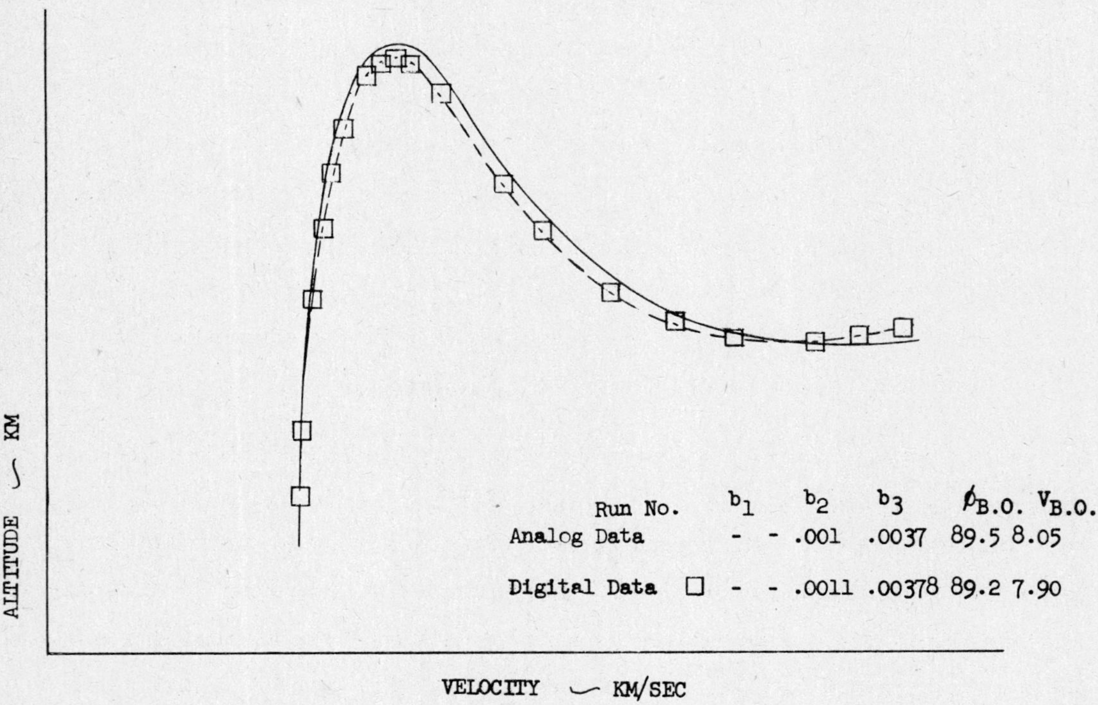

Fig 8 Saturn Dyna-Soar Combined Analog Digital Computer Simulation

RESULTS AND CONCLUSION

The Saturn Dyna-Soar trajectory optimization problem presented an excellent example to demonstrate the superiority of the combined analog-digital computer procedure. In the Saturn Dyna-Soar case, because of the restrictions placed on the ascent trajectory, it was necessary to initiate three separate tilting programs during the operation of the second stage to meet the required mission in the most efficient manner. The complexity and size of a general digital program which would anticipate and solve a problem of this nature without external direction would be very expensive and time consuming. If such a program were developed, it would most likely require modification for each new problem with its associated set of particular restrictions. Such modifications could conceivably result in a separate program, for each new problem which would become prohibitive from both a cost and time standpoint. The combination analog-digital computational procedure takes less total elapsed time, calendar time and provides a considerable reduction in cost over a straight digital computational technique. The continual observation advantage of the analog computer is very important in reducing the amount of data required to prescribe an optimum trajectory. Many times it is not necessary to complete a trajectory before a decision can be made regarding the optimum, or it may become evident from a few trajectories that a particular region is unrelated to the desired results. This enables large areas to be quickly eliminated without further computations. The introduction of the individual directly into the optimization process to contribute direction is an efficient procedure for the determination of optimum trajectories.

NOMENCLATURE

F_v Vacuum thrust

g_o Sea level gravitational acceleration

R_o Radius of the earth

R Distance from the center of the earth to the missile

h Altitude

t Flight time

V Missile velocity (space fixed)

W_0 Initial vehicle weight

W Instantaneous vehicle weight

$\dot{W}$ Propellant flow rate

a Coefficient determined by end conditions

α Angle between between missile axis and the velocity vector

Ω Angle between the launch vertical and the local vertical

ϕ Angle between the local vertical and the velocity vector

χ Angle between the launch vertical and the missile axis

REFERENCES

1. Dr. G. Leitmann, "Extremal Rocket Trajectories in Position and Time Dependent Force Fields," - Department of Mechanical Engineering, University of California, Berkeley. Paper presented at the 7th annual meeting of the American Astronautical Society, Dallas, Texas

2. Dr. Stanley Ross, "Composite Trajectories Yielding Maximum Coasting Apogee Velocity," ARS Journal, Vol. 29, Nov 1959

3. W. Carter and P. C. Krammer, "Escape of Manned Vehicles from Lunar Mission Ascent Trajectories," Astronautics Division, Chance Vought Aircraft Inc. Paper presented at the 7th annual meeting of the American Astronautical Society, Dallas, Texas

4. "Space Technology," Howard S. Seifert, John Wiley & Sons, Inc., 1959

INTERPLANETARY TRAJECTORIES*

F. G. Gravalos+

The aim of this work is to solve the two-point "limit problem" within the gravitational field of the solar system. To make possible a concise mathematical formulation of the problem, a method is presented for the definition of a physical model "ad hoc" to whatever interplanetary trajectories must be studied. Picard's iterative method for constructing solutions, with an essential modification found empirically, is given and numerical results systematically obtained are presented.

INTRODUCTION

In recent years, considerable work has been done on the problem of determining interplanetary trajectories, both in the United States and abroad. At home, various n-body programs, which are a mixture of astronomical and numerical integration techniques, were developed. These programs can be used directly for the solution of the initial boundary value problem: "Given the position and velocity at a given instant, determine the corresponding trajectory." By application of a trial and error procedure these programs have also been used to solve numerically the boundary value problem we call the "limit problem": "Determine the trajectory that goes through two given points at two given instants." The mathematical foundations that justify this type of approach were given at

---

*This paper was prepared with the collaboration of A. J. Dennison, Mathematician, Navigation and Control - Engineering, and M. McDevitt, Engineer, Flight Mechanics - Systems Engineering, Missile and Space Vehicle Department, General Electric Company, Philadelphia

+Consultant, Applied Mechanics - Systems Engineering, Missile and Space Vehicle Dept., General Electric Co., Philadelphia

the turn of the century by Paul Painleve. He showed that the singularities in the equations of celestial mechanics are always removable. From these thoughts sprang epoch-making papers by Levi-Civita and Sundman. Recently, Benedikt[1] has determined the technical value and limitations of Levi-Civita's work. My teacher, the late Professor G. D. Birkhoff,[2] elucidated Sundman's paper and made it accessible to a large public. However, from the point of view of scientific engineering, one is not concerned with "collision trajectories" in the sense of modern dynamics. Instead, it is the "limit problem" that has primary importance in technical applications.

The present work is divided into three parts. In Part 1 the physical model is defined. Although we have only dealt with trajectories from near the earth to the moon, the method outlined for the selection of the relevant planets, as well as the form of the corresponding differential equations, is completely general. This is, in spite of its simplicity, an essential novelty of this work.

Part 2 is divided into sections that, logically, are quite apart. Using ideas E. Picard[3] first published in 1893, the two-point limit problem is rigorously formulated in an integral form. Trying to use Picard's method of successive approximations for the actual construction of solutions, a cyclic process was devised - section 2. 2 - by means of which solutions were obtained even for cases when the original iterative process is divergent. This scheme, as presented here, has no scientific value. However, if numerical results are viewed as phenomena, this cyclic scheme becomes a very interesting experimental tool, by use of which solutions can be obtained and the nature of the specific problem at hand can be analyzed. This work is not concerned with the determination of the general conditions of applicability of this scheme; but, it is hoped, this and related questions will be studied in a forthcoming paper.

In Part 3, numerical results of high accuracy are presented. The numerical analysis that should precede, or accompany, the construction of solutions is outlined in section 3. 3.

1. PHYSICAL MODEL

1. 1 GENERAL CONCEPTS

For any problem involving the calculation of interplanetary trajectories, and independent of what is the particular problem at hand, the first questions to resolve are what physical model and what coordinate system of reference should be used. In this work, a system of Cartesian coordinates at absolute rest - a Galilean system* - is assumed with center at the Sun, with respect to which another Cartesian system rigidly attached to it - and, naturally, Galilean - is defined and used as a system of reference. All the planets and the Sun are considered particles - points with finite mass - and our differential equations define the motion of a negligible mass in the time dependent gravitational field created by certain planets, selected as bearing mensurable influence upon the motion in question, in their known motions within the solar system. Since we take the results of astronomical work as physical data, our equations are very much simpler than those usually used in the so-called n-body programs; on the other hand, a selection of the planets that determine the motion must be made "a priori."

Let these assumptions be examined more closely. The equations of free dynamical systems are the analytic expressions for the symbolic relation,

$$\text{Inertial Terms} = \text{Acting Forces} \tag{1}$$

Astronomers provide us with the positions of planets in the equatorial system - shown in Fig. 1 - versus time, with a given reference date. Since only gravitational forces are considered, and these are defined

*By definition, Galilean systems include also those moving without acceleration.

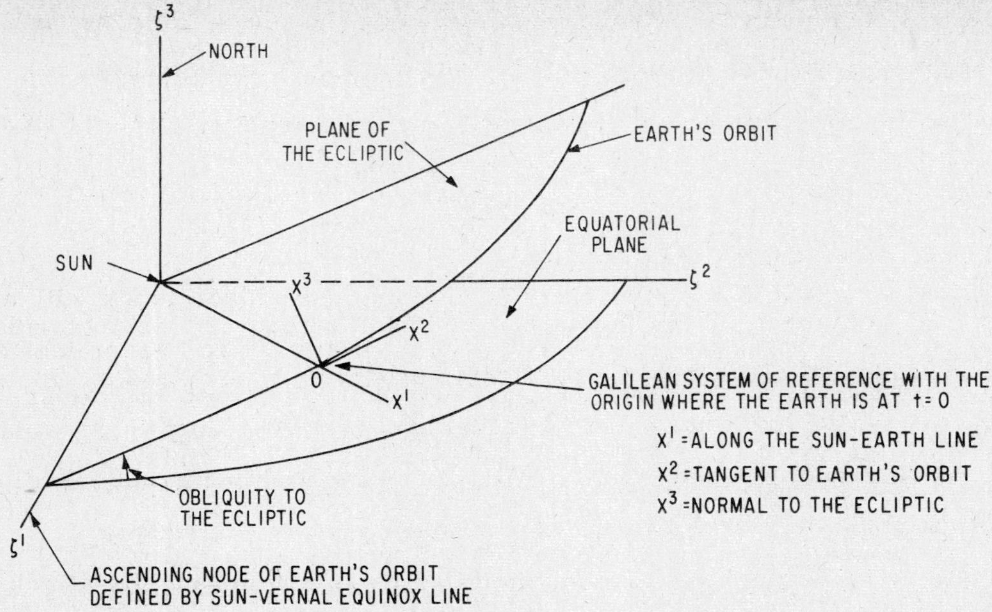

Fig.1 Definition of the Galilean System of Reference

completely by the relative position of the point masses, the right hand side of Eq. (1) will be written for our physical model with no "error", if the astronomical data is supposed to be "exact."

In Newtonian mechanics the existence of a system of coordinates is always assumed at absolute rest. Therefore, the error committed when writing the acceleration components on our Galilean system can only be examined with reference to motions known to exist and can only be determined with the accuracy with which those motions are quantitatively known. Let $\vec{\omega}_G$ and $\vec{\omega}_S$ be the angular velocity vectors of the motions of the center of mass, c. m., of the solar system about our galactic axis - the line at absolute rest - and of the Sun about the c. m. of the solar system, respectively, both $\vec{\omega}_G$ and $\vec{\omega}_S$ being obtained with the Sun as the reference point for the decomposition of the motions. If these motions are taken into account, the left-hand side of Eq. (1); i. e., the acceleration per unit mass, will be given by the vector.

$$\dot{\vec{V}} + (\vec{\omega}_G + \vec{\omega}_S) \times \vec{V}$$

where the local time-derivative, indicated by the "dot", is taken with respect to a system rigidly attached to the Sun.* Thence, the error committed on the left hand side of Eq. (1) due to the assumption that our system of reference is at absolute rest is given by

$$(\vec{\omega}_G + \vec{\omega}_S) \times \vec{V}$$

It is known that the solar system moves about our galactic axis with a periodic motion; let the period by $\bar{T}$. Neglecting $\vec{\omega}_S$, and assuming this motion rigid, for a vehicle moving at, say, 30, 000 ft/sec and for an optimum relative position of its velocity vector, $\vec{V}$, and $\vec{\omega}_G$, the error in position will be less than 4, 500 ft per year of flight time, for $\bar{T}$ equals 100 million years.

Let it be remarked that, although Einstein's gravitational tensor is independent of any specific system of coordinates, in order to write the differential equations of motion for the known integrable cases (Schwarzchild, Painlevé), a system of reference at absolute rest must also be explicitly assumed within the framework of the General Theory of Relativity.

1. 2 PHYSICAL MODEL FOR EARTH-MOON TRAJECTORIES

The motion of a space vehicle - as a point mass - will be studied in the gravitational field created by the Sun, the Earth, and the Moon. The qualitative reasoning for this selection runs as follows: Although, for relatively short intervals of total flight time, the influence of the Sun may be negligible, there will always be a portion of the flight under the almost exclusive influence of the Sun. Moreover, to account for the presence of the Sun does not bring any extraneous complexity in the computations, for the distances from the vehicle to the Earth and to the Moon must be calculated using their coordinates in the equatorial system.

An example of the quantitative analysis that should precede the selection of planets will show how insufficient a reasoning such as that in the

*In Newtonian mechanics, "all the clocks are synchronized"; thus, only the relative motions need be considered.

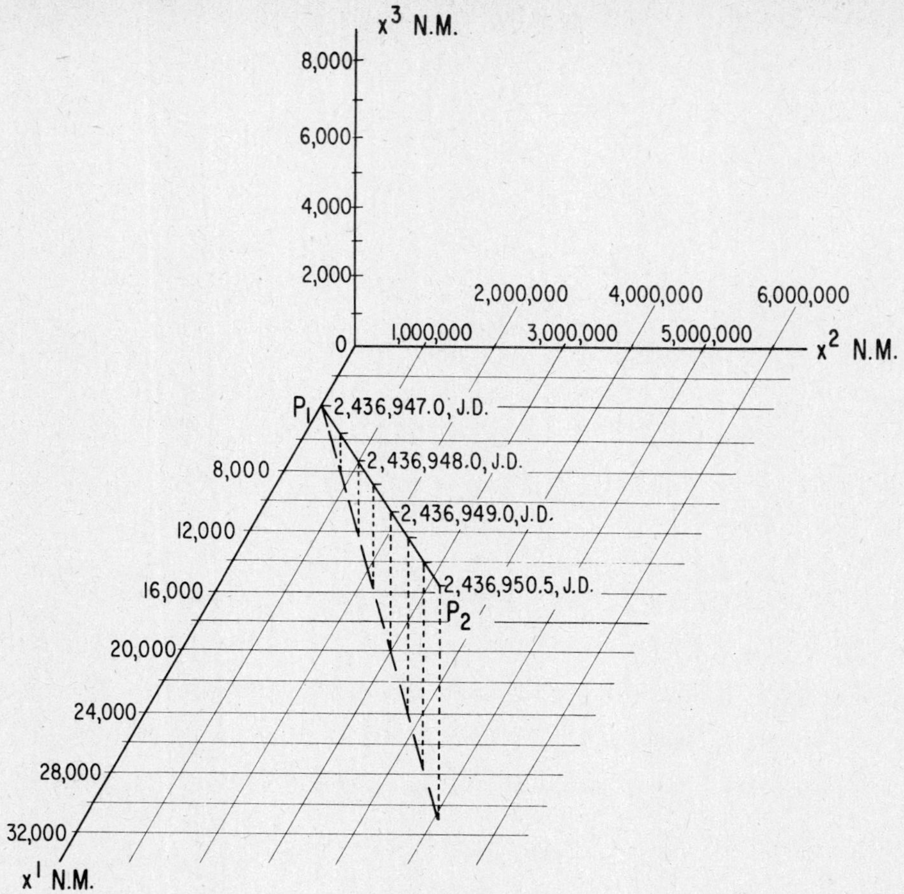

Fig. 2 "Initial Solution" for a Specific Time Interval

above paragraph could be. In Fig. 2, the motion from a point P_1 to another point P_2 is shown as being uniform, for given Julian dates. The forces per unit mass exerted upon this particular motion - the simplest that can be assumed - by different planets are given in g's in Fig. 3. From this graph an upper limit for the effect of neglecting, say, Jupiter is found to be about ten million feet in positional error for 75 hrs of flight time, a magnitude considerably larger than the numerical error of our solutions (see Part 3). It should, however, be noted that just the addition of another term in Eq. (3) would not avoid this error; to account for such effect, the summations in Eq. (10) must be adequately arranged.

By use of ephemeris tables, the coordinates of the Moon and the Earth, in our Galilean system, $0 - x^1$, x^2, x^3 at specific intervals of time, can be easily obtained. Although these data are given at discrete points, we shall write

$$X_S^i = d^i = \text{constant}$$

$$X_E^i = x_E^i(t) \qquad (i = 1, 2, 3) \tag{2}$$

$$X_M^i = x_M^i(t)$$

and suppose these functions to be - as they are - continuous in the interval of time in which the motion takes place. We shall see that, with our formulation of the problem and method of solution, the data can be treated as given by functional relations without direct reduction to analytic expressions.

For this physical model, the equations of motion of a particle of negligible mass - with respect to the gravitational masses involved - take in our system of coordinates the following simple form.

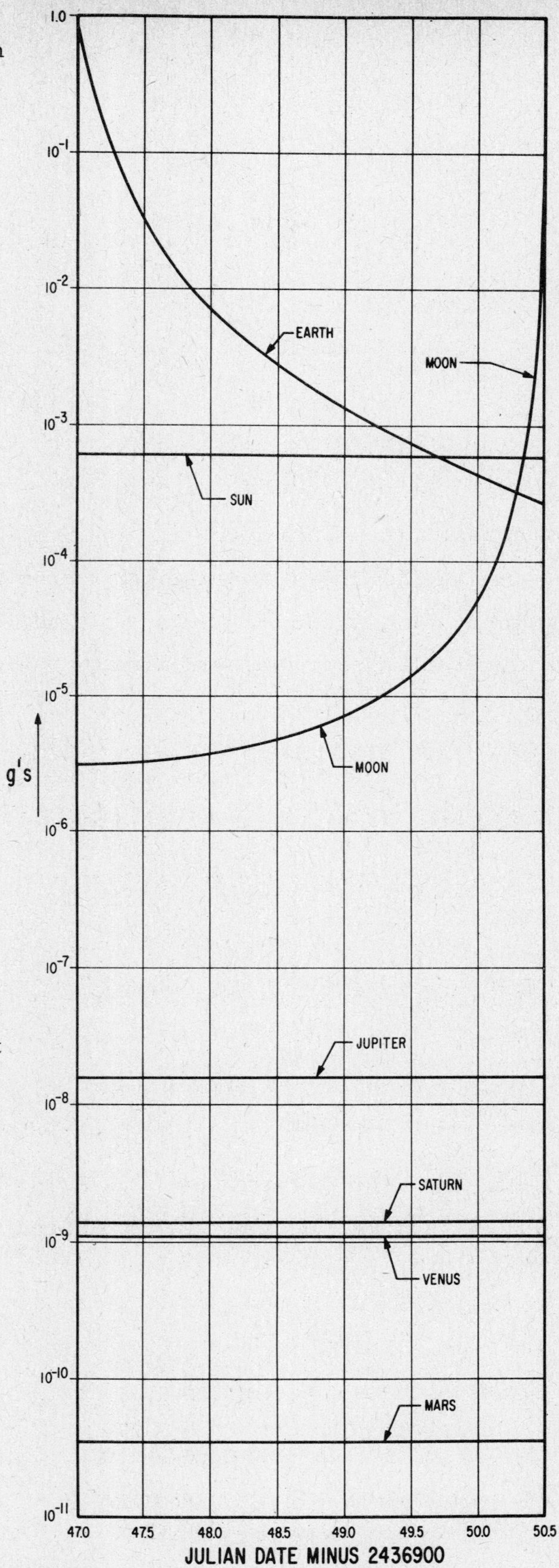

Fig. 3 Influence in g's of Different Planets Along the "Initial Solution"

$$\ddot{x}^i = K_S \left[d^i - x^i\right]\left\{\sum_{j=1}^{3}\left[d^j - x^j\right]^2\right\}^{-\frac{3}{2}}$$

$$+ K_E\left[x_E^i(t) - x^i\right]\left\{\sum_{j=1}^{3}\left[x_E^j(t) - x^j\right]^2\right\}^{-\frac{3}{2}} \tag{3}$$

$$+ K_M\left[x_M^i(t) - x^i\right]\left\{\sum_{j=1}^{3}\left[x_M^j(t) - x^j\right]^2\right\}^{-\frac{3}{2}}, \quad (i, j = 1, 2, 3)$$

where K_S, K_E, and K_M are the products of the universal gravitational constant times the masses of the Sun, the Earth, and the Moon, respectively. For other interplanetary problems that require taking into account the presence of other, or of additional planets, the corresponding differential equations will remain of the same general form.

2. THE LIMIT PROBLEM

2.1 MATHEMATICAL FORMULATION

Equations (3) are of the form

$$\ddot{x}^i = g^i(x^j, t) \quad (i, j = 1, 2, 3) \tag{4}$$

and a solution is sought such that

$$x^i(t_1) = a^i \tag{5}$$

$$x^i(t_1 + T) = b^i$$

i.e. the solution that goes from P_1: (a^i) to P_2: (b^i) in the interval of time T. The constant t_1 is an additive constant and will be always taken equal to zero.

If a solution, $x^i = x^i(a^j, \alpha^j, t)$, of the initial value problem

$$x^i(0) = a^i$$

$$\dot{x}^i(0) = \alpha^i$$

were obtainable in closed form and the equations

$$x^i(a^j, \alpha^j, T) = b^i$$

for α^i had a solution, the limit problem defined above would be solved. This approach is only possible in trivial cases.

To use Picard's thoughts for the solution of our limit problem, let the $g^i(x^j, t)$ of Eq. (4) be the functions in the right hand side of Eq. (3) and

$$x^i(t) \equiv x^i(a^j, b^j, t) \tag{6}$$

be a solution of (13) satisfying the conditions of Eq. (5). Then, the integral relation,

$$x^i(t) = a^i + \frac{t}{T}\left[(b^i - a^i) - \int_0^T (T - \tau) g^i(x^j(\tau), \tau)\, d\tau\right] + \int_0^t (t - \tau) g^i(x^j(\tau), \tau)\, d\tau, \tag{7}$$

will be satisfied identically in t and, conversely, any set of functions $x^i(t)$ satisfying Eq. (7) is a solution of Eq. (3) plus Eq. (5), as can be verified by derivation.

The right-hand side of Eq. (7) may be interpreted in a dual role: As a vector operator, $\Gamma_i\left[x^j(t)\right]$, that permits obtaining sequentially from any given set of functions $x^i_h(t)$ another set,

$$x^i_{h+1}(t) = \Gamma_i\left[x^j_h(t)\right] \tag{8}$$

that always verifies the conditions of Eq. (5), and as an operator that provides the means "to check" whether any given set of functions $\bar{x}^i(t)$ verifying Eq. (5) is a solution of Eq. (3). For, in that case,

$$\bar{x}^i(t) \equiv \Gamma_i\left[\bar{x}^j(t)\right]. \tag{9}$$

With $g^i(x^j_h(t), t) \equiv \bar{g}^i_h(t)$, the operator Γ_i may be written in the compact form,

$$x^i_{h+1}(t) = a^i + \frac{t}{T}\left[(b^i - a^i) - \int_0^T (T - r)\bar{g}^i_h(r)\,dr\right] + \int_0^t (t - r)\,\bar{g}^i_h(r)\,dr \equiv \Gamma_i\left[x^j(t)\right]. \tag{10}$$

The operator $\dot{\Gamma}_i$ *, giving the derivatives of $x^i_{h+1}(t)$ from $x^i_h(t)$, is defined by the similar formula,

$$\dot{x}^i_{h+1}(t) = \frac{1}{T}\left[(b^i - a^i) - \int_0^T (T - r)\bar{g}^i_h(r)\,dr\right] + \int_0^t \bar{g}^i_h(r)\,dr \equiv \dot{\Gamma}_i\left[x^i_h(t)\right]. \tag{11}$$

Both Γ_i and $\dot{\Gamma}_i$ may be obtained from Eq. (4) by formal integrations.

2.2 METHOD OF SOLUTION

Picard's iterative method - with some essential modifications - is used to construct solutions.

Take, as "initial guess" of the solution, an arbitrary set of functions $x^i_0(t)$, verifying the conditions of Eq. (5). Repeated use of the operator Γ_i, and of $\dot{\Gamma}_i$ yields the double sequence of functions

$$\Gamma_i\left[x^j_0(t)\right],\ \Gamma^2_i\left[x^j_0(t)\right],\ \text{-----}\ \Gamma^h_i\left[x^j_0(t)\right] = x^i_h(t) \tag{12}$$

$$\dot{\Gamma}_i\left[x^j_0(t)\right],\ \dot{\Gamma}_i\left[x^j_1(t)\right],\ \text{-----}\ \dot{\Gamma}_i\left[x^j_h(t)\right] \tag{13}$$

If the limit

$$\lim_{h\to\infty} \Gamma^h_i\left[x^j_0(t)\right] = \lim_{h\to\infty} x^i_h(t) = x^i(t) \tag{14}$$

exists, then

$$\dot{\Gamma}_i\left[\lim_{h\to\infty} x^j_h(t)\right] = \dot{x}^i(t) \tag{15}$$

---

*Although $\dot{\Gamma}_i\left[x^j(t)\right]$ is call d an operator, it is nothing but a convenient notation; for $\dot{\Gamma}^h_i\left[x^j_0(t)\right]$ is not well defined.

and $x^i(t)$ and $\dot{x}^i(t)$ are the coordinates and the velocity components of our original dynamical problem.

To obtain the conditions for convergence for a single differential equation, Picard wrote the sequences (12) and (13) in the form of series,

$$x_0^i(t) + \left[x_1^i(t) - x_0^i(t)\right] + \text{-----} \left[x_h^i(t) - x_{h-1}^i(t)\right] + \text{-----}$$

$$\dot{x}_0^i(t) + \left[\dot{x}_1^i(t) - \dot{x}_0^i(t)\right] + \text{-----} \left[\dot{x}_h^i(t) - \dot{x}_{h-1}^i(t)\right] + \text{-----}$$

and found that, if the right hand side of Eq. (3) (i= 1, only) verifies Lipschitz's condition and both T and the absolute value of the slope of the straight line joining the points P_1 and P_2 (in the x-t plane) are sufficiently small, a solution of Eq. (4) plus Eq. (5) exists and is unique. Instead of discussing these questions in general terms, something which always implies the "a priori" selection of dominant constants, we shall give numerical techniques that provide simultaneously the solution* and the analysis of the specific problem at hand.

In essence, these numerical techniques consist of the following steps:

i. - Divide the time interval (0, T) into n parts, which need not be equal.

ii. - Compute the values of $x_E^i(t)$ and $x_M^i(t)$ at those n values of t. (If, by electronic machines, these values are computed as needed, considerable storage space is saved, but the computational time is increased.)

iii. - Select an arbitrary set of functions $x_0^i(t)$ that verify conditions of Eq. (5)

iv. - Set an algorithm to carry out the operations in Eqs. (10) and (11); i. e. arrange the necessary calculations to obtain

---

*In all the cases studied solutions were obtained; these included trajectories for T = 84 hours.

$\Gamma_i\left[x_h^j(t)\right]$ and $\dot{\Gamma}_i\left[x_h^i(t)\right]$ from $x_h^i(t)$ at the selected values of t.

v. - Examine whether the differences

$$\left|\Gamma_i\left[x_h^j(t)\right] - x_h^i(t)\right| \tag{16}$$

$$\left|\dot{\Gamma}_i\left[x_h^j(t)\right] - \dot{x}_h^i(t)\right| \tag{17}$$

are, for all the n values of t, smaller than prescribed numbers - defining the accuracy of the iteration process - from a certain h onward.

vi. - The examination of numerical results shows that each of the sequences (12) and (13) separates into two distinct sequences,** corresponding to the odd and even iterations for each of the $x^i(t)$,

$$\begin{array}{l} x_1^i(t),\ x_3^i(t),\ \text{-----}\ x_{2h+1}^i(t)\ \text{-----} \\ x_2^i(t),\ x_4^i(t)\ \text{-----}\ x_{2h}^i(t)\ \text{-----} \end{array} \tag{18}$$

(and the same for their derivatives).

One observes further that these double sequences converge to a common limit, or diverge from each other, depending - as it should be - on the physical nature of the problem under scrutiny. In both cases, rates of convergence, or of divergence, of the sequences $x_{2h+1}^i(t)$ and $x_{2h}^i(t)$ are extremely small (very near unity) and almost identical. Therefore, the following simple, but essential, modification in the iteration process was introduced: After an arbitrarily selected number m of iterations, an average is taken of the last two, and the result,

$$x_{0,1}^i(t) \equiv \frac{x_m^i(t) + x_{m-1}^i(t)}{2}, \tag{19}$$

is used as a new initial guess, $x_{0,1}^i(t)$ of the solution. This mixed cycle of m-iterations and an average is repeatedly applied, and the resulting sequence for the $x_{0,k}^i(t)$ can be described by use of the operator Γ_i in

**The mathematical reasons for this phenomenon will be discussed in a forthcoming paper.

the following manner,

$$x^i_{0,1}(t) = \frac{\Gamma_i^{m-1}\left[x^j_0(t)\right] + \Gamma_i^m\left[x^j_0(t)\right]}{2} \tag{20}$$

$$x^i_{0,2}(t) = \frac{\Gamma_i^{m-1}\left[x^j_{0,1}(t)\right] + \Gamma_i^m\left[x^j_{0,1}(t)\right]}{2}$$

and so forth. To the sequence

$$\Gamma_i\left[x^j_{0,1}(t)\right],\ \Gamma_i\left[x^j_{0,2}(t)\right], \text{-----},\ \Gamma_i\left[x^j_{0,k}(t)\right] \text{-----} \tag{21}$$

there corresponds the sequence,

$$\dot{\Gamma}_i\left[x^j_{0,1}(t)\right],\ \dot{\Gamma}_i\left[x^j_{0,2}(t)\right], \text{-----},\ \dot{\Gamma}_i\left[x^j_{0,k}(t)\right] \text{-----} \tag{22}$$

for the $\dot{x}^i(t)$.

The examination of results is not, however, done on the sequences (20) and (21) directly. Using the role of Γ_i indicated by (9), the differences

$$\left|\Gamma_i\left[x^j_{0,k}(t)\right] - x^j_{0,k}(t)\right| \tag{23}$$

are examined to determine whether $x^i_{0,k}(t)$ is a solution of Eq. (4) and (5)*. It is of importance to examine also the double sequences,

$$\Gamma_i\left[x^j_{0,k}(t)\right],\ \Gamma_i^3\left[x^j_{0,k}(t)\right], \text{-----},\ \Gamma_i^{m-1}\left[x^j_{0,k}(t)\right] \tag{24}$$

$$\Gamma_i^2\left[x^j_{0,k}(t)\right],\ \Gamma_i^4\left[x^j_{0,k}(t)\right], \text{-----},\ \Gamma_i^m\left[x^j_{0,k}(t)\right] \tag{25}$$

(written for m even) and those corresponding to $\dot{x}^i(t)$, for the nature of the specific problem at hand is defined by the convergence, or divergence, of these sequences.

In general, a solution is attained after 5, 6, or 10 complete cycles - m equals 20 - even when the two original sequences (18) are divergent. These are truly remarkable numerical events, for a simple modification to Picard's method - the averaging process - permits construction of

---

*In this case, $\dot{\Gamma}_i\left[x^j_{0,k}(t)\right] = \frac{d}{dt}\Gamma_i\left[x^j_{0,k}(t)\right]$

solutions in cases where the results of Picard's straight iteration process tell us that the solution is not necessarily unique.

2.3 FACTORS AFFECTING THE ACCURACY AND THE RAPIDITY OF THE PROCESS

These factors are:

i. - The number and distribution of the n values of t, between O and T

ii. - The formula used for the numerical integrations in Γ_i and $\dot{\Gamma}_i$

iii. - The number of digits carried out in the computations

iv. - The initial guess of the solution

v. - The number m defining the cycle described in 2.2vi

These factors are related to each other and to the form in which astronomical data are available. If the n time-values are not coincident with those for which $x_E^i(t)$ and $x_M^i(t)$ are given, an interpolation formula will have to be used to acquire the needed values of x_E^i and x_M^i. This formula should fit consistently with the integration scheme for the calculation of Γ_i and $\dot{\Gamma}_i$; i.e., neither the interpolation nor the integration formula should imply an order of approximation which the other makes it impossible to reach. In this connection, some astronomical work comes into the picture, for the interpolation formula should be one of those used in astronomy.

With these ideas in mind, since computing machines have a limited storage capacity and to operate them is costly, the worker must make a compromise as to what computational arrangement is most adequate for his purpose.

Once these choices are made, and assuming that our process of constructing solutions is converging, the question arises of whether for different n's there will result different solutions; the optimum solution to the physical problem is given by that maximum division of the interval

(O, T) into $\bar{n}$ parts beyond which the accuracy of the solution cannot be improved. This, in turn, is related to the number of digits carried in the computations: The value of any term in the sums representing the integrals in Γ_i and $\dot{\Gamma}_i$ for $n = \bar{n}$ must always be such that, its own individual error is adequately bound, and large enough to alter at least the last significant digit of all possible partial sums. For any subdivision beyond such $\bar{n}$, the integration process will degenerate; this should be first detected by fluctuations in the values of $x^i(t)$ in some part of the trajectory.

The number of digits carried in the computations plays another, very important role.

The sequence of powers of the operator Γ_i applied to $x_0^i(t)$ represents successive corrections to the initial guess. Thence, it is clear that the number of digits carried in the computations determines the degree of accuracy - and, therefore, of speed - with which these corrections are performed, i. e. the number h for which

$$\left| \Gamma_i^{h+1}\left[x_0^j(t)\right] - \Gamma_i^h\left[x_0^j(t)\right]\right| < \epsilon$$

is a function of the number of digits carried in the arithmetic operations.

From the above, intuitive interpretation of the sequences Γ_i^h, it is also clear that the better the initial guess is, the faster will be the convergence to the solution. Since for the cases where the straight iteration process is not convergent it is not possible to guarantee the uniqueness of the solution, one should ask whether different choices of $x_0^i(t)$ would yield different, or differently converging solutions.

The same question is applicable to the choice of m: Does the repeated application of our complete cycle yield different solutions for different m's? The dependence on m of the value of k for which

$$\left| \Gamma_i^2\left[x_{0,k}^j(t)\right] - \Gamma_i\left[x_{0,k}^j(t)\right]\right| < \epsilon$$

from k on, is obvious.

3. NUMERICAL RESULTS

3.1 SIMPLIFIED PHYSICAL MODEL

Inasmuch as in the present work we are not concerned with the analysis of a specific mission, assumptions have been made that simplify considerably the computational work but leave unaltered the questions pertaining to the method.

These simplifications are:

i. - The Earth moves about the Sun in a perfect circle.

ii. - The Moon also moves about the Earth in a perfect circle.

iii. - The integrals in Γ_i and $\dot{\Gamma}_i$ are calculated by Simpson's rule.

To obtain the functions $x_E^i(t)$ and $x_M^i(t)$, to define the problem, and to present the results, the following systems of coordinates are used:

The Galilean system with center at the Sun, S - ζ^1, ζ^2, ζ^3

The fundamental system of reference, 0 - x^1, x^2, x^3

A system of coordinates with center at the Earth, E- ξ^1, ξ^2, ξ^3, and always parallel to 0 - x^1, x^2, x^3.

A system of coordinates with center at the Moon, M- η^1, η^2, η^3 and always parallel to 0-x^1, x^2, x^3

All these systems are well defined by Fig. 1. The data required in Eq. (3) can be written immediately.

For the Sun's coordinates,

$$d^1 = -R$$

$$d^2 = 0$$

$$d^3 = 0,$$

R = 80728492 N. M.

For the Earth's coordinates,

$$x_E^1(t) = R\left[\cos(\dot{\Omega}t) - 1\right]$$

$$x_E^2(t) = R\sin(\dot{\Omega}t)$$

$$x_E^3(t) = 0,$$

$$\dot{\Omega} = 7.1676658 \times 10^{-4} \text{ rad./hr.}$$

For the Moon's coordinates

$$x_M^1(t) = x_E^1(t) + r\left[\cos\phi\cos(\omega_0 + \dot{\omega}t) - \sin\phi\cos\psi\sin(\omega_0 + \dot{\omega}t)\right]$$

$$x_M^2(t) = x_E^2(t) + r\left[\sin\phi\cos(\omega_0 + \dot{\omega}t) + \cos\phi\cos\psi\sin(\omega_0 + \dot{\omega}t)\right]$$

$$x_M^3(t) = r\sin\psi\sin(\omega_0 + \dot{\omega}t)$$

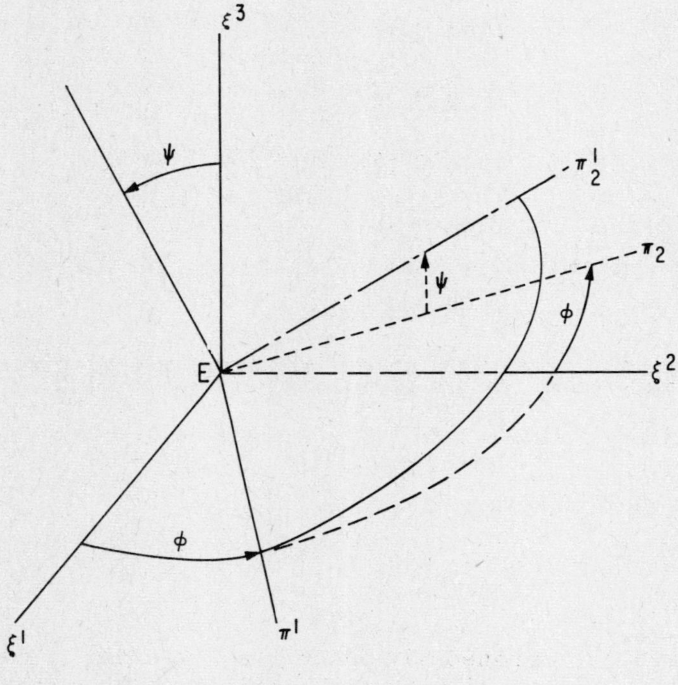

(For December 1961, the Moon's center moves roughly on this plane.)

Fig. 4 Location of Plane of the Idealized Moon's Motion

with r = 207561.40 N. M., $\dot{\omega}$ = 9.5823497 x 10^{-3} rad./hr., ϕ = 2.4434609 rad., ψ = 4.0944 x 10^{-2} rad, and ω_0 a constant value which is specified later on. Thence, the Moon moves in the plane E - π_1, π'_2, defined in the system E - ξ^1, ξ^2, ξ^3 as follows: π_1 and π_2 are the positions of ξ^1 and ξ^2 after a ϕ - rotation about ξ^3, and π'_2 is the position of π_2 after a ψ - rotation about π_1. (See Fig. 4)

3.2 DESCRIPTION AND SCOPE OF THE NUMERICAL EXAMPLES

The principal aim of the numerical work was to test, or to try Picard's iterative method. A systematic array of computations was prepared for the simplified model described in the preceding section.

The schedule for the set of runs, labeled Run I-1 is given below.

Runs I : T = 74 hrs.; $\overline{P_2M}$ = 1827 N. M.* at t = T

Run I-1 : $\overline{P_1E}$ = 10500 N. M. at t = 0.

Run I-2 : $\overline{P_1E}$ = 8000 N. M. " " " ".

Run I-3 : $\overline{P_1E}$ = 7000 N. M. " " " ".

Run I-4 : $\overline{P_1E}$ = 6000 N. M. " " " ".

Run I-5 : $\overline{P_1E}$ = 5000 N. M. " " " ".

Run I-6 : $\overline{P_1E}$ = 4000 N. M. " " " ".

Run I-7 : $\overline{P_1E}$ = 3700 N. M. " " " ".

These runs were started with an initial guess, $x_0^i(t)$, given by

$$(x_0^i - x_1^i)(t - T) - (x_0^i - x_2^i)\, t = \dot{0} \qquad (26)$$

This motion is that shown in Fig. 2

* An input card was erroneously punched at the onset of the numerical work and was kept uncorrected all through the computation. $\overline{P_2M}$ was intended to be such that the final distance to the moon's surface would be 1000 N. M.

By direct use of Picard's iterative process, solutions were obtained after 400 iterations for Runs I-1, I-2, but not for Run I-3; convergence in this case would have required something of the order of 5×10^5 iterations. Furthermore, for these two runs the convergence was achieved only to 4 or 5 digits. A change to the cyclic process described in Section 2. 2 was, therefore, indicated and the results presented in Tables I through VII were then obtained.

These trajectories represent a gradual approach to the Earth, carried as far as "practicable" (200 N. M.) from the surface of the actual Earth. It was intended to show, as it does, how the convergence process alters with the distance of P_1 to the "strong" singularity of the differential equations (3). The number of cycles required for convergency increases to 6 for $\overline{P_1E}$ = 5000 N. M. with m constant. For this value of $\overline{P_1E}$, the initial guess was changed from the arbitrary uniform motion of Eq. (26) to a more "educated" guess from the previous run; otherwise, the number of cycles required would have been considerably greater.

How the "odd" and "even" sequences begin to diverge from each other for $\overline{P_1E}$ = 6000 N. M. is shown in Fig. 5, for two values of t, t = 7. 4 hr. and t = 66. 6 hr.

Finally, a trajectory, Run II, Table VIII, from 200 N. M. from the surface of the earth to impact on the moon was computed. Again the numerical results corroborate the marked influence of the "weak" singularity, located at the Moon. Ten cycles were required to achieve a convergency of the same order as that of Tables I through VII. Since this run is obviously the most interesting, a graphical presentation of the results is given in Figs. 6 through 10.

From Runs I-1 through I-7 the two curves of Figs. 11a and 11b were obtained. For a systematic analysis of an interplanetary mission, these curves are most important; they define the requirements for launching at a certain altitude and give the necessary information to know what will be

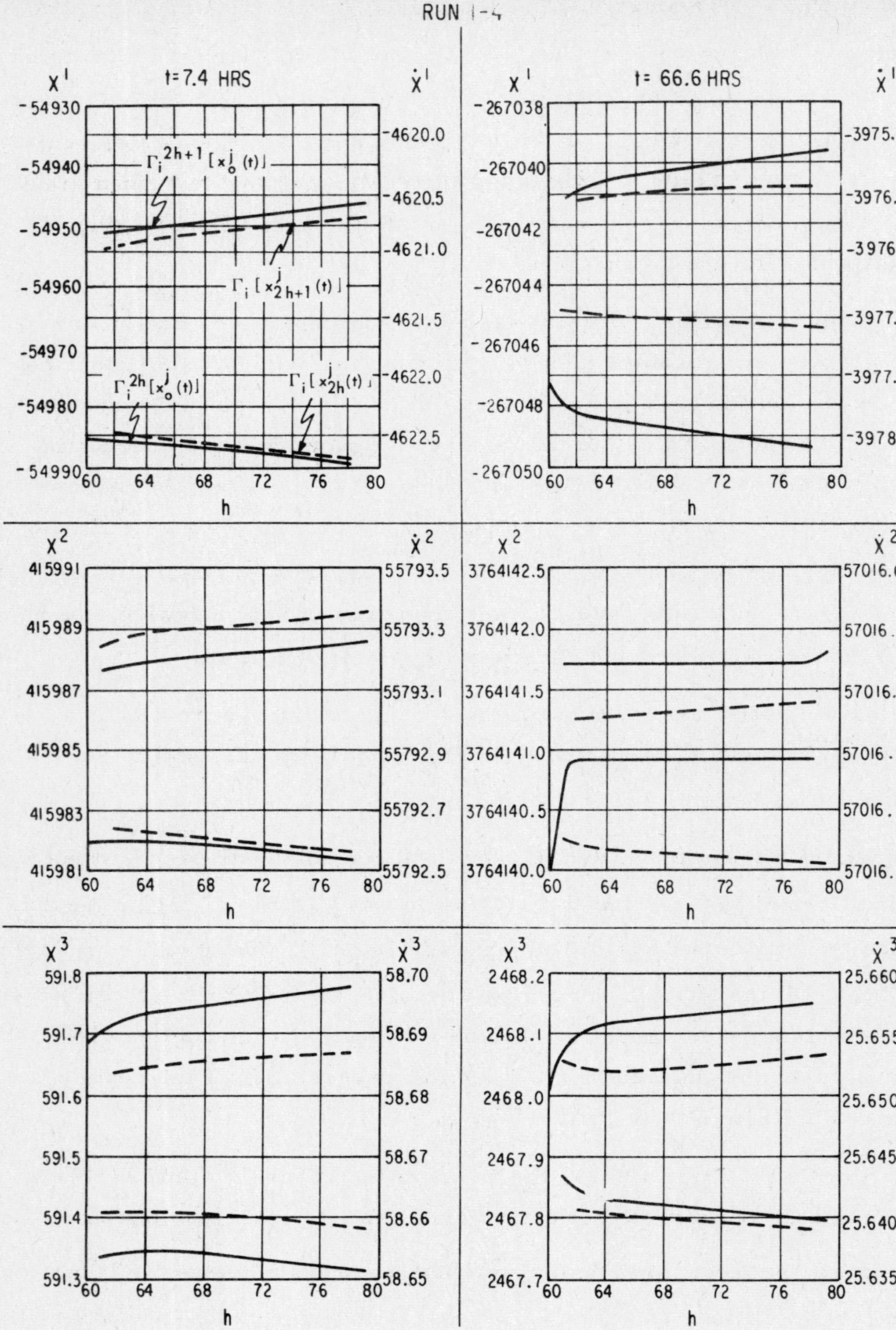

Fig. 5 SEQUENCES $\Gamma_i^{2h}[x_o^j(t)]$ and $\Gamma_i^{2h+1}[x_o^j(t)]$ and $\dot{\Gamma}_i[x_{2h}^j(t)]$ and $\dot{\Gamma}_i[x_{2h+1}^j(t)]$ vs h

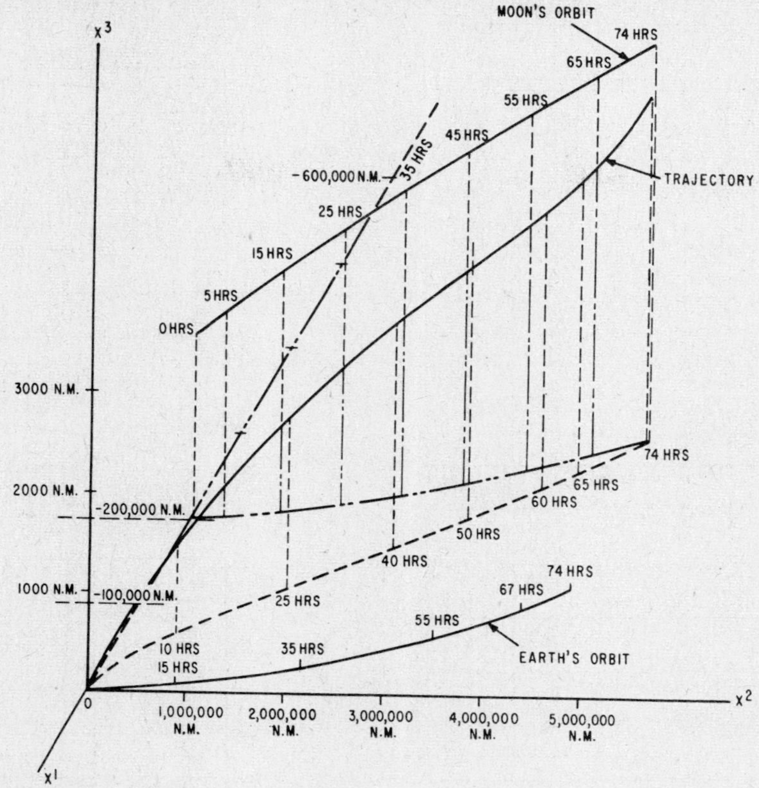

Fig. 6 Vehicle's, Earth's and Moon's Motions in the Galilean System of Reference

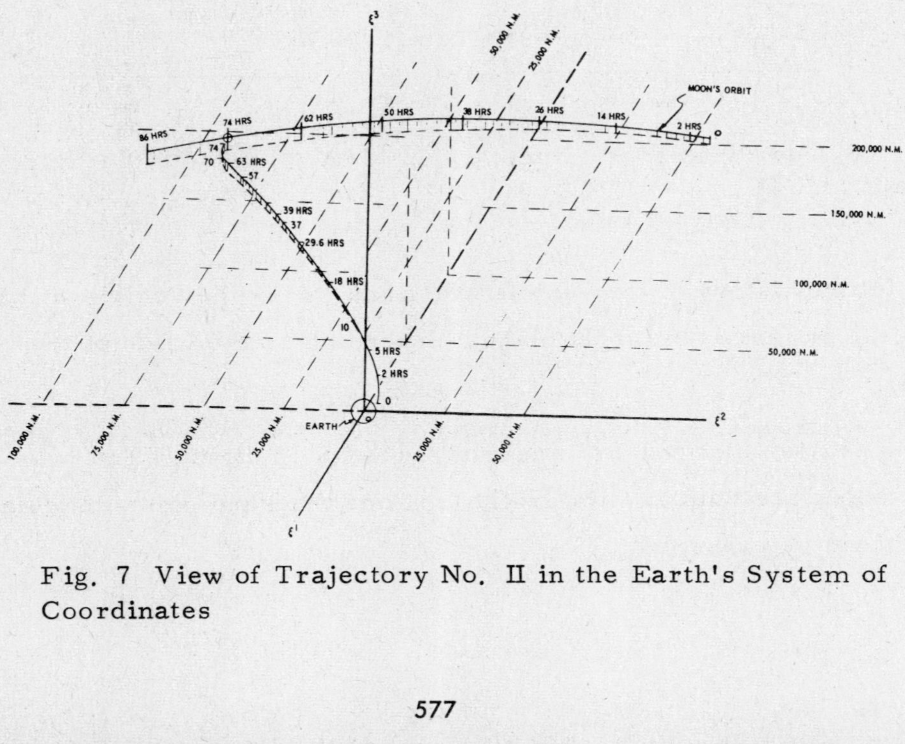

Fig. 7 View of Trajectory No. II in the Earth's System of Coordinates

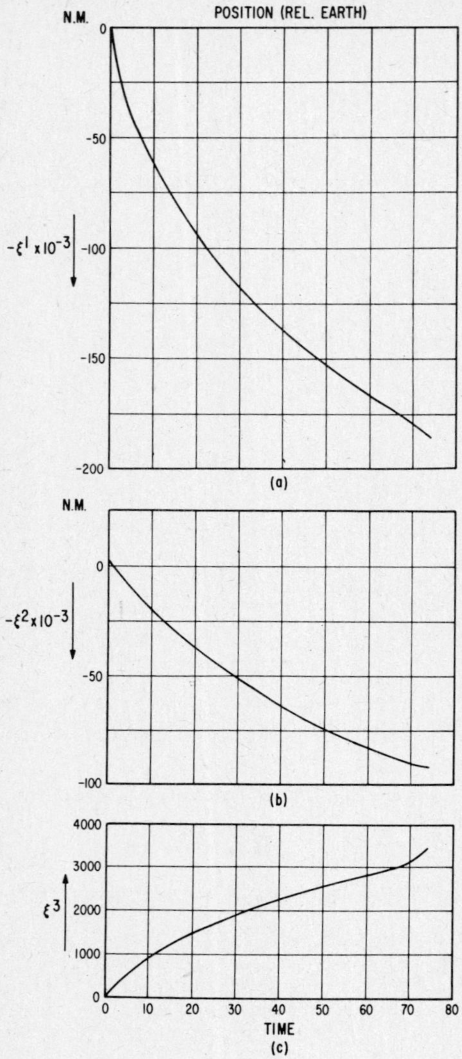

Fig. 8 Coordinates vs. Time, Trajectory II

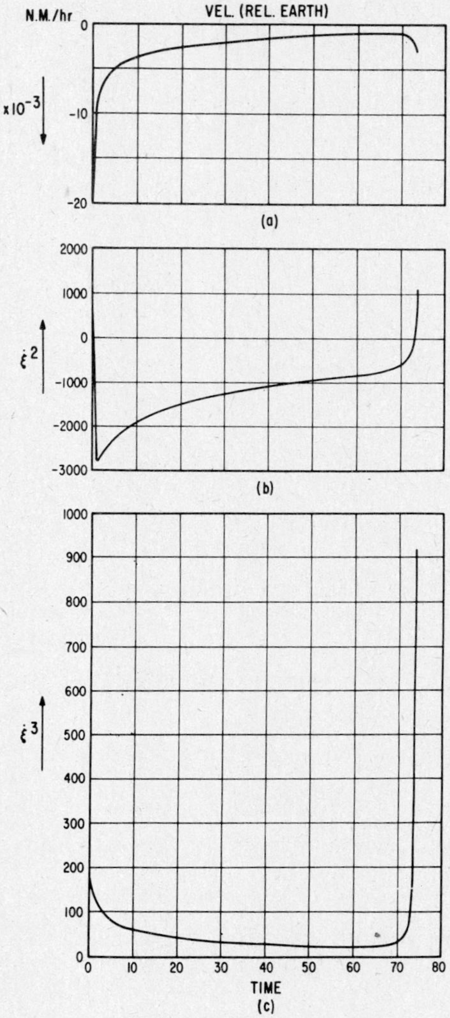

Fig. 9 Velocity Components vs Time, Trajectory II

the future history of the vehicle at P_2 and $t = T$, the vehicle will land on, circle, or leave the target planet, depending on the value of the velocity at P_2, $t = T$.

All these calculations were carried out with 12 digits. The values in the tables are typical. Naturally, the one thousand points calculated could not be presented.

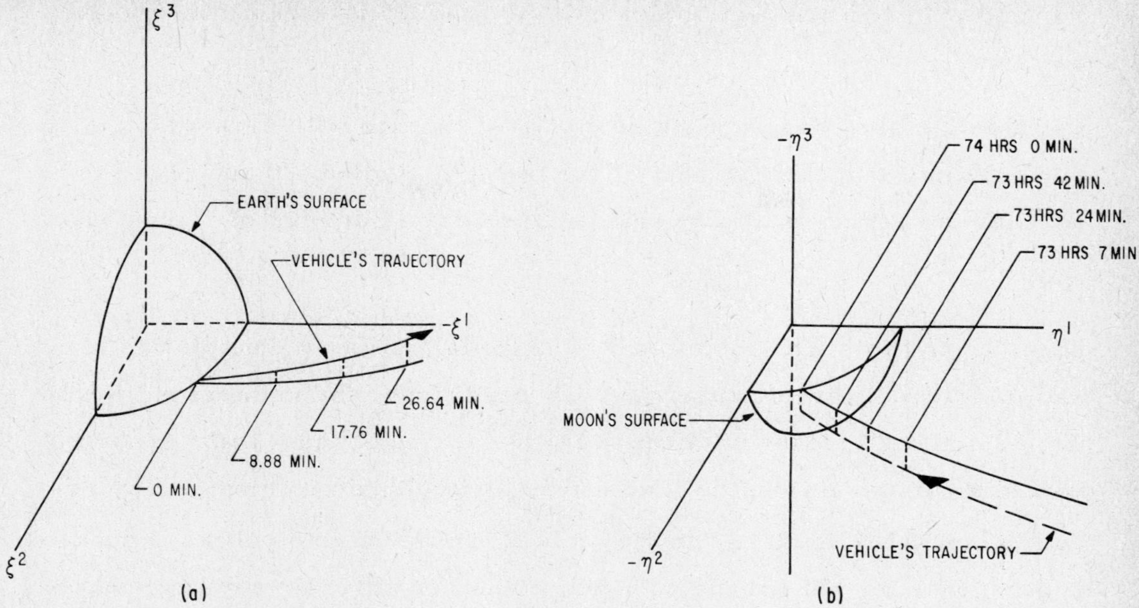

Fig. 10 Views of the Motion Near the Earth and Near the Moon

3.3 NUMERICAL ANALYSIS

The work of this section is an example of the numerical investigations that should be carried out to determine "the best" selection of n, m, and the initial guess. As noted previously, these inquiries are intimately connected with the formula used for the summations in Γ_i and $\dot{\Gamma}_i$, and to the number of digits carried in the computations. Therefore, the validity of these examples is limited to integrations by Simpson's rule, and to numerical operations with 12 digits.

The effect of the number of digits on the solution is shown in Table IX. The numbers printed are those to which the solution has converged after 80 iterations.

In Table X, the variation of the solution with n is clear. The printed numbers are those to which the solution has converged and it is evident that slightly different solutions result for different n's. Limitations in the memory capacity of the computer, for our special program, made it impossible to increase n beyond one thousand; therefore, it cannot be said that the "optimum" corresponds to the larger n tried. However, inspection of the numbers indicates that n = 1000 is very close to that optimum.

In Table XI, the change in the rate of convergence with m is shown, m = 6 being definitely the best of the two m's studied (20 and 6). An adequate selection of m may save a considerable amount of computational time.

Finally, in Table XII, the effect of the initial guess on the solution is shown. Two different initial guesses were tried, the uniform motion of Eq. (26) and a "guess" obtained from the previous run - that next and at a greater distance from the Earth - by a simple "stretching." The results show that the change in the "initial" guess affects only the rate of convergence to, but not the solution. Whether different solutions would would result for widely different "initial" guesses, cannot be ascertained.

Of course, all throughout these numerical investigations only one factor was changed at a time. The selection of initial conditions for testing the effect on the convergence process of varying m, n, the initial guess, and the number of digits was determined by purely logistic reasoning.

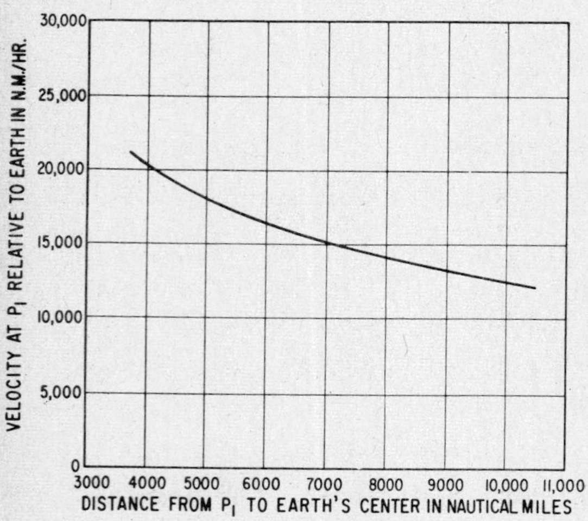

Fig. 11a - Initial Velocity vs Distance from Center of the Earth ($P_2^- M$ constant, T=74 hrs.)

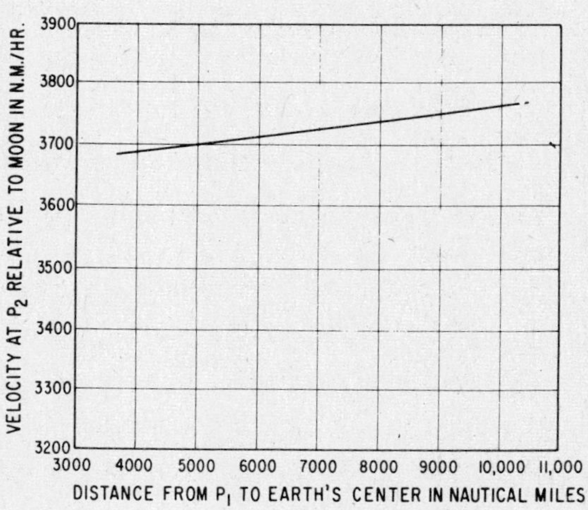

Fig. 11b - Final Velocity vs Initial Distance from Earth (P_2M constant, T=74 hrs.)

RUN I-1

TABLE I

P_1: (ξ = -8043, ξ^2 = 6749, ξ^3 = 0) at t = 0; P_2: (η^1 = 1037.82, η^2 = -1144.75, η^3 = -974.68) at t = T

$\overline{P_1E}$ = 10500, at t = 0; $\overline{P_2M}$ = 1827 at t = T. Lengths in N.M.

T = 74 hours n = 1001 m = 20 k = 5

x_o^i from $(x_o^i - x_1^i)(t - T) - (x_o^i - x_2^i)\, t = 0$. Input $\equiv x^i_{o,k}(t)$, Output $\equiv \Gamma_i\,[x^j_{o,k}(t)$.

t(hrs)		X^1(N.M.)	X^2(N.M.)	X^3(N.M.)	$\dot{X}^1$(N.M./hr)	$\dot{X}^2$(N.M./hr)	$\dot{X}^3$(N.M./hr)
0	INPUT	-8043.0000	6749.0000	0	-12180.889	58163.330	101.17823
	OUTPUT	-8043.0000	6749.0000	0	-12180.889	58163.330	101.17823
7.4	INPUT	-56352.105	421047.92	545.35016	-4632.8414	55858.609	57.463935
	OUTPUT	-56352.105	421047.92	545.35016	-4632.8414	55858.609	57.463950
14.8	INPUT	-86592.525	835300.81	917.06350	-3715.7929	56093.444	44.583728
	OUTPUT	-86592.525	835300.81	917.06351	-3715.7929	56093.444	44.583738
22.2	INPUT	-112728.36	1251090.1	1218.4498	-3400.4775	56273.905	37.404892
	OUTPUT	-112728.36	1251090.1	1218.4498	-3400.4775	56273.905	37.404897
29.6	INPUT	-137440.22	1668058.2	1476.1166	-3302.5710	56415.022	32.497369
	OUTPUT	-137440.22	1668058.2	1476.1166	-3302.5710	56415.022	32.497374
37.0	INPUT	-161874.11	2085970.7	1702.4135	-3314.9766	56530.916	28.825253
	OUTPUT	-161874.11	2085970.7	1702.4135	-3314.9766	56530.916	28.825263
44.4	INPUT	-186666.78	2504675.3	1904.7699	-3394.9517	56630.339	25.989683
	OUTPUT	-186666.78	2504675.3	1904.7699	-3394.9517	56630.339	25.989693
51.8	INPUT	-212238.78	2924075.3	2088.9079	-3523.7883	56719.896	23.916630
	OUTPUT	-212238.78	2924075.3	2088.9079	-3523.7883	56719.896	23.916640
59.2	INPUT	-238929.85	3344124.7	2261.5802	-3698.4268	56807.714	23.053020
	OUTPUT	-238929.85	3344124.7	2261.5802	-3698.4268	56807.714	23.053030
66.6	INPUT	-267154.74	3764877.4	2440.2742	-3950.7398	56917.831	26.947305
	OUTPUT	-267154.74	3764877.4	2440.2742	-3950.7398	56917.831	26.947315
74.0	INPUT	-298833.25	4187403.4	2999.9999	-5422.1597	57968.734	571.80040
	OUTPUT	-298833.25	4187403.4	2999.9999	-5422.1597	57968.734	571.79978

RUN I-2

TABLE II

P_1 : (ξ^1 = -6128, ξ^2 = 5142, ξ^3 = 0) at t = 0; P_2 : (η^1 = 1037.82, η^2 = - 1144.75, η^3 = -974.68) at t = T

$\overline{P_1E}$ = 8000, at t = 0; $\overline{P_2M}$ = 1827 at t = T. Lengths in N.M.

T = 74 hours n = 1001 m = 20 k = 5

x_o^i from $(x_o^i - x_1^i)(t - T) - (x_o^i - x_2^i)\, t = 0$ Input $\equiv x^i_{o,k}(t)$, Output $\equiv \Gamma_i\,[x^i_{o,k}(t)$]

t(hrs)		X^1(N.M.)	X^2(N.M.)	X^3(N.M.)	$\dot{X}^1$(N.M./hr)	$\dot{X}^2$(N.M./hr)	$\dot{X}^3$(N.M./hr)
0	INPUT	-6128.0000	5142.0000	0.	-14025.023	58416.614	115.05342
	OUTPUT	-6128.0000	5142.0000	0.	-14025.023	58416.614	115.05342
7.4	INPUT	-55605.452	418383.03	569.45450	-4631.1272	55823.074	58.150387
	OUTPUT	-55605.452	418383.03	569.45450	-4631.1271	55823.074	58.150382
14.8	INPUT	-85840.694	832606.95	943.59518	-3718.0885	56109.973	44.694204
	OUTPUT	-85840.694	832606.95	943.59519	-3718.0885	56109.973	44.694204
22.2	INPUT	-112017.13	1248593.0	1245.0647	-3408.8099	56308.069	37.342397
	OUTPUT	-112017.13	1248593.0	1245.0647	-3408.8099	56308.069	37.342397
29.6	INPUT	-136806.67	1665847.8	1501.9351	-3314.9355	56457.420	32.352519
	OUTPUT	-136806.67	1665847.8	1501.9351	-3314.9355	56457.420	32.352509
37.0	INPUT	-161342.21	2084091.8	1726.9583	-3329.8776	56577.716	28.628534
	OUTPUT	-161342.21	2084091.8	1726.9583	-3329.8776	56577.716	28.628534
44.4	INPUT	-186251.08	2503152.9	1927.6972	-3411.2943	56679.635	25.749204
	OUTPUT	-186251.08	2503152.9	1927.6972	-3411.2943	56679.635	25.749204
51.8	INPUT	-211946.46	2922923.4	2109.8748	-3540.6340	56770.620	23.623911
	OUTPUT	-211946.46	2922923.4	2109.8747	-3540.6341	56770.620	23.623911
59.2	INPUT	-238760.83	3343351.5	2280.0763	-3714.6837	56859.239	22.665166
	OUTPUT	-238760.83	3343351.5	2280.0763	-3714.6837	56859.239	22.665156
66.6	INPUT	-267097.94	3764487.4	2455.0451	-3964.2133	56969.870	26.254295
	OUTPUT	-267097.94	3764487.4	2455.0451	-3964.2133	56969.870	26.254295
74.0	INPUT	-298833.25	4187403.4	3000.0000	-5416.0586	58023.139	564.26813
	OUTPUT	-298833.25	4187403.4	3000.0000	-5416.0586	58023.139	564.26813

RUN I-3

TABLE III

P_1: (ξ^1 = -5362, ξ^2 = 4499.25, ξ^3 =0) at t = 0; P : (η^1 = 1037.82, η^2 = -1144.75, η^3 = -974.68) at t = T

$\overline{P_1 E}$ = 7000, at t = 0; $\overline{P_2 M}$ = 1827 at t = T. Lengths in N.M.

T = 74 hours n = 1001 m = 20 k = 5

x_0^i from $(x_0^i - x_1^i)(t - T) - (x_0^i - x_2^i)\, t = 0$. INPUT ≡ $x^i_{0.k}(t)$, OUTPUT ≡ $\Gamma_i\,[x^j_{0,k}(t)]$

t(hrs)		x^1(N.M.)	x^2(N.M.)	x^3(N.M.)	$\dot{x}^1$(N.M./hr)	$\dot{x}^2$(N.M./hr)	$\dot{x}^3$(N.M./hr)
0	INPUT	-5362.0000	4499.2500	0	-15027.142	58552.191	122.58228
	OUTPUT	-5362.0000	4499.2500	0	-15027.142	58552.191	122.58228
7.407	INPUT	-55328.336	417634.10	580.55973	-4625.7715	55808.623	58.393653
	OUTPUT	-55328.336	417634.10	580.55973	-4625.7715	55808.623	58.393653
14.815	INPUT	-85571.120	832266.95	955.83139	-3717.4297	56117.977	44.710857
	OUTPUT	-85571.120	832266.95	955.83139	-3717.4297	56117.977	44.710857
22.222	INPUT	-111783.06	1248760.5	1257.4364	-3411.7138	56323.587	37.290831
	OUTPUT	-111783.06	1248760.5	1257.4364	-3411.7138	56323.587	37.290831
29.630	INPUT	-136628.05	1666562.3	1514.0494	-3320.1429	56476.447	32.268353
	OUTPUT	-136628.05	1666562.3	1514.0494	-3320.1429	56476.446	32.268353
37.037	INPUT	-161232.78	2085373.6	1738.5932	-3336.6209	56598.627	28.523632
	OUTPUT	-161232.78	2085373.6	1738.5932	-3336.6209	56598.627	28.523632
44.444	INPUT	-186220.60	2505013.4	1938.6925	-3419.0649	56701.626	25.627369
	OUTPUT	-186220.60	2505013.4	1938.6925	-3419.0648	56701.626	25.627369
51.852	INPUT	-212001.86	2925369.6	2120.0834	-3549.0464	56793.257	23.483411
	OUTPUT	-212001.86	2925369.6	2120.0834	-3549.0464	56793.257	23.483411
59.259	INPUT	-238906.71	3346387.9	2289.3217	-3723.4007	56882.325	22.497339
	OUTPUT	-238906.71	3346387.9	2289.3217	-3723.4007	56882.325	22.497338
66.667	INPUT	-267337.20	3768118.6	2463.0866	-3973.1046	56993.743	26.056053
	OUTPUT	-267337.20	3768118.6	2463.0866	-3973.1045	56993.742	26.056052
74.000	INPUT	-298833.25	4187403.4	2999.9999	-5413.5971	58046.483	561.04537
	OUTPUT	-298833.25	4187403.4	2999.9999	-5413.5968	58046.483	561.04527

RUN I-4

TABLE IV

P_1: (ξ^1 = -4596, ξ^2 = 3856.5, ξ^3= 0) at t = 0; P : (η^1 = 1037.82, η^2 = -1144.75, η^3= -974.68) at t = T

$\overline{P_1 E}$ = 6000, at t = 0; $\overline{P_2 M}$ = 1827 at t = T. Lengths in N.M.

T = 74 hours n = 1001 m = 20 k = 5

x_0^i from $(x_0^i - x_1^i)(t - T) - (x_0^i - x_2^i)\, t = 0$. INPUT ≡ $x^i_{0,k}(t)$, OUTPUT ≡ $\Gamma_i\,[x^j_{0,k}(t)]$

t(hrs)		x^1(N.M.)	x^2(N.M.)	x^3(N.M.)	$\dot{x}^1$(N.M./hr)	$\dot{x}^2$(N.M./hr)	$\dot{x}^3$(N.M./hr)
0.0	INPUT	-4596.0000	3856.5000	0	-16270.940	58721.743	131.90968
	OUTPUT	-4596.0000	3856.5000	0	-16271.012	58721.750	131.91022
7.4	INPUT	-54968.276	415984.97	591.54361	-4621.6904	55792.961	58.672862
	OUTPUT	-54968.278	415984.97	591.54364	-4621.6991	55792.957	58.672989
14.8	INPUT	-85170.069	830192.68	967.44899	-3718.4104	56125.668	44.760118
	OUTPUT	-85170.069	830192.68	967.44903	-3718.4117	56125.666	44.760159
22.2	INPUT	-111375.65	1246359.6	1268.8179	-3415.7904	56339.059	37.269355
	OUTPUT	-111375.65	1246359.5	1268.8179	-3415.7889	56339.059	37.269361
29.6	INPUT	-136232.86	1663873.0	1524.8751	-3325.9130	56495.551	32.212148
	OUTPUT	-136232.86	1663873.0	1524.8752	-3325.9101	56495.551	32.212136
37.0	INPUT	-160859.33	2082414.7	1748.6905	-3343.2665	56619.671	28.444716
	OUTPUT	-160859.33	2082414.7	1748.6905	-3343.2627	56619.672	28.444693
44.4	INPUT	-185872.83	2501795.0	1947.9322	-3426.0311	56723.759	25.528422
	OUTPUT	-185872.82	2501795.0	1947.9322	-3426.0268	56723.760	25.528392
51.8	INPUT	-211679.57	2921897.0	2128.3212	-3555.8533	56815.976	23.358602
	OUTPUT	-211679.57	2921897.0	2128.3212	-3555.8486	56815.977	23.358568
59.2	INPUT	-238605.46	3342663.4	2296.2998	-3729.4117	56905.258	22.318764
	OUTPUT	-238605.46	3342663.4	2296.2998	-3729.4068	56905.259	22.318723
66.6	INPUT	-267044.55	3764141.3	2467.9732	-3976.5213	57016.240	25.646098
	OUTPUT	-267044.55	3764141.3	2467.9733	-3976.5161	57016.241	25.646036
74.0	INPUT	-298833.26	4187403.4	2999.9999	-5411.1277	58071.026	557.63267
	OUTPUT	-298833.26	4187403.4	2999.9999	-5411.1205	58071.024	557.63162

RUN I-5

TABLE V

P_1: ($\xi^1 = -3830$, $\xi^2 = 3213.75$, $\xi^3 = 0$; P_2: ($\eta^1 = 1037.82$, $\eta^2 = -1144.75$, $\eta^3 = -974.68$) at t = T

$\overline{P_1E} = 5000$, at t = 0; $\overline{P_2M} = 1827$ at t = T. Lengths in N.M.

T = 74 hours   n = 1001   m = 20   k = 6

$x_0^i(t)$ obtained by "stretching" solution I-4. INPUT $\equiv x^i_{0,k}(t)$, OUTPUT $\equiv \Gamma_i\,[x^j_{0,k}(t)]$

t(hrs)		x^1(N.M.)	x^2(N.M.)	x^3(N.M.)	$\dot{x}^1$(N.M./hr)	$\dot{x}^2$(N.M./hr)	$\dot{x}^3$(N.M./hr)
0	INPUT	-3830.0000	3213.7500	0	-17878.960	58948.287	143.87465
	OUTPUT	-3830.0000	3213.7500	0	-17878.977	58948.289	143.87478
7.407	INPUT	-54654.098	415068.97	604.30816	-4612.1998	55777.633	58.905854
	OUTPUT	-54654.100	415068.98	604.30819	-4612.2019	55777.632	58.905879
14.815	INPUT	-84849.252	829690.24	891.27299	-3717.1650	56135.209	44.766551
	OUTPUT	-84849.254	829690.24	891.27302	-3717.1652	56135.209	44.766579
22.222	INPUT	-111090.30	1246379.3	1282.6960	-3419.1074	56356.868	37.205978
	OUTPUT	-111090.30	1246379.3	1282.6961	-3419.1071	56356.868	37.205979
29.630	INPUT	-136007.87	1664458.5	1538.4011	-3331.9612	56517.227	32.114394
	OUTPUT	-136007.87	1664458.5	1538.4012	-3331.9606	56517.228	32.114391
37.037	INPUT	-160710.72	2083588.2	1761.6298	-3351.0650	56643.431	28.324726
	OUTPUT	-160710.73	2083588.2	1761.6298	-3351.0628	56643.431	28.324721
44.444	INPUT	-185811.53	2503569.1	1960.1128	-3434.9617	56748.711	25.389861
	OUTPUT	-185811.53	2503569.1	1960.1129	-3434.9608	56748.711	25.389855
51.852	INPUT	-211713.03	2924279.3	2139.5826	-3565.4633	56841.632	23.199300
	OUTPUT	-211713.03	2924279.3	2139.5827	-3565.4624	56841.633	23.199293
59.259	INPUT	-238738.33	3345658.7	2306.4433	-3739.2981	56931.380	22.127966
	OUTPUT	-238738.34	3345658.7	2306.4434	-3739.2971	56931.380	22.127957
66.667	INPUT	-267279.11	3767754.2	2476.6962	-3986.4200	57043.115	25.409632
	OUTPUT	-267279.11	3767754.2	2476.6963	-3986.4190	57043.115	25.408619
74.000	INPUT	-298833.26	4187403.4	2999.9998	-5408.7065	58097.143	554.01546
	OUTPUT	-298833.26	4187403.4	2999.9998	-5408.7049	58097.143	554.01521

RUN I-6

TABLE VI

P_1: ($\xi^1 = -3064$, $\xi^2 = 2571.2$, $\xi^3 = 0$) at t = 0, P_2: ($\eta^1 = 1037.82$, $\eta^2 = -1144.75$, $\eta^3 = -974.68$) at t = T

$\overline{P_1E} = 4000$, at t = 0; $\overline{P_2M} = 1827$ at t = T. Lengths in N.M.

T = 74 hours   n = 1001   m = 20   k = 6

$x_0^i(t)$ obtained by "stretching" solution I-4. INPUT $\equiv x^i_{0,k}(t)$, OUTPUT $\equiv \Gamma_i\,[x^j_{0,k}(t)]$

t(hrs)		x^1(N.M.)	x^2(N.M.)	x^3(N.M.)	$\dot{x}^1$(N.M./hr)	$\dot{x}^2$(N.M./hr)	$\dot{x}^3$(N.M./hr)
0	INPUT	-3064.0000	2571.2000	0	-20083.544	59286.342	159.97450
	OUTPUT	-3064.0000	2571.2000	0	-20087.140	59286.690	160.00239
7.407	INPUT	-54269.601	413615.14	617.80850	-4601.8372	55761.124	59.158968
	OUTPUT	-54270.141	413615.03	617.81460	-4602.1304	55760.993	59.163343
14.815	INPUT	-84429.298	828234.43	995.58461	-3716.6632	56145.298	44.787818
	OUTPUT	-84429.913	828234.28	995.59208	-3716.6973	56145.261	44.789030
22.222	INPUT	-110686.14	1245035.6	1296.8556	-3423.3642	56375.804	37.153809
	OUTPUT	-110686.74	1245035.4	1296.8634	-3423.3001	56375.810	37.153767
29.630	INPUT	-135645.84	1663272.2	1552.0254	-3338.8378	56540.314	32.025281
	OUTPUT	-135646.39	1663272.0	1552.0329	-3338.7239	56540.344	32.024592
37.037	INPUT	-160405.77	2082582.0	1774.4992	-3359.4594	56668.752	28.211181
	OUTPUT	-160406.25	2082581.8	1774.5062	-3359.3167	56668.796	28.210118
44.444	INPUT	-185572.20	2502755.6	1972.0629	-3444.1899	56775.300	25.255168
	OUTPUT	-185572.60	2502755.5	1972.0693	-3444.0296	56775.353	25.253871
51.852	INPUT	-211543.49	2923665.6	2150.4462	-3574.9909	56868.934	23.039082
	OUTPUT	-211543.80	2923665.5	2150.4519	-3574.8191	56868.991	23.037614
59.259	INPUT	-238638.74	3345248.8	2315.9709	-3748.5428	56959.044	21.921144
	OUTPUT	-238638.96	3345248.8	2315.9757	-3748.3624	56959.101	21.919473
66.667	INPUT	-267243.87	3767549.8	2484.2668	-3994.2339	57070.907	25.049298
	OUTPUT	-267243.98	3767549.7	2484.2705	-3994.0438	57070.957	25.046954
74.000	INPUT	-298833.28	4187403.3	2999.9997	-5406.3556	58125.353	550.07819
	OUTPUT	-298833.28	4187403.3	2999.9997	-5406.1118	58125.322	550.04768

RUN I-7

TABLE VII

P_1: (ξ^1 =-2840, ξ^2 = 2380, ξ^3 = 0) at t = 0; P_2; (η^1 = 1037.82, η^2 = -1144.75, η^3 = -974.68) at t = T

$\overline{P_1E}$ = 3700, at t = 0; $\overline{P_2M}$ = 1827 at t = T. Lengths in N.M.

T = 74 hours n = 1001 m = 6 k = 9

$x_0^i(t)$ obtained by "stretching" solution I-6. INPUT = $x_{0,k}^i(t)$, OUTPUT = $\Gamma_i\,[\,x_{0,k}^j(t)\,]$

t(hrs)		X^1(N.M.)	X^2(N.M.)	X^3(N.M.)	$\dot{X}^1$(N.M./hr)	$\dot{X}^2$(N.M./hr)	$\dot{X}^3$(N.M./hr)
0	INPUT	-2840.0000	2380.0000	0	-20910.594	59418.650	165.84133
	OUTPUT	-2840.0000	2380.0000	0	-20910.594	59418.650	165.84133
7.407	INPUT	-54146.154	413149.95	622.02038	-4598.1631	55756.013	59.232990
	OUTPUT	-54146.154	413149.95	622.02037	-4598.1631	55756.013	59.232990
14.815	INPUT	-84293.420	827769.31	1000.0331	-3716.4633	56148.571	44.793782
	OUTPUT	-84293.420	827769.31	1000.0331	-3716.4633	56148.571	44.793782
22.222	INPUT	-110555.27	1244606.5	1301.2564	-3424.7403	56381.872	37.137915
	OUTPUT	-110555.27	1244606.5	1301.2564	-3424.7403	56381.872	37.137915
29.630	INPUT	-135528.60	1662893.4	1556.2635	-3341.0647	56547.696	31.998239
	OUTPUT	-135528.60	1662893.4	1556.2635	-3341.0647	56547.696	31.998239
37.037	INPUT	-160307.01	2082260.8	1778.5081	-3362.1748	56676.843	28.176674
	OUTPUT	-160307.01	2082260.8	1778.5081	-3362.1748	56676.843	28.176674
44.444	INPUT	-185494.66	2502495.9	1975.7921	-3447.1718	56783.794	25.214130
	OUTPUT	-185494.66	2502495.9	1975.7921	-3447.1718	56783.794	25.214130
51.852	INPUT	-211488.49	2923469.8	2153.8439	-3578.0685	56877.654	22.990098
	OUTPUT	-211488.49	2923469.8	2153.8439	-3578.0685	56877.654	22.990098
59.259	INPUT	-238606.35	3345118.1	2318.9591	-3751.5310	56967.877	21.857579
	OUTPUT	-238606.35	3345118.1	2318.9591	-3751.5310	56967.877	21.857579
66.667	INPUT	-267232.30	3767484.6	2486.6506	-3996.7689	57079.773	24.937781
	OUTPUT	-267232.30	3767484.6	2486.6506	-3996.7689	57079.773	24.937781
74.000	INPUT	-298833.29	4187403.3	2999.9995	-5405.6538	58134.306	548.82671
	OUTPUT	-298833.29	4187403.3	2999.9995	-5405.6538	58134.306	548.82671

RUN II

TABLE VIII

P_1: (ξ^1 = -2840, ξ^2 = 2380, ξ^3 =0) at t = 0; P_2: (η^1 = 531.82596; η^2 = -589.74854, η^3 = -499.67821) at t = T

$\overline{P_1E}$ = 3700, at t = 0; $\overline{P_2M}$ = 940 at t = T. Lengths in N.M.

T = 74 hours n = 1001 m = 6 k = 10

x_0^i from $(x_0^i - x^i)\,(t - T) - (x_0^i - x_2^i)\,t = 0$, INPUT = $x_{0,k}^i(t)$, OUTPUT = $\Gamma_i\,[\,x_{0,k}^i(t)\,]$

t(hrs)		X^1(N.M.)	X^2(N.M.)	X^3(N.M.)	$\dot{X}^1$(N.M./hr)	$\dot{X}^2$(N.M./hr)	$\dot{X}^3$(N.M./hr)
0	INPUT	-2840.0000	2380.0000	0	-20908.927	59439.962	200.69884
	OUTPUT	-2840.0000	2380.0000	0	-20908.927	59439.962	200.69884
7.4	INPUT	-54138.279	412824.95	752.45050	-4603.1553	55763.857	71.736703
	OUTPUT	-54138.279	412824.95	752.45050	-4603.1553	55763.857	71.736703
14.8	INPUT	-84286.965	827079.05	1209.7826	-3720.1113	56154.359	54.242373
	OUTPUT	-84286.965	827079.05	1209.7826	-3720.1113	56154.359	54.242373
22.2	INPUT	-110545.94	1243538.0	1574.1153	-3427.4541	56386.604	44.952437
	OUTPUT	-110545.94	1243538.0	1574.1153	-3427.4541	56386.604	44.952437
29.6	INPUT	-135511.51	1661439.9	1882.3509	-3343.0244	56551.733	38.695650
	OUTPUT	-135511.51	1661439.9	1882.3509	-3343.0244	56551.733	38.695650
37.0	INPUT	-160277.20	2080414.8	2150.6238	-3363.4881	56680.363	34.009223
	OUTPUT	-160277.20	2080414.8	2150.6238	-3363.4881	56680.363	34.009223
44.4	INPUT	-185447.25	2500254.2	2388.0954	-3447.9241	56786.902	30.309783
	OUTPUT	-185447.25	2500254.2	2388.0954	-3447.9241	56786.902	30.309783
51.8	INPUT	-211418.88	2920828.8	2601.0686	-3578.3652	56880.414	27.374520
	OUTPUT	-211418.88	2920828.8	2601.0686	-3578.3652	56880.414	27.374520
59.2	INPUT	-238510.78	3342074.7	2795.4028	-3751.6145	56970.317	25.349829
	OUTPUT	-238510.78	3342074.7	2795.4028	-3751.6145	56970.317	25.349829
66.6	INPUT	-267110.87	3764035.7	2982.4325	-3997.9376	57081.885	26.203909
	OUTPUT	-267110.87	3764035.7	2982.4325	-3997.9376	57081.885	26.203909
74.0	INPUT	-299339.28	4187958.3	3474.9999	-6308.4840	58861.599	911.60593
	OUTPUT	-299339.28	4187958.3	3473.9999	-6308.4840	58861.599	911.60593

TABLE IX

P_1: (ξ^1 = -5362, ξ^2 = 4499.25 ξ^3 = 0) at t = 0; P_2: (η^1 = 1037.82, η^2 = -1144.75, η^3 = -974.68) at t = T

$\overline{P_1E}$ = 7000 at t = 0; $\overline{P_2M}$ = 1827 at t = T. Lengths in N.M.

T = 74 hours n = 1001 m = 20 k = 5 Number of Digits carried = 8, 12

x_0^i from $(x_0^i - x_1^i)$ (t - T) - $(x_0^i - x_2^i)$ t = 0

Time	No. of Digits	X^1(N.M.)	X^2(N.M.)	X^3(N.M.)	$\dot{X}^1$(N.M./hr)	$\dot{X}^2$(N.M./hr)	$\dot{X}^3$(N.M./hr)
0	8	-5362.0000	4499.2500	0	-15027.	58551.	122.65
	12	-5362.0000	4499.2500	0	-15027.	58552.	122.58
7.407	8	-55328.	417631.	580.9	-4625.6	55808.3	58.42
	12	-55328.	417634.	580.55	-4625.7	55808.62	58.393
14.815	8	-85570.	832262.	956.4	-3717.3	56117.78	44.737
	12	-85570.	832266.	955.83	-3717.42	56117.97	44.7108
22.222	8	-11178*.	124854.	1258.1	-3411.6	56323.42	37.313
	12	-11178*.	1248760.	1257.43	-3411.71	56323.587	37.2908
29.630	8	-13662*.	1666554.	1514.9	-3320.1	56476.30	32.288
	12	-136628.	1666562.	1514.04	-3320.1	56476.44	32.2683
37.037	8	-16123*.	2085364.	1739.6	-3336.6	56598.49	28.541
	12	-161232.	2085373.	1738.59	-3336.6	56598.62	28.5236
44.444	8	-186219.	2505001.	1939.8	-3419.0	56701.50	25.642
	12	-186220.	2505013.	1938.69	-3419.0	56701.62	25.627
51.852	8	-212001.	2925355.	2121.3	-3549.0	56793.13	23.496
	12	-212001.	2925369.6	2120.08	-3549.0	56793.25	23.483
59.259	8	-238906.	3346371.	2290.6	-3723.4	56882.21	22.507
	12	-238906.	3346387.9	2289.32	-3723.4	56882.32	22.497
66.667	8	-267337.	3768099.	2464.5	-3973	56993.65	26.056
	12	-267337.	3768118.6	2463.08	-3973.1	56993.74	26.056
74.000	8	-298833.25	4187403.4	2999.9999	-5407.	58045.6	556.
	12	-298833.25	4187403.4	2999.9999	-5413.6	58046.48	561.04

TABLE X

P_1: (ξ^1 = -4596, ξ^2 = 3856.5, ξ^3 = 0) at t = 0; P_2: (η^1 = 1037.82, η^2 = -1144.75, η^3 = -974.68) at t = T

$\overline{P_1E}$ = 6000, at t = 0; $\overline{P_2M}$ = 1827 at t = T. Lengths in N.M.

T = 74 hours n = 1001, 751, 501, 251 m = 20 k = 5

x_0^i from $(x_0^i - x_1^i)$ (t - T) - $(x_0^i - x_2^i)$ t = 0

Time	n	X^1(N.M.)	X^2(N.M.)	X^3(N.M.)	$\dot{X}^1$(N.M./hr)	$\dot{X}^2$(N.M./hr)	$\dot{X}^3$(N.M./hr)
	1001	-4596.0000	3856.5000	0	-16271.	58721.7	131.91
0	751	-4596.0000	3856.5000	0	-16294.	58751.1	131.84
	501	-4596.0000	3856.5000	0	-16360.	58833.3	131.66
	251	-4596.0000	3856.5000	0	-16693.	59241.0	130.82
	1001	-54968.2	415984.9	591.543	-4621.6	55792.9	58.672
7.4	751	-54965.2	415974.4	591.787	-4621.6	55792.8	58.690
	501	-54956.6	415945.2	592.459	-4621.6	55792.4	58.737
	251	-54913.2	415808.9	595.641	-4621.6	55790.7	58.968
	1001	-85170.	830192.6	967.449	-3718.41	56125.66	44.7601
14.81	751	-85166.	830182.1	967.793	-3718.42	56125.74	44.7716
	501	-85158.	830153.0	968.746	-3718.47	56125.95	44.8035
	251	-85113.	830016.4	973.285	-3718.78	56126.93	44.9581
	1001	-160869.	2082414.	1748.69	-3343.26	56619.67	28.4446
37.0	751	-160857.	2082407.	1749.21	-3343.33	56619.86	28.4508
	501	-160851.	2082387.	1750.68	-3343.52	56620.39	28.4679
	251	-160821.	2082295.	1757.69	-3344.45	56622.89	28.5505
	1001	-238905.4	3342663.4	2296.29	-3729.40	56905.25	22.3187
59.2	751	-238604.9	3342660.8	2296.92	-3729.48	56905.46	22.3114
	501	-238603.4	3342653.7	2298.67	-3729.68	56906.04	22.3289
	251	-238596.3	3342620.4	2307.04	-3730.68	56908.77	22.3658
	1001	-267044.5	3764141.3	2467.97	-3976.5	57016.24	25.646
66.6	751	-267044.5	3764140.3	2468.61	-3976.5	57016.45	25.645
	501	-267044.4	3764137.4	2470.38	-3976.77	57017.03	25.643
	251	-267044.4	3764124.2	2478.88	-3977.70	57019.79	25.635
	1001	-298833.26	4187403.4	2999.9999	-5411.1	58071.02	557.63
74.0	751	-298833.26	4187403.4	2999.9999	-5412.9	58073.93	560.44
	501	-298833.26	4187403.4	2999.9999	-5418.1	58082.10	568.35
	251	-298833.26	4187403.4	2999.9999	-5444.9	58123.45	608.22

TABLE XI

P_1: ($\xi^1 = -4596$, $\xi^2 = 3856.5$, $\xi^3 = 0$) at t = 0; P_2: ($\eta^1 = 1037.82$, $\eta^2 = -1144.75$, $\eta^3 = -974.68$) at t =T

$\overline{P_1E} = 6000$ at t = 0; $\overline{P_2M} = 1827$ at t = T. Lengths in N.M.

T = 74 hours n = 1001 m = 6, 20 k = 5

x_0^i from $(x_0^i - x_1^i)(t - T) - (x_0^i - x_2^i)\, t = 0$

Time	m	x^1 (N.M.)	x^2 (N.M.)	x^3 (N.M.)	$\dot{x}^1$ (N.M./hr)	$\dot{x}^2$ (N.M./hr)	$\dot{x}^3$ (N.M./hr)
0	20	-4596.0000	3856.5000	0	-16271.	58721.7	131.910
	6	-4596.0000	3856.5000	0	-16270.940	58721.742	131.90968
7.4	20	-54968.27	415984.97	591.5436	-4621.69	55792.95	58.672
	6	-54968.278	415984.97	591.5436	-4621.6904	55792.960	58.67286
14.8	20	-85170.07	830192.68	967.4490	-3718.41	56125.66	44.7601
	6	-85170.071	830192.68	967.4490	-3718.4103	56125.668	44.76011
22.2	20	-111375.65	1246359.	1268.8179	-3415.78	56339.059	37.26936
	6	-111375.65	1246359.5	1268.817	-3415.7903	56339.059	37.269355
37.0	20	-160859.33	2082414.7	1748.6905	-3343.26	56619.67	28.4446
	6	-160859.33	2082414.7	1748.690	-3343.2665	56619.671	28.444716
51.8	20	-211679.57	2921897.0	2128.3212	-3555.84	56815.97	23.3585
	6	-211679.57	2921897.0	2128.321	-3555.8532	56815.976	23.358602
59.2	20	-238605.46	3342663.4	2296.2998	-3729.40	56905.25	22.3187
	6	-238605.46	3342663.4	2296.299	-3729.4117	56905.258	22.318763
66.6	20	-267044.55	3764141.3	2467.973	-3976.51	57016.24	25.6460
	6	-267044.55	3764141.3	2467.973	-3976.5212	57016.240	25.646098
74.0	20	-298833.26	4187403.4	2999.9999	-5411.12	58971.02	557.63
	6	-298833.26	4187403.4	2999.9999	-5411.127	58071.025	557.6325

TABLE XII

P_1: ($\xi^1 = -3830$, $\xi^2 = 3213.75$, $\xi^3 = 0$) at t = 0; P_2: ($\eta^1 = 1037.82$, $\eta^2 = -1144.75$, $\eta^3 = -974.68$) at t = T

$\overline{P_1E} = 5000$, at t = 0; $\overline{P_2M} = 1827$ at t = T. Lengths in N.M.

T = 74 hours n = 1001 m = 20 k = 6

$x_0^i(t)$ obtained from $(x_0^i - x_1^i)(t - T) - (x_0^i - x_2^i)\, t = 0$ and by "stretching" solution I-4.

Time	INITIAL GUESS	x^1 (N.M.)	x^2 (N.M.)	x^3 (N.M.)	$\dot{x}^1$ (N.M./hr)	$\dot{x}^2$ (N.M./hr)	$\dot{x}^3$ (N.M./hr)
0	"6000 N.M."	-3830.0000	3213.7500	0	-17878.9	58948.28	143.874
	"ST. LINE"	-3830.0000	3213.7500	0	-18306.274	58993.237	147.14138
7.407	"6000 N.M."	-54654.10	415068.9	604.3081	-4612.20	55777.63	58.9058
	"ST. LINE"	-54688.602	415062.37	604.64098	-4655.3335	55759.239	59.535976
14.815	"6000 N.M."	-84849.25	829690.24	981.2730	-3717.165	56135.209	44.76657
	"ST. LINE"	-84887.414	829680.92	981.65608	-3722.9121	56129.499	44.954047
22.222	"6000 N.M."	-111090.30	1246379.3	1282.696	-3419.107	56356.868	37.20597
	"ST. LINE"	-111126.85	1246369.5	1283.0700	-3410.6489	56357.218	37.216000
29.630	"6000 N.M."	-136007.87	166458.5	1538.401	-3331.960	56517.22	32.11439
	"ST. LINE"	-136040.77	1664449.2	1538.7431	-3316.3015	56520.985	32.032863
37.037	"6000 N.M."	-160710.7	2083588.2	1761.6298	-3351.06	56643.431	28.32472
	"ST. LINE"	-160739.00	2083580.0	1761.9296	-3331.2594	56649.231	28.190319
44.444	"6000 N.M."	-185811.53	2503569.1	1960.112	-3434.96	56748.711	25.38985
	"ST. LINE"	-185834.65	2503562.4	1960.3656	-3412.5989	56755.715	25.222352
51.852	"6000 N.M."	-211713.03	2924279.3	2139.582	-3565.46	56841.63	23.1992
	"ST. LINE"	-211730.69	2924274.2	2139.7862	-3541.4248	56849.229	23.007417
59.259	"6000 N.M."	-238738.3	3345658.7	2306.443	-3739.29	56931.380	22.1279
	"ST. LINE"	-238750.35	3345655.3	2306.5968	-3714.0100	56938.972	21.906243
66.667	"6000 N.M."	-267279.11	3767754.2	2476.696	-3986.41	57043.115	25.4086
	"ST. LINE"	-267285.32	3767752.5	2476.7983	-3959.6398	57049.616	25.085702
74.000	"6000 N.M."	-298833.26	4187403.4	2999.9999	-5408.70	58097.143	554.015
	"ST. LINE"	-298833.26	4187403.4	2999.9999	-5373.3810	58091.865	549.38024

REFERENCES

1. E. T. Benedikt, "Collision Trajectories in the Three-Body Problem", Journal of the Astronautical Sciences, Vol VI, No. 2, 1959, pp 17-24

2. G. D. Birkhoff, "Dynamical Systems", American Mathematical Society, Colloquium Publications, Vol IX., New York, 1927

3. E. Picard, "Sur l' application des méthodes d' approximations successives à l'étude de certaines équations differentielles ordinaires", Journal de Math. Pures et Appliquées, Vol. IX, 1893, pp 217-291

ACKNOWLEDGMENT

To Mrs. Jean Kirk, who wrote the program for the electronic computer, and made the subsequent modifications as needed, many thanks are, indeed, due.

ROUND-TRIP TRAJECTORIES to MARS and VENUS *

Geza S. Gedeon **

Two types of round-trip trajectories are treated; "swing-around" trajectories, requiring no additional propulsion near the observation planet and "orbiting round-trip" trajectories, requiring propulsion to enter into a planetocentric orbit and to leave it after a specified waiting period. These problems are attacked by a new and very simple calculation procedure based on the "Lambertian Mechanics"[1] concept. The method presented is easily adaptable to an automatic digital computer, and thus it eliminates the tedious labor involved in constructing auxiliary charts and drawing envelopes to find optimum solutions.

Finally, in addition to the results in the usual form of graphs of total trip time versus characteristic velocity, the complete set of departure conditions is presented. These values can be used as good approximations to the initial conditions required by a precision orbit calculation routine.

INTRODUCTION

Exploration of outer space will undoubtedly start with inspection of the two nearest planets, Venus and Mars. The three steps probably required to accomplish this purpose are: (1) sending an instrumented package to these planets, (2) orbiting a crew around the planets, and (3) landing and returning a man from these planets. The second and third steps obviously involve round trips. However, it can be shown that "swing-around" trajectories offer many advantages over a one-way trajectory, even for the first phase. For example, the spacecraft would not be re-

---

*Presented at the eighth annual meeting of the American Astronautical Society, January 1962.

**Head of Astrodynamics, Northrop Space Laboratories, Hawthorne, California.

quired to carry long-distance communication systems. Long read-out times on a home orbit or recovery of observations are permitted, and the penalties associated with the extra propulsion requirements at the planet are removed. For these reasons it was deemed advisable to undertake an analysis of round-trip trajectories only.

Swing-around trajectories were first proposed by Ehricke et al.[2] Battin[3] investigated a class of trajectories having velocities of departure restricted to values producing total trip times in the two-to-three-year range. Johnson and Smith[4] have made the very interesting discovery that trajectories of 1-to 1.5-years duration can be realized with velocity requirements which are feasible with present day technology. This last publication is, to the writer's knowledge, the only one dealing comprehensively with swing-around trajectories, presenting both swing-in and swing-out trajectories for different values of closest planetary approach. Unfortunately, as confirmed by private communication with the author of that paper, a programming error has resulted in certain errors in the results presented.

The second class of trajectories, the so-called orbiting round trips, involve a stay on the planet or on a planetary orbit, i. e., the space vehicle moves in heliocentric space with the same speed as the planet for a prescribed waiting time. To accomplish this feat, it is necessary to apply propulsion at arrival in the planetary vicinity and at take-off from the planet or planetary orbit. Unlike the swing-around trips in which total trip time and closest approach completely define the trajectory, an infinite number of orbiting round-trip trajectories are possible for a given total trip time and orbiting altitudes. This being the case, it is logical to search for optimum solutions. Moeckel[5] made the first step towards optimization by employing transfer trajectories which are tangent to the heliocentric orbits at departure. He found that nonsymmetric trajectories yield considerably lower trip times than symmetric ones for equal fuel requirements. Johnson and Smith[4] took the next logical step by removing the tangency requirement. They utilized a set of working charts and a graphical optimization technique (envelope drawing) to obtain optimum orbiting round-trip trajectories to Mars. It is important to note that they realized that optimum round trips do not consist of two optimum one-way trajectories; rather, the optimization is utilized to determine the two segments into which the total trip should be divided. Johnson and Smith utilized a set of orbital equations developed by Moeckel (based on perihelion radii and velocities). Dugan[6] repeated the study of Martian round-trip trajectories using Vertregt's[7] method (based on eccentricity and semi-latus rectum) and extended it to cover Venutian[8] round trips. His method also utilized a set of working charts and graphical optimization.

In this paper, a new method of calculation based on the "Lambertian Mechanics"[1] is presented. The advantages of this method are twofold. First, the equations are continuous, i. e., they provide a smooth transition between the routes called, by the earlier authors, direct, perihelion, aphelion, and indirect. Second, the equations are so simple that no intermediate working charts are required, and the optimization can be readily performed by an automatic digital computer.

NOMENCLATURE

Symbols

μ	GM = force field constant
a	semimajor axis
e	eccentricity
r	radius
q	$\frac{r}{2a}$ = modulus
c	chord
s	$\frac{r_o + r + c}{2}$ = semiperimeter
w	$\pm\sqrt{1 - c/s}$ = shape factor
z	$\frac{s}{2a} = (\frac{s}{r})\, q$ = argument of universal time equation
V_h	$\sqrt{\frac{\mu_S}{r}}$ = heliocentric circular orbit velocity at radius r
V_p	$\sqrt{\frac{\mu_P}{r}}$ = planetocentric circular orbit velocity at radius r
K	$\frac{V}{V_h}$ or $\frac{V}{V_p}$ = heliocentric or planetocentric Kepler number
$K^{h.e.}$	hyperbolic excess Kepler number
ΔK	Kepler number increment
t	time, mean solar days
T	round-trip time, mean solar days
n	$\sqrt{\mu_S/r^3}$ = heliocentric angular velocity
θ	true anomaly
ϵ	transfer angle
α	Lambert's first angle
β	Lambert's second angle
ϕ	angle of elevation
γ	angle between chord and radius

σ	angle between hyperbolic excess velocity vector and the heliocentric velocity vector of the planet
κ	angular deflection at hyperbolic encounter
δ	angle measured from planet-sun line to planetary radius at departure from parking orbit, positive in direction of motion
A_n	$\frac{1.3.5.\ \ldots\ (2n-1)}{2.4.6.\ \ldots\ (2n)}$, $A_o = 1$ coefficients of the universal time equation
k	± 1 switching constant

Subscript

o	conditions at Earth
s	semiperimeter
S	Sun
E	Earth
h	heliocentric
p	planetary or planetocentric
w	waiting
1	outward leg
2	homeward leg
a	asymptote

ANALYSIS

Assumptions

The following assumptions are utilized in this analysis:

(1) The orbits of Earth, Mars, and Venus are circular and coplanar.

(2) The vehicle is attracted by only one inverse-square central force field at a time. No account is taken of transition from one sphere of influence to another in the vicinity of a planet, and the point at which the spacecraft enters or leaves the heliocentric force field is taken to be identical to the planet's position.

(3) The impulsive velocity-change concept is applied.

(4) Characteristic velocity rather than mass ratio is used as figure of merit.

Swing-Around Trajectories

A typical swing-in trajectory is shown in Fig 1. A trajectory cutting the planet sun line is designated a swing-out trajectory while one passing outside this line is designated a swing-in trajectory.

The analysis starts by assuming a heliocentric velocity vector at arrival to the vicinity of the target planet. The magnitude of the velocity is expressed by a unitless quantity, the Kepler number K_1. In Ref. 1, the source of many of the following equations, the Kepler number is defined as the actual velocity dividend by the local circular velocity. In this case, the latter corresponds to the heliocentric orbital velocity of the planet. The direction of the velocity vector is given by the angle of elevation ϕ_1.

Knowing the arrival velocity vector with magnitude and direction denoted by $[K_1, \phi_1]$ and the planet's assumed circular velocity vector $[1, 0]$, the hyperbolic excess velocity $K^{h.e.}$ can be calculated simply as the relative velocity with respect to the planet.

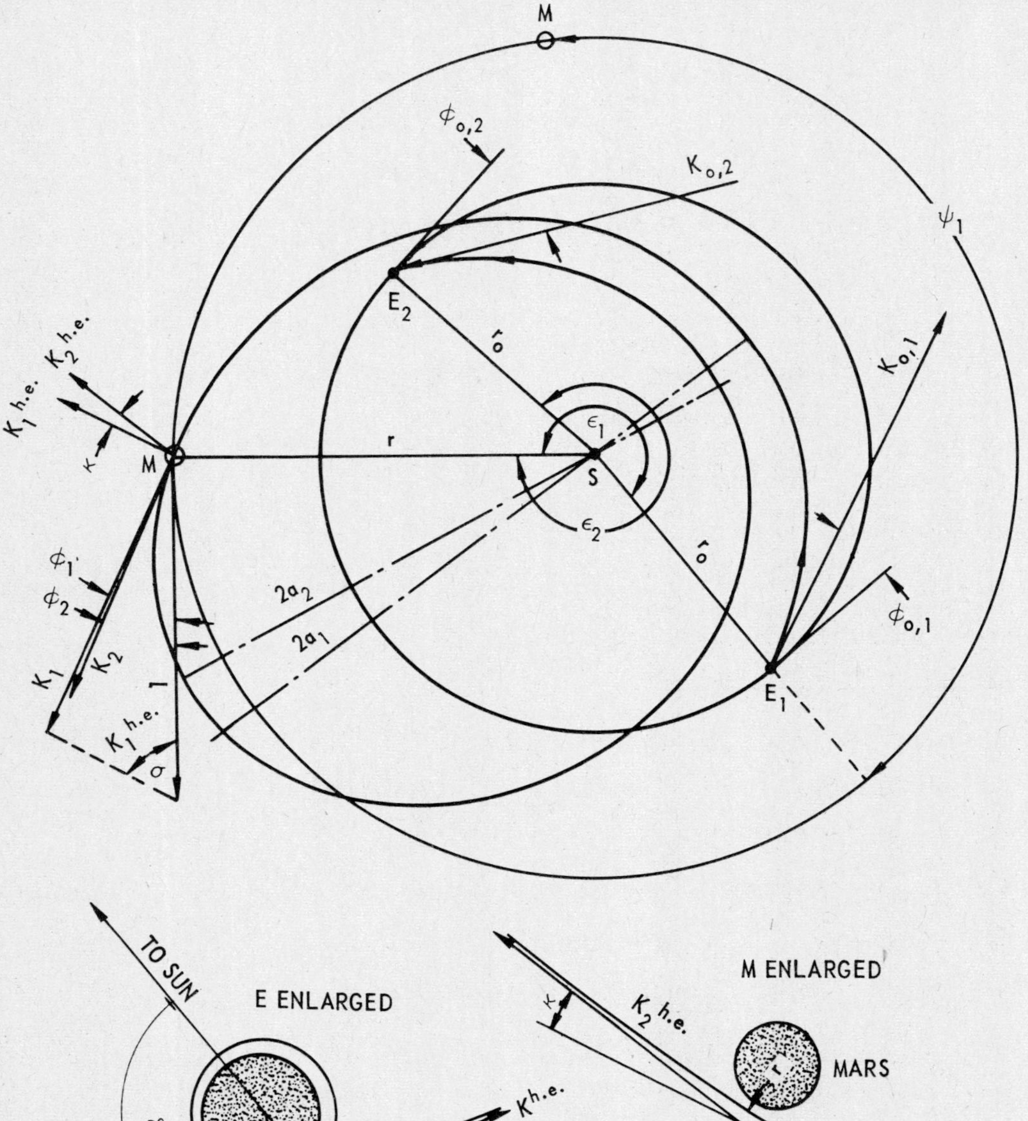

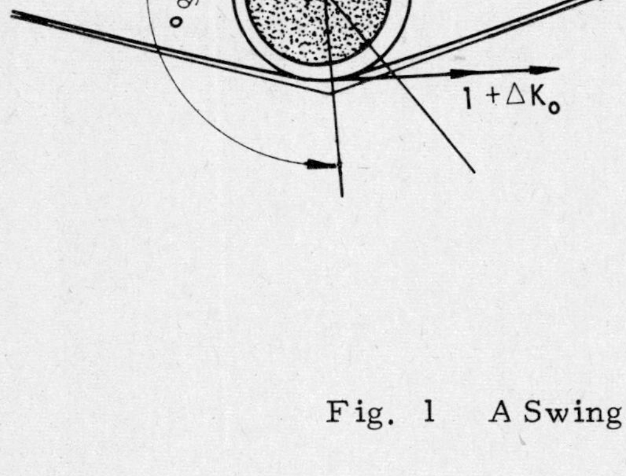

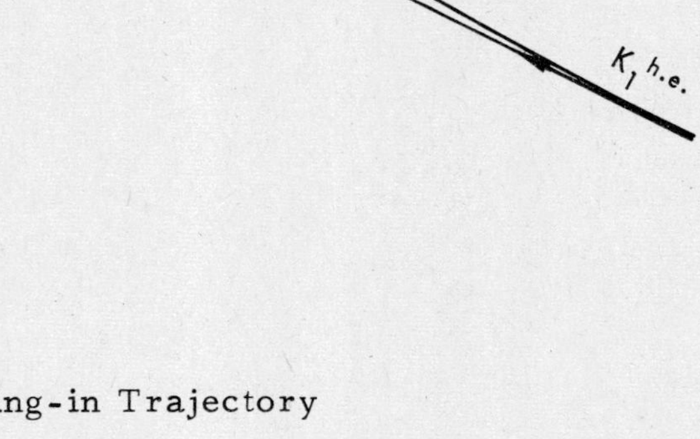

Fig. 1 A Swing-in Trajectory

$$K_1^{h.e.} = \sqrt{1 - 2K_1 \cos \phi_1 + K_1^2} \qquad (1)$$

On the enlarged view of point M it can be seen that $K_1^{h.e.}$ is the approach velocity on the asymptote of a planetocentric hyperbola. The spacecraft departs along the other asymptote with hyperbolic excess velocity, $K_2^{h.e.}$, having the same magnitude as $K_1^{h.e.}$

The gravitational attraction results in a deflection angle, , which is

$$\csc \kappa/2 = \left[\frac{K^{h.e.}}{V_p/V_h}\right]^2 + 1, \quad 0 \leq \frac{\kappa}{2} \leq \frac{\pi}{2} \qquad (2)$$

The above relationship is derived in Appendix A.

In this equation, V_p is the planetocentric circular orbit velocity at closest approach and V_h is the heliocentric velocity of the planet. The angle σ included between $K_1^{h.e.}$ and the heliocentric velocity vector of the planet is obtained from

$$\sin \sigma = \frac{K_1}{K_1^{h.e.}} \sin \phi, \quad \begin{matrix} 0 \leq |\sigma| \leq \frac{\pi}{2} \\ \frac{\pi}{2} \leq |\sigma| \leq \pi \end{matrix} \quad \text{if } K_1 \cos \phi_1 \begin{cases} < 1 \\ > 1 \end{cases} \qquad (3)$$

$K_2^{h.e.}$ makes an angle $(\sigma \pm \kappa)$ with the planets heliocentric velocity vector.

Now, the departure velocity vector can be calculated as the resultant of the heliocentric velocity of the planet and the hyperbolic excess velocity at departure by the relationships

$$K_2 = \sqrt{\left[1 - K_2^{h.e.} \cos(\sigma \pm \kappa)\right]^2 + \left[K_2^{h.e.} \sin(\sigma \pm \kappa)\right]^2} \qquad (4)$$

and

$$\tan \phi_2 = \frac{K_2^{h.e.} \sin(\sigma \pm \kappa)}{1 - K_2^{h.e.} \cos(\sigma \pm \kappa)}, \quad -\frac{\pi}{2} \leq \phi_2 \leq \frac{\pi}{2} \qquad (5)$$

Knowing the velocity vector at one end of the trajectory defines the orbital segment. Thus, according to Ref. 1, the two moduli q, q_o are given by

$$q = 1 - \frac{1}{2}K^2 , \quad q_o = \frac{q}{r/r_o} \tag{6}$$

and the transfer angle ϵ by

$$\tan \epsilon/2 = \frac{1 - q/q_o}{- \tan \phi \pm \frac{q}{q_o}\sqrt{\frac{1-q_o}{1-q} \frac{q_o}{q} \sec^2 \phi - 1}} \qquad 0 \leq \frac{\epsilon}{2} \leq \pi \tag{7}$$

The ± sign in the above equation is due to the fact that conics are symmetric around the line of apsides. In one case the trajectory contains one more apsis than in the other. Note that the above expression is based upon angle of elevation at the target planet, contrasted to the equation of Ref. 1, which was based on the initial angle of elevation. The difference between the two equations stems from the fact that the naught subscript is maintained, describing conditions at Earth. To use Eq. (7) for the second segment, the departure velocity vector is rotated through 180°, i. e., the sign of ϕ_2 is reversed. One thus has two orbital segments originating from the departure and arrival position of the Earth and meeting head-on at closest approach to the planet.

Next, according to Ref. 1, the transfer times t_1 and t_2 are calculated by first determining the chords c_1 and c_2 from

$$\frac{c}{r_o} = \sqrt{1 + (r/r_o)^2 - 2(r/r_o) \cos \epsilon} \tag{8}$$

the semiperimeters s_1 and s_2 from

$$s/r_o = \frac{1 + r/r_o + c/r_o}{2} \tag{9}$$

the shape factors w_1 and w_2 from

$$w = \pm \sqrt{1 - c/s} \qquad \begin{cases} + \text{ for } \epsilon < \pi \\ - \text{ for } \epsilon > \pi \end{cases}, \tag{10}$$

and, finally, the transfer time from

$$n_s t = (1 - k) \frac{\pi}{(2z)^{3/2}} + k\sqrt{2} \sum_{n=o}^{\infty} A_n \frac{1-kw^{2n+3}}{2n+3} z^n \quad * \tag{11}$$

where

$$k = \pm 1, \; n_s = \frac{n_E}{(s/r_o)^{3/2}}$$

n_E being the heliocentric angular velocity of Earth and

$$z = \frac{s}{2a} = \left(\frac{s}{r_o}\right) q_o \tag{12}$$

In Eq. (11), the ambiguity is hidden in the expression k which is restricted to +1 or -1. The plus sign applies when the empty focus falls outside the area enclosed by the segment and chord, the 'minus' sign when it falls inside. To determine the correct sign, the angle γ included between the chord and the radius to the planet is computed by

$$\tan \gamma = \frac{\sin}{r/r_o - \cos \epsilon} \qquad \begin{cases} \text{if in } 3^{rd} \text{ or } 4^{th} \text{ quadrant} \\ \text{subtract } 2\pi \text{ from } \gamma \end{cases} \tag{13}$$

and then compared with 2ϕ. The criterion used in Ref. 1 is adapted to the end conditions in the following manner,

$$\text{if} \quad \begin{matrix} \gamma > (-2\phi) \\ \gamma < (-2\phi) \end{matrix} \qquad \begin{cases} k = +1 \\ k = -1 \end{cases}$$

Since Eq. (7) is double valued and there are two segments, four individual trip time values are calculated: t_1', t_1'', t_2', t_2'', the first two

* See Ref. 1 about the convergence of above series.

corresponding to the first leg and the last two to the last leg. If any of the four combinations satisfy the requirement that

$$(t_1 + t_2)\, n_E = (\epsilon_1 + \epsilon_2)$$

where n_E is the angular velocity of Earth in the same units as t_1 then departing from Earth at the correct date, the returning space vehicle will intercept Earth at the end of the trip.

If none of the combinations satisfies Eq. (14), then one of the initial assumptions must be changed. K_1 can be kept fixed and ϕ_1 varied between arbitrary limits (high ϕ_1 means high propulsion requirement) or ϕ_1 fixed and K_1 varied. Both solutions are double valued in certain regions. Thus, care must be exerted to find both solutions.

If a solution to Eq. (14) is found, the constellation angle can be calculated from

$$\psi_1 = \epsilon_1 - n_p t_1 \tag{15}$$

where $n_p = n_E/(r/r_o)^{3/2}$ is the angular velocity of the planet. It is supposed that departure will be made from a geocentric parking orbit in the plane of the ecliptic. The amount of velocity increment necessary for departure can be calculated in the following manner.

First, the departure Kepler number is given by

$$K_o = \sqrt{2(1 - q_o)} \tag{16}$$

and the angle of elevation at departure by

$$\cos \phi_o = \sqrt{\frac{(1 - q)\, q}{(1 - q_o)\, q_o}} \cdot \cos \phi \tag{17}$$

Note that the sign of ϕ_o is undetermined; it agrees with ϕ if the number of apsides in the trajectory is even (0 or 2), and differs from ϕ if this number is odd. The number of apsides may be determined from the nature of the problem. This ambiguity can be avoided by using Eq. (27).

Next, the hyperbolic excess velocity $K_o^{h.e.}$ is obtained by Eq. (1) and then the heliocentric Kepler number increment ΔK_o from

$$\Delta K_o = \sqrt{(K_o^{h.e.})^2 + 2\left[\frac{V_p}{V_h}\right]_E^2} - \left[\frac{V_p}{V_h}\right]_E \tag{18}$$

where $(V_p)_E$ is the geocentric circular velocity on the parking orbit and $(V_h)_E$ is the heliocentric velocity of Earth. Thus, ΔK_o is normalized with respect to the Earth's heliocentric velocity (18.5 mi/sec or 29.8 km/sec).

The point of application of this impulse is assumed to be the vertex of the geocentric hyperbola of departure. To find the angle δ_o between the axis of this hyperbola and the Earth - Sun line, Eqs. (2) and (3) are employed again, using departure data (naught subscript). The angle δ_o is given by

$$\delta_o = (\sigma_o - \kappa_o/2) \tag{19}$$

The calculation of ΔK_o and δ_o can be repeated for the return point, and a characteristic velocity can be obtained by adding together the two impulses.

Orbiting Round Trip Trajectories

Figure 2 shows an orbiting round-trip trajectory. During the given waiting time t_w, the spacecraft moves through the angle ϵ_w with the angular velocity of the planet in heliocentric space where

$$\epsilon_w = n_p t_w \tag{20}$$

The total angle ϵ swept out by the Earth during the round trip is defined by the given total trip time t

$$\epsilon = \epsilon_1 + \epsilon_w + \epsilon_2 = n_E t \tag{21}$$

This last equation serves to define only the sum of the transfer angles ϵ_1 and ϵ_2, rather than the two angles individually. It is part of the opti-

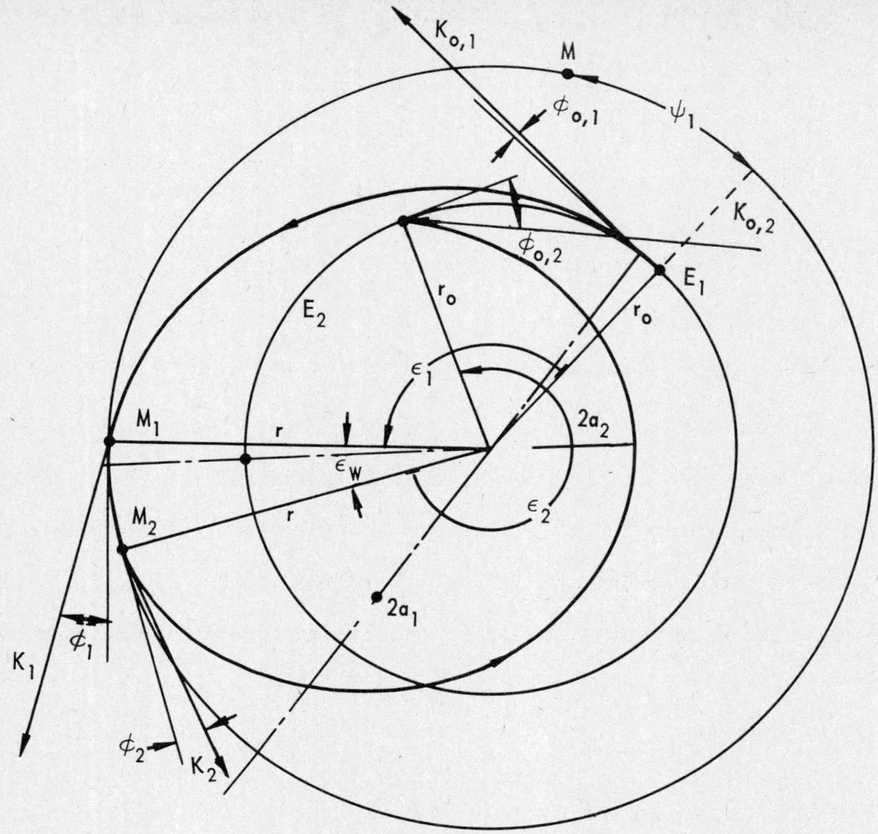

Fig. 2 An Orbiting Round Trip Trajectory

mization procedure to find each of the angles. Thus the calculation begins by assuming an ϵ_1 and a trajectory parameter of the first segment. This fixes the transfer time, and an optimization subroutine varies ϵ_1 and the trajectory parameter until minimum characteristic velocity is found.

The trajectory parameter selected can be either the modulus q or the argument z defined in Ref. 1. These quantities are interrelated by

$$z = \left(\frac{s}{r}\right) q \tag{22}$$

where (s/r) is a constant if ϵ is fixed. The choice of z is preferred since it has a fixed upper limit = + 1, which is a convenience in automatic machine calculations.

The calculation starts by computing the chords c_1 and c_2 by Eq. (8), the semiperimeters s_1 and s_2 by Eq. (9), and finally the shape factors w_1 and w_2 by Eq. (10).

The trip time t_1 can be calculated by Eq. (11). However, t_2 is defined by the relation

$$t_1 + t_w + t_2 = t \tag{23}$$

and therefore Eq. (11) should be used to calculate z_2. This appears to be a very cumbersome procedure. At the outset, it was hoped that a good approximation to the inverse of Eq. (11) could be found. Unfortunately this has not yet been realized. Therefore, a trial and error procedure has been used in programming this problem. For trial and error solution, Lambert's Equation in the form illustrated below was used in place of Eq. (11).

$$n_E t = \frac{(\alpha - \sin\alpha) - (\beta - \sin\beta)}{\left(\frac{2z}{s/r_o}\right)^{3/2}} \tag{24}$$

where

$$\alpha = (1-k)\pi + 2k \sin^{-1}\sqrt{z} \tag{25}$$

and

$$\beta = \pm \sin^{-1} w\sqrt{z} \qquad \begin{cases} + \text{ for } \epsilon < \pi \\ - \text{ for } \epsilon > \pi \end{cases} \tag{26}$$

Equations (24) through (26), along with Eq. (23), were used to solve directly for t_1 and by trial and error for z_2. The constant k_1 [k associated with the first leg] was an input, $k_1 = +1$ producing fast elliptic outward segments, and $k = -1$ allowing slow trajectories. The second constant k_2 [k associated with the second leg] was selected by the program. If no root has been found with $k_2 = +1$ then the machine switched automatically to $k_2 = -1$ to find z_2. No provision was made for parabolic or hyperbolic orbits since, in the total trip time range investigated, elliptic segment solutions were possible with considerable smaller fuel requirements.

As soon as z_2 was found, it was converted into the modulus q_2 by dividing by s/r. Then the Kepler numbers at the four endpoints of the two segments were calculated by Eq. (16) and the angles of elevation by the following equations

$$\tan \phi_o = \frac{1-q_o}{\tan \epsilon/2} - k \frac{\sqrt{(1-q_o)(q_o/q) - q_o \cos^2 \epsilon/2}}{\sin \epsilon/2} \tag{27}$$

and

$$\tan \phi = -\frac{1-q}{\tan \epsilon/2} + k \frac{\sqrt{(1-q)(q/q_o) - q(1-q \cos^2 \epsilon/2)}}{\sin \epsilon/2} \tag{28}$$

Note that the constant k must have the same value in these equations as used in Eq. (25).

Knowing the velocity vectors at the four endpoints (again both segments originate from the Earth and end at the planet at the arrival and departure position in heliocentric space), the hyperbolic excess velocities can be found by Eq. (1) and the Kepler number changes by Eq. (18).

Finally, the characteristic velocity (normalized to ecliptic speed) is

$$V_{CH}/(V_h)_E = \Delta K_{o,1} + \frac{\Delta K_1 + \Delta K_2}{\sqrt{(r/r_o)}} + \Delta K_{o,2} \tag{29}$$

To find minimum characteristic velocity, ϵ_1 and z_1 are changed by a subroutine utilizing a "gradient or steepest descent" method of optimization. When the minimum characteristic velocity is found, the departure conditions are calculated by the same procedure as in the case of swing-around trajectories, i. e., the configuration angle is obtained by Eq. (15), and the velocity increments for departures from an Earth orbit and from a planetary orbit by Eqs. (3) and (18).

RESULTS AND DISCUSSIONS

Solutions for Swing-Around Trajectories

Figures 3 and 4 illustrate velocity vectors (K, ϕ) at planetary encounter which produce round-trip trajectories originating and terminating at Earth under the assumption that the closest approach to the (center of the) planet is 1.1 planetary radius. Figures 5 and 6 illustrate several cases of particular interest.

Before attempting to explain the above solutions it is advisable to discuss the characteristics of those unperturbed trajectories which emanate from and return to Earth. One solution is obvious; all orbits having a semimajor axis identical to that of the Earth's heliocentric orbit (1 au) will intercept Earth a year later. Thus, the modulus of such an orbit at planetary encounter is one-half the mean radius of the planetary orbit expressed in au's. Then, Eq. (16) yields a heliocentric Kepler number at Mars of 0.6909 for all unperturbed one-year orbits. These orbits are represented by the vertical dotted line on Fig. 3. In the following sketches of unperturbed Martian orbits (Fig. 5), empty circles correspond to possible locations of massless planets. The filled circles correspond to planetary locations for which a perturbative analysis was carried out. The number 2 represents launch points for one-year orbits. The numbers 3 and 4 represent planetary locations at encounter. Number 5 designates an extra crossing of the Earth's orbit, and 6 the arrival position. For example, sketch (A) illustrates an orbit departing from Earth at point 2, passing by a massless Mars located at point 3/4, and returning to Earth at point 6 which is exactly coincident with the departure point.

However, one-year orbits do not constitute the only family of unperturbed orbits which intercept Earth at return. Martian orbits which overlap at perihelion can also be constrained to return to Earth by proper choice of the semimajor axis. The term 'overlapping' designates double

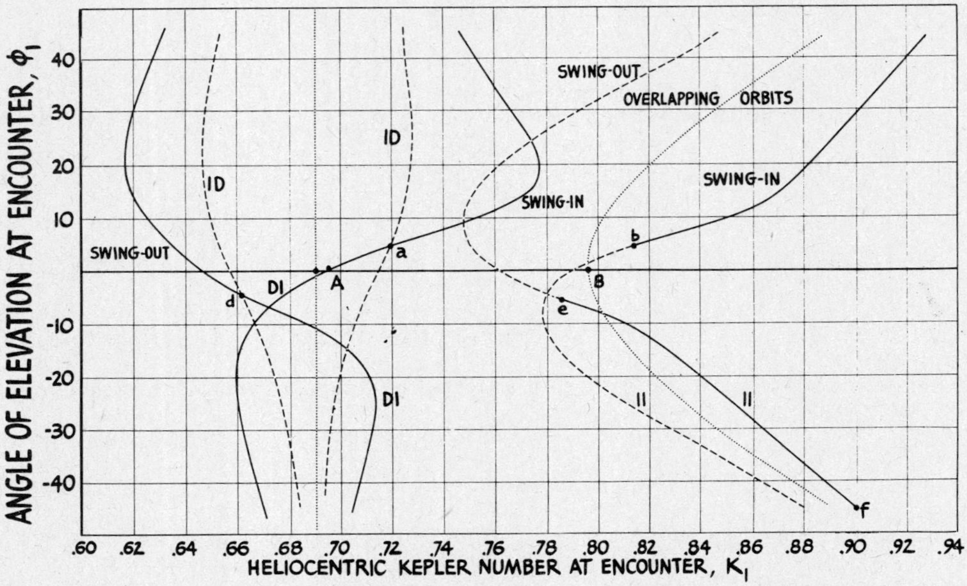

Fig. 3 Solutions to Swing-Around-Mars Trajectories. Closest approach 1.1 M.R.

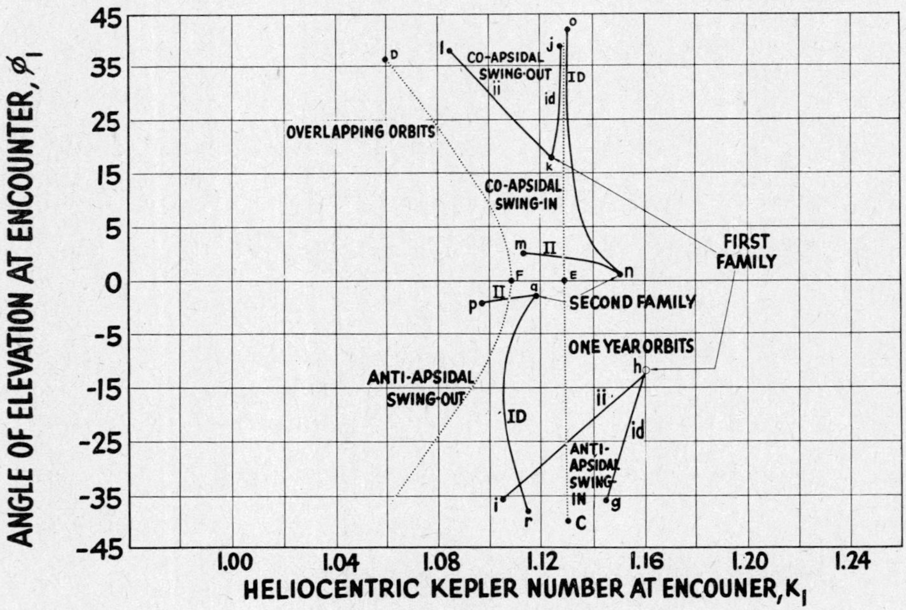

Fig. 4 Swing-Around-Mars Trajectories

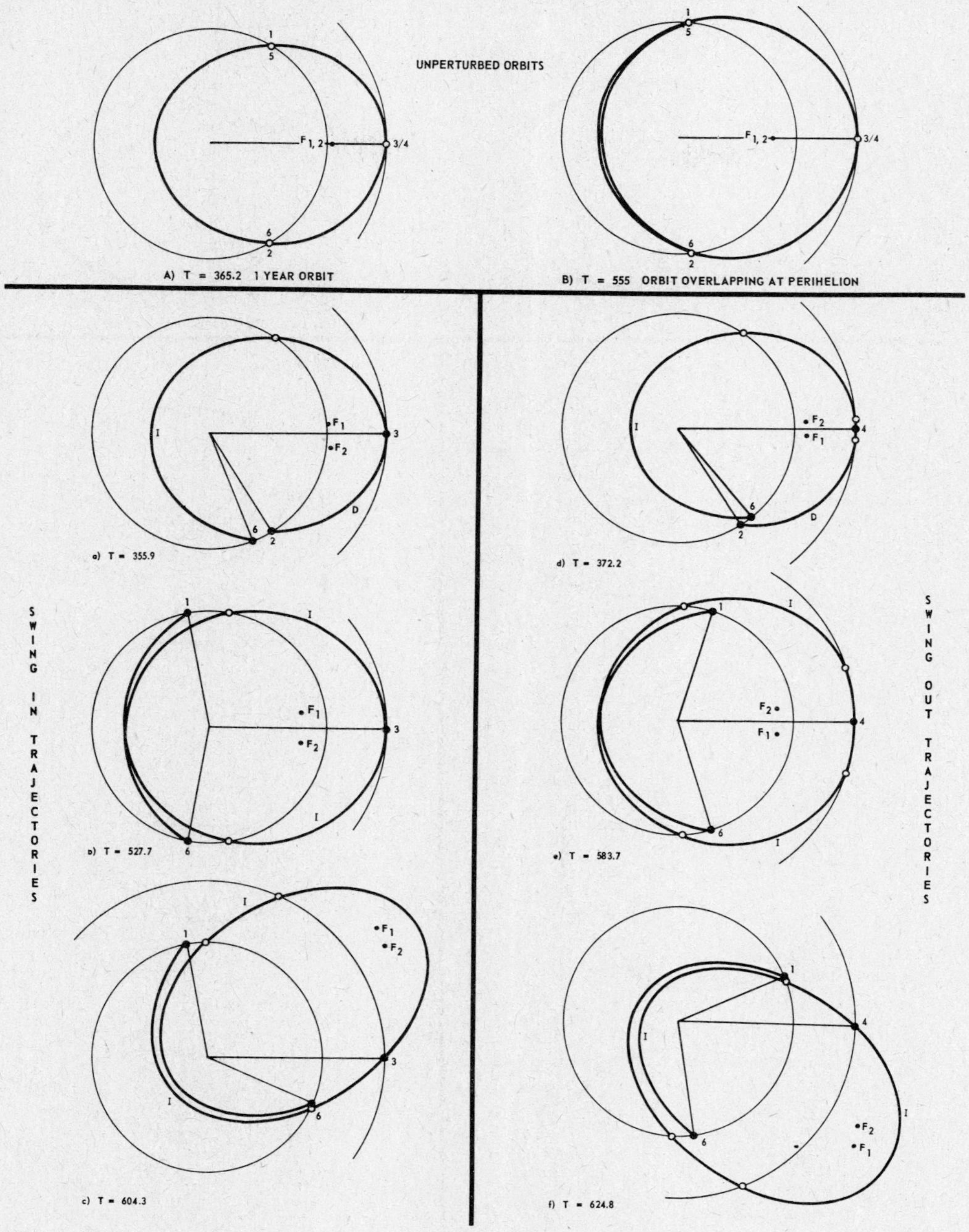

Fig. 5 Solutions to Swing-Around-Venus Trajectories. Closest approach 1.1 V.R. First Family: lower case letters, Second Family: capital letters (used for route designations)

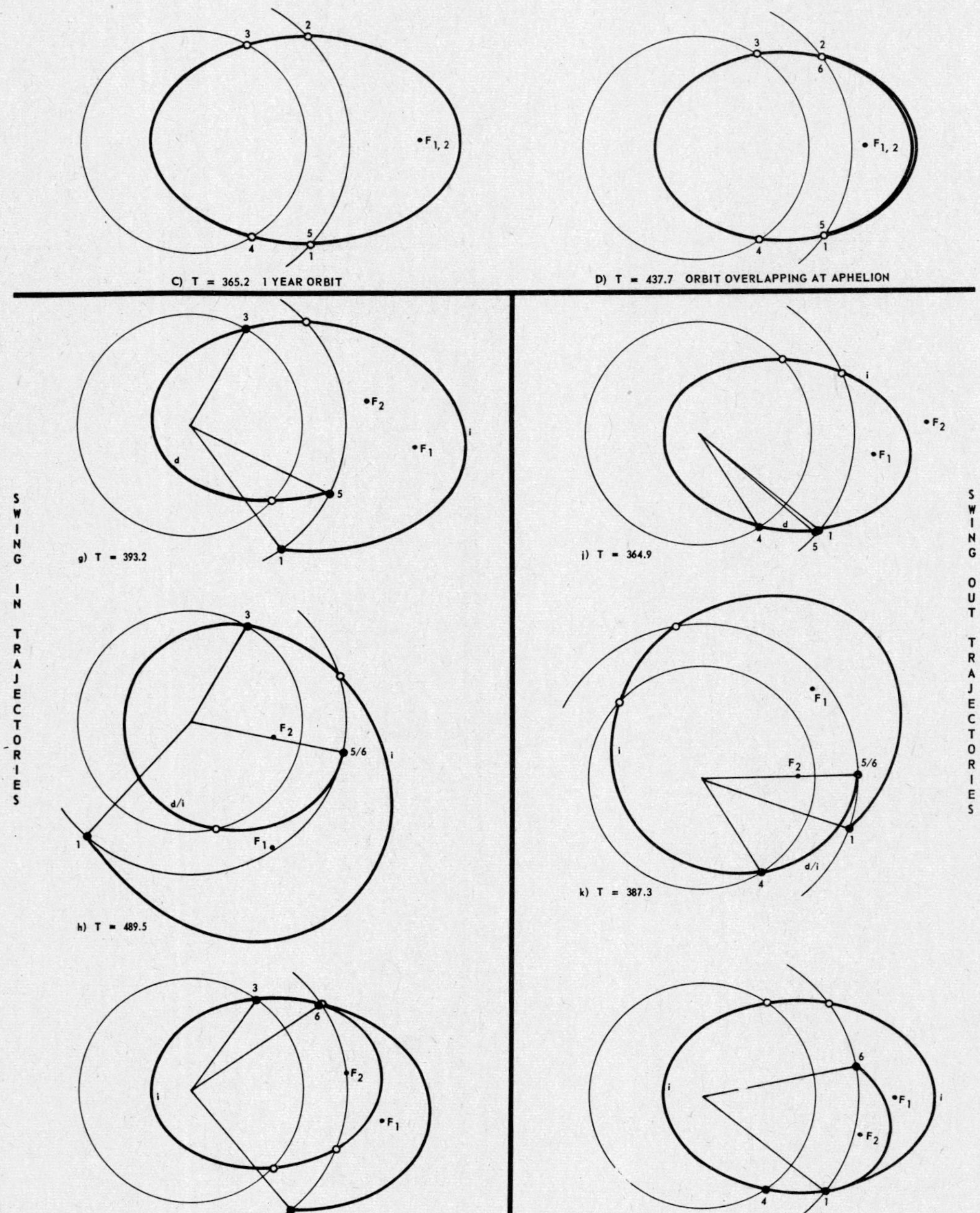

(a)

Fig. 6 Swing-Around-Venus Trajectories. a) First Family b) Second Family

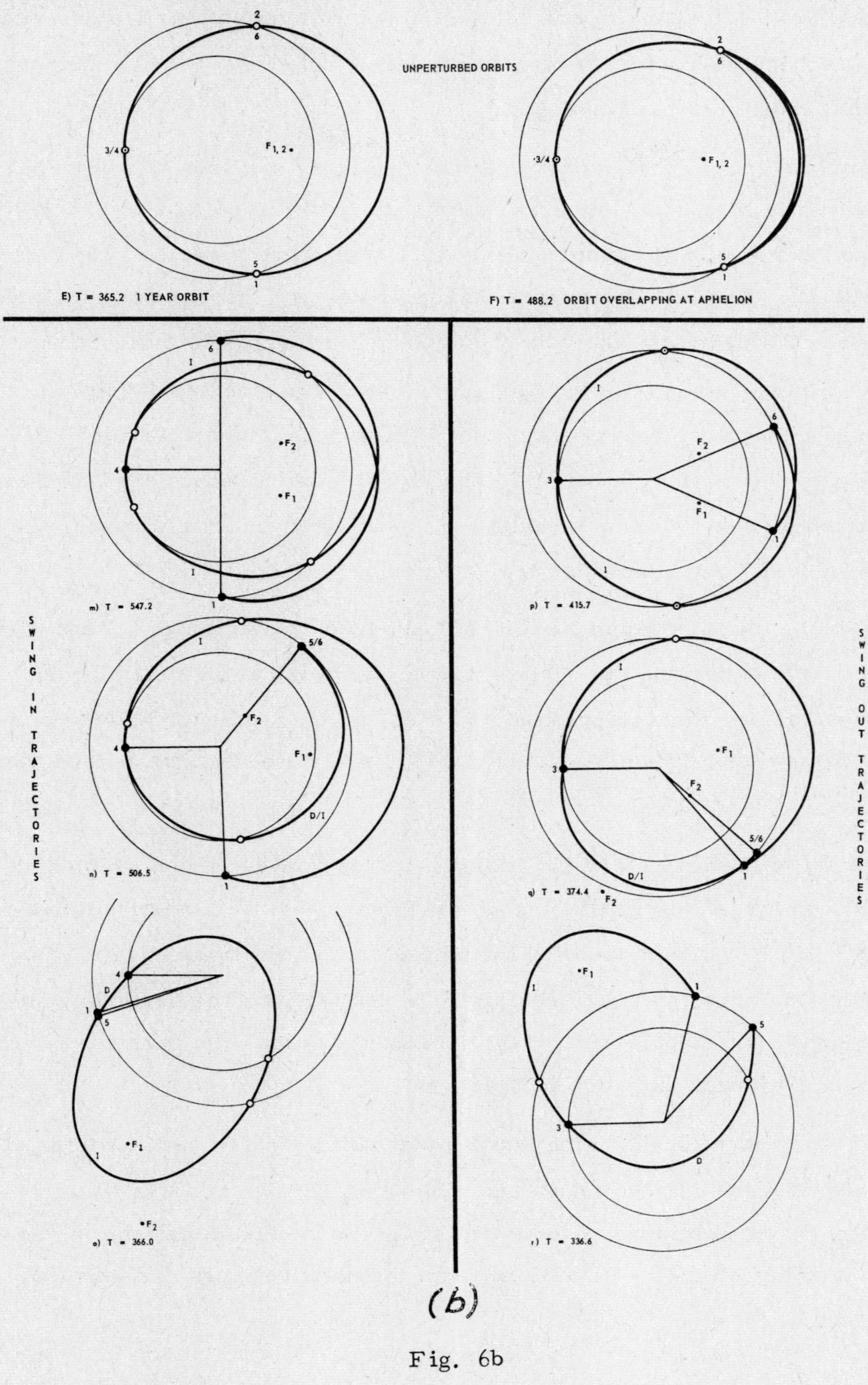

UNPERTURBED ORBITS
E) T = 365.2 1 YEAR ORBIT
F) T = 488.2 ORBIT OVERLAPPING AT APHELION
m) T = 547.2
n) T = 506.5
o) T = 366.0
p) T = 415.7
q) T = 374.4
r) T = 336.6
SWING IN TRAJECTORIES
SWING OUT TRAJECTORIES

(b)

Fig. 6b

passes through perihelion for Martian orbits. The dotted curve on Fig. 3 illustrates such a family, and sketch (B) shows a particular trajectory of longer than one-year duration. It should be noted that for overlapping orbits, departure occurs at point 1.

Should the space vehicle pass outside the planet-sun line, the one year orbit (A) is perturbed into a swing-in trajectory, sketch (A). Similarly, if the vehicle cuts the planet-sun line, a swing-out trajectory results (sketch D). The full line curves on Fig. 3 passing through points a and d were obtained from analysis of perturbed one-year nominal orbits having higher eccentricities than (A). The corresponding dashed line curves represent mirror image trajectories, i. e., trajectories originating at 6 and terminating at 2. Such orbits can be envisioned as those obtained by reversing both vehicle and planetary motions. Hence, they will be designated clockwise trajectories.

Perturbing the overlapping orbit (B), produces trajectories (c) and (e). Likewise, increasing the eccentricity of the original orbit results in curves passing through points c and e in Fig. 3. Although both sets of curves were not drawn beyond $\phi = \pm 45°$, solutions do exist beyond these values.

Each set of the trajectory is distinguished as between those which cross and do not cross the Earth's orbit; designated indirect and direct, respectively. It should be noted that these same adjectives have been used by other authors in a more restricted sense. Thus, the perturbed one-year orbits are considered direct-indirect, whereas the perturbed overlapping orbits are indirect-indirect.

Martian trajectories passing 2 and 5 planetary radii from the center of the planet were also investigated. Solutions similar to those illustrated in Fig. 3 were obtained. However, since the perturbation from a two-body orbit is smaller, deviations from the dotted curves are less pronounced.

Figure 4 illustrates the much more complicated swing-around Venus trajectories. The vertical dotted line corresponds to one-year orbits in a manner wholly analogous to the one-year Martian orbits illustrated in Fig. 3. As in the case of Martian orbits, empty circles in Fig. 6 correspond to possible locations of massless planets, and filled circles to planetary locations obtained from perturbative analysis. Sketches (C) and (E) represent minimum eccentricity (tangent to Venus' heliocentric orbit at fly-by), unperturbed one-year orbits. Sketches (D) and (F) illustrate minimum eccentricity overlapping orbits. For Venusian trajectories the term "overlapping" designates double passes through aphelion. For these trajectories, departure always occurs at point 1, extra-Earth orbital crossing at 2, planetary encounter at 3 or 4 and arrival at either 5 or 6. Should planetary encounter occur when the Earth and the planet at encounter are on the same side of the line of apsides of the first orbital segment, the perturbation is called co-apsidal and the number 4 is used. In the other case, the perturbation is called anti-apsidal, and the number 3 is used. The definitions of co-apsidal and anti-apsidal perturbations, as based on departure and encounter positions, hold true if vehicle and planetary motions are reversed, i. e., if departure occurs at 6 and arrival at 1 or 2. Unperturbed overlapping orbits of longer than one-year duration are represented by the dotted curve on Fig. 4. It should be noted that the vertex of this curve is much closer to the straight vertical line than in the Martian case.

Unlike Martian trajectories whose ϕ vs. K graphs were found to be continuous, discontinuous solutions were found for all perturbed Venusian trajectories (Fig. 4). These discontinuities were found to occur around small angles of elevation splitting the curves into two parts corresponding to co-apsidal and anti-apsidal perturbations. The upper and lower halves of Fig. 4 correspond respectively to co-apsidal and anti-apsidal perturbations for counterclockwise trajectories as illustrated on Fig. 6. The analysis of clockwise trajectories is omitted on Fig. 4 because of the obviously greater complexity of Venusian as compared with Martian trajectories. Considering first the anti-apsidal case, the curves $\widehat{gh}$ and $\widehat{qr}$ can be regarded respectively as swing-in and swing-out perturbations of one-year orbits. Similarly, $\widehat{ih}$ and $\widehat{pq}$ may be regarded as representations of perturbed overlapping orbits. In the case of co-apsidal perturbations, $\widehat{jk}$ and $\widehat{on}$ were obtained from perturbations of one-year orbits, and $\widehat{lk}$ and $\widehat{mn}$ from perturbations of the overlapping orbit. It should be noted that k, n, q and h, representing tangential Earth arrival, are common points for pairs of curves. Such points do not exist if the perburbation is small, i. e., they do not appear for a closest approach of 5 Venusian radii.

Examination of Fig. 4 and the subsequent charts reveals a similarity in appearance between co-apsidal swing-out and anti-apsidal swing-in trajectories. For this reason, both types of trajectories will be designated as the First Family of swing-around Venus trajectories. Similarly coapsidal swing-in and anti-apsidal swing-out trajectories are grouped together as the Second Family. Figure 6a presents the First Family of swing-around Venus trajectories. Sketch (C) illustrates an unperturbed highly eccentric one-year orbit with (g) and (j) representing trajectories obtained by anti-apsidal swing-in and co-apsidal swing-out, respectively. Highly eccentric overlapping orbits are perturbed into trajectories (i) and (l). Similarly, (h) and (k) are obtained from less eccentric one-year orbits.

Figure 6b presents the Second Family of swing-around-Venus trajectories. The symmetric trajectories (m) and (p) are obtained by perturbations of minimum eccentricity unperturbed orbits (E) and (F), respectively. Trajectories (n) and (q) represent cases when arrival is tangential to the Earth's orbit. Finally, (o) and (r) illustrate highly perturbed one-year nominal orbits.

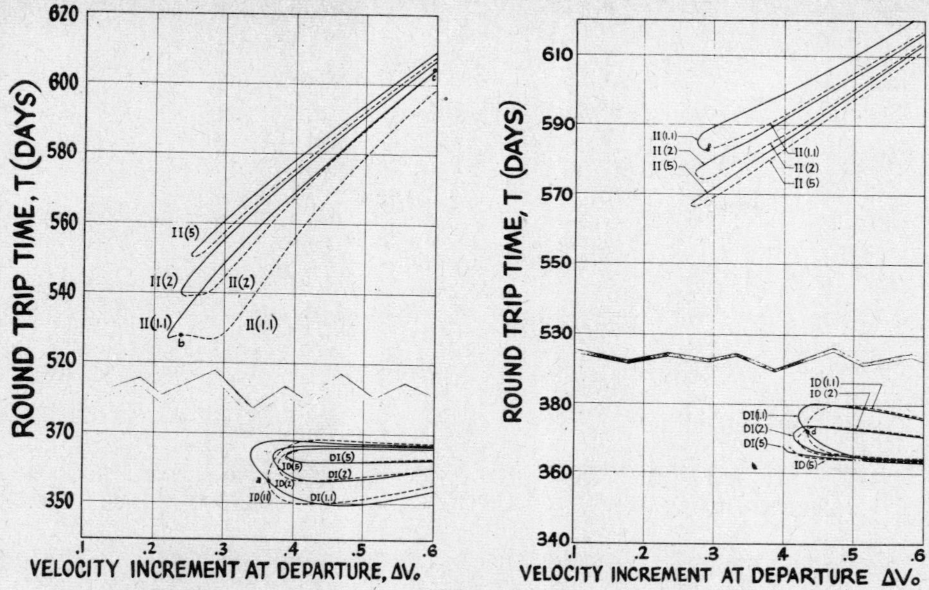

(a) (b)

Fig. 7 Round Trip Time vs. Velocity Increment (normalized with the heliocentric velocity of Earth) for Swing-Around-Mars Trajectories. Parking Orbit 1.1 E.R. Closest approach 1.1, 2, 5 M.R. a) Swing-in b) Swing-out

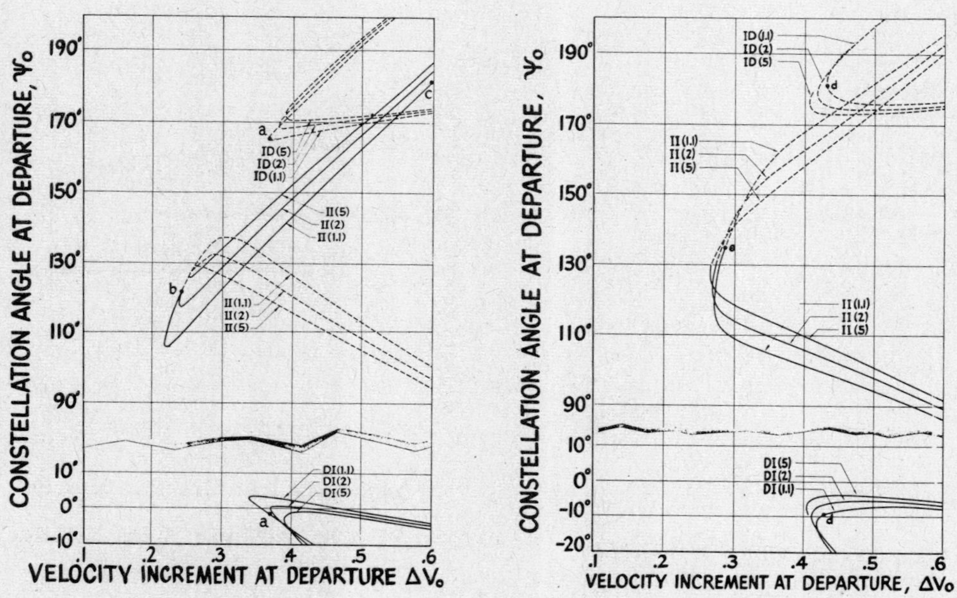

(a) (b)

Fig. 8 Departure Constellation Angle vs. Velocity Increment (normalized with the heliocentric velocity of Earth) for Swing-Around-Mars Trajectories. Parking Orbit 1.1 E.R. Closest Approach 1.1, 2, 5 M.R. a) Swing-in b) Swing-out

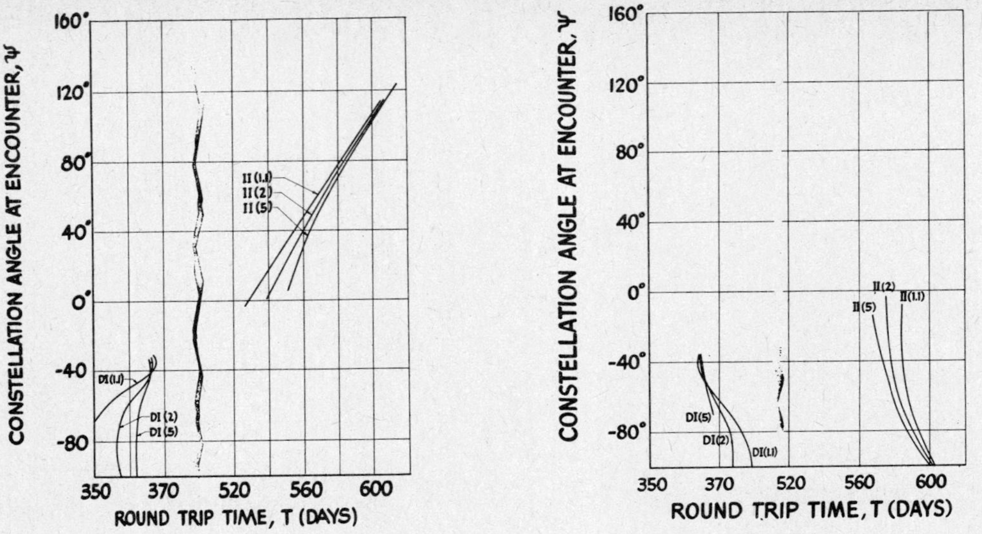

(a) (b)

Fig. 9 Encounter Constellation Angle vs. Total Trip Time for Swing-Around-Mars Trajectories. Parking Orbit 1. 1 E. R. Closest Approach 1. 1, 2, 5 M. R. a) Swing-in b) Swing-out

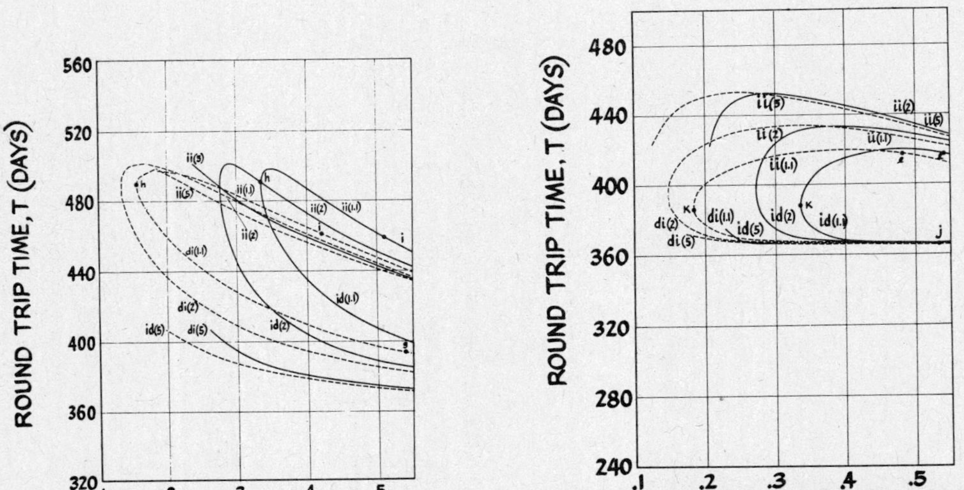

(a) (b)

Fig. 10 Total Trip Time vs. Velocity Increment (normalized with the heliocentric velocity of Earth) for Swing-Around-Venus Trajectories. Parking Orbit 1. 1 E. R. Closest Approach 1. 1, 2, 5 V. R. a) First Family Swing-in b) First Family Swing-out

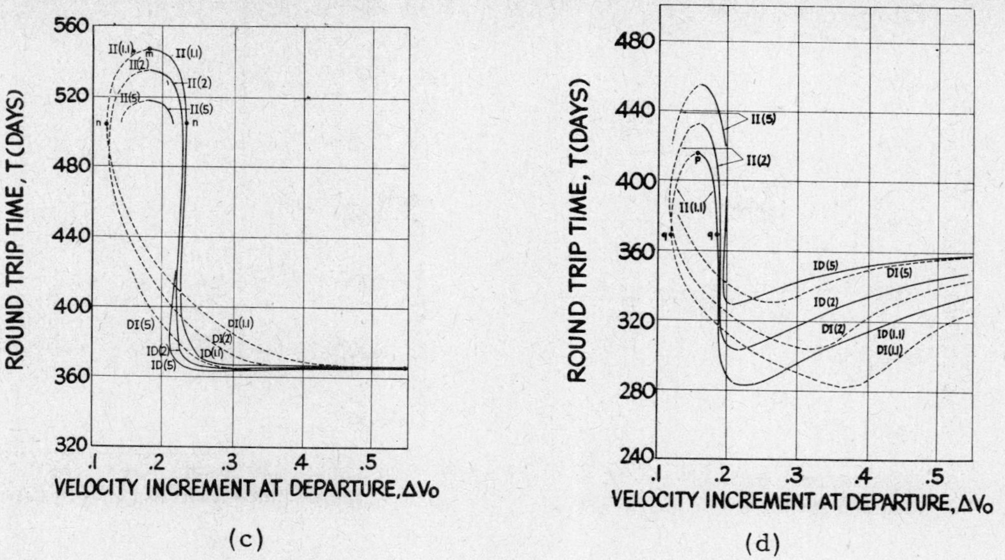

Fig. 10 Second Family Swing-in Second Family Swing-out

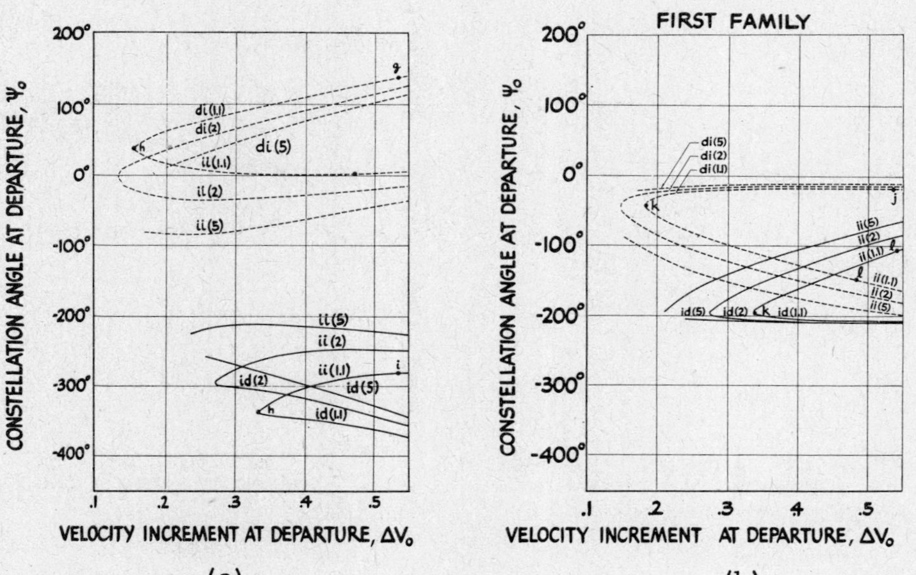

Fig. 11 Departure Constellation Angle vs. Velocity Increment (normalized with the heliocentric velocity of Earth) for Swing-Around-Venus Trajectories. Parking Orbit 1.1 E.R. Closest Approach 1.1, 2, 5 V.R. a) First Family Swing-in b) First Family Swing-out

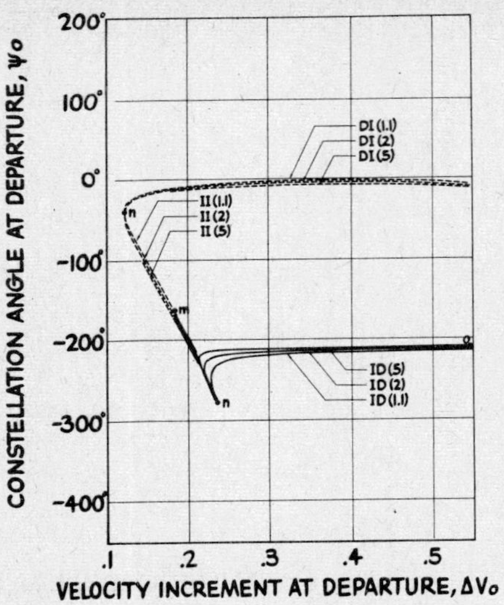

(c)

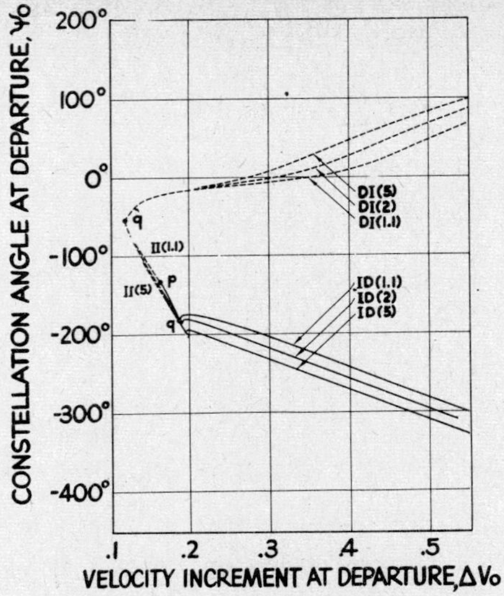

(d)

Fig. 11 Second Family Swing-in Second Family Swing-out

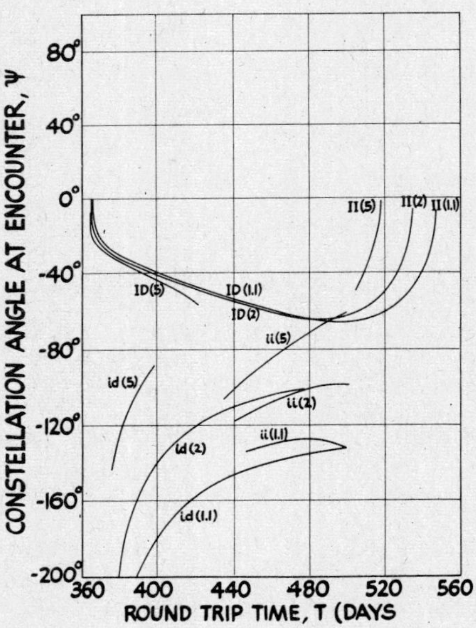

(a)

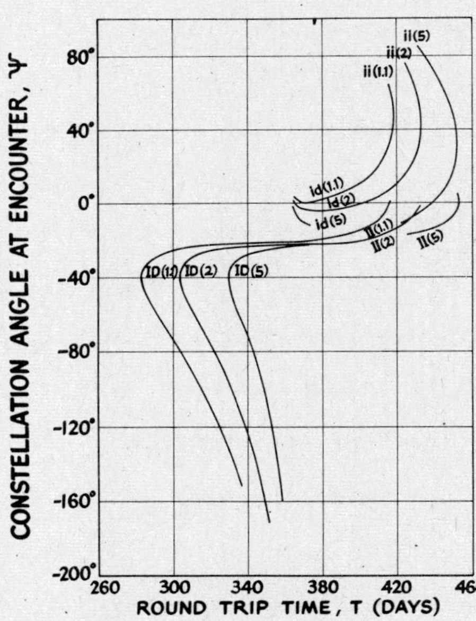

(b)

Fig. 12 Encounter Constellation Angle vs. Total Trip Time for Swing-Around-Venus Trajectories. Parking Orbit 1.1 E.R. Closest Approach 1.1, 2, 5 V.R. a) Swing-in b) Swing-out

Characteristics of Swing-Around Trajectories

Figures 7 through 12 illustrate the characteristics of swing-around Mars and swing-around Venus trajectories. These charts present four items of interest: total trip time, constellation angles at departure and encounter, and velocity increment required to depart from a geocentric circular parking orbit with radius equal to 1.1 Earth radii.

Figure 7a and 7b illustrate two classes of round-trip Martian trajectories; the lower curves indicating trajectories of about one-year duration, the upper curves trajectories of about 1 1/2 to 1-3/4 years. A swing-in trajectory of somewhat less than 1-1/2 year duration exhibits the minimum velocity requirement ($\sim$4 mi/sec). Figures 8a and 8b illustrate the constellation angles at departure. It is interesting to note that these angles are grouped into three different ranges, one extending between 90° and 180°, and two narrow ranges grouped around 0° and 180°. Therefore, departure is possible during one quarter of the synodic year when Mars leads Earth and during a short period of time near inferior conjunction of the planets. The lower velocity requirements are associated with Martian lead angles of 106° to 126°. Figures 9a and 9b present constellation angles at planetary encounter vs. round-trip time. Such information is important in determining communication distance between Earth and vehicle at planetary encounter. Comparison of Figs. 7a and 9a illustrates the fortunate happenstance of minimum communication distance (zero constellation angle) occurring with minimum velocity requirements and shortest total trip for the longer than one-year family. If one knows constellation angles at both launch and encounter, he can, by means of a very simple calculation, determine the transfer times for each leg. Each transfer time is merely equal to the quotient of the change of constellation angle and the difference of planetary mean motions.

$$t = \frac{\psi - \psi_o}{n - n_E}$$

For Venusian orbits, the second family of swing-out trajectories, Fig. 10d exhibits the most favorable combinations of trip time and velocity requirements. The duration of these trips is between 3/4 and 1-1/4 years with velocity requirements at departure as low as 2. 2 mi/sec. The range of constellation angles associated with these conditions lies between -50° and -200°. Thus, favorable departure is possible during almost half of the synodic year. But as can be seen from Figs. 11a through 11d, trips can be undertaken at any time if velocity increments up to 10 mi/sec are available. The comments regarding Figs. 7a and 9a for Martian trajectories apply also to Figs. 12a and 12b for Venusian voyages.

Initial Conditions for Swing-Around Trajectories

This section concerns itself with the initial conditions required as input for an n body precision orbit computation routine. Both initial and arrival conditions are needed, as the latter lead directly (by reversing the sign of the velocity vector) to initial conditions for clockwise trajectories. It was decided to present conditions at closest planetary approach rather than in terms of Earth departure and arrival conditions. This reduces the required amount of data by half, since the arrival and departure conditions at closest planetary approach are identical and both legs of the trajectory may be obtained by running forward and backward in time. Even more important, any errors resultant from the numerical calculations are now split into two segments rather than accumulating for the the entire duration of the mission. Therefore, the following data are presented: constellation angle at planetary encounter, location of closest approach (vertex of the planetocentric hyperbola) and velocity at this point expressed in units of planetocentric Kepler number. As explained above, these data can serve as initial conditions for both legs of the mission for precision calculation routines by running time in both positive and negative directions. If the resultant trajectory cuts the Earth's orbit either ahead or behind the Earth on both ends, the location of

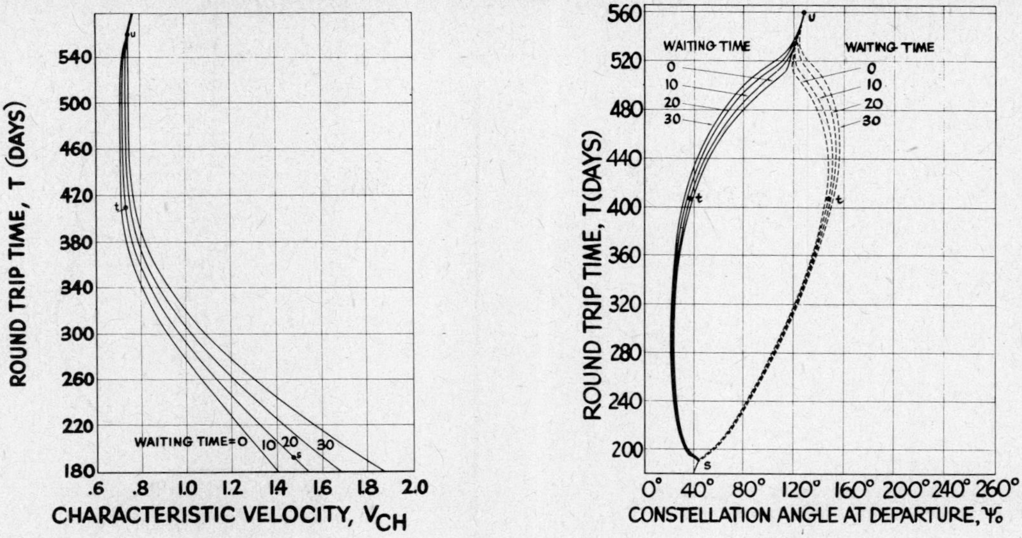

(a) (b)

Fig. 13 Optimum Martian Round-Trip Trajectories. Parking Orbits = 1.1 E.R. and 1.1 M.R. a) Round-Trip Time vs. Characteristic Velocity (normalized with the heliocentric velocity of Earth) b) Round-Trip Time vs. Departure Constellation Angle

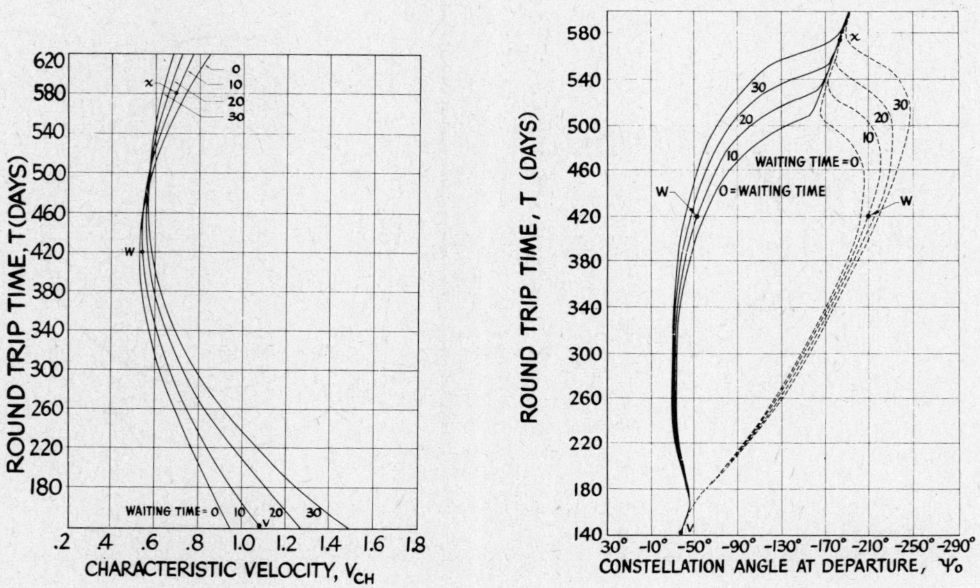

(a) (b)

Fig. 14 Optimum Venusian Round-Trip Trajectories. Parking Orbits = 1.1 E.R. and 1.1 V.R. a) Round Trip Time vs. Characteristic Velocity (normalized with the heliocentric velocity of Earth) b) Round-Trip Time vs. Departure Constellation Angle

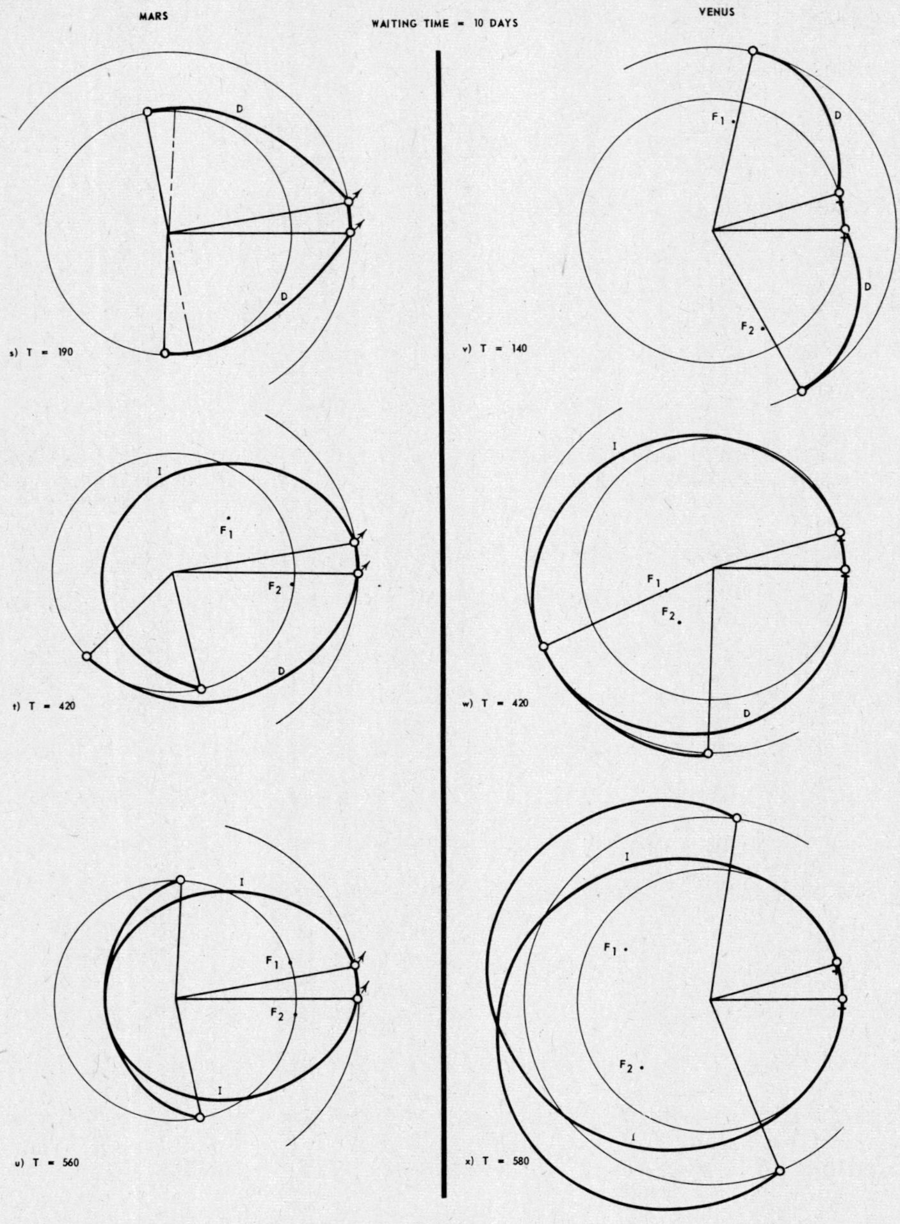

Fig 15 Orbiting Round-Trip Trajectories

closest approach should be adjusted. If, on the other hand, the trajectory branches "straddle" Earth, the velocity magnitude should be modified to achieve impacts.*

Characteristics of Optimum Orbiting Round-Trip Trajectories

Figures 13a through 14b present the characteristics of Optimum Round-Trip trajectories to Mars and Venus, involving waiting times of 0, 10, 20 and 30 days. During this waiting period, the spacecraft travels on a planetary parking orbit of 1.1 planetary radius. The orbital characteristics illustrated in these figures are: round-trip time, constellation angle at departure, and characteristic velocity (sun of the four impulses). Minimum characteristic velocities are obtained for zero waiting time for trip times of approximately 500 days for Martian and 420 days for Venusian trajectories. These velocity requirements are 13.1 mi/sec and 10.2 mi/sec for Martian and Venusian voyages, respectively. Waiting time increases the characteristic velocity requirements for Martian trajectories but does not affect the trip time associated with minimum characteristic velocity. For Venusian voyages, waiting increases the round-trip time associated with minimum characteristic velocity. However, the velocity requirement increases less severely than for Martian trajectories.

Figures 13b and 14b yield the interesting result that, for short and long round-trip times, the optimum orbiting round-trip trajectories consist of symmetrical segments resulting from the fact that the clockwise and counterclockwise segments are identical. This symmetry does not hold for trip times of intermediate duration. The constellation angle range is restricted to allow for departure during about 2/5ths of the synodic year for Martian and over 1/2 of the synodic year for Venusian trajectories. Figure 15 shows round-trip trajectories for Mars and Venus.

---

* For accurate charts, apply to Northrop Space Laboratories and cite NSL62-83.

Initial Conditions for Optimum Orbiting Round-Trip Trajectories

This section concerns itself with the initial conditions at departure from a 1.1 Earth radii parking orbit. The characteristics shown are: constellation angle, velocity increment, and departure point location. Full line curves represent initial conditions for counterclockwise trajectories, dashed curves for clockwise trajectories.*

Constants

The following constants were used in the trajectory calculations:

	Radius of Parking Orbit	Earth	Mars	Venus
R_P/R_E	----	1	1.52369	.72333
V_P/V_h	1.1	.25323	.14	.199
	2	----	.1039	.1476
	5	----	.0657	.0933
$1/n_E$		58.134778 days/radians		

Non-Coplanar Trajectories

In the body of this report the assumption was made that the trajectory of the vehicle lies in the plane of the planetary orbits which were, in turn, assumed to be coplanar. In Appendix B, the necessary modifications for adapting the initial conditions to the non-coplanar case will be discussed.

REFERENCES

1. G. S. Gedeon, "Lambertian Mechanics," Proc, XII Internat. Astrnaut. Cong. Also, "Orbital Segment Machanics," Norair report ASG-TM-61-43R

2. K. A. Ehricke, C. M. Whitlock, R. L. Chapman, and C. H. Purdy, "Calculations on a Manned Nuclear Propelled Space Vehicle", ARS

*See earlier footnote

Preprint 532-57

3. R. H. Battin, "The Determination of Round-Trip Planetary Reconnaissance Trajectories." Aero/Space Sci, Sept. 1959

4. P. G. Johnson, and R. L. Smith, "Round-Trip Trajectories for Mars Observation," Advances in the Astronaut. Sci, Vol. 5, Plenum Press, 1960

5. W. E. Moeckel, "Interplanetary Trajectories with Excess Energy," Proc. IX Internat Astronaut Cong, Springer Verlag, 1959

6. J. F. Dugan, Jr., "Analysis of Trajectory Parameters for Probe and Round-Trip Missions to Mars," NASA TN D-281, 1960

7. M. Vertregt, "Interplanetary Orbits" J. Brit. Interplanetary Soc. Mar-Apr 1958

8. J. F. Dugan, Jr., and C. R. Simsic, "Analysis of Trajectory Parameters for Probe and Round-Trip Missions to Venus," NASA TN D-470, 1960

9. G. S. Gedeon, "A Minimal Set of Orbital Equations," Norair report ASG-TM-61-63

APPENDIX A

HYPERBOLIC ENCOUNTER

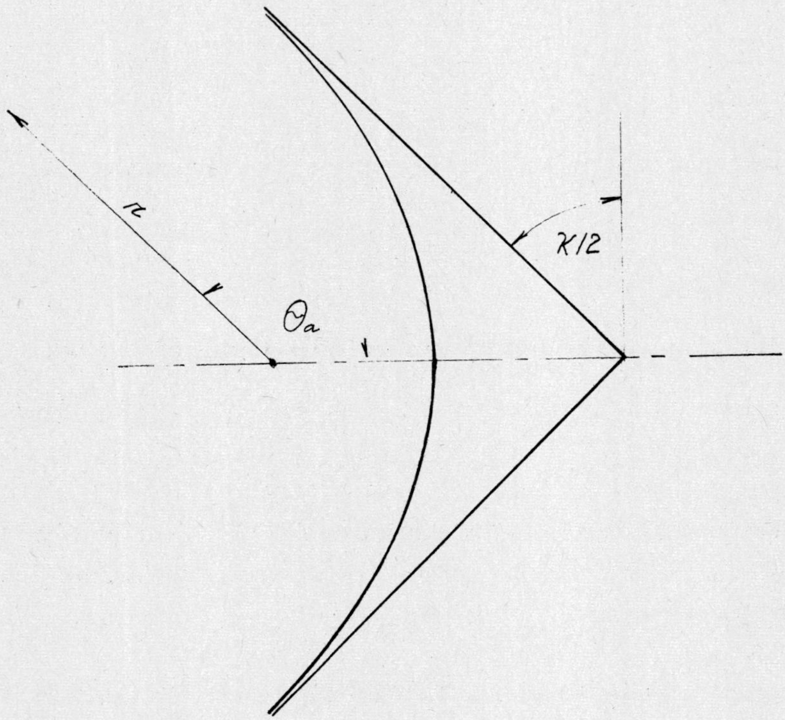

The true anomaly of the asymptote is found by equating the radius to infinity in the defining equation of a conic section

$$r = \frac{(1-e^2)\,a}{1 + e\cos\theta} \tag{1A}$$

yielding

$$\cos\theta_a = -\frac{1}{e} \tag{2A}$$

According to Ref. 9, the eccentricity is given by

$$e = \sqrt{1 - 4q(1-q)\cos^2\phi} \tag{3A}$$

where q is defined as r/2a and called the modulus. At periplanet, the closest approach to the planet, the angle of elevation is zero and therefore

$$e = 1 - 2q_p \tag{4A}$$

In the same reference the relation between the modulus and Kepler number is given as

$$K = \sqrt{2(1-q)} \tag{5A}$$

Thus, the eccentricity expressed by the periplanet Kepler number becomes

$$e = K_p^2 - 1 \tag{6A}$$

and, finally, the true anomaly of the asymptote is

$$\cos\theta_a = -\frac{1}{K^2 - 1} \tag{7A}$$

From the sketch, it can be seen that half of the deflection angle K is

$$K/2 = \theta_a - 90° \tag{8A}$$

Therefore,

$$\sin K/2 = -\cos\theta_o = \frac{1}{K_p^2 - 1} \qquad 0 \leq \frac{K}{2} \leq \frac{\pi}{2}. \tag{9A}$$

Now the vertex velocity is the square root of the sum of the square of the escape velocity and the square of the hyperbolic excess velocity. If unitless velocity expressions are used, the heliocentric hyperbolic excess Kepler number $K^{h.e.}$ must be converted into a planetocentric hyperbolic excess Kepler number. This is accomplished by applying the (V_p/V_h) conversion factor

$$K_p^2 = 2 + \left[\frac{K^{h.e.}}{V_p/V_h}\right]^2 \tag{10A}$$

Combining Eqs. (9A) and (10A) yields

$$\csc \kappa/2 = \left[\frac{K^{h.e.}}{V_p/V_h}\right]^2 + 1 \qquad (11A)$$

APPENDIX B

NON-COPLANAR ENCOUNTER

This appendix is concerned with obtaining initial conditions for non-planar spatial trajectories made necessary by the fact that the planetary orbits are not coplanar. Planetocentric hyperbolas which are "skewed" with respect to the planetary orbit plane will yield such type of trajectories. In order to utilize initial conditions obtained in the coplanar analysis, the following assumptions must be made:

1) The vertex of the planetocentric hyperbolas should be on the sun-planet line,

2) The earlier obtained $(K, \phi, \sigma)_{1, 2}$ sets are to be maintained intact. Utilizing above assumptions, one finds that the deflection angle κ' will be larger than the κ associated with the coplanar case if $(K, \phi, \sigma)_1$, and $(K, \phi, \sigma)_2$ are to be preserved. Simple vector algebra yields the relationship between κ' and κ as

$$\tan \kappa'/2 = \frac{\tan \kappa/2}{\cos i} \tag{1B}$$

Now Eq (9A) can be written as

$$\tan \kappa/2 = \frac{1}{K_p\sqrt{K_p^2 - 2}} \tag{2B}$$

Letting K'_p designate the planetocentric Kepler number at closest approach which produces the deflection angle κ', Eq (2B) can be rewritten for prime values and then combined with Eq (1B) to yield

$$K'_p\sqrt{K_p'^2 - 2} = K_p\sqrt{K_p^2 - 2} \quad \cos i \tag{3B}$$

or

$$K_p' = \left[1 + \sqrt{1 + K_p^{\,2}(K_p^{\,2} - 2)\cos^2 i}\right]^{1/2} \tag{4B}$$

Also, by combining Eqs (10A) and (4B) one obtains

$$(V_p/V_h) = \frac{K^{h.e.}}{\sqrt{K_p'^{\,2} - 2}} \tag{5B}$$

Therefore, the results of the coplanar investigations remain valid if K_p and the radius of closest approach are modified according to the last two equations (before inserting those as initial conditions into a precision trajectory calculation routine).

The plane changes i_1 and i_2 (in heliocentric space) with respect to the planetary orbit plane are obtained from the following relations:

$$K_1 \sin \varnothing_1 \sin i_1' = K^{h.e.} \cos K'/2 \ \sin i \ = K_2 \sin \phi_2 \sin i_2 \tag{6B}$$

From this last equation, it is evident that in general $i_1 + i_2$, and hence a net plane change was affected by the gravitational field of the planet without exterior propulsion.

AN ANALYSIS OF LUNAR INJECTION PARAMETERS AND THEIR EFFECT UPON THE CHARACTERISTICS OF ENTRY INTO THE EARTH'S ATMOSPHERE*

J.G. Swider** and R.D. Taylor**

Relationships between lunar injection parameters and the position of Earth return trajectories at entry into the Earth's atmosphere are presented. The effect of changing the longitude of the lunar injection radius vector upon the required injection energy as a function of return flight time is shown.
Emphasis is placed upon determining the inclination of the trajectory plane at entry into the Earth's atmosphere as a function of the instantaneous angle between the trajectory plane and lunar plane at injection. Also, the change in position of the Earth perigee vector as injection energy varies is examined. These relationships are used to determine the longitudinal range requirements of an entry vehicle for continuous monthly operation. All results have been obtained using an unrestrictive, four-body, integrating program on an IBM-704 computer.

---

* A complete version of this paper (AAS 62-26) may be obtained from the American Astronautical Society or by contacting the authors.

** Ling-Temco Vought, Inc., Astronautics Division, Dallas, Tex.

eighth
annual
meeting

AMERICAN ASTRONAUTICAL SOCIETY

THE NEED FOR ASTRONOMICAL TELESCOPES IN THE TRACKING OF SPACE PROBES*

J. C. Walker+

There exists a variety of astronomical telescopes whose limiting magnitudes are fainter than those of any existing tracking telescopes. Similarly, the angular accuracies of certain of the astronomical telescopes are more precise than those of tracking telescopes, at least for the observation of deep-space probes where time is not the limiting factor. In this paper, it is shown that probes equipped with simple visibility-enhancement devices can be tracked at interplanetary distances by existing astronomical telescopes. This paper shows that pointing information for the entire flight of an interplanetary vehicle may be derived merely from observations of its parking orbit. This would make possible the use of astronomical telescopes for observations of deep-space probes whose orbital elements have not been announced by the launching nation.

INTRODUCTION

The tracking task for deep-space probes of the National Aeronautics and Space Administration (NASA) consists primarily of keeping a precise track of their own space vehicles. These vehicles, in general, will transmit clear signals from positions that will be closely approximated in advance. The North American Air Defense Command (NORAD) has the responsibility of keeping track of all man-made objects in space. For this latter task, the requirements for accuracy may not always be

---

*Engineering Paper No. 1230 prepared under the sponsorship of the Douglas Aircraft Company Independent Research and Development Program, Account No. 9703-51741.

+Design Specialist, Aero/Astrodynamics Section, Missile and Space Systems Division, Douglas Aircraft Co., Inc.

as stringent as those of NASA, but there exists the problem of detecting probes which are less cooperative. These probes may transmit at times when they are least visible to NORAD's stations, at unknown times, and/or unknown frequencies. Of course, technical difficulties may also infrequently cause NASA to desire to track a probe that is not transmitting.

The advantages and limitations of astronomical telescopes for each of these purposes will be developed. These applications are explored after an evaluation and comparison of the tracking capabilities of the existing Baker-Nunn tracking network with existing astronomical telescopes. Also included are some launching-frequency estimates for both United States and Russian probes.

Throughout this paper, photographic recording without electronic enhancement is assumed. When an emulsion is specified, it is assumed to be Eastman Kodak 103a-0. Although it is known that electronic devices may be used to increase the limiting magnitude of telescopes, they were not considered because none exists which is suitable for use on existing telescopes in making the kinds of observations relevant to this paper.

TRACKING CAPABILITIES OF TELESCOPES

Comparison of Baker-Nunn and Astronomical Telescope Limiting Magnitudes

A comparison of the limiting performance of astronomical telescopes with the Baker-Nunn tracking telescopes for objects which produce streaked images (i. e., objects with motion imperfectly known or too rapid for the telescopes to track) is shown in Figs. 1 and 2. The calculations for these figures are made with reference to the work of Baum for the limiting magnitude (see Appendix). The telescope parameters used in the preparation of these figures were primarily obtained from Ref. 1. On these figures, curves are shown only for the Baker-Nunn telescope, the 200-inch Hale (which is the largest of the conventional

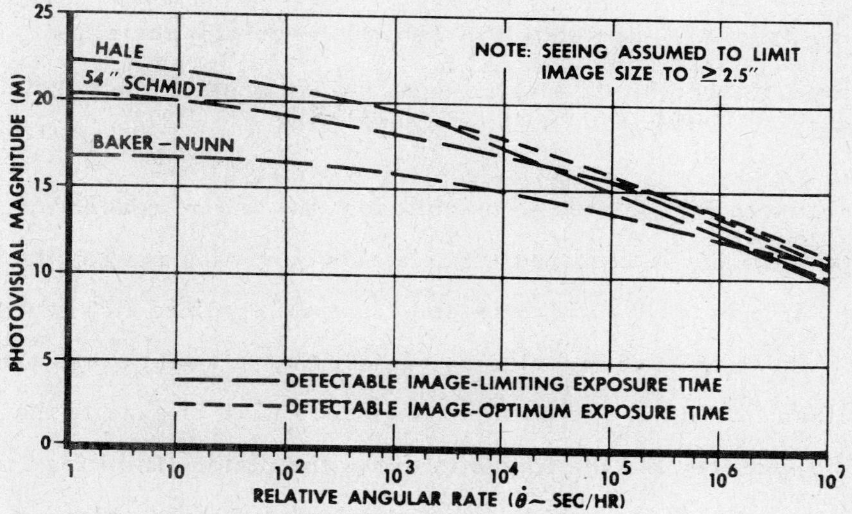

Fig. 1 Limiting Photovisual Magnitude as a Function of Object to Telescope Relative Angular Rate for Barely Detectable Images

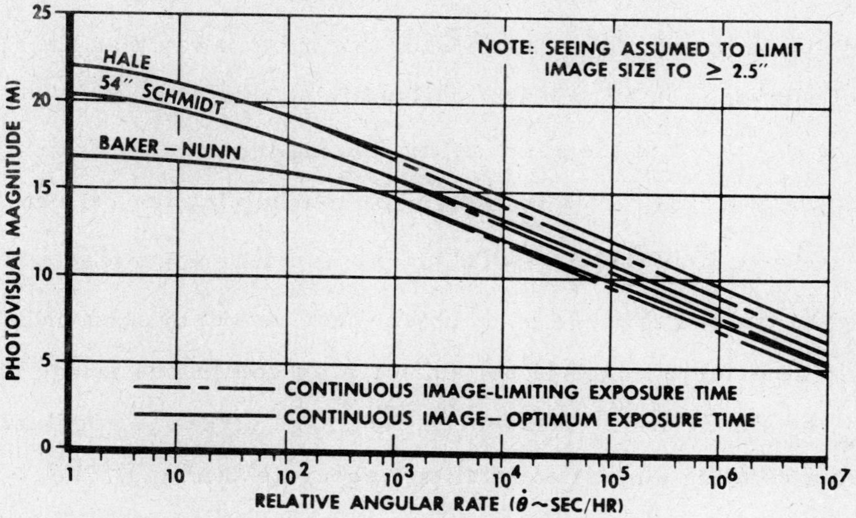

Fig. 2 Limiting Photovisual Magnitude as a Function of Object to Telescope Relative Angular Rate for Continuous Images

telescopes), and the 54-inch Schmidt (which is the largest of the Schmidt telescopes). For other astronomical telescopes, the limiting magnitudes, at one tracking rate only, are shown in Table I. Average seeing is assumed to limit image size to a value equal to, or greater than, 2.5 sec of arc.

The curves shown in Fig. 1 are the magnitudes that can be reached if only a detectable image is required. For a slowly moving image, the best exposure time is the limiting exposure time determined by sky fog. For an object whose image is moving so rapidly that it would cross the plate in less than this time, the optimum exposure time is exactly that which would be required for the image to cross the plate. Limiting magnitudes for both optimum and limiting exposure time are shown in Fig. 1.

The total spread of limiting magnitudes for all the telescopes studied is approximately 5.5. In relating the astronomer's magnitude scale to tracking needs, it must be remembered that magnitude is a logarithmic concept. This 5.5 magnitude difference means that the most powerful telescope can see the same object twelve times further away than the least powerful telescope. Other ways of visualizing this difference would be to say the effect is the same as increasing the diameter of each object in the sky by a factor of twelve, or multiplying the reflecting area by one hundred and forty-four.

To determine the time of passage of an object past a certain point on the plate, a barely detectable image is not sufficient; a continuous image is required. (In the Baker-Nunn telescopes, the image streak is regularly interrupted by a rotating shutter to facilitate accurate timing.) The faintest magnitude for which each telescope can provide such an image, if optimum exposure time is used, is shown by the solid lines in Fig. 2. For rapidly moving objects, this limiting magnitude is a very insensitive function of exposure time for times well beyond optimum. In fact, exposure

TABLE I
LIMITING MAGNITUDE OF SELECTED ASTRONOMICAL TELESCOPES
(CONTINUOUS IMAGE)

OBSERVATORY	LOCATION	LAT. (DEG)	LONG. (HRS)	TYPE	APERTURE (IN.)	FOCAL RATIO (f/)	LIMITING MAGNITUDE (AT 1"/HR)
TAUTENBURG	GERMANY	50:59	−0:47	SCHMIDT	54	3.0	20.5
MT. PALOMAR	CALIFORNIA	33:21	7:47	SCHMIDT	48	2.5	20.2
ROYAL	BELGIUM	50:48	−0:17	SCHMIDT	46	2.5	20.2
KVISTABERG	SWEDEN	59:30	−1:10	SCHMIDT	40	3.0	20.1
BOYDEN	SOUTH AFRICA	−29:02	−1:46	BAKER-SCHMIDT	32	3.7	20.0
STERNWARTE	GERMANY	53:29	−0:41	SCHMIDT	32	3.0	19.8
VATICAN	ITALY	41:45	−0:51	SCHMIDT	25	3.8	19.7
TONANTZINTLA	MEXICO	19:02	6:33	SCHMIDT	26	3.2	19.6
UNIVERSITY	BELGIUM	50:37	−0:22	SCHMIDT	24	3.75	19.6
HARVARD	MASSACHUSETTS	42:30	4:46	SCHMIDT	24	3.5	19.6
MT. STROMLO	AUSTRALIA	−35:20	−9:56	SCHMIDT	20	3.5	19.2
MT. PALOMAR	CALIFORNIA	33:21	7:47	REFLECTOR	200	3.65	22.1
LICK	CALIFORNIA	37:20	8:07	REFLECTOR	120	5.0	21.6
ASTROPHYSICAL	CRIMEA	44:44	−2:16	REFLECTOR	104	3.8	21.4
MT. WILSON	CALIFORNIA	34:13	7:42	REFLECTOR	100	5.0	21.4
McDONALD	TEXAS	30:40	6:56	REFLECTOR	82	4.0	21.1
SAINT MICHEL	FRANCE	43:56	−0:23	REFLECTOR	77	5.0	21.1
MT. STROMLO	AUSTRALIA	−35:20	−9:56	REFLECTOR	74	5.0	21.1
OKAYAMA	JAPAN	34:34	8:54	REFLECTOR	74	4.9	21.1
RADCLIFFE	SOUTH AFRICA	−25:47	−1:53	REFLECTOR	74	4.8	21.1
DOMINION	CANADA	48:31	8:14	REFLECTOR	73	5.0	21.1
PERKINS	OHIO	40:15	5:32	REFLECTOR	69	4.3	21.0
HARVARD	MASSACHUSETTS	42:30	4:46	REFLECTOR	61	5.1	20.9
NATIONAL	ARGENTINA	−31:36	4:18	REFLECTOR	61	5.0	20.9
PADUA UNIVERSITY	ITALY	45:52	−0:46	REFLECTOR	48	5.0	20.6
ROYAL	SOUTH AFRICA	−33:56	−1:14	REFLECTOR	40	4.5	20.3
LA PLATA	ARGENTINA	−34:54	3:52	REFLECTOR	33	18.0	19.8

times of the order of the time necessary for the image to cross its own diameter, and up to one thousand times as long, show no significant variation in limiting magnitude. However, as shown by the lower curve in Fig. 2 for each telescope, the limiting magnitude is fainter for the continuous image of a moving object, if the exposure is not extended to the limit determined by sky fog.

The limiting magnitude for a continuous image using the Baker-Nunn telescope will be observed to be brighter than that presented by Henize (Refs. 2 and 3). A further investigation is being made of this discrepancy, which has not yet been fully explained. It may be due largely to a different assumption concerning chemical fog and criterion for detectability. In any case, the author has been unable to find an observational verification of the Henize predictions, while Baum's results for limiting exposures have been substantiated with several telescopes. Due to reciprocity failure, effective exposure time varies from actual exposure time; therefore, actual exposure time should be used for calculating streak lengths. No distinction between the two different times was made in calculating these curves.

The breaks in the curves of Figs. 1 and 2 occur at the point where the streak length exceeds the plate size, assuming that the image moves along a diameter of the mirror. Errors in pointing will cause the image to move along a chord other than a diameter. This will have the effect of moving these breaks in the curves to the left. Conventional telescopes will be far more sensitive to such aiming errors than the Schmidt telescopes, because of their smaller fields of view. Limiting magnitudes are plotted versus relative angular rate on these figures. If an object could be tracked perfectly, this relative rate could be held to zero. However, as viewed from the earth, the moon is a rapidly moving object compared to other astronomical bodies. In the design of large astronomical telescopes, the astronomer's preoccupation with natural phenomena has usually resulted in tracking rates of the order of the moon's geocentric motion. Relatively speaking, the tracking rates of the Baker-Nunn telescopes are essentially infinite, and the design goal was the capability of tracking to an accuracy of 1 per cent. For low-altitude vehicles, which grossly exceed the tracking rates of astronomical telescopes, it is clear that the Baker-Nunn tracking telescopes can detect fainter objects, providing a prior estimate of the object's orbit is available. Thus, from a limiting magnitude point of view, the best telescope for a given task is dependent upon the angular rate of the object and the accuracy of the pointing and tracking information. There are many situations in which astronomical telescopes can do an appreciably better job of tracking than the existing tracking telescopes.

Of interest, then, is an examination of the abilities of telescopes, in general, as tracking instruments. First, it should be pointed out that the limiting magnitudes of Figs. 1 and 2 are calculated for observations made at the zenith, in a dark sky. Unfortunately, not all observations can be made under these conditions. For instance, an object near Venus can never get far enough from the sun to even approach these conditions. Second, it should be pointed out that for Figs. 1 and 2 average (actually average at Palomar) seeing conditions are assumed. Nights when the limits of the larger telescopes are a full magnitude fainter are common. Of course, on other nights, no observations are possible. Thus, considerable variation from the limiting magnitudes presented occurs. Third, a serious factor which has limited the use of the Baker-Nunn telescopes in detecting orbiting vehicles in the problem of earth shadow. This ceases to be a problem for observing objects far enough away to require the use of large astronomical telescopes. For instance, the infrequency of lunar eclipses shows that at lunar distances earth shadow is a minor problem.

Photovisual Magnitude to Object Size Relationship

To determine what can and cannot be seen in specific situations, some very general plots may be drawn. Figure 3 provides a means of estimating the magnitudes of perfectly-reflecting specular and diffuse spheres at various positions. The use of Fig. 3 is shown by the following example. Assume that a diffuse reflecting sphere 10 ft in diameter is 5 million miles from the earth, at a direction of 135 deg from the sun. Starting from the left, draw a line from 10, on Scale A, through 5×10^6 on Scale B to Scale C. The value read on Scale C is the magnitude of a 10-ft specularly reflecting sphere 5 million miles from the earth, directly opposite and at a distance of 1 astronomical unit from the sun. (This geometry would, of course, be impossible.) To correct this magnitude to a diffusely reflecting sphere at the proper phase and proper distance from the sun, begin at 5×10^6 miles on the horizontal scale at the right, read up to the dashed line for 135 degrees, and left to the magnitude correction scale, Scale E. Connect this point with the previously obtained point on Scale C. The desired magnitude is then read on Scale D as 22.75.

The observation angle effect on specular spheres is due to the choice of distance from the earth and observational angle as independent variables. This choice causes the variation in magnitude due to the distance of the sphere from the sun to appear as an observation angle effect. Figure 3, which is based on the work of Tousey (Ref. 4), gives the magnitudes of spherical objects at the zenith. The equations for the calculation of this nomograph are given in the Appendix.

In addition to illustrating the use of the plot, the previous example illustrates that even a vehicle large by present-day standards cannot be optically tracked to interplanetary distances without some sort of visibility enhancement. One method of enhancement would be to use large spheres of a construction similar to the Echo passive communication satellite.

However, the use of an Echo-type object 100 ft in diameter could extend the range of the larger Schmidt telescopes to only about 15 million miles. This method is simple, but the cost, in terms of weight for even this extension, is high.

The California Institute of Technology's Jet Propulsion Laboratory (JPL) reports that the orientation of U. S. interplanetary probes will, in some cases, be controlled to within 0.05 deg (Ref. 5), and that the probes will have both directional antennas pointed at the earth, and sun sensors. Figure 4 shows the photovisual magnitudes of a plane circular perfectly-reflecting mirror which is oriented to reflect sunlight to the earth; i. e., the line normal to the mirror's surface bisects the earth-sun angle. For a mirror at a distance of 1 astronomical unit from the sun, this requires a pointing accuracy of ± 1/8 deg. Figure 4 is calculated by the method of Ref. (6) (see Appendix), and is read in the same manner as Fig. 3. This type of visibility enhancement is more complicated in that it requires orientation, but such mirrors could offer great weight savings compared to spherical reflectors, particularly if they could be constructed, for example, of an inflatable torus with a sheet of aluminized Mylar stretched over the hole. As shown in Fig. 5, at the orbit of Venus and at an angle from the sun of 30 deg, a mirror 3 ft in diameter could easily be photographed by the large Schmidt telescopes. The curves shown in this figure are derived from the data presented in Figs. 3 and 4.

By orientating to a lesser accuracy a small zone, of one base, of a large specular sphere, the magnitude of the entire sphere is essentially obtained. Specifically, if a is the angular diameter of this zone, as subtended at the center of the sphere, then at 1 astronomical unit, where the sun subtends 32 min of arc, the sphere need be oriented only within a tolerance of $\pm 1/2\,(\alpha - 16')$. See Fig. 6. For example, a zone subtending 2 deg, 16 mins at the center of the sphere will have the magnitude of the entire sphere, if orientated within ± 1 deg. For a 100-ft diameter sphere,

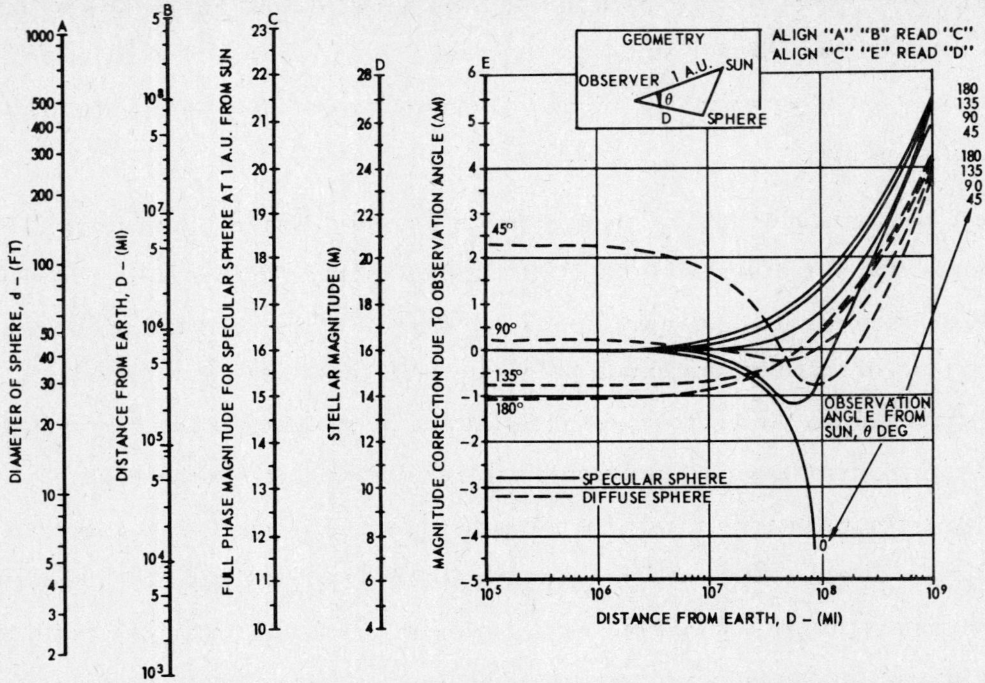

Fig. 3 Nomograph: Stellar Magnitudes--Perfectly Reflecting Spheres

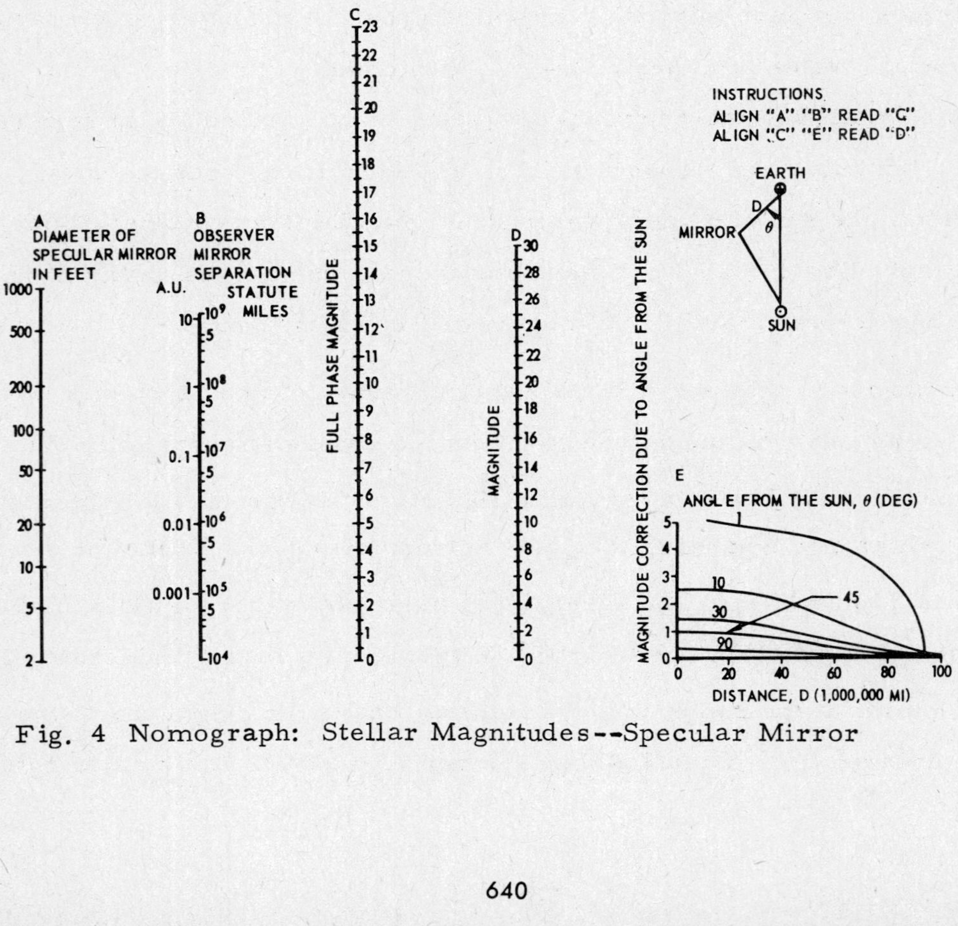

Fig. 4 Nomograph: Stellar Magnitudes--Specular Mirror

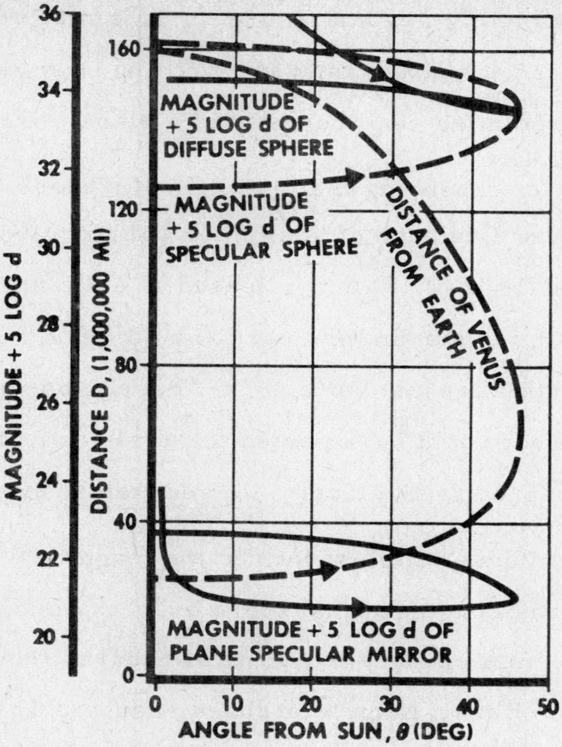

Note:

d = diameter of object (ft)

Fig. 5 Stellar Magnitudes of Various Reflecting Surfaces at the Orbit of Venus

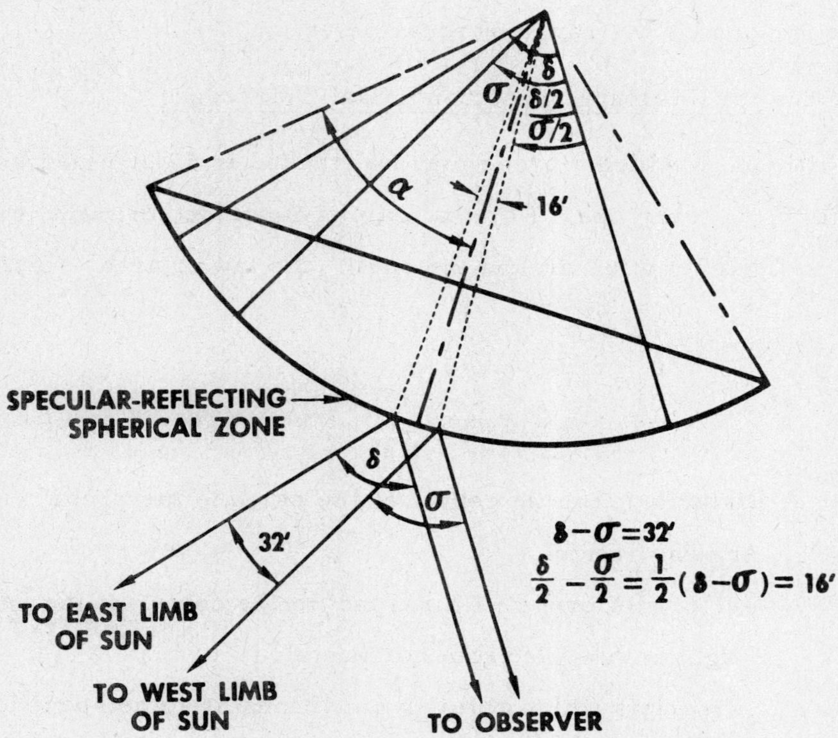

Fig. 6 Specular Reflection of Sun from Spherical Zone

such a zone would be approximately 2 ft in diameter. This method may provide a compromise between the disadvantages resulting from the weight of the sphere and the pointing requirements of a plane mirror.

There is also the possibility of enhancing the visibility of a space vehicle for a short period of time by the release of sodium or lithium clouds. In Ref. 7 it is calculated that 10 lb of lithium released at a distance of 1 astronomical unit from both the sun and the earth would, in 2.5 sec, become a star-like object with a magnitude of 16.5. Based upon Soviet experience (Ref. 8), this object could be expected to persist for at least 4 min, and could be photographed by the larger Schmidt telescopes.

Comparing the limiting magnitudes of telescopes to the magnitude of objects versus their size, it is concluded that there exist astronomical telescopes which could track most probes for several hundred thousand miles beyond the range of the Baker-Nunn telescopes, and could track suitably equipped probes throughout the solar system. It would also appear that telescopes provide some hope for detecting non-transmitting probes that are far beyond the range of radar detection.

Minimum Altitudes for Tracking by Astronomical Telescopes

The minimum altitude at which a probe may be tracked is determined by the angular rate of the telescope. For that part of its trajectory where a probe executes Kepler motion around the earth, the law of areas states:

$$r^2\dot{\theta} = r_o V_o \cos y_o$$

where:

r	=	Distance from the center of the earth to the probe
$\dot{\theta}$	=	Angular rate of r
V	=	Velocity of probe with respect to the center of the earth
γ	=	Flight path angle from horizontal
o	=	Subscripts which refer to conditions at some specific time

The highest performance probe considered possible is assumed to describe a hyperbola with a perigee altitude of 100 nm and a velocity of 100, 000 fps. From the above equation, assuming a maximum tracking rate of 3/4 deg per hour, it follows that an astronomical telescope would not be able to track such a probe nearer than 126, 000 nm from the center of the earth. Slower probes may be tracked at smaller distances; e. g., for a parabolic orbit with a perigee altitude of 100 nm, the lower limit of geocentric distance is 76, 000 nm.

Given perigee altitude and velocity, or eccentricity and semi-major axis (semi-latus rectum for a parabola) of the escape conic, Fig. 7 enables the calculation of angular rate versus radius. As an example, this figure shows that Pioneer V could not be tracked with a telescope with a tracking rate of 3/4 deg per hour at a radial distance of less than 78, 000 nm, which Pioneer V attained approximately 10 hours after launch.

At distances where vehicles can be tracked by astronomical telescopes, precise timing should not be as important as it is for near-earth satellites. At a tracking rate of 3/4 deg per hour, a timing error of over 1 sec would become the predominant error. (The Baker-Nunn telescopes provide timing to within approximately 0. 001 sec.) However, at Venus's inferior conjunction, a vehicle approaching Venus at a velocity of even 100, 000 fps would have an angular rate of only about 0. 2 deg per hour. For Mars, the corresponding angular rate is 0. 1 deg per hour at opposition.

APPLICATIONS

Finding and Tracking Probes Which Cannot be Detected by The Deep Space Net

The North American Air Defense Command (NORAD) is charged with the responsibility of maintaining up-to-date ephemerides of all man-made objects in space. Probes launched by the U. S. will be equipped with either a beacon or a transponder which will enable the deep-space radars of the Jet Propulsion Laboratory to track them to interplanetary distances.

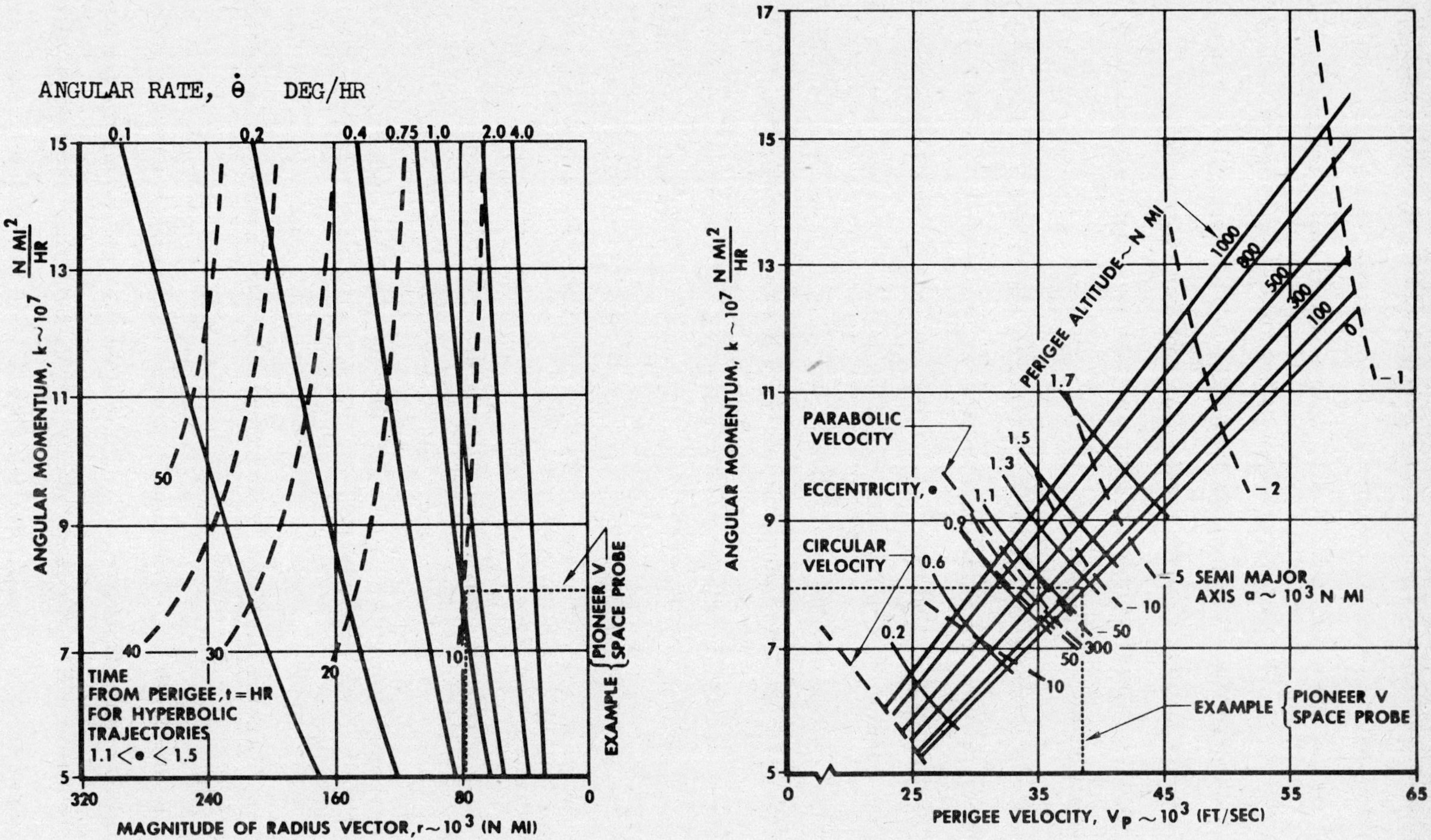

Fig. 7 Nomograph: Angular Rate for Geocentric Orbits

The transmission frequencies of probes launched by the U. S. S. R. are not designed for compatibility with the deep-space net. Furthermore, they will probably transmit only while over the U. S. S. R., which is a poor location for detection by the U. S. deep-space net. At present, skin-track radar and optical techniques are the only established methods of finding and tracking non-transmitting probes. The capability of radar techniques decreases as the fourth power of distance, whereas optical capabilities decrease as the second power of distance. Therefore, for deep-space tracking, only optical surveillance appears feasible. This fact must be contrasted with the problem of surveillance of near-earth satellites. Because of the earth's shadow, these objects will generally be unobservable by optical means; however, with space probes this quickly ceases to be a problem. Of course, as a deep-space probe recedes from the earth, its apparent magnitude at any phase may be expected to become fainter.

Many problems need to be solved, however, if astronomical telescopes are to contribute to the solution of this problem. One is the problem of availability. There is already a high utilization of large astronomical telescopes with the problems of astronomy. To obtain the use of the telescope, particularly on an immediate emergency basis as might be necessary for an unannounced Soviet space probe, is almost certain to disturb the astronomical work which is really the prime interest of the observatories. The cooperation of the observatories can be expected only if every effort is made to minimize the disturbance caused by such a request. A second problem is the obtaining of adequate pointing information. Except for the Schmidt-type telescopes, astronomical telescopes have a very limited field of view and, as shown above, their effectiveness can be greatly increased if the approximate angular rate of the vehicle is known. In some applications, this information can be supplied by analytic studies with only limited observational data.

Prediction, years in advance, of likely launching dates for interplanetary vehicles to within a period of a few months is relatively simple. In fact, the most likely dates can be estimated to within a few weeks. Once the orbit of an earth satellite has been observed, it can be inferred whether or not it is a possible parking orbit for flight to another planet. This may assist in short-term scheduling and in detecting errors in the transmittal of pointing information. The method of making both long-and short-term predictions is explained below.

In Fig. 8 are plotted contours of the transit time that would be required to reach Venus, given a fixed launch velocity capability, beginning at various launch dates in 1960 and 1961. This pattern of launch velocities is repeated, except for changes near the 180-deg transfer line, at periods equaling the Venus synodic year. The slight variations in this pattern are due to variations in the relative position of the nodes, since they depend on Venus's period rather than on its synodic year. As may be seen from this figure, the Hohman orbit which requires minimum energy for transfer between coplanar orbits (Ref. 9) is not a practical trajectory for orbits that are even slightly non-coplanar. The reasons for this circumstance are discussed in Ref. 10. Plots similar to Fig. 8 can be drawn for flight to Mars.

The use of the plot in Fig. 8 for long-term predictions is as follows: The achievement of velocities beyond approximately 40, 000 fps is very expensive in terms of payload. Restricting the velocity to 40, 000 fps or less still leaves a fairly large choice of launch dates; thus, the region inside this velocity contour represents a preferred period for interplanetary launch. It is shown in Ref. 10 that a velocity of at least a few hundred feet per second above the absolute minimum is a necessity, from a guidance-sensitivity standpoint; thus, the region inside such a velocity contour should be avoided. Another consideration relates to communication problems at the time the probe reaches Venus. Obviously, for

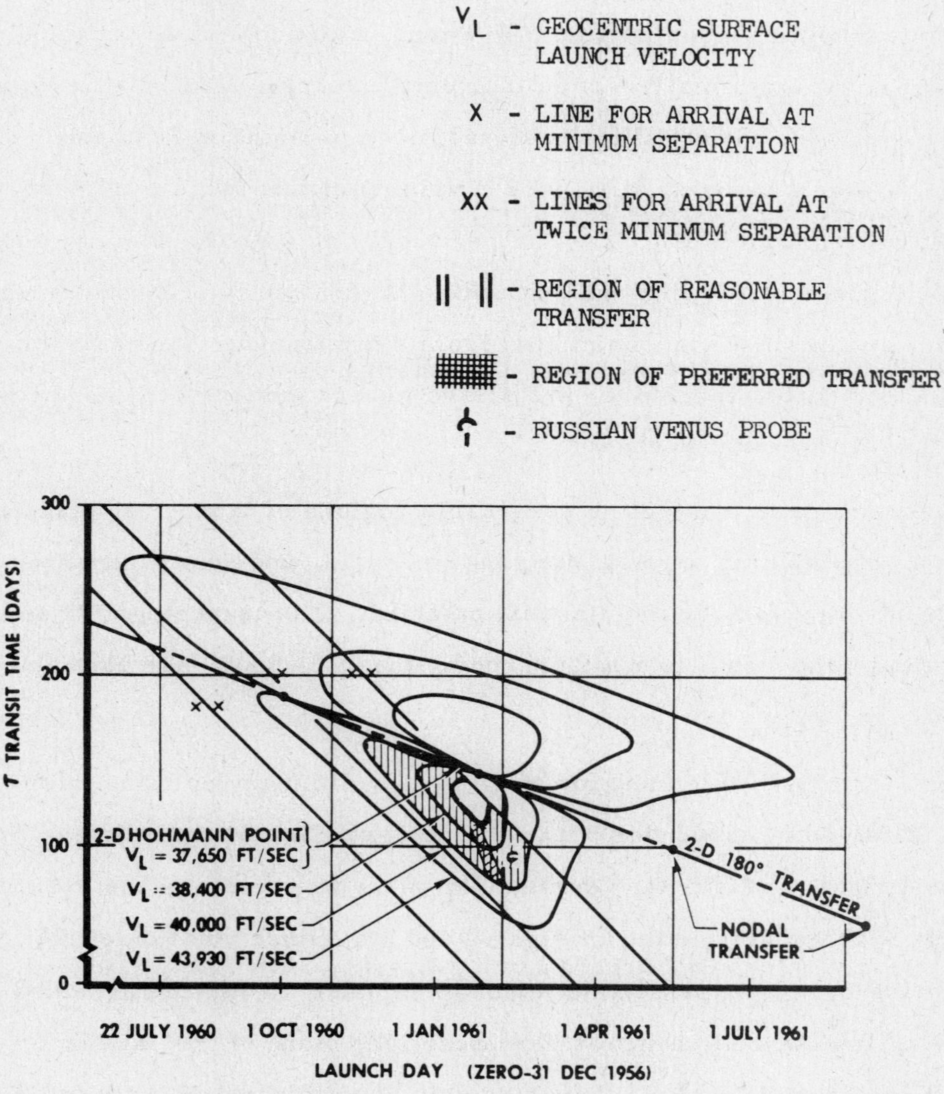

Fig. 8 Earth-to-Venus Transit Time (τ) as a Function of the Launch Day for Constant Launch Velocities (V_L)

optical tracking purposes, the worst time for viewing a probe near Venus is when Venus is at inferior conjunction, because of the angular proximity of the probe to the sun. It is shown in Fig. 6 that observation angles up to 40 deg result in no loss of visibility due to increasing range. Noise from the sun would also make difficult radio communication with a probe near Venus at inferior conjunction. However, more than twice minimum Earth-Venus separation at the time the vehicle reaches Venus is not desirable either, due to the difficulty in radio communication resulting from the increase in distance. Lines of minimum-and twice minimum-separation are shown in Fig. 8. Furthermore, for a given velocity, the choice of a launch date which will minimize flight time is preferable, not only from a reliability standpoint, but from a propaganda standpoint as well. Launch dates, therefore, are biased toward the line running through the minima of the velocity curves.

The above considerations define reasonable regions of launch for planetary probes as shown by the vertical hatching in Fig. 8, and sub-regions of preferred launching times by the diagonal hatching. The launching date and flight time for the Venus probe launched by the U. S. S. R. are also shown on this figure.

For short-term prediction and for providing preliminary pointing information, it should be noted that with each point on Fig. 8, there is an associated right ascension and declination of the asymptote of the escape hyperbola. These are plotted in Figs. 9 and 10. Since rotation of the orbital plane is very expensive in terms of impulse, it may be assumed that this asymptote will be in, or very near, the plane of the parking orbit. If on a probable probe launching date, as defined in Fig. 8, a new object is observed in orbit, then a simple check could be made to determine if this orbit contains any of the suitable escape vectors for that day, as obtained from Figs. 9 and 10. This will enable the likelihood that this is a parking orbit of an interplanetary vehicle to be determined.

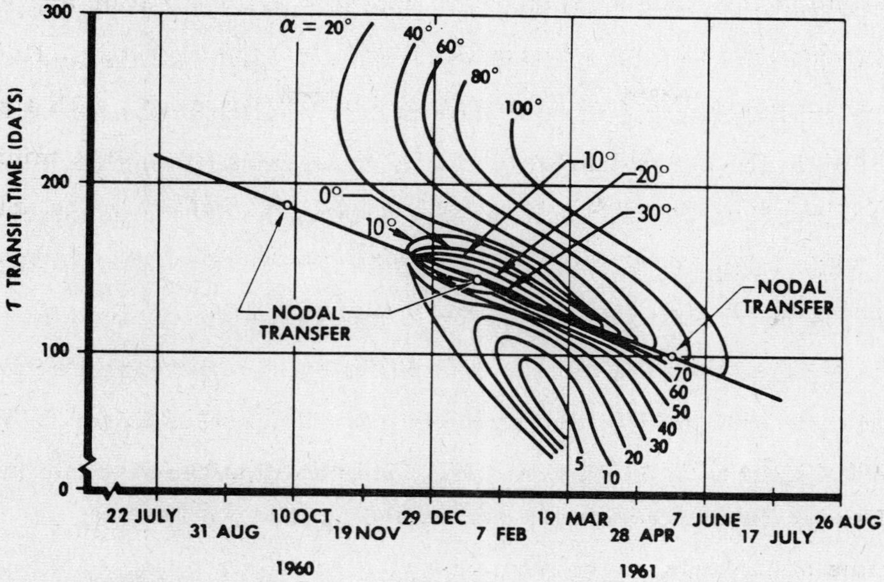

Fig. 9 Transit Time (τ) as a Function of the Launch Day for Constant Right Ascension (α)

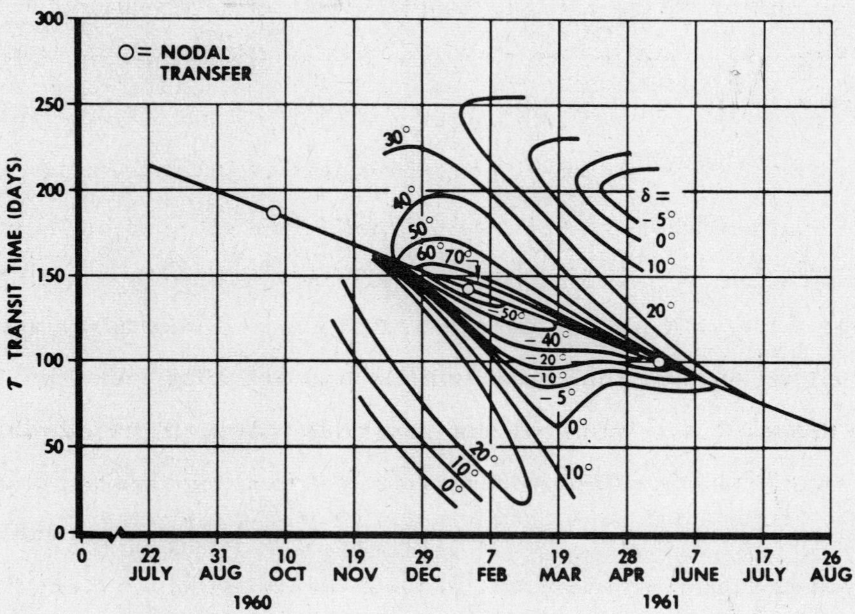

Fig. 10 Transit Time (τ) as a Function of the Launch Day for Constant Declination (δ)

(It should be noted that low orbits are the most efficient parking orbits for all vehicles, except perhaps those with a very high thrust-to-weight ratio. Therefore, vehicles will not be left in these orbits very long because of the resulting high drag losses.) If it is assumed that the impulse to put the vehicle into the escape hyperbola is applied instantaneously and tangentially to the parking orbit (which is the most efficient way), then the escape hyperbola from each passage around the orbit is uniquely defined by Figs. 8 through 10 and the ephemeris of the earth satellite. If the object is observed to disappear or to modify its appearance on a particular cycle, then the telescope pointing information may be derived from Figs. 8 through 11 to determine if this was, as suspected, a planetary probe. A check of Figs. 9 and 10 indicate that the satellite launched on February 4, 1961 by the U. S. S. R. was not intended as a Venus probe, but was probably a test of the first two stages of the probe launched February 12, 1961.

These plots can be prepared in advance, thereby minimizing the communication problem during the critical time immediately after a new object has been discovered. After the first few days of flight, the two-body theory will no longer suffice, but by then regular teletype ephemerides should be available.

The above considerations, of course, do not define all possible launching times for all types of missions, but considering the present state of Russian technology, it may be expected that they will take advantage of each of the next several such opportunities that arise for launching interplanetary probes. Astronomical observatories may well provide the only observation of these Russian probes, since the existing tracking facilities available as recently as last February did not provide any observations of the Venus probe launched by the U. S. S. R.

Similar plots can be drawn for the moon, except with the whole cycle taking place each month instead of each synodic year. Dates for launching to the moon extend over a period of approximately five consecutive days each month. Which months are chosen by the Russians will depend as much upon the purpose of the mission as on kinematical considerations.

The plots used as examples have been derived from an approximate theory, but they could be prepared precisely for each interplanetary launch period illustrated in Fig. 12. These would be sufficient to supply, merely from information about the parking orbit, no only scheduling data but pointing information for the Baker-Nunn's 5 x 30 deg field of view or probably the field of any Schmidt. They might also, on occasion, provide data accurate enough for use by conventional astronomical telescopes. Whether lunar flight could be predicted to the required degree of precision is subject matter for a further study.

Because of the availability problems associated with conventional telescopes, it would appear that NORAD's surveillance capabilities would be enhanced by the addition of an optical net of several large telescopes. The use of astronomical telescopes could provide data for interim emergencies and, it is hoped, provide experimental or developmental knowledge in optical surveillance of deep space. From the above study, it is inferred that by careful preparation, confusion may be minimized, the observatories may be assured that careful use of all the existing tracking facilities has been made, and the probability of obtaining useful surveillance data may be maximized.

Increased Accuracy in Tracking Transmitting Probes

The present primary system for tracking space probes is the Jet Propulsion Laboratory's deep-space tracking network. This network will consist of three large radars, located in California, South Africa, and Australia. During the flight of the Pioneer IV probe, the Goldstone, California radar antenna demonstrated a tracking accuracy of ± 1 min of the arc at lunar distance. The transmitter on this probe transmitted 200 mw of power (Ref. 11). A data rate of one point each 10 sec was

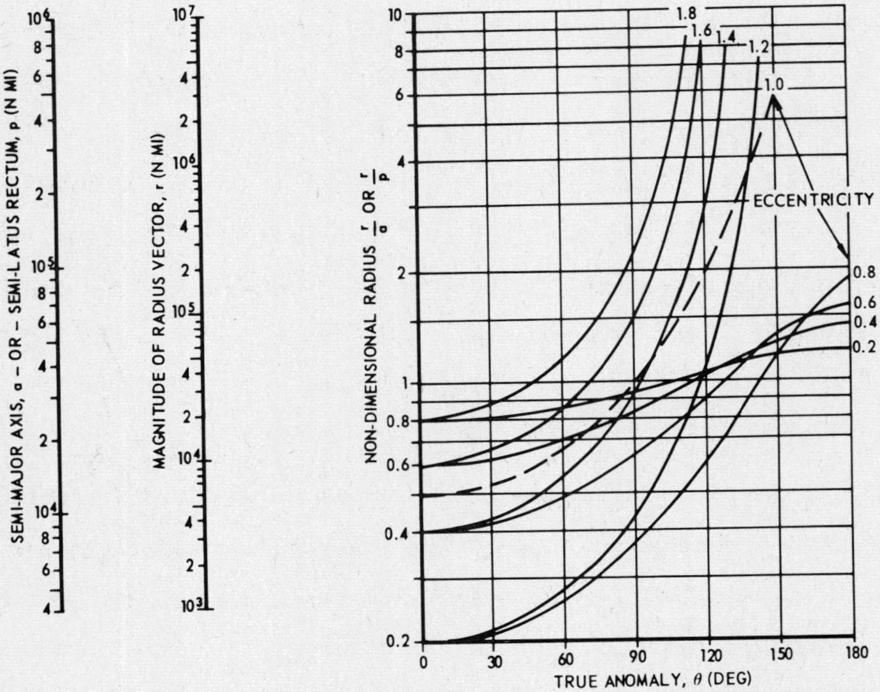

Fig. 11 Nomograph: True Anomaly vs. Radius and Orbit Parameters

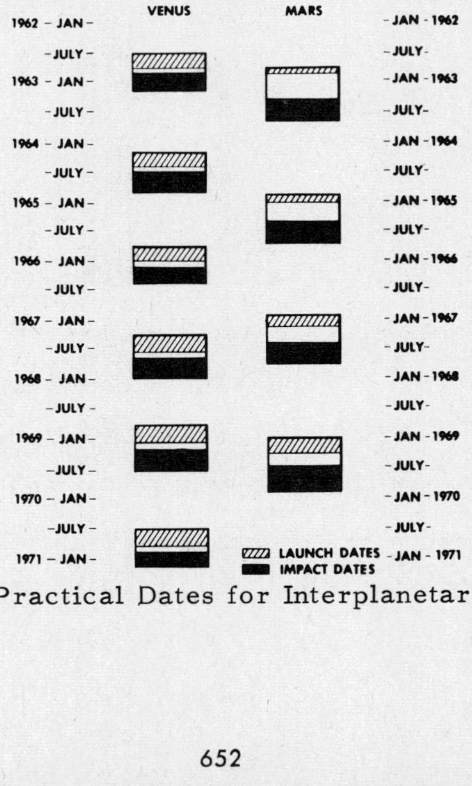

Fig. 12 Practical Dates for Interplanetary Transfer

maintained for 9 hours each day.

With this network, orbit prediction is accomplished by using a sophisticated differential correction statistical computer program. (Refs. 12 and 13.) According to Ref. 11, during the tracking of the Pioneer IV probe "all predictions made after the first day of tracking, for periods of one day later, were found to agree with the observations to within 0.05 deg". The basic procedure, at present, is to assume that all dimensions and mass ratios of the solar system are known. The trajectory of the probe, considered as an $\underline{n}$ body problem corrected for the oblateness of the earth, may be specified by six initial conditions at the time of final burnout.

By the statistical method of maximum likelihood, the orbit-prediction program repeatedly solves for the set of initial conditions most closely compatible with the observed data. Simultaneously, the program attempts to improve the knowledge of the bias and noise in the data received from each observational station.

Reference 11 also states: "The most serious unresolved problem in the tracking program is a proper determination of the degree of degeneracy of the matrix involved in the solution for the changes in initial conditions." The procedure used during the tracking of the Pioneer IV probe was to perform several dry-runs prior to actual tracking, in order to determine the regions of singularity. The number of quantities considered as unknown was reduced during the sensitive periods, and the remaining equations were sometimes solved in an alternating fashion, rather than simultaneously. The orbit prediction program now has an option of comparing the calculated correction to the calculated standard deviation, and thus automatically determines which quantities cannot be improved from the data available, and hence, how much the order of the matrix needs to be reduced. This method, although automatic, still "solves" the problem of matrix singularity by abandoning the attempt to improve some of the parameters in question.

Some work on obtaining a best approximate solution for an actually singular matrix has been done, for instance, by Hestenes of the University of California at Los Angeles (UCLA). However, even for a near-singular matrix, a simple example given by Milne (Ref. 14) reminds one that there are limits to the improvements that can be obtained by numerical analysis. He presents the equations:

a. $x - y = 1$

b. $x - 0.99999y = 0$

These have the solutions: $y = -100,000$, and $x = -99,999$. He then points out that if the coefficient of y in equation (b) had an uncertainty of 0.00002 so that equation (b) should actually read:

$$x - 1.00001\,y = 0$$

then the correct answer would be $y = 100,000$ and $x = 100,001$. It is obvious that with this degree of uncertainty in the coefficients, numerical analysis cannot reduce the uncertainty of the result. However, if the same coefficient were known to two more decimal places, at the worst, equation (b) should read:

$$x - 0.9999902\,y = 0$$

then the results would have a maximum error of approximately two per cent. This example provides an indication of how the precision of a derived quantity can be vastly improved by increasing the precision of the data. Two orders of magnitude is approximately the angular accuracy improvement to be expected from using large astronomical telescopes, compared to the present Jet Propulsion Laboratory tracking network.

In tracking at interplanetary distances, certain difficulties, beyond those which have arisen in the Pioneer IV experiment at lunar distances, may be expected to arise which could adversely affect the results of the orbit prediction program. These difficulties result from uncertainties in our knowledge of solar system constants.

As regards the astronomical unit, de Vaucouleurs (Ref. 15) has taken weighted means of all the measurements made by each of eight methods. The averages vary from 149, 524, 000 km, for trigonometric methods, to 149, 556, 000 km for measurements of the parallactic inequality of the moon. He adopts a weighted mean value of 149, 536, 000 ± 3000 km. Makemson, Baker, and Westrom (Ref. 16) adopt for astronomical work Rabe's value of 149, 520, 000 ± 34, 000 km "with a larger uncertainty to partially encompass the (radar) values." New radar measurements by the Jet Propulsion Laboratory, substantiated by Jodrell Bank and Lincoln Laboratories, agree on about 149, 599, 000 ± 500 km. Without arguing with either the dynamical or radar methods, it must be remembered that the vehicle orbit prediction is based on a dynamical argument, but using radar data. The reason for the discrepancy between the two methods has not yet been explained. Therefore, there would appear to be some risk in assuming an uncertainty of less than 60, 000 km. In fact, Rabe's dynamical uncertainty of 6, 797 km, or even the Jet Propulsion Laboratory's uncertainty of ± 500 km would present great difficulties for a probe as it neared a planet.

The mass ratios of the planets, particularly of planets such as Venus which have no natural satellites, are also subject to considerable uncertainty. Rabe (Ref. 17) estimates the mass ratio of Venus to have an uncertainty of 0. 5 percent.

The present Jet Propulsion Laboratory orbit prediction program attempts to provide ephemerides for probes launched by the U. S. by ascertaining the initial conditions and instrument biases. From the above discussion, it is concluded that the program could be certain of providing accurate ephemerides near another planet only if it was modified to also consider as unknowns certain of the constants of the solar system. This increase in the size of the matrix should contribute as much to the singularity of the matrix as should the additional distance.

In view of the above, it is very uncertain whether improved knowledge of the solar system constants could be obtained by tracking a probe with present tracking accuracies. Dr. G. M. Clemence of the U. S. Naval Observatory, has expressed the belief that with the presently planned deep-space missions, even the accuracy of astronomical telescopes would be only marginal for improving the measurement of the astronomical unit from dynamical considerations.

At this point, it should be noted that the evaluation of the deep-space network capability has been based on the tracking of the Pioneer IV probe. At the time Pioneer IV was tracked, only the receiver at Goldstone and a portable receiver at Puerto Rico were operating; thus, no range data were available. Range data will be available on the very first planetary probes. Certain errors in both angular and range-rate data that were observed at the time of the Pioneer IV flight have been corrected. Thus, order-of-magnitude improvements by telescopes over this early tracking are necessary in order to indicate that optical methods will be significantly more accurate than the present deep-space network. However, even for the largest radio telescopes planned, angular accuracy will probably not exceed ± 30 sec of arc.

It is not suggested here that data obtained by optical means ever be used alone, but that it be properly weighted and then combined with radar range, angle, and range-rate data. If a standard deviation for optical data of one one-hundredth of that of radar angular data is assumed, then the linearized maximum likelihood equations would weight one optical measurement the same as ten thousand radar angular measurements.

CONCLUSION

Large astronomical telescope will be capable of photographing space probes of the expected order of size by reflected sunlight out to distances of a few hundred thousand miles. It would be quite feasible to extend this range, at least to several astronomical units, for probes which are equipped with specular reflectors, or which liberate atomic scatters. For non-cooperating probes, the astronomical observations may be the only observations possible. Whenever telescope data are available, they will be much more precise than radio angular data.

As to what volume of observations may be required, we note that Fig. 12 indicates that it is possible some interplanetary probes may be in transit 68 percent of the time. However, several considerations indicate that optical surveillance throughout all flights will not be required. First, unless the U. S. S. R. provides continuous visibility enhancement of their probes or provides the rest of the world with a schedule of events, such as the release of sodium clouds, we shall be able to observe their probes only for a short period after launch. Secondly, it would be hoped that the knowledge of solar system constants necessary for future orbit prediction could be obtained from tracking only a very few early probes launched by the U. S. Optical tracking of U. S. vehicles should then be necessary only in regions where the maximum likelihood matrix of the Jet Propulsion Laboratory's orbit prediction program is very nearly singular. When and if such times will exist might be determined, as in the past, by supplying simulated trajectories, together with noise and generous errors in the constants of the solar system, to the Jet Propulsion Laboratory's program and observing the accuracy of the result. Nevertheless, as indicated above, the U. S. plans deep-space probes with goals other than interplanetary transfer. The U. S. also may be expected to launch at least 12 lunar probes in this decade, and the U. S. S. R. will certainly launch several. Previous lunar probes have soon after launching become fainter than the limiting magnitude of the Baker-Nunn network.

If the national space objectives require regular optical coverage of this wide assortment of missions, then an extension of the present optical network is indicated. Astronomical telescopes could offer an interim extension of this coverage and, even if they were to be used only sparingly on these problems, they would provide experience which would be of value in the design of suitable new instruments. A carefully planned program would probably obtain the cooperation of the observatories.

ACKNOWLEDGMENTS

The author wishes to acknowledge the many important contributions by D. C. Fisher, J. T. Cragin, and B. Shields of the Communications Section; W. K. Ming, P. L. Cohen, R. T. Giuli, L. L. Lakamp, and G. A. Weissenberger of the Aero/Astrodynamics Section; N. Xerikos of the Missiles Aerodynamics Section ; R. L. Warren of the Product Support Section; and the patience and assistance of G. C. Goldbaum, Chief of the Aero/Astrodynamics Section, Douglas Aircraft Company, Inc.

APPENDIX

Figs. 1 and 2, "Limiting Photovisual Magnitude as a Function of Object to Telescope Relative Angular Rate," are obtained by the following modifications to the method of Ref. 18. For the computations, various numerical constants are utilized, and these are listed together with their source.

SYMBOLS AND DEFINITIONS USED IN FIGS. 1 and 2

D	Aperture of telescope
E	Maximum number of statistically effective photo-events which can populate a unit area at the primary focus. For the Eastman Kodak 103a-0 plate, $E = 5 \times 10^6 \frac{\text{granulations}}{\text{cm}^2}$
f	Focal length of telescope
k	Coefficient of certainty which relates the actual threshold of recognition to the statistical uncertainty expressed as a standard deviation. (k = 4 is used.)
M	Magnitude of object
N	Average number of eligible photons per second received per unit area at the surface of the earth from a unit solid angle of sky background
n	Average number of eligible unscattered photons per second received per unit area at the surface of the earth from the object
q	Average quantum efficiency in terms of independent registered events per eligible photon received during an exposure. ($q = 10^{-3}$ is used.)
t	Effective integration time
α	Angular width of image

γ Angular length of image = $\dot{\theta} t + \alpha$

$\dot{\theta}$ Angular rate of image relative to telescope

ψ Number of photo-events per unit solid angle of background events due to chemical fog

EXPRESSIONS OF INTEREST

The total number of photo-events due to the object is: nD^2qt. Generalizing from Ref. 18, one finds that for the case of appreciable chemical fog, the average number of background events is: $(ND^2qt + \Psi)$ (angular area), where ψ accounts for the chemical fog. For a barely detectable image, the threshold values of nD^2qt is

$$nD^2qt = k \left[2 \ (ND^2\,qt + \psi)\,\alpha\gamma + nD^2\,qt\right]^{1/2}$$

Solving for n, one obtains

$$n = \frac{k}{D^2\,qt}\left[2 + \sqrt{2}\ \sqrt{2 + (ND^2\,qt + \psi)\,\alpha\gamma}\right]$$

Upon differentation of the above expression for n with respect to t, the optimum exposure time is found to equal the limiting time for film saturation.

For a continuous measurable image, it is required that each point image, of area α^2, be detectable. The light from the object per area α^2 is given by $\frac{nD^2qt}{\alpha\gamma}\alpha^2$, and the background, per area α^2, by $\alpha^2(ND^2qt + \psi)$. Thus,

$$nD^2\,qt\ \frac{\alpha}{\nu} = k \left[2\,\alpha^2\,(ND^2qt + \gamma)\ + nD^2\,qt\ \frac{\alpha}{\gamma}\right]^{1/2}$$

Therefore,

$$n = \frac{8\gamma}{\delta t}\left[1 + \{1 + \frac{\alpha^2}{2}\left(\frac{N\delta t}{\alpha} + \psi\right)\}^{1/2}\right]$$

and optimum exposure time is given by

$$\dot{\theta} = \frac{4B^{1/2}\,[1 + B^{1/2}] - \delta N\,\alpha t}{\delta N\,t^2}$$

where

$$\left.\begin{aligned} k &= 4 \\ y &= \dot{\theta}t + a \end{aligned}\right\} \text{ defined previously}$$

$$\delta = D^2 q a$$

$$B^{1/2} = \left[1 + \frac{a^2}{2}\left(\frac{N\delta t}{a} + \psi\right)\right]^{1/2}$$

The magnitude of an object whose image is at the threshold is given by

$$M = -2.5 \log \frac{n}{n_o}$$

where n = n of the object

n_o = n of a zero magnitude star (given in Item 3 below)

Constants

1. The chemical fog coefficient, ψ, is assumed to be $\frac{E}{3}$. For the Eastman Kodak 103a-0 plate, $\psi' = 1.67 \times 10^6 \frac{\text{granulation}}{\text{cm}^2}$. To use this number in the above equations, a conversion to angular area must be made utilizing the telescope scale factors given by s. f. $= \frac{206,265}{f\ (\text{cm})}$. For example, ψ for the Hale telescope is

$$\psi = \frac{\psi'}{\text{s.f.}} = \frac{\left(1.67 \times 10^6 \frac{\text{gran}}{\text{cm}^2}\right)}{\left(110 \frac{\text{sec}}{\text{cm}}\right)^2} \cdot \left(3600 \frac{\text{sec}}{\text{deg}}\right)^2 \left(57.3 \frac{\text{deg}}{\text{rad}}\right)^2 = 5.86 \times 10^{12} \frac{\text{gran}}{\text{rad}^2}$$

2. N is obtained from the expression for the saturated value of the average number of background events as given in Ref. 2. For example,

$$ND^2 q a^2 t = a^2 f^2 E$$

After substitution of the following values for the Hale telescope,

$$D = 508 \text{ cm}$$

$$\alpha = 1.21 \times 10^{-5}$$

$$f = 1855 \text{ cm}$$

$$q = 10^{-3}$$

$$t = 1800 \text{ sec}$$

the following value is obtained:

$$N = 3.7 \times 10^{7} \frac{\text{photons}}{\text{cm}^2 \text{ sec steradian}}$$

3. The illuminance for a zero-magnitude star is obtained from the following values given in Ref. 19. The illuminance of a zero-magnitude star outside the earth's atmosphere is 2.43×10^{-6} meter-candle. Considering the atmospheric attenuation factor 0.83, the corresponding value inside the atmosphere is

$$n_o = (0.83)(2.43 \times 10^{-6}) = 2.02 \times 10^{-6} \text{ meter-candle}$$

4. The conversion from meter-candles to $\frac{\text{photons}}{\text{cm}^2 \text{ sec}}$ is obtained as follows:

According to Ref. 18, the magnitude of the sky background, M_B, is -4.4. Substitution of this number, together with the value obtained in Item 3, into the magnitude equation results in the following:

$$M = M_B - 2.5 \log \frac{n_o}{n_{-4.4}} = 0$$

$$\therefore n_{-4.4} = 1.16 \times 10^{-4} \frac{\text{meter-candle}}{\text{steradian}}$$

Since the value obtained from Item 2 is equal to this number, the following is obtained:

$$1 \text{ meter-candle} = \frac{3.7 \times 10^{7}}{1.16 \times 10^{-4}} = 3.19 \times 10^{11} \frac{\text{photons}}{\text{cm}^2 \text{ sec}}$$

Therefore, $n_o = 6.45 \times 10^{5} \frac{\text{photons}}{\text{cm}^2 \text{ sec}}$, which corresponds to 4440 Å light.

Figure 3, "Stellar Magnitudes--Perfectly Reflecting Spheres," is an extension of the nomograph given in Ref. 4.

SYMBOLS AND DEFINITIONS USED IN FIG. 3

D	=	Distance from satellite to observer (ft)
M	=	Magnitude
M'	=	Full phase magnitude at 1 a. u. from the sun
R	=	Radius of sphere (ft)
r_c	=	Distance from vehicle to sun (mi)
r_c'	=	$D \cos \phi \pm \sqrt{1 - D^2 \sin^2 \phi}$ (a. u.) (derived from law of cosines)
ϕ	=	Phase angle (rad)

The equations used in the calculation of Fig. 3 are as follows:

$$M = -2.5 \log E - 16.45$$

$$E_{(specular)} = 3170 \left(\frac{R}{D}\right)^2 e^{-0.117 \left(\frac{92,914,900}{r_c}\right)^2}$$

$$E_{(diffuse)} = 8470 \left(\frac{R}{D}\right)^2 e^{-0.117 \left(\frac{92,914,900}{r_c}\right)^2} \left[\frac{\sin \phi + (\pi-\phi) \cos \phi}{\pi}\right]$$

Converting the above equations to the form of magnitude corrections (ΔM) which are functions of r_c and Θ, the Sun-observer-object angle (see Fig. 13), the following equations are obtained:

$$M'_{(specular)} = -2.5 \log \left[3170 \left(\frac{R}{D}\right)^2 e^{-0.117}\right] -16.45$$

$$\Delta M_{(specular)} = 5 \log r_c'$$

$$\Delta M_{(diffuse)} = 5 \log r_c' - 2.5 \log \left[\frac{\sin \phi (\pi-\phi) \cos \phi}{\pi}\right] - 1.06$$

To obtain ϕ as a function of Θ, one used the law of sines as follows:

$$\sin \phi = \frac{\sin \phi}{r_c}$$

Therefore, the final magnitude equations are

$$M_{(specular)} = M'_{(specular)} + \Delta M_{(specular)}$$

$$M_{(diffuse)} = M'_{(specular)} + \Delta M_{(diffuse)}$$

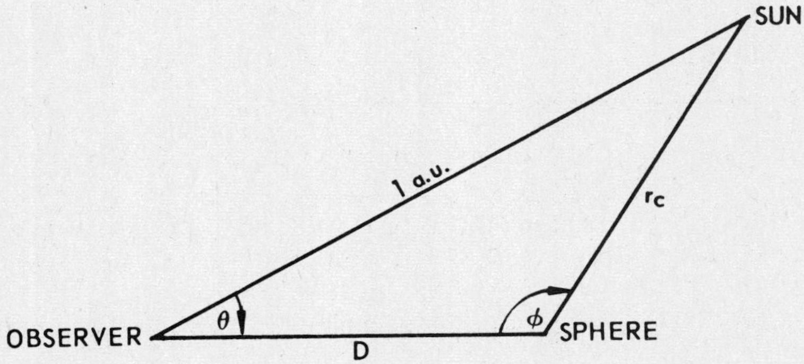

Fig. 13 Viewing Geometry--Spherical Mirror

Figure 4, "Stellar Magnitudes--Specular Mirror," is obtained by a method similar to the one previously described for Fig. 3.

SYMBOLS AND DEFINITIONS USED IN FIG. 4 (see Fig. 14)

D	=	Distance from satellite to observer (a. u.)
D'	=	Distance from satellite to observer (ft)
d	=	Diameter of mirror (ft)
M	=	Magnitude
Θ	=	Elongation angle
$\emptyset$	=	Phase angle

The equations used in the calculation of Fig. 4 are as follows:

$$M_1 = 5 \log \left[\frac{0.0093D'}{d}\right] - 26.7 \text{ (full phase magnitude)}$$

$$\sin \emptyset = \frac{\sin \Theta}{\sqrt{D^2 + 1 - 2D \cos \Theta}}$$

$$M = M_1 - 2.5 \log (\cos \frac{\emptyset}{2})$$

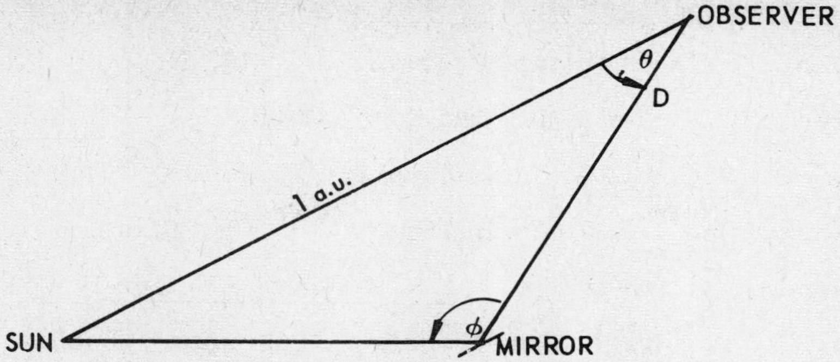

Fig. 14 Viewing Geometry--Flat Mirror

REFERENCES

1. G. P. Kuiper, and B. M. Middlehurst, Telescopes, Chicago: University of Chicago Press, 1940.

2. K. G. Henize, "The Baker-Nunn Satellite-Tracking Camera," Sky and Telescope, Vol. 16 (January, 1957), pp. 108-111.

3. K. G. Henize, "Tracking Artificial Satellites and Space Vehicles," Advances in Space Sciences, Vol. 2. New York: Academic Press, 1960.

4. R. Tousey "The Visibility of an Earth Satellite," Astronautica Acta, Vol. II (1956), pp. 101-112.

5. "JPL Industry Conference Proceedings," Jet Propulsion Laboratory, October 26, 1960.

6. F. L. Whipple, "Orbit Accuracy and Ranges From Ground-Based Optical Tracking," Vistas in Astronautics, Vol. 1. New York: Pergamon Press, (1957), pp. 258-262.

7. Interplanetary Midcourse Guidance Study, Second Interim Report, Goodyear Aircraft Corp., Report No. GER9482, October 31, 1959. (Secret)

8. I. S. Shklovskii, V. E. Esipov, V. G. Kurt, V. I. Moroz, and P. V. Shcheglov, "An Artificial Comet," Soviet Astronomy AJ, Vol. 3, No. 6, (May-June, 1960).

9. Walter Hohmann, The Attainability of Heavenly Bodies. Translation

of Die Erreichbarkeit der Himmelskörper, Oldenbourg, R. Munich-Berlin: 1925. NASA Technical Translation F-44, November, 1960.

10. R. J. Gunkel, D. N. Lascody, and D. S. Merrilees, Impulsive Midcourse Correction of an Interplanetary Transfer, Douglas Aircraft Company, Inc., Engineering Paper No. 804, September, 1960.

11. M. Eimer, and Y. Hiroshige. Evaluation of Pioneer IV Orbit Determination Program, Jet Propulsion Laboratory Technical Release No. 34-26, February 22, 1960.

12. J. Lorell, R. E. Carr, and R. H. Hudson, The Jet Propulsion Laboratory Lunar Probe Tracking and Orbit-Determination Program, Jet Propulsion Laboratory Technical Release No. 34-16, March 10, 1960.

13. R. E. Carr, and R. H. Hudson, The JPL Tracking and Orbit-Determination Program, Jet Propulsion Laboratory Technical Release No. 32-7, February 22, 1960.

14. W. B. Milne, Numerical Calculus, Princeton, Princeton University Press, 1949.

15. G. de Vaucouleurs, The Astronomical Unit of Distance, Geophysics Corporation of America Report, January, 1961.

16. M. W. Makemson, R. M. L. Baker, Jr., and G. B. Westrom. "Analysis and Standardization of Astrodynamic Constants," Journal of the Astronautical Sciences, Vol. VIII, No. 1, (Spring 1961), pp. 1-13.

17. E. Rabe, "Derivation of Fundamental Astronmonical Constants From the Observations of Eros During 1926-1945," The Astronomical Journal, No. 1184 (1949), p. 112.

18. W. A. Baum, "The Detection of Faint Images Against the Sky Background," Transactions at the International Astronomical Union, Vol. IX (1957), pp. 681-687.

19. C. W. Allen, Astrophysical Quantities, London: University of London, Athlone Press, 1955.

A PROBABILITY MODEL FOR DESIGNING MILITARY SATELLITE SYSTEMS

Joseph A. Joseph *

A mathematical model is developed for use in surveillance and other problems, which can be used in the design of military space systems. It gives results in expected coverage and probability of observing. Several examples are described to show the use and value of the model. Tradeoffs are obtained between number of satellites in the system, orbit paramenters, and sensor radius.

It is assumed that the initial injection into orbit of these satellites is random and the density of the subsatellite positions has a Poisson distribution. It follows directly from this assumption that if several satellites are in orbit in or very near to a common plane, the distribution of the magnitude of the spaces between the subsatellite positions will be exponential. From these considerations, several useful results are obtained. For sensor-bearing satellites having a specified radius of coverage on the Earth's surface, tradeoff relations are obtained relating the number of satellites, radius of the sensor coverage, and probability of coverage. Typical of the analytical results obtained are: (1) fraction of the Earth's surface covered by a number of randomly distributed, sensor-bearing satellites; (2) the required number of satellites for no gaps in the surveillance net for a given probability of assurance; (3) the required number of satellites to assure, at a given probability level, that at least one satellite is above the horizon; (4) tradeoffs between latitude of surveillance station, radius of coverage of sensor, angle of inclination of orbital plan of satellites, number of satellites in the plane and probability of coverage.

---

* International Business Machines Corp., Federal Systems Division, Rockville, Maryland

INTRODUCTION

In this paper, a probability model is developed that can be applied to Earth-satellite surveillance problems to obtain useful results in problems of coverage (as in satellite-to-Earth communications), interception (Earth-to-space), collision (satellite-to-satellite), and so on in terms of probabilities. A few applications will be given, but the main purpose is to describe the model for general use. A surface model will be described first in which it is assumed that a large number of sensor-bearing satellites are randomly distributed about the Earth. Then, only those satellites that lie in (or are very close to) a common plane in inertial space and remain for many days will be considered. The assumption of random distribution of the satellites in a plane leads to interesting and useful results.

THE GENERAL SPACE PROBLEM

Let there by N satellites in orbit around the Earth. Each satellite has been launched at random intervals with different initial velocity, and is subject to a unique set of unpredictable perturbing forces, including drag. At any instant of time, the subsatellite (ss.) point of the ith satellite is specified by the longitude and latitude (θ_i, ϕ_i) where $i = 1, 2, \ldots N$. It is assumed that at any instant the ss. points are randomly distributed over the surface of the Earth.

Let us divide the surface of the Earth into M equal and non-overlapping areas of size A*. The average number of ss. points per square at any instant is $\lambda = N/M$. A Poisson distribution of the number of ss. points per area A is assumed. That is, the probability of k ss. points in an area at any instant is given by

(1) $$P_k = \frac{e^{-\lambda} \lambda^k}{k!}$$

* The model here is an area distribution model similar to the "bacteria and blood count" problem described by Feller, Probability Theory and Its Applications, 1st ed., pp. 122-3

or, in other words, the number of areas that contain exactly k ss. points is

$$N_k = M \frac{e^{-\lambda}\lambda^k}{k!} , \tag{2}$$

For example, if N = 100 and M = 50, then $\lambda = 2$ and

$$N_k = \frac{50\, e^{-2} 2^k}{k!}$$

so that

k	N_k
0	6.8
1	13.5
2	13.5
3	9.0
4	4.5
5	1.8
6	.6
7	.2
8	0.
	50

This says that 6.8 (i.e., 7) of the 50 areas are expected to contain no satellites, 13.5 contain 1 satellite, 9 contain 3 satellites, etc.

If A is taken to be the area of coverage (communications, infra-red, photography, etc.) for each of N satellites, and M = number of areas of size A making up the entire surface of the earth, than an estimate of the fraction of the earth's surface that will not be covered at any instant is, from Equation (1),

$$P_o = e^{-(N/4)\,(r/R)^2} \tag{3}$$

where r = radius of coverage of the sensing device on each of the N satellites and R is the radius of the earth, both in the same units. A plot of this relation is shown in Figure 1. The quality of the estimate decreases as r becomes large in comparison to R.

Another estimate that would be of interest is the fraction of the earth's surface that does not have simultaneous coverage from two or more satellites. All we can say is that this fraction, F_1, is greater than the per cent of empty cells plus the per cent of cells that contain one ss. point.

$$(4) \qquad F_1 = P_0 + P_1 = e^{-(N/4)\,(r/R)^2}\left[1 + (N/4)r^2/R^2\right]$$

This is a weak estimate. The quality assumes that all ss. points are at the center of their area.

The weakness of the above model is that the ss. points are not completely uncorrelated. Any two satellites with ss. points (θ_1, ϕ_1) and (θ_2, ϕ_2) will maintain a constant difference in longitude, $\theta_1 - \theta_2$ for many days, except for small perturbations. The longitudes of all satellites are correlated by their common cause, the earth's rotation. The "degree" of correlation between satellites from plane to plane is greater than that between two satellites in the same plane. Furthermore, for large orbital inclinations the number of satellites in the surface model will be greater near the poles than near the equator. This is contrary to the assumption of random distribution over the surface of the earth. It is for these reasons that a more realistic model concerns itself with only those satellites that lie in or close to a common inertial plane. It is assumed that the positions of the subsatellite points in the common plane are randomly distributed. A more precise statement and a development of the planar concept is given next.

THE PLANAR SET OF SATELLITES[1]

Let us examine a set, n, of random satellites that lie in or very close to a common plane.

We define a subsatellite (ss.) point as the point in the plane on the line from the focus (earth's center) to the satellite at a distance R from the focus; R is the radius of the earth. The ss. point can be specified by a clock angle in the plane of rotation of the satellites. The satellites will

---

[1] See Bibliography for two papers on this same subject.

continue to occupy the same plane in inertial space. Consider any half line, fixed in inertial space, radiating out in a unique direction from the common focus of all the orbits. Each satellite will transit this line once during its period, and an infinite set of such transit times form a chronological set (T_j) regardless of which satellite does the crossing and regardless of the direction of crossing (clockwise or counter-clockwise). The set of successive differences of these times ($t_j = T_{j+1} - T_j$), from j = 1, 2, . . . , is of interest as a stochastic variable. It is possible that after a sufficiently long time, the t_j will repeat, but this is assumed to be a long time, and the size of the set, up to that time, is assumed large. Let the average period of all the n satellites be

$$\bar{P} = \frac{\sum_i^n P_i}{n}$$

The average t_j is approximately $\bar{P}/n$, if the variation in P_i is small.

Consider a small interval of time, Δ. The probability that at least one of the n satellites crosses the radial line in time Δ is

$$p = 1 - (1 - \frac{\Delta}{\bar{P}})^n$$

If Δ is small compared with $\bar{P}$, then

$$p \cong \frac{n\Delta}{\bar{P}} .$$

The probability of a t_j between $k\Delta$ and $(k + 1)\Delta$ is

$$P[k\Delta \leq t_j \leq (k + 1)\Delta] = pq^k \text{ for } k = 0, 1, 2, \ldots .$$

where $q = 1 - p$. This gives us the distribution function of the set (t_j).

Choosing any time to initiate a look, the probability that a satellite will cross the line within time $t = (a + \epsilon)\Delta$, where a is a non-negative integer and $\epsilon < 1$ and positive, is

$$P[\text{time} < t] = \sum_{k=0}^{a} pq^k = p\left(\frac{1 - q^{a+1}}{1 - q}\right) = 1 - (1 - p)^{a+1}$$

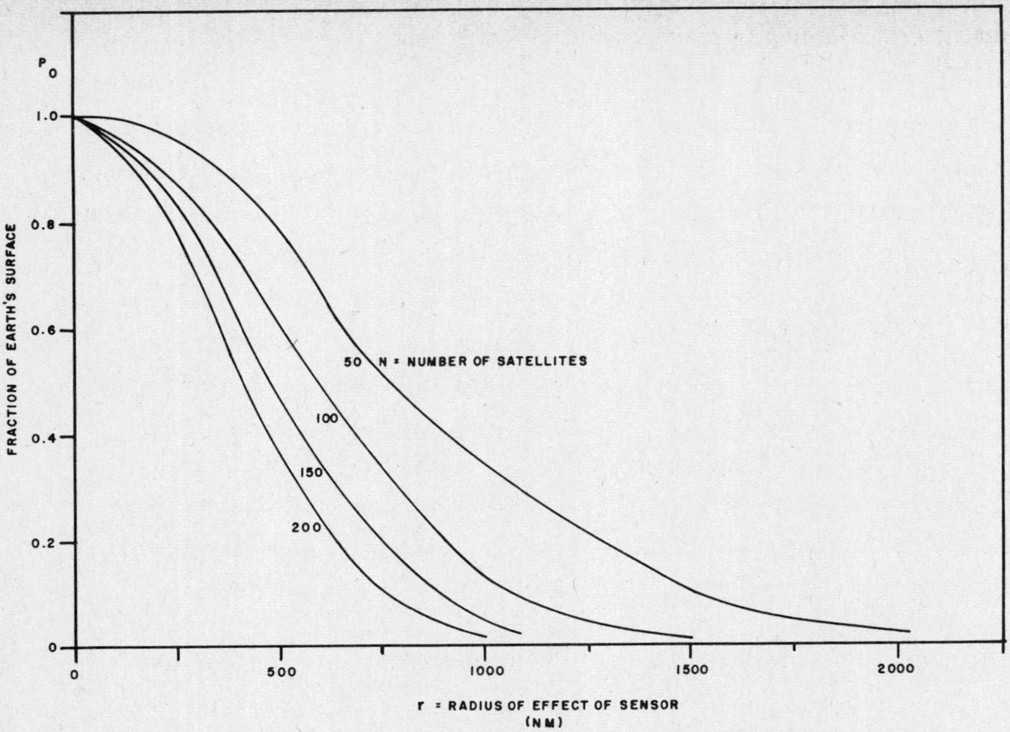

Fig. 1 Fraction of Earth's Surface not Covered by N Randomly Distributed Satellites

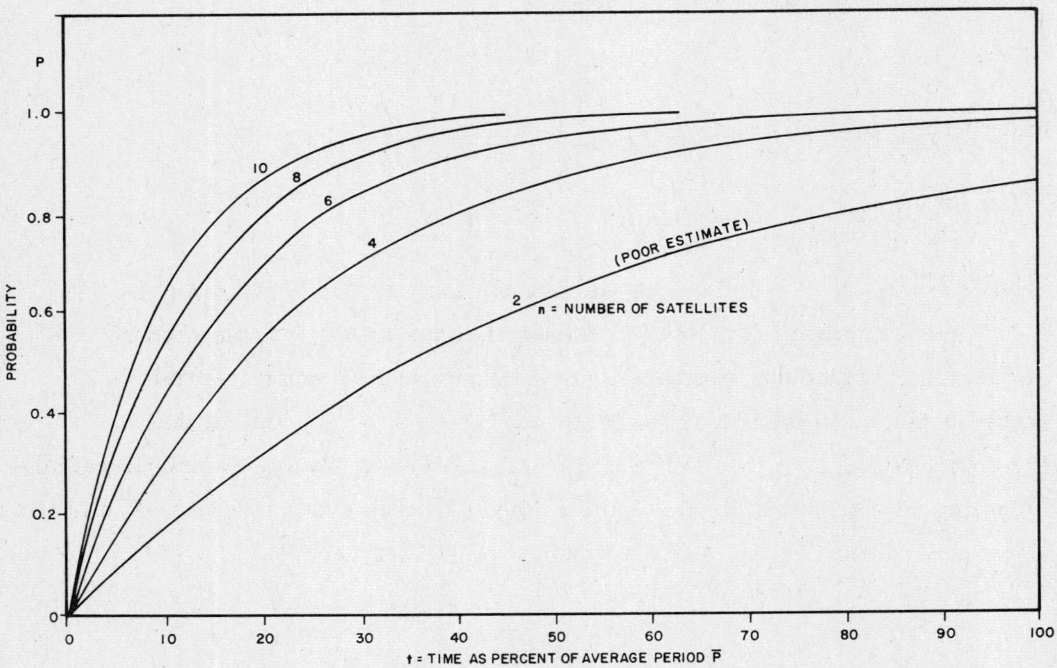

Fig. 2 Probability of a Time Interval Less than t for n Satellites in a Plane

Putting in the values of p and a in terms of Δ will give

$$P[\text{time} < t] = 1 - \left(1 - \frac{n\Delta}{\bar{P}}\right)^{t/\Delta + 1 - \epsilon}$$

$$= 1 - \left[\left(1 - \frac{n\Delta}{\bar{P}}\right)^{\frac{\bar{P}}{n\Delta}}\right]^{n\Delta/\bar{P}\,[(t/\Delta) + 1 - \epsilon]}$$

For very small Δ this becomes

$$(5) \qquad P[\text{time} < t] = 1 - e^{-nt/\bar{P}}$$

This is plotted in Figure 2. The expected value of t is $\bar{P}/n$.

Instead of thinking in terms of spaces between satellites, one can think in terms of number of satellites in an interval of size D radians. The probability of at least one satellite in D is, as we saw before,

$$1 - \left(1 - \frac{D}{2\pi}\right)^n$$

The probability of no ss. points in D is $(1 - D/2\pi)^n$.
This is approximately equal to $e^{-nD/2\pi}$ for small D. By induction, it can be shown[2] that the probability of exactly k ss. points in D is

$$(6) \qquad P_k = \frac{e^{-\lambda}\lambda^k}{k!}$$

where $$\lambda = \frac{nD}{2\pi}$$

namely the Poisson Distribution.

APPLICATIONS

Surveillance Gaps.

Consider n satellites having almost the same altitude and therefore period, P, in a circular orbit, polar or inclined, in a common inertial plane. They have been fired randomly into this common plane so that the time spacing ($t_j = Ps_j/2\pi$) is also random, where s_j is the distance in radians between successive subsatellite points. The expected spacing is $E(t) = P/n$ in time

[2] Feller, 1st Edition, pp 142-143.

and E (s) = $2\pi/n$ in radians. Let a surveillance sensor have a span of coverage of r nautical miles from the subsatellite point, so that the unsurveyed gap is $G_j = s_j - 2r/R$, where R is the radius of the earth in nm (see Figure 3). Then

(7) $$\overline{G} = \text{the expected gap} = \sum_{(2r/R)+1}^{\infty} pq^k (kD - 2r/R)$$

$$= \frac{2\pi}{n} - \frac{2r}{R} \quad \text{radians.}$$

The probability of a gap being greater than some value G is

(8) $$P[\text{gap} > G] = P[\text{space } s > G + 2r/R]$$

$$= P\left[k > \frac{G + 2r/R}{D}\right]$$

$$= \sum_{\frac{G+2r/R}{\Delta}}^{\infty} pq^k$$

$$= (1 - nD/2\pi)^{\frac{[(G+2r)/R]}{D}}$$

$$\cong e^{-(G + 2r/R)\, n/2\pi}$$

for small D. The case G = 0 is of interest. It says the probability of any randomly selected gap being greater than zero is

(9) $$P[\text{a gap} \geq 0] = e^{-nr/\pi R}$$

and the probability that a randomly chosen space has no gap is

(10) $$P_r[\text{a gap} \leq 0] = 1 - e^{-nr/\pi R}$$

Solving for n we get

(11) $$n = -\frac{\pi R}{r} \log_e (1 - P_r),$$

which gives the required number of satellites with sensor radius r to assure no gap on the first random choice at level of assurance P_r. This is plotted in Figure 4.

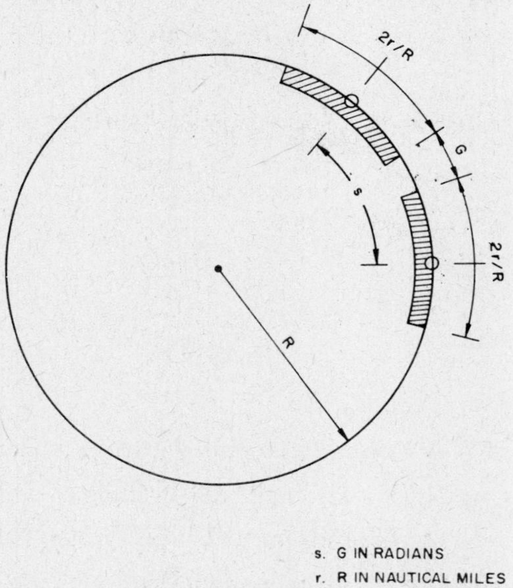

Fig. 3 Typical Spacings and Gap G

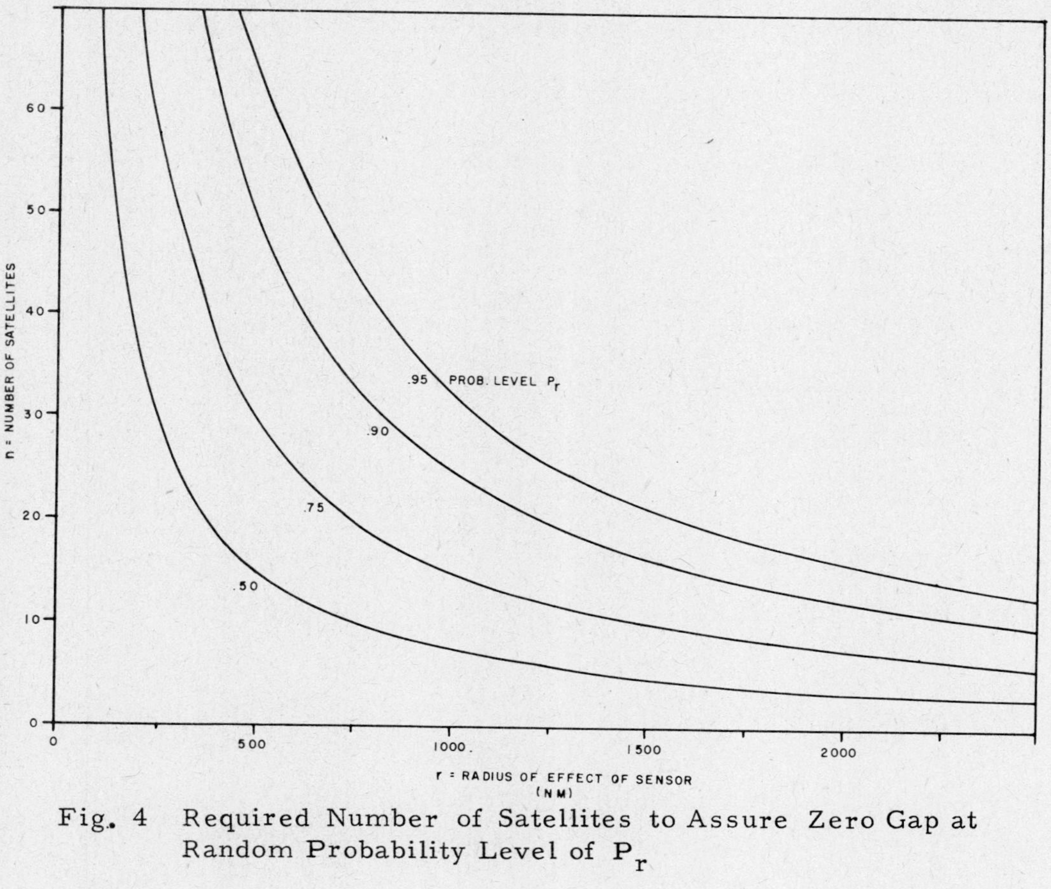

Fig. 4 Required Number of Satellites to Assure Zero Gap at Random Probability Level of P_r

The probability of no gap around the whole circle is $\left(1 - e^{-nr/\pi R}\right)^n$ since there are n gaps in sequence to consider.

To get an expected gap of zero, $\overline{G} = 0$, we see from Eqn. (7) that $n = \pi R/r$.

This is different from getting no gaps on n successive selections which require that

$$r = \frac{\pi R}{n} \left[\log n - \log (1 - P) \right] \tag{12}$$

This is plotted in Figure 5.

Horizon Problem

Of n satellites in a plane of inclination i, the probability of at least one satellite above the horizon, relative to some fixed latitude ϕ on the earth's surface (see Figure 6) is the same as the probability of a time interval greater than the time for a satellite to go from one horizon to the other. The time to transit from one horizon to the other is

$$t = \frac{\overline{P}}{2\pi} \left(2 \cos^{-1} \frac{R'}{R' + h'} \right) \tag{13}$$

where

$$R' = \frac{R}{\cos (\phi - i)}$$

$$h' = R \left[1 - \sec (\phi - i) \right] + h.$$

$$P \left[\text{at least one above the horizon} \right] \tag{14}$$

$$= P \left[\text{time interval} < \left(\overline{P}/\pi \right) \cos^{-1} \frac{R'}{R' + h'} \right]$$

$$= 1 - e^{-n/\pi} \cos^{-1} \frac{R'}{R' + h'}$$

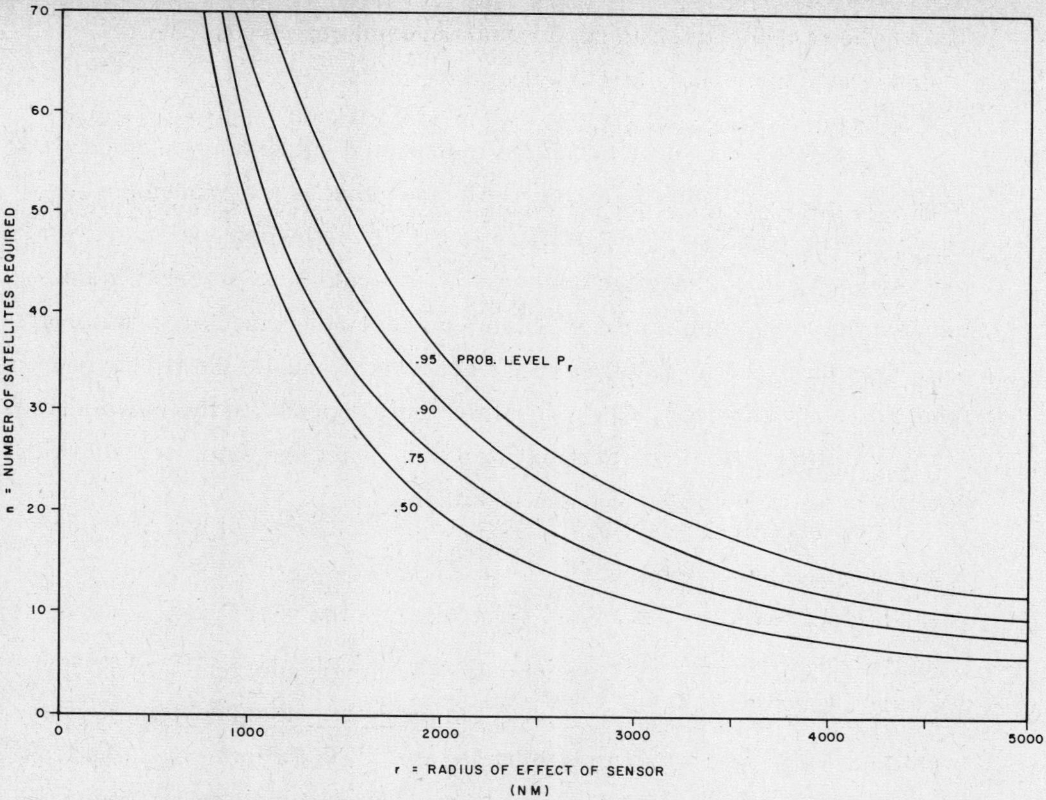

Fig. 5 Number of Satellites Required to Assure No Gaps for One Period at Probability Level P_r

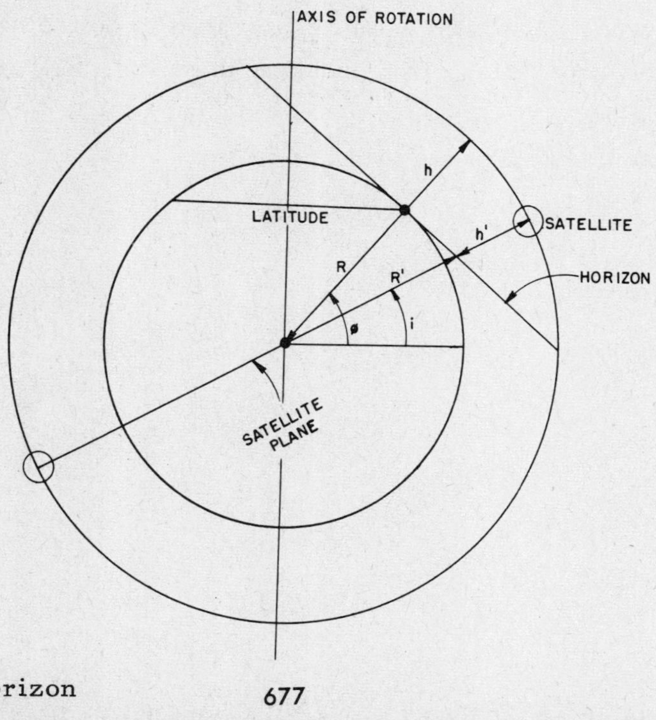

Fig. 6 Horizon

The equation is plotted in Figure 7 for a special case.

Expected Value Problems

A set of interesting problems in expected value can be solved. For example, for n satellites in an inertial plane, what is the expected wait and the expected number of satellites to pass over before one can have at least two satellites above the horizon? This would be of interest in communications. To simplify the problem, we assume a circular, equatorial orbit with the receiving station on the equator. Compute the time gap t, from horizon to horizon, Eqn. (13), which only depends on the mean altitude of the satellites. Then the probability p of at least two satellites above the horizon is the probability of a gap less than t.

$$p = 1 - e^{-nt/P}, \tag{15}$$

where P is the period. The probability of a gap greater than t is $q = 1 - p$. Table 1 shows the details for solving the expected value problem E_i is the event that i gaps greater than t go by and the ith is followed by a gap less than t. The probability of this event is $q^i p$. For example in E_0, the probability that the first gap over is less than t is p. E_1, the probability that the first gap is more than t and the second is less than t is qp, and so on. These are shown in the first two columns of the table. The third column shows the minimum wait for each event E_i. The fourth column shows, for each event e_i, how many satellites, N_i, went by before two appeared simultaneously above the horizon, including those two. Now we can compute the expectations.

The expected minimum wait is

$$\overline{W}_{min.} = \sum_{0}^{\infty} q^i p \cdot it \tag{16}$$

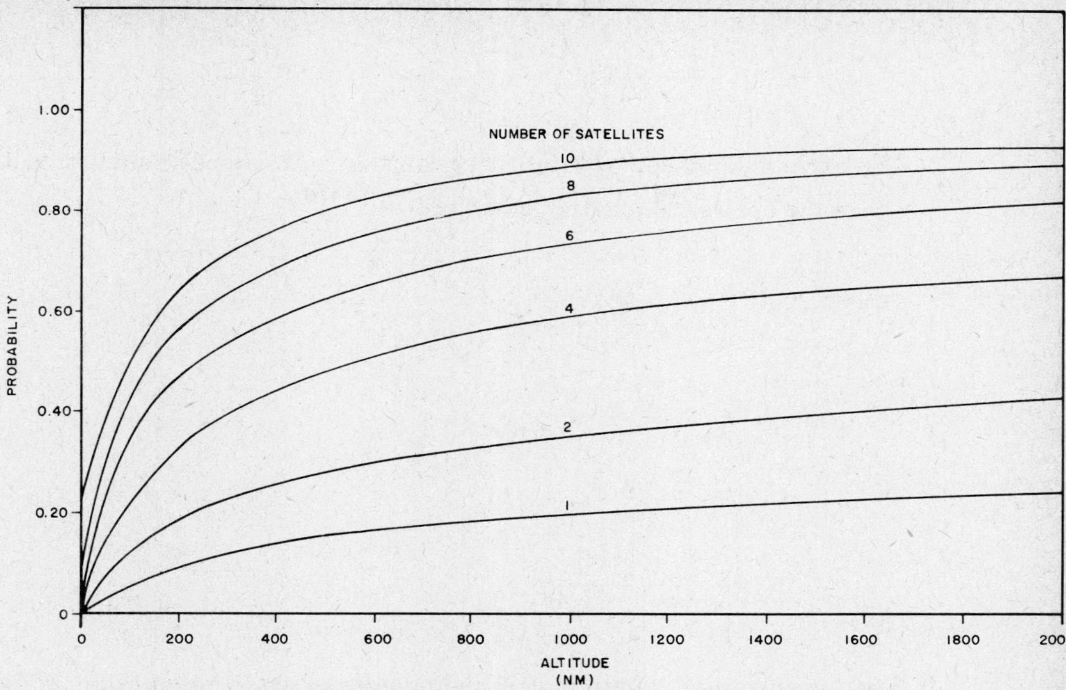

Fig. 7 Probability of at Least One Satellite Above the Horizon - Simple Case: $\phi = i = 90^{o}$

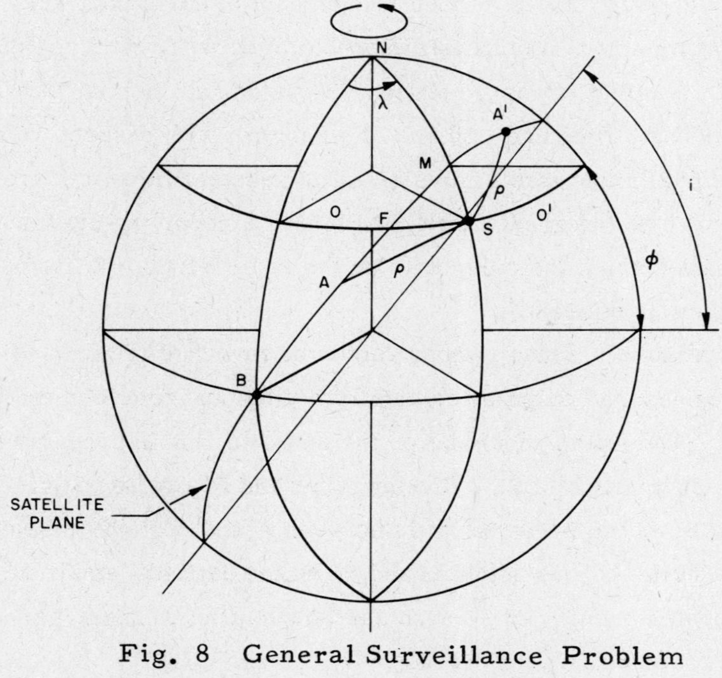

Fig. 8 General Surveillance Problem

$$= tp \cdot q/p^2$$

$$= t\left(\frac{e^{-nt/P}}{1 - e^{-nt/P}}\right)$$

The expected number of satellites over before two appear simultaneously above the horizon (including those two) is

$$(17) \qquad \bar{N} = \sum_{0}^{\infty} q^i p\,(i + 2)$$

$$= p\left(\frac{q}{p^2} + \frac{2}{p}\right)$$

$$= \frac{1 + p}{p}$$

$$= \frac{2 - e^{-nt/P}}{1 - e^{-nt/P}} \quad .$$

The formulae will help to select the number of satellites n required in orbit for the expectation to be less than some specified value. The method outlined is typical of several interesting applications.

Expected Coverage of a Station

Previous applications have been independent of earth rotation. This application (of which we only state the problem) directly requires the consideration of the effect of earth rotation. Consider an inertial plane containing several satellites and inclined at an angle i with respect to the earth's equatorial plane (see Figure 8). The subsatellite points are randomly positioned on the great circle BAFMA'. A receiving station S with latitude ϕ rotates about the polar axis on the small circle FS; its position in inertial space is defined by λ .

At the station is a sensor whose radius of coverage is given as an arc of ρ radians measured on great circles radiating out from S on the surface of the earth. (The sensor could be on the satellite if ρ is properly interpreted.) In any position of S, defined by λ , A and A' are ss. points at distance ρ from S; then if an ss. point is between A and A', it is in sensor contact with S. There are two points O and O' on the latitude small circle FS and on opposite sides of F such that the station S must be on the arc OFO' to be able to "observe" the satellite. So two conditions for "observation" are simultaneously essential: S must be in OFO', and the ss. point must be in AA'. A similar configuration occurs in the opposite hemisphere.

There are six distinct, though related configurations to consider according to the relative magnitudes of ϕ , ρ, and i. These are shown schematically in Figure 9.

The geometrical relations on the surface of the sphere and the probability considerations necessary to give the probability of a station "seeing" one or more satellite is too detailed to be included in this paper, in which we only wish to develop the model and show its usefulness. It will be presented in another report.

Monte Carlo Model for Satellites in a Plane

The model described in the section on The Planar Set of Satellites can be used with a high-speed computer to solve simulation problems using Monte Carlo techniques and involving the position of subsatellite points of a family of n co-planar satellites. The time spacing of successive satellites would occur according to the probability distribution.

(18) $$y = p[\text{time} < T] = 1 - e^{-nT/\bar{P}}.$$

The technique of selecting successive time spaces is as follows:

(a) Sample y from a uniform rectangular distribution of two-digit decimals from zero to one. Use a table of random numbers. (Ignore values of y which give $x > n$, since $T \geq \bar{P}$.)

(b) From Table 2, select the x corresponding to the randomly selected y.

(c) Compute the time T by the relation

(19) $$T = \bar{P}x/n.$$

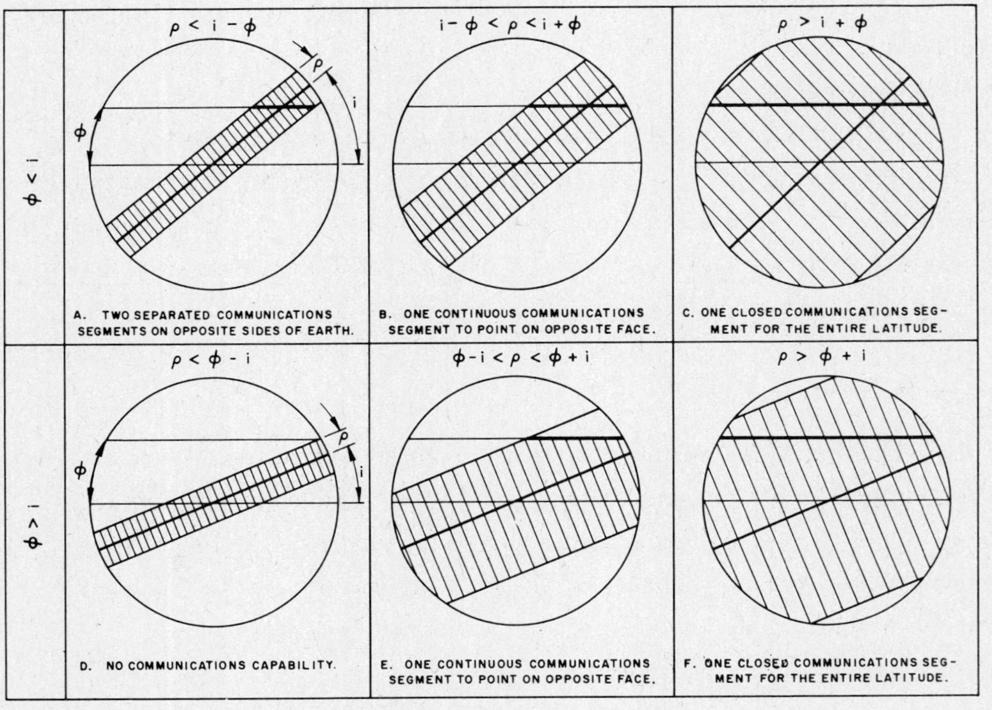

Fig. 9 Communications Segments

Table 1
EXPECTED VALUE CALCULATION

E_i	$P(E_i)$	Minimum W_i	N_i
E_o	p	0	2
E_1	qp	t	3
E_2	q2p	2t	4
E_3	q3p	3t	5
.	.	.	.
.	.	.	.
.	.	.	.
E_i	q^ip	it	i+2
.	.	.	.
.	.	.	.
.	.	.	.

Table 2
TABLE FOR SAMPLING
$y = 1 - e^{-x}$

If Sampled y is	Choose x	If Sampled y is	Choose x
.00 - .04	.0	.81 - .82	1.7
.05 - .13	.1	.83 - .84	1.8
.14 - .21	.2	.85	1.9
.22 - .29	.3	.86	2.0
.30 - .35	.4	.87	2.1
.36 - .41	.5	.88	2.2
.42 - .47	.6	.89	2.3
.48 - .52	.7	.90	2.4
.53 - .56	.8	.91	2.5
.57 - .60	.9	.92	2.7
.61 - .64	1.0	.93	2.8
.65 - .67	1.1	.94	3.0
.68 - .70	1.2	.95	3.2
.71 - .73	1.3	.96	3.5
.74 - .76	1.4	.97	3.9
.77 - .78	1.5	.98	4.6
.79 - .80	1.6	.99	4.7

This is the time interval to the arrival of the next satellite. (Because of the truncation in (a), maximum T = $\bar{P}$.)

(d) Sample a new y as in (a). The set of values of T so selected will be distributed according to the random relation given above.

BIBLIOGRAPHY

1. F.W. Bennett, T.L. Coleman, and J.C. Houbolt, Determination of the Required Number of Randomly Spaced Communications Satellites. Tech Note No. D-619, NASA, Jan 1961

2. W. Feller, An Introduction to Probability Theory and Its Applications, 1st Ed., vol. 1, John Wiley and Sons, New York, 1950

3. R. David Luders, "Satellite Networks for Continuous Zonal Coverage," ARS Journal, vol. 31, No. 2, Feb 1961, pp. 179-184

MILITARY APPLICATIONS OF SPACE SYSTEMS

Maj. Gen. C. H. Mitchell*

I am happy to be here today and take part in this program of the American Astronautical Society. It is always a pleasure to meet with those who are actively interested in space and have made significant contributions to the science of astronautics.

We in the military are interested in space for one primary reason--our survival as free men may very well depend on what we as a nation achieve in space. For many years our country could be adequately defended by forces that could fight on land and at sea. But the invention of the airplane added a decisive new dimension to warfare, and the advent of the Space Age has brought the possibility that this dimension may be extended outward hundreds or thousands of miles from the surface of the earth.

This means we must achieve the capability to prevent space from being used as an avenue of aggression. Naturally, as a nation, we would prefer to see space used only for purely scientific experimentation and eventual economic exploitation.

In an effort to achieve this end, the United States has presented a detailed program to the United Nations, proposing a number of specific steps to be taken toward achievement of international cooperation in the peaceful use of outer space.

At the same time, we have been forced to recognize that some nations do not endorse our national policy for the peaceful use of space, any more than they share our aspirations for peace and freedom on earth. To quote the words used by Ambassador Stevenson at the UN last month,

---

*Vice Commander, Air Force Systems Command, Andrews Air Force Base, Washington, D. C.

"The military questions of space are closely entangled with the military questions of earth."

The military questions on earth today are quite clear. We face adversaries in world affairs who are dedicated to a continuing drive for world domination, and who have never hesitated to make use of force, terror, subversion, and deception as expedient to attain their ends.

In the words of the official Program of the Communist Party of the Soviet Union, which was adopted last fall in Moscow, "The success of the struggle of the working class for the victory of the revolution will depend on how well it and its party master all forms of struggle--peaceful and nonpeaceful, parliamentary and extraparliamentary--and how well they are prepared to switch from one form of struggle to another as quickly and unexpectedly as possible."

The seriousness of this challenge is shown by Communist actions in Korea, Hungary, Tibet, Laos, Cuba, Berlin, South Vietnam, and other trouble spots in the world. The Soviets have shown conclusively that they are restrained neither by moral considerations nor by world opinion. Their aggressive actions have been stopped only when they have been opposed with determination by the Free World, backed up by adequate military strength.

The responsibility of the armed forces today is to create and maintain this strength, both for today and for the years to come. It is obvious that this strength will have served its true purpose only if it never has to be used. But we must build it, nevertheless.

In this connection I am reminded of the story of the Civil War veteran from Georgia who was recalling the military prowess of the Confederate armies. "Why shucks," he boasted, "we could have beat them Yankees with cornstalks." Then, after a pause, he added sadly, "Trouble is, they wouldn't fight with cornstalks."

The point of the story is this: We are engaged in a competition that is unique in history, and our opponents will not necessarily use weapons of our choosing. In fact, they will make every effort to achieve a decisive technological advantage, and one of the areas that looks most promising to them today is space. In this realm we must not be taken by surprise.

For this reason, we need to be realistic about the possible military uses the Soviets may make of space. In the past we have often been slow to see the military application of new inventions. The airplane, for instance, is an American invention, but its military potential was not realized very rapidly in this country. American airmen flew in combat during World War I, but the aircraft they flew were all of European design.

We showed a similar lack of imagination in regard to another American invention--the liquid-fueled rocket. Professor Goddard had successfully demonstrated his rocket in 1926, but there was little American interest in the military application of this invention until the beginning of World War II. And it was not until the German V-2's began falling on London that many people came to see the potential of the liquid-fueled rocket as a weapon.

Today the potential military threat posed by Soviet space achievements is becoming increasingly evident. Last year the Soviets placed Major Gagarin and Major Titov in orbit around the earth. Since their flights these two Soviet Air Force officers have continually been exhibited in uniform and wearing their military medals. Moreover, Mr. Khrushchev has threatened that rockets similar to those that carried Gagarin and Titov could be used to carry 100 megaton bombs.

From these achievements and these threats, one thing is clear. We cannot permit space to be dominated by those who have shown themselves to be the enemies of freedom. For the nation that dominates space might --if it chose--be able to dominate the world. I am convinced that we must be prepared to operate in space in order to preserve peace on earth.

To operate in space, we must first be able to reach space. The way has been provided by the intercontinental ballistic missile, which travels outside the earth's atmosphere during a large part of its trajectory. Today our strategic missiles provide most of the boosters for both military and civilian space programs. They have attained a quite satisfactory reliability, and that reliability is still improving. Both liquid-and solid-fueled missiles will serve as an increasingly important part of our nation's strategic forces.

To operate effectively in space we will need a mix of unmanned and manned space vehicles. Some of the potential uses of unmanned earth satellites have already been demonstrated. Satellite systems have shown their feasibility for such operations as communications, navigation, and weather prediction, and their potential use for a variety of information-gathering tasks.

But, in spite of the many capabilities of satellites and other unmanned systems, we will also need a capability for manned military operations in space, because machines--no matter how sophisticated--cannot completely replace man.

Man is superior to a machine in a number of ways. Perhaps the basic difference is one of motivation. You cannot design courage or a love of freedom into a computer. In addition, man has several capabilities that machines do not possess. For instance, in the processing of information man can do superbly two things no machine can do as well: he can discriminate between important and incidental information, and he can make correct decisions from inadequate or even incorrect data.

Moreover, unlike machines, man can use judgment and exercise control based on that judgment. He is unique in his ability to cope with unexpected situations. For these reasons man will be an essential element in many future space systems.

Studies have shown a number of functions in space that could best be performed by man. Among these are flight control, in-flight maintenance, and control of vehicles in rendezvous and docking operations.

Man also has a potential role in satellite and missile observation systems. His ability to observe, evaluate and make judgments could provide an information filter that would reduce the requirement for complex electronic data storage and transmission equipment.

In addition, man's ability to diagnose malfunctions and make repairs could be a major factor contributing to the prolonged operational life of military space vehicles.

These are but a few of the potential requirements we now see for military man in space. Their importance requires that we make a serious study of the problems of transporting, supporting, and utilizing man in space on a routine, day-to-day basis.

The transportation of man into space and back to earth requires the development of a number of basic capabilities. Among these are the ability to put large payloads into space in a reliable and reasonably economic manner; the ability to maneuver, to achieve rendezvous, to dock with other space vehicles, and to transfer personnel and materiel from one space vehicle to another; and finally, the ability to re-enter the atmosphere from both high and low orbit, and to make controlled horizontal landings.

The support of man in space requires that we develop means of providing protection against such hazards or potential hazards as radiation, weightlessness, isolation. We must also perfect systems to provide food, water and fresh air, and to dispose of waste products.

But it is not enough merely to send man into space and keep him alive there. He must also be able to function in space and he must be able to leave his vehicle safely for a variety of purposes, such as emergency escape, vehicle repair, transfer to another vehicle, and assembly of space stations.

In outlining some of the potential uses of man in military space operations, I do not mean to imply that we have defined specific military missions in space. It is probable that the evolution of space craft will follow somewhat the same pattern as the evolution of aircraft.

When the first U. S. military airplane was purchased, more than 50 years ago, there existed no set of specified missions for military aviation. First of all it was necessary to build a machine that would fly. Then it had to be improved, in terms of altitude, speed, range, and carrying capacity. When we began to understand the aircraft itself, we could begin to work out the possible tasks that it could perform.

I believe the same sequence will occur as we develop the basic technology for unmanned and manned space flight. As our knowledge and understanding increase, we shall be able to design systems for specific missions, just as our specialized fighters, interceptors, bombers, and transports were all developed after we had begun by building a machine that would fly.

This means that the Air Force will be serving a double function in space.

In developing the basic techniques of space flight it will be actively supporting the National Aeronautics and Space Administration in carrying out the national manned lunar landing program. This will be a continuation of the support the Air Force has given to other NASA programs in the past and is currently giving to the NASA Mercury program. At the same time the Air Force is responsible for developing the operational capabilities that will be needed to prevent the exploitation of space as an avenue of aggression.

The goal of our efforts in space should not differ from our national goal on earth--the preservation of peace and freedom in the world. Ideally, man should go into space as scientist or explorer. Realistically, as long as our nation faces a potential Soviet threat in space, he must go there prepared to play a military role as well.

Through this preparation, space can be used to help preserve the peace. I am convinced that as we work toward our national goals with wisdom and courage, we will see space flight--like aviation--become a source of great benefit to all the peoples of the earth.

A POSSIBLE MILITARY APPLICATION OF A CIS-MARTIAN ASTEROID*

Dandridge M. Cole*+

One of the close-approach or cis-Martian asteroids could conceivably be diverted from its orbit in such a manner as to impact on a preselected earth target area.

This could be done by employing a Nova-class rocket vehicle to carry a crew of men or remotely controlled devices to such an object and applying a velocity increment through explosion of hydrogen bombs on the surface.

The energy release at impact on the earth would be on the order of one thousand times as great as the energy of a multi-megaton bomb which could be carried by the same Nova-class vehicle on an ICBM trajectory to the same target.

This unusual weapon system shares with the biological weapons the possibility for masquerading as a natural catastrophe. The attacker could hope to escape blame and retribution from surviving retaliation forces.

While such an irreverent meddling with cosmic forces may seem highly implausible or even impossible on superficial examination, the need for careful consideration of such a system is supported by the highly favorable energy ratio and strategic advantages which it could supply.

---

* A complete version of this paper (AAS 62-32) may be obtained from the American Astronautical Society. Some of the material contained in this paper and additional material on asteroids has appeared as follows:

Dandridge M. Cole, "Asteroids Stir Growing Interest," Missiles and Rockets, 25 Feb 1963, pp.43, 46, 48, 50, 52, 53, 54.

Dandridge M. Cole, "Capturing the Asteroid," Astronautics and Aerospace Engineering, Mar 1963, pp. 88-93.

+ General Electric, Missile and Space Vehicle Dept., Philadelphia, Pa.

Rabe, E. (1950). Derivation of fundamental astronomical constants from the observations of Eros during 1926-1945. Astron. J. 55, 112.

Brown, H. (1960). The density and mass distribution of meteoritic bodies in the neighborhood of the Earth's orbit. J. Geophys. Research 65, 1679.

Institute of Theoretical Astronomy. (1959). "Ephemerides of the Minor Planets for 1960". Academy of Sciences, Moscow.

Alter, D. (1949). The importance and nature of the asteroids. Griffith Observer 13, 26.

Kuiper, G. P. (1953). Note on the origin of the asteroids. Proc. Natl. Acad. Sci. U. S. 39, 1159.

Hirayama, K. (1918). Groups of asteroids probably of common origin. Astron. J. 31, 185.

Hirayama, K. (1933). Present state of the families of asteroids. Proc. Imp. Acad. Tokyo, 9, 482.

Brouwer, D. (1951). Secular variations of the orbital elements of minor planets. Astron. J. 56, 9.

Rabe, E. (1956). On the origin of the Kirkwood gaps and the minor planet families. Z. Astrophys. 40, 107.

Kuiper, G. P., Fujita, Y., Gehrels, T., Groeneveld, I., Kent, J., Van Biesbroeck, G., and van Houten, D. J. (1958). Survey of asteroids. Astrophys. J. Suppl. Ser. 3, 289.

Gehrels, T. (1957). Photometric studies of asteroids. VI. Photographic magnitudes. Astrophys. J. 125, 550.

Gehrels T., Carstens, J. R., Peters, J. G., and Peters, L. S. (1960). Photometric studies of asteroids. VIII. Additional photographic magnitudes. Astrophys. J. 132, 232.

Groeneveld, I., and Kuiper, G. P. (1954). Photometric studies of asteroids. I. Astrophys. J. 120, 200.

Groeneveld, I., and Kuiper, G. P. (1954). Photometric studies of Asteroids. II. Astrophys. J. 120, 529.

Shatzel, A. V. (1954). Photometric studies of asteroids. III. The light-curve of 44 Nysa. Astrophys. J. 120, 547.

Ahmad, I. I. (1954). Photometric studies of asteroids. IV. The light curves of Ceres, Hebe, Flora, and Kalliope. Astrophys. J. 120, 551.

Gehrels, T. (1956). Photometric studies of asteroids. V. The light-curve and phase function of 20 Massalia. Astrophys. J. 123, 331.

van Houten-Groeneveld, I., and van Houten, C. J. (1958). Photometric studies of asteroids. VII. Astrophys. J. 127, 253.

Urey, H. C. (1959). Primary and secondary objects. J. Geophys. Research 64, 1721.

Fish, R. A., Goles, G. G. and Anders, E. (1960). The record in the meteorites. III. On the development of meteorites in asteroidal bodies. Astrophys. J. 132, 243.

Herrick, S. (1953). Icarus and the variation of parameters. Astron. J. 58, 156.

Gilvarry, J. J. (1953). Relativity precession of the asteroid Icarus. Phys. Rev. 89, 1046.

Urey, H. C. (1956). Diamonds, meteorites, and the origin of the solar system. Astrophys. J. 124, 623.

Urey, H. C. and Donn, B. "Chemical Heating for Meteorites" Astrophysical J. Vol. 124, July - Nov., 1956.

Kornhauser, Murray, "The Meteorite Danger to Earth" G. E. -MSVD. The American Astronautical Society Summer Meeting, San Francisco, Aug., 1961.

CIVILIAN APPLICATION OF ASTRONAUTICAL SYSTEMS*

Donald N. Michael+

*This paper (AAS 62-34) may be available by contacting the author.

+Peace Research Institute, 1329 18th St. N.W., Washington 6, D. C.

eighth
annual
meeting

AMERICAN ASTRONAUTICAL SOCIETY

ABORT DURING MANNED ASCENT INTO SPACE*

R. L. Gervais and M. C. Johnson+

The objectives of this paper are to define the abort problem throughout an entire ascent into space and then establish abort procedures that will provide for the safe return of man to the earth's surface. Environmental restrictions are determined and imposed throughout the entire analysis. System malfunctions are defined and abort procedures that alleviate the hazards created are advanced. Two particular malfunctions appear to influence abort design conditions since their resulting hazards approach several of the environmental restrictions established. They are: abort from a launch pad booster explosion, and abort at high altitudes. Consequently, an explosion analysis is conducted as well as a flight mechanics management study. Results show that a successful abort is possible in nearly all phases of an ascent with only minimum compromise to mission performance, while methods of increasing abort system effectiveness are also advanced.

INTRODUCTION

Since we are presently experiencing the advent of manned space flights, techniques which will insure the safety of personnel participating in such flights must be thoroughly explored. After these techniques have been evaluated, they must then be correlated with presently accepted aircraft safety standards to determine the degree of emphasis that must be given the required safety measures. The purpose of this study, then, is to first determine and evaluate the types of failures that may occur during manned space flights, and then present techniques which will

---

*Engineering paper No. 1243 prepared under the sponsorship of the Douglas Aircraft Co., Independent Research and Development Program, Account No. 9703-51733

+Aero/Astrodynamics Section, Missile & Space Systems Division, Douglas Aircraft Company, Inc., Santa Monica, Calif.

minimize the hazards resulting from such failures. In particular, failure modes of the booster system and space vehicle will be investigated during the lift-off, atmospheric, and vacuum phases of the ascent, as well as during a cislunar transfer phase. These potential failures will also be evaluated to determine their effect on total mission reliability. Finally, corrective action, in the form of propulsion and flight kinematics management, will be recommended to reduce the hazards resulting from these failures.

As the state of the art advances and contemplated missions become more ambitious, the scope of such an abort evaluation must be increased. Since increased mission capability will be directly reflected in the size of the booster required, the hazards associated with a booster system malfunction become greater. Because of this increased booster requirement, emphasis is placed upon resolution of the lift-off and low altitude booster explosion problem for boost system lift-off weights up to 10 million pounds.

In an effort to emphasize the need for such an abort evaluation, Table I presents an example of the reliability anticipated from future space systems where the boost system exhibits a reliability of only 90 per cent (Ref. 1). When compared to the fatality rate of current "century series" fighter aircraft of two fatalities per 10, 000 flights, it can be seen that safety methods for manned space flights must be improved considerably.

ABORT CRITERIA

Before recovery techniques applicable to abort conditions are discussed, a definition of some of the malfunctions requiring an aborted flight will be made, and the methods of their detection will be reviewed.

Possible Malfunctions

The most catastrophic type of malfunction is a booster explosion. The probability and magnitude of such an explosion are dependent on the type

TABLE I
PROJECTED FAILURE PROBABILITY
(AFTER GARNETT)

NOTE: ONE PASS ORBITAL MISSION

TOTAL SYSTEM FAILURE PROBABILITY = 12.85 PERCENT

	PERCENTAGE OF SUBSYSTEM FAILURES (PERCENT)
NAVIGATION GUIDANCE	7.8
FLIGHT CONTROLS	5.7
GLIDER POWER SYSTEMS	3.6
CREW ENVIRONMENT	1.3
OTHER SYSTEMS	5.1
BOOSTER SYSTEM	76.5
TOTAL	100.0 PERCENT

of propellant used. Propellant properties, and hence explosion probabilities, are discussed in Ref. 2. The magnitude of the resulting explosion will depend upon the amount of propellant available; the position of the fuel and oxidizer tanks relative to each other; the particular fuel and oxidizer utilized; the mixing rate of the reactants; and the ignition delay.

Less severe types of failure are loss of thrust or failure to ignite. These types of failure usually will occur during upper stage operation, but will still require an abort action. While there are many possible modes of failure, those of primary significance are guidance and/or control failure, and crew cabin malfunction.

Methods of Detection

There appears to be no difficulty in detecting any of the types of malfunctions mentioned above. The speed with which the malfunction can be detected is of the utmost importance, however, because an escape system requires the maximum warning time available from any given detection system. Detectors or sensors that could be used to detect an impending booster explosion are pressure sensors in the propellant tanks, or a turbine speed measuring sensor. Considerable study should be devoted to the design of crew cabin sensors due to the immediate effect of crew cabin malfunctions upon the crew, but these sensors will probably not be as demanding in reaction time as some of the boost system sensors.

Recovery Modes and Constraints

Once a boost system or space vehicle malfunction is detected, an abort action sequencing must necessarily follow. If a boost system malfunction occurs, an automatic sequencing is required because of the response times involved; however, if the failure is confined to the space vehicle itself, situations may arise where manually initiated, rather than automatic, abort is desirable.

When a boost system explodes, large amounts of energy are released, accompanied by a characteristic propagating shock wave. If the space vehicle is still attached to the booster, the impulse of the shock wave and/or overpressure will damage it, and fatally injure its occupants. The intensity of this shock wave will decrease as its distance from the explosion center increases. Since space vehicles are designed to withstand only a certain value of overpressure, the vehicle must be removed from the booster, in the event a boost system explosion occurs, in order to preserve the structural integrity of the vehicle and prevent injury to its occupants.

As the booster gains altitude, the explosion problem becomes less acute, due to decreased atmospheric pressure and available propellant, and only less severe malfunctions, such as thrust loss, guidance and/or control loss, etc., are likely to occur. All of these malfunctions will require separation of the space vehicle from the booster, the only difference being the rate of separation required. Due to this difference in rate of separation, recovery constraints based upon environmental limitations must now be introduced. From an ecological standpoint, the rate of acceleration onset and offset, and the duration and magnitude of acceleration onset and offset, and the duration and magnitude of acceleration must be considered. From an aerodynamic standpoint, the space or escape vehicle must possess aerodynamic characteristics compatible with such a maneuver. From a structural standpoint, a compromise must be made; the vehicle must be stressed either to accommodate the high accelerations of separation or to resist the higher overpressures encountered. An ecological-structural compromise must also be made, i. e., the high accelerations required to outrun a shock wave of a defined lower overpressure must be weighed against improving the structural integrity of the vehicle so that the vehicle is capable of withstanding high overpressures.

Once the escape vehicle has separated from the booster, other limitations must be considered. For extremely low altitude aborts, landing is of immediate concern and landing speed and/or rate of descent must be maintained within reasonable limits. However, as the altitude at which the abort occurs increases, the resulting descent flight path becomes of main concern because such a descent could compromise either the psychophysiological integrity of the crew or the thermal integrity of the escape vehicle. Aborts occurring at increasingly higher altitudes approach other ecological limitations, such as passage through the radiation belts, and the time required to return to earth now becomes of major concern.

Thought must also be given to the area in which a vehicle, descending from an aborted flight, will land. If present launch complex locations are retained, 95 per cent of all aborted flights will touch down on water (Ref. 1); hence, the recovery constraint of buoyancy is introduced. In addition, the question of economics must also be considered. For example, specified retrieval areas must be designated or else the cost of any retrieval operation will become exorbitant. This retrieval constraint also imposes a maneuverability requirement on the escape vehicle. Consequently, the escape vehicle must possess inherent aerodynamic maneuverability and/or the capability of increasing its velocity to permit maneuverability.

Space Vehicle Design Requirements

Since abort is obviously not the prime objective of the mission, abort recovery accommodations should be designed to minimize any resulting decrease in the mission capabilities of the space vehicle. Accordingly, the design of the space vehicle is almost totally influenced by the prime goal of the mission, that of placing man at his destination in space and then returning him safely to earth. Since re-entry presents one of the most formidable problems in the accomplishment of this mission, design of the space craft should reflect the conditions that will be encountered during such a re-entry.

After a review of the criteria that affect such a design (Ref. 3), a vehicle capable of aerodynamic maneuverability was selected for analysis in this study. The specific characteristics of this vehicle are as follows: a maximum hypersonic lift-to-drag ratio of one, a wing loading of 24 lb/ft^2, and a vehicle weight of 10, 000 lb. These aerodynamic characteristics, in addition to conforming to the re-entry requirements, also coincide with the maneuverability requirements of an abort retrieval operation. An illustration of this vehicle separating from an exploding booster is shown in Fig. 1.

BOOSTER EXPLOSIONS

As has been indicated in the previous paragraphs, possible abort situations resulting from booster explosions exist from launch pad to final stage burnout. These conditions will be analyzed in chronological order;

Fig. 1 Space Vehicle Separating From Exploding Booster

hence, launch pad malfunctions become the initial area of investigation.

Detonation

In order for the space or escape vehicle to retain its structural integrity, it must immediately be removed from the area of the explosion. In order to accomplish this removal, an impulse must be imparted to the abort recovery system. The magnitude of this impulse is determined by the amount of overpressure the vehicle is designed to withstand and the magnitude of the energy released in the detonation.

If a fundamental approach to this explosion problem is taken, conditions at and inside the detonation front must be known. Since data regarding these conditions are non-existent, the substitution of an equivalent amount of TNT as the explosive source has been used in this study. This equivalency is established by choosing the correct weight of TNT, so that the total energy available is equal to that of a particular rocket fuel explosion. This assumption, however, is not acceptable by itself because the presence of the booster structure will affect the rate of energy release. Consequently, a more realistic theory which allows for a finite rate of energy release is necessary (Ref. 2). This is accomplished by introducing new initial conditions at the boundary of the explosion source.

It is believed the assumption that the various propellants be rated according to their TNT equivalence is closer to reality than the empirical nuclear bomb data contained in Ref. 4. The basis for this belief is the difference in reaction rates between TNT and nuclear explosions and the variance in the respective pressure-time data. Therefore, the mathematical model used in analyzing a booster explosion involves a numerical solution of the partial differential equations of hydrodynamic motion for the case of a center-detonated spherical charge of TNT (Ref. 5). The assumption of symmetry is not valid near the explosive center, but as the shock wave propagates outward it will approach a symmetrical state. Solution of these hydrodynamic equations for various TNT equivalent weights yields the sea-level overpressure-distance relationships presented in Fig. 2. This same overpressure-distance data is then related to an over-pressure-time relationship as shown in Fig. 3. Of particular note is the time required for the shock wave to propagate to lower energy levels. This small interval of time will subsequently be shown to have a direct bearing on the specifications for the malfunction-detecting system previously mentioned. It should be noted that image effects are inherent in the calculations of these curves and thus are considered in the results derived.

Now that these overpressure relationships have been established for various TNT equivalent weights, the problem that arises is equating them to particular propellant weights. In order for an exact solution to

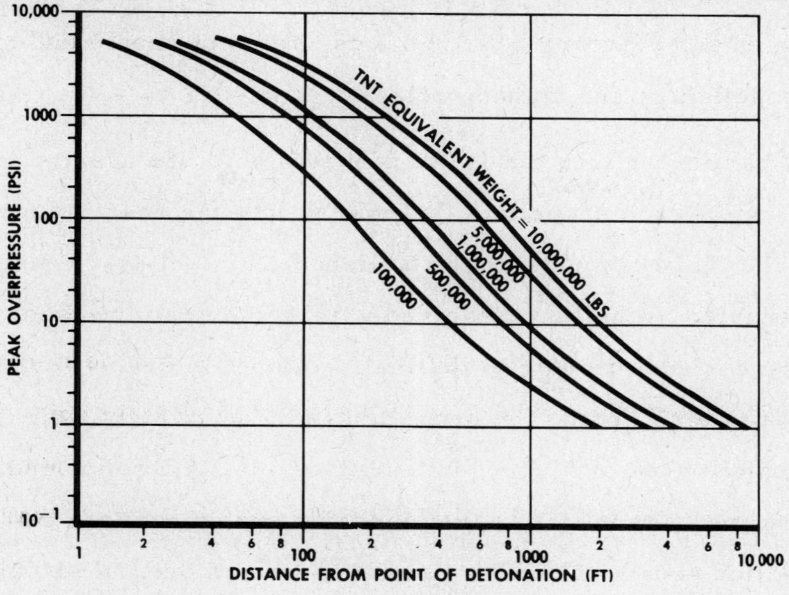

Fig. 2 Overpressure Distance Relationship (Sea Level Explosion)

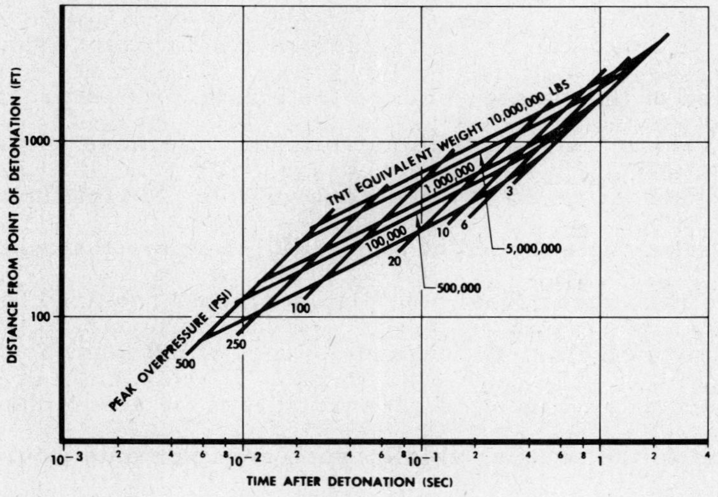

Fig. 3 Launch Pad Explosion Criteria

be obtained, the following criteria must be known: type and extent of tank rupture, temperature of propellants, rates of vaporization, mixing rate of propellants, and ignition delay.

Since this particular solution is not now available, an assumption of the exact amount of each propellant that enters into the reaction must be made. Once these amounts are established, the fuel-to-oxidizer ratios of the particular propellants involved will then determine the total amount of propellant on board prior to the malfunction. In order to present results in parametric form, the propellant-TNT equivalent weight relationship was established at 50 percent. That is, if 2 million pounds of propellant are on board initially, the explosion energy release can be equated to 1 million pounds of TNT. However, when a detonation takes place on or near the surface of the earth, an image or Mach stem effect becomes evident. In order to account for this image effect in the energy released in the detonation, twice the actual amount of TNT is presumed. However, this presumption is inherent in the calculations and therefore should not be taken into additional account when using Figs. 2 and 3.

Abort Vehicle Separation

Before Figs. 2 and 3 can be used to determine the exact separation distance required of the escape vehicle, the maximum overpressure capability of the vehicle must first be established. Because of localized weakness in several areas of the escape vehicle, the maximum tolerable overpressure was established at 6 psi. Fig. 3 shows that where overpressures do not exceed 6 psi, the displacements required for large TNT equivalent weights are thousands of feet which must be traversed in tenths of seconds. Since the separation is to be accomplished by imparting thrust to the escape vehicle, such requirements would be unrealistic because of the excessively high thrust-to-weight ratios necessary. In order that this separation time be made more tolerable, the malfunction detection system must provide a warning time in advance of detonation. Included in this warning time must be allowances for system sensing and reaction time. It is in this area of malfunction detection that emphasis should be placed, in order to provide as much warning time as possible.

Fig. 4 presents distances and burning times required for various TNT equivalent weights as a function of both escape vehicle thrust-to-weight ratios and warning times so that a peak overpressure of 6 psi is never exceeded. In order to obtain a parameterized approach, an initial displacement of 100 feet is assumed between the explosion center and the escape vehicle. The specific example illustrated is indicative of a system required to place a space vehicle of 10, 000 pounds on a circumlunar trajectory; i. e., 5 million pounds of propellant or 2 1/2 million pounds of TNT. If a warning time of 2. 0 seconds is available, a thrust-to-weight ratio of 20 is capable of moving the escape vehicle the required distance in the required time. Fig. 5 gives a synopsis of the thrust-to-weight ratio required for various propellant loadings as a function of both warning and burn times. Of particular interest here is the length of burn times associated with various thrust-to-weight ratios. For example, a 10 million pound propellant explosion with a 1. 5 second warning time and 50 per cent of propellant reacting requires a thrust-to-weight ratio of 26. 8 for approximately 2 seconds of burn time.

The data presented in Figs. 4 and 5 were accumulated on a high speed digital computer utilizing a six degree of freedom dynamical solution. Because a lateral as well as a vertical displacement is required between the escape vehicle and the booster during an atmospheric abort, a thrust deflection of 20 degrees was used in the computations.

Launch Pad Constraints

Obviously, several of the constraints mentioned previously might have to be compromised in order to assure that the vehicle would not be subjected

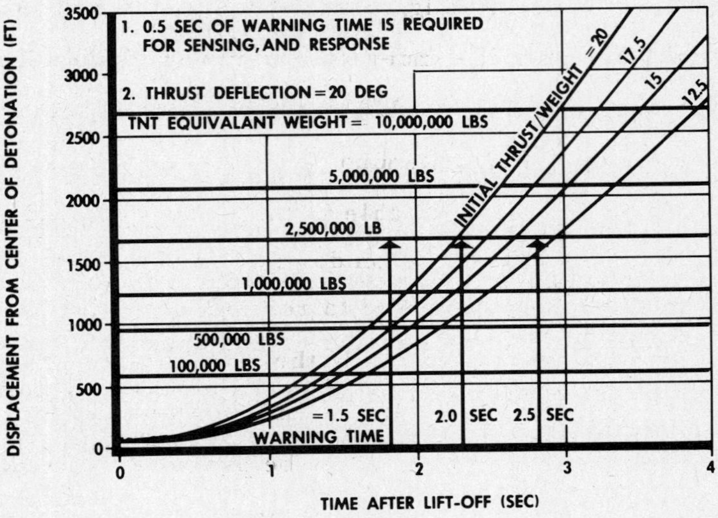

Fig. 4 Displacement vs Time from Center of Detonation as a Function of Thrust/Weight for a 6 PSI Overpressure at Sea Level

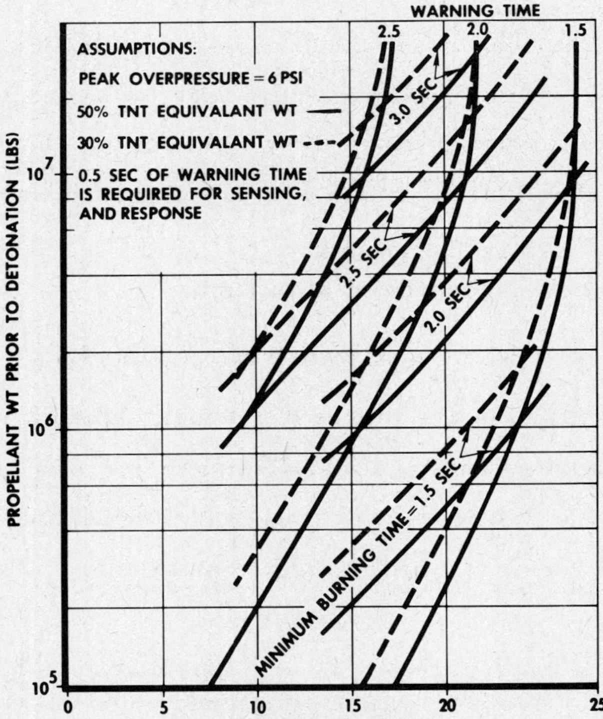

Fig. 5 Thrust/Weight Ratio Required for Adequate Displacements

to overpressures in excess of 6 psi. Of prime concern is the safety of the crew, and in that regard Fig. 6 is introduced. From a chronological standpoint, the first area of concern is the rate of onset of the acceleration. As can be seen in Fig. 6, 1,000 g's per second appear tolerable (Ref. 6). From a designer's standpoint, this limitation presents no problems because of tolerances available in thrust buildup. However, the acceptable durations of acceleration do present obstacles, because durations of 2 to 3 seconds are required to restrict the overpressure level (Fig. 5). The next area of concern is the rate of offset that would follow such an impulse; in some instances, a differential of 25 to 30 g's might be experienced. However, this too can be provided for by the designer by the use of multiple, step, or regressing rockets.

Since an ecological restraint appears to require compromise, several procedures which possibly could decrease the thrust-to-weight ratio requirement were investigated. One area that appeared promising was an

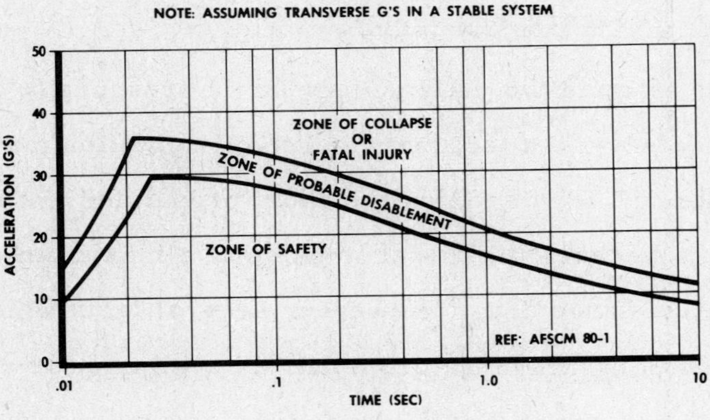

Fig. 6 "Hiad" Acceleration Limitations

automatic destruct system which would be actuated if it was determined that the booster was about to explode. By means of such a predetermined destruct pattern the amount of propellant reacting could be reduced, thereby reducing the energy released in the explosion. However, Fig. 5 shows that reacting only 30 percent of the propellant on board does not reduce the acceleration requirements sufficiently to justify compromising the safety margin through the activation of an automatic destruct system. Another approach investigated was to increase the tolerable overpressure by enhancing the structural integrity of the escape vehicle. Unfortunately, this method not only does not appreciably reduce the accelerations required; it also increases overall system weight, thereby reducing mission performance. The only remaining method of reducing the accelerations imposed upon the crew is acceleration attenuation through the use of seating restraints; e.g., seat positioning, contour seating, acceleration absorbers, and customized apparel. It has been revealed (Refs. 7 and 8) that the high accelerations mentioned above are tolerable only when such restraint apparatus is provided. Thus, it appears that real, but not insurmountable, ecological obstacles exist during launch-pad abort. However, the question still to be answered is: what booster explosion condition is most critical from a thrust requirement standpoint?

Maximum Dynamic Pressure Explosions

The decision to investigate the maximum dynamic pressure (q) explosion condition is predicated on the fact that the lift-off condition may not be the design condition, as far as sizing the abort propulsion system is concerned. This is because the thrust impulse of a maximum q abort is countered by a rather large drag. However, several facts which tend to counterbalance this point are: less propellant is available for detonation; there is no surface on which a shock wave might reflect and cause a fused shock; shock propagation rates are, in general, reduced by decreased atmospheric pressure; the initial speed of the escape vehicle is probably transonic at the moment of detonation, and perhaps only a portion of the escape vehicle may require separation during launch pad abort.

Solution of the hydrodynamic equations at the altitude of maximum q condition (31, 800 ft) yield the overpressure time-distance relationships presented in Fig. 7. When these results are compared with the sea level explosion criteria shown in Fig. 3, the effects of the image concept and altitude can be readily seen. In order to finalize abort propulsion design, the example presented previously requiring 5 million pounds of propellant will be pursued at the maximum q condition. Since the propellant remaining at this point in the boost trajectory is now only 3. 25 million pounds, the TNT equivalent is reduced to 1. 625 million pounds. Figure 7 shows that the displacement required for this TNT equivalent weight at an overpressure of 6 psi is only 953 ft. Figure 8 graphically illustrates how these requirements fall short of the launch pad requirements. Initially, the escape vehicle and the propellant center of gravity are assumed to have a displacement of 50 ft. Also, two seconds of warning time are available, 1/2 second of which is utilized by the detection and activation systems. In addition, launch pad values of a thrust-to-weight ratio of 20 and a thrust deflection angle of 20 degrees are used. Thus, with all comparisons established, it can be seen that the launch-pad requirements insure a displacement of 1240 ft in the required time interval, thereby establishing the launch pad requirements as the design condition. The above illustration would not have proved as convincing if the drag were allowed to increase over the value used. Hence, drag must be kept to a minimum (approximately zero degree angle of attack) by designing into the vehicle the capability of trimming at zero degree angle of attack. Since other phases of the mission, besides abort, require angles of attack much greater than zero, the control system must be designed to be compatible with these requirements.

As the vehicle continues its ascent into space after experiencing the maximum q condition, the dangers of a propagating shock wave due to booster explosion are considerably reduced, due to an ever decreasing propellant

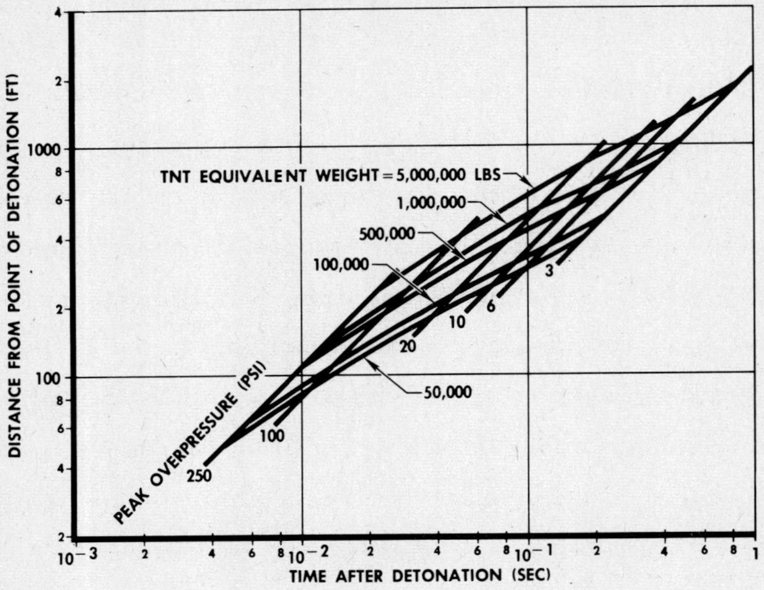

Fig. 7 Explosion Criteria for Maximum 'q' Conditions

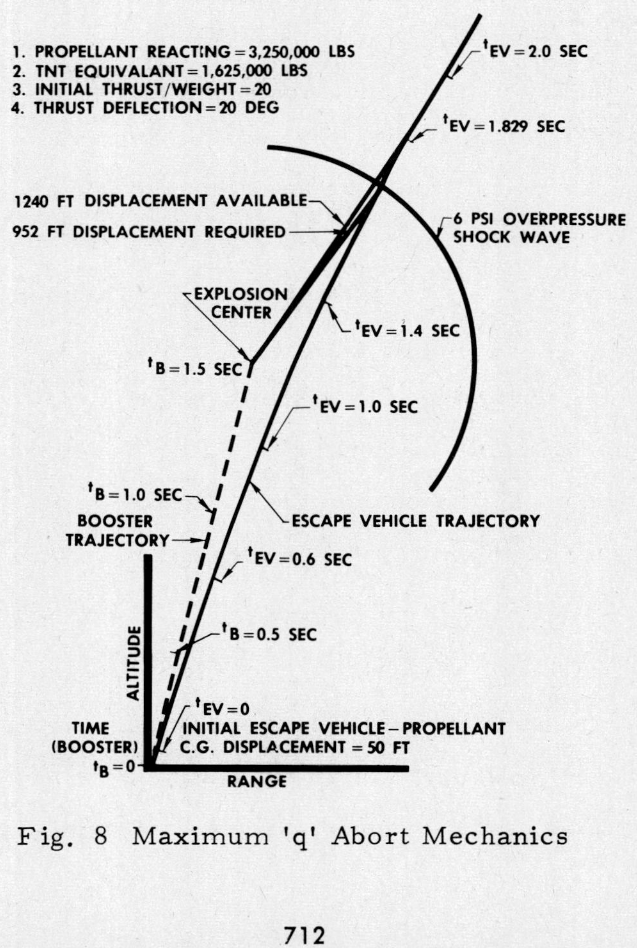

Fig. 8 Maximum 'q' Abort Mechanics

weight and the decreasing ability of the shock wave to propagate at extreme altitudes.

ATMOSPHERIC ABORTS

Aborts resulting from malfunctions other than booster explosions exhibit a common problem, that of excessive deceleration. Low altitude aborts exhibit this tendency at the moment of escape vehicle separation, while high altitude aborts experience decelerations in the descent following the initial separation. Before permissible abort conditions can be established, the endurance limits of the manned payload must first be defined since they determine in part the limits of these conditions.

Acceleration Requirements

Just as material strength, modulus, and temperature limit the structure, the manned payload places a precise design limit on a space vehicle. However, man's limitation is more absolute because the addition of weight seems always to heal the ailments of hardware. The tolerance requirements of man are dictated for the most part by the "g" environment to which he is subjected. The tolerance "limit" indicates the point at which an individual will be unable to withstand continued acceleration. Within the tolerance, the subject would be expected to see, think, and function adequately. The main characteristics of the acceleration environment that must be considered are the magnitude of the acceleration, the time duration, the rate of onset, and/or offset, and the direction through which the acceleration force acts. This last element provides the greatest opportunity for the designer to offset the relative effects of acceleration; i. e., seating can be arranged to minimize the effects of acceleration.

The endurance level is dependent upon the physical and mental condition of the individual at the time he encounters the high acceleration environment. Previous exposure experience and individual motivation are also considerations affecting endurance level. While a great deal is yet to be learned about these interactions, it is safe to say that the deceleration encountered during re-entry from a space mission or extra atmospheric abort must be considered separately from take-off exit accelerations and atmospheric abort decelerations. Mental tension, exhaustion-deceleration levels for exit and atmospheric abort cases, as well as an estimate of the reduced endurance level during the re-entry phase that has followed a typical space mission or an extra atmosphere abort, are shown as the allowable acceleration tolerance in Fig. 9 (Ref. 8).

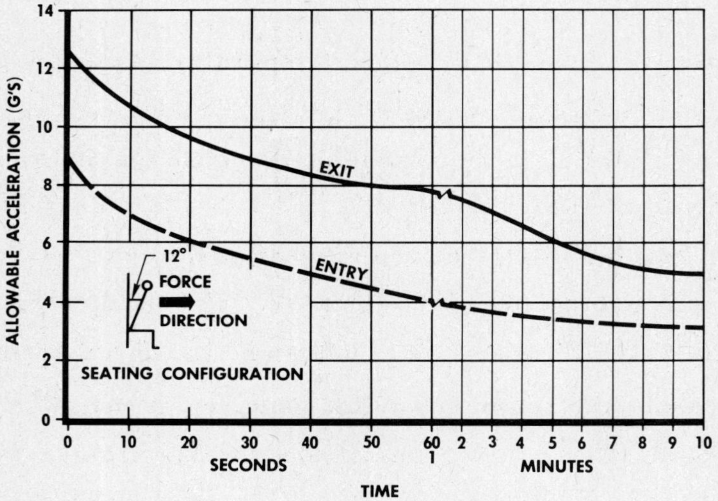

Fig. 9 Allowable Acceleration Tolerance

Low Altitude Aborts

In order to comprehend more fully the low altitude abort situation, the approach taken will be that only a minute impulse is imparted to the escape system at moment of separation. Although this has been, and will subsequently be shown to be, an unrealistic assumption, it does portray more vividly the actual mechanics of such an abort. The decelerations encountered at the moment of separation for various altitude-velocity combinations at a zero degree angle of attack are shown in Fig. 10. These results are quite unexpected, when compared with booster accelerations at the moment prior to escape vehicle separation. This deviation is due to the ballistic parameter (W/C_DA) difference between the booster-space vehicle system and the space vehicle alone. These results are indications, however, of the thrust required to attain certain separation displacements at low-altitude conditions for avoidance of booster-escape vehicle collision, etc. The results of Fig. 10 are characteristic of instantaneous conditions 'and hence' are not dependent on flight path angle. When flight path angle is introduced, a deceleration-time dependence becomes evident where, for a given altitude-velocity condition, the deceleration decay is greater for larger flight-path angles. After the booster has attained an altitude of approximately 50, 000 feet, the effect of the atmosphere starts noticeably to decrease the decelerations encountered. Also, the problem of shock wave propagation caused by explosion is diminished due to increased altitude. Consequently, the need for the abort rocket thrust decreases. This suggests the use of multiple rockets capable of being jettisoned, since a booster performance penalty is paid as long as they are retained. Once the vehicle has ascended above the high drag regions, the only reason for retaining some abort-rocket thrust capability is to avoid collision by the booster after separation and to avoid being struck by fragments in the event of an upper atmosphere explosion. Consideration has also been given to the possibility of the booster tumbling before escape vehicle separation. However, excessive booster divergence rates have been inherently restricted by booster design.

It has therefore been shown that abort rockets are required for launch pad and low altitude explosions, and are also required for booster-escape vehicle separation because of drag effect. In addition, they are required

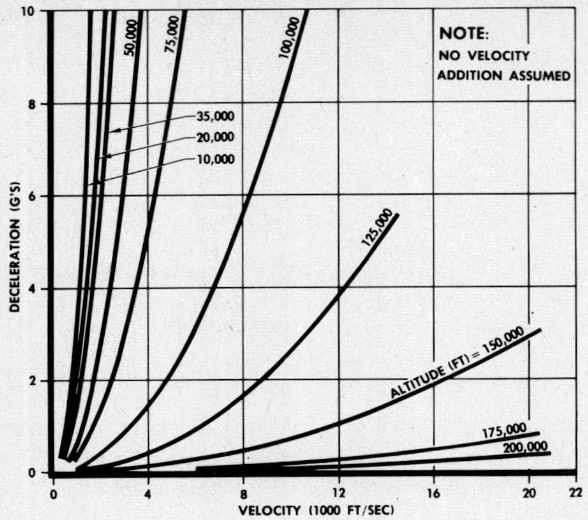

Fig. 10 Instantaneous Decelerations of the Escape Vehicle after Separation from the Booster

to reduce the rate of deceleration offset after a period of acceleration or deceleration. It has also been shown that they should be multiple, step, and/or regressing, since jettisoning them after their usefulness ends increases performance in some cases. Exceptions are those cases where their prime purpose is not abort but some other mission requirement; e. g., space maneuvering.

High Altitude Abort

With low-altitude aborts, the effects of deceleration are encountered at the moment of separation and not during descent. Exactly the reverse is true during a high-altitude abort; i. e., separation effects are small, but considerable high descent decelerations occur.

In an effort to control these decelerations, several approaches are considered. The first includes a controlling of the flight corridor so that a certain maximum deceleration resulting from an abort is never exceeded. Included in the boundaries of this flight corridor are the thermodynamic aspects of boost and re-entry, and the anticipated apogee conditions. Thermodynamics exert a strong influence on flight corridor boundary conditions, depending upon the thermal protection system; e.g., a particular thermal protection system cannot function above its design thermal environment, hence will limit the flight corridor to include its capabilities.

Since the vehicle is designed to trim at a high angle of attack during the re-entry phase of a space mission, it will also trim at this angle of attack during the high-altitude abort situation. Hence, the hypersonic lift-to-drag ratio during the descent from abort is 0.7, where the lifting parameter ($W/C_L S$) is 40 lb/ft^2. Refs. 3 and 9 present the thermodynamic arguments for the high angle of attack trim condition. While these particular aerodynamic characteristics are sufficient to maintain the re-entry decelerations experienced during descent from a standard orbital or space mission to the levels indicated in Fig. 9 (Ref. 10), such is not the case for suborbital re-entry. Figure 11 illustrates this situation where maximum deceleration is presented as a function of abort velocity at 400,000-ft altitude for a constant abort-flight-path angle. The point to be made here is that such a sub-orbital entry exceeds deceleration limitations, and that any artificial reduction of velocity does not improve the situation.

The results of controlling the flight corridor so that a maximum deceleration of 10 g's is not exceeded during descent are shown in Fig. 12 where all calculations were made on a high speed digital computer. In Fig. 12, the region of 10 g's or less lies to the left of and below each specific altitude curve. Limiting the flight corridor at velocities greater than satellite velocity are the characteristic skip or grazing-decay limits, where a skip limit is defined as the minimum abort-flight-path angle which will insure single pass re-entry. Also shown in Fig. 12 are stagnation-point equilibrium temperatures resulting from aborts at specific altitudes, velocities, and flight-path angles corresponding to the 10 g limitation. For abort altitudes less than 250,000 ft, the vehicle

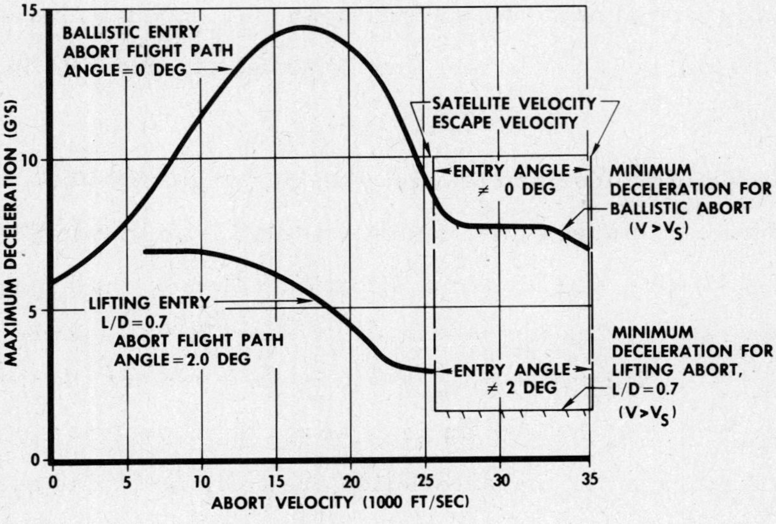

Fig. 11 Maximum Deceleration vs Abort Velocity at an Altitude of 400, 000 ft

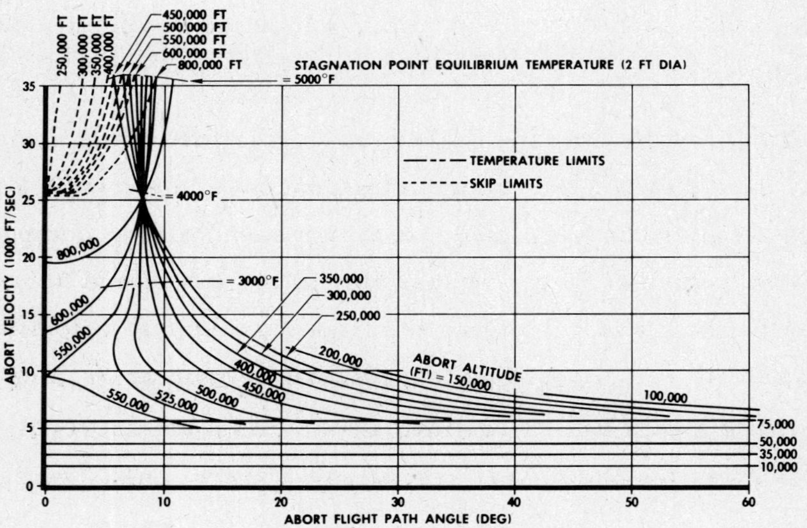

Fig. 12 Flight Corridor Corresponding to 10 G's

is flown to apogee in a minimum drag orientation, in order to minimize the decelerations and heat rates encountered. If the vehicle is flown to apogee in a lifting attitude from these intermediate altitudes, the ensuing trajectory will exhibit high rates of climb, resulting in higher apogee altitudes, and hence, higher descent decelerations. Apogee conditions in or above the inner Van Allen Belt are not noted, but in all conditions analyzed, abort velocities greater than satellite velocity result in entry into the radiation belts.

The extremely low-altitude aborts represented on Fig. 12 are indicative of the decelerations experienced at the moment of separation, and not during descent; however, they are the maximum decelerations encountered during that particular trajectory, and therefore can be justified for inclusion in this particular illustration. Since several types of boost trajectories exhibit negative flight path angles during ascent, provisions for this occurrence can be made in the computation of these flight corridors. For altitudes in excess of 300, 000 feet, the illustrated flight corridors are also applicable to negative flight path angles by virtue of elliptical symmetry.

Figures 13, 14, and 15 illustrate how the flight corridor becomes restricted when the tolerable decelerations are reduced from 10 to 8, to 6, and finally, to 4 g's. All of the clarification offered in conjunction with Fig. 12 is also applicable to Figs. 13, 14, and 15. As will be subsequently shown in the paragraph on boost trajectory compatibility, conforming to the more restrictive flight corridors adversely affects booster performance. Therefore, other methods of restricting deceleration that do not restrict ascent envelopes must be investigated.

Velocity Addition

Another method of restricting the decelerations created by an abort is the application of a velocity addition. Through this addition, flight-path angles and/or velocity can be so constrained that a specified maximum

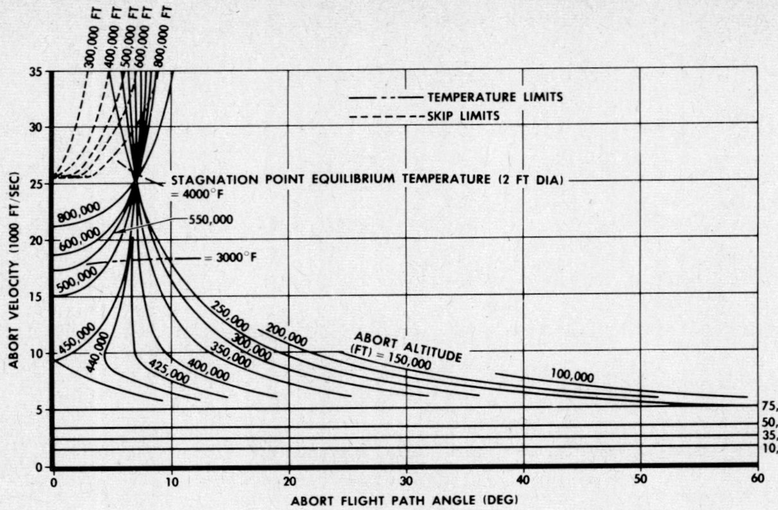

Fig. 13 Flight Corridor Corresponding to 8 G's

deceleration is never exceeded. Before such a problem can be analyzed, two fundamental questions must be answered; namely, when and how is this impulse imparted to the vehicle? Entire abort trajectories were analyzed by a high-speed digital computer in an effort to ascertain where this correction should be made. Conclusions reached agreed with the work of Slye (Ref. 11), where a condition just prior to the entry of the vehicle into the sensible atmosphere (300, 000 ft) was chosen. This conclusion is predicated on the fact that greater accuracy can be derived at low altitudes. Correction within the atmosphere was not considered, since the required thrust direction will result in aerodynamic moments which require excessive control torques. How this impulse or thrust is to be applied can be answered by defining the thrust direction and magnitude.

The easiest definition concerns thrust direction, since it is not system dependent. The definition is simply the direction which eventually results in the greatest reduction in deceleration, or, to put it another way,

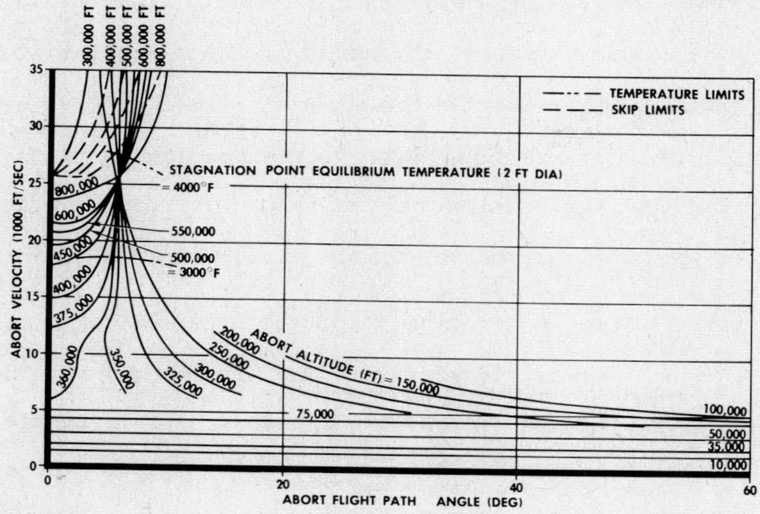

Fig. 14 Flight Corridor Corresponding to 6 G's

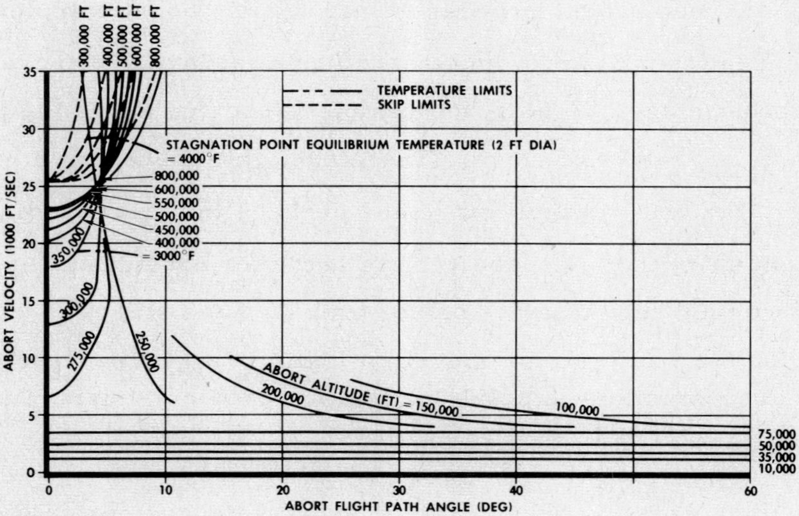

Fig. 15 Flight Corridor Corresponding to 4 G's

the direction which results in the greatest increase in the flight corridor, while still maintaining a specified maximum deceleration. In the majority of the cases analyzed, the optimum thrust vector was found to be greater than 90 degrees from the velocity vector, or, to be more precise, approximately perpendicular to the resulting velocity vector. This result conforms with the conclusions drawn from Fig. 11, that flight-path angle exerts a greater effect on deceleration than velocity.

Two approaches to the thrust impulse problem can be taken. These are: how great a thrust impulse is required, and/or how great a thrust impulse is available. The more feasible approach is to evaluate the thrust impulse or impulsive velocity that is required throughout other portions of the mission profile, and then determine whether it is applicable to an abort situation. Upon investigation of the least ambitious manned space mission, impulsive velocities of one or two thousand feet per second for correction purposes is available. Obviously, as more ambitious missions are conceived, larger impulsive velocities become available for abort correction. A typical example of how such a velocity can be attained 'without sacrifice of mission performance' can be derived from a presently contemplated multi-stage lunar landing boost complex. This design concept consists of a lunar-landing and lunar-takeoff stage above its main boost stages. The space vehicle is located above the lunar-landing and lunar-takeoff stage. Therefore, in the event of a high altitude abort, the lunar-takeoff stage and perhaps the lunar landing stage would separate with the escape vehicle and later provide the velocity correction required. Whether one or both of these stages is used will depend upon the ability of the stages to ignite in a gravity-free environment. If the propellant is solid, no ignition problem exists; but if it is liquid, ullage thrust must be provided.

With this formulation, flight-corridor expansions are performed with varying amounts of velocity correction, so that a maximum deceleration of 4 g's is never exceeded. Figure 16 presents these results for velocity additions of 2,000, 6,000, and 10,000 ft/sec. When these velocity-enlarged flight corridors are compared to the deceleration-monitored flight corridors shown in Figs. 12 through 15, it is obvious that velocity addition, rather than deceleration control, is the approach that should be pursued, as shown in Fig. 17. However, when the problems of retrieval of an aborted manned payload are introduced, the velocity expenditure method discussed previously becomes increasingly complex. That is, if a launch complex is situated at one of our existing missile ranges, the probability of landing an aborted flight on water has already been established at 95 per cent. From a safety aspect, it would be desirable to establish numerous pickup or retrieval stations; however, even with the aid of aerodynamic maneuvering, this might prove prohibitive from an economic standpoint. Therefore, the only method of keeping these pickup stations to a minimum is to use velocity for maneuvering, in addition to the already presented aerodynamic capabilities. It would appear that either higher decelerations must be tolerated, or mission performance must be compromised by non-useful abort velocity, in order to keep the system cost within reason.

EXTRA-ATMOSPHERIC ABORTS

The possibility of a malfunction occurring during the latter stages of ascent is small, because the boost system velocity has either already been attained or is approaching attainment. Therefore, the most probable failure requiring an extra-atmospheric abort would be one occurring within the space vehicle. In general, a failure occurring in proximity to the manned payload will require immediate action and a minimum-time return flight to earth. Because of the minimum time requirement, this extra-atmospheric abort area can extend as far down as the 30,000 ft/sec velocity level on the boost trajectory. This is due to the fact that an unaltered abort trajectory, initiated at velocities below escape velocity, but in excess of 30,000 ft/sec, would require many hours, and in some cases, days to return to earth.

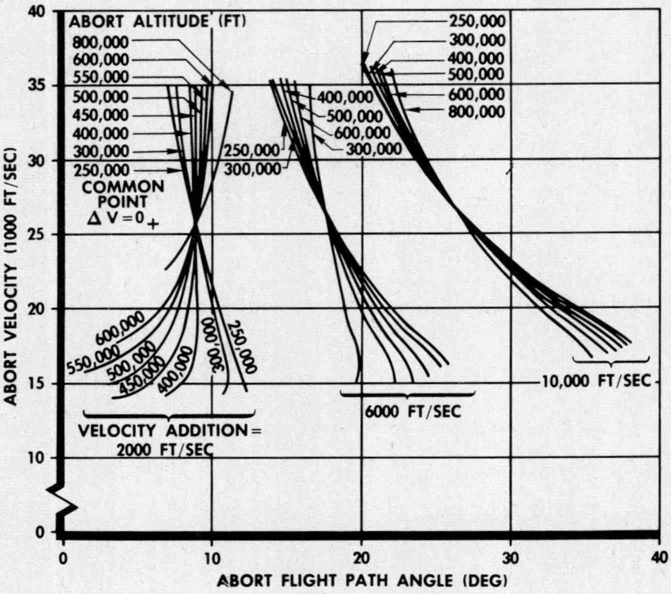

Fig. 16 Effect of Velocity Addition on Flight Corridor

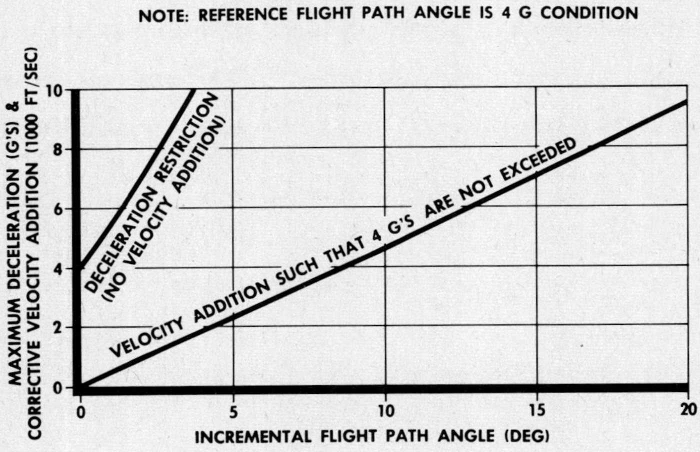

Fig. 17 Comparison of Abort Flight Path Angle Increase at Satellite Velocity

As a vehicle escapes the earth's gravitational field and ventures further away from earth, it becomes obvious that a velocity addition is the only means by which the vehicle may return directly to earth. An abort analysis for a lunar impact transfer was conducted therefore, where the assumed burnout conditions are: parabolic velocity; a flight path angle of 3.16 degrees; and an altitude of 70 nautical miles. These burnout conditions are compatible with the boost trajectory that will subsequently be shown to be optimum for a lunar landing mission. A single acceptable entry angle of -6 degrees is assumed, as is a single impulse abort maneuver. The results of this analysis are presented in Fig. 18, where the corrective velocity needed to return to earth is presented as a function of the position of the vehicle on the transfer trajectory (altitude), and time required to return to entry altitude. At specified altitudes corresponding to the nominal trajectory, the true anomaly, velocity, and flight-path angle are known. At each of the altitudes indicated, various injection velocities are applied at injection angles that will result in minimum return-time trajectories which enter the atmosphere at -6 degrees. In this manner, the velocity and time required to return to earth from an aborted lunar transfer can be obtained. As can be seen from Fig. 18, the time required to return to earth decreases as injection velocity increases, hence, if a system is velocity-limited, a minimum time to return to earth from a given abort location is established. However, too large a velocity addition may not be practical, because the resultant atmospheric entry velocities may become large enough to subsequently compromise the thermal integrity of the space vehicle.

It has been shown that propulsion is required for safe abort during all phases of an ascent into space. The phases that require propulsion, and the amounts of propulsion required are:

a. Launch pad to maximum dynamic pressure region - High thrust levels for short periods of time to outrun shock-wave propagation and/or overcome separation drag.

b. Maximum dynamic pressure region to upper limit of the atmosphere

- Reduced thrust levels for short periods to separate from low-energy explosions and overcome drag upon separation.

c. High-altitude aborts - Low thrust levels to separate from booster, but large velocity additions derived from existing mission propulsion systems to monitor deceleration.

d. Lunar or planetary transfers - Large impulsive velocities to obtain minimum-time returns to earth.

EFFECT OF AERODYNAMIC CHARACTERISTICS

The introduction of lift decreases the decelerations experienced by man re-entering the atmosphere. Figure 19 illustrates this decrease in deceleration at the most optimistic conditions possible; i. e., where entry angle guidance tolerances are of the highest order. This figure indicates that extrapolation of data derived from this study to other lift-to-drag ratios can be justified, provided that the extrapolation is restricted to lift-to-drag ratios of 0. 5 and 1. 0 and compatible wing loadings.

BOOST TRAJECTORY COMPATIBILITY

The flight mechanics studies of the boost phase of the trajectory are performed assuming a "typical" three stage vehicle. The characteristics of the vehicle are obtained from generalized boost vehicle studies being conducted by the Douglas Aircraft Co., Inc., and represent a reasonably efficient vehicle, although no attempt is made to optimize the configuration with respect to the payload-to-gross weight ratio. The pertinent characteristics are as follows:

	First Stage	Second Stage	Third Stage
Initial thrust-to-weight ratio	1. 25	1. 5	. 42
Propellant-to-gross weight ratio	. 49	. 24	. 11
Vacuum specific impulse-seconds	300	420	420

The trajectories investigated are characterized by a zero-lift gravity

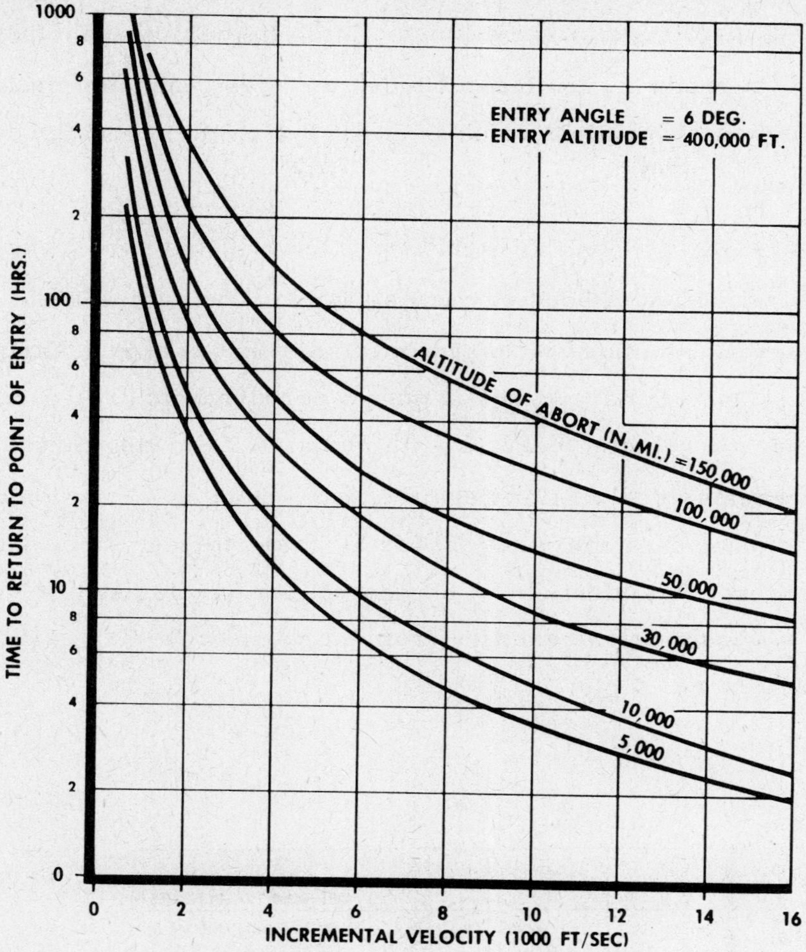

Fig. 18 Trans-Lunar Abort Requirements

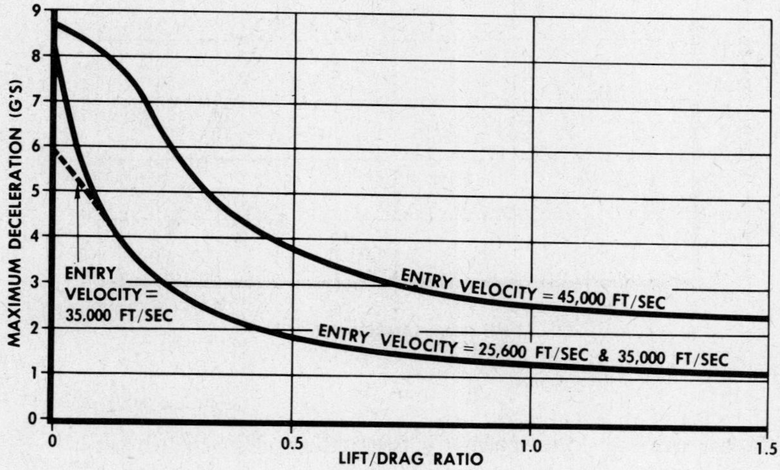

Fig. 19 Maximum Deceleration vs Lift/Drag Ratio
Entry Altitude = 400,000 ft

turn during first stage burning. The upper stage trajectories are computed utilizing a calculus of variations solution to maximize the payload at third stage burnout. Terminal altitudes of 70, 85, and 100 nautical miles are studied. Terminal velocity is parabolic at each of the designated altitudes.

The maximum deceleration that would exist under abort conditions is evaluated as a function of booster burning time. The nature of the resulting deceleration can be discussed in two distinct phases. Early in the trajectory, the primary effect is that of aerodynamic drag produced by booster-escape vehicle separation immediately following thrust loss. Figure 20 presents deceleration resulting from drag as a function of boost time. These data are identical for all trajectories, since the first stage tilt program was identical in all cases. As can be seen in Fig. 20, the maximum deceleration resulting from an abort occurring early in the flight does not exceed 4 g's.

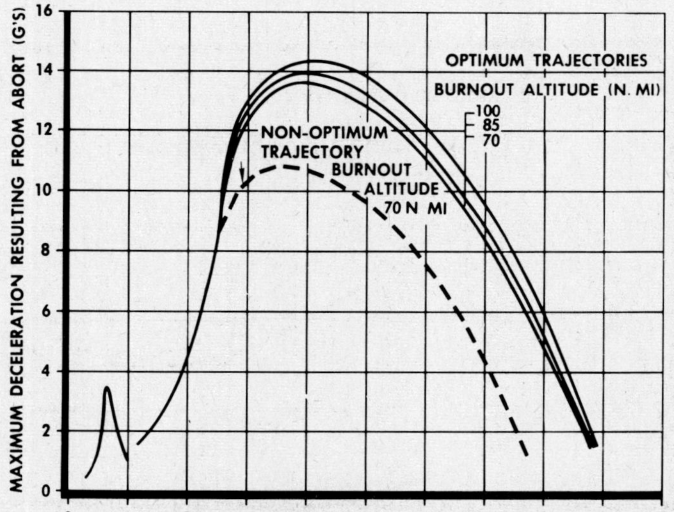

Fig. 20 Maximum Decelerations Resulting From Aborts

If an abort does not occur until after the maximum dynamic pressure region of the boost trajectory, the maximum deceleration of the space or escape vehicle is experienced upon re-entry into the earth's atmosphere under conditions determined at the time of abort. These loadings are also shown in Fig. 20 for the three terminal altitudes considered. Although the maximum g's decrease with decreasing burnout altitude, the reduction is not significant.

Analysis of the boost trajectories indicates that the peak altitude during third stage burning is approximately 80 miles above the burnout altitude. Because of the low initial thrust-to-weight ratio of the third stage, the trajectory is lofted during second-stage burning. An abort during this period will therefore result in relatively large flight-path angles at entry, with resulting high decelerations.

To reduce the maximum deceleration, the vehicle characteristics and trajectory tilt program are modified to a slightly non-optimum situation.

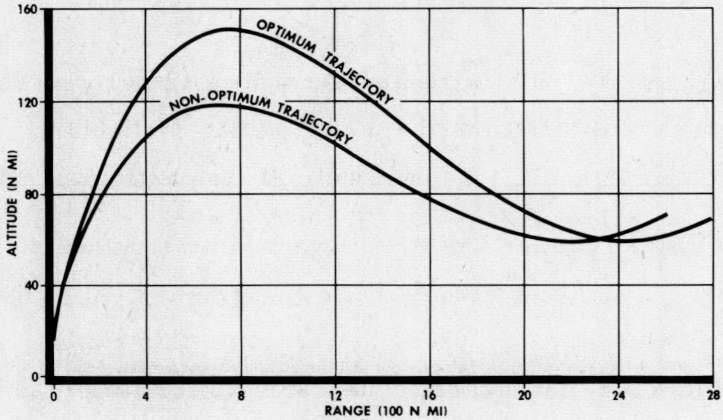

Fig. 21 Boost Trajectory Histories

The thrust-to-weight ratio of the third stage is increased slightly by off-loading propellant, and the trajectory is forced to deviate so that the lofting is reduced. These perturbations have a significiant effect on reducing the maximum loading, as shown in Fig. 20. The resulting payload penalty is only a 1.0 per cent reduction in payload capability. The trajectory profiles of the optimum and non-optimum trajectories are shown in Fig. 21.

Since velocity addition has already been shown to be the most efficient method of reducing decelerations resulting from an abort, analysis of the optimum boost trajectory, utilizing velocity addition, is warranted. Results of this investigation show that a velocity addition of only 2,500 ft/sec is required to reduce the magnitude of the deceleration from 14 g's to 4 g's.

After evaluating the results of the velocity addition study, it was concluded that a more detailed investigation of configuration and boost trajectory perturbations was not required of this study. The results of this brief boost trajectory analysis do indicate, however, that significant gains can be made in reducing the peak acceleration resulting from an abort occurring outside the atmosphere with relatively little loss in payload capability.

CONCLUSIONS

An abort recovery system is warranted because a 12 per cent failure probability is predicted for manned space flights. Methods of reducing this probability have been discussed in relation to particular malfunction:

The most critical malfunction has been shown to be a launch pad explosion where abort rockets are necessary to insure survival of the manned payload. The accelerations required to overcome this explosion probler increase as vehicle size increases; consequently, for large launch vehicles, the magnitude of these required accelerations might compromise man's psycho-physiological integrity. In an effort to reduce the degree of this compromise, advances in the state-of-the-art will have to be made in malfunction detection systems, as well as advances in acceleration attenuation devices.

Control of entry decelerations resulting from an abort was also investigated, and was found to be practicable with little or no compromise in booster performance capability. However, if sufficient velocity capability for trajectory control is not present in the booster system, then controlling entry deceleration, either through boost trajectory shaping or velocity addition will degrade payload capability. An escape vehicle possessing lifting capabilities was found to be advantageous in reducing these decelerations.

After evaluation of the various malfunctions contributing to the 12 per cent failure probability, it is shown in Fig. 22 that corrective techniques can reduce this probability to 2 per cent (Ref. 1). Unfortunately the failure probability for manned space vehicles is still orders of magnitude above current fighter aircraft failure probability and, hence, this area should receive increased effort as our space missions become more ambitious.

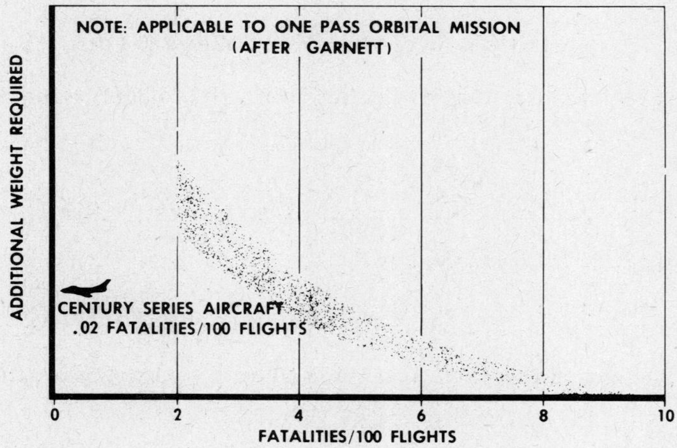

Fig. 22 Hazard Elevation

ACKNOWLEDGMENT

The authors wish to express their appreciation to Messrs. R. L. Warren and H. Wash for their assistance in the preparation of this paper.

REFERENCES

1. R. J. Garnett, "The Selection of an Escape System for a Boost Glide or Orbital Re-entry Vehicle," American Rocket Society Preprint No. 2213-61, presented at the American Rocket Society Space Flight Report to the Nation, New York, October 9-15, 1961.

2. J. J. Oslake, R. J. Getz, R. A. Romine, and K. SooHoo. "Explosive Hazards of Rocket Launchings," Aeronutronic (Ford Motor Company) Technical Report No. U-108:98, November 30, 1960,

3. R. L. Gervais, A. A. duPont, and R. E. Lowe. "A Practical Manned Entry System," Douglas Aircraft Company Engineering Paper No. 1095, presented at the American Rocket Society Lifting Re-entry Vehicles, Structures, Materials and Design Conference, Palm Springs, California, April 4, 1961.

4. Samuel Gladstone. The Effects of Nuclear Weapons, U.S. Atomic Energy Commission, June 1957.

5. H. L. Brode, "A Calculation of the Blast Wave From a Spherical Charge of TNT," RAND Corp. Report No. RM-1965, August 21, 1957.

6. "Handbook of Instruction for Aircraft Designers," Air Force Systems Command Manual 80-1, May 1961.

7. Personal communication with Col. H. Blackshear, USAF, Chief, Aeromedical Field Laboratory, Air Force Systems Command, Aero-Space Medical Laboratory, Holloman AFB.

8. Personal communication with Konecci, E. B., and B. A. Olson, Douglas Aircraft Company, Inc.

9. R. E. Lowe, R. B. Greene, and R. L. Gervais. "The Effect of Trajectory and Aerodynamic Characteristics on Entry Heating of Lifting Spacecraft," Douglas Aircraft Company Engineering Paper No. 1238, presented at the Annual Meeting of the Institute of Aerospace Sciences, New York, January 23, 1962.

10. R. E. Lowe, and R. L. Gervais. "Manned Entry Mission to Mars and Venus," Douglas Aircraft Company Engineering Paper No. 1223, presented at American Rocket Society Space Flight Report to the Nation, New York, October 9-15, 1961.

11. R. E. Slye, "Velocity Requirements for Abort From the Boost Trajectory of a Manned Lunar Mission," NASA TN-D-1038, July 1961.

RADIATION CONSIDERATIONS FOR LUNAR MISSIONS

J. Abel*

The hazard of the Van Allen belt and solar flare radiations are evaluated for a lunar mission. The implications with respect to trajectory selection and vehicle design are discussed. Van Allen belt doses are determined for lunar ascent trajectories and emergency returns therefrom. A method of analyzing a space vehicle design from the point of view of radiation shielding is described. The solar flare problem is discussed from a probability point of view, and the results are integrated with vehicle shielding characteristics to define the relationship between radiation safety and weight penalty.

INTRODUCTION

Van Allen belt and solar flare radiations represent a formidable problem in spacecraft design. This paper discusses these hazards with regard to their implications to vehicle design, and presents a method of characterizing the relationship between radiation safety and spacecraft weight penalty. Reduction of the problem to one of reliability versus weight penalty permits the spacecraft designer to treat the radiation hazard in the manner in which other system risks are treated. One underlying assumption in this treatment is that a biological radiation dose limit is established for the crew of the lunar vehicle, which in this case will be taken as 100 Rem. In addition, a few remarks will be made on some of the outstanding areas of uncertainty which bear on the problem and which necessarily qualify the conclusions expressed here.

Van Allen-Integrated Dose

For this aspect of the radiation problem, two classes of trajectories are of interest:

Lunar Ascent

Emergency Return

* General Electric Company, Missile and Space Vehicles Dept., Philadelphia

A number of examples of each type were analyzed; the integrated dose and dose as a function of time were determined for each trajectory. The digital computer program written for this analysis accepts initial position and velocity information and generates a point by point two-body solution of the trajectory equation. A model of the radiation belts is stored in memory in matrix form giving dose in roentgens/hr as a function of geomagnetic latitude and distance from the midpoint of the earth's magnetic dipole, (Fig. 1). The program records the dose rate at each point and integrates the maximum value of dose rate in each time interval, thereby yielding an upper estimate of the integrated dose. The model of the Van Allen belts is based on the data from the Minnesota Experiment in Explorer VI, passes 1-14 August 7-14, 1959. The data are extrapolated by assuming symmetry with respect to the axis of the geomagnetic dipole and with respect to the geomagnetic equator. Interpolation in intensity is taken in accord with counting rate data obtained by instruments in the Pioneer III and Explorer IV satellites. The ionization data from the Explorer VI correspond to a measurement taken behind approximately 1 gm/cm^2 of shielding and as such provide a conservatively high estimate of the level encountered by a man in a space vehicle. Particularly, the dose levels in the inner Van Allen belt would be considerably reduced by the nominal amount of shielding 2-6 gms/cm^2 provided by the vehicle structure [1]. To simulate cosmic ray background radiation a value of .0006 R/hr was employed. Spatial variation of this radiation was considered to be of negligible effect and the constant value representing the peak cosmic ray dose rate of .0006 R/hr was taken at all points outside the radiation belts.

For purposes of evaluating the biological dose in REM from the ionization data, a ratio of one Rem per roentgen will be approximately correct. The principal error incurred in this approximation arises when considering the dose contribution from the inner Van Allen belt protons. In this case an appreciable number of the incident protons are in the energy range in which the RBE is greater than one. However, as indicated, the conservative estimate of the effective shielding inherent in the data compensates for this error. In the outer belt the assumption is subject to the variation in the energy absorption of tissue (nominally 93 ergs/gm per roentgen) over the range of photon energies 10 Kev - 800 Kev[2]. This variation is small and well within the uncertainties inherent in the experimental data.

A group of lunar ascent trajectories were run using initial conditions representative of burn-out data on launch from Cape Canaveral. In each case, the trajectories were run until the position of the vehicle was well outside the limits of the Van Allen belts. The total transit time for passage through the belts was approximately 2.45 hrs for all cases. Table I gives the input conditions and total dose after 5 hrs.

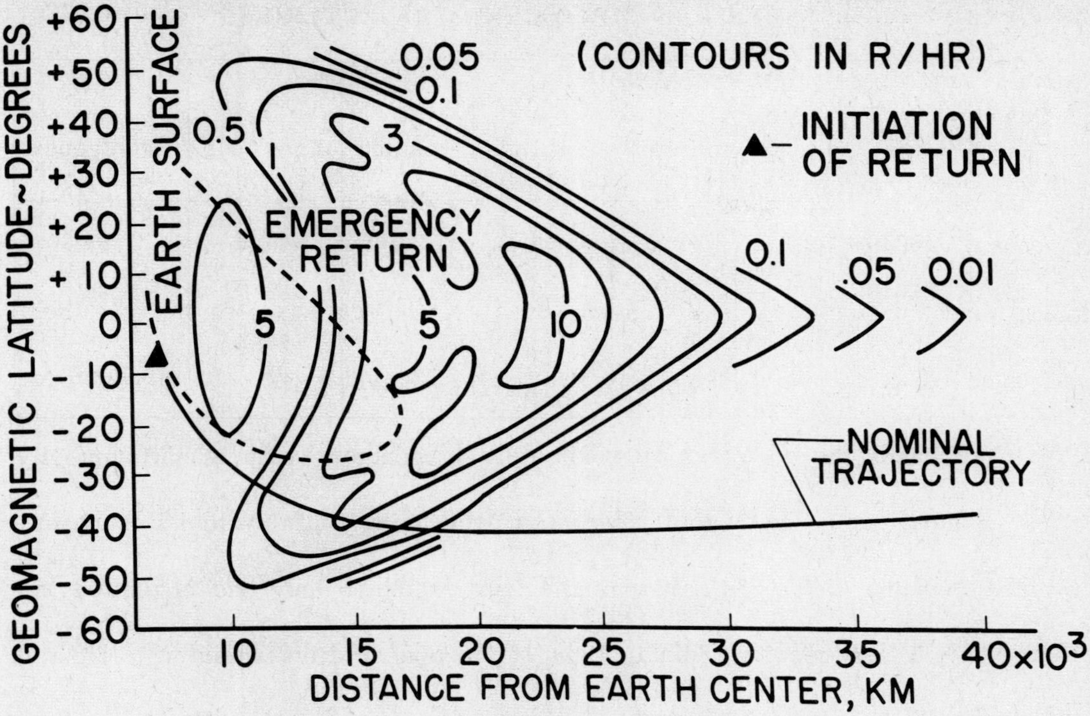

Fig. 1 Van Allen Belt Trajectories

Fig. 2
Integrated Dose
vs. Time Lunar
Ascent

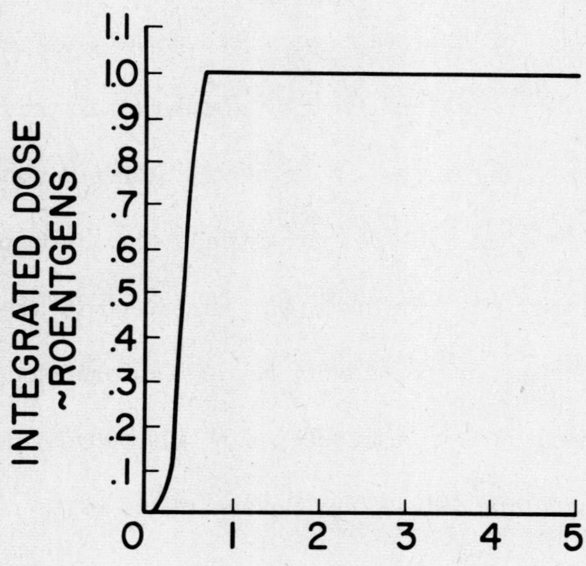

TABLE I LUNAR ASCENT TRAJECTORIES

Case	Velocity V fps	Flight Path Angle γ	Altitude r (st. mi)	β Latitude	ψ Heading	λ Longitude	Dose R_T Roentgens
A	36240	0	4019	9.045	127.0	-48.3	1.01
B	36019	0	4019	9.298	125.1	-45.6	1.09
C	36141	0	4019	9.045	127.0	-48.3	1.06
D	36341	0	4019	9.045	127.0	-45.6	1.01

In all cases the dose was approximately 1 roentgen which is satisfactorily low. A small increase in total dose with decreasing initial velocity is observable. Because of the similarity of the four examples only one is plotted in Figure 2. These results taken in light of the conservative shielding estimate, indicate that, for a nominal direct-ascent trajectory, the Van Allen belts do not constitute a primary hazard.

The question of the dose encountered on an emergency return from an outbound trajectory was examined by simulating returns to earth from three different points along the ascent trajectory (Case B above). The normal Case B ascent was run until the desired altitude was attained. At that point a change in velocity and path angle was introduced to bring the vehicle back to earth. The change in the velocity vector was precalculated to produce acceptable re-entry conditions at an altitude of 400,000 ft. Return from an altitude of 3,998 statute miles above the surface of the earth resulted in a relatively high dose of 22 roentgens. Two alternate velocity vector changes were simulated in an effort to determine if a return from that altitude could be achieved with a lower total dose. The desired result was achieved as indicated in Table II.

Definition of symbols in Table II:

h_a = altitude above earth at initiation of return

V = resultant velocity

γ = resultant path angle

T_a = time to reach h_a

T_e = time to return to earth

R_T = integrated radiation dose in roentgens

TABLE II EMERGENCY RETURN TRAJECTORIES

Case	h_a	V	γ	T_a	T_e	R_T Approach	R_T Return
I	345.5	29219	12.0	0.10	3.20	0.00	11.63
II_A	3998.2	20565	36.4	0.45	6.35	0.33	21.82
II_B	3998.2	23856	42.3	0.45	23.5	0.33	8.22
II_C	3998.2	22167	39.6	0.45	11.4	0.33	12.00
III	11641.1	12953	47.6	1.15	9.95	1.09	10.09

Referrring to Table II it may be observed that an acceptably low dose is encountered in return from the three altitudes shown. Case II_A resulted in a total dose of 22.2 roentgens. The alternate velocity changes represented by cases II_B and II_C gave a 50 percent decrease in total dose. The trajectory traces for the three returns from h_a = 3998.2 st. miles permit a qualitative explanation of the result. In case II_A the vehicle path traverses the center of both Van Allen belts and does so at low geomagnetic latitudes, thereby resulting in what is probably an upper limit to the dose which might be encountered in an emergency return. The trace for Case II_B indicates that the trajectory does not pass throught the highest intensity portions of either belt, and in fact avoids the inner belt entirely. For Case II_C the dose is increased due to traversal of the center of the outer belt at the geomagnetic equator. It may be concluded that for any given altitude a safe emergency return can be effected. The selection of velocity increment and resultant path angle, however, will affect the total radiation dose and must be chosen subject to that constraint.

The results obtained indicate that for the two classes of trajectories studied, the integrated dose absorbed in passing through the Van Allen belts will not constitute a hazard. In some instances a small variation in the controlled variables of a flight will be necessitated to reduce the dose to a very low value as in the emergency return, IIA. The analyses, subject to the assumptions stated above, give a conservatively high estimate of the integrated dose. This serves to increase the confidence in the assertion that the safety of the crew would not be endangered by traversal of the trapped radiation belts in a normal lunar mission or in a prematurely terminated mission of either type.

SOLAR FLARE RADIATION

Because of the apparent randomness of the occurrence of solar flares, the problem of encountering these streams of high energy protons is best approached on a probability basis. The probability of encountering a solar flare is naturally a function of frequency of occurrence and mission duration. Figure 3 shows encounter probability as a function of mission duration for flare frequencies of 9 per year, 3 per year, and 1/4 per year. These frequencies correspond to an estimate of the peak rate of occurrence of three types of flares classified by energy spectra as follows:

Spectrum	Example obtained on -	Frequency Per Year
Low Energy High Intensity	22 August 1958	9
Low Energy Extreme Intensity	10 May 1959	3
High Energy High Intensity	23 February 1956	1/4

The flare occurrence is taken to be random. While there are data which indicate a tendency for flares to cluster, and perhaps even some seasonal variations in rate of occurrence, not enough is known to permit more sophisticated descriptions of the behavior.

The model of the relationship between frequency of occurrence and flare type underlies the analysis of the probable dose function to be described later. The probabilities of encounter for a 1 - week mission are, respectively, 16 percent, 5.8 percent, and 0.3 percent. Thus for a lunar mission of a week or longer the solar-flare hazard is of sufficiently high probability to merit further analysis. Foelsche[3] has given the characteristics for the variation of dose with shielding for some solar proton events. These characteristics are summarized in Table III. It must be mentioned here that the doses given are for leak-through protons only, and that secondary neutrons and protons have been neglected.

The error introduced with this assumption is small for thin shields perhaps 5 to 10 percent, but may be as large as 50 percent for thick shields of 30 gm/cm^2 or more.

TABLE III SHIELDING REQUIRED TO LIMIT DOSE S~gm/cm^2 H_2O

Dose Rem	Flare		
	Aug. '58	May '59	Feb. '56
1	8 gm/cm^2	---	---
5	4.2	---	---
10	3.5	34	---
15	3.3	28	---
20	3.0	27	36
25	2.8	25	32
50	2.5	19	21
100	2.0	15	13

STRUCTURAL ANALYSIS

The high probability of encounter associated with the solar flare events leads us to analyze the spacecraft design from the point of view of radiation shielding. In the analysis of spacecraft structure three principal characteristics are sought:

1) Because of the isotropic nature of the solar flare radiation complete continuous enclosure of the crew compartment by the structural element is essential.

2) The rapid decrease of dose with increased shielding (approximately inversely as the cube of shielding) places emphasis on the amount of mass per unit area.

3) Where possible low atomic number material is desired to minimize the production of secondary radiation.

Figure 4 shows a reentry vehicle with details of two areas of the structure. It may be seen that the elements shown fulfill the three requirements set forth above. Complete enclosure of the crew compartment is afforded by the ablation material and substructure. The values of 2 gm/cm^2 and 5 gm/cm^2 represent appreciable contributions to the radiation protection. The ablation material being a hydrocarbon, perhaps phenolic nylon, is characterized by low atomic number, and high stopping power. Similarly the aluminum substructure is a low atomic number material.

Applying a similar analysis to the total spacecraft yields a representation of the vehicle shielding. This plot termed the Vehicle Mass Distribution, (Figure 5), describes completely the shielding characteristics of the vehicle relative to a point at the mass center of the crew. On the mass distribution the radius is proportional to the shielding in gm/cm^2, and the angle is equal

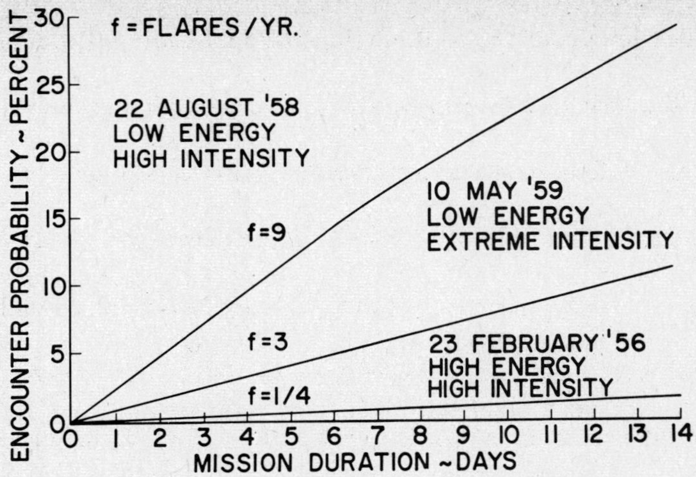

Fig. 3 Encounter Probability vs. Mission Duration

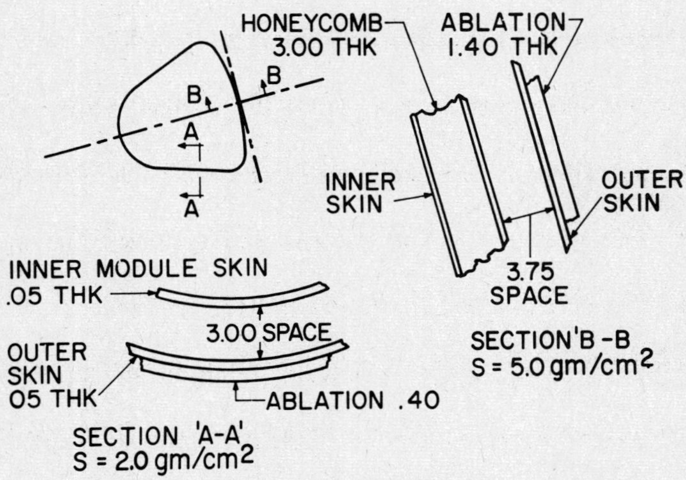

Fig. 4 Structure Shielding Analysis

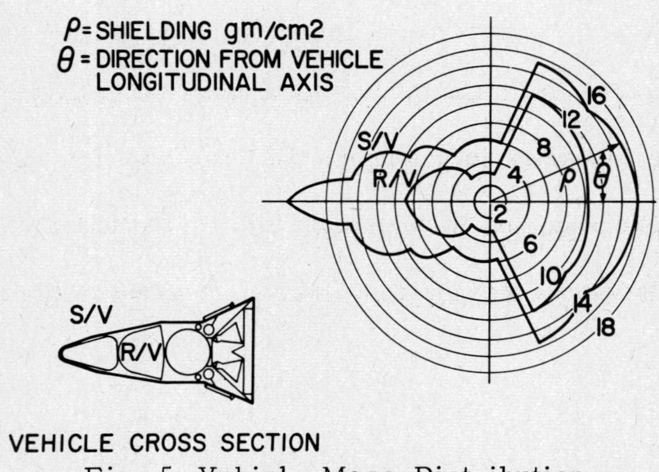

Fig. 5 Vehicle Mass Distribution

to the direction from the vehicle longitudinal axis. Circles are loci of constant shielding thickness, and the radius of the largest circle wholly contained within the vehicle contour gives the minimum value of all-over protection. As the radius increases we see larger and larger portions of the circle lying outside the contour, thereby defining so-called Vulnerable Solid Angles. If the locus of constant shielding thickness intersects the vehicle contour at angle θ, and an arc of the circle equal to ψ degrees lies outside the contour then the vulnerable angle is given by:

$$\theta_V = 2\pi \left[\cos\theta - \cos(\theta + \psi) \right]$$

The relationship θ_V as a function of S for the sample vehicle is shown in Figure 6. If it is desired to achieve a certain minimum value of all-over shielding, then supplemental shielding must be provided in the vulnerable areas. This plot then provides the basis for determining the weight penalty incurred as a function of the desired augmented shielding, a basic input to the weight penalty vs. mission safety tradeoff. In figure 5 we have two concentric contours representing the reentry vehicle alone (R/V) and the entire space vehicle (S/V) configuration in assembly. For any given value of shielding, the complete spacecraft configuration affords more complete protection.

In order to evaluate the weight penalty associated with a given value of total shielding, it is necessary to determine the supplemental weight required to increase the vehicle shielding in vulnerable areas. To facilitate this computation, it may be assumed that the additional mass added to the structure

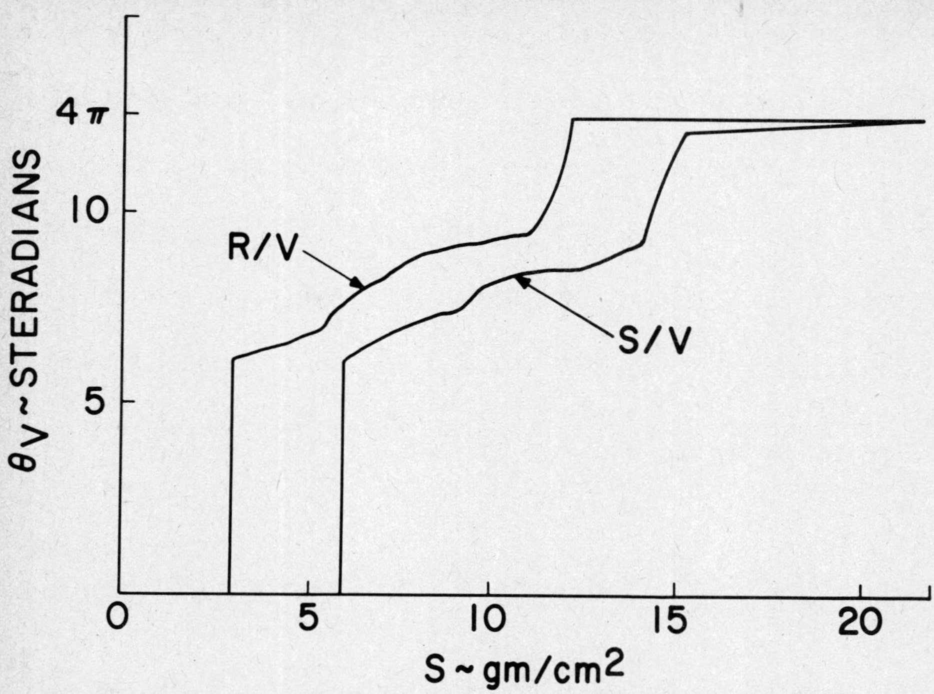

Fig. 6 Vulnerable Solid Angle vs. Shielding Thickness

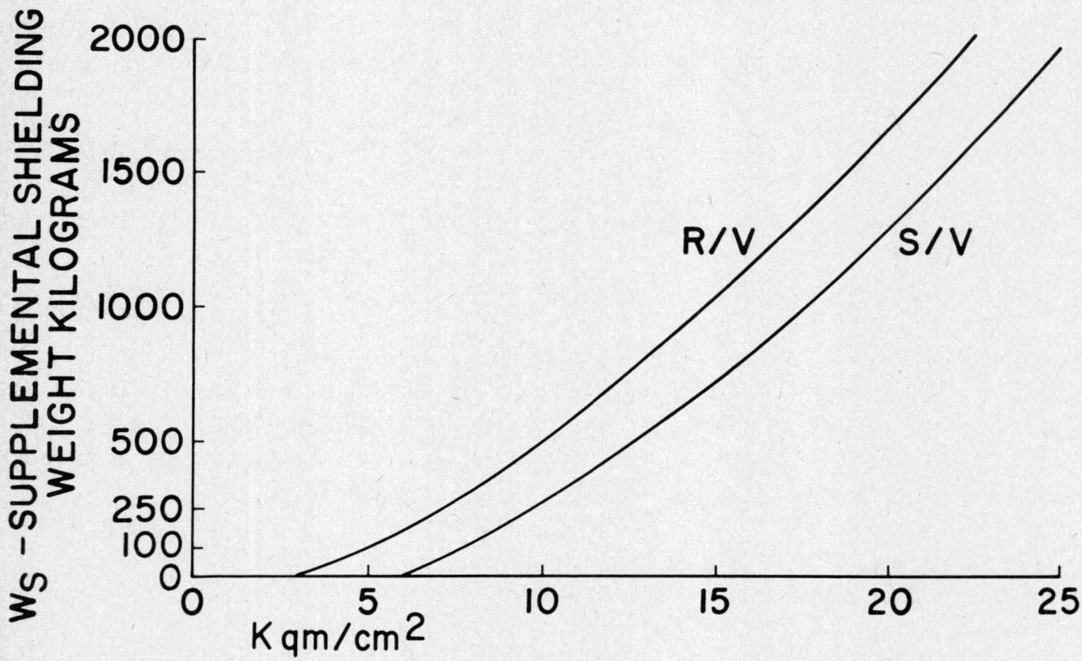

Fig. 7 Supplemental Shielding Weight vs. Minimum Shielding Thickness

will be in the form of a convenient geometrical form. The simplest such form, of course, is that of a segment of a sphere. For the vehicle being considered, a sphere seven feet in diameter is assumed. To achieve a given value of shielding K gm/cm^2, each increment of area possessing mass, S, must be supplemented by (K-S) gm/cm^2. Expressed in equation form this yields:

$$\Delta W = r^2 (K-S) \Delta \theta_v$$

Summing the above for all values of S less than K and passing to the limit yields:

$$W \text{ supp.} = r^2 \int_{\theta_v (S \text{ min.})}^{\theta_v (S \text{ max.})} (K-S) \, d\theta_v$$

In this expression, S is implicitly taken as a function of θ, the inverse of the relationship plotted in Figure 6. The expression for W supp. was evaluated as a function of K, minimum shielding thickness, and the results are shown in Figure 7.

PROBABLE DOSE

By combining the encounter probability data and the dose vs. shielding characteristics of solar flares a probable dose distribution may be derived for a mission of specified duration. By considering the probability of encountering one, two or three flares of the same or mixed types we exhaust the modes of encounter having probabilities in excess of 0.1%. This produces a probability density distribution consisting of a finite number of points which

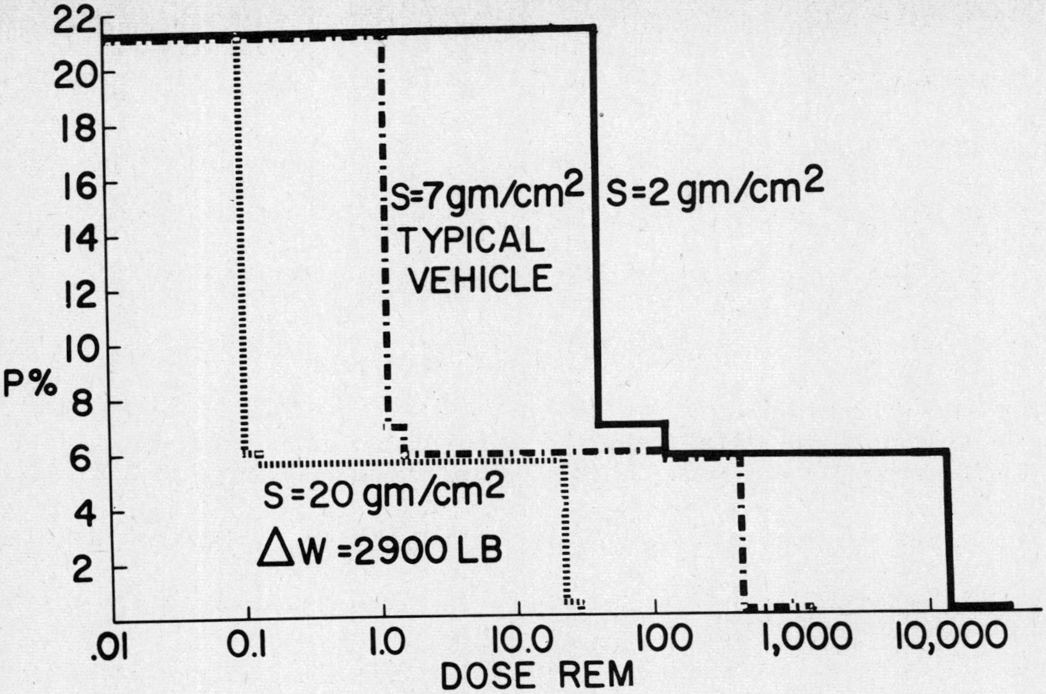

Fig. 8 Probability of Exceeding Dose 7 Day Mission

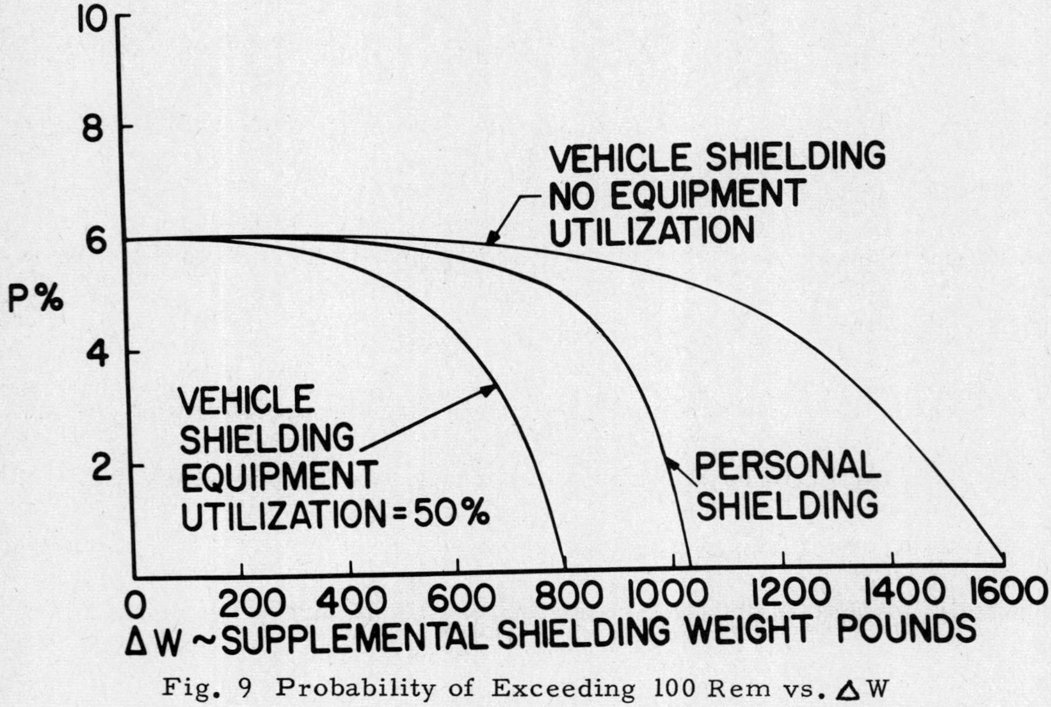

Fig. 9 Probability of Exceeding 100 Rem vs. ΔW

leads to an integral distribution composed of a series of steps. This unfortunate discreteness in the integral distribution could only be eliminated by constructing a continuous relationship between flare type and frequency of occurrence which the present scarcity of data does not permit. In Figure 8 we have the probability of exceeding a given dose for a one week mission. The "curves" are given for three values of shielding. The value of 2 gm/cm^2 represents a lightly shielded vehicle, or perhaps a man outside a vehicle in a pressure suit. The 7 gm/cm^2 corresponds to the effective shielding inherent in the example spacecraft structure with little equipment utilization. By incurring a weight penalty of 2900 pounds, the vehicle may be supplemented to achieve an effective value of shielding of 20 gm/cm^2. The decrease in risk with shielding is shown by considering the probability of exceeding 100 Rem on a seven day mission. The probability varies from 7 to 6 to virtually zero percent as the effective shielding is increased. We see here the basis for a tradeoff wherein the probability of exceeding the design dose limit, 100 Rem, may be decreased by incurring a weight penalty. Thus, the probable dose determination leads to the relationship between Weight Penalty and Mission Safety, defining the "price" which must be paid to ensure a given level of reliability. This then characterizes the radiation hazard in a manner whereby it may be treated along with the other system risks, and wherein it may be handled as a fundamental input to the analysis of the overall system reliability.

In Figure 9 curves of Probability of Exceeding 100 Rem versus Supplemental Shielding Weight, show the relationship for three modes of shielding augmentation. The curve on the right represents vehicle shielding wherein supplemental material is added to the structure in vulnerable areas. The center curve represents the case in which supplemental material is placed in close proximity to each crew member in the form of a garment or small enclosure. The curve on the left represents the case which 50% of the supplemental shielding is developed by appropriate positioning of vehicle equipment and supplies, which is an estimate of what might be achieved in an optimized design. To reduce the probability of exceeding the design dose to less than 0.1%, the required supplemental shielding weight is for the three methods 1600 lbs., 1030 lbs., and 800 lbs. The choice of shielding mode will of course be affected by factors relating to crew comfort and performance, structural design and materials selection which might override certain weight savings.

On the basis of these analyses we are able to state these conclusions: With respect to the Van Allen belts:

a) For a direct ascent trajectory the Van Allen belts do not constitute a primary hazard.

b) Consideration of Emergency Return points up a need for trajectory shaping on premature termination of the mission.

For solar flares we have established the risk of exceeding the dose limit. We see that a significantly high probability of encounter exists and this places emphasis on the shielding characteristics of the spacecraft under consideration. For a spacecraft with shielding of 7.0 gm/cm^2 the probability of exceeding the 100 Rem dose is 6%. The probability may be decreased to a prescribed level of less than 1% by augmenting the shielding of the spacecraft considered with attendant weight penalties of 800-1600 lbs.

The final resolution of the radiation problem requires the establishment of a dose limit and desired confidence level. It is then incumbent upon the spacecraft designer to provide sufficient shielding to attain this reliability and to tradeoff the attendant weight penalty with other means of improving system reliability.

REFERENCES

1. Schaefer, H. J., "Dosimetry of Proton Radiation in Space" - U. S. Naval School of Aviation Medicine, June 6, 1961, Report 19.

2. Hine and Brownell, "Radiation Dosimetry" - Academic Press - New York, 1956.

3. Foelsche, T., "Protection Against Solar Flare Protons" - American Astronautical Society, January 16, 1961.

HEAT TRANSFER TO POROUS NOSE CAPS AND LEADING EDGES OF HYPERSONIC VEHICLES AND REENTRY AIDS

Harold G. Lorsch*

This paper demonstrates the considerable increase in heating rates connected with the use of porous instead of solid surfaces at locations near the stagnation point. The results of calculations based on laminar continuum flow are applied to show that the shocks from the individual filaments of a porous cloth weave will not interact, if the weave distance is about the filament diameter D, and the shock detachment distance is small compared to D. Equations and graphs are given which evaluate quantitatively the resultant, significant heat transfer increase both at the stagnation point and over the forward half of a leading edge and a nose cap. The equilibrium temperatures for porous cloths are sufficiently higher than those for solid surfaces to make their use in hypersonic vehicles and reentry aids highly inadvisable. If the filament diameter is calculated which, for a cloth of zero porosity, makes the heating rate equal to that for free molecular flow, it is virtually identical to the free molecular path length. This may be interpreted as a proof of the validity of this approach to the heating of porous surfaces.

---

* General Electric Company. Manager, Advanced Structures and Materials, Missile and Space Vehicle Department, Philadelphia

INTRODUCTION

The results obtained elsewhere [1, 2, 3] show that the heat generated at the nose cap or the leading edge of a vehicle in hypersonic flight is inversely proportional to the square root of the radius of curvature of such a nose cap or leading edge. The use of large radii of curvature has, therefore, become a design requirement for these parts of currently contemplated hypersonic vehicles (Dyna-soar, Apollo, and others). The heat generated at corresponding locations of a re-entry aid, such as a paraglider [4], similarly depends on the radius of curvature at the leading edge.

Several writers have recommended porous drag or lifting re-entry aids (Fig. 1) because of the advantages which they offer from the point of view of flexibility, packaging, manufacture, aerodynamic stability, etc. [4, 5]. To the author's knowledge, the serious disadvantage of increased heat transfer from the surrounding air to the porous vehicle skin, however, has not received sufficient consideration. This paper compares the heating rates and radiation equilibrium temperatures of porous and solid nose caps and leading edges of hypersonic vehicles in an attempt to draw attention to this serious design problem.

HEAT TRANSFER

Stagnation Point

Scala and others [1, 2, 3] have calculated laminar heat transfer rates at the nose (Fig. 2) and leading edge (Fig. 3) of a hypersonic vehicle for the assumption of equilibrium binary mixture dissociated air in continuum flow. Fig. 4 shows graphs of these heating rates at the nose cap, multiplied by the square root of the nose radius, as a function of flight speed and altitude based on the ARDC 1956 model atmosphere.

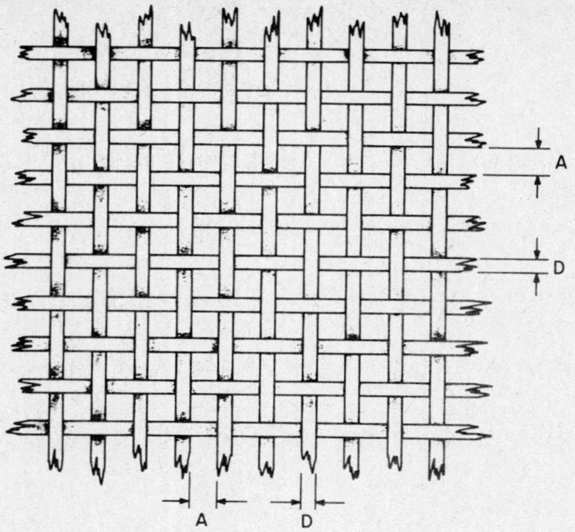

Fig. 1 Porous Cloth Weave

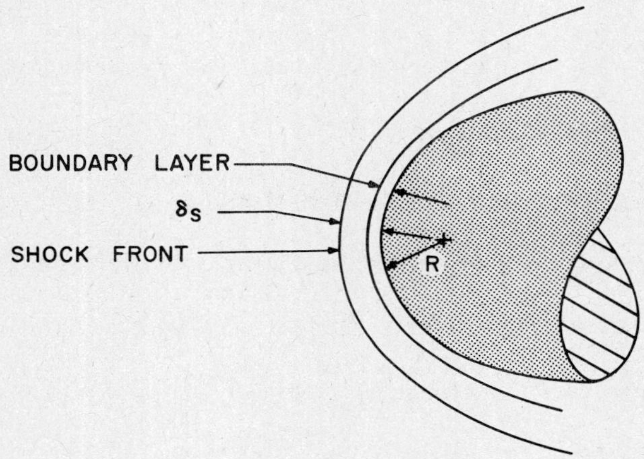

Fig. 2 Axially Symmetric Flow (Nose Cap)

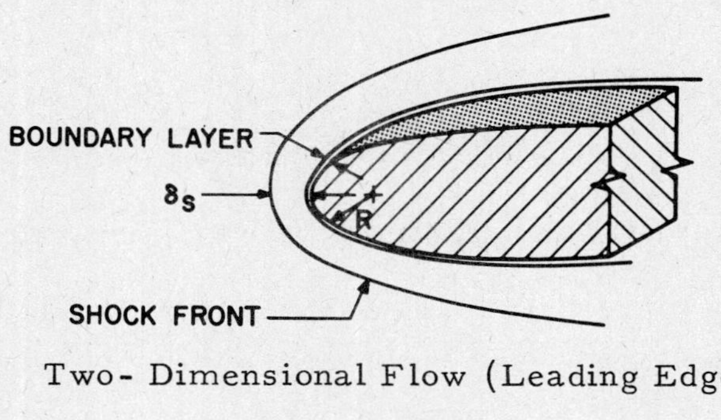

Fig. 3 Two-Dimensional Flow (Leading Edge)

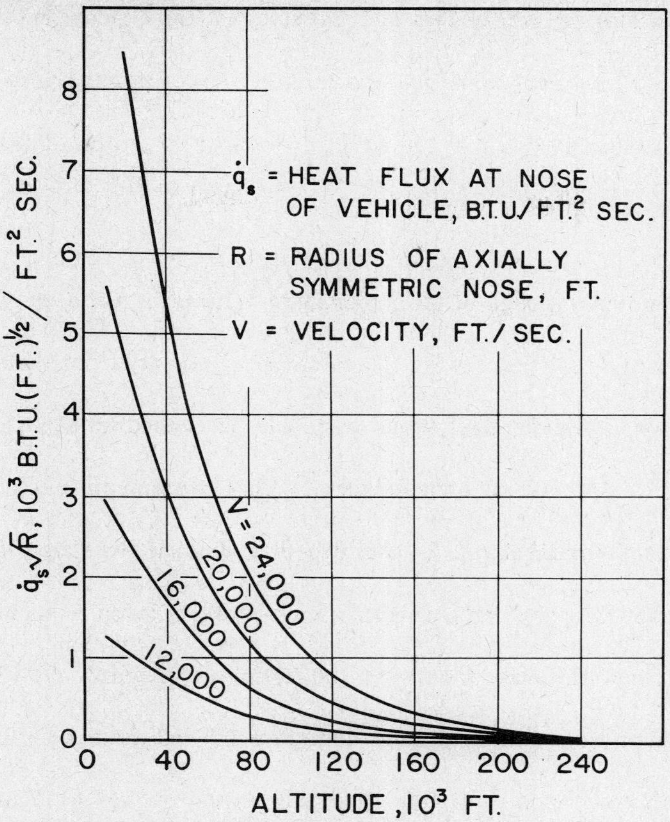

Fig. 4 Nose Heating Rate for Hypersonic Vehicle

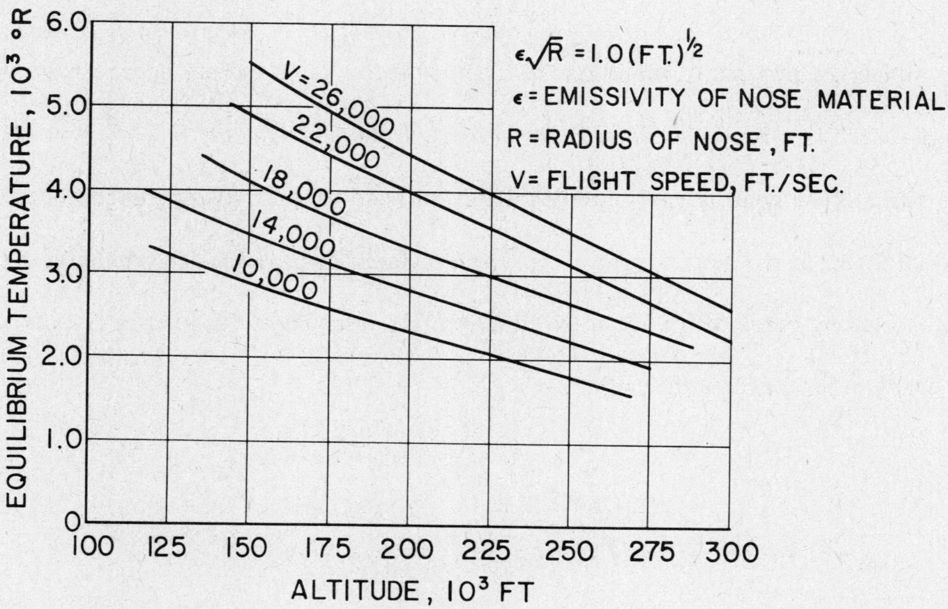

Fig. 5 Equilibrium Temperature of Vehicle Nose Material

For a nose cap of one foot (1 foot) radius this figure gives the heating rate directly. Fig. 5 shows graphs of the corresponding equilibrium temperatures based on normal shock tables[6] and a value of $\epsilon \sqrt{R} = 1.0 \text{ (ft.)}^{1/2}$ where ϵ is the emissivity of the material and R the nose radius in feet.

If the nose cap or leading edge of the hypersonic vehicle is made up of a porous weave consisting of individual round wire filaments (Fig. 1), it is concluded from the results of Ref. (7) that an independent shock wave can be caused by each filament for weaves having porosities as low as a few percent. This is apparent from Fig. 6 which shows the shock detachment distance δ_s for two-dimensional (leading edge) and axi-symmetrical (nose cap) flow conditions. These curves were taken from Refs. (2) and (7). As long as the clear distance A between the filaments of the weave (Fig. 1) is of the same order of magnitude as the filament diameter D, and the shock detachment distance δ_s is small compared to D, the shocks from the individual filaments will not interact and form a single shock front. Therefore, the total heat transfer to such a woven cloth is equal to the sum of the heat transfers to the individual filaments.

Since the product of the stagnation point heating rate and the square root of the nose radius is a function of the flight speed and the density only[1, 2, 3]: $\dot{q} \sqrt{R} = f(V, \rho)$, the ratio of the heating rates for two different nose caps or leading edges is inversely proportional to the square root of the ratio of their nose radii. In particular, the ratio of the heating rates of porous to solid materials is equal to the square root of the ratio of nose radius to filament radius.

$$\frac{\dot{q}_p}{\dot{q}_s} = \sqrt{\frac{2R}{D}}$$

Here $\dot{q}$ is the stagnation point heating rate, R the nose radius, and the subscripts p and s denote porous and solid materials, respectively.

This relation has been plotted in Fig. 7 for single-weave cloths of zero porosity. For a cloth of non-zero porosity, the ordinates of Fig. 7 must be multiplied by the solidity ratio which is the ratio of the exposed material in elevation to the total area of the leading edge in elevation. (The solidity ratio thus equals unity minus the porosity.)

Fig. 7 clearly illustrates the enormous heat transfer penalty introduced by using porous cloths. For example, a weave consiting of .006 inch diameter filaments, 100 to the inch in each direction (solidity ratio 0.84, porosity 0.16), receives 53 times as much heat as a solid material of 1 foot radius located at the same stagnation point position. If the diameter of the filaments is reduced to .0005 inch,and the weave has 1500 filaments to the inch each way (solidity ratio 0.9375, porosity 0.0625), this factor increases to 205.

Leading Edge

Comparing heat transfers over the forward half of a leading edge rather than at the stagnation point only, the porous cloth suffers from an additional disadvantage. The heat transfer to this half of the solid leading edge is approximately equal to

$$\dot{Q}_s = \int_0^{\pi/2} 2R\,\dot{q}_s \cos\theta\, d\theta = 2R\,\dot{q}_s$$

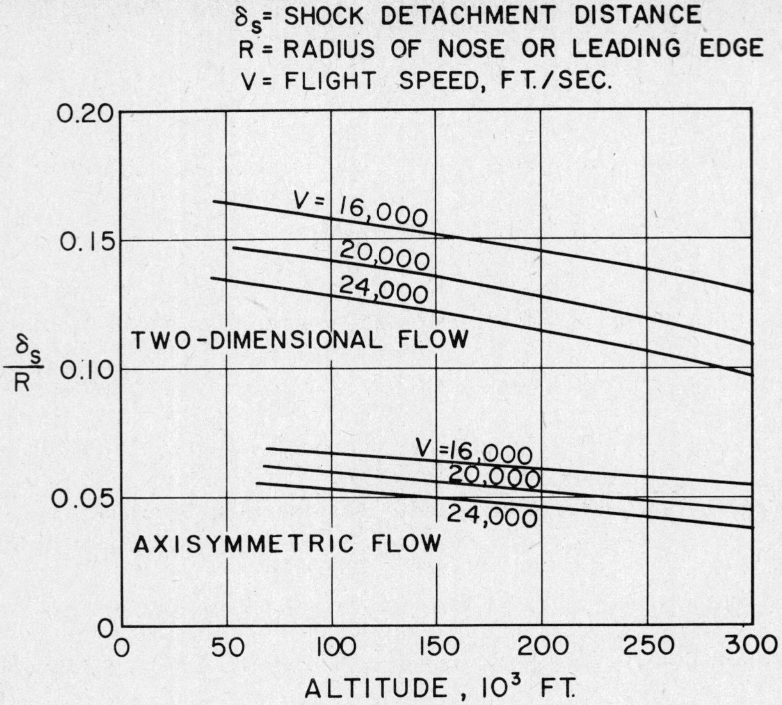

Fig. 6 Shock Detachment Distance

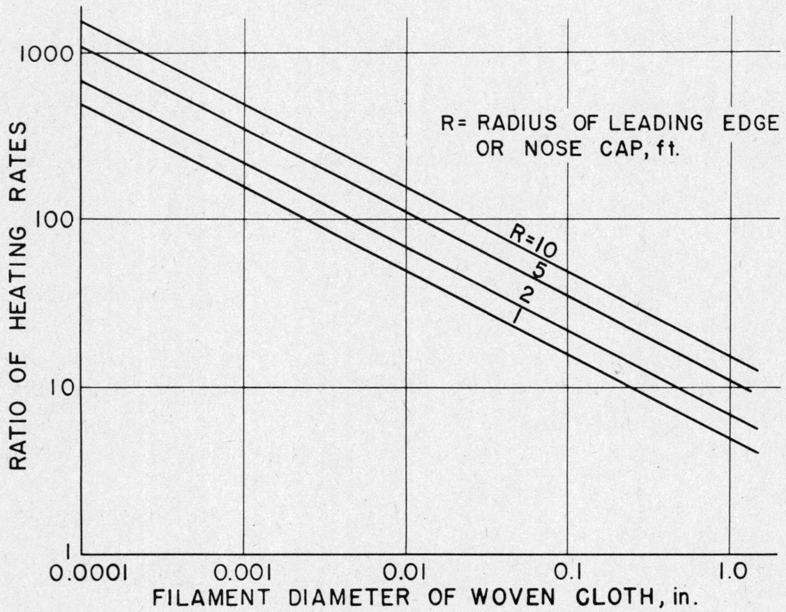

Fig. 7 Ratio of Leading Edge or Nose Cap Heating Rates of Zero Porosity Woven Cloth to Solid Material

while the heat transfer to the porous cloth (Fig. 8) is approximately given by

$$\dot{Q}_p = \int_{o}^{(\pi/2)-\phi} 2\,R\,\dot{q}_p\,d\theta + \int_{(\pi/2)-\phi}^{\pi/2} 2\,R\,\dot{q}_s \cos\theta\,d\theta$$

$$\dot{Q}_p = 2\,R\,\dot{q}_p\left(\frac{\pi}{2} - \phi\right) + 2\,R\,\dot{q}_s\left(1 - \sin\left[\frac{\pi}{2} - \phi\right]\right)$$

$$\dot{Q}_p \cong 2\,R\,\dot{q}_p\left(\frac{\pi}{2} - \phi\right) + 2\,R\,\dot{q}_s\left(\frac{\phi^2}{2}\right).$$

Here $\dot{Q}$ is the total heat transfer over the front half of the leading edge, and ϕ is the angle beyond which the porous material is no longer effectively porous due to the blocking action of the individual shock waves which tend to overlap (Fig. 8). This angle depends on the porosity of the cloth, the altitude, and the flight speed of the vehicle.

The ratio between the above quantities is, therefore,

$$\frac{\dot{Q}_p}{\dot{Q}_s} \cong \frac{\dot{q}_p}{\dot{q}_s}\left(\frac{\pi}{2} - \phi\right) + \frac{\phi^2}{2}$$

where the second term on the right hand side can be neglected by comparison with the first term. One of the plots of Fig. 9 shows leading edge heating ratios compared to stagnation point heating ratios with the latter term omitted.

Nose Cap

If similar calculations are performed for the total heating over the front half of a spherical nose cap, the following results are obtained.

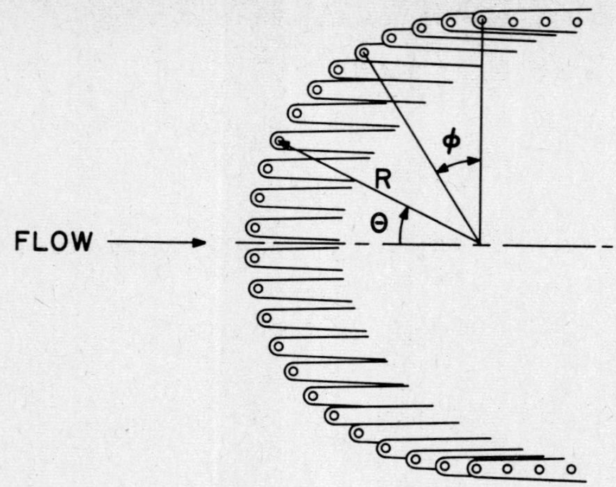

Fig. 8 Merging of Shock Waves From Filaments of Porous Leading Edge or Nose Cap

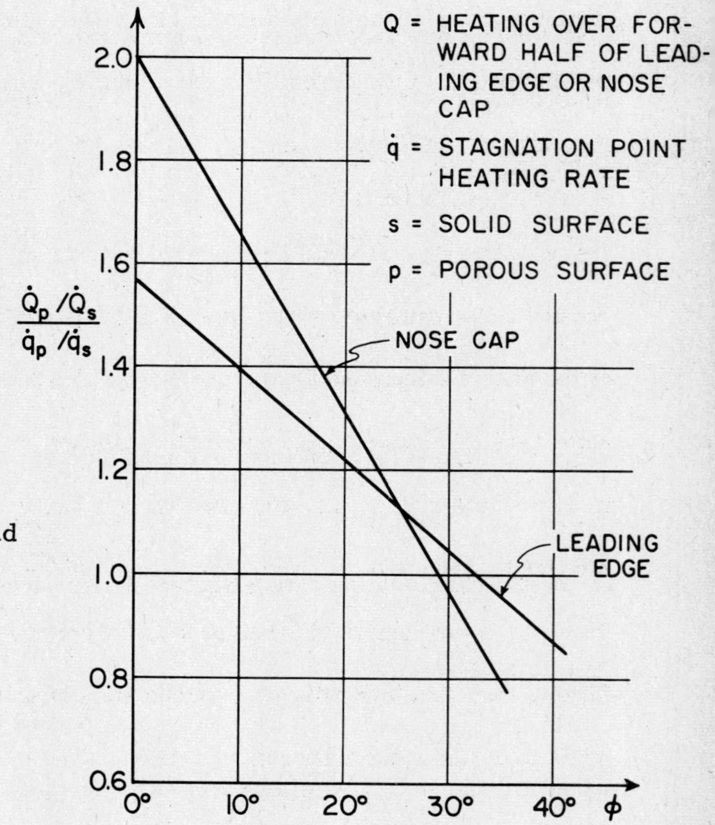

Fig. 9 Ratio of Leading Edge and Nose Cap Heating to Stagnation Point Heating Rates

$$\dot{Q}_s = \int_o^{\pi/2} 2\pi R \, \dot{q}_s \, \sin\theta \, \cos\theta \, R \, d\theta = \pi R^2 \dot{q}_s$$

$$\dot{Q}_p = \int_o^{(\pi/2)-\phi} 2\pi R \, \dot{q}_p \, \sin\theta \, R \, d\theta + \int_{(\pi/2)-\phi}^{\pi/2} 2\pi R \, \dot{q}_s \, \sin\theta \cos\theta \, R \, d\theta$$

$$\dot{Q}_p = 2\pi R^2 \dot{q}_p (1 - \sin\phi) + \pi R^2 \dot{q}_s \sin^2\phi$$

$$\frac{\dot{Q}_p}{\dot{Q}_s} \cong \frac{\dot{q}_p}{\dot{q}_s} \, 2 \left(1 - \phi + \frac{\phi^3}{6}\right) + \phi^2$$

The last equation with the non-linear terms neglected has also been plotted in Fig. 9. This figure thus gives a heat transfer factor to be used in addition to the factor obtained from Fig. 7 if total heating rather than stagnation point heating is critical.

Equilibrium Temperature

The increase in equilibrium temperature is, fortunately, less drastic than the increase in heating rate, because the rate of re-radiation increases with the fourth power of the absolute temperature[8], and woven cloths have higher emissivities than solid materials. This is easily seen if one considers the effect of the great number of cavities in a porous weave on the reflection of radiation. These cavities tend to act in a manner similar to a black body, hence they increase the absorptivity above that of the solid material. Since emissivity and absorptivity are equal under equilibrium conditions, it follows that the emissivity of a porous structure is greater than that of a solid structure made from the same material.

If a solid leading edge, with a nose radius R equal to one foot and an emissivity ϵ_s equal to 0.5, is compared to weaves having various filament diameters and having an emissivity ϵ_p equal to 0.95, the ratios of the stagnation point equilibrium temperatures are shown in Fig. 10. For the previous example of a weave having 100 .006-inch diameter filaments to the inch both ways, the equilibrium temperature is multiplied by 1.8 while the heat transfer rate is multiplied by 53. For the weave having 1500 .0005-inch diameter filaments to the inch both ways, the equilibrium temperature is multiplied by 2.4 and the heat transfer rate by 205. The ordinates of Fig. 10 remain unchanged for different porosities, because the effect of porosity is presumed to be included in the determination of the emissivity of the woven cloth.

Effect of Porosity

If the number of weaves per inch is increased and the filament size held constant, the porosity decreases. It is apparent that decreasing the porosity and decreasing the filament diameter will increase the heating rate. The maximum possible heating rate, however, is the heating rate for free molecular flow $\dot{Q}_f$[9].

$$\dot{Q}_f = \frac{1}{2} \rho V^3$$

Here ρ is the density of the surrounding medium and V the flight speed of the vehicle. The filament diameter was determined which, for a cloth of zero porosity, makes the heating rate equal to that for free molecular flow. This diameter, which is a function of altitude but almost independent of flight speed, was plotted on the free molecular path length[7] in Fig. 11. The figure shows that the two are almost identical. This can be interpreted as proof that the procedures used are valid for conditions for which the mean path length is considerably smaller than any geometric dimensions of the vehicular surface.

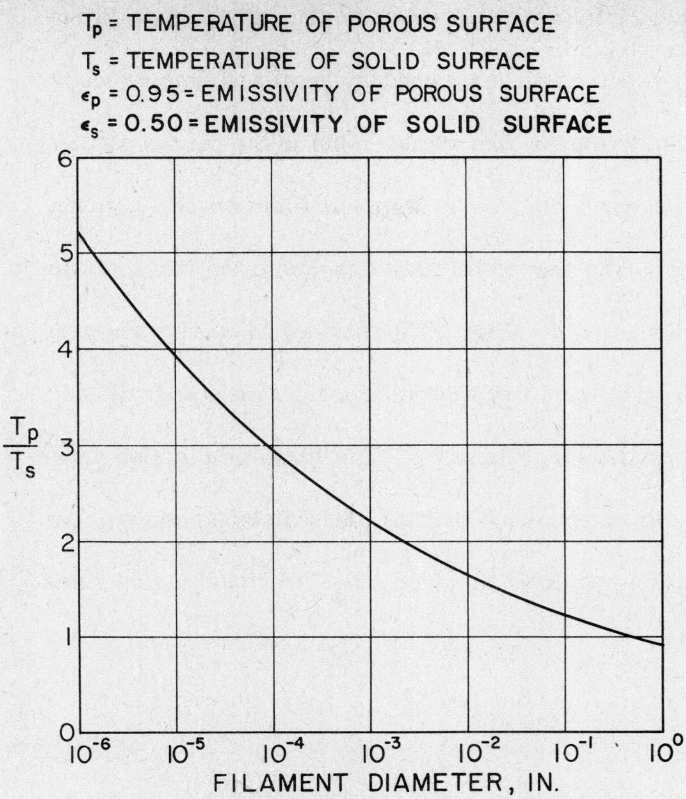

Fig. 10 Ratio of Stagnation Point Equilibrium Temperatures

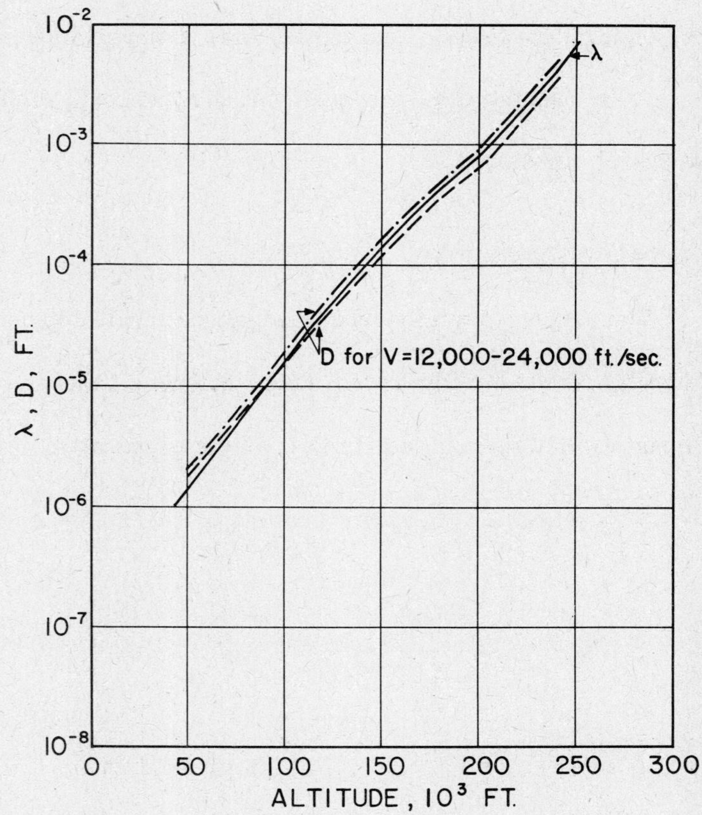

Fig. 11 Comparison of Free Molecular Path Length, λ, and Equivalent Filament Diameter, D

This approach to calculating the heating rates of porous surfaces is subject to two limitations: One is the transition between continuum boundary layer and free molecular flows, the other is the ability of the air to pass through the voids in the porous cloth. The first limitation is clearly spelled out in Ref. (7) in terms of flight speed, altitude, Reynolds number or Knudsen number. The second limitation is due to the fact that the porosity of the woven cloth for angles $\theta < \phi$ (Fig. 8) is assumed to be high enough to let all air pass through without a significant increase in pressure ratio and without interaction of the flows around two neighboring filaments. This limitation is also apparent from[7] and requires that a cloth be chosen such that the clear distance between the filaments A is a multiple of the shock detachment distance δ_s. Obviously, this rules out cloths of zero porosity.

CONCLUSION

Studies to determine the minimum porosity required for this analysis to be valid are continuing. The results, presented here, however, indicate porous cloths to suffer such severe heating penalties that their use in hypersonic vehicles and re-entry aids must be seriously questioned.

ACKNOWLEDGMENT

The author gratefully acknowledges the invaluable help of Dr. S.M. Scala, Manager of High Altitude Aerodynamics, Space Sciences Laboratory, Missile and Space Vehicle Department, General Electric Co., in the preparation of this paper.

NOTATION

A	clear distance between filaments of a porous weave
D	filament diameter
$\dot{Q}$	heating rate of forward half of leading edge or nose cap
$\dot{q}$	heating rate at stagnation point
R	radius of curvature of leading edge or nose cap
V	flight speed
δ_s	shock detachment distance
ϵ	emissivity
θ	angular variable of integration
ϱ	density
ϕ	limiting angle of effective porosity

Subscripts

p	porous
s	solid
f	free molecular flow

REFERENCES

1. S.M. Scala and C. Baulknight, "Transport and Thermodynamic Properties in a Hypersonic Laminar Boundary Layer", Part I, Properties of the Pure Species, Journal ARS, vol. 29, No. 1, January 1959.

2. Ibid, Part II, Applications, Journal ARS, vol. 30, No. 4, April 1960.

3. T.Y. Li and R. Geiger, "Stagnation Point of a Blunt Body in Hypersonic Flow", J. Aero Sci., vol. 24, No. 1, January 1957.

4. F.M. Rogallo, J.G. Lowry, D.R. Croom, and R.T. Taylor, Preliminary Investigation of a Paraglider, NASA TN D-443, August 1960.

5. Arthur D. Little, Inc., Candidate Materials for High Temperature Fabrics, WADC TR 59-155, September 1959.

6. P.W. Huber, Tables and Graphs of Normal Shock Parameters at Hypersonic Mach Numbers and Selected Altitudes, NACA TN-4352, September 1958.

7. S.M. Scala, "Hypersonic Viscous Shock Layer", Journal ARS, vol. 29, No. 7, July 1959.

8. M. Zemansky, Heat and Thermodynamics, McGraw - Hill Book Co., 4th ed. 1957, p. 106.

9. S. Schaaf and P. Chambré, Fundamentals of Gas Dynamics, Princeton High Speed Aerodynamics and Jet Propulsion Series, 1958, vol. 3, p. 707.

AERODYNAMIC HEATING FOR LUNAR REENTRY VEHICLES

B. H. Little, Jr.*

A survey is presented of problem areas involved in the prediction of aerodynamic heating for a lunar mission lifting reentry vehicle. A reentry corridor and configurations for vehicles which might fly this corridor are presented. The flow field around the vehicle in such a corridor is discussed with the purpose of defining significant flow phenomena. Methods and charts for predicting aerodynamic heating in the reentry corridor are given, and the optimum nose radius for minimizing total heat transfer rate is discussed. Thermal protection system requirements are discussed with attention directed to the nature and magnitude of the uncertainties involved in predicting heating loads.

INTRODUCTION

Methods for predicting aerodynamic heating for lifting orbital reentry vehicles have been reasonably well established by both theory and experiment. For vehicles reentering the earth's atmosphere at escape speed, however, much uncertainty still exists. This uncertainty may be partly attributed to the lack of knowledge of the exact condition of shock layer gases at the extreme temperatures encountered in atmospheric flight at velocities approaching 36,000 feet per second. It is also attributed to the lack of knowledge concerning the effects of thermal non-equilibrium and ionization on the heat transfer processes in hypervelocity flow. For lifting reentry at orbital

*Aerospace Technology Dept., Operations Research Division, Lockheed Georgia Company, Marietta, Georgia

velocity (about 26,000 feet per second) these effects are small and can, in general, be neglected as insignificant. The extrapolation of the methods used for orbital reentry to higher velocities, however, cannot be done without a careful look at the possible consequences.

This paper is intended to present a survey of the problems involved in the prediction of aerodynamic heating for a manned lunar mission reentry vehicle. It is also intended to provide equations and charts by which estimates can be made of the significant heating parameters for lunar reentry. The original work upon which this paper was based was part of an overall study at the Lockheed-Georgia Company of manned lifting vehicles for reentry from a lunar mission. This study consisted of many parts - the selection of reentry technique and flight corridor, the selection of vehicle configurations capable of performing the lunar reentry maneuvers, the selection of methods for predicting aerodynamic heating, and an examination of methods for providing thermal protection against the aerodynamic heating encountered. This paper draws from portions of this overall investigation other than that specifically concerned with aerodynamic heating in order that a proper perspective may be gained of such problems as follow:

1. The analysis of the flow physics typical of a lifting reentry vehicle flight
2. The selection of typical configurations for which heating predictions are required
3. The determination of regions of the vehicle which must be protected against aerodynamic heating.

The author, therefore, acknowledges the valuable contributions of a number of other members of the AeroSpace Technology Department who by their specialized efforts contributed portions of the information presented in this paper.

FLIGHT CORRIDOR AND VEHICLE CONFIGURATION

This analysis of aerodynamic heating was focussed on reentry conditions for a "realistic" manned lunar mission. Consequently, attention was directed to the determination of a vehicle capable of containing three men and supporting equipment and the corridor through which such a vehicle would re-enter the earth's atmosphere.

In the following paragraphs, a brief description is given of the ways in which reentry corridor and vehicle shape were derived. Subsequently, the analysis of reentry heating will be based on flow physics in such a corridor and around such a vehicle.

Studies of the reentry flight corridor for the lunar mission vehicle were made using a 2-degree-of-freedom automatic computer program designed specifically for such purposes. It was assumed that reentry must be made in a single pass and that, since the vehicle was to be manned, the maximum allowable deceleration must nominally be 10 g's. These assumptions fixed the upper and lower limits of the flight corridor. It was further assumed that, by way of reentry technique, the vehicle would be capable of (1) maintaining constant L/D until the point of peak deceleration was passed, and (2) then producing negative lift to prevent skipping out of the earth's atmosphere. Under these conditions it was found that the widest reentry corridor was obtained for a vehicle possessing an $L/D = 0.8$ and $W/C_DS = 150$ lbs/ft.* This corridor, in

* Mathematical symbols and their definitions appear at the end of this paper.

terms of entry angle at 400,000 feet altitude, lay between γ_0 = -5.5° and γ_0 = -8.25°. By sacrificing about 10 percent of this corridor width, however, the reentry corridor would accommodate vehicles with L/D values ranging from 0.5 to 1.3 and W/C_DS values ranging from 50 to 300. The corridor thus defined was chosen for this study. It is shown in terms of velocity and altitude in Fig. 1. The boundaries shown represent the loci of extreme velocity—altitude plots with the exception that skip spikes along the upper boundary are not shown.

Also shown in Fig. 1 is a computed trajectory for a typical reentry vehicle. It has the characteristics of W/C_DS = 125, and L/D = 0.5. The trajectory shown is for an atmosphere entry angle (at 400,000 feet altitude) of -7 degrees and an initial load factor of 5.0. This trajectory will be used hereinafter as a typical or reference lunar reentry trajectory and is called "Trajectory A."

Vehicles possessing aerodynamic coefficients within the range of interest (L/D = 0.5 to 1.3, W/C_DS = 50 to 300) may assume a wide variety of shapes. Two vehicles which were studied in this general program and which qualify aerodynamically for successful lunar reentry are shown in Fig. 2. A detailed discussion of the aerodynamic characteristics of these vehicles is given in Reference 1. Both configurations of Fig. 2 are sized to house three astronauts and their necessary equipment. The base diameter was chosen for suitability to mating with a Saturn-sized lunar rocket.

Configuration 1 of Fig. 2 is designed to achieve the necessary L/D characteristics only in the angle of attack range from + 25° to + 45°. It

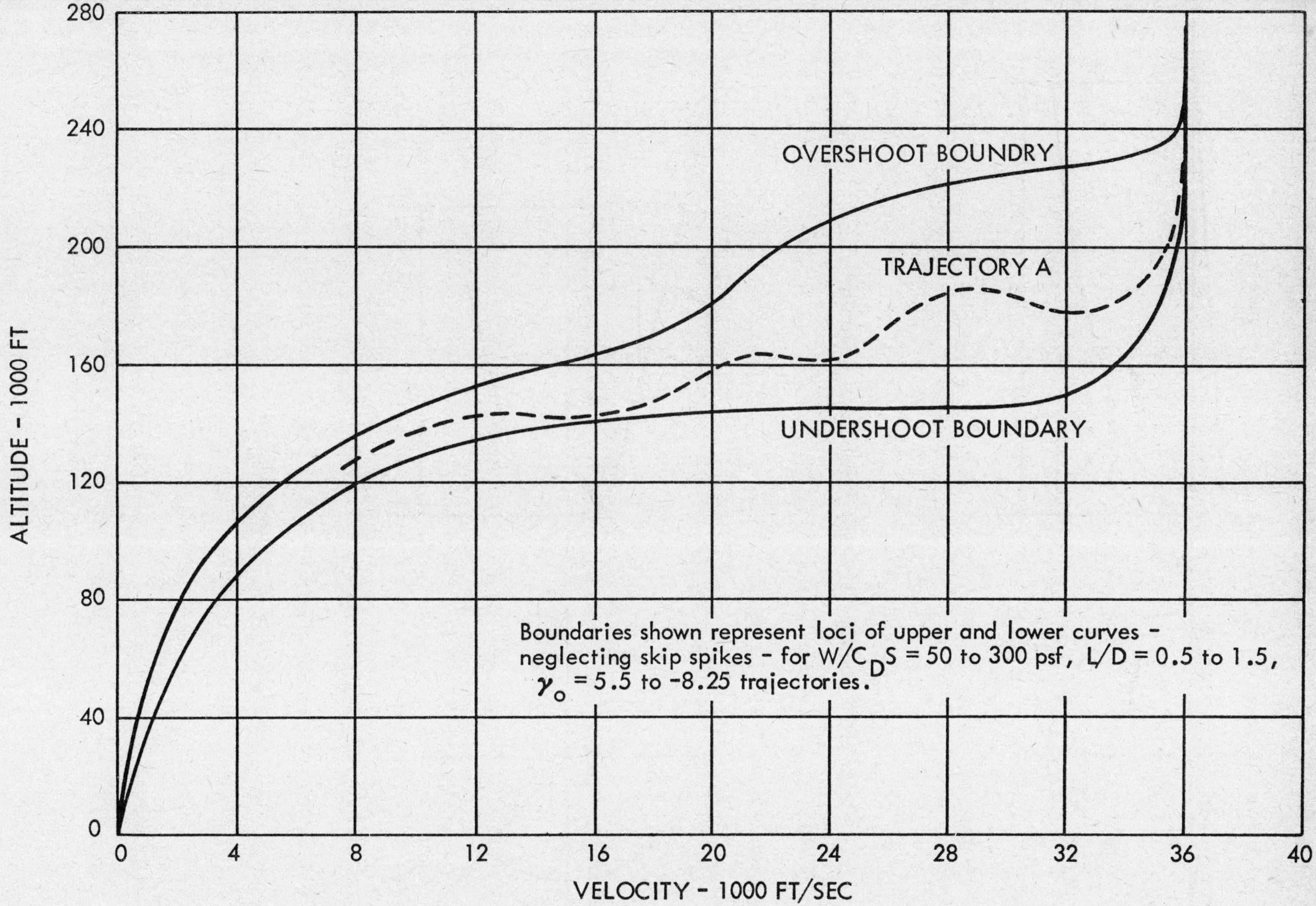

Fig. 1 Lunar Mission Re-entry Corridor

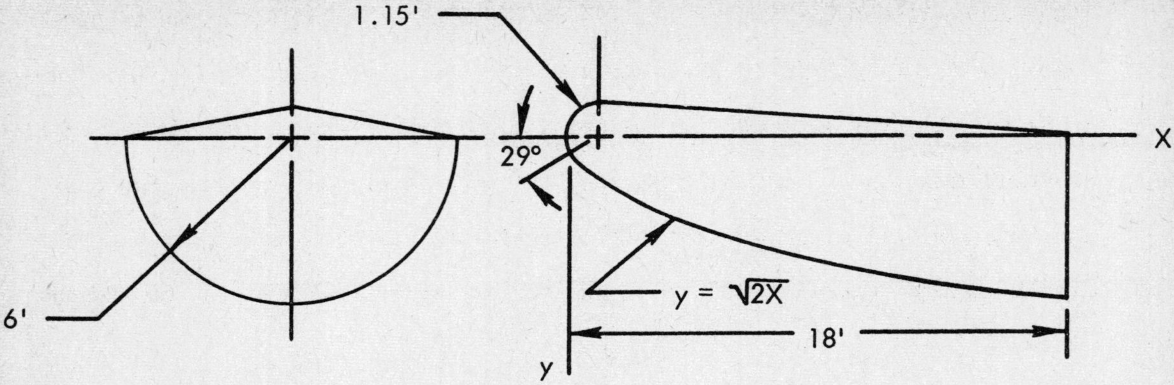

Base Area = 56.5 ft^2
Volume ≅ 510 ft^3
Configuration 1 - Half Parabola of Revolution

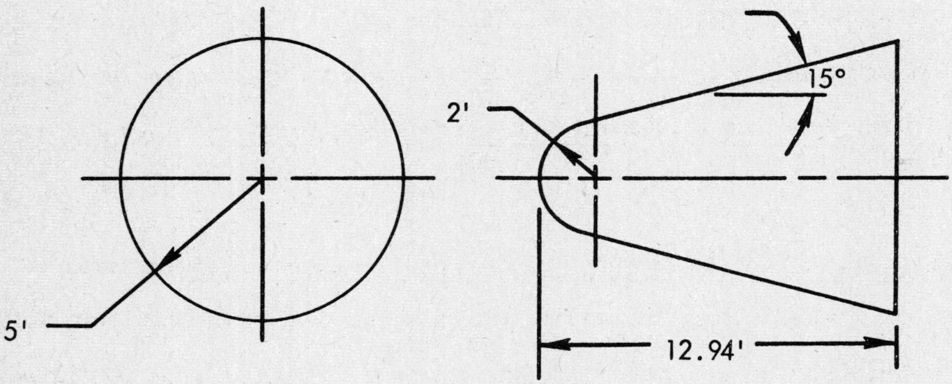

Base Area = 78.5 ft^2
Volume = 472 ft^3
Configuration 2 - Blunted Cone

Fig. 2 Geometry used in Aerodynamic Evaluation of Re-entry Vehicles

thus accomplishes the desired reentry requirement of negative lift after peak deceleration by rolling over on its back when negative lift is required. Such a vehicle has the advantage of maintaining loadings always in the same direction. Heating also can be confined primarily to one side of the vehicle. Maximum heating would occur on the nose and along the bottom surface which would be inclined over the major portion of the surface at angles of 40° to 60° to the flow.

Configuration 2 of Fig. 2 achieves the necessary L/D characteristics at angles of attack in the $\pm 5^\circ$ to 40°-range. It can navigate the desired reentry corridor by use of pitch control, or as in the case of Configuration 1, by roll control. Heating again over a large portion of the vehicle surface would be that for surface inclinations up to 60° relative to the flow. A number of shapes other than those shown might be used. These two, however, will be satisfactory for demonstrating the heat shield requirements for lunar reentry and for determining characteristic dimensions needed for flow evaluations.

Summarizing, it can be stated that for vehicles reentering the earth's atmosphere in a single pass from escape velocity the flow around the body will be that associated with the reentry flight corridor of Fig. 1. The configuration will fly with large portions of the vehicle surface inclined to the flow at angles up to about 60°. The maximum vehicle dimension, as shown in Fig. 2, will be about 20 feet.

FLOW PHYSICS

In this section the flight corridor defined in Fig. 1 is examined to determine the significant flow phenomena which the lunar reentry vehicle encounters. The results of the examination guide in the selection of methods for estimating aerodynamic heating.

Real Gas Effects

The limitations of the assumption of continuum flow are best defined by the Knudsen number concept.[2-5] This concept states that when the Knudsen number (or ratio of the mean free path length λ to a characteristic flow dimension) is much less than 1, the flow is dominated by intermolecular collisions and may be considered a continuum. Conversely, when the Knudsen number is much greater than 1, the flow is dominated by molecule—body surface collision and must be considered a free molecule flow.

For a blunt axisymmetric vehicle, Hayes and Probstein[2] state that the significant dimensions to be used in the Knudsen number application for stagnation point flow examination are the mean free path length behind the shock λ_s and the shock layer thickness Δ . While the vehicle of interest here is not axisymmetric in reentry attitude, the departure should not introduce significant error into the flow field analysis. Reference 2 also states that the ratio λ_s/Δ is of the order of M_∞/R_∞ . Since velocity cancels out of the above expression, the flow field for blunt reentry vehicles may be described in terms of altitude alone. Fig. 3 presents an altitude-velocity plot showing the position of the lines of constant value of the ratio M_∞/R_∞. It can be seen that for stagnation region flow considerations all portions of the reentry trajectory below about 270,000 feet altitude can be considered in the continuum regime. The significant peak heating portion of the lunar reentry trajectory lies well within this continuum regime. Only for detailed analyses of reentry heating of vehicles entering at shallow reentry angles is heating outside this regime significant.

Boundary Layer Transition

Considerable doubt exists as to the Reynolds number for transition in hypervelocity flows. Experimental evidence tends to place the occurrence of transition in the range of Reynolds numbers between $5x10^5$ to about $2x10^7$ based on a body station dimension. A commonly used value (Reference 6) is $5x10^6$. Fig. 3 shows lines of constant Reynolds number per foot plotted on the velocity-altitude chart with the reentry corridor superimposed for comparison. Since characteristic lengths for stagnation region heating will be on the order of one foot or less, the stagnation heating flow will always be laminar in nature. Using the value of critical Reynolds number of $5x10^6$ just mentioned, the reentry vehicle of interest with a maximum body dimension of about 20 feet would be expected to maintain laminar flow overall for altitudes above about 160,000 feet. It is safe, therefore, for this analysis to assume laminar flow for first order estimates of reentry heating.

Dissociation and Ionization - The thermodynamic and chemical states of the fluid have a significant bearing on the nature and magnitude of aerodynamic heating. It is important then to determine the nature of the fluid in the stagnation region of the lunar reentry vehicle. Attention is first directed to Fig. 4 where equilibrium shock layer temperatures and pressures are presented on the velocity altitude chart. This figure is reproduced from Reference 7. It can be seen that the lunar reentry vehicle flies through regions where the shock layer temperatues approach 12,000°K. Orbital reentry vehicles by way of contrast encounter temperatures of only about 8000°K. Lees[8], Kivel[9], and others point out that at temperatures above 8000°K the emissivity of heated air becomes significant so that radiative heating from the shock layer becomes important for the lunar reentry vehicle.

The high temperature - low pressure shock layer of the lunar reentry also changes the convective heat transfer picture. This is primarily a result of ionization of atoms at the high temperatures shown in Fig. 4. This can be seen more clearly in Fig. 5 where the regions in which dissociation and ionization are significant are shown on the velocity-altitude chart. For the lunar reentry trajectory, the region of peak heat transfer lies well within the ionization zone.

One of the major effects of ionization of air is to increase the conductivity, as shown by Hansen[7]. This will have to be considered in the analysis of convective heat transfer.

Equilibrium - The finite time interval required for a gas to react chemically to changes in temperature, though small, are often significant when considered with respect to the times required for the gas particles to move through the shock layer of a hypervelocity vehicle. Thus, it is possible for the shock layer to be in a state of non-equilibrium under certain conditions of temperature and pressure. Fay and Riddell[10] in their treatment of laminar stagnation point heating showed the effects of non-equilibrium recombination to be a reduction in the convective heat rates if the body surface were non-catalytic to recombination. On the other hand, non-equilibrium dissociation

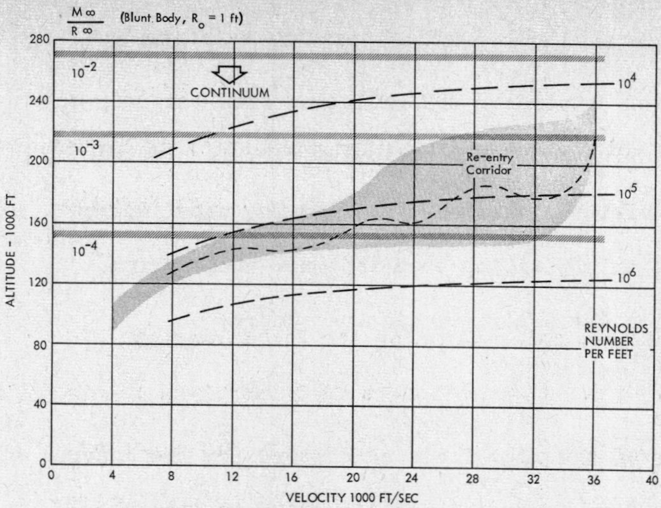

Fig. 3 Knudsen and Reynolds Numbers in Flight Spectrum

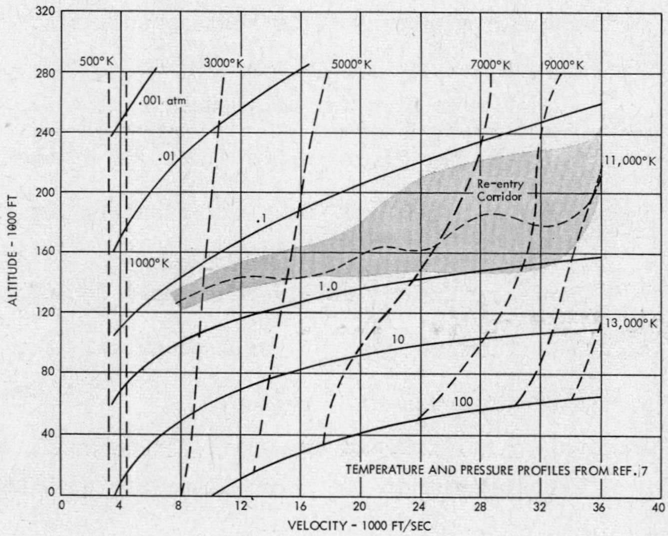

Fig. 4 Pressure and Equilibrium Temperature Profiles in Flight Spectrum

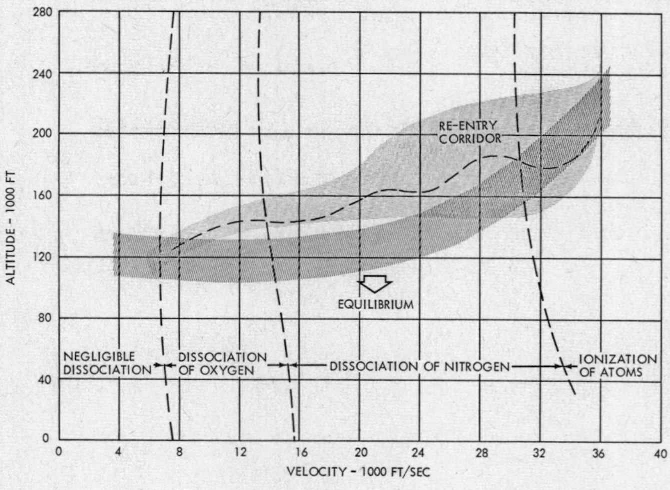

Fig. 5 Flow Regions in the Flight Spectrum

and ionization at the shock front result in temperatures much higher than those for equilibrium, and consequently increase the radiation from the heated layer. The boundary between equilibrium and non-equilibrium flow is plotted as a shaded region on the velocity-altitude chart of Fig. 5. The superimposed reentry corridor can be seen to lie in a position such that non-equilibrium effects can be expected in the region where maximum heating occurs.

<u>Flow Physics Summary</u>

From the foregoing examination of the flow field encountered by the lunar reentry vehicle, some conclusions may be drawn as to the conditions imposed on the lunar reentry vehicle heating analysis.

A continuum flow may be assumed except for detailed analyses. This assumption is valid for flight below altitudes of about 270,000 feet, and this is the region in which most of the heating occurs for the single pass reentry vehicle.

Laminar flow may safely be assumed in and near the stagnation regions. Turbulent flow, if present, will occur only for body stations somewhat downstream of the stagnation region.

Temperatures in the shock layer are such as to introduce two factors into the heating analysis which could be neglected for reentry at orbital velocity. These factors are (1) radiation from the shock layer to the body; and (2) the influence of ionization on the convective heating.

A departure from chemical equilibrium in the shock layer may be expected and should be considered in analyses of reentry heating.

AERODYNAMIC HEATING

For the lunar mission reentry vehicle, primary attention herein is focused on the determination of heat transfer in the stagnation region. It is in this region that the effects of ionization and non-equilibrium flow will be most severe. The stagnation region heating will also define the most severe requirements for the thermal protection system.

Convective Heating

Laminar Stagnation Point - A number of methods have been published in the literature [8, 10, 11, 12] for calculating laminar stagnation point convective heating. The effects of dissociation and non-equilibrium were treated in some detail by Fay and Riddell[10] and Kemp, Rose, and Detra[13]. Some of the methods cited are too unwieldy for approximate heating analyses. A noteworthy exception is the approximate method of Lees[8] which presents the heat transfer rate equation as follows:

$$\dot{q}_c = 15.5 \frac{2^{k/2}}{\sqrt{R_o}} \sqrt{\rho_\infty} \; U_\infty^3 (1 - hw/hs) \times 10^{-9} \text{ Btu/ft}^2\text{-sec} \qquad (1)$$

For heat transfer to a body of revolution, and the assumption of a cold wall ($hw/hs \ll 1.0$), this equation reduces to the following:

$$\dot{q}_c \sqrt{R_o} = 20.9 \sqrt{\rho_\infty} \; U_\infty^3 \times 10^{-9} \qquad (2)$$

In Reference 6, a number of methods for computing laminar stagnation point heating were compared and it was found that Equation (1) produced results close to the numerical mean of all the methods examined. Equation (1) is therefore widely accepted as valid for dissociating gases. It may not be adequate, however, for ionized flows.

The results presented in Fig. 5 indicate that peak heating for the lunar reentry corridor will occur in a region where ionization is significant. Adams[14] presented an approximate analysis in which the effects of ionization for reentry at escape velocity were estimated to result in an increase in heat transfer rates of about 30 percent or less over the values predicted by methods such as the Lees equation. Results from this analysis are shown in Fig. 6. Recently it was reported[15] that a more detailed analysis shows the correction for ionization to be more nearly a factor of 2.0 to 2.5. However, the exact conditions under which such large corrections might be required have not yet been generally published. The analysis herein was made, therefore, using a correction derived from the Adams analysis. Some remarks are made in the thermal protection discussion about the possible effects of larger convective heating rates.

While the calculations of Adams for the ionization correction are somewhat involved, the final results of this analysis show a correction with velocity for altitudes ranging from sea level to 250,000 feet in the band shown in Fig. 6. For estimates of the heating rates of this analysis a linear correction was applied to the Lees equation for all velocities greater than 23,000 feet per second in the form:

$$\dot{q}_c'/\dot{q}_c = 1 + .015\ (u_\infty - 23{,}000)$$

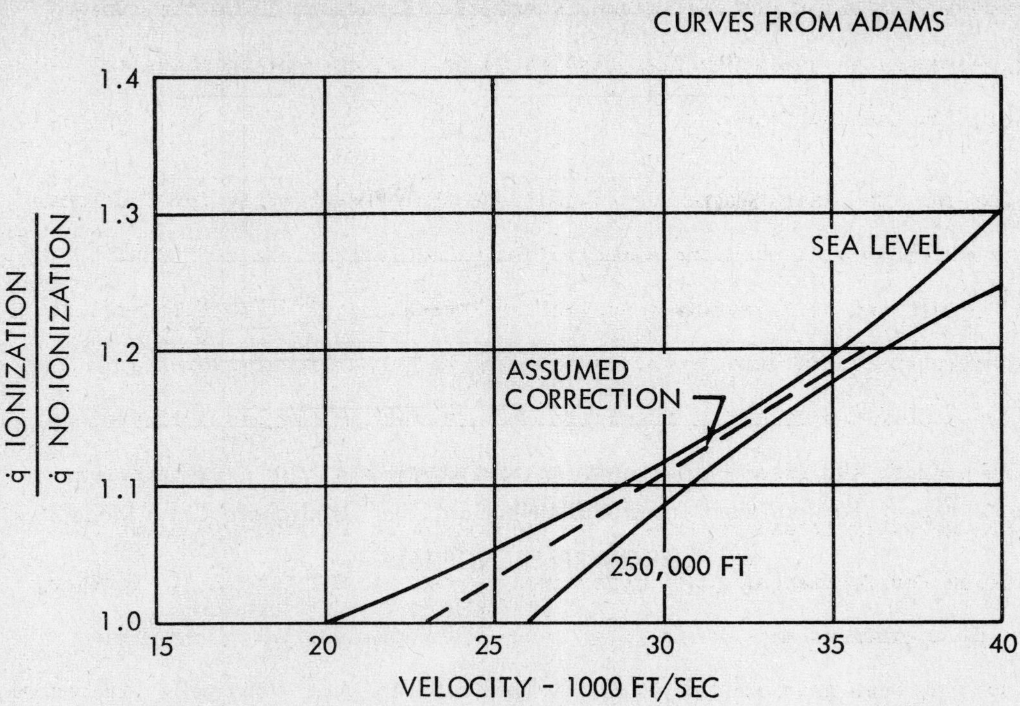

Fig. 6 Correction to Convective Heating Due to Ionization

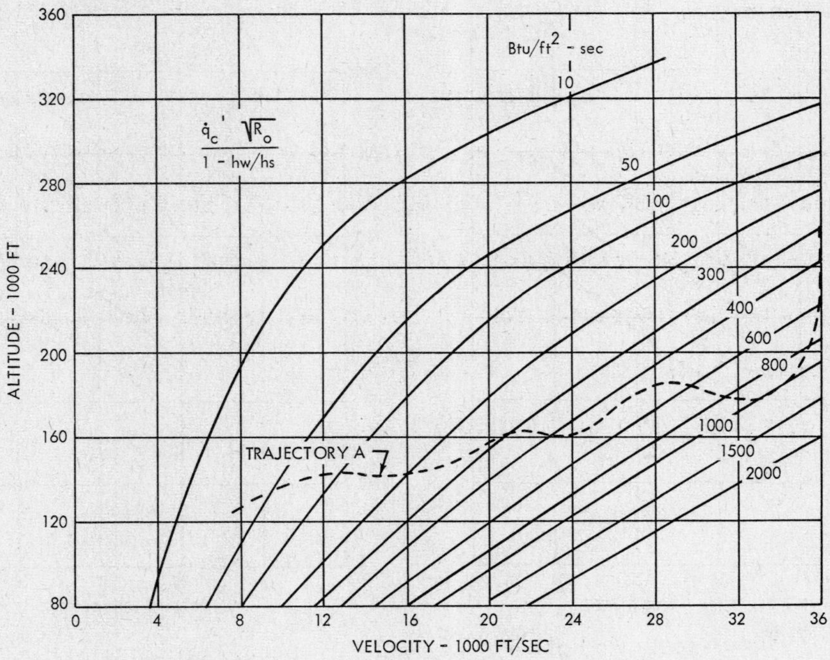

Fig. 7 Laminar Stagnation Point Heat Transfer Rate

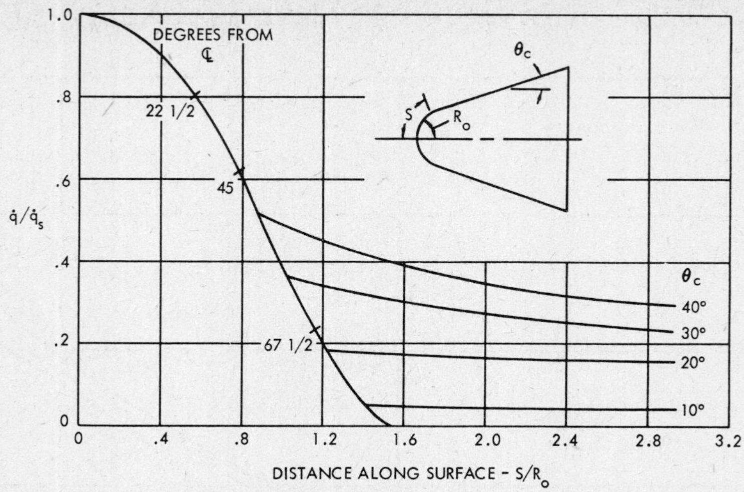

Fig. 8 Laminar Heat Transfer Distribution Over Hemisphere-Cone Models
(FROM REFERENCE 16)

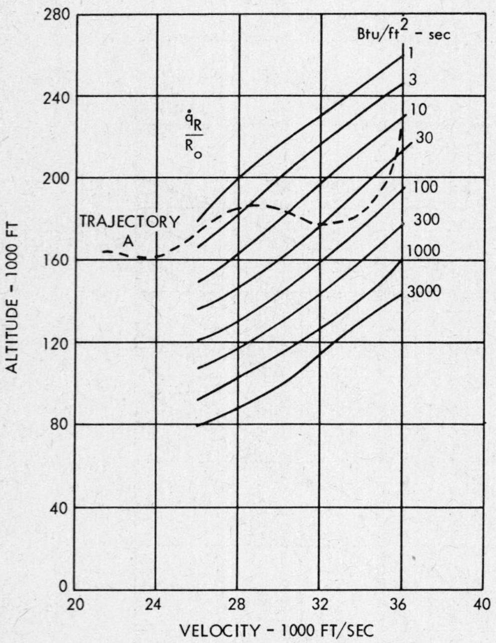

Fig. 9 Shock Layer Radiative Heat Transfer Rates

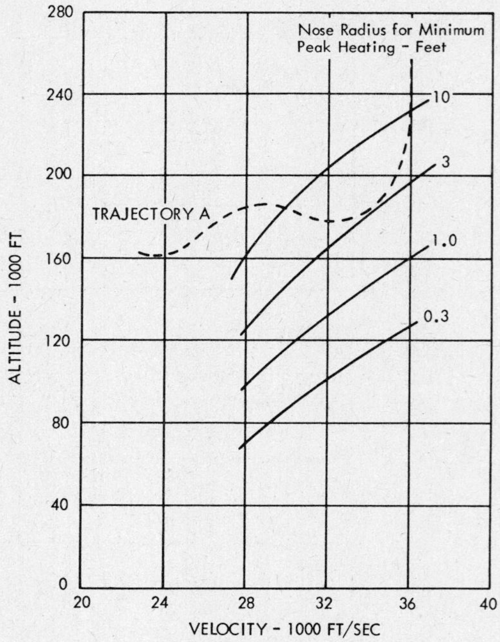

Fig. 10 Nose Radii which Minimize Peak Stagnation Point Heating Rate

This assumed correction factor is also plotted in Fig. 6.

Since the heat transfer rate of Equation (1) depends on nose radius R_o and wall temperature or enthalpy h_w it can be made more useful for general application by writing it:

$$\dot{q}_c \sqrt{R_o}/(1 - h_w/h_s) = 20.9 \sqrt{\rho_\infty} \quad u_\infty^3 \times 10^{-9} \tag{4}$$

In this form with the correction of equation 3 values of $\dot{q}_c \sqrt{R_o}/(1 - h_w/h_s)$ were calculated and are plotted on the velocity-altitude chart in Fig. 7. For any trajectory, then, it is possible from this figure to obtain heat transfer rates directly for a nose radius of one foot and the "cold wall" condition.

The lunar reentry trajectory A is superimposed on the curves of Fig. 7. It can be seen that for the $W/C_DS = 125$, $L/D = .5$ trajectory, the peak heating rate $\dot{q}_c \sqrt{R_o}$ is about 1000 Btu/Ft^2 sec. For a nose radius, $R_o = 2$ ft, the stagnation point heat rate is about 700 Btu/Ft^2-sec.

Convective Heating To Areas Downstream Of The Stagnation Point

Laminar convective heat transfer rates to areas downstream of the stagnation point may be predicted by the methods of Lees[16]. Good agreement of these methods with experiment has been obtained so long as pressure distributions over the body can be adequately predicted. Curves are presented in Fig. 8 which show the distribution of heat transfer rate around hemisphere-cone models. Such curves can be readily used for first order estimates of convective heat transfer rate distribution over a reentry body. For ellipsoidal nose shapes, Detra and Hidalgo[17] and Vinokur[18] present curves for computing local heat transfer rates.

It can be seen in Fig. 8 that for the reentry bodies of interest which will fly with surface regions inclined at angles up to about 60° to the flow that laminar heat transfer rates will range from about 50 percent down to about 30 percent of the stagnation point values. On the nose, the heating rate drops to 90 percent of the stagnation point value at points $22\frac{1}{2}^\circ$ from the stagnation point and down to 60 percent at 45°.

Turbulent Heat Transfer

Turbulent heat transfer rates were not computed since it was felt that transition would be far enough removed from the stagnation point to reduce such considerations to a secondary nature. A good discussion of turbulent heat transfer through a highly cooled, partly dissociated boundary layer is given, however, in the report of Rose, Probstein, and Adams[19]. It was found that turbulent heat transfer for the conditions cited did not depart appreciably from values predicted by incompressible flow theory. A correction factor for the effects of dissociation was presented for application to the incompressible theory.

Heat Transfer In Rarefied Gas Flows

While peak heating rates occur for the lunar reentry in flow regions where the assumption of a continuum is valid, it sometimes is necessary to make detailed analyses of trajectories near the upper boundary of the corridor in order to investigate rarefied gas flows. A number of complex interactions may take place between shock layer and boundary layer in the region between continuum and free-molecule flow. The reader is referred to the work of Hayes and Probstein[2] for more detailed discussion of this region. For approximate analyses, a conservative estimate of heat transfer in this region may be obtained by computing the free-molecule value by methods such as those of Reference 2.

Shock Layer Radiation - It can be seen in Fig. 4 that shock layer temperatures for the lunar reentry reach values of about 12,000°K for equilibrium flow. For non-equilibrium conditions, temperatures may be considerably higher. Even at the equilibrium temperatures, however, radiation from this gas layer to the reentry body nose reaches proportions which may be significant with respect to the convective heating. This problem has been studied in some detail, and results are presented in References 8, 9, and others. For non-equilibrium radiation, the lack of sufficient fundamental data makes this problem a serious one. Analyses by Camm, Kivel, Taylor, and Teare[20], for velocities up to satellite velocity, indicate that non-equilibrium radiation may be one or more orders of magnitude higher than that predicted for equilibrium flow. For the present study, however, the calculations shown are those for equilibrium radiation. Some consideration is given in the following section to the possible effects of non-equilibrium radiation on the total heating loads.

Reference 21 presents the following equation for calculation of shock layer radiation heat transfer rates:

$$q_R = R_o A \left(\frac{\rho_\infty}{\rho_o}\right)^{1.78} \left(\frac{u_\infty}{10^4}\right)^8$$

where

$A = 0 \; : \; B = 0 \qquad u_\infty < 25{,}000$

$A = 6.8; \; B = 12.5 \qquad 25{,}000 < u_\infty < 30{,}000$

$A = 0.003; \; B = 19.5 \qquad 30{,}000 < u_\infty < 35{,}000$

$A = 20.4; \; B = 12.5 \qquad 35{,}000 < u_\infty$

Equation (5) may also be made more useful for general analyses by writing it in the form of $\dot{q}_R/R_o$.

Equation (5) was used to compute lines of constant $\dot{q}_R/R_o$ which are plotted on the velocity-altitude chart of Fig. 9. It can be seen that for Trajectory "A", the shock layer radiative heat transfer rates for $R_o = 1$ are of the order of 10 percent of the convective heat transfer rates. Peak values shown for $\dot{q}_R/R_o$ are about 200 Btu/ft^2-sec for the reentry corridor of interest. If, however, non-equilibrium effects at escape velocity are as large as those at orbital velocity the radiative heat transfer will reach values of the same order of magnitude as those for convective heat transfer.

Total Aerodynamic Heat Transfer - The preceding paragraphs present charts by means of which the heating rates at the stagnation point of a lunar reentry vehicle may be predicted. These charts were presented in terms of $\dot{q}$ and R_o so that a nose radius must be selected before the actual heat transfer rates can be calculated. Since convective heating rates vary proportionally to $R_o^{-\frac{1}{2}}$ and radiative heating rates vary proportionally to R_o there exists a nose radius for any given trajectory which minimizes the peak total heating rate. While peak heating rate is not the only criterion for selection of nose radius, it is an important one and should be considered. It is convenient, therefore, to determine the nose radius which minimizes the sum of convective and radiative heat transfer. This may be done in the following way:

From equations (1) and (5) we may write the total heat transfer rate as follows:

$$\dot{q}_T = \dot{q}_R + \dot{q}_c = A\, R_o \left(\frac{\rho_\infty}{\rho_o}\right)^{1.78} \left(\frac{u_\infty}{10^4}\right)^B + 1.02 \times 10^3\ 1/\sqrt{R_o} \left(\frac{\rho_\infty}{\rho_o}\right)^{1/2} \left(\frac{u_\infty}{10^4}\right)^3 \quad (6)$$

By differentiating this expression with respect to R_o, setting the result equal to zero, and solving for R_o, the following is obtained:

$$R_o = \left[\frac{5.1 \times 10^2}{A\left(\rho_\infty/\rho_o\right)^{1.28}\left(u_\infty/10^4\right)^{B-3}}\right]^{2/3} \quad (7)$$

Equation (7) provides a convenient way of determining optimum values of R_o for flight at any velocity and altitude.

Optimum R_o values, computed by means of equation (7) are presented on a velocity-altitude chart in Fig. 10. A word of caution should be inserted here. The values of optimum nose radius shown are strongly dependent on the equations used for determining convective and radiative heat transfer rate. A major change in the magnitude of either mode of aerodynamic heating will change the position of the curves shown.

The reentry trajectories, when superimposed on the curves of Fig. 10, indicate that the total of convective and radiative heat rates at peak heating is minimized with a nose radius of about 3 feet. If non-equilibrium effects increase the radiative heating significantly, the optimum radius curves will shift upwards in altitude to indicate a smaller optimum radius. If ionization effects increase the convective heating significantly, the curves will shift downwards to indicate a larger optimum radius.

Aerodynamic Heating Summary

Convective and radiative heat transfer rates to the stagnation point of a lunar reentry vehicle may be estimated by means of the charts shown in Fig. 7 and Fig. 9 respectively.

The convective heat transfer rates (Fig. 7) are uncertain at velocities above about 32,000 ft/sec since the effects of ionization are not precisely known. Recent releases of an analysis of these effects states that ionization may increase the non-ionized convective heat rates by a factor of 2 to 2.5.

The radiative heat transfer rates (Fig. 9) are presented for equilibrium flow. These are also uncertain since non-equilibrium effects may be expected in the peak heating region. Information on this effect at orbital velocities show the non-equilibrium radiation to be an order of magnitude higher than equilibrium radiation.

THERMAL PROTECTION SYSTEM REQUIREMENTS

Design Requirements

As was pointed out in previous discussions, the uncertainties which exist in prediction of aerodynamic heating for the lunar mission reentry vehicle are large and may have considerable effect on the design of the thermal protection system. The magnitude of the uncertainties is indicated in Fig. 11 where stagnation point heat rate is plotted as a function of time. The solid line curves show the results of computations of heat rate for a vehicle of nose radius R_o = 2 ft, L/D = 0.5, W/C_oS = 125, γ_o = -7 degrees, and an initial load factor of 5.0. These computations were made by the use of equations (1), (3), and (5) for convective and radiative heating. The lower curve shows the convective heat rate (equations (1) and (3)) and the upper curve shows the total heat rate. It can be seen that the peak stagnation point heat rate is about 800 Btu/ft^2-sec.

The dashed line curves of Fig. 11 are plotted to indicate the nature of the heating problem if arbitary correction factors are applied to the convective and radiative heat rates to account for uncertainties due to the effects of ionization and non-equilibrium respectively. The factors applied were 2.0 for the convective heat rate and 10.0 for the radiative heat rate. The net effect is to increase the peak heating rate to about 2900 Btu/ft^2-sec. The value R_o = 2 ft for nose radius is slightly off the optimum value for minimum peak heating but, even with optimum radius, the peak heating rate would still be about 2,500 Btu/ft^2-sec.

The total heat to the stagnation point during reentry is plotted in Fig. 12. The computed heat transfer for Trajectory "A" is about 60,000 Btu/ft^2 as shown by the solid line curve. With the assumed adjustments for the effects of ionization and non-equilibrium the heat transfer is about 120,000 Btu/ft^2.

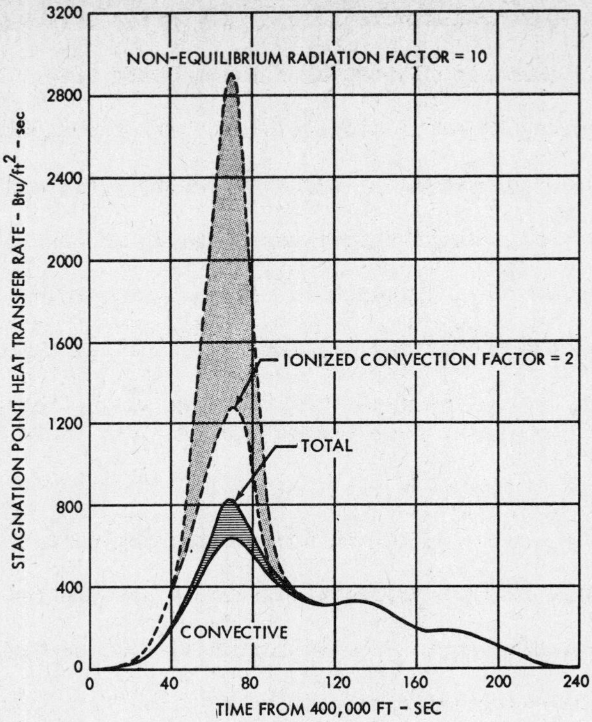

Fig. 11 Heating Rate Time History - Trajectory A

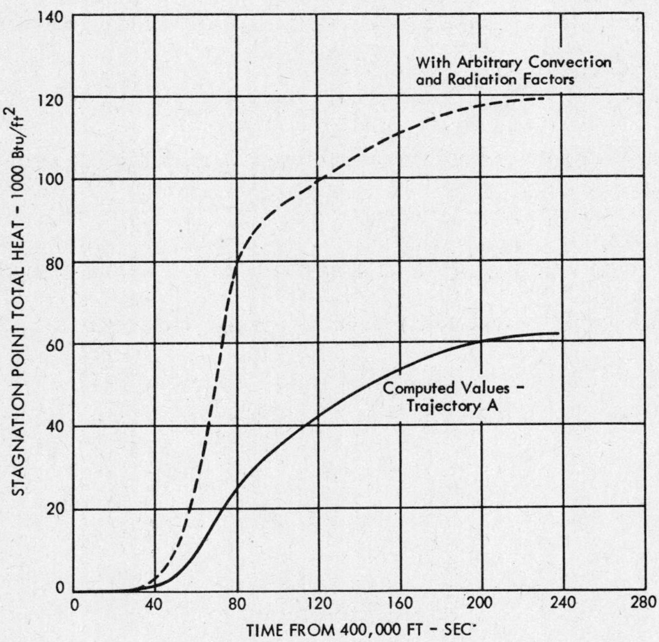

Fig. 12 Total Stagnation Point Heat Load - Trajectory A

The uncertainties discussed herein apply primarily to flow in the stagnation region of the vehicle. Ionization effects which are strongly temperature dependent will probably not be significant outside the stagnation region. Non-equilibrium in itself is more dependent on pressure than temperature but radiation from the shock layer, the heating mechanism which it affects, is strongly temperature dependent. Thus, the non-equilibrium problem also is confined to the stagnation region.

The areas receiving heat transfer loads as great as 50 percent of the stagnation heat transfer loads for this investigation were never more than about 10 percent of the total surface area excluding the base area. Thus the heat transfer problems discussed affect only a small part of the total exposed area. This should not detract, however, from the fact that the thermal protection requirements at the nose are quite severe.

Possible Thermal Protection Systems

Thermal protection systems were examined on the basis of the following:

1. Ability to meet design requirements
2. Reliability
3. Efficiency in terms of heat absorption per pound of installed weight.

Attention was confined again to the severe requirements in the stagnation region.

Radiation cooling is certainly attractive from the standpoint of reliability and efficiency but is marginal at best for the computed heat loads. It is not able to approach the shielding requirements if ionization and non-equilibrium effects are large. Pyrolytic graphite with a sublimation temperature of about 6,500°F can radiate heat at the rate of about 800 Btu/ft^2-sec and appears as the best possibility for a radiative heat shield. At present, however, it has a number of drawbacks which must be overcome before its application becomes practical. These include difficulties of fabrication, low working stress levels, and attachment problems.

Ablative heat shields have demonstrated capability to absorb high heat rates and presumably can withstand even those which may result from the uncertainties in ionization and non-equilibrium effects. They are time-limited, however, and at the present state of the art are not able to withstand the prolonged heating of the lunar mission reentry.

The best possibility, at present, appears to be a combination system employing ablation to absorb the initial heat rates, and radiation cooling for the longer heating period thereafter. This might be accomplished by the use of an ablating layer which burns off to leave a radiating surface or, perhaps, by the development of a new material which, in ablating, leaves a layer capable of radiating at the required heat rates.

CONCLUSION

The present state of the art of predicting aerodynamic heating is such that large uncertainties exist as to the exact magnitude of the heat loads for a lunar mission reentry vehicle. These uncertainties are great enough that the peak heating rates may be increased by a factor of about 3 over the values obtained with the best methods known today. The total heat loads to the stagnation point may be increased by a factor of about 2. These uncertainties arise due to a lack of knowledge as to the effects of ionization and non-equilibrium on the heating rates at velocities of the order of escape velocity and higher.

Fortunately, the effects of ionization and non-equilibrium are confined primarily to the stagnation region and thus affect a relatively small portion of the vehicle surface. They are, nonetheless, quite important since they are capable of producing unusually severe requirements for the thermal protection system in the regions of maximum heating.

NOTATION

L/D - lift to drag ratio

W/C_DS - ballistic parameter

M_∞ - freestream Mach number

R_∞ - Reynolds number based on freestream conditions and R_o

R_o - vehicle nose radius

$\dot{q}_c$ - convective heat transfer rate

$\dot{q}_c'$ - convective heat transfer rate with ionization factor

$\dot{q}_R$ - shock layer radiative heat transfer rate

$\dot{q}_T$ - total heat transfer rate

u_∞ - freestream velocity

k - geometric index in convective heating equation

h_w - flow enthalpy evaluated at wall temperature

h_s - freestream total enthalpy

γ_o - trajectory entry angle

Δ - shock layer thickness

λ_s - mean free path in the shock layer

ρ_∞ - freestream density

REFERENCES

1. Yoder, J. F. Aerodynamic Pressure, Force, and Heating Data for Lunar Re-entry Vehicles. ORD Report MR 75, Lockheed Aircraft Corporation, Marietta, Georgia. 1 January 1961.
2. Hayes, Wallace D. and Probstein, Ronald T. Hypersonic Flow Theory. Academic Press. 1959.
3. Tsien, H. S. Superaerodynamics, Mechanics of Rarefied Gases. Jour. Aero. Sci., Vol. 13, pp. 653-664, 1946.
4. Schaaf, S. A., and Chambre, P. L.: Flow of Rarefied Gases, in Fundamentals of Gas Dynamics (H. W. Emmonds, ed.) Sect. H. pp 687-739 (Vol. III of High Speed Aerodynamics and Jet Propulsion). Princeton Univ. Press, Princeton, 1958.
5. Lange, R. H., Little, B. H., Jr., Mylrea, B. W., and Jacobs, W. F.: Survey and Analysis of Hypersonic and Re-Entry Vehicles. Lockheed-Georgia Company, ARL-62.
6. Hankey, Wilbur L., Jr.; Neumann, Richard D.; and Flinn, Evard H.: Design Procedures For Computing Aerodynamic Heating at Hypersonic Speeds. WADC Technical Report 59-610. ASTIA AD 239943. June 1960.
7. Hansen, C. Frederick: Approximations for the Thermodynamic and Transport Properties of High Temperature Air. NASA TN 4150. March 1958.
8. Lees, Lester: Heat Transfer at Hypersonic Speeds in a Planetary Atmosphere. Notes from lecture series, Space Technology, University of California. 1959.

9. Kivel, Bennett: Radiation from Hot Air and Stagnation Heating. Avco Everett Research Laboratory Report 79. October 1959.

10. Fay, J. A., and Riddell, R. R. Theory of Stagnation Point Heat Transfer in Dissociated Air, Jour. Aero. Sci. Vol. 25, No. 2. February, 1958.

11. Sibulkin, M. J. Heat Transfer Near the Forward Stagnation Point of a Body of Revolution, Jour. Aero. Sci. Vol. 19, No. 8, August 1952.

12. Cohen, C. B., and Reshotko, E. Heat Transfer at the Forward Stagnation Point of Blunt Bodies. NACA TN 3513, July 1955.

13. Kemp, Nelson H., Rose, Peter H., and Detra, Ralph W.: Laminar Heat Transfer Around Blunt Bodies in Dissociated Air. Jour. Aero/Space Sci., Vol. 26, No. 7, July 1959.

14. Adams, Mac C.: A Look at the Heat Transfer Problem at Super Satellite Speeds. Avco Everett Research Laboratory AMP 53, December 1960.

15. Aviation Week. November 20, 1961, p 29.

16. Lees, Lester: Laminar Heat Transfer over Blunt Nosed Bodies at Hypersonic Flight Speeds. Jet Propulsion Volume 26, No. 4. April 1956.

17. Detra, R. W. and Hidalgo, H. Generalized Heat Transfer Formulae and Graphs. Avco Everett Research Laboratory Report 72. March 1960.

18. Vinokur, Marcel: Laminar Heat Transfer Distribution on Oblate Ellipsoidal Noses in Hypersonic Flow. Lockheed Aircraft Corporation. Missiles and Space Division Report LMSC 894001. April 1961.

19. Rose, P. H., Probstein, R. F., and Adams, M. C. Turbulent Heat Transfer Through a Highly Cooled Partially Dissociated Boundary Layer. Avco Research Laboratory Research Report 14, January 1958.

20. Comm, J. C., Kivel, B., Taylor, R. L., and Teare, J. D.: Absolute Intensity of Non-Equilibrium Radiation in Air and Stagnation Heating at High Altitudes. Avco Everett Research Laboratory Report 93, December 1959.

21. Perrine, C. Estimation of Maximum Heat Rates and Total Heat Load During Supercircular Re-Entry. MLV TN-10, The Martin Company, August 1960.

REENTRY/RECOVERY DESIGN CONSIDERATIONS OF A LUNAR RETURN DATA VEHICLE *

Donald M. Waltz+

The design considerations pertinent to the reentry and recovery system of a vehicle to bring lunar data from the Moon to the Earth are presented. The data vehicle is defined as a component of a lunar return spacecraft and is that section which will separate from the return spacecraft prior to entry into the Earth's atmosphere. The lunar return data vehicle configuration is discussed from the standpoint of trajectory analysis, Earth atmospheric entry techniques, deceleration load factor, aerodynamic heating, stability, and dispersion. Parameter analysis is performed to establish the data vehicle configuration, gross weight, W/C_DA value, entry angle limitations, design criteria for control of entry heating, deceleration load factor limits, recovery methods, and retrieval techniques. Two recovery and retrieval concepts are suggested: A parachute landing system with retrieval by surface vessel after water impact, and a parachute/balloon buoyancy system with retrieval by aircraft pickup after the data vehicle has stabilized at a design buoyancy altitude.

* A complete version of this paper (AAS 62-40) may be available from the American Astronautical Society or by contacting the author.

+ Northrop Corporation, Ventura Division, Conejo, Calif.